Wave Dispersion Characteristics of Continuous Mechanical Systems

Wave Dispersion Characteristics of Continuous Mechanical Systems provides a mechanical engineering-based analysis of wave dispersion response in various structures created from different materials. Looking at materials including strengthened nanocomposites, functionally graded materials, metal foams, and anisotropic materials, it uses analytical solution methods to solve typical problems in the framework of a micromechanics approach.

Nanocomposites are a novel type of composite materials, fabricated by dispersing nanosized reinforcements in a matrix to combine the material properties of the matrix with the improved properties of nanosized elements. This book enables readers to learn about the theory and practical applications of this rapidly evolving field. Practically minded, the book investigates the impact of employing various nanofillers and demonstrates how this augments stiffness within the nanocomposite. Topics covered include agglomeration and waviness of nanofillers, porosity, elastic mediums, fluid flow, and the impact of the thermal environment on a propagated wave. Using mathematical formulations to solve wave dispersion characteristics of structures including beams, plates, and shells, the book obtains equations of structures using first- and higher-order shear deformation theories.

This book will be of interest to professional engineers working in material and mechanical engineering, nanocomposites, nanofillers, and micromechanics. It will also be of interest to students in these fields.

Wave Dispersion Characteristics of Continuous Mechanical Systems

Farzad Ebrahimi

CRC Press
Taylor & Francis Group
Boca Raton London New York

CRC Press is an imprint of the
Taylor & Francis Group, an **informa** business

First edition published 2024
by CRC Press
2385 NW Executive Center Drive, Suite 320, Boca Raton FL 33431

and by CRC Press
4 Park Square, Milton Park, Abingdon, Oxon, OX14 4RN

CRC Press is an imprint of Taylor & Francis Group, LLC

ISBN: 978-1-032-21834-2 (hbk)
ISBN: 978-1-032-63597-2 (pbk)
ISBN: 978-1-003-27026-3 (ebk)

DOI: 10.1201/9781003270263

Typeset in Times
by SPi Technologies India Pvt Ltd (Straive)

Contents

Preface

The dispersion of elastic waves within a continuous mechanical system can provide significant insights into the properties of a specimen. Such information may be challenging to obtain through alternative methods or may not be obtainable at all. An exemplary illustration of the utility of the wave propagation technique is its efficacy in prognosticating imperfections in solids. This methodology may be employed in instances where it is not feasible to conduct nondestructive testing of the specimen being scrutinized.

The primary objective of this study is to address the issue of insufficient analytical examinations pertaining to the dispersion reactions exhibited by diverse continuous mechanical systems that employ varying material characteristics. The initial two chapters of the book focus on presenting fundamental assumptions and definitions, as well as mathematical techniques that are essential for a mechanical engineer to manipulate the constitutive equations of composites and nanocomposites. Additionally, these chapters provide the necessary tools to solve the wave propagation problem of a continuous mechanical system, assuming that the governing equations have already been established. The initial chapter provides an introduction to composites and nanocomposites by offering a concise overview of their fundamental concepts. The chapter concludes by outlining the diverse range of applications for these innovative materials in contemporary industries such as automotive, aerospace, naval and marine, sensors and actuators, civil structures, and biomedical fields. Subsequently, Chapter 2 will explicate the mathematical methodologies essential for conducting the analyses delineated in the ensuing sections of the manuscript. This chapter will provide an overview of the elasticity relations and various theories pertaining to shells, beams, and plates. Subsequently, Chapters 3–7 are dedicated to the examination of wave dispersion in diverse continuous mechanical systems composed of a variety of materials. Chapter 3 is dedicated to deriving the equations of motion for wave propagation in functionally graded structures, encompassing beams, plates, and shells that are exposed to diverse mechanical, thermal, and magnetic loads. Chapter 4 provides an overview of wave propagation in fluid-conveying structures. In contrast, Chapter 5 focuses on the wave dispersion characteristics of various graphene-based structures, including single- and double-layered graphene sheets, graphene oxide–based materials, graphene platelet–reinforced beams, and graphene foam structures. Chapter 6 pertains to the analysis of wave dispersion in nanocomposites, while Chapter 7 is dedicated to examining the wave dispersion characteristics of metal foam structures.

The present text endeavors to convey mathematical formulations in a self-contained manner, thereby, obviating the need for readers to consult external sources. Furthermore, a variation-based approach will be utilized to articulate the kinematic equations governing the dynamics of continuous mechanical systems characterized by beam, plate, and shell geometries. Classical and shear-deformable kinematic theories are employed for the purpose of facilitating the analysis of continuous mechanical systems, which may be thin or thick and composed of diverse materials. Moreover, it is feasible for any individual with a rudimentary understanding of engineering mathematics to derive all of the formulations that have been presented. This literary work may prove to be a valuable resource for scholars pursuing advanced degrees who are interested in examining the mechanical properties of continuous mechanical structures. The author endeavored to produce a comprehensive and inclusive text with the aim of facilitating readers' comprehension of the wave dispersion properties of continuous mechanical systems. The author earnestly requests that all readers provide their thoughts and comments without reservation. Constructive feedback from a discerning reader can undoubtedly aid the author in enhancing the caliber of this written work in subsequent revisions.

F. Ebrahimi
Imam Khomeini International University, Qazvin, Iran

Author's Biography

Farzad Ebrahimi is an associate professor in the Department of Mechanical Engineering, IKIU, Qazvin, Iran. His research interests include mechanical behaviors of nanoengineered systems, mechanics of composites and nanocomposites, functionally graded materials, viscoelasticity, and smart materials and structures. Dr. Ebrahimi has authored more than 400 high-quality peer-reviewed research articles in his fields of interest. He has also edited and authored multiple books for well-known publishers. He is an associate editor of the journal *Shock and Vibration*, an Editorial Board member of the *Journal of Computational Applied Mechanics*, and a distinguished reviewer, whose expertise helps the editors of prestigious journals judge research articles.

Abbreviations

Acronym	Definition
CF	carbon fiber
CFRP	carbon fiber–reinforced polymer
GFR	glass fiber-reinforced
MMCs	metal–matrix composites
FRPs	fiber–reinforced plastics
CNTs	carbon nanotubes
PNC	polymer matrix nanocomposites
MD	molecular dynamics
PV	photovoltaic
SWCNT	single-walled carbon nanotubes
BNNTs	boron nitride nanotubes
NSGT	nonlocal strain gradient theory
EBB	Euler–Bernoulli beam
TB	Timoshenko beam
CS	cylindrical shell
DLGSs	double-layered graphene sheets
MHCR	multi-hybrid nanocomposite reinforcement
ER	electro-rheological
MR	magneto-rheological
MEE	magneto-thermoelastic
NET	nonlocal elasticity theory
TWBNNT	triple-walled boron nitride nanotubes
GPLR	graphene platelet-reinforced
GDQM	generalized differential quadrature method
GNPRC	graphene nanoplatelet-reinforced composite
GF	graphene foam
GOP	graphene oxide powder
PGF	porous graphene foam
FSDT	first-order shear deformation theory
HSDT	higher-order shear deformation theory
FGM	functionally graded material
CNCs	carbon nanocones
CFR	carbon fiber–reinforced
CPT	classical plate theory

Symbols

Symbol	Definition
U	Displacement vector in the x direction
V	Displacement vector in the y direction
W	Displacement vector in the z direction
ϕ_x	Rotation in the x direction
ϕ_y	Rotation in the y direction
u_0	Initial displacement in the x direction
v_0	Initial displacement in the x direction
w_0	Initial displacement in the x direction
a	Length
b	Width
h_c	The thickness of the magnetostrictive layer
h_f	The thickness of the FG layer
K_c	Coil constant
e_{ij}	Magnetic modules
N_{xx}	Normal forces
M_{xx}	Bending moment
ξ	The amount of porosity
k	Gradient index
ε_{ij}	Strain vector
σ_{ij}	Stress vector
μ	Nonlocal parameter
λ	Length scale parameter
$[C]$	Damping matrix
$[K]$	Stiffness matrix
$[M]$	Mass matrix
I_i	Moment of inertia
E_f	Young's modulus of FG materials
E_m	Young's modulus of magnetostrictive materials
ρ_f	Density of FG materials
ρ_m	Density of magnetostrictive materials
ω	Frequency
Ω	Non-dimensional frequency
P	Critical buckling load
H_{zz}	Magnetic field
∇^2	Laplacian operator
X_m	Admissible Galerkin functions along the x direction
Y_n	Admissible Galerkin functions along the y direction
∂T	Variation of strain energy
∂U	Variation of potential energy
∂W	Variation of external work

z_0	Neutral location
k_s	Shear correction factor
δ	Variation
ε	Strain
σ	Stress
K_w	Winkler coefficient
K_p	Pasternak coefficient
I_i	Mass moment of inertia
C_{ijkl}	Elastic coefficient
t	Time

1 Introduction to Composites and Nanocomposites

1.1 BACKGROUND

This chapter will be dedicated to discussing composites and nanocomposites. In fact, a qualitative perspective will be observed in the majority of sections of this chapter, and a quantitative perspective will be avoided. As the introduction to this chapter, a broad overview of heterogeneous materials will be provided. The main reason for our brief discussion in this subsection is that this book focuses on the mechanical characteristics of nanocomposites. In other words, the innovation of the book will be compromised if we insist on using standard literature concepts of composites. However, due to the fact that nanocomposites are a unique type of composites, a concise overview of composites is necessary. After a comprehensive examination of composites, we will introduce the genesis of nanocomposites to enter the nanocomposites environment. In this subsection, the prevalent issues in nanocomposites will be discussed. Various classifications of nanocomposites will be presented, as well as a discussion of the various nanoscale reinforcing agents that will be utilized in the production of nanocomposites. This section will be followed by a discussion of the most prevalent manufacturing techniques used to produce nanocomposites. In a subsequent subsection, emphasis will be placed on the experimental, analytical, and numerical methods used to analyze nanocomposites using various techniques. Once the abovementioned universal discussions are finished, a wide subsection will be allocated to the investigation of the polymer-reinforced nanocomposites.

1.2 A BRIEF REVIEW OF COMPOSITES

1.2.1 GENERAL CONCEPTS

For the past many years, people have been accustomed to being acquainted with the term "composite." Composites are one of the most well-known engineering materials because they are used in so many different kinds of engineering. These days, the use of composite materials on the exteriors of buildings is not at all unusual. Additionally, many people are acquainted with composites as a result of having visited an orthodontist at some point in their lives. You are aware that dental implants that will substitute for an injured tooth should be made of composite materials, which are the best option available. This growing predilection for composites in the real world is primarily due to the fact that composites have superior

DOI: 10.1201/9781003270263-1

characteristics when compared to conventional kinds of homogeneous materials. To put it another way, the development of composites was a direct consequence of the capacity of such materials to capitalize on the benefits offered by the constituents while compensating for the drawbacks posed by those constituents. Consider the concept of producing strengthened concrete as a straightforward demonstration of this possibility (reinforced concrete). Because it is a malleable solid, steel is able to withstand tension loads but not compression loads, as you may remember from the course on the mechanics of materials that you took. Concrete, on the other hand, is a fragile substance that can perform excellently when submitted to compressive pressure, but cannot reliably demonstrate tension resilience. Because of this, thin steel bars were used along with concrete to produce a novel substance that exhibited improved behavior in both the tension and compression phases [1]. The fundamental idea behind composites can be understood with the help of an illustration such as this one. When discussing combinations, the term "matrix" is generally reserved for the component that accounts for the greatest proportion of the total capacity. Additionally, typical composites have a high strength-to-weight ratio, which is an essential quality for engineering designs to possess. In point of fact, as the inventor, you will be granted permission to make use of lightweight components that are able to support substantial weights. In actual practice, the volume fraction of the host polymer and the reinforcing part will be personalized [1] based on the designers' understanding of the required material behavior. This is because in actuality the volume fraction of the host polymer is typically larger than that of the reinforcing part. The three most prevalent types of composites are fiber-reinforced, particulate, and laminated composites. Fiber-reinforced composites are the most common type of composites. The matrix of fiber-reinforced composites will have a predetermined amount of fiber-like strengthening agent components distributed throughout it in a predetermined proportion. The inclination angle at which the fibers in the matrix are oriented is what determines which direction in these composites has the highest strength. When looking at things from a very broad perspective, the selection of fiber-like strengthening agents is based on the direct association between the reduction in size of the materials and the characteristics they possess. To put it another way, the mechanical, thermal, and electrical performance of minute sections of any substance can be improved in comparison to the performance of mass materials [2]. According to accounts, the matrix of a fiber-reinforced composite is responsible for transferring the load from the environment to the fibers, as well as maintaining the orientation of the fibers and preventing the fibers from deviating in different directions [1, 2]. The performance of the second type of composites, known as particulate composites, can be enhanced by fusing together the host substance and a collection of strengthening element particles. By stacking plies with the appropriate orientation angles, one can generate laminates, the ultimate form of composites. Laminates can be used in a variety of applications. This category has the potential to incorporate the two combination categories that came before it. In fact, some of the plies can be manufactured from particulate composites, while others can be manufactured from fiber-reinforced composites. This is determined by the specifications of the design. Changing the orientation angles of the constituent plies can obviously change the mechanical reaction of laminated composites to external excitations [1]. This is something that can be done in a number of different ways. It is essential to remember that short and long strands are both viable options for use in the production of fiber-reinforced composite materials. When compared to long fibers, short fibers have superior material characteristics, so it is recommended that they be used whenever possible. Long fibers, on the other hand, should be avoided whenever possible. Additionally, it is interesting to note that there are four primary types of fiber-reinforced composites that can be manufactured. Unidirectional, bidirectional, discontinuous, and braided configurations are all possible for these fiber-reinforced composites. In this book, we will not enter into the intricacies of these four categories of fiber-reinforced composites; however, participants are strongly encouraged to conduct research on the aforementioned information in the composite literature [2]. In conclusion, it is important to point out that the use of composites comes with a number of drawbacks in addition to the benefits it provides. In fact, the book that came before this one concentrated on the desirable properties of composites but failed to provide any information about the detrimental behaviors of composites. The delamination phenomenon, also known as the detachment of the stacking plies from each other, is one of the most significant and ubiquitous phenomena in laminated composites. It is also one of the most common phenomena. An instantaneous change in shear stress distribution at the intersection of neighboring plies is what causes delamination in laminates [2]. Rather than the so-called comparable single-layer approach, the concept of the layer-wise theory is preferable for addressing the occurrence of delamination when working with laminated composites. This is because the layer-wise theory takes into account all of the layers in the laminated structure. Another aspect that needs to be taken into consideration is the possibility that the composite will contain openings and holes as a direct consequence of the standard defects that occur during the manufacturing process. After the selection of fiber-reinforced composites, it is necessary to conduct research into the actual connection condition that exists between the fiber and the matrix. In fact, the presumption of flawless bonding between the fiber and matrix is not always accurate, and deboning can sometimes occur, which has the potential to unfavorably affect the development of composite materials [1].

1.2.2 APPLICATIONS

Some clarifications about the common implementations of composites in day-to-day living were presented at the commencement of the segment that came before this one. Now, the writer plans to talk about some of the most spectacular implementations of the various kinds of composites used in a broad variety of different industries. As a result, the implementations of composite materials in various fields of engineering, including automobile, marine, space and aeronautical, thermal, electronic, municipal, and fusion engineering, will be addressed in this subsection of the book. In addition to the applications that have been discussed thus far, we will also discuss the applications of composites in the fields of medicine and the bio-environment. In order to keep things as straightforward as possible moving forward, each of the previously mentioned applications will be dissected within the context of their own distinct segment.

1.2.2.1 Composites in the Automotive Industry

Possibly everyone is aware of the extensive use of composite materials in automobile manufacturing. In this section of the book, we will examine the conceptual basis for this trend. According to our fundamental understanding of material science, the common fibers utilized in the production of composites have a very high stiffness-to-weight ratio. Consider the comparison between carbon fiber (CF) and steel, for instance. Low-stiffness CFs can provide a higher Young's modulus than steel, while their mass density is approximately one-fourth that of steel. Consequently, the use of carbon fiber-reinforced polymer (CFRP) in the automotive sector can be an excellent option, as the utilized composite can be at least as effective as the conventionally used steel alloys. Once metallic alloys are replaced with CFRPs, however, the total weight of the designed segment will decrease significantly. Ford Corporation used CFRPs in the prototype of one of its automobiles at the end of the 1970s for this primary reason [3]. They replaced steel with carbon fiber-reinforced plastic in certain automotive components. Front end, frame, doors, bumpers, driveshaft, and body-in-white were among these components. This material substitution reduced the car's total weight by approximately 33%. However, the replacement had no negative effect on the automobile's performance, and the manufactured car's performance was at worst identical to that of the previously metal-only vehicle [3].

Even though the above-said example demonstrated the performance suitability of CFRPs in the automotive industry, such composites cannot be extensively utilized in the production of automobiles. As you may have predicted, the primary issue with this method is the high cost of using CFRPs in the manufacturing process. Researchers made numerous endeavors to reduce the expense of utilizing these materials. However, the ultimate cost for the fibers was approximately $8 per pound, which is inappropriate for this industry because the consumer will be unable to afford the manufactured automobile. Due to this limitation, glass fibers (GFs) were chosen for use in the automotive industry in an effort to reduce costs. As indicated in Section 1.1, there are various categories of glass fiber-reinforced (GFR) composites. In reality, both fiber-reinforced and particulate composites can be utilized in the automotive industry. In common applications, thermoset polymers such as polyester and vinyl-ester are used to host the distributed particles or fibers of the reinforcing glasses. Nevertheless, thermoplastics can be utilized in these composites, albeit at a lower efficiency. In fact, the use of thermoplastics as resin will reduce thermal resistance, which is not a desirable design outcome. Therefore, it is preferable to use thermosets in composites reinforced with glass for automobile manufacturing [3].

Automobile devices must also be capable of ideal energy absorption from themselves, a factor that is of paramount importance in their design. In previously extensively used metallic materials, such as steel alloys or aluminum, energy absorption will be carried out by the element via plastic deformation. In fact, the absorption procedure will be satisfactory to the ductile nature of these metallic materials, which can withstand a certain amount of plastic strain before they fail. Therefore, the phenomenon of absorption can be linked to the stress–strain behavior of such materials. However, there is no similarity between the stress–strain behavior of fiber-reinforced plastics (FRPs) and that of metals [3]. The stress–strain curves of these materials are identical to that of brittle materials. Although it may appear that this difference makes it impossible for FRPs to serve as an excellent energy absorber, it should be noted that these materials are superior to metallic ones in absorbing used energy. Indeed, the mechanism of energy absorption in FRPs differs from that of metallic materials. It has been demonstrated that brittle materials can absorb the high energy of a desired projectile by dispersing its initial energy in the form of conical surface fragmentation [3]. Consequently, only the method of energy absorption in FRPs differs from that of metals. In addition, it is demonstrated that low-quality composites have a greater specific energy absorption than metals. It is only natural that this specific value could be enhanced if FRPs of superior quality were used instead of those of inferior quality [3].

In contrast, rigidity of the implemented materials in the automotive industry is one of the most influential factors on the minds of designers. A designer's general belief is that the optimal material has the highest potential rigidity. Moreover, it is evident that the rigidity of FRPs is significantly lower than that of metals. However, the stiffness-to-weight ratio of these composites is at least equivalent to that of metals. In addition, it must be remembered that, due to the lower weight of FRPs in comparison to metals, their rigidity can be increased by increasing the material thickness [3]. In addition, it is important to note that the use of composites can result in an increase in the structural rigidity of the elements by eliminating the connections that cannot be disregarded when metals are employed. In other words, some of the junctions that enhance the system's flexibility (i.e., correspond to a reduction in structural rigidity) will be eliminated as a result of the components integration alternative that can be implemented once composites are utilized in the design of automobile parts [3]. In addition to the issue of rigidity, composites can also result in a superior suspension mechanism because FRPs have greater internal damping than metallic materials. Customers in the automotive industry can be more satisfied with their comfort if the vehicle has a superior suspension system, and this issue can be directly related to the damping ability of the material used. As a consequence, the use of composites can result in a higher stiffness-to-weight ratio for the designed elements of a car, followed by a more effective dissipation of oscillation [3].

It has also been demonstrated that the implementation of lightweight textile-reinforced composites housed by a thermoplastic polymer can result in mechanical performance that is superior to that of previously known composites with a reduced reinforcing phase content [4]. Indeed, researchers demonstrated that such composites can exhibit improved impact resistance, enhanced rheological behavior, and acceptable rigidity due to the lattice structure of the textile. According to reference [3], the aforementioned architecture resulted in a remarkable improvement in the energy absorption capability of the composite, which is of great significance in the automotive industry.

Although the use of GFs in the framework of FRPs in the automotive industry has resulted in a number of benefits for manufacturers and consumers, the implementation of such composites cannot be disregarded without qualification. For instance, the poor machinability potential, low recyclability, and hazards to the living environment of FRPs prompted engineers and scientists to search for an alternative composite material to replace FRPs [5]. Researchers discovered composites made from natural fibers to be suitable for this purpose. Natural fiber-reinforced composites can reduce weight, increase the likelihood of recycling, and reduce dependence on hydrocarbon resources. On the premise of European Union sanctions, car manufacturers in Europe were required to produce a certain percentage of recyclable vehicles. This quantity, which was determined as a percentage of the vehicle's total weight, was approximately 80% in 2006 and 85% in 2015. This decision led to the production of less-polluting automobiles of comparable quality through the substitution of GFs with natural fibers such as kenaf, hemp, flax, jute, and sisal [5]. In response to these advantages, many of the world's leading automakers, including Mercedes-Benz, Ford, Audi, and Toyota, have begun using natural textiles in their productions. From now on, we will refer to these composites as green composites because they assist the environment by conserving hydrocarbon resources and reducing pollutants. In the following paragraph, we will examine the constituents of green composites in greater detail. Among all the natural fibers utilized in the production of green composites, jute is the most well-known due to its abundance worldwide. In addition, flax and ramie are popular green reinforcements among producers. These fibers are capable of exhibiting exceptional mechanical performance on their own [2]. Among all varieties of natural fibers, ramie possesses the highest stiffness-to-weight ratio, followed by hemp, kenaf, flax, abaca, curaua, jute, and sisal. It is fascinating to note that the stiffness-to-weight ratio of ramie, hemp, and kenaf is greater than that of E-GF, demonstrating the satisfactory mechanical performance of green composites [2].

In addition to polymer-based composites, metal–matrix composites play an enormous role in the automotive industry. Different filaments or particulates can be dispersed in a metallic matrix to create these composites. Metal-based composites can have a high strength-to-weight ratio, a high stiffness-to-weight ratio, excellent damping performance, machinability, enhanced wear, corrosion, and friction resistance, and a decreased coefficient of thermal expansion [6] depending on the selected constituents and their utilized content. Each of these characteristics is pertinent to one of the car's components. For instance, the wear resistance, thermal performance, and outstanding fatigue life of metallic composites are logical justifications for their widespread use in the piston crown and valves of automobile engines. The low weight and resistance to friction of these composites make them appropriate for use in the production of bearings for automobile engines. As another example, metal-based composites are utilized extensively in the production of cylinder blocks for automobile engines due to their excellent corrosion and wear resistance, as well as their low friction and low weight. In addition to using metal-based composites in engine components, the designers also incorporated these materials in brakes, transmission and differential bearings, and drivetrain gears and shift forks due to their low weight and resistance to wear. Refer to reference [6] for more information regarding the widespread applications of metallic composites in automobiles.

1.2.2.2 Naval and Marine Applications

Ship engineering and naval area is an additional engineering discipline that has a close relationship with the application of composites. From a historical perspective, composites were only used in noncritical components of naval devices because their material properties have not been established to be effective enough for use in critical components as well. After the Second World War, however, composites are widely used in the fabrication of naval and marine components. The use of composites in US Navy patrol vessels and corvettes, for example, is documented in the open literature. It is noteworthy that composites have been used as lightweight materials in the construction of patrol vessels since the 1960s. Due to the limited rigidity of the hull girder, the maximum length of these vessels is generally capped at 10 meters. In such vessels, GF-reinforced composites are commonly used. Following the successful production of patrol boats by the US Navy, other navies throughout the globe began utilizing composites for the production of patrol boats of various configurations. Examining the review published in the early 2000s will provide additional historical information on this topic. Moreover, composites have also caught the attention of the designers of mine-countermeasure vessels. Conventionally, such vessels were constructed from wood due to its independence from magnetic signals, which allowed the ships to navigate in the direction of magnetic explosives. However, the scarcity of timber in the world following the Second World War compelled engineers to find an alternative material for this application. In contrast, the replacement of wood with composites resulted in a significant reduction in the maintenance-related through-life costs of mine-countermeasure vessels made of wood. The designers of mine-countermeasures deemed the use of composites enticing enough to develop entirely composite ones in the 1970s [7]. Corvettes are another naval application of composite materials. These massive vessels are designed for a variety of missions, including surveillance, combat, mine-laying, mine-countermeasures, and anti-submarine warfare operations. Composites play a crucial role in the manufacturing of such enormous naval components. Sweden began manufacturing composite corvettes under the YS-2000 project before any other nation in the world [7]; it was the world leader in this field. Engineers utilized hybrid composites reinforced with carbon and glass fibers in accordance with reference [7] in order to achieve the designed properties in practice. It must be stated that composites are not used in the naval industry solely due to their increased stiffness/strength or mine-clearing capability. These materials can be utilized as air ducts in military ships for air conditioning and heating purposes. In addition to the aforementioned examples, composites are widely used in other sections of naval and marine instruments, such as superstructures, masts, and propulsion, whose detailed explanations can be found in the open literature [7].

Polymeric composites can demonstrate remarkable resistance to corrosion and also meet the lightweight criterion, which are critical factors in the design of naval instruments. Additionally, the use of this class of sophisticated materials can eliminate the magnetic-pulse dependence of previously employed metallic materials. Nevertheless, the issue of utilizing optimal thermal barriers in naval facilities is one of the most crucial design criteria in their design process. In fact, the high thermomechanical working conditions and potential for conflagration in large ships necessitate the use of fire-resistant materials in the manufacture of naval devices. In addition to the aforementioned requirements, a massive ship's components must be robust enough to provide acceptable strength and rigidity while maintaining a low weight. Consequently, fire-resistant composites are among the most promising materials whose use in this sector can result in intriguing performance. In order to satisfy the aforementioned requirements, several varieties of composites were examined experimentally for their thermal behavior to determine if they are suitable for use in naval applications. It was determined that using intumescent coatings alone or in combination with ceramic coatings can satisfy the American Society for Testing and Materials (ASTM)-issued design criterion. It was demonstrated that the flame spread index and heat release of such composites were suitable for use in naval applications. In a subsequent experiment, it was determined that the use of a modified variety of halogenated composites will result in enhanced thermal performance. Using matrices modified with aluminum trihydrate (ATH) in the manufacturing process has also been shown to increase ignition time while decreasing heat release rate. Researchers found it intriguing at the end of the 1990s to use brominated fire-resistant composites in marine applications to improve thermal barriers in naval instruments. Under cone calorimeter testing, it has been reported that ATH composites can achieve a maximum 20–25% reduction in the apex of the release rate [8].

1.2.2.3   Aerospace Applications

It has been known for a number of years that the local temperature in some space activities around Earth and Mars exceeds approximately 1800 Kelvin, which corresponds to a substantial thermal burden on the device responsible for functioning under the aforementioned working conditions. In addition, many structural elements in aerospace applications require high specific stiffness and strength, so fibrous composites are widely used in these applications. Therefore, the structural elements used in the fabrication of such devices must withstand such severe thermal loading, which cannot be satisfied unless a material with virtually zero coefficient of thermal expansion (CTE) is used as the device's constituent material. Inspired by this reality, metal matrix composites (MMCs) have played an outstanding role in aerospace engineering. In the aforementioned applications, it is common to implement MMC tube, plate, and panel structures in critical conditions. Despite the fact that MMCs have been demonstrated to be effective options for space and aerospace applications, it is important to note that their production encounters certain barriers, such as formation and manufacturing costs. NASA was one of the first organizations to successfully use boron–aluminum (B/Al) MMC as the truss lattice structures in the mid-fuselage section of the Space Shuttle Orbiter. It is worth noting that this effort resulted in a weight reduction of 45% compared to the case where complete aluminum rods were used as the supports of the under-observation truss lattice. The Hubble Space Telescope's antenna mast is made of graphite-reinforced aluminum-based (Gr/Al) MMC, which was used in another effective project. The latter resulted in the certification of the requisite rigidity with a very low CTE, which is one of the most important design criteria in this field of study. Due to the superior electrical conductivity of the constructed MMC, it was possible to efficiently receive electrical signals, which was another accomplishment of this material choice. In addition, other varieties of MMCs with similar multitasking capabilities as Gr/Al can be found. For instance, graphite-reinforced Copper (Gr/Cu) can provide increased stiffness along with high thermo-electrical conductivity, which is suitable for aerospace applications [9].

In contrast, discontinuously reinforced aluminum MMCs are suitable for use in a variety of aerospace applications, such as joints, attachment fittings of truss lattices, longerons, electronic packaging, thermal planes, and bearings. In such cases, either particulate or whisker-type reinforcing agents will be used to produce the composite [9]. Referring to the author's previous book [1] appears useful for obtaining additional information about various composites. To return to our primary topic, it must be considered that, among all of the potential applications of discontinuously reinforced aluminum (DRAs), such MMCs exhibit their finest functionality in electronic packaging and thermal management due to their exceptional thermo-electrical conductivity and tunable CTE. For instance, particulate DRAs are widely used in communication satellites and global positioning system satellites. It is also reported that using DRAs for thermal management in spacecraft power semiconductor modules will result in a weight reduction of more than 80%.

In addition to MMCs, polymer matrix composites (PMCs) can meet many aerospace design criteria. PMCs, for instance, can exhibit adequate specific rigidity coupled with an acceptable interlaminar shear strength (ILSS). In addition, the effective bonding between the fiber and polymeric matrix is typically a strong one that prevents failure due to the debonding phenomenon. Even though such composites have a number of advantages that can be useful in aerospace applications, specimens made from certain resins, such as epoxy, have a high risk of ignition, which is not negotiable in the aforementioned application. Therefore, these composites must be modified from a combustion retardant standpoint prior to use in critical positions. PMCs hosted by a shape memory polymer (SMP) matrix, also known as SMP composites (SMPCs), are a subcategory of PMCs with unique material properties that are in high demand in the space and aerospace industries. Before discussing the potential applications of SMPCs, it would be prudent to be familiar with the general characteristics of SMPs. These polymers are a distinct class of polymeric substances that can respond to external excitations by modifying their shape and pigment, followed by a recovery procedure. SMPs are lightweight, inexpensive, and simple-to-produce materials with remarkable bio-degradability and tunable glass transition temperature. However, these polymers cannot demonstrate adequate stiffness on their own. To remedy this deficiency in these materials, they will be utilized as host matrices for composite materials reinforced with high-rigidity fibers whose stiffness is many times that of the SMPs. Therefore, the obtained SMPCs are applicable for space and aerospace applications due to their wide spectrum of qualities. The mechanical behaviors of SMPCs can be influenced by factors such as high vacuum, severe thermal loading, and ultraviolet radiation, which cannot be disregarded in space and aerospace engineering [10].

The shape memory property of SMPCs has enabled the aerospace industry to incorporate such advanced composites as hinges. Implementation of rubber-based SMPCs in tape hinges can result in enhanced impact control and increased strain energy storage. Late in the 21st century, it was reported that carbon fiber-reinforced (CFR) SMPCs could be viable candidates for being used as tape hinges, whose recovery is crucial in aerospace applications. On the other hand, SMPCs can also be utilized as satellite platforms. Implementation of SMPCs as booms of such sophisticated instruments will result in a notable reduction in boom weight compared to the case where metallic materials are used instead. According to reports, accurate procurement of the wondrous properties of SMPCs results in the fabrication of various types of devices for space satellites. In addition, as you may be aware, conventional metallic booms consist of tubular, extendable booms and collapsible truss booms whose manipulation requires a motor. However, once the intelligent behaviors of SMPCs are customized, the desired elongation and compression can be attained without the use of additional mechanical elements. Different varieties of booms, such as collapsible truss boom, coilable truss boom, and storable tubular extendible member boom, can be manufactured based on the advanced characteristics of SMPCs. US Air Force Academy (USAFA) allocated funding for the FalconSat-3 mission to produce foldable SMPC truss platforms with two or three longerons, Commonly, SMPCs are incorporated into the longerons of coilable booms to facilitate the extension process. Clearly, foldable and coilable platforms are not the only possible implementations of SMPCs. In addition to the positions enumerated above, SMPCs will be increasingly utilized in various space and aerospace devices, such as deployable reflectors, morphing structures, and expandable lunar habitats [10].

1.2.2.4   Composites in Thermal Environments

Unquestionably, one of the most important engineering challenges is to control the enormous thermal gradient that occurs under various industrial working conditions. Despite the fact that numerous researchers have attempted to develop metallic super alloys with acceptable thermal stability, inherent limitations make it impossible to enhance the thermal properties of such alloys beyond a certain limit. In numerous high-temperature situations, the local temperature exceeds approximately 1500 Kelvin, which corresponds to approximately 80% of the melting temperature of the majority of metallic alloys. Therefore, in such extreme thermal conditions, metallic elements cannot independently demonstrate adequate rigidity and creep resistance. Ceramic matrix composites (CMCs) have garnered the attention of engineers due to the aforementioned facts and, in their opinion, are intriguing enough to be utilized in such working conditions, such as in engines. In fact, the use of CMCs ensures thermal durability for temperatures up to approximately 1800 Kelvin. Due to the enhanced thermal stability of CMCs, this exemplary manifest is not at all surprising. According to the literature, the aforementioned variety of composite materials can withstand temperatures as high as 2300 Kelvin. The designers are now able to use these composites in aircraft engines and gas turbines due to this accomplishment. Notably, individual ceramics cannot be used in the aforementioned critical applications due to their brittleness, which results in poor tensile properties.

The only applications of composites in engineering disciplines involving high-temperature working conditions cannot be summed up as described in the preceding paragraph. In other words, in certain circumstances, it is essential to create conditions that accelerate heat transfer [1]. The light-emitting diode (LED) industry is one of the most glaring examples of the aforementioned cases. Due to their extended lifespan compared to conventional illumination, LEDs are utilized in a vast array of situations. As a result of the elevated heat discharge generated within these elements, however, their performance may be negatively impacted. Consequently, it is crucial to facilitate heat transfer from the interior of LEDs to their surrounding atmosphere. To achieve this objective, materials with high thermal conductivity are required, and PMCs in general satisfy this requirement [1]. In addition, once the electronic packaging of electrical components is targeted, the vital function of high-range thermal conductivity composites can be perceived once more. In other words, the significance of rapid cooling in the components of electrical devices has made it necessary to maximize the benefits of highly thermally conductive materials. For instance, the chilling procedure in single-chip containers must be carried out using thermally conductive composites that provide adequate adhesion in order to be effective. Moreover, PMCs are ideally suited for electrical control unit (ECU) applications in the automotive sector. In this instance, thermally conductive, low-rigidity, and viscoelastic materials are required, and all of these design requirements can be met with PMCs. According to the literature, silicone PMCs are among the most effective candidates for the situation described above. The battery industry is also interested in thermally conductive PMCs for the purpose of satisfying the process of rapid heat transfer. If the batteries operate at a high charge/discharge rate, the rate of heat production will exceed its rate of attenuation. In order to prevent the advent of a battery-burning phenomenon, composites with exceptional thermal conductivity are required. Lastly, it is noteworthy that composites are also utilized in the solar cell industry. The impact of an increase in working temperature on the energy conversion efficacy of solar cells can be mitigated by incorporating a heat dissipation mechanism into such systems. However, photovoltaic (PV) materials used in the fabrication of solar cells cannot be efficiently chilled, and this shortcoming is typically mitigated for by electrically insulating additives whose constituents are PMCs in general [11].

1.2.2.5   Composites in Sensors and Actuators

The concept of energy conversion, particularly from electrical to mechanical and vice versa, is utilized in a great number of modern multitasking advanced devices. Intelligent piezoelectric transducers are increasingly utilized in these high-tech applications. Sensors and actuators are the two most common varieties of transducers that are utilized. In the first, a mechanical input (i.e., strain or tension) must be converted into an electrically measurable form by utilizing the piezoelectricity phenomenon. In contrast, the latter variety of transducers must demonstrate the effect of an initial electrical signal on a mechanical sign. Notably, sensors and actuators are called direct (passive sound receiver or hydrophone) and indirect (active sound transmitter or loudspeaker) transducers, respectively. For instance, hydrophones are typically expected to detect relatively low-frequency underwater sounds (i.e., waves with frequencies inside 40 kHz). Due to the increased length of the sound wave in comparison to the dimensions of the sensor, it is evident that a hydrostatic stress state exists in such cases. In the aforementioned conditions, the sensitivity of the hydrophone will be determined by measuring the output voltage produced by the hydrostatic pressure. In addition, the voltage coefficient, which relates the hydrostatic strain coefficient to the permittivity of the utilized material, will be measured to determine the applicability of the piezoelectric for such applications. In addition to the electric perspective, the mechanical flexibility, impact resistance, and formability of the sensor are of utmost significance for the application in question [12].

The thickness of the electromechanical coupling of the transducers reveals the transducer's efficacy in converting mechanical and electrical energies to one another. In addition, the mechanical quality factor will be used to evaluate the mechanical loss (i.e., acoustic type) of the transducers. Clearly, optimizing this factor can result in a significant reduction in acoustic losses. This parameter, however, cannot be increased indefinitely because its increase corresponds to low-resolution images in applications such as intelligent medical imaging. Therefore, a mid-range must be chosen in order to meet all of the intended design specifications. According to reports, qualitative factors ranging from 2 to 10 can satisfy the intended objective. Smart piezocomposites appear to be effective enough for use in the transducers industry based on the aforementioned design criteria. Using such intelligent composites can satisfy the discussed material properties in addition to allowing the manufacturer to form the transducer into any arbitrary shape required for a specific application [12].

Previously, one of the materials used in the transducers was also known as lead zirconium titanate (PZT). This group of piezoelectric materials, however, is unsuitable for use as hydrostatic transducers, as their low hydrostatic piezoelectric coefficient will result in a low voltage coefficient, which is undesirable in such applications. According to reports, the phenomenon described above is caused by the coupling between and coefficients of PZT. Intelligent PMCs with piezoelectric properties can be utilized effectively to address this issue. In the latter, the coefficients are decoupled from one another, and the permittivity of the enriched piezocomposites is decreased. As a result of the inverse relationship between permittivity and voltage coefficient, the secondary consequence of the aforementioned material selection amplifies the voltage coefficient. In addition to their light weight and formability, silicone rubber piezocomposites demonstrated exceptional sensitivity to low-frequency sound during experimental testing, making them suitable for use as transducers. The combination of polymeric matrices and piezoelectric ceramics can generate clever piezocomposites whose applications as electrical elements vary as the ceramic–polymer connectivity varies. For instance, one to three piezocomposites can be utilized as transducers in certain applications where simultaneous sensing and actuation procedures are required. Consider, for the sake of clarification, a pulse-echo medical ultrasonic imaging instrument in which passive and active conversions must occur sequentially [12].

Although the previously mentioned piezocomposites have a lower acoustic impedance than conventional ones, their acoustic impedance is significantly greater than that of human tissue, which has a negative impact on medical ultrasound imaging. Using an interlayer connective piezoelectric between the original one and tissue, it is reported that this issue can be resolved. However, there is still a problem, which is the dependence of the system's sensitivity on the layer thickness. To this end, a superior alternative is described, namely the use of solitary crystals of the solid solution of, also known as PZN, and, also known as PT. The required electromechanical coupling and low-range acoustic impedance can be readily met using this method. In a similar experiment, the same behavior was observed in one to three PMCs that have been reinforced with, also known as PMN-PT [12]. In conclusion, PZN-PT and PMN-PT piezocomposites are viable options for ultrasonic transducer applications [1].

1.2.2.6 Composites in Civil Structures

FRPs possess remarkable properties that can satisfy a wide variety of civil engineering material requirements. The most important characteristics are their eye-catching specific rigidity and strength, light weight, and durability. FRPs have recently supplanted conventional metal and concrete components. In general, the applications of composites in civil structures can be divided into two main categories: rehabilitation and new construction. In the first category, FRPs will be used to restore an aging structure by either reinforcing it or upgrading it against external environmental excitations such as seismic. In the latter phase, however, FRPs, either alone or in combination with concrete, will be used to construct new structures. When utilized in the construction of bridges, hybrid FRP/concrete composites, for instance, demonstrated exceptional performance. In the mentioned application, such lightweight hybrid composites demonstrated remarkable durability. In an experiment conducted by University of California, San Diego researchers, a five-story masonry structure was reinforced with FRPs containing carbon fibers. Once the pseudo-dynamic seismic excitation was applied to the structure, improved retrofitting was observed, which can be ascribed to the material properties of the FRP [13].

Wall tensile strength enhancement is another aspect of civil structures that can be addressed with composites. To achieve this, FRP wall overlays will be applied to the existing wall. Due to the fact that the mode of failure in this condition corresponds to the diagonal direction, the reinforcing fibers are oriented at a 45-degree angle along the longitudinal direction. FRPs can also be utilized to improve the flexural resistance of civil structures. As the primary objective in such situations is to improve buckling performance, it must be considered that the aforementioned requirement cannot be met unless there is an optimal load transmission from the original structure to the overlay. Furthermore, CFRPs have been demonstrated to be effective options for reinforcing surfaces. In this regard, an experiment was conducted to compare the outcomes of strengthened and unreinforced slabs subjected to three- and four-point deformation tests. It is observed that slab deflection will be significantly reduced if CFRP reinforcers are utilized. According to the findings of this research, the ultimate strength of the reinforced slabs is at least twice that of the individual slabs [13].

In contrast, since the earthquakes of the 1980s and 1990s in the state of California, retrofitting columns of suspended structures such as bridges has received more attention. In order to retrofit the columns against seismic stimulations, steel and FRP casings were utilized. Reportedly, a noticeable enhancement in the structure's dynamic resistance was observed. Despite the fact that both steel and FRP were capable of enhancing the dynamic resistance of the under-observation structures, it was determined that FRP jackets were the superior option. In fact, a greater improvement was observed when steel vests were utilized, which has been demonstrated to be superfluous under the aforementioned working conditions. Moreover, FRPs with low maintenance costs and low weight are reportedly suitable for use as bridge deck components. The use of FRPs in bridge platforms enables the construction of long, lightweight bridges capable of supporting tremendous live loads. Even though FRPs have desirable characteristics for civil applications, it is well-known that their manufacturing and curing processes limit their use. Using microwave-assisted curing techniques is reportedly one of the finest methods for overcoming this shortcoming. Following this procedure, the mechanical strength and storage modulus of FRPs will be significantly increased [13].

Hybrid FRP–concrete composites will be used as an efficient material in civil structures for another form of application. The primary reason for selecting such a material is to compensate for the poor tensile performance of concrete. In actuality, a novel variety of reinforced concrete (RC) that is significantly lighter than conventional RCs will be created in this manner. During the implementation of FRP/concrete composite structures, the thermal durability of the resulting hybrid composite is an important consideration. Because of the thermally graded working conditions that the composite structure must endure, phenomena such as composite freezing and thawing can occur. The failure possibility is attributed to the difference in properties between concrete and FRP, as stated in reference [13].

On the other hand, there is no doubt that environmental preservation is one of the hottest topics worldwide. Obviously, one of the engineering disciplines closely related to this issue is civil engineering. The compatibility of civil structures with the living environment is, in truth, one of the most important topics. Since the late 1970s, a transformation in the constituent materials of civil structures has begun on the basis of this logic. In this regard, civil engineers did their best to find eco-friendly materials to lessen the impact of civilization on the environment while also constructing modern, energy-efficient buildings. Engineers were attracted to cellulosic fiber-reinforced composites as one of the environmentally friendly materials. When used as reinforcing agents in cement- or concrete-based composites, these natural fibers demonstrated satisfactory mechanical performance. In fact, it has been demonstrated that cellulosic cementitious composites possess remarkable specific rigidity and strength as well as enhanced tensile, flexural, and impact properties that have never been observed in individual concrete. Furthermore, these composites are eco-friendly materials with notable biodegradability, renewability, cost-effectiveness, and light weight. According to reports, these materials can mitigate the vulnerabilities of cement and concrete. Cellulosic composites hosted by concrete, for instance, can exhibit enhanced resistance to fracture propagation due to the fact that cellulosic fibers function as a bond to retain the matrix's cracks. Flax, sisal, jute, hemp, coir, hibiscus, cannabinus, eucalyptus, malva, ramie, pineapple leaf, kenaf bast, *Sansevieria* leaf, abaca leaf, vakka, bamboo, banana, and palm are the most well-known cellulosic fibers that can be used in such composites. Despite the fact that cellulosic cement- and concrete-based composites are excellent options for use in civil structures, it is reported that three main considerations must be made before their industrial fabrication can begin. The first concern is that there are no reliable data regarding the durability of such composites in the open literature. In addition, the possibility of cavities in the composite and insufficient distribution of reinforcing natural fibers in various composite segments have been reported as potential outcomes of using a high content of reinforcing fibers in these composites. Changes in the chemical composition, geometrical dimensions, and surface texture of cellulosic fibers are responsible for the final defect. According to reports, these variable properties are crucial factors negatively influencing the performance of the natural composite. It has been demonstrated, however, that all of the aforementioned potential drawbacks can be mitigated by using woven composites of such cellulosic fibers as the reinforcing agent of the concrete in civil structures [14].

In addition to their extensive applications in civil structures, composites are also suitable candidates for use as infrastructure constituent materials. In reality, the superior properties of FRPs cannot be summed up by their specific strength and rigidity. Such materials are capable of exhibiting exceptional resistance to corrosion, humidity, and ultraviolet radiation, as well as promoting an appropriate response to fatigue loading. FRPs are utilized as connective connections, bridge piers, bumper systems, frame structures, truss bridges, and solar panel supports due to their aforementioned properties [14].

1.2.2.7   Biomedical Applications

In the coming years, applications in the medical field will make use of the more sophisticated properties that composite materials offer. For the purpose of manufacturing mechanical ventricular assist-type muscles, for instance, such materials are utilized. These artificial composite muscles have the potential to be transplanted into the bodies of heart patients in order to fortify them against irregularities that take place in the heart as a consequence of the weakening of the cardiac muscles. It is important to note that the management of tachycardia will be achieved through the implementation of composites in the applications described above. When this occurs, ionic polymeric metal composite sensors and controllers will be implanted in the bodies of the patients in a manner that does not interfere with the process of blood circulation. Because the prosthetic muscle will be in contact with the heart ventricle, it is essential that it be flexible enough to prevent the patient's heart from becoming injured during the compression process. In addition, the composite material needs to be one that is intelligent and has strong electrical properties. It is important to note that the battery of the prosthetic muscle will be recharged by coming into contact with the patient's epidermis, as this will allow the battery to draw energy from the patient. Ionic polymeric metal composites are used in another biological application to generate artificial smooth muscle actuators. These artificial smooth muscle actuators have the potential to be extensively employed as the constituents of artificial sections of human bodies. A significant number of portions that are connected to one another can be arranged in this manner to generate such components. Once again, the intelligent nature of composite materials will be utilized in this instance to initiate any section followed by another one through the use of a controlled framework. After initiation, a wave-based motion begins, and thanks to the intelligent electromechanical connection that is made available in the composite, the artificial smooth muscle will be allowed to move and will also be able to perform the task that was prescribed for it. The purpose of utilizing these intelligent manufactured devices is to achieve the objective of increasing the motional portions of human beings. In order to accomplish this, a wide variety of artificial components, such as portable, electronically self-powered, exoskeletal prosthetics, orthoses, and integrated muscle fabric system constituents such as jackets, trousers, mittens, and footwear, can be put into use. The rectification of refractive defects in human vision is yet another biological application that could be addressed by the aforementioned type of composites. In this essential scenario, many challenges pertaining to human eyesight, such as astigmatism, are able to be conquered through the utilization of intelligent composites. In addition to the applications described above, ionic polymeric metal composites may also be utilized in a variety of other contexts, as is necessary to point out here. In other words, these types of smart composites can be utilized in a variety of assistive devices, such as those for incontinence, peristaltic pumps, laparoscopic instruments, and so on [15].

In addition to this, polymeric composites that are strengthened with natural fibers are of great magnificence in the context of their use in biological applications. PMCs that have been strengthened with silk strands, for instance, can be utilized in the manufacturing of wound closures. In this kind of situation, the knot strength and handling characteristic of the silk strands sound like they could be fascinating tools for physicians to use when they are suturing the tissues, such as cardiovascular tissues. The same composites can be helpful in other contexts, such as when they are utilized as substrates for damaged tissue. The sluggish deterioration rate of such materials is the primary reason why these composites are suitable for use in the circumstances that were described. This allows for the damaged tissue to be restored over a comparatively extended period of time, which is one of the reasons why these composites are useful. Additionally, the inventor is able to change the disintegration time of silk fiber-reinforced composites to correspond with the approximate amount of time that is required for the formation of the neotissue. Damage to the tissue of the anterior cruciate ligament, more commonly referred to as ACL, is one of the most well-known areas of the human body that can be restored using composites that have been strengthened with silk strands. In addition, it has been observed that the manufacturing of bio-composites with a broad variety of applications can be achieved by combining natural fibers with biopolymers such as polyvinyl alcohol, which are examples of biopolymers. On the other hand, there are many other kinds of composites that have the potential to be utilized in a wide variety of applications as manufactured biological instruments. For example, functionally graded (FG) composites, which comprise at least two distinct stages, are of utmost significance in the field of biological applications. Additionally, metamaterials that exhibit certain mechanical characteristics that are contingent on their geometries are playing an increasingly important role in the applications of biological engineering. The academic community has shown a greater interest in metamaterials with auxetic lattices than in any other kind [15]. This is true across the board for all kinds of metamaterials.

1.3 MICROMECHANICAL HOMOGENIZATION OF COMPOSITES AND NANOCOMPOSITES

In this section, the micromechanical methods will be used to obtain the equivalent material properties of composites and nanocomposites. These properties will be compared and contrasted. It is planned to make an effort to demonstrate, using a variety of approaches, how the various approximations that are produced by the various approaches differ from one another. In order to accomplish this goal, the most fundamental technique of standardization, known as the rule of the mixture, will be presented first, followed by a discussion of more complicated models in greater depth later on.

1.3.1 RULE OF THE MIXTURE

A certain volume will be designated to each of the constituent phases (carbon nanotubes (CNTs) and polymer matrix) of the polymer nanoparticle *composites* (PNC) in accordance with this method. Next, the effective material properties of the PNC will be determined by summing the quantity of each phase's desired property multiplied by the phase's volume fraction. In this method, it is supposed that the bonding between nanofiller and matrix is of the ideal type, and nanotubes are regarded as aligned, straight reinforcing elements. In contrast to NCs, however, the aforementioned assumptions hold true for conventional composites. A common method for compensating for this disparity is to combine the basic form of the rule of the mixture with atomistic molecular dynamics (MD) simulations. To this end, the material properties of the carbon nanotube-reinforced (CNTR) PNC will be calculated using MD, and some efficiency coefficients will be added to the mixture formula. The efficiency coefficients will be determined by configuring the predicted MD-assisted response with the expressions derived from the rule of the mixture. After that, the efficiency coefficients will be implemented for further estimations to reduce the disparity between the theoretical and atomistic modeling-derived material properties.

Using the following definitions [1], Young's and shear modules of the CNTR PNC material can be calculated using the aforementioned instructions:

$$E_{11} = \eta_1 V_r E_{11}^r + V_m E_m \tag{1.1}$$

$$\frac{\eta_2}{E_{22}} = \frac{V_r}{E_{22}^r} + \frac{V_m}{E_m} \tag{1.2}$$

$$\frac{\eta_3}{G_{12}} = \frac{V_r}{G_{12}^r} + \frac{V_m}{G_m} \tag{1.3}$$

where E_{11} and E_{22} are the longitudinal and transverse Young's modules of the CNTR PNC and G_{12} denotes the shear modulus of the same material in the plane constructed from the intersection of longitudinal and transverse directions. In the above relations, the η_i's ($i = 1, 2, 3$) are the efficiency coefficients that must be determined from the comparison of theoretical and MD answers. Also, the terms V_r and V_m are the volume fractions of CNT and matrix, respectively. Clearly, the sum of the aforementioned volume fractions must equal one. On the basis of this method, the shear modulus in the plane between longitudinal and transverse orientations will be calculated, and the shear modulus in all other planes will be regarded as half of the value obtained for the G_{12} [1].

The CNTR PNC material's mass density, Poisson's ratio between longitudinal and transverse dimensions, and longitudinal and transverse CTEs can be determined using the following formulas:

$$\rho = \rho_r V_r + \rho_m V_m \tag{1.4}$$

$$\nu_{12} = \nu_{12}^r V_r + \nu_m V_m \tag{1.5}$$

$$\alpha_{11} = \alpha_{11}^r V_r + \alpha_m V_m \tag{1.6}$$

$$\alpha_{22} = \left(1+v_{12}^r\right)\alpha_{22}^r V_r + \left(1+v_m\right)\alpha_m V_m - v_{12}\alpha_{11} \tag{1.7}$$

in which ρ_r and v_{12}^r are density and Poisson's ratio of the CNT, respectively. The longitudinal and transverse CTEs of the CNT are respectively shown as α_{11}^r and α_{22}^r. Besides, the density, Poisson's ratio, and CTE of the matrix are shown as ρ_m, v_m, and α_m, respectively. It must be pointed out that the volume fraction of the CNTs in Eqs. (1.1)–(1.7) can be computed according to the following relation:

$$V_r = \frac{W_r}{W_r + \dfrac{\rho_r}{\rho_m}\left(1-W_r\right)} \tag{1.8}$$

where W_r is the mass fraction of the CNTs.

1.3.2 Halpin–Tsai Method

Using the Halpin–Tsai method, the material properties of the CNTR PNC will be determined here. On the basis of this method, a parameter will be used to capture the effects of the inclusion's geometrical characteristics and loading conditions on the estimated property values of any desired form of composite. This enables the user to analyze various categories of materials by employing the aforementioned parameter's appropriate expression. Similar to the rule of the mixture, it is presumed that the bonding conditions between nanofiller and matrix are optimal. On the basis of these assumptions and the geometry of the reinforcing CNTs, Young's modulus of a CNTR PNC can be derived from the following definition [1]:

$$E = \left[\frac{5}{8}\left(\frac{1+2\xi V_r}{1-\xi V_r}\right) + \frac{3}{8}\left(\frac{1+2\{l_r/d_r\}\zeta V_r}{1-\zeta V_r}\right)\right]E_m \tag{1.9}$$

in which

$$\xi = \frac{\left(E_r/E_m\right)-\left(d_r/4t_r\right)}{\left(E_r/E_m\right)+\left(d_r/2t_r\right)} \tag{1.10}$$

$$\zeta = \frac{\left(E_r/E_m\right)-\left(d_r/4t_r\right)}{\left(E_r/E_m\right)+\left(l_r/2t_r\right)} \tag{1.11}$$

In Eqs. (1.9)–(1.11), the length, outer diameter, wall thickness, and Young's modulus of the CNTs are shown as l_r, d_r, t_r, and E_r, respectively. Obviously, E_m stands for Young's modulus of the matrix. It is noteworthy that ξ and ζ are dimensionless parameters provided to be in charge of controlling the effect of using nanofillers for reinforcement. It is interesting to mention that the expression of these dimensionless terms can be varied if the nanofiller is changed and another element is employed to reinforce the polymer. This is the way that the impact of different types of reinforcements on the mechanical properties of the under-observation inhomogeneous material.

In Eq. (1.9), the volume fraction of the CNTs (V_r) can be calculated by using the definition previously presented in Eq. (1.8). It is worth mentioning that the equivalent mass density and Poisson's ratio of the CNTR PNC can be attained with the aid of the rule of the mixture as formerly discussed in Section 1.1.1 (see Eqs. (1.4) and (1.5)). However, it must be noticed that the obtained Poisson's ratio will be the general Poisson's ratio of the CNTR PNC material because of the fact that the whole nanocomposite (NC) is assumed to be an isotropic linearly elastic solid in this type of homogenization. Furthermore, the CTE of the CNTR PNC material can be obtained by means of the Halpin–Tsai method in the following form [1]:

$$ \alpha = \frac{1}{2}\left[\left(\frac{\alpha_r E_r V_r + \alpha_m E_m V_m}{E_r V_r + E_m V_m} \right)(1-v) + (1+v_m)\alpha_m V_m + (1+v_r)\alpha_r V_r \right] \quad (1.12) $$

where v is the Poisson's ratio of the CNTR PNC. Also, the CTEs corresponding to matrix and CNT are shown as α_m and α_r, respectively [1].

Again, it can be demonstrated that the modulus of the polymer can be significantly increased by the addition of the reinforcing phase. The cause of this problem is described in Section 1.1.1. In conjunction to this pattern, the figure reveals that incorporating longer CNTs to reinforce the polymer can result in an appreciable increase in the nanomaterial's rigidity. Despite the fact that the aforementioned issue tends to promote its impact on the modulus of the CNTR PNC as an entirely increasing one, it must be noted that the use of thin nanofillers can raise the probability of the existence of nonstraight shapes in CNTs that cannot be accounted for by the Halpin–Tsai method.

1.3.3   MODIFIED HALPIN–TSAI METHOD

Using the Halpin–Tsai model, Section 1.3.2 determined the material properties of PNC materials reinforced with straight-shaped CNTs. Although the geometrical characteristics of the reinforcing nanofiller are considered in the preceding method, the existence of a curve in the CNTs' structure was disregarded. The necessity of incorporating the effect of the waviness effect on the material properties of CNTR PNCs can be better perceived when it is considered that chemical vapor deposition (CVD) is the most widely used synthesis technique for the production of CNTs. In this technique, the likelihood of CNTs containing defects is extremely high. Undoubtedly, the existence of curves in the CNT's structure is one of the most likely defects to manifest.

According to the modified Halpin–Tsai method, the CNTR PNC will be assumed to be a linearly elastic isotropic solid in order to determine its properties.

In this method, the impact of a 2D or 3D random distribution of reinforcing elements in the PNC will also be considered. In order to accomplish this, an alignment factor will be introduced and multiplied by the modulus of the CNTs prior to the homogenization procedure. Notably, the present model is valid for analyzing PNCs with reinforcement contents of less than 2 weight percentage. The reason is the nanomaterial's nonlinear behavior at higher CNT concentrations, which is caused by the presence of CNT agglomerates in such instances.

The waviness coefficient is known, so the Young's modulus of the PNC reinforced with CNTs can be calculated using the modified Halpin–Tsai equation [1] shown below.

$$E = \frac{1 + C\eta V_r}{1 - \eta V_r} E_m \tag{1.13}$$

where

$$\eta = \frac{\alpha C_w \left(E_r / E_m \right) - 1}{\alpha C_w \left(E_r / E_m \right) + C} \tag{1.14}$$

in which α stands for the orientation factor which is provided to account for the type of the random distribution of the nanofillers in the PNC. Commonly, the orientation factor of $\alpha = 1/3$ is employed for the cases where the length of the CNTs is greater than the dimensions of the media and 2D case has happened. Whereas, $\alpha = 1/6$ can be used whenever the length of the CNTs is smaller than the geometrical dimensions of the under-observation material. Therefore, implementation of the latter case for the analyses of the nanocomposite structures that possess macroscopic dimensions can be a reasonable choice. In addition, C is a coefficient that is utilized in order to account for the geometrical specifics of the CNT and is equal to two times the length-to-diameter ratio of the CNT ($C = 2l_r/d_r$).

1.3.4 MORI–TANAKA METHOD

Utilizing the well-known Mori–Tanaka method, the equivalent properties of CNTR PNCs will be derived here. In this technique, it is supposed that the CNTs utilized reinforce the polymer to have a perfect linear shape. In addition, the case in which CNTs are perfectly aligned will be considered. In accordance with the Mori–Tanaka approach, any preferred addition in the PNC will be accompanied by matrix that may be attracted by tension or strain. The above stresses and strains must represent the average of the stresses and strains applied to the matrix. By assuming that all inclusions are identical and applying the Mori–Tanaka homogenization algorithm, it is possible to formulate the following equation for the elasticity tensor of PNCs reinforced with oriented, straight CNTs [1]:

$$\mathbf{C} = \left(V_m \mathbf{C}_m + V_r \mathbf{C}_r : \mathbf{A} \right) : \left(V_m \mathbf{I} + V_r \mathbf{A} \right)^{-1} \tag{1.15}$$

It is important to emphasize the fact that the tensors of the second and fourth orders are responsible for the boldface indications. In this instance, the tensors of flexibility of the matrix and inclusion (i.e., CNTs) are represented by $\mathbf{C}_m$ and $\mathbf{C}_r$, respectively.

Besides, $\mathbf{I}$ stands for the fourth-order identity tensor. In Eq. (1.15), the tensor $\mathbf{A}$, also known as dilute tensor, is provided to relate the average strains of the matrix and CNT together as below [1]:

$$\varepsilon_r = \mathbf{A} : \varepsilon_m \tag{1.16}$$

where

$$\mathbf{A} = \left[\mathbf{I} + \mathbf{S} : \mathbf{C}_m^{-1} : \left(\mathbf{C}_r - \mathbf{C}_m\right)\right]^{-1} \tag{1.17}$$

in which $\mathbf{S}$ is the Eshelby tensor. The nonzero arrays of the Eshelby tensor for the case of analyzing aligned, straight CNTs are in the following form [1]:

$$S_{1111} = S_{3333} = \frac{5 - 4v_m}{8\left(1 - v_m\right)}, \quad S_{1122} = S_{3322} = \frac{v_m}{2\left(1 - v_m\right)},$$

$$S_{1133} = S_{3311} = \frac{4v_m - 1}{8\left(1 - v_m\right)}, \quad S_{1313} = \frac{3 - 4v_m}{8\left(1 - v_m\right)}, \quad S_{2323} = S_{1212} = 0.25, \tag{1.18}$$

Now what needs to be done is to demonstrate the stress–strain relationship of the CNTR PNC. If we assume that the polymeric matrix is a linearly isotropic solid, with E_m and v_m, representing its Young's modulus and Poisson's ratio, respectively, and if we further assume that the PNC contains aligned, straight CNTs, then the relation between stress and strain can be presented in the following form [1]:

$$\begin{Bmatrix} \sigma_{11} \\ \sigma_{22} \\ \sigma_{33} \\ \sigma_{23} \\ \sigma_{13} \\ \sigma_{12} \end{Bmatrix} = \begin{bmatrix} k+m & l & k-m & 0 & 0 & 0 \\ l & n & l & 0 & 0 & 0 \\ k-m & l & k+m & 0 & 0 & 0 \\ 0 & 0 & 0 & p & 0 & 0 \\ 0 & 0 & 0 & 0 & m & 0 \\ 0 & 0 & 0 & 0 & 0 & p \end{bmatrix} \begin{Bmatrix} \varepsilon_{11} \\ \varepsilon_{22} \\ \varepsilon_{33} \\ 2\varepsilon_{23} \\ 2\varepsilon_{13} \\ 2\varepsilon_{12} \end{Bmatrix} \tag{1.19}$$

In the above identity, k, n, and l are plane-strain bulk modulus perpendicular to the direction of the CNTs' alignment, uniaxial tensile modulus parallel with the direction of the CNTs' alignment, and associated cross modulus, respectively. Also, the shear modules in the planes normal and parallel to the direction of the

CNTs' alignment are respectively shown with m and p. It is noteworthy that the above terms are called the elastic constants of Hill.

To calculate Hill's elastic constants, the nonzero parts of the Eshelby tensor, which were presented in Eq. (2.29), must be implanted into the description of the Dilute tensor in order to identify the nonzero arrays of this tensor. This can be done by inserting the nonzero elements of the Eshelby tensor into the description of the dilute tensor. When you make the aforementioned change, you will arrive at [1]:

$$A_{1111} = A_{3333} = -\frac{a_3}{a_1 a_2}, \quad A_{1133} = A_{3311} = \frac{a_4}{a_1 a_2},$$

$$A_{1122} = A_{3322} = \frac{l_r \left(1 - v_m - 2v_m^2\right) - E_m v_m}{a_1}, \quad A_{2222} = 1, \tag{1.20}$$

$$A_{2323} = A_{1212} = \frac{E_m}{E_m + 2p_r \left(1 + v_m\right)}, \quad A_{1313} = \frac{2E_m \left(1 - v_m\right)}{a_2}$$

where

$$a_1 = \left(2v_m - 1\right)\left[E_m + 2k_r \left(1 + v_m\right)\right],$$

$$a_2 = E_m + 2m_r \left(3 - v_m - 4v_m^2\right),$$

$$a_3 = E_m \left(1 - v_m\right)\left\{E_m \left(3 - 4v_m\right) + 2\left(1 + v_m\right)\left[m_r \left(3 - 4v_m\right) + k_r \left(2 - 4v_m\right)\right]\right\}, \tag{1.21}$$

$$a_4 = E_m \left(1 - v_m\right)\left\{E_m \left(1 - 4v_m\right) + 2\left(1 + v_m\right)\left[m_r \left(3 - 4v_m\right) + k_r \left(2 - 4v_m\right)\right]\right\}$$

in which k_r, l_r, m_r, n_r, and p_r are Hill's elastic constants of the CNTs. Now, by using Eqs. (1.20) and (1.15), the Hill's elastic constants of the PNC can be extracted and written in the below form:

$$k = \frac{E_m \left\{E_m V_m + 2k_r \left(1 + v_m\right)\left[1 + V_r \left(1 - 2v_m\right)\right]\right\}}{2\left(1 + v_m\right)\left[E_m \left(1 + V_r - 2v_m\right) + 2V_m k_r \left(1 - v_m - 2v_m^2\right)\right]} \tag{1.22}$$

$$l = \frac{E_m \left\{v_m V_m \left[E_m + 2k_r \left(1 + v_m\right)\right] + 2l_r V_r \left(1 - v_m^2\right)\right\}}{\left(1 + v_m\right)\left[2k_r V_m \left(1 - v_m - 2v_m^2\right) + E_m \left(1 + V_r - 2v_m\right)\right]} \tag{1.23}$$

$$m = \frac{E_m \left[E_m V_m + 2m_r \left(1 + v_m\right)\left(3 + V_r - 4v_m\right)\right]}{2\left(1 + v_m\right)\left\{E_m \left[V_m + 4V_r \left(1 - v_m\right)\right] + 2V_m m_r \left(3 - v_m - 4v_m^2\right)\right\}} \tag{1.24}$$

$$n = \frac{E_m{}^2 V_m \left(1 + V_r - V_m v_m\right) + 2 V_r V_m \left(k_r n_r - l_r{}^2\right)\left(1 + v_m\right)^2 \left(1 - 2 v_m\right)}{\left(1 + v_m\right)\left\{2 V_m k_r \left(1 - v_m - 2 v_m{}^2\right) + E_m \left(1 + V_r - 2 v_m\right)\right\}}$$
$$+ \frac{E_m \left[2 V_m{}^2 k_r \left(1 - v_m\right) + V_r n_r \left(1 - 2 v_m + V_r\right) - 4 V_m l_r v_m\right]}{2 V_m k_r \left(1 - v_m - 2 v_m{}^2\right) + E_m \left(1 + V_r - 2 v_m\right)} \tag{1.25}$$

$$p = \frac{E_m \left[E_m V_m + 2 \left(1 + V_r\right) p_r \left(1 + v_m\right)\right]}{2 \left(1 + v_m\right)\left[E_m \left(1 + V_r\right) + 2 V_m p_r \left(1 + v_m\right)\right]} \tag{1.26}$$

By substituting for Hill's elastic constants of the CNTR PNC from Eqs. (1.22)–(1.26) in Eq. (1.19), Young's modules of the PNC in both the longitudinal and transverse directions can be gathered as below:

$$E_L = n - \frac{l^2}{k}, \quad E_T = \frac{4m\left(kn - l^2\right)}{kn - l^2 + mn} \tag{1.27}$$

Using the material properties of the polymer matrix introduced in Section 1.1.2 incorporated with Hill's elastic constants of the single-walled carbon nanotubes (SWCNT) (10,10) tabulated in Table 1.1, the longitudinal and transverse Young's modules of the CNTR PNC can be monitored.

1.3.5   ESHELBY–MORI–TANAKA METHOD

Through the utilization of the Eshelby–Mori–Tanaka technique, the purpose of the current section is to investigate the spectacular influence that the aggregation of the strengthening CNTs has on the material characteristics of the CNTR PNC materials. It is important to keep in mind that the foundation of this model is built on the presumption that the nanofillers for reinforcement are in the form of straight-shaped CNTs. Therefore, in the framework of the Eshelby–Mori–Tanaka algorithm, the waviness issue, which was earlier addressed in Section 1.1.3, will be disregarded as irrelevant. With this method, the CNTs that are accessible in the PNC will be segmented into two separate categories. The first group is dedicated

TABLE 1.1
The Values of Hill's Elastic Constants of the SWCNT

k_r [GPa]	l_r [GPa]	m_r [GPa]	n_r [GPa]	p_r [GPa]
271	88	17	1089	442

to those of the CNTs that are distributed in the polymer matrix, and the second group matches with those that are accumulated in randomized inclusions inside the PNC to create CNT agglomerates. Both groups are considered to be part of the CNTs. Now, assuming that the overall volume of the CNTs is equal to, the mathematical reduction that is presented below can be taken into consideration [1]:

$$W_r = W_r^{in} + W_r^m \tag{1.28}$$

In the above identity, W_r^{in} and W_r^m are the volumes of the CNTs entangled in the inclusions and dispersed in the matrix, respectively. By considering the fact that the total volume of the structure can be obtained by addition of the volume of the matrix (W_m) to the volume of the reinforcing phase, one can write [1]:

$$W = W_r + W_m \tag{1.29}$$

If both sides of the above identity are divided by the left-hand side of it, the volume fractions of the reinforcing and host phases will be obtained in the following form:

$$V_r = \frac{W_r}{W}, \\ V_m = \frac{W_m}{W} \tag{1.30}$$

The scenario of having a consistent number for the volume percentage of the CNTs will be evaluated in the majority of the studies that are being conducted. A more practical analysis of the agglomeration phenomenon can be achieved, however, by taking into account the progressively changing volume percentage of the strengthening phase. For example, the volume proportion of the CNTs present in a PNC can be represented using the following expression [1, 2]:

$$V_r(z) = \left[\frac{\rho_r}{w_r \rho_m} - \frac{\rho_r}{\rho_m} + 1 \right]^{-1} F(z) \tag{1.31}$$

where w_r denotes the mass fraction of the CNTs and can be defined as below:

$$w_r = \frac{M_r}{M_r + M_m} \tag{1.32}$$

In the above definition, M_r and M_m are the mass of the reinforcing phase (CNTs) and matrix, respectively. In Eq. (1.31), $F(z)$ is a geometry-dependent function that is provided to govern the continuous variation of the volume fraction across the thickness direction. Generally, two types of functions will be utilized in the open literature. The difference between these models is about the location of the polymeric matrix and CNT agglomerates through the thickness direction. In one of them, the matrix will be presumed to be at the bottom surface of the material and the top surface will be dedicated to the reinforcing phase. Meanwhile, the position of the aforesaid phases will be substituted with each other in the other type of modeling. Based on the above explanations, the below expressions can be assigned to $F(z)$ [1]:

According to the preceding description, M_r and M_m represent, respectively, the mass values of the strengthening phase (CNTs) and the matrix. In Eq. (2.42), $F(z)$ is a function that is geometry dependent and supplied to control the variability of the volume fraction across the thickness direction. This variation is governed by the fact that the function is provided. In most cases, the accessible literature will make use of one of two distinct categories of functions. The position of the polymeric matrix and CNT agglomerates along the thickness direction is what differentiates these two versions from one another. In one of them, it is going to be assumed that the matrix is on the underside of the substance, and the upper surface is going to be reserved for the strengthening phase. In the meantime, the location of the aforementioned stages will be interchanged with one another in the modeling approach that uses the other type. On the basis of the justifications provided above, the following expressions can be ascribed to $F(z)$ [1]:

$$F(z) = \begin{cases} \left(\dfrac{1}{2} + \dfrac{z}{h} \right)^p \\ \left(\dfrac{1}{2} - \dfrac{z}{h} \right)^p \end{cases} \tag{1.33}$$

where p is the gradient index, which is responsible for determining the dispersion of the nanofillers along the thickness direction and is represented by the symbol. In the event that the polymer is located on the lowermost surface, the first variety of the aforementioned options for $F(z)$ must be chosen. It should come as no surprise that the second instance is connected to the scenario in which the polymer matrix is located on the upper surface of the PNC. h is the overall thickness of the substance and is used in Eq. (2.44).

Now, in order to make it feasible to determine the stiffness of the CNTR PNCs despite the presence of consolidated CNTs, two other variables need to be brought into the equation. These parameters, which are also referred to as *agglomeration parameters*, are able to be specified with the assistance of the volume of the accessible inclusions and the volume of the CNTs contained within the inclusions in the following manner:

$$\mu = \frac{W_{in}}{W},$$
$$\eta = \frac{W_r^{in}}{W_r} \tag{1.34}$$

According to the previous description, the volume fraction of the inclusions (clusters) is denoted by μ, and the volume fraction of the CNTs contained within the inclusions is denoted by η. The respective values of the agglomeration parameters allow for the possibility of three different situations, each of which will be addressed in more detail in the following paragraphs.

Combining Eqs. (2.41) and (2.45) and performing various mathematical operations on the combined data will result in the following expressions being derived:

$$\frac{W_r^{in}}{W_{in}} = \frac{V_r \eta}{\mu} \tag{1.35}$$

$$\frac{W_r^{m}}{W - W_{in}} = \frac{V_r (1-\eta)}{1-\mu} \tag{1.36}$$

The following relationships can be derived for the bulk and shear modules of the areas where the strengthening CNTs are inside the inserts using the well-known Eshelby–Mori–Tanaka [1] method:

$$K_{in}(z) = K_m + \frac{V_r \eta (\delta_r - 3K_m \alpha_r)}{3(\mu - V_r \eta + V_r \eta \alpha_r)} \tag{1.37}$$

$$G_{in}(z) = G_m + \frac{V_r \eta (\eta_r - 2G_m \beta_r)}{2(\mu - V_r \eta + V_r \eta \beta_r)} \tag{1.38}$$

where K_m and G_m represent the matrix's bulk and shear modules, accordingly. Using the same technique, the bulk and shear modules of the remainder of the PNC can be described as follows [1]:

$$K_{out}(z) = K_m + \frac{V_r (1-\eta)(\delta_r - 3K_m \alpha_r)}{3\left[1 - \mu + V_r (1-\eta)(\alpha_r - 1)\right]} \tag{1.39}$$

$$G_{out}(z) = G_M + \frac{V_r (1-\eta)(\eta_r - 2G_M \beta_r)}{3\left[1 - \mu + V_r (1-\eta)(\beta_r - 1)\right]} \tag{1.40}$$

In Eqs. (1.37)–(1.40), stiff terms α_r, β_r, δ_r, and η_r are in the following form:

$$\alpha_r = \frac{3(K_m + G_m) + k_r + l_r}{3(G_m + k_r)} \tag{1.41}$$

$$\beta_r = \frac{1}{5}\left[\frac{4G_m + 2k_r + l_r}{3(G_m + k_r)} + \frac{4G_m}{G_m + p_r} + \frac{2(G_m[3K_m + G_m] + G_m[3K_m + 7G_m])}{G_m(3K_m + G_m) + m_r(3K_m + 7G_m)}\right]$$

(1.42)

$$\delta_r = \frac{1}{3}\left[n_r + 2l_r + \frac{(2k_r + l_r)(3K_m + G_m - l_r)}{G_m + k_r}\right]$$

(1.43)

$$\eta_r = \frac{1}{5}\left[\begin{array}{c} \dfrac{2}{3}(n_r - l_r) + \dfrac{8G_m p_r}{G_m + p_r} + \dfrac{(2k_r - l_r)(2G_m + l_r)}{3(G_m + k_r)} \\[3mm] + \dfrac{8m_r G_m(3K_m + 4G_m)}{3K_m(m_r + G_m) + G_m(7m_r + G_m)} \end{array}\right]$$

(1.44)

In Eqs. (1.41)–(1.44), k_r, l_r, m_r, n_r, and p_r are the stiff Hill's constants of the reinforcing phase. A list of Hill's constants for the SWCNT (10,10) can be found in Section 1.1.4.

The effective mass and shear modules of the PNC substance can reportedly be displayed in the following manner, as stated in [1]:

$$K(z) = K_{out}\left[1 + \frac{\mu\left(\dfrac{K_{in}}{K_{out}} - 1\right)}{1 + (1 - \mu)\left(\dfrac{K_{in}}{K_{out}} - 1\right)\dfrac{1 + v_{out}}{3(1 - v_{out})}}\right]$$

(1.45)

$$G(z) = G_{out}\left[1 + \frac{\mu\left(\dfrac{G_{in}}{G_{out}} - 1\right)}{1 + (1 - \mu)\left(\dfrac{G_{in}}{G_{out}} - 1\right)\dfrac{8 - 10v_{out}}{15(1 - v_{out})}}\right]$$

(1.46)

where

$$v_{out} = \frac{3K_{out} - 2G_{out}}{6K_{out} + 2G_{out}}$$

(1.47)

The fundamental equations of solid mechanics can be used in this context to derive the following form for expressing Young's modulus and Poisson's ratio of the CNTR PNC material:

$$E(z) = \frac{9K(z)G(z)}{3K(z)+G(z)} \tag{1.48}$$

$$v(z) = \frac{3K(z)-2G(z)}{6K(z)+2G(z)} \tag{1.49}$$

For the purpose of determining the relative density of the CNTR PNC substance, one may make use of the traditional formula of the mixture, which states as follows:

$$\rho(z) = (\rho_r - \rho_m)V_r + \rho_m \tag{1.50}$$

1.3.6   3D Mori–Tanaka Method

In this subsection, we will collect the material characteristics of the PNCs that were strengthened via unidirectionally positioned CNTs by taking into consideration the matrix's random orientation of the CNTs. For the purposes of this form of simulation, the complete PNC substance will be presumed to operate in the same way as objects that are isotropic in the horizontal direction. It is important to observe that all of the strengthening CNTs will be regarded to be straight, slender nanofillers that have high ratios of length to thickness. This number will be shown to have a significant influence on the corresponding characteristics of the PNC, as will be shown in the following demonstration. In order to discover the characteristics of the nanomaterial under examination, there are five distinguishing variables that need to be identified first. This is because the nanomaterial possesses a transversely isotropic character.

These variables are longitudinal Young's modulus E_{11}, transverse Young's modulus E_{22}, in-plane shear modulus G_{12}, out-of-plane shear modulus G_{23}, and plane-strain bulk modulus K_{23}. Based upon the 3D Mori–Tanaka method [1], the above variants can be formulated in the following form [1]:

$$\frac{E_{11}}{E_m} = \frac{1}{1 + V_r \left(A_1 + 2v_m A_2 \right)/A} \tag{1.51}$$

$$\frac{E_{22}}{E_m} = \frac{1}{1 + V_{CNT}\left[\left(1-v_m\right)A_4 - 2v_m A_3 + \left(1+v_m\right)A_5 A \right]/2A} \tag{1.52}$$

$$\frac{G_{12}}{G_m} = 1 + \frac{V_r}{G_m/\Delta G + 2\left(1-V_r\right)S_{1212}} \tag{1.53}$$

$$\frac{G_{23}}{G_m} = 1 + \frac{V_r}{G_m / \Delta G + 2(1 - V_r) S_{2323}} \qquad (1.54)$$

$$\frac{K_{23}}{\overline{K}_0} = \frac{(1 + v_m)(1 - 2v_m)}{1 - v_m (1 + 2v_{12}) + V_r \left\{ 2(v_{12} - v_m) A_3 + \left[1 - v_m (1 + 2v_{12}) \right] A_4 \right\} / A} \qquad (1.55)$$

where V_r is the volume fraction of reinforcing phase and can be calculated using Eq. (1.8). In addition, E_m, G_m, and v_m are respectively Young's modulus, shear modulus, and Poisson's ratio of the matrix. ΔG stands for the difference between the shear modules of the CNT and matrix ($\Delta G = G_r - G_m$). Also, v_{12} is the Poisson's ratio of the matrix in the 1–2 plane, which can be assumed to be identical with the matrix's Poisson's ratio due to the isotropic nature of the matrix. The term $\overline{K}_0$ is the bulk modulus of the matrix under plane–strain condition and can be considered as $\overline{K}_0 = \lambda_m + G_m$. The A_i's utilized in Eqs. (1.51)–(1.55) can be defined in the below form [1]:

$$\begin{aligned}
A_1 &= D_1 (B_4 + B_5) - 2B_2, \\
A_2 &= (1 + D_1) B_2 - (B_4 + B_5), \\
A_3 &= B_1 - D_1 B_3, \\
A_4 &= (1 + D_1) B_1 - 2B_3, \\
A_5 &= \frac{1 - D_1}{B_4 - B_5}, \\
A &= 2B_2 B_3 - B_1 (B_4 + B_5)
\end{aligned} \qquad (1.56)$$

in which

$$\begin{aligned}
B_1 &= V_r D_1 + D_2 + (1 - V_r)(D_1 S_{1111} + 2S_{2211}), \\
B_2 &= V_r + D_3 + (1 - V_r)(D_1 S_{1122} + S_{2222} + S_{2233}), \\
B_3 &= V_r + D_3 + (1 - V_r)\left[S_{1111} + (1 + D_1) S_{2211} \right], \\
B_4 &= V_r D_1 + D_2 + (1 - V_r)(S_{1122} + D_1 S_{2222} + S_{2233}), \\
B_5 &= V_r + D_3 + (1 - V_r)(S_{1122} + S_{2222} + D_1 S_{2233})
\end{aligned} \qquad (1.57)$$

In the above relations, S_{ijkl}'s indicate the components of the fourth-order Eshelby tensor and can be computed using the below definitions [1]:

$$S_{1111} = \frac{1}{2(1 - v_m)} \left[1 - 2v_m + \frac{3\alpha^2 - 1}{\alpha^2 - 1} - g\left(1 - 2\mu_m + \frac{3\alpha^2}{\alpha^2 - 1} \right) \right] \qquad (1.58)$$

$$S_{2222} = S_{3333} = \frac{3}{8(1-v_m)} \frac{\alpha^2}{\alpha^2-1} + \frac{1}{4(1-v_m)}\left(1-2v_m - \frac{9}{4(\alpha^2-1)}\right)g \quad (1.59)$$

$$S_{2233} = S_{3322} = \frac{1}{4(1-v_m)}\left[\frac{\alpha^2}{2(\alpha^2-1)} - \left(1-2v_m - \frac{3}{4(\alpha^2-1)}\right)g\right] \quad (1.60)$$

$$S_{2211} = S_{3311} = -\frac{1}{2(1-v_m)}\frac{\alpha^2}{\alpha^2-1} + \frac{1}{4(1-v_m)}\left[\frac{3\alpha^2}{\alpha^2-1} - (1-2v_m)\right]g \quad (1.61)$$

$$S_{1122} = S_{1133} = -\frac{1}{2(1-v_m)}\left(1-2v_m+\frac{1}{\alpha^2-1}\right) + \frac{1}{2(1-v_m)}\left(1-2v_m+\frac{3}{2(\alpha^2-1)}\right)g$$

$$(1.62)$$

$$S_{2323} = S_{3232} = \frac{1}{4(1-v_m)}\left[\frac{\alpha^2}{2(\alpha^2-1)} + \left(1-2v_m - \frac{3}{4(\alpha^2-1)}\right)g\right] \quad (1.63)$$

$$S_{1212} = S_{1313} = \frac{1}{4(1-v_m)}\left[1-2v_m - \frac{\alpha^2+1}{\alpha^2-1} - 0.5\left[1-2v_m - \frac{3(\alpha^2+1)}{\alpha^2-1}\right]g\right] \quad (1.64)$$

where

$$g = \frac{\alpha^2\sqrt{\alpha^2-1} - \alpha\cosh^{-1}\alpha}{(\alpha^2-1)\sqrt{\alpha^2-1}} \quad (1.65)$$

In Eqs. (1.58)–(1.65), α is the length-to-diameter ratio of the CNTs. In the definitions provided in Eqs. (1.56) and (1.57), D_i's can be obtained by:

$$D_1 = 1 + \frac{2\Delta G}{\Delta\lambda},$$

$$D_2 = \frac{\lambda_m + 2G_m}{\Delta\lambda}, \quad (1.66)$$

$$D_3 = \frac{\lambda_m}{\Delta\lambda}$$

in which λ_m is the Lame's constant of the matrix and $\Delta\lambda$ is the difference between the Lame's constants of the CNT and matrix ($\Delta\lambda = \lambda_r - \lambda_m$). The Lame's constants corresponding with matrix and nanotube can be defined as follows:

$$\lambda_m = \frac{v_m E_m}{(1-2v_m)(1+v_m)}, \quad \lambda_r = \frac{v_r E_r}{(1-2v_r)(1+v_r)} \tag{1.67}$$

Using the relation between Young's modulus, shear modulus, and Poisson's ratio of the linearly elastic isotropic solids, the shear modules of the matrix and CNT can be expressed in the following form:

$$G_m = \frac{E_m}{2(1+v_m)}, \quad G_r = \frac{E_r}{2(1+v_r)} \tag{1.68}$$

Now, the equivalent bulk and shear modules of the CNTR PNC material can be obtained as below [1]:

$$K = \frac{E_{11} + 4(1+v_{12})^2 K_{23}}{9} \tag{1.69}$$

$$G = \frac{E_{11} + (1-2v_{12})^2 K_{23} + 6(G_{12} + G_{23})}{15} \tag{1.70}$$

With the aid of the existing relationships between shear and bulk modules of a desired material with its Young's modulus and Poisson's ratio, one can write:

$$E = \frac{9KG}{3K+G} \tag{1.71}$$

$$v = \frac{E}{2G} - 1 \tag{1.72}$$

It is worth regarding that the effective mass density of the CNTR PNC can be attained by the means of the so-called rule of the mixture (see Eq. (1.50)).

1.3.7 HOMOGENIZATION OF MULTISCALE NANOCOMPOSITES

Here, the homogenization method will be completed by introducing the equivalent properties of PNC materials reinforced with nanoscale CNTs and macrofibers. In Section 1.1, numerous homogenization techniques for the purpose of determining the material properties of CNTR PNCs were discussed in depth. The properties obtained in the Section 1.3.6 will be considered as input values in this segment. In other words, the characteristics of the CNTR PNC will serve as the material

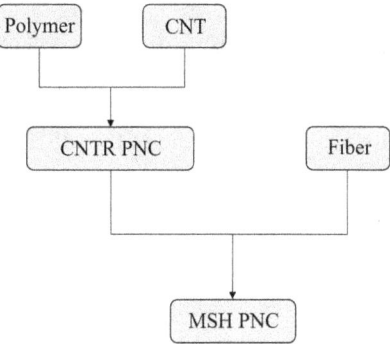

FIGURE 1.1 Hierarchical stages diagrammed for extracting MSH PNC material characteristics. The initial step is to collect data on the characteristics of CNTR PNC. Then, the
calculated values will serve as inputs for the second stage, which will result in the recovery
of the MSH PNC's equivalence characteristics.

properties of the matrix that will be strengthened by the addition of macrofibers.
Figure 1.1 illustrates the sequential phases of homogenization [1].

The "NCM" subscript and superscript will be used throughout the rest of this
section to denote the various material characteristics of the CNTR PNC. In addition, the structural characteristics of the multi-scale hybrid (MSH) PNC substance
will be discussed for the most general scenario. To put it another way, a formulation that can be used to calculate the CTE of the composite nanomaterial will be
supplied. If, on the other hand, determining the CTE is the goal, it is important to
keep in mind that the modified Halpin–Tsai, Mori–Tanaka, Eshelby–Mori–
Tanaka, and 3D Mori–Tanaka algorithms cannot be utilized.

Following the determination of the normal and shear stiffness of the MSH
PNC, the following equations need to be employed in order to deduce the longitudinal Young's modulus, the transverse Young's modulus, and the shear modulus
of the MSH nanomaterial [1]:

$$E_{11} = V_F E_{11}^F + V_{NCM} E_{11}^{NCM} \tag{1.73}$$

$$\frac{1}{E_{22}} = \frac{1}{E_{22}^F} + \frac{V_{NCM}}{E_{22}^{NCM}} - V_F V_{NCM} - \frac{v_F{}^2 E_{22}^{NCM} / E_{22}^F + v_{NCM}{}^2 E_{22}^F / E_{22}^{NCM} - 2v_F v_{NCM}}{V_F E_{22}^F + V_{NCM} E_{22}^{NCM}} \tag{1.74}$$

$$\frac{1}{G_{12}} = \frac{V_F}{G_{12}^F} + \frac{V_{NCM}}{G_{12}^{NCM}} \tag{1.75}$$

The longitudinal and transverse Young's modules of the CNTR material (E_{11}^{NCM}
and E_{22}^{NCM}) must be considered equal to the obtained Young's modulus in situations where the properties obtained indicate isotropic behavior for the CNTR
PNC. This declaration must be made if properties promote isotropic behavior.

In an analogous manner, when conditions such as these exist, the in-plane shear modulus of the CNTR PNC must be considered equivalent to the value obtained for the shear modulus.

On the other hand, the following formula [1] can be used to determine the corresponding mass density and Poisson's ratio of the MSH PNC:

$$\rho = V_F\rho_F + V_{NCM}\rho_{NCM} \tag{1.76}$$

$$\nu_{12} = V_F\nu_F + V_{NCM}\nu_{NCM} \tag{1.77}$$

where ρ_F and ν_F are the mass density and Poisson's ratio of the fiber, respectively. In addition, the CTE of the MSH PNCs can be derived using the below formulas [1]:

$$\alpha_{11} = \frac{V_F E_{11}^F \alpha_{11}^F + V_{NCM} E_{11}^{NCM} \alpha_{11}^{NCM}}{V_F E_{11}^F + V_{NCM} E_{11}^{NCM}} \tag{1.78}$$

$$\alpha_{22} = \left(1+\nu_F\right)V_F\alpha_{22}^F + \left(1+\nu_{NCM}\right)V_{NCM}\alpha_{22}^{NCM} - \nu_{12}\alpha_{11} \tag{1.79}$$

In Eqs. (2.89) and (2.90), α_{11}^F and α_{22}^F represent the longitudinal and transverse CTEs of the reinforcement fiber, respectively. Each of the aforementioned relationships represents the volume fraction of the fibers used as macrosize reinforcement. Clearly, the volume fractions of the components must equal one when added together.

REFERENCES

[1] Ebrahimi, F., & Dabbagh, A. (2022). *Mechanics of multiscale hybrid nanocomposites* (1st ed.), Elsevier.
[2] Ebrahimi, F., & Dabbagh, A. (2020). *Mechanics of Nanocomposites: Homogenization and Analysis* (1st ed.). Boca Raton, FL: CRC Press. https://doi.org/10.1201/9780429316791
[3] Beardmore, P., & Johnson, C.F. (1986). The potential for composites in structural automotive applications. *Composites Science and Technology*, 26(4), 251–281. https://doi.org/10.1016/0266-3538(86)90002-3
[4] Hufenbach, W., et al. (2011). Polypropylene/glass fibre 3D-textile reinforced composites for automotive applications. *Materials & Design*, 32(3), 1468–1476. https://doi.org/10.1016/j.matdes.2010.08.049
[5] Holbery, J., & Houston, D. (2006). Natural-fiber-reinforced polymer composites in automotive applications. *JOM*, 58(11), 80–86. https://doi.org/10.1007/s11837-006-0234-2
[6] Rohatgi, P. (1991). Cast aluminum-matrix composites for automotive applications. *JOM*, 43(4), 10–15. https://doi.org/10.1007/BF03220538
[7] A.P. Mouritz, et al. (2001). Review of advanced composite structures for naval ships and submarines. *Composite Structures*, 53(1), 21–42. https://doi.org/10.1016/S0263-8223(00)00175-6

[8] Le Lay, F., & Gutierrez, J. (1999). Improvement of the fire behaviour of composite materials for naval application. *Polymer Degradation and Stability*, *64*(3), 397–401. https://doi.org/10.1016/S0141-3910(98)00140-2

[9] Rawal, S.P. (2001). Metal-matrix composites for space applications. *JOM*, *53*(4), 14–17. https://doi.org/10.1007/s11837-001-0139-z

[10] Liu, Y., et al. (2014). Shape memory polymers and their composites in aerospace applications: a review. *Smart Materials and Structures*, *23*(2), 023001. https://doi.org/10.1088/0964-1726/23/2/023001

[11] Chen, H., et al. (2016). Thermal conductivity of polymer-based composites: Fundamentals and applications. *Progress in Polymer Science*, *59*, 41–85. https://doi.org/10.1016/j.progpolymsci.2016.03.001

[12] Akdogan, E.K., Allahverdi, M., & Safari, A. (2005). Piezoelectric composites for sensor and actuator applications. *IEEE Transactions on Ultrasonics, Ferroelectrics, and Frequency Control*, *52*(5), 746–775. https://doi.org/10.1109/TUFFC.2005.1503962

[13] Van Den Einde, L., Zhao, L., & Seible, F. (2003). Use of FRP composites in civil structural applications. *Construction and Building Materials*, *17*(6), 389–403. https://doi.org/10.1016/S0950-0618(03)00040-0

[14] Yan, L., Kasal, B., & Huang, L. (2016). A review of recent research on the use of cellulosic fibres, their fibre fabric reinforced cementitious, geo-polymer and polymer composites in civil engineering. *Composites Part B: Engineering*, *92*, 94–132. https://doi.org/10.1016/j.compositesb.2016.02.002

[15] Shahinpoor, M., & Kim, K.J. (2004). Ionic polymer–metal composites: IV. Industrial and medical applications. *Smart materials and Structures*, *14*(1), 197. https://doi.org/10.1088/0964-1726/14/1/020

2 Kinematic Relations for Continuous Systems

2.1 BACKGROUND

In this chapter, the primary emphasis will be placed on the presentation and application of a variety of kinematic theorems to characterize the mechanical properties of continuous systems. These theorems will be used throughout. There have been many different kinds of kinematic theories established up until this point, each of which has a different level of precision in predicting the accurate reaction of the system. The effect that the presence of shear displacement has on the mechanical characteristics of the continua is ignored in traditional theories [1, 2]. The reason for this problem is that it was believed that the shear deformation of the system will be too minor compared to the deformation that will be induced by bending, and because of this, it can be disregarded. In general, such theories can be believed when applied to elements with narrow walls but not when applied to elements with substantial walls. A theory that has been slightly altered is referred to as first-order shear deformation theory (FSDT). In this theory, a constant approximation for shear deformation is recommended all over the thickness of the structure. This gives the theory the ability to provide more accurate responses than the classical model that was introduced earlier [1–3]. In addition, more modern higher-order shear deformation theory (HSDTs) give us the ability to take into consideration the impact of shear displacement on the mechanical reaction of the structures. This is done by approximating the distortion of the cross-section of the continuous system using a shape function [1–3]. A strategy that is based on energy will be used throughout the entirety of this chapter as a replacement for the kinematic theory, which is typically applied in order to characterize the mechanical characteristics of continuous systems. The concept of virtual labor is based on the use of this technique. Calculating the fluctuations of the total tension energy of the system and the work done on it by external pressure, followed by totaling the aforementioned energy functional, and finally setting the solution to zero will suffice when dealing with situations that are stagnant. The following expression is capable of being expressed in the mathematical form as [1–3]:

$$\delta \left(U + V \right) = 0 \qquad (2.1)$$

in which the strain energy and the work done by the strain energy and work done are denoted by U and V correspondingly. Even though the preceding identification only applies to static problems, the principle behind it can easily be expanded to dynamic issues as well. In point of fact, according to this principle, the fluctuation in the total energy of a continuous system must be equal to zero in order to

DOI: 10.1201/9781003270263-2

realize the motion equations of the system. When dealing with changing issues, a different form of energy will manifest itself. In other words, the manifestation of kinetic energy is a direct consequence of the time dependency that exists. As a result, it is possible to deduce that in this scenario, the fluctuation of the total energy of the system in any given time interval that was chosen must be equal to zero. With the assistance of the dynamic form of the virtual work's principle, also known as Hamilton's principle, this physical occurrence is able to be represented in the form of a mathematical equation. The following formula can be used to express this principle [1]:

$$\int_{t_1}^{t_2} \delta \left(U + V - T \right) dt = 0 \qquad (2.2)$$

where T is the kinetic energy of the system. Both the strain energy and the work done by exterior pressure are identical to those that were introduced before.

2.2 KINEMATIC RELATIONS OF BEAMS

2.2.1 EULER–BERNOULLI BEAM THEORY

This section will begin with an introduction to the Euler–Bernoulli beam theory, also referred to as the classical beam theory. According to this theory, the effects of the shear mode on structural displacement will be considered to be negligible enough to the point where they can be ignored. This is because the twisting of the structure can be generally attributed to the displacement of the structure in slender structures (e.g., beams with slenderness ratios that are greater than 50). This is the reason why this is the case. Assuming this to be true, the presumption predicts that there will be no deformation in the cross-section of the beam. In addition, it is going to be assumed that any line that is requested and that is normal to the cross-section of the structure will continue to be normal to the cross-section after the structure has been loaded. Even though this theory seems to defy reasoning, it is important to note that the reaction that can be obtained by using this theory can be relied upon in thin-type beams. This is something that should be kept in mind. In the framework of this theory, the displacement field of the beam can be represented using the following form [1, 2]:

$$u_x \left(x, z, t \right) = u \left(x, t \right) - \left(z - \overline{z} \right) \frac{\partial w \left(x, t \right)}{\partial x},$$
$$u_z \left(x, z, t \right) = w \left(x, t \right) \qquad (2.3)$$

Herein, $u(x, t)$ and $w(x, t)$ represent the longitudinal displacement of the neural axis of the beam and the beam's bending deflection, respectively. In the preceding definition, the effect of time on the displacement field is considered for purposes

of generality. In static problems, however, this term may be omitted. In the preceding definition, $\bar{z}$ denotes the precise position of the neutral axis of the beam
along the direction of its thickness. Consideration of the precise position of the
beam's neutral axis enables simpler decoupling of the problem's motion equations by eliminating the stretching-bending coupling rigidity from the calcula-
tions. Following the below-described definition, it is simple to calculate $\bar{z}$ [1, 2]:

$$
\bar{z} = \frac{\int_{-h/2}^{h/2} zE(z)\,dz}{\int_{-h/2}^{h/2} E(z)\,dz}
\tag{2.4}
$$

It is presumed, for the sake of thoroughness, that Young's modulus of the under-
observation beam is a function of its thickness. This value is denoted by $E(z)$.

The only nonzero element of the strain tensor of Euler–Bernoulli beams can be
represented as follows [1, 2], in accordance with the von Kármán nonlinear strain-
displacement relations. In the text that is to follow, an attempt will be made to
articulate the problems of nonlinear oscillations and post-buckling. Given that
there is a temporal domain involved in the dynamic analyses, it is abundantly
obvious that the technique for deriving the solutions to the aforementioned problems is not the same. However, in order to solve either of these challenges, it is
necessary to determine how the overall tension energy of the beam varies. The
following is a formulation that can be used to describe the fluctuation of the strain
energy for beams that have been described using the Euler–Bernoulli theorem:

$$
\delta U = \int_0^L \left\{ N\left[\frac{\partial \delta u}{\partial x} + \frac{\partial w}{\partial x}\frac{\partial \delta w}{\partial x} \right] - M\frac{\partial^2 \delta w}{\partial x^2} \right\} dx
\tag{2.5}
$$

According to the description given above, N and M refer, respectively, to the normal force and the bending moment. These stress-causing factors can be described
in the following ways:

$$
\begin{Bmatrix} N \\ M \end{Bmatrix} = \int_0^b \int_{-h/2}^{h/2} \begin{Bmatrix} 1 \\ z - \bar{z} \end{Bmatrix} \sigma_{xx}\,dz\,dy
\tag{2.6}
$$

2.2.1.1 Wave Propagation Problem

This section's primary purpose is to provide an analytical solution to the problem
of wave propagation in thin-walled beams. This issue will be investigated in the
linear domain, which must be accounted for. The strain-displacement relations of
the beam must therefore be regenerated. The beam's geometrical characteristics

are comparable to those described in earlier sections. In addition, the wave media are assumed to be resting on a three-parameter visco-Pasternak medium with Winkler (k_W), Pasternak (k_P), and damping (c_d) and to be subjected to a $f(t)$ magnitude of uniform transverse loading.

Consider the following form for Euler–Bernoulli beam displacement fields [1, 2]:

$$u_x\left(x,z,t\right) = -z\,\frac{\partial w\left(x,t\right)}{\partial x},$$

$$u_z\left(x,z,t\right) = w\left(x,t\right)$$

(2.7a)

in which $w(x, t)$ indicates the beam's deflection due to the structure's deformation. In the above definition, the effect of the axial displacement of the beam's mid-axis is not accounted for. Using the continuum mechanics definition of the linearized strain tensor, the only nonzero component of the beam's strain tensor can be expressed as follows:

$$\varepsilon_{xx} = -z\,\frac{\partial^2 w}{\partial x^2}$$

(2.7b)

Therefore, it can be asserted that the axial strain of the beam in this condition is caused by the curvature that develops in the structure's geometry under load. Using the definition of the beam's strain energy variation, one can derive the following relationship:

$$\delta U = -\int_0^L M\,\frac{\partial^2 \delta w}{\partial x^2}\,dx$$

(2.8a)

in which M is the bending moment which can be defined as below:

$$M = \int_0^b \int_{-h/2}^{h/2} z\sigma_{xx}\,dz\,dy$$

(2.8b)

As mentioned above, the beam will be considered to be rested on a viscoelastic foundation. So, the variation of work done on the beam by the surrounding substrate can be formulated in the following form [1, 2]:

$$\delta V = \int_0^L \left[k_W w - k_P\,\frac{\partial^2 w}{\partial x^2} + c_d\,\frac{\partial w}{\partial t} + f\left(t\right) \right] \delta w\,dx$$

(2.9)

Also, the variation of the kinetic energy of the beam can be extracted as follows:

$$\delta T = \int_{-h/2}^{h/2} \left[I_0 \frac{\partial w}{\partial t} \frac{\partial \delta w}{\partial t} + I_2 \frac{\partial^2 w}{\partial x \partial t} \frac{\partial^2 \delta w}{\partial x \partial t} \right] dx \qquad (2.10a)$$

in which

$$\begin{Bmatrix} I_0 \\ I_2 \end{Bmatrix} = \int_0^b \int_{-h/2}^{h/2} \begin{Bmatrix} 1 \\ z^2 \end{Bmatrix} \rho(z) \, dz \, dy \qquad (2.10b)$$

Now, all of the material needed to find the motion equation of the beam are in hand. To complete the derivation procedure, it is enough to substitute Eqs. (2.8), (2.9), and (2.10) into Eq. (2.2). Then, if the nontrivial response of the obtained equation is chosen, the below Euler–Lagrange equation can be written:

$$\frac{\partial^2 M}{\partial x^2} - k_W w + k_P \frac{\partial^2 w}{\partial x^2} - c_d \frac{\partial w}{\partial t} + f(t) = I_0 \frac{\partial^2 w}{\partial t^2} - I_2 \frac{\partial^4 w}{\partial t^2 \partial x^2} \qquad (2.11)$$

Now, the above equation must be rewritten in terms of the displacement field's variables to attain the governing equation of the problem. To this purpose, the 1D Hook's law ($\sigma_{xx} = E(z)\varepsilon_{xx}$) must be considered and integrated over the cross-section area of the beam with regard to the definition of the bending moment expressed in Eq. (2.8b) Therefore, the bending moment can be rewritten as below:

$$M = -D_{xx} \frac{\partial^2 w}{\partial x^2} \qquad (2.12)$$

In the above definition, the cross-sectional bending rigidity of the plate can be defined in the following form:

$$D_{xx} = \int_0^b \int_{-h/2}^{h/2} z^2 E(z) \, dz \, dy \qquad (2.13)$$

Substitution of Eq. (2.12) into Eq. (2.11) gives:

$$D_{xx} \frac{\partial^4 w}{\partial x^4} + k_W w - k_P \frac{\partial^2 w}{\partial x^2} + c_d \frac{\partial w}{\partial t} + I_0 \frac{\partial^2 w}{\partial t^2} - I_2 \frac{\partial^4 w}{\partial t^2 \partial x^2} = f(t) \qquad (2.14)$$

Once the above equation is solved, the wave response of the system will be obtained. In common, an exponential solution will be considered to derive the natural frequency, phase, and group velocities of the scattered waves. The general guidelines of implementation of this method can be found by referring to the authors' previous textbooks [1–3]. Herein, the main concept of the exponential method, that is, the expression of the spatial part of the solution will be kept and a different point of view for solving the problem in the time domain will be introduced. Indeed, the solution of the problem will be considered in the following general form:

$$w(x,t) = W(t)\exp(i\beta x) \tag{2.15}$$

In the above solution, β denotes the wave number and $W(t)$ is considered as the amplitude of the dispersed waves. In this solution, the amplitude will be a function of time. By simple substitution of the above solution in Eq. (2.14), one can reach the below expression:

$$M\ddot{W}(t) + C\dot{W}(t) + KW(t) = f(t) \tag{2.16}$$

where M, C, and K are respectively mass, damping, and stiffness of the under-observation system and can be calculated as below:

$$M = I_0 + I_2\beta^2, \ C = c_d, \ K = D_{xx}\beta^4 + k_W + k_P\beta^2 \tag{2.17}$$

The above formula describes the damped fluctuation of a single degree of freedom (DOF) whose schematic view can be observed in Figure 2.1.

In what follows, the time-dependent response of the preceding problem will be analyzed. In order to achieve this objective, the domain of the problem will be transformed from time to frequency using Laplace's technique. The transferred equation can then be written as follows:

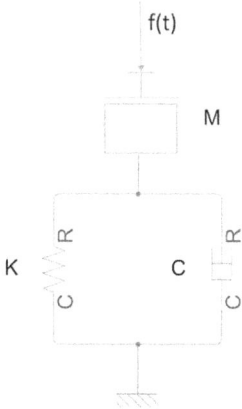

FIGURE 2.1 This schematic depicts a non-conservative oscillating system with a single degree of freedom.

$$\left[Ms^2 + Cs + K\right]\bar{W} = \bar{F} \tag{2.18}$$

In the above equation, $\bar{W}$ and $\bar{F}$ represent the frequency domain–transformed deflection amplitude and force, respectively. After performing algebraic operations on the preceding equation, the transformed deflection can be enhanced. The obtained expression must then be transferred to the time domain using a straightforward Laplace inversion in the region of system convergence. After the preceding stages have been completed, the amplitude of the wave's motion as a function of time can be determined.

To elucidate the preceding procedure, suppose that the deflection amplitude of the wave in the frequency domain is given as follows:

$$\bar{W} = \frac{1}{Ms^2 + Cs + K}\bar{F} \tag{2.19}$$

The amplitude of deflection in the time domain can now be determined. Take the following general form for the frequency domain deflection:

$$\bar{W} = \frac{N(s)}{D(s)} = \frac{a_k s^k + a_{k-1}s^{k-1} + \ldots + a_1 s + a_0}{b_l s^k + b_{l-1}s^{k-1} + \ldots + b_1 s + b_0} \tag{2.20}$$

$N(s)$ and $D(s)$ are polynomial functions with no common zeros. The preceding expression has straightforward type poles. In other words, the zeros of function $D(s)$ are straightforward roots. Therefore, these roots do not attribute a value of "0" to the function's derivative $D(s)$. Typically, these origins may have real or complex values. If the number of real and complex poles are denoted by n_r and n_c, respectively, the expression below can be used to determine the inverse Laplace transformation of $\bar{W}$:

$$w(t) = L^{-1}\bar{W} = \mathrm{Re}\left\{\sum_{i=1}^{n_c}\frac{N(c_i)}{E(c_i)}\exp(c_i t)\right\} + \sum_{j=1}^{n_r}\frac{N(r_j)}{E(r_j)}\exp(r_j t) \tag{2.21}$$

in which c_i and r_j stand for the i-th complex root and j-th real root, respectively. Also, the function $E(s)$ denotes the first derivative of $D(s)$ with respect to s. Following the above method yields the dispersed wave deviation magnitude in time.

A control engineering-assisted structure will now watch for issue stability. First, explain the analogous electrical circuit of the issue. Figure 2.2 shows this circuit. Kirchhoff's circuit law (KCL) for the given electrical system yields the following property [1–3]:

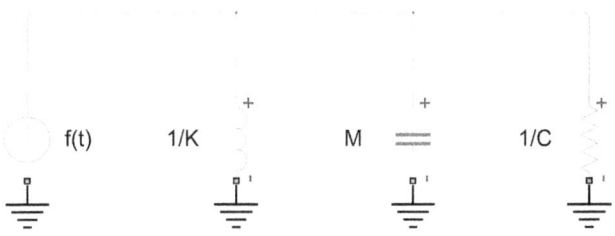

FIGURE 2.2 Schematic illustration of the analogous electrical circuit for the issue shown mechanically in Figure 2.1.

$$\frac{s\overline{W}}{1/Ms} + \frac{s\overline{W}}{s/K} + \frac{s\overline{W}}{1/C} = \overline{F} \tag{2.22}$$

By doing simple mathematical manipulations, Eq. (2.22) can be rewritten in the following form:

$$\frac{\overline{W}}{\overline{F}} = \frac{1}{Ms^2 + Cs + K} \tag{2.23}$$

The obtained relation is indeed a new version of Eq. (2.18), and it gives the transfer function of the system, that is, identical with the ratio between the transformed input and output of the system in the frequency domain. The above transfer function corresponds with the following characteristic equation of a typical second-order system [1–3]:

$$s^2 + 2\xi\omega_n s + \omega_n^2 = 0 \tag{2.24}$$

in which ξ is the damping ratio and ω_n is the natural frequency of the undamped (conservative) system. In order to track the stability of the system in the framework of the well-known stability criteria in the control's open literature, the following variables must be known [1–3]:

$$\omega_d = \omega_n\sqrt{1-\xi^2} \tag{2.25}$$

$$\omega_r = \omega_n\sqrt{1-2\xi^2} \tag{2.26}$$

$$\alpha = \xi\omega_n \tag{2.27}$$

in which ω_d and ω_r are the damped and resonance frequencies of the second-order equation. Also, the term α stands for the damping coefficient of the mentioned system. As the final variable important for us, the percentage of overshoot (PO) of the introduced system can be calculated with the aid of the below formula [1–3]:

$$PO(\%) = \exp\left(-\frac{\xi\pi}{\sqrt{1-\xi^2}}\right) \tag{2.28}$$

Using the preceding definitions and the Bode stability criterion, it is straightforward to monitor the system's stability under any condition.

2.2.1.2 Refined Higher-order Beam Theory

Once the beam has been loaded, it will be presumed for the purposes of the Euler–Bernoulli theory that the cross-section of the beam has not been altered in any way as a result of the loading. The oversimplification that results from making this presupposition makes this theory inadequate for the analysis of components of the thick type. In point of fact, the bending mode–induced deformation contributes a disproportionately large amount of weight to the structure's overall displacement; as a result, this impact only manifests itself in a feeble form in structures with thin walls. A problem of this nature cannot be tolerated in components with thick walls because the dimension relation between the length and the thickness of the construction is relatively low. Because of this deficiency in the traditional theory of beams, thickness-dependent shape functions were incorporated into the theory in order to make it possible to approximate the profile of the distorted cross-section of the beam. This was done in order to compensate for the aforementioned deficiency. As a consequence of this, higher-order theories came into existence so that the shear-induced displacement of the beams could be accounted for in the computations. The rotation of the beam's cross-section will be a new variable in conventional theories, in addition to the previously accessible longitudinal displacement and bending deformation of the structure. This is because the beam's cross-section rotates as it bends. However, in a different category of higher-order theories known as refined HSDTs, the revolution of the cross-section is not taken into account. Instead, the overall displacement of the continuous system is broken up into two portions that are based on the bending and shearing modes. The shape function can also be found in refined-type theories, just like it can be found in conventional HSDTs.

In this part of the book, the improved beam theories will be discussed for time-dependent issues in their broadest possible context. After that, it will be described, with the assistance of this kind of beam theory, how to handle the problems of bending, buckling, and free oscillation associated with the beams. In this section, the previously stated mathematical approximations regarding the geometric variables are going to be maintained. To put it another way, it is going to be believed that the beam has dimensions of length L, width b, and thickness h. It is possible to think of the displacement field of an improved beam theory as having the following shape [1–3]:

$$u_x(x,z,t) = u(x,t) - z\frac{\partial w_b(x,t)}{\partial x} - f(z)\frac{\partial w_s(x,t)}{\partial x},$$

$$u_z(x,z,t) = w_b(x,t) + w_s(x,t)$$

(2.29)

where the longitudinal displacement of the neutral axis is represented by $u(x, t)$ and the bending and shear deflections of the beam are represented by $w_b(x, t)$ and $w_s(x, t)$, respectively. According to the preceding description, the shape function $f(z)$ is the component that is accountable for regulating the distribution of shear strain as well as stress all the way through the thickness of the beam. As a result of the fact that the stress-free condition on the upper and lower margins of the construction will be fulfilled, such theories do not require the addition of an additional shear adjustment coefficient. In the following paragraphs, the following geometry functions will be used [1–3]:

$$f(z) = \frac{he^z}{h^2 + \pi^2}\left[\pi\sin\left(\frac{\pi z}{h}\right) + h\cos\left(\frac{\pi z}{h}\right)\right] - \frac{h^2}{h^2 + \pi^2}$$

(2.30)

It is now possible to display the nonzero components of the strain tensor of the continuous system in the following form:

$$\varepsilon_{xx} = \frac{\partial u}{\partial x} - z\frac{\partial^2 w_b}{\partial x^2} - f(z)\frac{\partial^2 w_s}{\partial x^2},$$

$$\gamma_{xz} = 2\varepsilon_{xz} = g(z)\frac{\partial w_s}{\partial x}$$

(2.31)

in which

$$g(z) = 1 - \frac{df(z)}{dz}$$

(2.32)

With the aid of refined HSDT of beams, we will formulate a variety of problems in the linear domain in this section. Beams are susceptible to static deformation, static buckle, and unconstrained vibration. Among all of these issues, the variation of the system's strain energy is unique. Consequently, this energy functional will be expressed in mathematical language prior to the subdivision of the current section into specific subsections. The continuous system's strain energy variation can be expressed as follows:

$$\delta U = \int_0^L \left\{ N\frac{\partial\delta u}{\partial x} - M_b\frac{\partial^2\delta w_b}{\partial x^2} - M_s\frac{\partial^2\delta w_s}{\partial x^2} + Q\frac{\partial\delta w_s}{\partial x}\right\} dx$$

(2.33)

in which

$$
\begin{Bmatrix} N \\ M_b \\ M_s \end{Bmatrix} = \int_0^b \int_{-h/2}^{h/2} \begin{Bmatrix} 1 \\ z \\ f(z) \end{Bmatrix} \sigma_{xx} dz dy, \quad Q = \int_0^b \int_{-h/2}^{h/2} g(z) \sigma_{xz} dz dy \tag{2.34}
$$

2.3 KINEMATIC RELATIONS OF PLATES

2.3.1 KIRCHHOFF–LOVE PLATE THEORY

This section will discuss the kinematic relationships of the Kirchhoff–Love plates. Similar to Euler–Bernoulli beams, the classical theory of plates disregards the influence of shear deformation due to the assumption that it is negligible in comparison to bending deformation. Indeed, the distortions will be disregarded, and so thick-walled plates cannot be analyzed using this theory. The results of this theory can be relied upon if the under-observation structure has narrow walls and a length-to-thickness ratio that is greater than 100, for instance. This theory states that the displacement field of a plate-type element can be represented as follows [1–3]:

$$
u_x(x,y,z,t) = u(x,y,t) - (z - \bar{z})\frac{\partial w(x,y,t)}{\partial x},
$$
$$
u_y(x,y,z,t) = v(x,y,t) - (z - \bar{z})\frac{\partial w(x,y,t)}{\partial y}, \tag{2.35}
$$
$$
u_z(x,y,z,t) = w(x,y,t)
$$

in which the components of the displacement in the longitudinal and transverse dimensions are denoted by $u(x, y, t)$ and $v(x, y, t)$, respectively. In addition to this, there is the plate's bending deformation. According to the aforementioned description, the $\bar{z}$ refers to the precise location of the neutral surface of the plate. After beginning the process of deriving the governing equation for the plate's nonlinear forced vibration problem, the magnitude of the problem of determining the precise location of the plate's neutral surface will become more apparent to the observer. In point of fact, this will make it possible for us to separate the pressures and seconds, which is something that will be advantageous for us. In addition to this, the problem that has been formulated will be one that is more applicable to real life as a result of the consideration, given to the precise position of the neutral surface on the plate.

In accordance with von Kármán's nonlinear strain-displacement equations, the nonzero components of the strain tensor of the rectangular plates can be represented as follows [1–3]:

$$\varepsilon_{xx} = \frac{\partial u}{\partial x} + \frac{1}{2}\left(\frac{\partial w}{\partial x}\right)^2 - (z-\bar{z})\frac{\partial^2 w}{\partial x^2},$$

$$\varepsilon_{yy} = \frac{\partial v}{\partial y} + \frac{1}{2}\left(\frac{\partial w}{\partial y}\right)^2 - (z-\bar{z})\frac{\partial^2 w}{\partial y^2}, \qquad (2.36)$$

$$\gamma_{xx} = \frac{\partial u}{\partial y} + \frac{\partial v}{\partial x} + \frac{\partial w}{\partial x}\frac{\partial w}{\partial y} - 2(z-\bar{z})\frac{\partial^2 w}{\partial x \partial y}$$

2.3.1.1 Wave Propagation Problem

In this subparagraph, the problem of transversely loaded rectangular plates impacted by elastic wave dispersion will be expressed mathematically. The geometries of the plate will be evaluated similarly to those described in the preceding subsection. Consider the rectangular plate to be embedded in a medium with both rigid and viscous properties. Two Winkler and Pasternak coefficients will be responsible for simulating the substrate's linear and shear springs. In addition, the damper will be evaluated using the damping coefficient. As indicated previously, it will be assumed that the plate is subject to a transverse load of amplitude $f(t)$. In this instance, the following displacement field form will be considered [1–3]:

$$u_x(x,y,z,t) = -z\frac{\partial w(x,y,t)}{\partial x},$$

$$u_y(x,y,z,t) = -z\frac{\partial w(x,y,t)}{\partial y}, \qquad (2.37)$$

$$u_z(x,y,z,t) = w(x,y,t)$$

in which the bending displacement of the plate is represented by the symbol $w(x, y, t)$. The following expressions can be written for the nonzero components of the strain tensor, based on the displacement field that was discussed earlier, as well as with regard to the definition of the linear strain tensor in the field of continuum mechanics:

$$\varepsilon_{xx} = -z\frac{\partial^2 w}{\partial x^2}, \; \varepsilon_{yy} = -z\frac{\partial^2 w}{\partial y^2}, \; \gamma_{xy} = -2z\frac{\partial^2 w}{\partial x \partial y} \qquad (2.38)$$

Therefore, the difference in the plate's strain energy can now be articulated by using the stresses that have been discussed thus far. The following shape can be taken into consideration for a linearly elastic plate:

$$\delta U = -\int_0^b\int_0^a\left[M_{xx}\frac{\partial^2 \delta w}{\partial x^2} + 2M_{xy}\frac{\partial^2 \delta w}{\partial x \partial y} + M_{yy}\frac{\partial^2 \delta w}{\partial y^2}\right]dxdy \qquad (2.39)$$

In the above definition, the bending-induced moments can be calculated via:

$$
\begin{Bmatrix} M_{xx} \\ M_{yy} \\ M_{xy} \end{Bmatrix} = \int_{-h/2}^{h/2} z \begin{Bmatrix} \sigma_{xx} \\ \sigma_{yy} \\ \sigma_{xy} \end{Bmatrix} dz
\tag{2.40}
$$

It is possible to describe the variation in the amount of work done by the visco-elastic substructure and the horizontal loading on the plate-type element using the following form refs. [1–3]:

$$
\delta V = \int_0^b \int_0^a \left[k_W w - k_P \left(\frac{\partial^2 w}{\partial x^2} + \frac{\partial^2 w}{\partial y^2} \right) + c_d \frac{\partial w}{\partial t} + f(t) \right] \delta w\, dx\, dy
\tag{2.41}
$$

In regard to the description of the variation of the kinetic energy, the following formulation for the variation of the plate's kinetic energy can be collected by sub-stituting the displacement field introduced in Eq. (2.2) into it:

$$
\delta T = \int_0^b \int_0^a \left[I_0 \frac{\partial w}{\partial t} \frac{\partial \delta w}{\partial t} + I_2 \left(\frac{\partial^2 w}{\partial x \partial t} \frac{\partial^2 \delta w}{\partial x \partial t} + \frac{\partial^2 w}{\partial y \partial t} \frac{\partial^2 \delta w}{\partial y \partial t} \right) \right] dx\, dy
\tag{2.42}
$$

in which the mass moments of inertia used in the above relation can be defined via:

$$
\begin{Bmatrix} I_0 \\ I_2 \end{Bmatrix} = \int_{-h/2}^{h/2} \begin{Bmatrix} 1 \\ z^2 \end{Bmatrix} \rho(z)\, dz
\tag{2.43}
$$

Now, the equation that describes the motion of the plate can be obtained in a straightforward manner. In order to achieve this objective, it is sufficient to substitute for the variations of strain energy, work done on the plate, and kinetic energy from Eqs. (2.39), (2.41), and (2.42) into Hamilton's principle and choose the non-trivial response of the problem, which corresponds with setting the coefficient to zero. This will accomplish the desired result. This brings us to:

$$
\frac{\partial^2 M_{xx}}{\partial x^2} + 2\frac{\partial^2 M_{xy}}{\partial x \partial y} + \frac{\partial^2 M_{yy}}{\partial y^2} - k_W w + k_P \left(\frac{\partial^2 w}{\partial x^2} + \frac{\partial^2 w}{\partial y^2} \right) - c_d \frac{\partial w}{\partial t}
$$
$$
= I_0 \frac{\partial^2 w}{\partial t^2} - I_2 \left(\frac{\partial^4 w}{\partial x^2 \partial t^2} + \frac{\partial^4 w}{\partial y^2 \partial t^2} \right)
\tag{2.44}
$$

It is necessary to describe the bending moments in terms of the plate's displacement in order to be able to determine the ultimate governing equation of the problem in terms of the bending deflection of the plate. This will allow you to solve the problem in terms of the bending deflection of the plate. In order to accomplish this, it is necessary to integrate Eq. (2.42) with respect to the thickness direction, taking into account the specifications presented in Eq. (2.40). When this process is finished, a connection that is analogous to the one that we presented earlier in Eq. (2.40) will have been attained. Nevertheless, the through-the-thickness bending rigidities of the plate need to be specified in the following manner:

$$
\begin{Bmatrix} D_{11} \\ D_{12} \\ D_{22} \\ D_{66} \end{Bmatrix} = \int_{-h/2}^{h/2} z^2 \begin{Bmatrix} Q_{11} \\ Q_{12} \\ Q_{22} \\ Q_{66} \end{Bmatrix} dz \tag{2.45}
$$

Now, the problem's governing equation can be readily derived by substituting the bending moments from Eq. (2.40) into Eq. (2.44). Therefore, the problem's governing equation is:

$$
D_{11}\frac{\partial^4 w}{\partial x^4} + 2\left(D_{12} + 2D_{66}\right)\frac{\partial^4 w}{\partial x^2 \partial y^2} + D_{22}\frac{\partial^4 w}{\partial y^4} + k_W w - k_P\left(\frac{\partial^2 w}{\partial x^2} + \frac{\partial^2 w}{\partial y^2}\right)
$$
$$
+ c_d\frac{\partial w}{\partial t} + I_0\frac{\partial^2 w}{\partial t^2} - I_2\left(\frac{\partial^4 w}{\partial x^2 \partial t^2} + \frac{\partial^4 w}{\partial y^2 \partial t^2}\right) = f\left(t\right) \tag{2.46}
$$

The solution to the problem can now be found by working through the equation that was presented earlier. In order to find a solution for this equation, we will make use of the well-known wave propagation technique that was described in Section 2.1.1.1. However, it is important to take note that the geographic solution to the current problem is not the same as the solution to the problem that we were required to solve in Section 2.1.1.1. Due to the fact that this particular problem must be handled in the plane, it is necessary to determine two wave numbers. It is best to get started on the process of solving the problem as soon as possible so that the described statements can be clarified. The following equations can be used to describe the solution to the problem of the plate curving and deflecting [1–3]:

$$
w\left(x, y, t\right) = W\left(t\right)\exp\left\{i\left(\beta_1 x + \beta_2 y\right)\right\} \tag{2.47}
$$

where β_1 and β_2 are the wave numbers for the longitudinal and transverse components of the wave, respectively. Again, the magnitude of the system's displacement is thought of as having a time-varying quality to it. By substituting the newly developed solution function into the problem's governing equation (Eq. (2.46)), one will be able to obtain an equation that is analogous to Eq. (2.16), the mass, damping, and stiffness of which can be expressed in the following manner [1, 2]:

$$M = I_0 + I_2 \left(\beta_1^2 + \beta_2^2 \right), \quad C = c_d,$$

$$K = D_{11}\beta_1^4 + 2 \left(D_{12} + 2D_{66} \right) \beta_1^2 \beta_2^2 + D_{22}\beta_2^4 + k_W + k_P \left(\beta_1^2 + \beta_2^2 \right)$$

(2.48)

It will not be necessary to reiterate any of the information contained in Section 2.1.1.1 because the remaining steps of the solution are identical to those described in that section. Therefore, in order to continue working through the issue with the plate, the viewers should return to Section 2.1.1.1.

2.3.2   Refined Higher-order Plate Theory

In the Section 2.3.1.1, the fundamental concept and mechanical connections of the classical theory of the rectangular plates were addressed in both linear and nonlinear situations. This section will continue the discussion begun in the previous section. The deformation of the continuous system's cross-section is not taken into account by traditional theories, as was previously mentioned. According to this problem, the Kirchhoff–Love theorem can be used to investigate only plates with thin walls; it cannot be applied to investigate plates that have walls that are either relatively thick or thick. In order to accomplish this goal, a plate theory was developed that was analogous to the Timoshenko beam theory. This theory was able to determine the shear displacement as a constant number throughout the entire thickness domain. However, this does not necessarily guarantee that the shear-induced displacement of the plates will be included in its entirety in the mechanical evaluations. As a result, HSDTs were developed in order to approximate the profile of the shear strain that runs along the thickness of the plate using a contour function that depends on the thickness. However, using the HSDT of plates rather than the Kirchhoff–Love hypothesis results in an expansion of the final eigenvalue problem. Specifically, the three governing equations are substituted with five equations as a direct consequence of this change. However, by using more sophisticated HSDTs in place of more conventional ones, this problem can be addressed in a less harsh manner. In these kinds of theories, there won't be a variable that's specifically devoted to the movements of the cross-section area around the longitudinal or transverse axis. According to these hypotheses, the complete plate's displacement can be broken down into two primary categories of displacement. The plate's bending deflection will be caused by the bigger portion, while the shear deflection will be caused by the smaller section. Both deflections will occur simultaneously. The improved plate theories predict that the displacement field of the plate will take the shape shown below [1–3]:

$$u_x \left(x, y, z, t \right) = u \left(x, y, t \right) - z \frac{\partial w_b \left(x, y, t \right)}{\partial x} - f(z) \frac{\partial w_s \left(x, y, t \right)}{\partial x},$$

$$u_y \left(x, y, z, t \right) = v \left(x, y, t \right) - z \frac{\partial w_b \left(x, y, t \right)}{\partial y} - f(z) \frac{\partial w_s \left(x, y, t \right)}{\partial y},$$

$$u_z \left(x, y, z, t \right) = w_b \left(x, y, t \right) + w_s \left(x, y, t \right)$$

(2.49)

where the displacements of the plate's mid-surface in the longitudinal and transverse dimensions are indicated by $u(x, y, t)$ and $v(x, y, t)$ correspondingly and
represent the bending and shearing deflections, respectively. In a manner that is
analogous to refined higher-order theorems that describe the kinematic behaviors
of beams, the function is inserted into the displacement field of the plate in the
current problem. This is done with the intention of simulating the distorted shape
of the cross-section of the plate by employing an expression that is dependent on
the plate's thickness. In a manner analogous to that described in Section 2.1.2, the
contour function will be analyzed using Eq. (2.30). The following expression can
be used to describe the linear strains that are present in the plate [1–3]:

$$
\begin{aligned}
\varepsilon_{xx} &= \frac{\partial u}{\partial x} - z\frac{\partial^2 w_b}{\partial x^2} - f(z)\frac{\partial^2 w_s}{\partial x^2}, \\
\varepsilon_{yy} &= \frac{\partial v}{\partial y} - z\frac{\partial^2 w_b}{\partial y^2} - f(z)\frac{\partial^2 w_s}{\partial y^2}, \\
\gamma_{xy} &= \frac{\partial u}{\partial y} + \frac{\partial v}{\partial x} - 2z\frac{\partial^2 w_b}{\partial x\partial y} - 2f(z)\frac{\partial^2 w_s}{\partial x\partial y}, \\
\gamma_{xz} &= g(z)\frac{\partial w_s}{\partial x}, \\
\gamma_{yz} &= 2\varepsilon_{yz} = g(z)\frac{\partial w_s}{\partial y}
\end{aligned}
\tag{2.50}
$$

in which $g(z) = 1 - df(z)/dz$. The issues of static bending, buckling, and free oscillations of the rectangular plates are going to be examined in the subsequent stages.
The computation of the difference of the plate's strain energy functional serves
as the intermediary between all of the aforementioned challenges. The fluctuation
of the plate's strain energy can be formulated as follows if we make the presumption that we are investigating a linearly elastic solid plate in order to arrive at this
conclusion:

$$
\delta U = \int_0^b\int_0^a
\begin{bmatrix}
N_{xx}\dfrac{\partial\delta u}{\partial x} - M_{xx}^b\dfrac{\partial^2\delta w_b}{\partial x^2} - M_{xx}^s\dfrac{\partial^2\delta w_s}{\partial x^2} + N_{yy}\dfrac{\partial\delta v}{\partial y} \\[2mm]
- M_{yy}^b\dfrac{\partial^2\delta w_b}{\partial y^2} - M_{yy}^s\dfrac{\partial^2\delta w_s}{\partial y^2} + N_{xy}\left(\dfrac{\partial\delta u}{\partial y} + \dfrac{\partial\delta v}{\partial x}\right) \\[2mm]
- 2M_{xy}^b\dfrac{\partial^2\delta w_b}{\partial x\partial y} - 2M_{xy}^s\dfrac{\partial^2\delta w_s}{\partial x\partial y} + Q_{xz}\dfrac{\partial\delta w_s}{\partial x} + Q_{yz}\dfrac{\partial\delta w_s}{\partial y}
\end{bmatrix} dxdy \tag{2.51}
$$

In the above definition, the stress resultants can be achieved using the below
formula:

$$
\begin{Bmatrix} N_{xx} & M_{xx}^b & M_{xx}^s \\ N_{yy} & M_{yy}^b & M_{yy}^s \\ N_{xy} & M_{xy}^b & M_{xy}^s \end{Bmatrix} = \int_{-h/2}^{h/2} \begin{Bmatrix} \sigma_{xx} & z\sigma_{xx} & f(z)\sigma_{xx} \\ \sigma_{yy} & z\sigma_{yy} & f(z)\sigma_{yy} \\ \sigma_{xy} & z\sigma_{xy} & f(z)\sigma_{xy} \end{Bmatrix} dz,
\tag{2.52}
$$

$$
\begin{Bmatrix} Q_{xz} \\ Q_{yz} \end{Bmatrix} = \int_{-h/2}^{h/2} g(z) \begin{Bmatrix} \sigma_{xz} \\ \sigma_{yz} \end{Bmatrix} dz
$$

2.4 KINEMATIC RELATIONS OF SHELLS

2.4.1 CLASSICAL SHELL THEORY

To characterize the motion of truncated conical shells, we present the classical theory here. Imagine the under-observation structure as an inclined line of finite length rotating around an axis parallel to the longitudinal direction of the inclined line. Thus, and due to the inclination of the line, a shell-type element will be produced with cross-sections fashioned like circles with differing diameters. In order to provide accurate results, the conical shell will be modeled as a structure with narrow walls whose thickness is many times less than its length and diameter. In addition, it is supposed that the shell is enclosed by linear and shear springs with stiffnesses, respectively. On the basis of the aforementioned assumptions, the displacement field of this type of projectile can be represented as follows [1–3]:

$$
u_x(x,\theta,z,t) = -z\frac{\partial w(x,\theta,t)}{\partial x},
$$

$$
u_\theta(x,\theta,z,t) = -\frac{z}{x\sin\alpha}\frac{\partial w(x,\theta,t)}{\partial\theta},
\tag{2.53}
$$

$$
u_z(x,\theta,z,t) = w(x,\theta,t)
$$

where $w(x, \theta, t)$ represents the shell's bending deflection and is the semi-vertex cone angle, which is the angle between each shell edge and its longitudinal axis. To use the concept of the linear strain tensor of a continuous system in the polar coordinate system, it is possible to derive the nonzero strains of the conical shell using the following definitions [1–3]:

$$
\varepsilon_{xx} = \frac{\partial u_x}{\partial x},
$$

$$
\varepsilon_{\theta\theta} = \frac{1}{x\sin\alpha\left(1+\dfrac{z}{x}\tan\alpha\right)}\left(\frac{\partial u_\theta}{\partial\theta}+u_x\sin\alpha+u_z\cos\alpha\right),
$$

$$
\gamma_{x\theta} = \frac{1}{x\sin\alpha\left(1+\dfrac{z}{x}\tan\alpha\right)}\left(\frac{\partial u_x}{\partial\theta}+x\sin\alpha\left(1+z/x\tan\alpha\right)\frac{\partial u_\theta}{\partial x}-u_\theta\sin\alpha\right)
\tag{2.54}
$$

In light of the fact that we are about to conduct an investigation into shells with thin walls, and by inserting Eq. (2.53) into the formulation presented earlier, the nonzero linear strains that are present in a cylindrical shell can be expressed in the following form:

$$\varepsilon_{xx} = -z \frac{\partial^2 w}{\partial x^2},$$

$$\varepsilon_{\theta\theta} = \frac{1}{x \sin \alpha} \left(-\frac{z}{x \sin \alpha} \frac{\partial^2 w}{\partial \theta^2} - z \sin \alpha \frac{\partial w}{\partial x} + w \cos \alpha \right), \qquad (2.55)$$

$$\gamma_{x\theta} = \frac{1}{x \sin \alpha} \left(-2z \frac{\partial^2 w}{\partial x \partial \theta} + \frac{z}{x} \frac{\partial w}{\partial \theta} \right)$$

Now, it is time to discover an expression for the variation of the shell's strain energy. To achieve this objective, it is necessary to first define this energy functional. This capability can be derived by:

$$\delta U = \int_0^{2\pi} \int_{x_0}^{x_0+L} \int_{-h/2}^{h/2} \left[\sigma_{xx} \delta \varepsilon_{xx} + \sigma_{\theta\theta} \delta \varepsilon_{\theta\theta} + \sigma_{x\theta} \delta \gamma_{x\theta} \right] x \sin \alpha \, dz dx d\theta \qquad (2.56)$$

Substitution of Eq. (2.55) in Eq. (2.56) gives:

$$\delta U = \int_0^{2\pi} \int_{x_0}^{x_0+L} \left[\begin{array}{c} -M_{xx} x \sin \alpha \dfrac{\partial^2 \delta w}{\partial x^2} - \dfrac{M_{\theta\theta}}{x \sin \alpha} \dfrac{\partial^2 \delta w}{\partial \theta^2} - M_{\theta\theta} \sin \alpha \dfrac{\partial \delta w}{\partial x} \\[2mm] + N_{\theta\theta} \cos \alpha \delta w - 2M_{x\theta} \dfrac{\partial^2 \delta w}{\partial x \partial \theta} + \dfrac{M_{x\theta}}{x} \dfrac{\partial \delta w}{\partial \theta} \end{array} \right] dx d\theta \qquad (2.57)$$

in which the forces and moments utilized in above relation can be introduced in the following form:

$$N_{\theta\theta} = \int_{-h/2}^{h/2} \sigma_{\theta\theta} dz,$$

$$\left\{ \begin{array}{c} M_{xx} \\ M_{\theta\theta} \\ M_{x\theta} \end{array} \right\} = \int_{-h/2}^{h/2} z \left\{ \begin{array}{c} \sigma_{xx} \\ \sigma_{\theta\theta} \\ \sigma_{x\theta} \end{array} \right\} dz \qquad (2.58)$$

The conical shell is assumed to be rested on a two-parameter stiff foundation. So, the work done by the elastic medium on the shell can be expressed as follows:

$$\delta V = \int_0^{2\pi} \int_{x_0}^{x_0+L} \int_{-h/2}^{h/2} \left[k_w x \sin \alpha + k_p \left(\frac{\partial^2 w}{\partial x^2} + \frac{1}{x \sin \alpha} \frac{\partial^2 w}{\partial \theta^2} \right) \right] \delta w dz dx d\theta \qquad (2.59)$$

Consider the variation in the kinetic energy of the conical shells to be as [1–3]:

$$\delta T = \int_0^{2\pi} \int_{x_0}^{x_0+L} \int_{-h/2}^{h/2} \rho(z) \left[\frac{\partial u_x}{\partial t} \frac{\partial \delta u_x}{\partial t} + \frac{\partial u_\theta}{\partial t} \frac{\partial \delta u_\theta}{\partial t} + \frac{\partial u_z}{\partial t} \frac{\partial \delta u_z}{\partial t} \right] x \sin\alpha \, dz dx d\theta \qquad (2.60)$$

Now, the variation in the kinetic energy of the shell can be rewritten in the following form if Eq. (2.53) is inserted into Eq. (2.60):

$$\delta T = \int_0^{2\pi} \int_{x_0}^{x_0+L} \left[I_0 x \sin\alpha \frac{\partial w}{\partial t} \frac{\partial \delta w}{\partial t} - I_2 x \sin\alpha \frac{\partial^2 w}{\partial x \partial t} \frac{\partial^2 \delta w}{\partial x \partial t} \right.$$

$$\left. + \frac{I_2}{x \sin\alpha} \frac{\partial^2 w}{\partial\theta \partial t} \frac{\partial^2 \delta w}{\partial\theta \partial t} \right] dx d\theta \qquad (2.61)$$

in which the mass moments of inertia can be defined as:

$$\begin{Bmatrix} I_0 \\ I_2 \end{Bmatrix} = \int_{-h/2}^{h/2} \rho(z) \begin{Bmatrix} 1 \\ z^2 \end{Bmatrix} dz \qquad (2.62)$$

Now, it is sufficient to substitute the system's energy functionals into the definition of Hamilton's principle to obtain the motion equation of the conical shell. By substituting Equations (2.57), (2.59), and (2.61) into Eq. (2.2), we obtain:

$$\frac{\partial^2 (x M_{xx})}{\partial x^2} \sin\alpha + \frac{1}{x \sin\alpha} \frac{\partial^2 M_{\theta\theta}}{\partial\theta^2} + \frac{\partial M_{\theta\theta}}{\partial x} \sin\alpha - N_{\theta\theta} \cos\alpha$$

$$+ 2 \frac{\partial^2 M_{x\theta}}{\partial x \partial\theta} - \frac{1}{x} \frac{\partial M_{x\theta}}{\partial\theta} - k_W w x \sin\alpha - k_P \left(\frac{\partial^2 w}{\partial x^2} + \frac{1}{x \sin\alpha} \frac{\partial^2 w}{\partial\theta^2} \right) \qquad (2.63)$$

$$+ I_2 x \sin\alpha \frac{\partial^4 w}{\partial x^2 \partial t^2} + \frac{I_2}{x \sin\alpha} \frac{\partial^4 w}{\partial\theta^2 \partial t^2} + I_0 x \sin\alpha \frac{\partial^2 w}{\partial t^2} = 0$$

To determine the problem's governing equation, the forces and moments in the preceding equation must be expressed in terms of the bending deflection of the conical shell. To accomplish this, it is necessary to first introduce the constitutive equation of the conical shell. Consider the following Hook's law variant [1–3]:

$$\begin{Bmatrix} \sigma_{xx} \\ \sigma_{\theta\theta} \\ \sigma_{x\theta} \end{Bmatrix} = \begin{bmatrix} Q_{11} & Q_{12} & 0 \\ Q_{12} & Q_{22} & 0 \\ 0 & 0 & Q_{66} \end{bmatrix} \begin{Bmatrix} \varepsilon_{xx} \\ \varepsilon_{\theta\theta} \\ \gamma_{x\theta} \end{Bmatrix} \qquad (2.64)$$

Integrating from the above relation over the thickness of the conical shell and using the definitions introduced in Eq. (0.68), the below expressions can be extracted:

$$N_{\theta\theta} = \frac{A_{22}}{x \tan \alpha} w - B_{12} \frac{\partial^2 w}{\partial x^2} - \frac{B_{22}}{x^2 \sin^2 \alpha} \frac{\partial^2 w}{\partial \theta^2} - \frac{B_{22}}{x} \frac{\partial w}{\partial x} \tag{2.65}$$

$$M_{xx} = \frac{D_{12}}{x \tan \alpha} w - D_{11} \frac{\partial^2 w}{\partial x^2} - \frac{D_{12}}{x^2 \sin^2 \alpha} \frac{\partial^2 w}{\partial \theta^2} - \frac{D_{12}}{x} \frac{\partial w}{\partial x} \tag{2.66}$$

$$M_{\theta\theta} = \frac{B_{12}}{x \tan \alpha} w - D_{12} \frac{\partial^2 w}{\partial x^2} - \frac{D_{12}}{x^2 \sin^2 \alpha} \frac{\partial^2 w}{\partial \theta^2} - \frac{D_{22}}{x} \frac{\partial w}{\partial x} \tag{2.67}$$

$$M_{x\theta} = \frac{D_{66}}{x^2 \sin \alpha} \frac{\partial w}{\partial \theta} - \frac{2D_{66}}{x \sin \alpha} \frac{\partial^2 w}{\partial x \partial \theta} \tag{2.68}$$

In the above relations, the through-the-thickness rigidities of the conical shell can be calculated using the following formulas:

$$\begin{Bmatrix} A_{22} \\ B_{22} \\ D_{22} \end{Bmatrix} = \int_{-h/2}^{h/2} Q_{22} \begin{Bmatrix} 1 \\ z \\ z^2 \end{Bmatrix} dz, \quad \begin{Bmatrix} B_{12} \\ D_{12} \end{Bmatrix} = \int_{-h/2}^{h/2} Q_{12} \begin{Bmatrix} z \\ z^2 \end{Bmatrix} dz, \quad \begin{Bmatrix} D_{11} \\ D_{66} \end{Bmatrix} = \int_{-h/2}^{h/2} z^2 \begin{Bmatrix} Q_{11} \\ Q_{66} \end{Bmatrix} dz \tag{2.69}$$

Now, the governing equation can be achieved by substituting for the forces and moments from Eqs. (2.65)–(2.68) in Eq. (2.63). Doing so, the below governing equation will be enhanced:

$$-\left[\frac{B_{22} \sin \alpha}{x^2 \tan \alpha} + \frac{A_{22} \cos \alpha}{x \tan \alpha} \right] w + \left[\frac{D_{22} \sin \alpha}{x^2} + \frac{2B_{22} \sin \alpha}{x \tan \alpha} \right] \frac{\partial w}{\partial x}$$

$$+ \left[2B_{12} \cos \alpha - \frac{D_{22} \sin \alpha}{x} + k_P \right] \frac{\partial^2 w}{\partial x^2} - \left[2D_{11} \sin \alpha + \frac{D_{12}}{x \sin \alpha} + D_{12} \sin \alpha \right] \frac{\partial^3 w}{\partial x^3}$$

$$-D_{11} x \sin \alpha \frac{\partial^4 w}{\partial x^4} + \left[\frac{2(D_{12} - D_{22} + 2D_{66})}{x^2 \sin \alpha} \right] \frac{\partial^3 w}{\partial x \partial \theta^2} - \left[\frac{2(D_{12} + 2D_{66})}{x \sin \alpha} \right] \frac{\partial^4 w}{\partial x^2 \partial \theta^2}$$

$$+ \left[\frac{2B_{22}}{x^2 \sin \alpha \tan \alpha} + \frac{2(D_{22} - D_{12} - 2D_{66})}{x^3 \sin \alpha} + \frac{k_p}{x \sin \alpha} \right] \frac{\partial^2 w}{\partial \theta^2} - \left[\frac{D_{22}}{x^3 \sin^3 \alpha} \right] \frac{\partial^4 w}{\partial \theta^4}$$

$$+ k_W w x \sin \alpha + I_0 x \sin \alpha \frac{\partial^2 w}{\partial t^2} + \frac{I_2}{x \sin \alpha} \frac{\partial^4 w}{\partial \theta^2 \partial t^2} + I_2 x \sin \alpha \frac{\partial^4 w}{\partial x^2 \partial t^2} = 0$$

$$\tag{2.70}$$

We can now calculate the system's inherent frequency through free oscillations by solving the above equation. This will be accomplished by addressing the issue independently in the time and space domains. Let's start by assuming the following broad shape for the answer to the problem:

$$w(x,\theta,t) = W(x)\sin(m\theta)\exp(i\omega t) \tag{2.71}$$

in which m and ω stand for circumferential wave number and natural frequency, respectively. Also, the influence of the variations over the longitudinal direction will be captured by the geometry-dependent function $W(x)$. If the above solution function is inserted into Eq. (2.70), the following expression can be achieved:

$$
\begin{aligned}
&-\left[\frac{B_{22}\sin\alpha}{x^2\tan\alpha} + \frac{A_{22}\cos\alpha}{x\tan\alpha}\right]W(x) + \left[\frac{D_{22}\sin\alpha}{x^2} + \frac{2B_{22}\sin\alpha}{x\tan\alpha}\right]\frac{\partial W(x)}{\partial x} \\
&+\left[2B_{12}\cos\alpha - \frac{D_{22}\sin\alpha}{x} + k_p\right]\frac{\partial^2 W(x)}{\partial x^2} \\
&-\left[2D_{11}\sin\alpha + \frac{D_{12}}{x\sin\alpha} + D_{12}\sin\alpha\right]\frac{\partial^3 W(x)}{\partial x^3} - D_{11}x\sin\alpha\frac{\partial^4 W(x)}{\partial x^4} \\
&-\frac{2m^2(D_{12}-D_{22}+2D_{66})}{x^2\sin\alpha}\frac{\partial W(x)}{\partial x} + \frac{2m^2(D_{12}+2D_{66})}{x\sin\alpha}\frac{\partial^2 W(x)}{\partial x^2} \\
&-\left[\frac{2m^2 B_{22}}{x^2\sin\alpha\tan\alpha} + \frac{2m^2(D_{22}-D_{12}-2D_{66})}{x^3\sin\alpha} + \frac{m^2 k_p}{x\sin\alpha}\right]W(x) - \frac{m^4 D_{22}}{x^3\sin^3\alpha}W(x) \\
&+k_w x\sin\alpha - I_0 x\omega^2\sin\alpha W(x) + \frac{I_2 m^2\omega^2}{x\sin\alpha}W(x) - I_2 x\omega^2\sin\alpha\frac{\partial^2 W(x)}{\partial x^2} = 0
\end{aligned}
\tag{2.72}
$$

As a 1D equation, the aforementioned relationship can be solved analytically or via computational techniques. To address this issue, this part opts for the tried-and-true generalized differential quadrature method (GDQM). An overview of the final answer must be given before the process can be considered complete. A discretization matrix with a fixed number of points is selected for a numerical solution in GDQM. Using the roots of Chebyshev polynomials as grid locations is a common GDQM-assisted method. For this reason, we can derive the grid points by:

$$x_i = \frac{L}{2}\left[1-\cos\left(\frac{i-1}{N-1}\pi\right)\right], \quad i = 1, 2, \cdots, N-1 \tag{2.73}$$

By the means of the sample points gathered from the above definition, the derivatives of any arbitrary function $f(x)$ with respect to x can be calculated via:

$$F^{(n)}\left(x_i\right) = \sum_{j=1}^{N} C_{ij}^{(n)} f\left(x_j\right), \; n = 1, 2, \ldots, N-1 \tag{2.74}$$

in which the weighting coefficients corresponding with the first-order derivative of the original function can be attained by:

$$C_{ij}^{(1)} = \frac{M\left(x_i\right)}{\left(x_i - x_j\right) M\left(x_j\right)}, \; i, j = 1, 2, \ldots, N \text{ and } j \neq i \tag{2.75}$$

in which

$$M\left(x_i\right) = \prod_{\substack{j=1 \\ j \neq i}}^{N} \left(x_i - x_j\right) \tag{2.76}$$

In addition, the higher-order weighting coefficients can be achieved using the below formulas:

$$C_{ij}^{(n)} = n\left(C_{ii}^{(n-1)} C_{ij}^{(1)} - \frac{C_{ij}^{(n-1)}}{x_i - x_j} \right), \; i, j = 1, 2, \ldots, N \text{ and } j \neq i \tag{2.77}$$

$$C_{ii}^{(n)} = -\sum_{\substack{j=1 \\ j \neq i}}^{N} C_{ij}^{(n)}, \; \begin{cases} i = 1, 2, \ldots, N \\ n = 1, 2, \ldots, N-1 \end{cases} \tag{2.78}$$

Once the above instructions are followed in order to compute the derivatives of the bending deflection of the conical shell with respect to x and inserting them into Eq. (2.72), the following relation will be obtained:

$$\begin{bmatrix} A_{bb} & A_{bd} \\ A_{db} & A_{dd} \end{bmatrix} \begin{Bmatrix} x_b \\ x_d \end{Bmatrix} = \omega^2 \begin{bmatrix} 0 & 0 \\ B_{db} & B_{dd} \end{bmatrix} \begin{Bmatrix} x_b \\ x_d \end{Bmatrix} \tag{2.79}$$

in which A and B are from types of stiffness and mass, respectively. The subscripts "b" and "d" indicate the boundary and domain points, respectively. In order to find the final relation, it is needed to omit the boundary points by inserting the boundary conditions (BCs) corresponding with the ends of the truncated conical shell into problem. Once the above manipulation is completely done, the below eigenvalue problem can be gathered:

$$\left[\mathbf{K} - \omega^2 \mathbf{M} \right] \mathbf{x_d} = \mathbf{0} \tag{2.80}$$

in which $\mathbf{K}$ and $\mathbf{M}$ denote stiffness and mass matrices, respectively. Also, $\mathbf{x_d}$ reveals the amplitude vector, which is constructed from the domain grid points. Solving the above eigenvalue equation for ω, the natural frequency of the conical shell can be achieved. It is worth noting that the BCs must be applied on the points $x = x_0$ and $x = x_0 + L$. For simply supported edges, the below mathematical constraint must be applied:

$$w = 0 \quad \text{and} \quad \frac{\partial^2 w}{\partial x^2} = 0 \qquad (2.81)$$

Also, the below constraints can satisfy the clamped condition:

$$w = 0 \quad \text{and} \quad \frac{\partial w}{\partial x} = 0 \qquad (2.82)$$

2.4.2 FIRST-ORDER SHEAR DEFORMATION SHELL THEORY

Here we introduce the FSDT of shells. Overall, FSDTs excel over traditional ones because they account for shear deformations up to the first degree, which the latter does not. Although this estimate lacks completeness, it does provide better solutions than the situation of using the conventional theory and permits the user to explore denser structures. Using this theory, we can account for two rotation factors in addition to the shell-type components' longitudinal, circular, and bending deformations. This theory predicts that the structure's displacement field can be represented as follows [1].

$$
\begin{aligned}
u_x\left(x,\theta,z,t\right) &= u\left(x,\theta,t\right) + z\varphi_x\left(x,\theta,t\right), \\
u_\theta\left(x,\theta,z,t\right) &= v\left(x,\theta,t\right) + z\varphi_\theta\left(x,\theta,t\right), \\
u_z\left(x,\theta,z,t\right) &= w\left(x,\theta,t\right)
\end{aligned}
\qquad (2.83)
$$

where axial, circumferential, and lateral deformations are represented by $u(x, \theta, t)$, $v(x, \theta, t)$, and $w(x, \theta, t)$, correspondingly. Additionally, $\varphi_x(x, \theta, t)$ and $\varphi_\theta(x, \theta, t)$ represent revolutions about the axial and circular axes, respectively. In order to analyze static issues, it is necessary to remove the time component from the preceding relations, as they are represented in the dynamic state. The subsequent subsections will concentrate primarily on the linear characteristics of these spheres; therefore, the components of the strain tensor will be determined for the linear case in this part. Sections devoted specifically to discussing nonlinear strain-displacement correlations will be included. The nonzero components of the strain tensor of a sphere can be expressed as follows [1–3] when the aforementioned polar coordinate displacement field is taken into account.

$$\varepsilon_{xx} = \frac{\partial u}{\partial x} + z\frac{\partial \varphi_x}{\partial x},$$

$$\varepsilon_{\theta\theta} = \frac{1}{R}\left(\frac{\partial v}{\partial \theta} + z\frac{\partial \varphi_\theta}{\partial \theta} + w\right),$$

$$\gamma_{x\theta} = \frac{1}{R}\frac{\partial u}{\partial \theta} + \frac{\partial v}{\partial x} + \frac{z}{R}\frac{\partial \varphi_x}{\partial \theta} + z\frac{\partial \varphi_\theta}{\partial x},$$

$$\gamma_{xz} = \varphi_x + \frac{\partial w}{\partial x}, \quad \gamma_{\theta z} = \varphi_\theta + \frac{1}{R}\frac{\partial w}{\partial \theta} - \frac{v}{R}$$

(2.84)

This equation for the change of the shell's strain energy can be expressed assuming that the shell is linearly elastic.

$$\delta U = \int_0^L\int_0^{2\pi}\left[\begin{array}{l} N_{xx}\dfrac{\partial\delta u}{\partial x} + M_{xx}\dfrac{\partial\delta\varphi_x}{\partial x} + \dfrac{N_{\theta\theta}}{R}\left(\dfrac{\partial\delta v}{\partial\theta}+\delta w\right) + \dfrac{M_{\theta\theta}}{R}\dfrac{\partial\delta\varphi_\theta}{\partial\theta} \\[2mm] +N_{x\theta}\left(\dfrac{1}{R}\dfrac{\partial\delta u}{\partial\theta}+\dfrac{\partial\delta v}{\partial x}\right) + M_{x\theta}\left(\dfrac{1}{R}\dfrac{\partial\delta\varphi_x}{\partial\theta}+\dfrac{\partial\delta\varphi_\theta}{\partial x}\right) + \\[2mm] Q_{xz}\left(\delta\varphi_x+\dfrac{\partial\delta w}{\partial x}\right) + Q_{\theta z}\left(\delta\varphi_\theta+\dfrac{1}{R}\dfrac{\partial\delta w}{\partial\theta}-\dfrac{\delta v}{R}\right) \end{array}\right] R\,d\theta\,dx \quad (2.85)$$

The forces and moments arising from the aforementioned relationship can be written as follows:

$$\begin{Bmatrix} N_{xx} & M_{xx} \\ N_{\theta\theta} & M_{\theta\theta} \\ N_{x\theta} & M_{x\theta} \end{Bmatrix} = \int_{-h/2}^{h/2} \begin{Bmatrix} \sigma_{xx} & z\sigma_{xx} \\ \sigma_{\theta\theta} & z\sigma_{\theta\theta} \\ \sigma_{x\theta} & z\sigma_{x\theta} \end{Bmatrix} dz,$$

$$\begin{Bmatrix} Q_{xz} \\ Q_{z\theta} \end{Bmatrix} = \kappa_s \int_{-h/2}^{h/2} \begin{Bmatrix} \sigma_{xz} \\ \sigma_{z\theta} \end{Bmatrix} dz$$

(2.86)

in which κ_s is the shear correction factor and can be considered to be $\kappa_s = 5/6$.

2.4.2.1 Wave Propagation Problem

The FSDT will be used to describe the issue of elastic wave distribution in circular shells in the following part. We already know the changes in strain energy and work done by the elastic base, two of the shell energy functionals necessary to construct the aforementioned issue. Therefore, all that needs to be done to finish the deduction is to deduce a suitable formula for the change of the shell's kinetic energy. The primary meaning of $x - \theta - z$ coordinate system kinetic energy variance [1–3] is

$$\delta T = \int_0^L \int_0^{2\pi} \int_{-h/2}^{h/2} \rho(z) \left[\frac{\partial u_x}{\partial t} \frac{\partial \delta u_x}{\partial t} + \frac{\partial u_\theta}{\partial t} \frac{\partial \delta u_\theta}{\partial t} + \frac{\partial u_z}{\partial t} \frac{\partial \delta u_z}{\partial t} \right] R dz d\theta dx \quad (2.87)$$

In Eq. (2.399), the shell-type elements' kinetic energy variant can be rewritten as follows:

$$\delta T = \int_0^L \int_0^{2\pi} \left[\begin{array}{c} I_0 \left(\dfrac{\partial u}{\partial t} \dfrac{\partial \delta u}{\partial t} + \dfrac{\partial v}{\partial t} \dfrac{\partial \delta v}{\partial t} + \dfrac{\partial w}{\partial t} \dfrac{\partial \delta w}{\partial t} \right) + \\[2mm] I_1 \left(\dfrac{\partial \varphi_x}{\partial t} \dfrac{\partial \delta u}{\partial t} + \dfrac{\partial u}{\partial t} \dfrac{\partial \delta \varphi_x}{\partial t} + \dfrac{\partial \varphi_\theta}{\partial t} \dfrac{\partial \delta v}{\partial t} + \dfrac{\partial v}{\partial t} \dfrac{\partial \delta \varphi_\theta}{\partial t} \right) \\[2mm] + I_2 \left(\dfrac{\partial \varphi_x}{\partial t} \dfrac{\partial \delta \varphi_x}{\partial t} + \dfrac{\partial \varphi_\theta}{\partial t} \dfrac{\partial \delta \varphi_\theta}{\partial t} \right) \end{array} \right] R d\theta dx \quad (2.88)$$

In the above relation, the mass moments of inertia can be defined as below:

$$\left\{ \begin{array}{c} I_0 \\ I_1 \\ I_2 \end{array} \right\} = \int_{-h/2}^{h/2} \rho(z) \left\{ \begin{array}{c} 1 \\ z \\ z^2 \end{array} \right\} dz \quad (2.89)$$

It is important to note that there is no axial compression acting on the shell in this instance. Consequently, the variation of the work done by external loading presented in Eq. (2.59) must be reduced to the expressions below [1–3]:

$$\delta V = \int_0^L \int_0^{2\pi} \int_{-h/2}^{h/2} \left[k_w w - k_p \left(\frac{\partial^2 w}{\partial x^2} + \frac{1}{R^2} \frac{\partial^2 w}{\partial \theta^2} \right) \right] R dz d\theta dx \quad (2.90)$$

Now, the Euler–Lagrange equations can be gathered by inserting Eqs. (2.87), (2.88), and (2.89) into Eq. (2.80). After doing the above substitution, the motion equations of the shell in the dynamic form can be expressed as below:

$$\frac{\partial N_{xx}}{\partial x} + \frac{1}{R} \frac{\partial N_{x\theta}}{\partial \theta} = I_0 \frac{\partial^2 u}{\partial t^2} + I_1 \frac{\partial^2 \varphi_x}{\partial t^2} \quad (2.91)$$

$$\frac{\partial N_{x\theta}}{\partial x} + \frac{1}{R} \frac{\partial N_{\theta\theta}}{\partial \theta} - \frac{1}{R} Q_{z\theta} = I_0 \frac{\partial^2 v}{\partial t^2} + I_1 \frac{\partial^2 \varphi_\theta}{\partial t^2} \quad (2.92)$$

$$\frac{\partial Q_{xz}}{\partial x} + \frac{1}{R} \frac{\partial Q_{z\theta}}{\partial \theta} + \frac{N_{\theta\theta}}{R} - k_w w + k_p \left(\frac{\partial^2 w}{\partial x^2} + \frac{1}{R^2} \frac{\partial^2 w}{\partial \theta^2} \right) = I_0 \frac{\partial^2 w}{\partial t^2} \quad (2.93)$$

$$\frac{\partial M_{xx}}{\partial x} + \frac{1}{R}\frac{\partial M_{x\theta}}{\partial \theta} + Q_{xz} = I_1\frac{\partial^2 u}{\partial t^2} + I_2\frac{\partial^2 \varphi_x}{\partial t^2} \tag{2.94}$$

$$\frac{\partial M_{x\theta}}{\partial x} + \frac{1}{R}\frac{\partial M_{\theta\theta}}{\partial \theta} + Q_{z\theta} = I_1\frac{\partial^2 v}{\partial t^2} + I_2\frac{\partial^2 \varphi_\theta}{\partial t^2} \tag{2.95}$$

The stress resultants in Eqs. (2.91)–(2.95) must be replaced with their analogous formulas in terms of the shell's displacement field to find the final governing equations. Thus, Eqs. (2.87)–(2.90) must be replaced in (2.2). After substituting, the shell's governing formulae are:

$$A_{11}\frac{\partial^2 u}{\partial x^2} + \frac{A_{66}}{R^2}\frac{\partial^2 u}{\partial \theta^2} + \frac{A_{12}+A_{66}}{R}\frac{\partial^2 v}{\partial x\partial\theta} + \frac{A_{12}}{R}\frac{\partial w}{\partial x} + B_{11}\frac{\partial^2 \varphi_x}{\partial x^2}$$
$$+ \frac{B_{66}}{R^2}\frac{\partial^2 \varphi_x}{\partial \theta^2} + \frac{B_{12}+B_{66}}{R}\frac{\partial^2 \varphi_\theta}{\partial x\partial\theta} = I_0\frac{\partial^2 u}{\partial t^2} + I_1\frac{\partial^2 \varphi_x}{\partial t^2} \tag{2.96}$$

$$\frac{A_{12}+A_{66}}{R}\frac{\partial^2 u}{\partial x\partial\theta} + A_{66}\frac{\partial^2 v}{\partial x^2} + \frac{A_{11}}{R^2}\frac{\partial^2 v}{\partial \theta^2} + A_{55}^s\frac{v}{R^2} + \frac{A_{11}-A_{55}^s}{R^2}\frac{\partial w}{\partial\theta}$$
$$+ \frac{B_{12}+B_{66}}{R}\frac{\partial^2 \varphi_x}{\partial x\partial\theta} + B_{66}\frac{\partial^2 \varphi_\theta}{\partial x^2} + \frac{B_{11}}{R^2}\frac{\partial^2 \varphi_\theta}{\partial \theta^2} - \frac{A_{55}^s}{R}\varphi_\theta = I_0\frac{\partial^2 v}{\partial t^2} + I_1\frac{\partial^2 \varphi_\theta}{\partial t^2} \tag{2.97}$$

$$\frac{A_{12}}{R}\frac{\partial u}{\partial x} + \frac{A_{11}-A_{55}^s}{R^2}\frac{\partial v}{\partial\theta} + \frac{A_{11}}{R^2}w + A_{55}^s\left(\frac{\partial^2 w}{\partial x^2} + \frac{1}{R^2}\frac{\partial^2 w}{\partial\theta^2}\right) - k_w w$$
$$+ k_p\left(\frac{\partial^2 w}{\partial x^2} + \frac{1}{R^2}\frac{\partial^2 w}{\partial\theta^2}\right) + \left(A_{55}^s + \frac{B_{12}}{R}\right)\frac{\partial\varphi_x}{\partial x} + \left(\frac{B_{11}}{R^2} + \frac{A_{55}^s}{R}\right)\frac{\partial\varphi_\theta}{\partial\theta} = I_0\frac{\partial^2 w}{\partial t^2} \tag{2.98}$$

$$B_{11}\frac{\partial^2 u}{\partial x^2} + \frac{B_{66}}{R^2}\frac{\partial^2 u}{\partial \theta^2} + \frac{B_{12}+B_{66}}{R}\frac{\partial^2 v}{\partial x\partial\theta} + \left(\frac{B_{12}}{R}+A_{55}^s\right)\frac{\partial w}{\partial x} + D_{11}\frac{\partial^2 \varphi_x}{\partial x^2}$$
$$+ \frac{D_{66}}{R^2}\frac{\partial^2 \varphi_x}{\partial \theta^2} + A_{55}^s\varphi_x + \frac{D_{12}+D_{66}}{R}\frac{\partial^2 \varphi_\theta}{\partial x\partial\theta} = I_1\frac{\partial^2 u}{\partial t^2} + I_2\frac{\partial^2 \varphi_x}{\partial t^2} \tag{2.99}$$

$$\frac{B_{12}+B_{66}}{R}\frac{\partial^2 u}{\partial x\partial\theta} + B_{66}\frac{\partial^2 v}{\partial x^2} + \frac{B_{11}}{R^2}\frac{\partial^2 v}{\partial \theta^2} - \frac{A_{55}^s}{R}v + \left(\frac{B_{11}}{R^2}+\frac{A_{55}^s}{R}\right)\frac{\partial w}{\partial\theta}$$
$$+ \frac{D_{12}+D_{66}}{R}\frac{\partial^2 \varphi_x}{\partial x\partial\theta} + D_{66}\frac{\partial^2 \varphi_\theta}{\partial x^2} + \frac{D_{11}}{R^2}\frac{\partial^2 \varphi_\theta}{\partial \theta^2} + A_{55}^s\varphi_\theta = I_1\frac{\partial^2 v}{\partial t^2} + I_2\frac{\partial^2 \varphi_\theta}{\partial t^2} \tag{2.100}$$

To determine dispersed wave dynamics, solve the above-linked governing equations. The suggested method for shell displacement field components is [1–3]:

$$u(x,\theta,t) = U \exp\left[i\left(\beta x + m\theta - \omega t\right)\right],$$
$$v(x,\theta,t) = V \exp\left[i\left(\beta x + m\theta - \omega t\right)\right],$$
$$w(x,\theta,t) = W \exp\left[i\left(\beta x + m\theta - \omega t\right)\right],$$
$$\varphi_x(x,\theta,t) = \phi_x \exp\left[i\left(\beta x + m\theta - \omega t\right)\right],$$
$$\varphi_\theta(x,\theta,t) = \phi_\theta \exp\left[i\left(\beta x + m\theta - \omega t\right)\right] \tag{2.101}$$

Axial and circular wave values are β and m. Also, ω denotes the frequency of the propagated waves. Inserting analytical answers of Eq. (2.101) into Eqs. (2.96)–(2.100) yields the following eigenvalue equation:

$$\left[\mathbf{K} - \omega^2 \mathbf{M}\right] \mathbf{d} = 0 \tag{2.102}$$

where $\mathbf{K}$ and $\mathbf{M}$ are the rigidity and mass symmetric matrices, respectively. In addition, $\mathbf{d}$ is a column vector containing the unknown displacement amplitudes. By setting the determinant of the coefficient behind the amplitude vector to zero, it is possible to determine the frequency of the propagating waves:

$$\left|\mathbf{K} - \omega^2 \mathbf{M}\right| = 0 \tag{2.103}$$

Since Eq. (2.102) introduces stiffness and mass matrices, their nonzero components are as follows [1–3]:

$$k_{11} = -\left(A_{11}\beta^2 + \frac{A_{66}}{R^2}m^2\right), \quad k_{12} = -\frac{A_{12} + A_{66}}{R}m\beta,$$

$$k_{13} = i\beta\frac{A_{12}}{R}, \quad k_{14} = -\left(B_{11}\beta^2 + \frac{B_{66}}{R^2}m^2\right),$$

$$k_{15} = -\frac{B_{12} + B_{66}}{R}m\beta, \quad k_{22} = -\left(A_{66}\beta^2 + \frac{A_{11}}{R^2}m^2 - \frac{A_{55}^s}{R^2}\right),$$

$$k_{23} = im\frac{A_{11} - A_{55}^s}{R^2}, \quad k_{24} = -\frac{B_{12} + B_{66}}{R}m\beta,$$

$$k_{25} = -\left(B_{66}\beta^2 + \frac{B_{11}}{R^2}m^2 + \frac{A_{55}^s}{R^2}\right),$$

$$k_{33} = \frac{A_{11}}{R^2} - A_{55}^s\left(\beta^2 + \frac{m^2}{R^2}\right) - k_W - k_P\left(\beta^2 + \frac{m^2}{R^2}\right), \quad k_{34} = i\beta\left(A_{55}^s + \frac{B_{12}}{R}\right),$$

$$k_{35} = im\left(\frac{B_{11}}{R^2} + \frac{A_{55}^s}{R}\right), \quad k_{44} = -\left(D_{11}\beta^2 + \frac{D_{66}}{R^2}m^2 - A_{55}^s\right), \qquad (2.104)$$

$$k_{45} = -\frac{D_{12} + D_{66}}{R}m\beta, \quad k_{55} = -\left(D_{66}\beta^2 + \frac{D_{11}}{R^2}m^2 - A_{55}^s\right)$$

and

$$m_{11} = m_{22} = m_{33} = -I_0, \ m_{14} = m_{25} = -I_1, \ m_{44} = m_{55} = -I_2 \qquad (2.105)$$

REFERENCES

[1] Ebrahimi, F., & Dabbagh, A. (2022). *Mechanics of multiscale hybrid nanocomposites* (1st ed.) Elsevier.

[2] Ebrahimi, F., & Dabbagh, A. (2020). *Mechanics of nanocomposites: Homogenization and analysis* (1st ed.) CRC Press. https://doi.org/10.1201/9780429316791

[3] Ebrahimi, F., & Dabbagh, A. (2019). *Wave propagation analysis of smart nanostructures*. CRC Press.

3 Wave Dispersion Characteristics of Functionally Graded Structures

3.1 BACKGROUND

In this chapter, the wave dispersion properties of a variety of functionally graded (FG) structures, such as FG hygrothermal loading on beams, plates, and shells, are discussed. Furthermore, the power law formulas are assumed to be used while calculating the effective material characteristics of functionally graded nanoplate. Functionally graded materials of both elastic and viscoelastic types are considered. For both thin and moderately thick nanoshells, the issue is formulated using first-order shear deformation theory (FSDT) along with Hamilton's principle. An improved higher-order plate theory is provided to improve the given model's ability to forecast the consequences of shear deformation. Utilizing an improved sinusoidal shear deformation beam theory, we additionally take into account the impact of shear deformation. The scattered waves' phase velocity is determined using an analytical solution. In the end, the findings are graphically represented to show how factors such as wave number, nonlocality, the length scale parameter, magnetic potential, electric voltage, gradient index, and angular velocity influence the wave frequency, phase velocity, and escape frequency of the structures.

3.1.1 A Review of Wave Dispersion Analysis of FG Structures

Scientists now have access to a wider range of collections from which to choose the constitutive material in their new designs because of the improvements in material qualities presented in functionally graded materials (FGMs). Using FGMs makes it far more likely that a mechanical design will be free from the stress concentration of its planned elements, which is one of the most significant qualities of a well-executed mechanical design. These modern materials really reduce stress concentration. However, owing to its great thermal resistance, this cutting-edge composite material type can withstand substantial thermal loadings. The advantage of such composites over traditional ones is that they are often built from two completely separate phases, and the distribution of each phase can be readily regulated. A team of Japanese researchers created FGMs for the first time in the 1980s as part of an aeronautical project [1]. Many writers have used FGMs because of their usefulness in the research process.

DOI: 10.1201/9781003270263-3

When carbon nanotubes (CNTs) were originally discovered in the early 1990s, they revealed hitherto unseen properties [1]. Because of their unique properties, nanostructures are finding more and more uses in a wide variety of nano-electro-mechanical systems (NEMSs). Because the classical continuum theory lacks a scale parameter, it is unable to predict the mechanical reactions of nanoscale structures, even though the differences in mechanical behavior between nano and macro scales are obvious to everyone. To account for scale effects in the mechanical properties of nanostructures, we must use size-dependent continuum theories. The Eringen nonlocal elasticity theory [2–3] is widely used for studying nanostructures' vibration, bending, buckling, and wave propagation properties. For instance, Narendar and Gopalakrishnan [4] analyzed the wave propagation in multi-walled CNTs, taking into account the tiny size. Wang et al. [5] used nonlocal elasticity to study the wave dispersion characteristics of double-layered nanoplates. Narendar and Gopalakrishnan [6] examined the wave propagation characteristics of nanoplates, paying special attention to the effects of thermal loading in addition to the size dependence of the structure. Ghadiri and Shafiei [7] analyzed the nonlinear bending vibration of spinning nanobeams using Eringen's elasticity theory. Ebrahimi and Hosseini [8] study the thermomechanical nonlinear vibration characteristics of size-dependent nanoplates. Vibration analysis of orthotropic single-layered graphene sheets using nonlocal elasticity was shown by Ebrahimi and Shafiei [9]. Although nonlocal elasticity has been widely used to analyze the mechanical behavior of nanostructures, recent research has shown that the theory is limited in its ability to account for the size dependence of nanostructures. The modified pair stress hypothesis [10] proposes that nanoscale components have a stiffness-enhancing impact, whereas nonlocal elasticity can only indicate the opposite. As an alternative to treating nanoscale materials as a collection of neighboring points, the strain gradient theory of elasticity by Lam et al. [11] suggests treating them as atoms with a higher-order deformation architecture. Thus, it is shown that nanostructures behave differently according to nonlocal elasticity and strain gradient theory. In addition, Lim et al. [12] create the nonlocal strain gradient theory (NSGT) to encapsulate these impacts in a single theory. In NSGT, we take into account not just the nonlocal elastic stress field but also the strain gradient stress field. There has been a recent influx of writers trying to do research on nanostructures inside the NSGT framework. Li and Hu [13] provided evidence that NSGT may affect the buckling behaviors of size-dependent beams. When evaluating the buckling reactions of embedded nanoplates, Farajpour et al. [14] used an NSGT to emphasize the diminishing and growing impacts of nonlocal and length scale parameters. Recently, the longitudinal vibrational behaviors of nanorods using NSGT have been shown by Li et al. [15].

However, in recent years, the intelligent behavior of mechanical components has played a significant part in the scientific explorations of numerous writers. The piezoelectric properties, through which one kind of mechanical or electrical energy may be converted into another, were formerly more well recognized as the defining feature in question. However, in current engineering designs, the interplay between the magneto-electro-mechanical properties of materials is of utmost relevance. Magneto-electro-elastic materials (MEEMs) have the potential to convert mechanical, electronic, or magnetic stimulations into the corresponding magnetic, electronic, or mechanical stimulations. Such materials must be smart enough to change the mechanical phenomena of electromagnetic loadings, especially elongation, into magnetic or electric fields when they are loaded along an axis. Because of these qualities, the use of such materials in NEMSs as actuators, controllers, and sensors is progressing quickly. However, a thorough understanding of how MEEMs react to varied external loadings is essential for successful design. Ramirez et al. [16] investigated the free vibration parameters of magneto-electro-elastic (MEE)-laminated plates in two dimensions. For instance, Ebrahimi and Dabbagh [17] presented a mathematical framework for the wave propagation problem of small-scale nanobeams and nanoplates made from different materials, such as smart piezoelectric materials, smart magneto-electro-elastic materials, smart magnetostrictive materials, and porous materials, for their analysis of wave propagation in smart nanostructures. In their book, they use the well-known Hamilton's concept to express the wave propagation issue utilizing both the classical and refined higher-order shear deformation beam and plate assumptions. The mechanical properties of nanostructures were analyzed using nonlocal elasticity and nonlocal strain gradient elasticity, with a focus on the impact of nanobeams. The dispersion curves of nanostructures were incorporated into multiple examples to demonstrate the impact of different factors, including elastic springs of an elastic foundation, damping coefficient of a viscoelastic substrate, various types of temperature changes, applied electric voltage and magnetic potential, and intensity of an external magnetic field. Furthermore, the Nonlinear Schrödinger–Gross–Pitaevskii equation was employed to monitor the responses of FGM nanoplates toward the transmission of thermo-acoustic waves, as reported in reference [18]. The study conducted in reference [19] investigated the dispersion properties of thermomechanical waves in double-nanobeams made of FGMs under the influence of a magnetic field. The task was achieved by utilizing the Euler–Bernoulli theorem in combination with the NSGT. Only two published works have utilized temporal nonlocality. In a previous study [20], an initial effort was made to investigate the propagation of waves in FGM nanobeams with viscoelastic damping using a framework that incorporates fractional time-space nonlocal strain gradient viscoelasticity. This was done in an effort to try to slow down the wave propagation. Later on, in reference [21], the same conceptual framework was used in order to explore the dynamic behaviors of waves that were dispersed across viscoelastic FGM nanoplates. In addition, Ke et al. [22] investigated the effects of vibration on MEE nanoplates.

In addition, FGMs have several enhanced material qualities in contrast to conventional laminated composites; they were first suggested for use in ambivalent thermomechanical situations [23]. In fact, these materials often include completely distinct top and bottom surfaces created from separate materials. Some of the advantages of these innovative composites include increased heat resistance, greater stress distribution, and reduced intensity factor. Given these advantages, the authors find it worthwhile to use FGMs in studies of the mechanical behavior of structures. The dynamic properties of beams were analyzed by Eltaher et al. [24] using FGMs and the finite element approach. Thai and Vo [25] employed an analytical solution to investigate many higher-order beam theories' predictions for the bending and free vibration behavior of FG beams. Choi and Thai [26] conduct a study of FG plates' free vibrations. Ebrahimi [27] discusses and presents the thermo-electro-mechanical static and dynamic reactions of circular FG plates. Nonlinear thermal buckling and post-buckling analysis of embedded FG beams have also been studied by Esfahani et al. [28] using FGMs. Nonlinear stability of FG beams under thermo-electrical stresses was addressed in detail by Kargani et al. [29]. The temperature dependence of the thermal stability findings of an FG beam was shown by Eslami et al. [30]. Ebrahimi et al. [31] show how the vibration parameters of FG temperature-dependent beams are affected by thermal loading in addition to porosity effects. Recently, the FG porous plates' coupled electromechanical vibration behaviors are studied by Barati et al. [32]. The bending behavior of FG plates under electrical influence has been explored by Barati et al. [33], who also took into account the effects of porosity.

The thermo-electro-mechanical properties of FG nanostructures as a function of size is also a hot focus of study. This assertion is supported by an orderly overview of related prior efforts.

Ebrahimi and Barati [34] use a Reddy beam to examine the influence of different temperature distributions on the thermal vibration parameters of FG-MEE nanobeams. Ebrahimi and Barati [35] also looked into the effects of humidity and temperature on the vibrational properties of FG nanobeams. The ripple effect of revising our understanding of shear deformation is also laid bare here. Ebrahimi and Barati [36] have created a nonlocal, higher-order, revised magneto-electro-viscoelastic model for studying the vibrations of smart nanobeams. Taking into account nonlinear thermal loading, Ebrahimi et al. [37] propose a precise method for analyzing the wave dispersion responses of FG spinning nanobeams. Recently, Ebrahimi and Dabbagh [38] used NSGT to examine the reactions of MEE-FG nanoplates to the propagation of flexural waves. The effects of porosity volume fraction on the free vibration characteristics of FG size-dependent embedded plates were analyzed by Ameri et al. [39]. Ebrahimi and Salari [40] compile a Timoshenko beam theory to analyze FG nanobeams' temperature-dependent buckling and free vibration. Additionally, Ebrahimi and Salari [41] used a semi-analytical differential transformation technique to investigate the flexural vibration characteristics of FG nanobeams. The buckling reactions of FG nanobeams were studied by Ebrahimi and Salari [42] using thermo-electrical actuation. Ebrahimi and Barati [43] provide the vibration responses of FG nanobeams within the context of a higher-order beam theory. Ebrahimi and Barati [44], using a higher-order beam theory, investigated the effects of different thermal loadings on the vibrational responses of FG nanobeams. Ebrahimi and Barati [45] look at the free vibration behavior of size-dependent FG beams under hygrothermal loadings. Recently, the piezo-thermo-electrical vibration capabilities of FG nanosize beams have been shown by Ebrahimi and Salari [46]. Vibration analysis of METE-FG (magneto-electro-thermo-elastic functionally graded) beams with size dependence with respect to shear deformation effects was also accomplished by Ebrahimi and Barati [47]. In addition, using an improved sinusoidal beam theory, Ebrahimi and Barati [48] have investigated the vibrational responses of magneto-electro-viscoelastic FG nanobeams. They have also taken into account the effects of different boundary conditions. Magnetic field effects on thermally influenced propagation were studied by Ebrahimi and Dabbagh [49]. Buckling responses of FG piezoelectric embedded nanobeams were modeled by Ebrahimi and Barati [50] using Reddy beam theory. Within the context of an improved inverse cotangential theory, Barati et al. [51] investigated the thermomechanical buckling behaviors of FG nanoplates. The buckling behavior of FG nanobeams was explored by Ebrahimi and Barati [52] using the third-order shear deformation beam theory, continuing the topic of buckling analysis of size-dependent beams. Ebrahimi and Barati [53] use METE-FG nanobeams and a higher-order shear deformation beam theory to conduct a precise investigation of temperature effects. Recently, Ebrahimi and Barati [54] provide a nonlocal strain gradient (NSG) beam model to account for size effects in a buckling analysis of curved FG nanobeams. In addition, Ebrahimi and Barati [55] use an improved higher-order beam theory to analyze the vibration of FG nanobeams in terms of hygrothermal effects. In this article, NSGT was utilized to look at how size and scale affected nanobeam. The study also takes into account the neutral surface orientation.

In addition to the aforementioned study, the analysis of wave propagation in nanostructures is becoming an increasingly important topic in the scientific community. Using nonlocal elasticity, Narendar and Gopalakrishnan [56] demonstrate the precise wave propagation responses of spinning CNTs. Using nonlocal strain gradient elasticity theory, Ebrahimi and Dabbagh [57] described the wave dispersion characteristics of rotating heterogeneous magneto-electro-elastic nanobeams. Nonlocal strain gradient-based wave dispersion behavior of spinning smart magneto-electro-elastic nanoplates was also tried to be shown [58]. Wave behaviors in an FG nanobeam were investigated by Zeng et al. [59], who looked at the combined impacts of thermo-piezoelectricity and surface elasticity. In their discussion of the wave propagation properties of MEE nanoplates, Zhang et al. [60] accounted for surface effects. Ebrahimi and Barati [61] use the NSGT to examine the wave dispersion characteristics of FG nanobeams in thermal situations where material properties change with temperature. In addition, Ebrahimi et al. [62] demonstrated how FG nanoplates' wave propagation properties change under nonlinear thermal stress. Ghorbanpour Arani et al. [63] use an accurate sinusoidal plate theory to investigate the surface impacts on the electro-magneto wave propagation characteristics of viscoelastic sandwich nanoplates. Ebrahimi and Barati [64] use NSGT to investigate the effects of a longitudinal magnetic field on the propagation of transverse waves in a sigmoid functionally graded (SFG) nanobeam. Ke et al. [65] provided two nonlocal models to investigate the wave propagation responses of MEE nanobeams within the context of Euler–Bernoulli and Timoshenko beam theories. Ebrahimi et al. [66] have provided a comprehensive discussion of the wave dispersion characteristics of FG-MEE axially stimulated nanobeams. To the authors' knowledge, the issue of wave propagation responses of an FG-MEE rotating size-dependent beam has never before been examined using an NSGT, despite attempts to investigate the wave propagation analysis of nanostructures.

3.2 FUNCTIONALLY GRADED BEAMS: WAVE DISPERSION CHARACTERISTICS

Dispersion properties of waves traveling through a rotating FG beam are discussed in this section [64]. Within the context of an improved sinusoidal beam theory, shear deformation is also considered, with no shear correction factor assumed. In addition, as can be seen in Figure 3.1, the beam's rotational velocity around the z-axis acts as an external axial load. After deriving the nonlocal governing equations, an analytical solution will be applied, and the resulting data will be precisely evaluated to reveal how different factors affect the wave propagation parameters of rotating FG-MEE beams of varying sizes.

A magneto-electro-elastic FG nanobeam with length L and thickness h, as shown in Figure 3.1, is assumed to be subjected to an electric potential $\Phi(x, z, t)$ and magnetic potential $\Upsilon(x, z, t)$ here. The nanobeam is made of $BaTiO_3$ and $CoFe_2O_4$ with the properties established in Table 3.1 [64].

According to the modified power law form, the porosity volume fraction is thought to smoothly change throughout the thickness of the porous FG piezoelectric nanobeam, as shown below.

$$P_f = P_c V_c + P_m V_m + \frac{\alpha}{2}\left(P_c - P_m\right) \tag{3.1a}$$

In Eq. (3.1a), P_m and P_c denote metal and ceramic material properties, respectively, and α stands for the volume fraction of porosity generated by defects during the fabrication procedure. Moreover, V_c and V_m are the volume fractions of

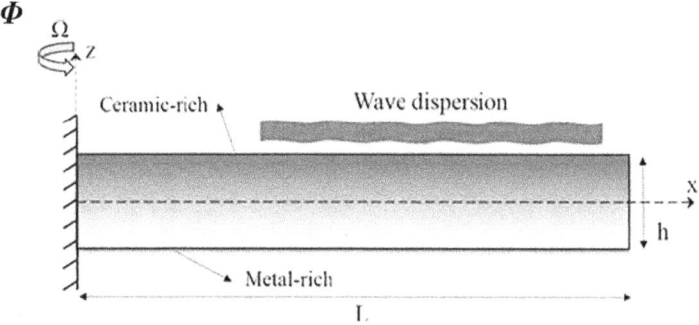

FIGURE 3.1   Geometry of a functionally graded nanobeam.

TABLE 3.1

Magneto-Electro-Elastic Coefficients of Material Properties

Properties	BaTiO$_3$	CoFe$_2$O$_4$
$c_{11} = c_{22}$ (GPa)	166	286
c_{33}	162	269.5
$c_{13} = c_{23}$	78	170.5
c_{12}	77	173
c_{55}	43	45.3
c_{66}	44.5	56.5
e_{31} (Cm^{-2})	−4.4	0
e_{33}	18.6	0
e_{15}	11.6	0
q_{31} (N/Am)	0	580.3
q_{33}	0	699.7
q_{15}	0	550
s_{11} (10^{-9} C^{2}m^{-2}N^{-1})	11.2	0.08
s_{13}	12.6	0.093
χ_{11} (Ns^{2}C^{-2}/2)	5	−590
χ_{33}	10	157
$d_{11} = d_{22} = d_{33}$	0	0
ρ (kg m^{-3})	5800	5300

ceramic and metal phases, respectively, which can be related to each other by the following equation:

$$V_c + V_m = 1 \qquad (3.1b)$$

In any thickness, the following formula may be used to determine the volume fraction of the ceramic phase:

$$V_c = \left(\frac{z}{h} + \frac{1}{2} \right)^p \qquad (3.1c)$$

Here p is the gradient index, which has the role of governing the smooth distribution of material through the thickness of the beam and z is the distance calculated from the neutral plane of the FG nanobeam. Now, substituting Eqs. (3.1b) and (3.1c) in Eq. (3.1a) results in an equation for equivalent material properties of porous FG beam:

$$P(z) = (P_c - P_m)\left(\frac{z}{h} + \frac{1}{2} \right)^p + P_m + \frac{\alpha}{2}(P_c - P_m) \qquad (3.1d)$$

It shall be considered that the top surface of porous FG nanosize beam $\left(z = \dfrac{h}{2} \right)$ is fully CoFe$_2$O$_4$, whereas the bottom surface $\left(z = -\dfrac{h}{2} \right)$ is fully BaTiO$_3$.

Also, the Mori–Tanaka homogenization model represents the effective material properties including effective local bulk modulus K_e and shear modulus μ_e in the following form by Ebrahimi and Barati [47]:

$$\frac{K_e - K_m}{K_c - K_m} = \frac{V_c}{1 + V_m \left(K_c - K_m\right)/\left(K_m + 4\mu_m/3\right)} \tag{3.2a}$$

$$\frac{\mu_e - \mu_m}{\mu_c - \mu_m} = \frac{V_c}{1 + V_m \left(\mu_c - \mu_m\right)/\left[\left(\mu_m + \mu_m\left(9K_m + 8\mu_m\right)\right)/\left(6\left(K_m + 2\mu_m\right)\right)\right]} \tag{3.2b}$$

Finally, the effective Young's modulus (E), Poisson's ratio (v), and mass density (ρ) can be expressed by:

$$E(z) = \frac{9K_e\mu_e}{3K_e + \mu_e} \tag{3.2c}$$

$$v(z) = \frac{3K_e - 2\mu_e}{6K_e + 2\mu_e} \tag{3.2d}$$

$$\rho(z) = \rho_c V_c + \rho_m V_m \tag{3.2e}$$

The thermal expansion coefficient (α) and thermal conductivity (κ) may be expressed by

$$\frac{\alpha_e - \alpha_m}{\alpha_c - \alpha_m} = \frac{\dfrac{1}{K_e} - \dfrac{1}{K_m}}{\dfrac{1}{K_c} - \dfrac{1}{K_m}} \tag{3.2f}$$

$$\frac{\kappa_e - \kappa_m}{\kappa_c - \kappa_m} = \frac{V_c}{1 + V_m \dfrac{\left(\kappa_c - \kappa_m\right)}{3\kappa_m}} \tag{3.2g}$$

The temperature-dependent behavior of the parameter values of material phases can be represented by the relation below:

$$P = P_0 \left(P_{-1} T^{-1} + 1 + P_1 T + P_2 T^2 + P_3 T^3\right) \tag{3.3}$$

where P_0, P_{-1}, P_1, P_2 and P_3 are the temperature-dependent constants that are tabulated in Table 3.2 for Si_3N_4 and SUS 304. The bottom and top surfaces of FG nanobeam are fully metal (SUS 304) and fully ceramic (Si_3N_4), respectively.

TABLE 3.2

Temperature-Dependent Coefficients for Si$_3$N$_4$ and SUS 304

Material	Properties	P_0	P_{-1}	P_1	P_2	P_3
Si$_3$N$_4$	E(Pa)	348.43e+9	0	−3.070e-4	2.160e-7	-8.946e-11
	α(K^{-1})	5.8723e-6	0	9.095e-4	0	0
	ρ(Kg/m^3)	2370	0	0	0	0
	κ(W/mK)	13.723	0	−1.032e-3	5.466e-7	-7.876e-11
	ν	0.24	0	0	0	0
SUS 304	E(Pa)	201.04e+9	0	3.079e-4	-6.534e-7	0
	α(K^{-1})	12.330e-6	0	8.086e-4	0	0
	ρ(Kg/m^3)	8166	0	0	0	0
	κ(W/mK)	15.379	0	−1.264e-3	2.092e-6	-7.223e-10
	ν	0.3262	0	−2.002e-4	3.797e-7	0

The temperature shows nonlinear behavior throughout the thickness in this investigation. The temperature distribution is obtained by determining the steady-state heat conduction problem with boundary conditions on the bottom and upper surfaces of the nanobeam along its thickness.

$$-\frac{d}{dz}\left(\kappa\left(z,\,T\right)\frac{dT}{dz}\right)=0 \tag{3.4a}$$

Considering the boundary conditions

$$T\left(\frac{h}{2}\right)=T_c,\;\;T\left(-\frac{h}{2}\right)=T_m \tag{3.4b}$$

The solution of Eq. (3.4a) is:

$$T=T_m+\left(T_c-T_m\right)\frac{\displaystyle\int_{-\frac{h}{2}}^{z}\frac{1}{\kappa\left(z,T\right)}dz}{\displaystyle\int_{-\frac{h}{2}}^{\frac{h}{2}}\frac{1}{\kappa\left(z,T\right)}dz} \tag{3.4c}$$

where, $\Delta T = T_c - T_m$ is the temperature change. Herein, the displacement fields of an MEE-FG rotary nanobeam are propounded employing a refined sinusoidal beam theory (SBT) as follows:

$$u_x\left(x,z,t\right)=u\left(x,z,t\right)-z\frac{\partial w_b}{\partial x}-\mathrm{f}\left(z\right)\frac{\partial w_s}{\partial x} \tag{3.5}$$

$$u_z\left(x,z,t\right) = w_b\left(x,y,t\right) + w_s\left(x,y,t\right) \tag{3.6}$$

In Eqs. (3.5) and (3.6), u is the longitudinal displacement of the mid-plane of the beam, and w_b, w_s are bending and shear deflections, respectively. In addition, $f(z)$ is a shape function, which estimates the shear stress through the thickness, and can be defined as follows for SBT [48]:

$$f\left(z\right) = z - \frac{\sin\left(\xi z\right)}{\xi} \tag{3.7}$$

The external electric and magnetic potentials are represented as a function of linear and cosine fluctuations [52, 53] to meet Maxwell's equation in the quasi-static approximation.

$$\Phi\left(x,z,t\right) = -\cos\left(\xi z\right)\phi\left(x,z,t\right) + \frac{2z}{h}V \tag{3.8}$$

$$Y\left(x,z,t\right) = -\cos\left(\xi z\right)\gamma\left(x,z,t\right) + \frac{2z}{h}\Omega \tag{3.9}$$

It must be mentioned that in Eqs. (3.7)–(3.9), $\xi = \frac{\pi}{h}$. V and Ω are external applied electric voltage and magnetic potential, respectively. Then, the nonzero strains of the MEE nanobeam can be written in the following form:

$$\varepsilon_{xx} = \frac{\partial u}{\partial x} - z\frac{\partial^2 w_b}{\partial x^2} - f\left(z\right)\frac{\partial^2 w_s}{\partial x^2} \tag{3.10}$$

$$\varepsilon_{xz} = \left(1 - \frac{df\left(z\right)}{dz}\right)\frac{\partial w_s}{\partial x} \tag{3.11}$$

Now, on the basis of Eq. (3.8), the electric field (E_x, E_z) and electric potential (Φ) can be related to each other as noted below:

$$E_x = -\frac{\partial \Phi}{\partial x} = \cos\left(\xi z\right)\frac{\partial \phi}{\partial x}$$

$$E_z = -\frac{\partial \Phi}{\partial z} = -\xi \sin\left(\xi z\right)\phi - \frac{2V}{h} \tag{3.12}$$

Furthermore, Eq. (3.9) reveals the relations between the magnetic field (H_x, H_z) and magnetic potential (Y) as follows:

$$H_x = -\frac{\partial \Upsilon}{\partial x} = \cos\left(\xi z\right)\frac{\partial \gamma}{\partial x}$$

$$H_z = -\frac{\partial \Upsilon}{\partial z} = -\xi \sin\left(\xi z\right)\phi - \frac{2\Omega}{h}$$

(3.13)

In this case, we use an expanded version of Hamilton's principle to derive the Euler–Lagrange equations for MEE-FG nanobeams, and this form of Hamilton's principle looks like this:

$$\int_0^t \delta\left(\Pi_S - \Pi_K + \Pi_W\right)dt = 0$$

(3.14)

where, Π_S is strain energy, Π_K is kinetic energy, and Π_W is the work done by external energy. The variation in strain energy is expressed as:

$$\delta \Pi_S = \int_0^L \int_A \left(\sigma_{xx}\delta\varepsilon_{xx} + \sigma_{xz}\delta\varepsilon_{xz} - D_x\delta E_x - D_z\delta E_z - B_x\delta H_x - B_z\delta H_z\right)dAdx$$

(3.15)

By plugging in the solutions to Eqs. (3.10) through (3.13) into Eq. (3.15), we get the following expression for the strain energy variation:

$$\delta \Pi_S = \int_0^L \left(N\left(\frac{\partial \delta u}{\partial x}\right) - M_{xx}^b\left(\frac{\partial^2\delta w_b}{\partial x^2}\right) - M_{xx}^s\left(\frac{\partial^2\delta w_s}{\partial x^2}\right) + Q_{xz}\left(\frac{\partial \delta w_s}{\partial x}\right)\right)dx$$

$$+ \int_0^L \int_A \left(-D_x\cos\left(\xi z\right)\left(\frac{\partial \delta\phi}{\partial x}\right) + D_z\xi\sin\left(\xi z\right)\delta\phi\right.$$

$$\left. - B_x\cos\left(\xi z\right)\left(\frac{\partial \delta\gamma}{\partial x}\right) + B_z\xi\sin\left(\xi z\right)\delta\gamma\right)dAdx$$

(3.16)

where the axial forces and bending moments utilized in the preceding equation represent the stress resultants.

$$N = \int_A \sigma_{xx}dA$$

$$\left(M_{xx}^b, M_{xx}^s\right) = \int_A \left(z, f\left(z\right)\right)\sigma_{xx}dA$$

(3.17)

$$Q_{xz} = \int_A \left(1 - \frac{df\left(z\right)}{dz}\right)\sigma_{xz}dA$$

The following expression may be used to indicate the varying amount of virtual work done by external forces:

$$\delta \Pi_W = \int_0^L \left(\left(N^E + N^H + N^R \right) \frac{\partial \left(w_b + w_s \right)}{\partial x} \frac{\partial \delta \left(w_b + w_s \right)}{\partial x} \right) dx \qquad (3.18)$$

where N^E, N^H, and N^R are normal in-plane forces generated due to electric voltage, magnetic potential, and beam's angular velocity around the z-axis, respectively, and can be defined in the following form:

$$N^E = \int_A \tilde{e}_{31} \frac{2V}{h} dA; N^H = \int_A \tilde{q}_{31} \frac{2\Omega}{h} dA; N^R = \int_x \int_A \left(\rho(z) A\Phi^2 x \right) dA dx \qquad (3.19)$$

where Φ denotes the angular velocity of the nanobeam. In this research, the maximum amount of the axial force generated by the rotation of the beam is considered as follows [56]:

$$N_{\max}^R = \int_0^L \int_A \left(\rho(z) A\Phi^2 x \right) dA dx \qquad (3.20)$$

The first variation in kinetic energy may be represented by the expression:

$$\delta \Pi_K = \int_V \left(\dot{u}_x \delta \dot{u}_x + \dot{u}_z \delta \dot{u}_z \right) \rho(z) dV = \int_0^L \left(I_0 \left(\frac{\partial u}{\partial t} \frac{\partial \delta u}{\partial t} + \frac{\partial \left(w_b + w_s \right)}{\partial t} \frac{\partial \delta \left(w_b + w_s \right)}{\partial t} \right) \right.$$

$$-I_1 \left(\frac{\partial u}{\partial t} \frac{\partial^2 \delta w_b}{\partial t \partial x} + \frac{\partial^2 w_b}{\partial t \partial x} \frac{\partial \delta u}{\partial t} \right) + I_2 \left(\frac{\partial^2 w_b}{\partial t \partial x} \frac{\partial^2 \delta w_b}{\partial t \partial x} \right)$$

$$-J_1 \left(\frac{\partial u}{\partial t} \frac{\partial^2 \delta w_s}{\partial t \partial x} + \frac{\partial^2 w_s}{\partial t \partial x} \frac{\partial \delta u}{\partial t} \right) + K_2 \left(\frac{\partial^2 w_s}{\partial t \partial x} \frac{\partial^2 \delta w_s}{\partial t \partial x} \right)$$

$$\left. +J_2 \left(\frac{\partial^2 w_b}{\partial t \partial x} \frac{\partial^2 \delta w_s}{\partial t \partial x} + \frac{\partial^2 \delta w_b}{\partial t \partial x} \frac{\partial^2 w_s}{\partial t \partial x} \right) \right) dx$$

$$(3.21)$$

Differentiation with respect to time is indicated by the dot-superscript in all equations; the mass inertias employed in the aforementioned equations should have the following form.

$$\left(I_0, I_1, J_1, I_2, J_2, K_2 \right) = \int_A \left(1, z, f(z), z^2, z f(z), f^2(z) \right) \rho(z) dA \qquad (3.22)$$

By substituting Eqs. (3.16), (3.18), and (3.21) into Eq. (3.14) and setting the coefficients of δu, δw_b, δw_s, $\delta\phi$, and $\delta\gamma$ to zero, the Euler–Lagrange equations of MEE-FG rotary nanobeam can be written as:

$$\frac{\partial N}{\partial x} = I_0\ddot{u}_0 - I_1\frac{\partial\ddot{w}_b}{\partial x} - J_1\frac{\partial\ddot{w}_s}{\partial x} \tag{3.23}$$

$$\frac{\partial^2 M_x^b}{\partial x^2} = \left(N^E + N^H + N_{\max}^R\right)\nabla^2\left(w_b + w_s\right) + I_0\left(\ddot{w}_b + \ddot{w}_s\right)$$
$$+ I_1\left(\frac{\partial\ddot{u}}{\partial x}\right) - I_2\nabla^2\ddot{w}_b - J_2\nabla^2\ddot{w}_s \tag{3.24}$$

$$\frac{\partial^2 M_x^s}{\partial x^2} + \frac{\partial Q_{xz}}{\partial x} = \left(N^E + N^H + N_{\max}^R\right)\nabla^2\left(w_b + w_s\right) + I_0\left(\ddot{w}_b + \ddot{w}_s\right)$$
$$+ J_1\left(\frac{\partial\ddot{u}_0}{\partial x}\right) - J_2\nabla^2\ddot{w}_b - K_2\nabla^2\ddot{w}_s \tag{3.25}$$

$$\int_A \left(\cos\left(\xi z\right)\frac{\partial D_x}{\partial x} + \xi\sin\left(\xi z\right)D_z\right)dA = 0 \tag{3.26}$$

$$\int_A \left(\cos\left(\xi z\right)\frac{\partial B_x}{\partial x} + \xi\sin\left(\xi z\right)B_z\right)dA = 0 \tag{3.27}$$

The stress field, in accordance with the nonlocal strain gradient theory, accounts for the influence of both strain gradient stress field and nonlocal elastic stress field. For magneto-electro-elastic solids, this leads us to the following theoretical expression [66]:

$$\left[1-\left(e_1 a\right)^2\nabla^2\right]\left[1-\left(e_0 a\right)^2\nabla^2\right]\sigma_{ij} = \left[1-\left(e_1 a\right)^2\nabla^2\right]\left(C_{ijkl}\varepsilon_{kl} - e_{mij}E_m - q_{nij}H_n\right)$$
$$- l^2\left[1-\left(e_0 a\right)^2\nabla^2\right]\nabla^2\left(C_{ijkl}\varepsilon_{kl} - e_{mij}E_m - q_{nij}H_n\right) \tag{3.28}$$

$$\left[1-\left(e_1 a\right)^2\nabla^2\right]\left[1-\left(e_0 a\right)^2\nabla^2\right]D_i = \left[1-\left(e_1 a\right)^2\nabla^2\right]\left(e_{ikl}\varepsilon_{kl} + s_{im}E_m + d_{in}H_n\right)$$
$$- l^2\left[1-\left(e_0 a\right)^2\nabla^2\right]\nabla^2\left(e_{ikl}\varepsilon_{kl} + s_{im}E_m + d_{in}H_n\right) \tag{3.29}$$

$$\left[1-\left(e_1 a\right)^2\nabla^2\right]\left[1-\left(e_0 a\right)^2\nabla^2\right]B_i = \left[1-\left(e_1 a\right)^2\nabla^2\right]\left(q_{ijk}\varepsilon_{kl} + d_{im}E_m + \chi_{in}H_n\right)$$
$$- l^2\left[1-\left(e_0 a\right)^2\nabla^2\right]\nabla^2\left(q_{ijk}\varepsilon_{kl} + d_{im}E_m + \chi_{in}H_n\right) \tag{3.30}$$

By discarding the terms of order $O(\nabla^2)$ and also assuming $e = e_0 = e_1$, and defining the Laplacian operator as $\nabla^2 = \dfrac{\partial^2}{\partial x^2}$, the simplified constitutive relation can be written as follows:

$$\left(1 - \mu \frac{\partial^2}{\partial x^2}\right)\sigma_{ij} = \left(1 - \eta \frac{\partial^2}{\partial x^2}\right)\left(C_{ijkl}\varepsilon_{kl} - e_{mij}E_m - q_{nij}H_n\right) \qquad (3.31)$$

$$\left(1 - \mu \frac{\partial^2}{\partial x^2}\right)D_i = \left(1 - \eta \frac{\partial^2}{\partial x^2}\right)\left(e_{ikl}\varepsilon_{kl} + s_{im}E_m + d_{in}H_n\right) \qquad (3.32)$$

$$\left(1 - \mu \frac{\partial^2}{\partial x^2}\right)B_i = \left(1 - \eta \frac{\partial^2}{\partial x^2}\right)\left(q_{ijk}\varepsilon_{kl} + d_{im}E_m + \chi_{in}H_n\right) \qquad (3.33)$$

Therefore, we may formulate the following stress-strain equations:

$$\left(1 - \mu \frac{\partial^2}{\partial x^2}\right)\sigma_{xx} = \left(1 - \eta \frac{\partial^2}{\partial x^2}\right)\left(\tilde{c}_{11}\varepsilon_{xx} - \tilde{e}_{31}E_z - \tilde{q}_{31}H_z\right) \qquad (3.34)$$

$$\left(1 - \mu \frac{\partial^2}{\partial x^2}\right)\sigma_{xz} = \left(1 - \eta \frac{\partial^2}{\partial x^2}\right)\left(\tilde{c}_{55}\gamma_{xz} - \tilde{e}_{15}E_x - \tilde{q}_{15}H_x\right) \qquad (3.35)$$

$$\left(1 - \mu \frac{\partial^2}{\partial x^2}\right)D_x = \left(1 - \eta \frac{\partial^2}{\partial x^2}\right)\left(\tilde{e}_{15}\gamma_{xz} + \tilde{s}_{11}E_x + \tilde{d}_{11}H_x\right) \qquad (3.36)$$

$$\left(1 - \mu \frac{\partial^2}{\partial x^2}\right)D_z = \left(1 - \eta \frac{\partial^2}{\partial x^2}\right)\left(\tilde{e}_{31}\varepsilon_x + \tilde{s}_{33}E_z + \tilde{d}_{33}H_z\right) \qquad (3.37)$$

$$\left(1 - \mu \frac{\partial^2}{\partial x^2}\right)B_x = \left(1 - \eta \frac{\partial^2}{\partial x^2}\right)\left(\tilde{q}_{15}\gamma_{xz} + \tilde{d}_{11}E_x + \tilde{\chi}_{11}H_x\right) \qquad (3.38)$$

$$\left(1 - \mu \frac{\partial^2}{\partial x^2}\right)B_z = \left(1 - \eta \frac{\partial^2}{\partial x^2}\right)\left(\tilde{q}_{31}\varepsilon_x + \tilde{d}_{33}E_z + \tilde{\chi}_{33}H_z\right) \qquad (3.39a)$$

where $\mu = (ea)^2$, $\eta = l^2$ are nonlocal and length scale parameters, respectively. Also, $\tilde{c}_{ij}$, $\tilde{d}_{ij}$, $\tilde{e}_{ij}$, $\tilde{s}_{ij}$, $\tilde{q}_{ij}$ and $\tilde{\chi}_{ij}$ are reduced coefficients of an MEE-FG nanobeam when it is subjected to a plane stress state [66]:

$$\tilde{c}_{11} = c_{11} - \frac{c_{13}^2}{c_{33}}, \; \tilde{s}_{11} = s_{11}$$

$$\tilde{e}_{31} = e_{31} - \frac{c_{13}e_{33}}{c_{33}}, \; \tilde{q}_{31} = q_{31} - \frac{c_{13}q_{33}}{c_{33}}, \; \tilde{d}_{11} = d_{11}, \tilde{\chi}_{11} = \chi_{11} \qquad (3.39b)$$

$$\tilde{d}_{33} = d_{33} - \frac{e_{33}q_{33}}{c_{33}}, \tilde{s}_{33} = s_{33} - \frac{e_{33}^2}{c_{33}}, \tilde{\chi}_{33} = \chi_{33} - \frac{q_{33}^2}{c_{33}}$$

The force-strain and moment-strain formulae of the nonlocal refined FG beam across its cross-section can be obtained by integrating Eqs. (3.34)–(3.39).

$$\left(1 - \mu \frac{\partial^2}{\partial x^2}\right) N = \left(1 - \eta \frac{\partial^2}{\partial x^2}\right)\left(A_{xx}\frac{\partial u}{\partial x} - B_{xx}\frac{\partial^2 w_b}{\partial x^2} - B_{xx}^s\frac{\partial^2 w_s}{\partial x^2} + A_{31}^e\phi + A_{31}^m\gamma\right)$$
$$- \left(N^E + N^E + N_{\max}^R\right) \qquad (3.40)$$

$$\left(1 - \mu \frac{\partial^2}{\partial x^2}\right) M_x^b = \left(1 - \eta \frac{\partial^2}{\partial x^2}\right)\left(B_{xx}\frac{\partial u}{\partial x} - D_{xx}\frac{\partial^2 w_b}{\partial x^2} - D_{xx}^s\frac{\partial^2 w_s}{\partial x^2} + E_{31}^e\phi + E_{31}^m\gamma\right)$$
$$- \left(M_{bx}^E + M_{bx}^H + M_{bx}^R\right) \qquad (3.41)$$

$$\left(1 - \mu \frac{\partial^2}{\partial x^2}\right) M_x^s = \left(1 - \eta \frac{\partial^2}{\partial x^2}\right)\left(B_{xx}^s\frac{\partial u}{\partial x} - D_{xx}^s\frac{\partial^2 w_b}{\partial x^2}\right.$$
$$\left. - H_{xx}^s\frac{\partial^2 w_s}{\partial x^2} + F_{31}^e\phi + F_{31}^m\gamma\right) - \left(M_{sx}^E + M_{sx}^H + M_{sx}^R\right) \qquad (3.42)$$

$$\left(1 - \mu \frac{\partial^2}{\partial x^2}\right) Q_{xz} = \left(1 - \eta \frac{\partial^2}{\partial x^2}\right)\left(A_{xz}^s\frac{\partial w_s}{\partial x} - A_{15}^e\frac{\partial \phi}{\partial x} - A_{15}^m\frac{\partial \gamma}{\partial x}\right) \qquad (3.43)$$

$$\left(1 - \mu \frac{\partial^2}{\partial x^2}\right)\int_A \left(D_x \cos(\xi z)\right) dA = \left(1 - \eta \frac{\partial^2}{\partial x^2}\right)\left(E_{15}^e\frac{\partial w_s}{\partial x} + F_{11}^e\frac{\partial \phi}{\partial x} + F_{11}^m\frac{\partial \gamma}{\partial x}\right) \qquad (3.44)$$

$$\left(1 - \mu \frac{\partial^2}{\partial x^2}\right)\int_A \left(D_z\xi \sin(\xi z)\right) dA = \left(1 - \eta \frac{\partial^2}{\partial x^2}\right)\left(A_{31}^e\frac{\partial u}{\partial x} - E_{31}^e\frac{\partial^2 w_b}{\partial x^2}\right.$$
$$\left. - F_{31}^e\frac{\partial^2 w_s}{\partial x^2} - F_{33}^e\phi + F_{33}^m\gamma\right) \qquad (3.45)$$

$$\left(1-\mu\frac{\partial^2}{\partial x^2}\right)\int_A \left(B_x \cos(\xi z)\right)dA = \left(1-\eta\frac{\partial^2}{\partial x^2}\right)\left(E_{15}^m\frac{\partial w_s}{\partial x} + F_{11}^m\frac{\partial \phi}{\partial x} + X_{11}^m\frac{\partial \gamma}{\partial x}\right) \quad (3.46)$$

$$\left(1-\mu\frac{\partial^2}{\partial x^2}\right)\int_A \left(B_z\xi \sin(\xi z)\right)dA = \left(1-\eta\frac{\partial^2}{\partial x^2}\right)\left(A_{31}^m\frac{\partial u}{\partial x}\right.$$
$$\left. - E_{31}^m\frac{\partial^2 w_b}{\partial x^2} - F_{31}^m\frac{\partial^2 w_s}{\partial x^2} - F_{33}^m\phi + X_{33}^m\gamma\right) \quad (3.47)$$

T sectional rigidities are simply defined in Eqs. (3.40)–(3.47) as follows:

$$\left(A_{xx}, B_{xx}, B_{xx}^s, D_{xx}, D_{xx}^s, H_{xx}\right) = \int_A \tilde{c}_{11}\left(1, z, f(z), z^2, zf(z), (f(z))^2\right)dA \quad (3.48)$$

$$A_{xz}^s = \int_A \tilde{c}_{55}\left(1-\frac{df(z)}{dz}\right)^2 dA \quad (3.49)$$

$$\left(A_{31}^e, E_{31}^e, F_{31}^e\right) = \int_A \tilde{e}_{31}\xi \sin(\xi z)\left(1, z, f(z)\right)dA \quad (3.50)$$

$$\left(A_{31}^m, E_{31}^m, F_{31}^m\right) = \int_A \tilde{q}_{31}\xi \sin(\xi z)\left(1, z, f(z)\right)dA \quad (3.51)$$

$$\left(A_{15}^e, E_{15}^e\right) = \int_A \tilde{e}_{15}\cos(\xi z)\left(1, \left(1-\frac{df(z)}{dz}\right)\right)dA \quad (3.52)$$

$$\left(A_{15}^m, E_{15}^m\right) = \int_A \tilde{q}_{15}\cos(\xi z)\left(1, \left(1-\frac{df(z)}{dz}\right)\right)dA \quad (3.53)$$

$$\left(F_{11}^e, F_{33}^e\right) = \int_A \left(\tilde{s}_{11}\cos^2(\xi z), \tilde{s}_{33}\xi^2 \sin^2(\xi z)\right)dA \quad (3.54)$$

$$\left(F_{11}^m, F_{33}^m\right) = \int_A \left(\tilde{d}_{11}\cos^2(\xi z), \tilde{d}_{33}\xi^2 \sin^2(\xi z)\right)dA \quad (3.55)$$

$$\left(X_{11}^m, X_{33}^m\right) = \int_A \left(\tilde{\chi}_{11}\cos^2(\xi z), \tilde{\chi}_{33}\xi^2 \sin^2(\xi z)\right)dA \quad (3.56)$$

Furthermore, the angular velocity of the beam in addition to the magnetic and electric fields' normal bending and shear moments may be written as:

$$\left(M_{bx}^{E}, M_{sx}^{E}\right) = -\int_{A} \tilde{e}_{31} \frac{2V}{h} \left(z, f\left(z\right)\right) dA$$

$$\left(M_{bx}^{H}, M_{sx}^{H}\right) = -\int_{A} \tilde{q}_{31} \frac{2V}{h} \left(z, f\left(z\right)\right) dA$$

$$\left(M_{bx}^{R}, M_{sx}^{R}\right) = \int_{0}^{L}\int_{A} \left(\rho\left(z\right) A\Phi^{2}x\right)\left(z, f\left(z\right)\right) dA dx$$

(3.57)

Here, we show that the displacement fields may be used to directly rewrite the nonlocal governing equations for MEE-FG nanosize beams.

$$\left(1-\eta\frac{\partial^2}{\partial x^2}\right)\left(A_{xx}\frac{\partial^2 u}{\partial x^2} - B_{xx}\frac{\partial^3 w_b}{\partial x^3} - B_{xx}^s\frac{\partial^3 w_s}{\partial x^3} + A_{31}^e\frac{\partial\phi}{\partial x} + A_{31}^m\frac{\partial\gamma}{\partial x}\right)$$

$$+\left(1-\mu\frac{\partial^2}{\partial x^2}\right)\left(-I_0\ddot{u} + I_1\frac{\partial\ddot{w}_b}{\partial x} + J_1\frac{\partial\ddot{w}_s}{\partial x}\right) = 0$$

(3.58)

$$\left(1-\eta\frac{\partial^2}{\partial x^2}\right)\left(B_{xx}\frac{\partial^3 u}{\partial x^3} - D_{xx}\frac{\partial^4 w_b}{\partial x^4} - D_{xx}^s\frac{\partial^4 w_s}{\partial x^4} + \left(E_{31}^e - A_{15}^e\right)\frac{\partial^2\phi}{\partial x^2}\right)$$

$$+\left(E_{31}^m - A_{15}^m\right)\frac{\partial^2\gamma}{\partial x^2} + \left(1-\mu\frac{\partial^2}{\partial x^2}\right)\left(-I_0\left(\ddot{w}_b + \ddot{w}_s\right) - I_1\frac{\partial\ddot{u}}{\partial x} + I_2\frac{\partial^2\ddot{w}_b}{\partial x^2}\right)$$

(3.59)

$$+J_2\frac{\partial^2\ddot{w}_s}{\partial x^2} - \left(N_{max}^R + N^H + N^E\right)\frac{\partial^2\left(w_b + w_s\right)}{\partial x^2} = 0$$

$$\left(1-\eta\frac{\partial^2}{\partial x^2}\right)\left(B_{xx}^s\frac{\partial^3 u}{\partial x^3} - D_{xx}^s\frac{\partial^4 w_b}{\partial x^4} - H_{xx}^s\frac{\partial^4 w_s}{\partial x^4} + \left(F_{31}^e - E_{15}^e\right)\frac{\partial^2\phi}{\partial x^2}\right)$$

$$+\left(F_{31}^m - E_{15}^m\right)\frac{\partial^2\gamma}{\partial x^2} + \left(1-\mu\frac{\partial^2}{\partial x^2}\right)\left(-I_0\left(\ddot{w}_b + \ddot{w}_b\right) - J_1\frac{\partial\ddot{u}}{\partial x}\right)$$

(3.60)

$$+J_2\frac{\partial^2\ddot{w}_b}{\partial x^2} + K_2\frac{\partial^2\bar{H}}{\partial x^2} - \left(N_{max}^R + N^H + N^E\right)\frac{\partial^2\left(w_b + w_s\right)}{\partial x^2} = 0$$

$$\left(1-\eta\frac{\partial^2}{\partial x^2}\right)\left(A_{31}^e\frac{\partial u}{\partial x}-E_{31}^e\frac{\partial^2 w_b}{\partial x^2}-\left(F_{31}^e-E_{15}^e\right)\frac{\partial^2 w_s}{\partial x^2}+F_{11}^e\frac{\partial^2\phi}{\partial x^2}\right.$$
$$\left.+F_{11}^m\frac{\partial^2\gamma}{\partial x^2}-F_{33}^e\phi-F_{33}^m\gamma\right)=0 \tag{3.61}$$

$$\left(1-\eta\frac{\partial^2}{\partial x^2}\right)\left(A_{31}^m\frac{\partial u}{\partial x}-E_{31}^m\frac{\partial^2 w_b}{\partial x^2}-\left(F_{31}^m-E_{15}^m\right)\frac{\partial^2 w_s}{\partial x^2}+F_{11}^m\frac{\partial^2\phi}{\partial x^2}\right.$$
$$\left.+X_{11}^m\frac{\partial^2\gamma}{\partial x^2}-F_{33}^m\phi-X_{33}^m\gamma\right)=0 \tag{3.62}$$

By using the same substitutions as in reference [49], we can also arrive at the final governing equations of the FG-double-nanobeam system.

$$\left(1-\lambda^2\nabla^2\right)\left(A\frac{\partial^2 u_1}{\partial x^2}-B\frac{\partial^3 w_1}{\partial x^3}\right)+\left(1-\mu^2\nabla^2\right)\left(-I_0\frac{\partial^2 u_1}{\partial t^2}+I_1\frac{\partial^3 w_1}{\partial x\partial t^2}\right)=0 \tag{3.63}$$

$$\left(1-\lambda^2\nabla^2\right)\left(B\frac{\partial^3 u_1}{\partial x^3}-D\frac{\partial^4 w_1}{\partial x^4}\right)$$
$$+\left(1-\mu^2\nabla^2\right)\left(\begin{array}{c}-I_0\frac{\partial^2 w_1}{\partial t^2}-I_1\left(\frac{\partial^3 u_1}{\partial x\partial t^2}\right)+I_2\nabla^2\frac{\partial^2 w_1}{\partial t^2}+\eta AH_x^2\frac{\partial^2 w_1}{\partial x^2}\\-k_w w_1+k_p\nabla^2 w_1-\left(N^T+N^R\right)\nabla^2 w_1-K_0\left(w_1-w_2\right)\end{array}\right)=0 \tag{3.64}$$

$$\left(1-\lambda^2\nabla^2\right)\left(A\frac{\partial^2 u_2}{\partial x^2}-B\frac{\partial^3 w_2}{\partial x^3}\right)+\left(1-\mu^2\nabla^2\right)\left(-I_0\frac{\partial^2 u_2}{\partial t^2}+I_1\frac{\partial^3 w_2}{\partial x\partial t^2}\right)=0 \tag{3.65}$$

$$\left(1-\lambda^2\nabla^2\right)\left(B\frac{\partial^3 u_2}{\partial x^3}-D\frac{\partial^4 w_2}{\partial x^4}\right)$$
$$+\left(1-\mu^2\nabla^2\right)\left(\begin{array}{c}-I_0\frac{\partial^2 w_2}{\partial t^2}-I_1\left(\frac{\partial^3 u_2}{\partial x\partial t^2}\right)+I_2\nabla^2\frac{\partial^2 w_2}{\partial t^2}+\eta AH_x^2\frac{\partial^2 w_2}{\partial x^2}\\-k_w w_2+k_p\nabla^2 w_2-\left(N^T+N^R\right)\nabla^2 w_2-K_0\left(w_2-w_1\right)\end{array}\right)=0 \tag{3.66}$$

In Eqs. (3.63)–(3.66), K_0 is the interlayer stiffness that couples the motion of nanobeams. Here we solve analytically the nonlocal governing equations developed in the preceding section. Below, we shall define the displacement fields on the assumption that they are exponential:

$$\begin{Bmatrix} u(x,z,t) \\ w_b(x,z,t) \\ w_s(x,z,t) \\ \phi(x,z,t) \\ \gamma(x,z,t) \end{Bmatrix} = \begin{Bmatrix} U\exp\left[i(\beta x - \omega t)\right] \\ W_b\exp\left[i(\beta x - \omega t)\right] \\ W_s\exp\left[i(\beta x - \omega t)\right] \\ \Phi\exp\left[i(\beta x - \omega t)\right] \\ Y\exp\left[i(\beta x - \omega t)\right] \end{Bmatrix} \qquad (3.67)$$

where U, W_b, W_s, Φ and Y are the unknown coefficients. Also, β is wave number of propagating waves, and finally ω is wave's angular frequency. Now, substituting Eq. (3.67) to Eqs. (3.58)–(3.62) reveals:

$$\left([K]_{5\times5} - \omega^2[M]_{5\times5}\right)\begin{Bmatrix} U \\ W_b \\ W_s \\ \Phi \\ Y \end{Bmatrix} = \{0\} \qquad (3.68)$$

where the corresponding k_{ij}, m_{ij} are presented in Appendix A [57]. In order to attain wave frequency, the determinant of the left-hand side of Eq. (3.68) must be set to zero:

$$\left|[K] - \omega^2[M]\right| = 0 \qquad (3.69)$$

Solving the obtained equation for ω, the wave's angular frequency of MEE-FG rotary nanobeam can be expressed as below:

$$\omega_i = M_i(k), i = (0,1,2,3,4) \qquad (3.70)$$

The angular frequencies of the wave's first five natural modes are given in Eq. (3.70). Each mode's phase velocity may be calculated by dividing its frequency by its wave number, as follows:

$$c_i = \frac{M_i(k)}{k}, (i = 0,1,2,3,4) \qquad (3.71)$$

When the wave number is allowed to tend to infinity, the escape frequency of the MEE-FG rotating nanobeam is calculated.

$$\tilde{\omega}_i = \lim_{k\to\infty}\frac{\omega_i}{2\pi}, i = (0,1,2,3,4) \qquad (3.72)$$

Specifically, there are three types of motion that may occur in a double-nanobeam system.

- Out-of-phase ($w_{rel} = w_1 - w_2 \neq 0$).
- In-phase ($w_{rel} = w_1 - w_2 = 0$).
- One nanobeam fixed ($w_{rel} = w_1 = 0$).

In fact, in-phase motion occurs anytime there is synchronous motion between two or more nanobeams. However, the out-of-phase condition arises when the nanobeams' relative velocity is not synchronized. In the end, the motion of a single-layered nanobeam is obtained if one of the nanobeams is kept still.

3.2.1 MAGNETIC FIELD EFFECTS

In this subsection, the wave propagation characteristics of an MEE-FG rotary nanobeam made of $BaTiO_3$ and $CoFe_2O_4$ is analyzed in detail once subjected to an external electric voltage and magnetic potential. This part consists of plenty of diagrams, which are plotted in order to investigate the variations of wave frequency and phase velocity and also to escape frequency of MEE-FG rotary nanobeam. The effect of wave number, nonlocality, length scale parameter, material distribution, electric voltage, magnetic potential, and angular velocity on wave characteristics of MEE-FG rotary nanobeam is studied in reference [64]. In Figure 3.2 the

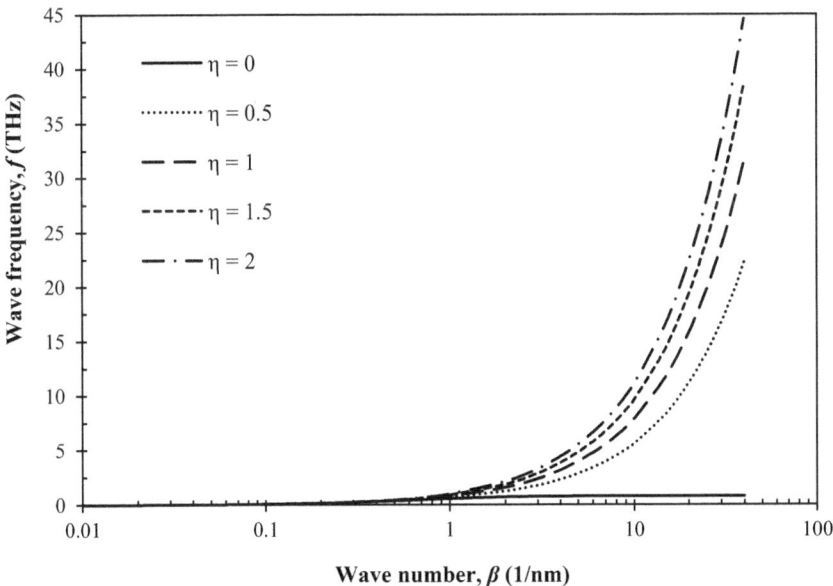

FIGURE 3.2 Variation in the wave frequency of FG rotary nanobeam versus wave number for various length scale parameters ($p = 1$, $V = \Omega = 0$, $\mu = 1$ nm^2, $\Phi = 1$ Grad/s).

variations in wave frequency versus wave number are studied using a large variety
of different length scale parameters at $p = 1$, $V = \Omega = 0$, $\mu = 1$ nm², $\Phi = 1$ Grad/s.
It is clear that length scale parameter has a hardening influence on the wave frequency of FG nanobeam, which is not rendered in nonlocal elasticity. Obviously,
an increasing trend can be seen for the wave frequency whenever the length scale
parameter increases. Another interpretation can be devoted to the different behavior of wave frequency once the length scale parameter possesses various amounts.
In other words, wave frequency approximately remains constant if the length scale
parameter is set to zero ($\eta = 0$). On the other hand, wave frequency tends to infinity
once the length scale parameter has a nonzero amount ($\eta \neq 0$).

Furthermore, for $p = 1$, $V = \Omega = 0$, $\mu = 1$ nm², and $\Phi = 1$ Grad/s, the effects of
different length scale factors on the phase velocity of FG rotational nanoscale
beams are explored. The length scale parameter's influence on stiffness-hardening
may be classified into three subtypes. When the length scale parameter is less than
the nonlocal parameter ($\eta < \mu$), the phase velocity rises until it reaches a peak and
then it drops to a minimum. For each given value of the wave number, the phase
velocity initially grows to a maximum value, where it then stays constant regardless of the nonlocal or length scale parameters ($\eta = \mu$). Finally, remember that once
the phase velocity has reached its limit magnitude, it cannot increase any further
regardless of the wave number. This is because the phase velocity continues to
increase even after it has reached its initial relative maximum value. Consequently,
it is clear that the phase velocity of FG rotational nanobeams may be affected by
the wave number and the length scale parameter in a nonincreasing way [64].

When the length scale parameter is assumed to be smaller than the nonlocal
parameter, equal to, and larger than it at $p = 1$, $V = \Omega = 0$, $\mu = 1$ nm², $\Phi = 1$ Grad/s,

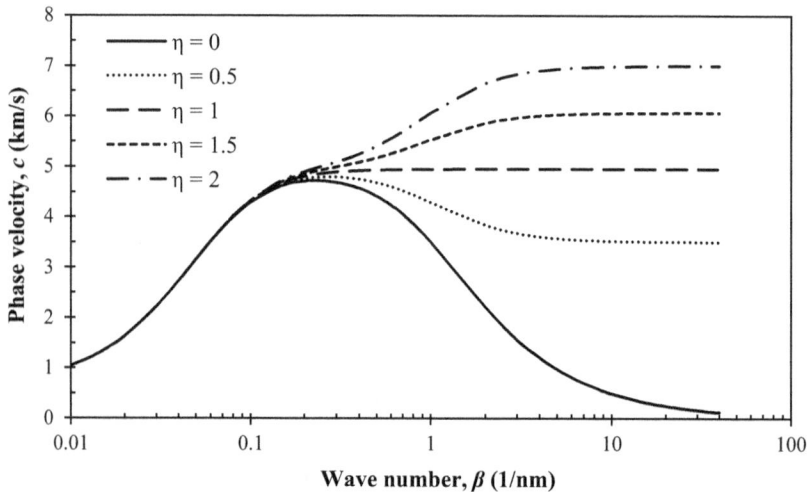

FIGURE 3.3 Variation in the phase velocity of FG rotary nanobeam versus wave number
for various length scale parameters ($p = 1$, $V = \Omega = 0$, $\mu = 1$ nm², $\Phi = 1$ Grad/s).

the smart magneto-electric characteristics of waves propagating across a nano-beam can be studied by plotting the variations of phase velocity versus wave frequency for different magnetic potentials and electric voltages (see Figures 3.4 and 3.5). According to the diagrams, as the frequency of the waves rises, so does the speed at which the phases pass. In fact, it is evident that a rise in the wave frequency may also result in an increase in the amount of phase velocity, and this is true irrespective of the fraction of the length scale parameter to the nonlocal parameter. However, phase velocity does not exhibit uniformity among the three cases. In other words, when the length scale parameter is less than the nonlocal parameter, the phase velocity decreases as the wave frequency increases, but it

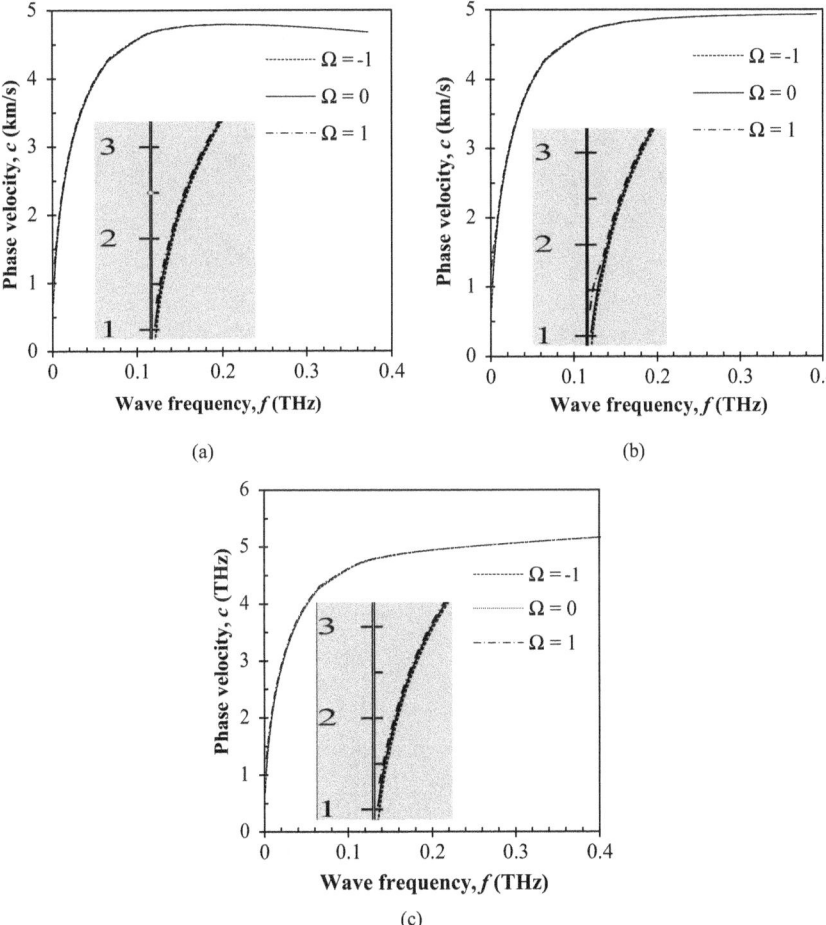

FIGURE 3.4 Variation in the phase velocity of FG rotary nanobeam versus wave frequency for various amounts of magnetic potential for (a) $\eta < \mu$, (b) $\eta = \mu$, and (c) $\eta > \mu$ ($p = 1$, $V = 0$, $\mu = 1$ nm^2, $\Phi = 1$ Grad/s).

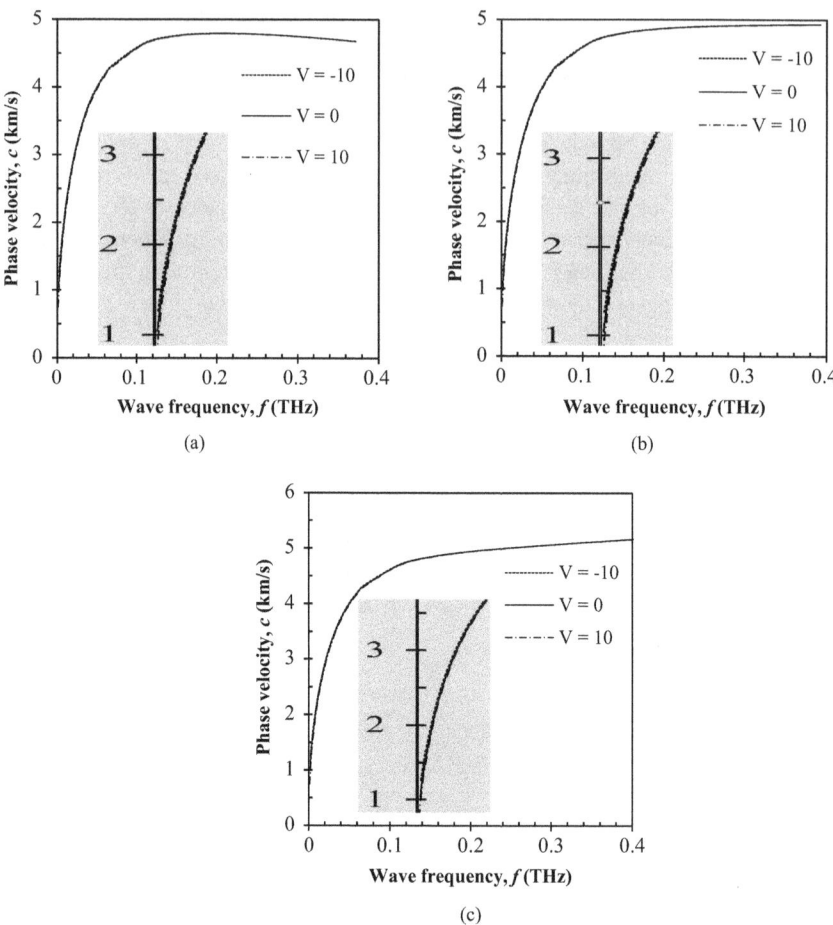

FIGURE 3.5 Variation in the phase velocity of FG rotary nanobeam versus wave frequency for various amounts of electric voltage for (a) $\eta < \mu$, (b) $\eta = \mu$, and (c) $\eta > \mu$ ($p = 1$, $\Omega = 0$, $\mu = 1$ nm^2, $\Phi = 1$ Grad/s).

doesn't change much when the two are equal. If the length scale parameter is bigger than the nonlocality parameter ($\eta > \mu$), then the phase velocity continues becoming larger as the wave frequency grows. Similarities between the three kinds allow us to deduce that a larger magnetic potential may lead to a higher phase velocity value. Furthermore, the impact of magnetic potential is quite distinct from the effects of applied electric voltage. Increasing the electric voltage results in a reduction in the quantity of phase velocity for FG nanobeam, proving the inverse relationship between the two. The phase velocity of propagating waves is shown to change as a function of the applied electric voltage and magnetic potential in these two graphs.

Furthermore, in Figures 3.6 and 3.7, the effects of magnetic potential and electric voltage on the phase velocity of FG nanobeams versus wave number are shown, providing a brief summary of the intelligent properties of propagating waves. The effects of the length scale parameter are also taken into account by splitting the inquiry into three distinct groups, as shown in Figures 3.4 and 3.5. When the length scale variable is less than the nonlocal parameter, a decrease in the phase velocity can be observed after it reaches its maximum value. At the point where the length scale parameter equals the nonlocal parameter, a specific wave number (approximately $\beta = 0.3 \times 10^9$) corresponds to the maximum phase

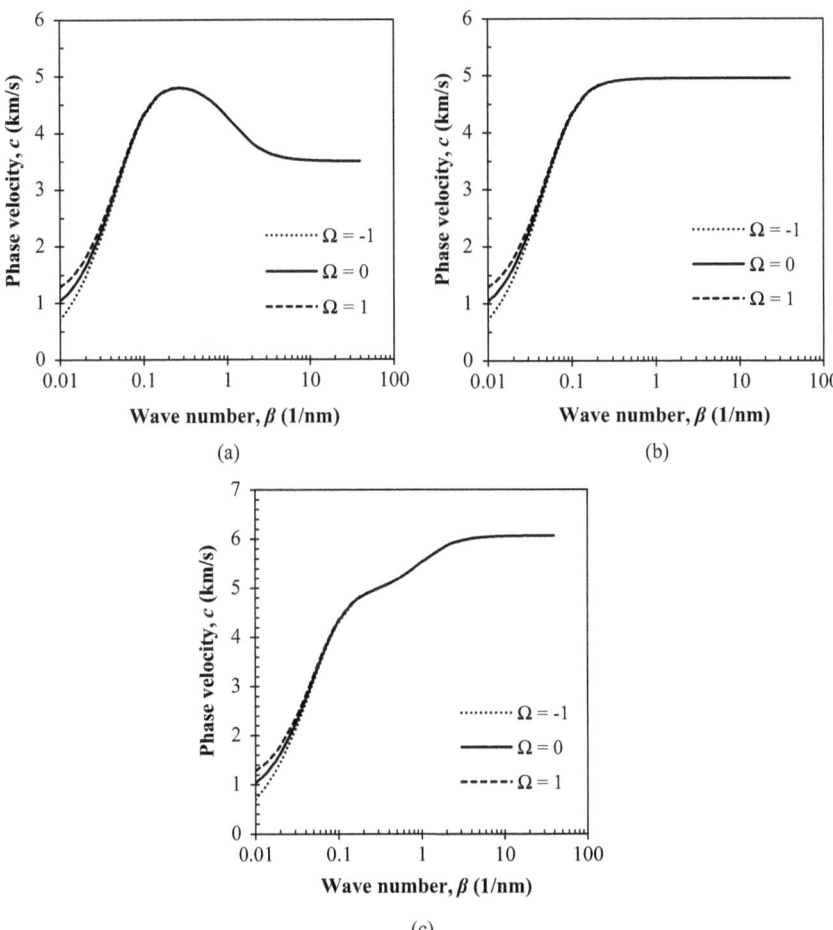

FIGURE 3.6   Variation in the phase velocity of FG rotary nanobeam versus wave number for various amounts of magnetic potential for (a) $\eta < \mu$, (b) $\eta = \mu$, and (c) $\eta > \mu$ ($p = 1$, $V = 0$, $\mu = 1$ nm², $\Phi = 1$ Grad/s).

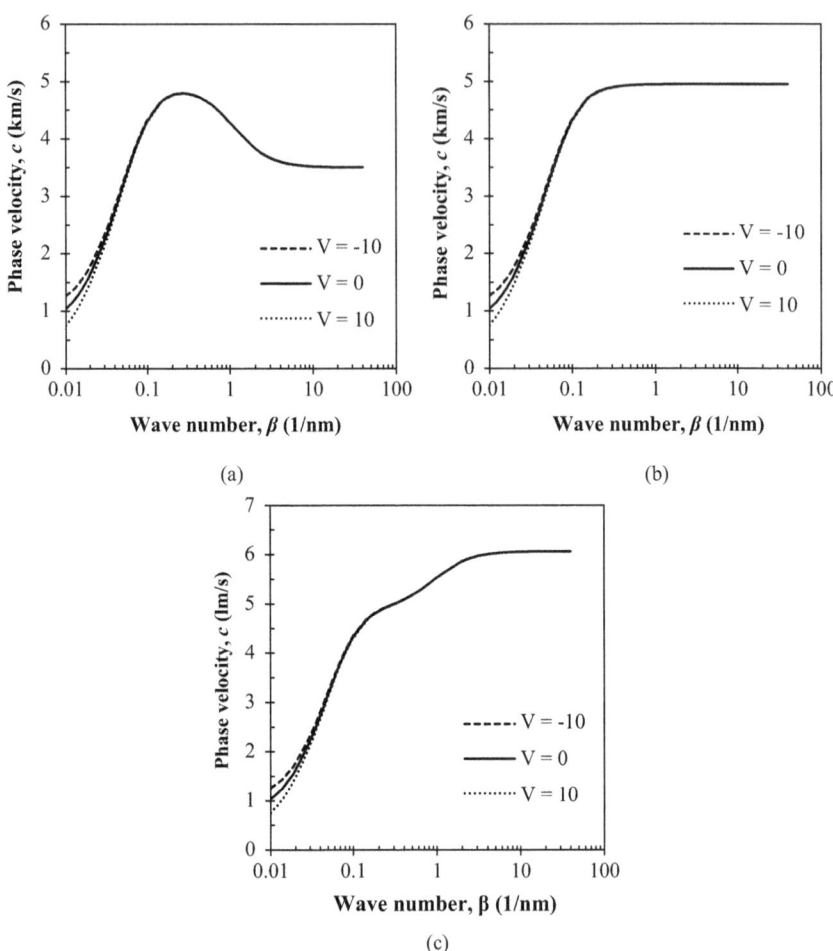

FIGURE 3.7 Variation in the phase velocity of FG rotary nanobeam versus wave number for various amounts of electric voltage for (a) $\eta < \mu$, (b) $\eta = \mu$, and (c) $\eta > \mu$ ($p = 1$, $\Omega = 0$, $\mu = 1$ nm^2, $\Phi = 1$ Grad/s).

velocity. As the wave number approaches almost $\beta = 5 \times 10^9$, the phase velocity continues to increase until it reaches its maximum value, after which it remains constant. The smart behavior of nanobeam is important, in addition to the stated effects. This phenomenon is a localized effect that becomes readily apparent below the wave number $\beta = 0.05 \times 10^9$. In the range of wave numbers stated, increasing the phase velocity may be achieved by adjusting the magnetic potential and electric voltage. In other words, the smart interaction of the waves propagating in an FG beam may be viewed as a rise in the amount of phase velocity whenever magnetic potential is raised or the amount of applied electric voltage is lowered. The phase velocity of an FG nanobeam is insensitive to variations in magnetic potential or electric voltage at high wave numbers.

Now that we have covered the clever behaviors of nanobeams, we can shift our attention to the finer points of angular velocity. At $p = 1$, $V = \Omega = 0$, and $\mu = 1$ nm^2, Figure 3.8 displays the phase velocity of an FG rotating nanobeam as a function of wave number over a range of angular velocities. This graphic clearly shows how the phase velocity of FG nanobeams is affected by the beam's rotational velocity. The addition of the angular velocity of the beam clearly magnifies the phase velocity across a very narrow range of wave numbers. There are practical limits to this enhancement; specifically, the range of wave numbers for which an increase in angular velocity may increase the phase velocity of scattered waves is generally bounded by $\beta = 0.2 \times 10^9$. This graphic shows the differing behavior of size-dependent beams when the length scale parameter is less than, equal to, or greater than the nonlocal parameter, in addition to the effects of beam's rotation.

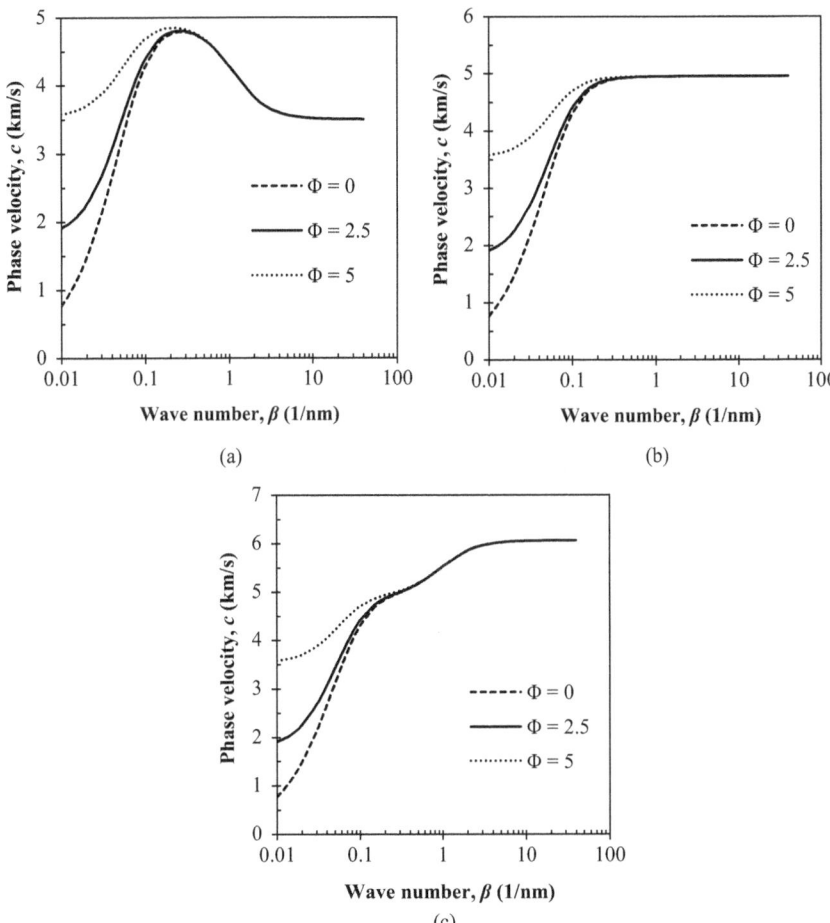

FIGURE 3.8 Variation in phase velocity versus wave number for various angular velocities for (a) $\eta < \mu$, (b) $\eta = \mu$, and (c) $\eta > \mu$ ($p = 1$, $V = \Omega = 0$, $\mu = 1$ nm^2).

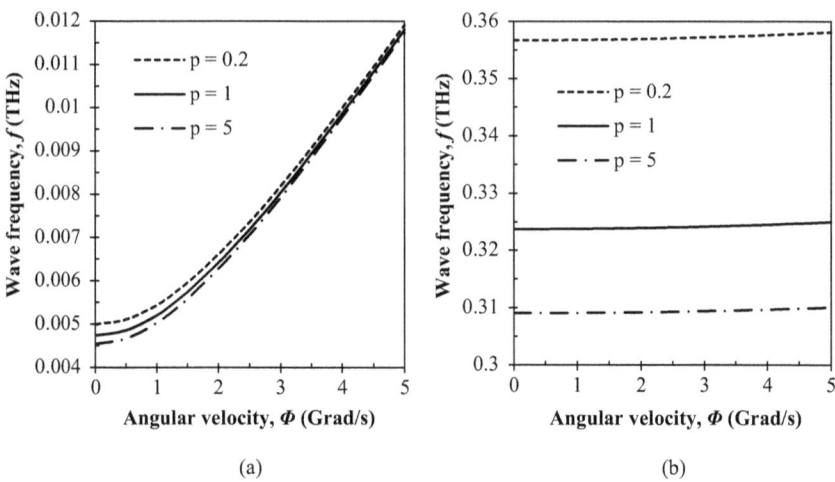

FIGURE 3.9 Variation in wave frequency versus angular velocity for various values of power law exponent for (a) $\beta = 0.02 \times 10^9$ and (b) $\beta = 0.4 \times 10^9$ (V = Ω = 0, μ = 1 nm^2, η = 1.5 nm^2).

For three distinct wave numbers (V = Ω = 0, μ = 1 nm^2, and η = 1.5 nm^2), Figure 3.9 considers the effects of the beam's rotational velocity and gradient index on the FG nanobeam's wave frequency. It is understood that the frequency of a wave is more sensitive to changes in angular velocity for smaller wave numbers than for larger ones. Increasing the beam's rotational velocity may raise the frequency of the reflected waves; however, this effect is much more pronounced at low wave numbers and disappears at higher ones. The graphic also takes into account the impact of the parameter of material distribution (the power law exponent). When a power law exponent is applied, the frequency of the wave decreases for any given rotational velocity and wave number, as described. Therefore, it is concluded that a larger angular velocity or a smaller power law exponent may enhance the wave frequency of FG nanobeams.

The influence of material distribution and angular velocity on the phase velocity of FG nanobeams at V = Ω = 0, μ = 1 nm^2, and η = 1.5 nm^2 are also explored in Figure 3.10. When the wave number is low, the relationship between angular velocity and phase velocity becomes more obvious, as seen by this graphic and the one before it. Nonetheless, for sufficiently high wave numbers, one may get an approximation of phase velocity independent of angular velocity. However, when the volume proportion of ceramic and metal phases is altered, a consistent pattern emerges. It is true that the phase velocity of the FG nanobeam decreases with increasing gradient index. It is important to note that the effect of the gradient index is not constant for low wave numbers. For example, once Φ = 1 Grad/s, the change in phase velocity and wave frequency may be more significant than Φ = 5 Grad/s if the power law exponent were altered from p = 0.2 to p = 5. If the power

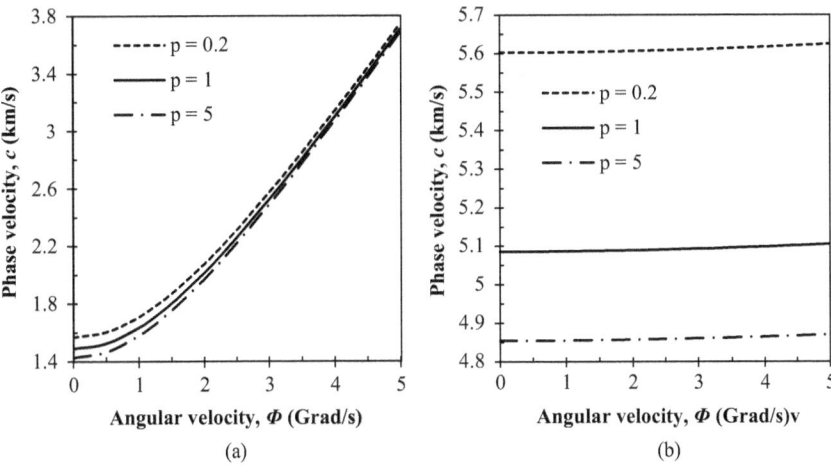

FIGURE 3.10 Variation in phase velocity versus angular velocity for various values of power law exponent for (a) $\beta = 0.02 \times 10^9$ and (b) $\beta = 0.4 \times 10^9$ (V = Ω = 0, μ = 1 nm^2, $\eta = 1.5$ nm^2).

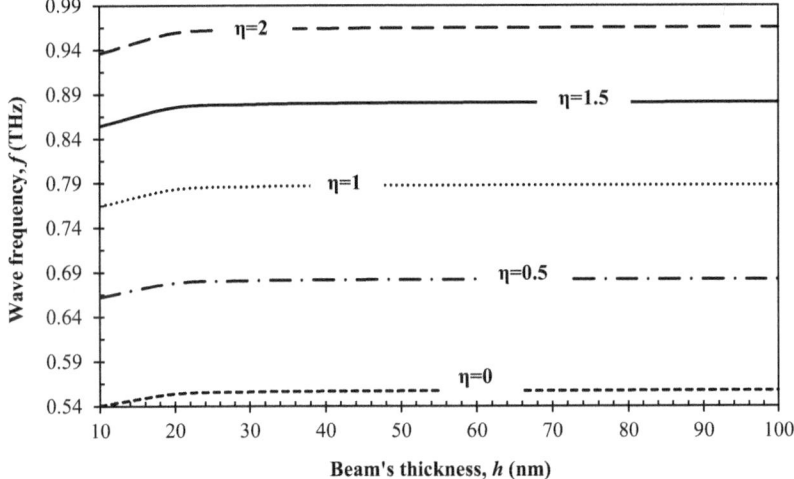

FIGURE 3.11 Variation in wave frequency versus beam's thickness for various length scale parameters ($p = 1, \mu = 1$ nm^2, V = Ω = 0, $\beta = 1 \times 10^9$, $\Phi = 1$ Grad/s).

law exponent is reduced or the angular velocity of the beam is raised, then higher phase velocities become available.

When $p = 1$, $\mu = 1$ nm^2, V = Ω = 0, $\beta = 1 \times 10^9$, and $\Phi = 1$ Grad/s, the effects of the length scale parameter and the beam's thickness on the FG nanobeam's wave frequency are displayed in Figure 3.11. In each specified thickness, the length scale parameter serves to amplify the wave frequency. It is also evident that the quantity of frequency of a wave may be increased by increasing the value of the beam's thickness. Thinner sections reveal this amplifying impact more clearly than thicker sections do. Beam thickness is no longer a viable option for increasing the wave frequency after $h = 20$ nm.

In addition, Figure 3.12 shows the results of an investigation into the relationship between the applied electric voltage and the phase velocity over a range of magnetic potentials and angular velocities: $p = 1$, $\mu = 1$ nm², $\eta = 1.5$ nm², and $\beta = 0.02 \times 10^9$. In reality, a recap of earlier diagrams may be seen in this one. It has been shown that the phase velocity value decreases noticeably as the applied electric voltage rises. When the angular velocity is zero, the effect becomes more apparent than with other values. Note that by adding angular velocity, phase velocity may be increased. Nanobeams' intelligent magnetic behaviors may also be seen once again. Increasing the magnetic potential is therefore another means of accelerating the phase transition.

In addition, the behavior of the escape frequency of scattered waves is investigated for different values of the expected parameters in Figures 3.13 and 3.14. At $\mu = 1$ nm², $V = \Omega = 0$, and $\Phi = 1$ Grad/s, all plausible patterns are shown again in Figure 3.13, which plots changes in escape frequency versus gradient index for different length scale values. According to the diagram, the quantity of escape frequency may be increased by adding the length scale parameter. Parameters like applied electric voltage and magnetic potential that affect the intelligent behavior of FG nanobeam are assumed to be zero in this diagram because they cannot stimulate wave frequency in large wave numbers and escape frequency of nanobeam is obtained by tending wave number to infinity. Additionally, the effects of

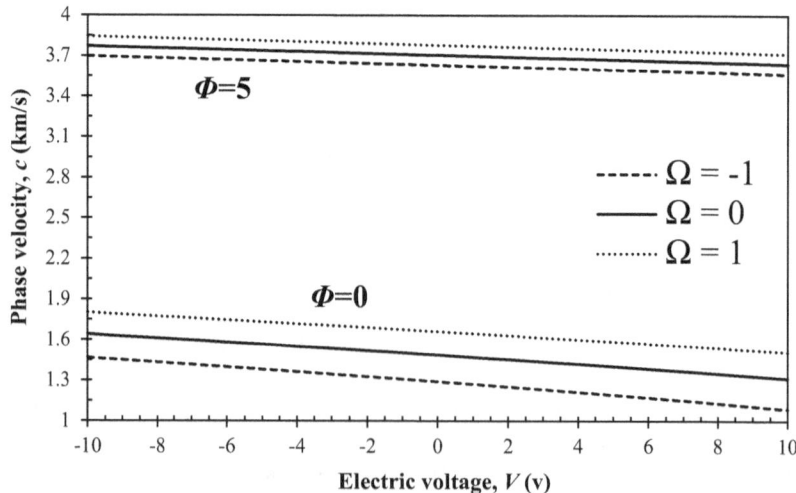

FIGURE 3.12 Variation in phase velocity versus electric voltage for various magnetic potentials ($p = 1$, $\mu = 1$ nm², $\eta = 1.5$ nm², $\beta = 0.02 \times 10^9$).

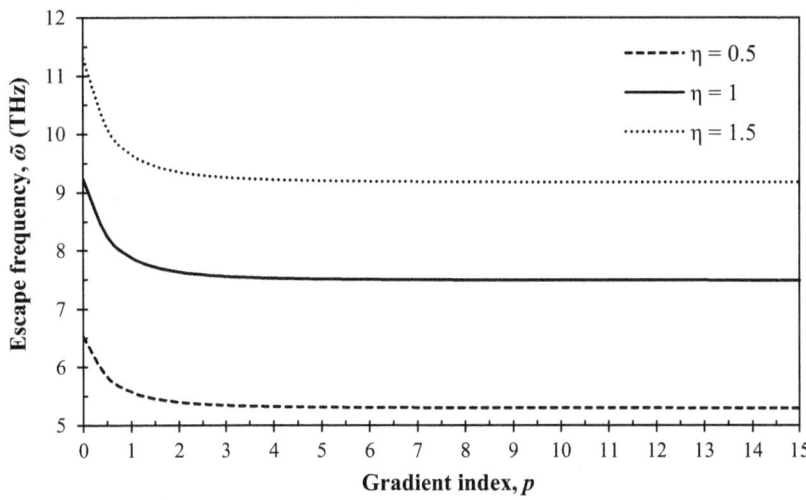

FIGURE 3.13 Variation in escape frequency versus gradient index for various length scale parameters ($\mu = 1$ nm², V $= \Omega = 0$, $\Phi = 1$ Grad/s).

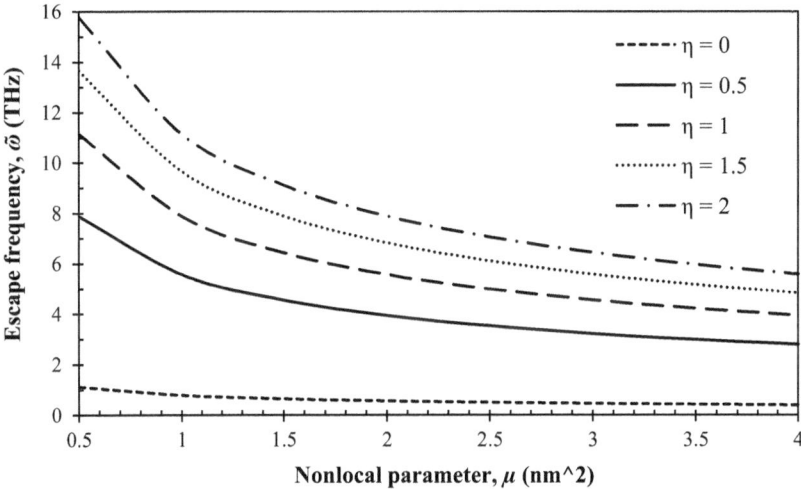

FIGURE 3.14 Variation in escape frequency versus nonlocal parameter for various length scale parameters ($p = 1$, V $= \Omega = 0$, $\Phi = 1$ Grad/s).

material dispersion on the FG nanobeam's escape frequency are taken into account. It is evident that the amplitude of the escape frequency decreases when the volume fraction of the metal phase is raised, meaning the gradient index is added, at low gradient indices. However, if the underlying gradient index is assumed to be greater than around $p = 3$, then the escape frequency is independent of the underlying gradient index.

Finally, the effects of scale are presented in Figure 3.14 for the FG nanobeam escape frequency of $p = 1$, $V = \Omega = 0$, and $\Phi = 1$ Grad/s. The escape frequency of an FG nanobeam is shown to be very sensitive to the nonlocal parameter. In fact, for any value of the length scale parameter, the escape frequency goes to infinity after nonlocality is disregarded. There are a number of ways in which the escape frequency may be modified by nonlocal and length scale characteristics. In other words, the escape frequency of an FG nanobeam decreases when nonlocality is increased, whereas it increases proportionally with the magnitude of the length scale parameter. To be more exact, it can be noted that when the nonlocal parameter is adjusted from 0 to $\mu = 0.5$ nm^2, the resulting softening effect is more pronounced than when the value is changed from $\mu = 3.5$ nm^2 to $\mu = 4$ nm^2.

3.2.2 THERMAL FIELD EFFECTS

Figure 3.15 shows the relationship between the wave number (β) and the frequency ($f = \omega/2\pi$) of a temperature-dependent FG nanobeam for different wave modes, with the gradient index (p) and nonlocality parameter ($\mu = 1$ nm^2) held constant

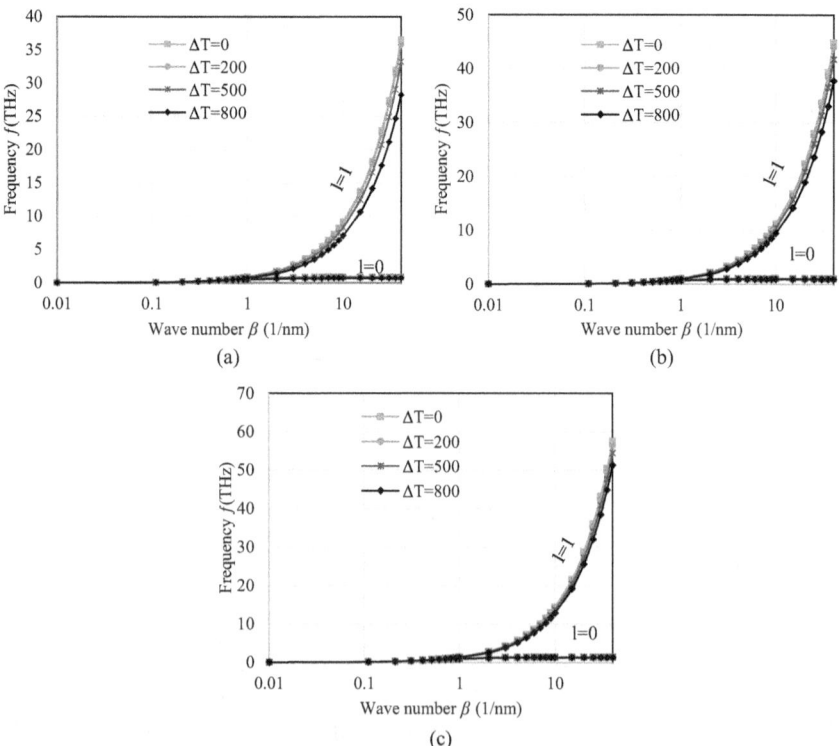

FIGURE 3.15 Effect of nonlinear temperature and length scale parameter on the frequency of temperature-dependent FG nanobeam for various wave modes ($\mu = 1$, $p = 1$). (a) M_0 mode, (b) M_1 mode, and (c) M_2 mode.

at 1. For small values of β (less than or equal to 0.1), it is clear that temperature variations have no discernible impact on the wave frequency. However, this is not the case for larger wave numbers. Specifically, greater levels of wave number see a decrease in frequency as the temperature rises. Therefore, the value of the wave number determines the impact of temperature change on the frequency of the wave. In addition, the notion of nonlocal strain gradients ($l = 1$ nm^2) states that greater numbers of waves result in higher frequencies. Furthermore, when $l = 0$, the wave number has no appreciable influence on the frequency of waves [61, 64].

Figure 3.16 depicts the correlation between the wave number and phase velocity of a temperature-dependent (ΔT) FG nanobeam, while maintaining a constant gradient index ($p = 1$) and nonlocality variable ($\mu = 1$nm^2). The phase velocity remains unaffected by the modification of the length scale parameter for $\beta \leq 0.1$. However, for higher wave numbers, the impact of the length scale parameter on the phase velocity becomes more prominent while maintaining a constant temperature. It can be inferred that the phase velocity remains relatively constant as the wave number increases. Regardless of the length scale value, an increase in

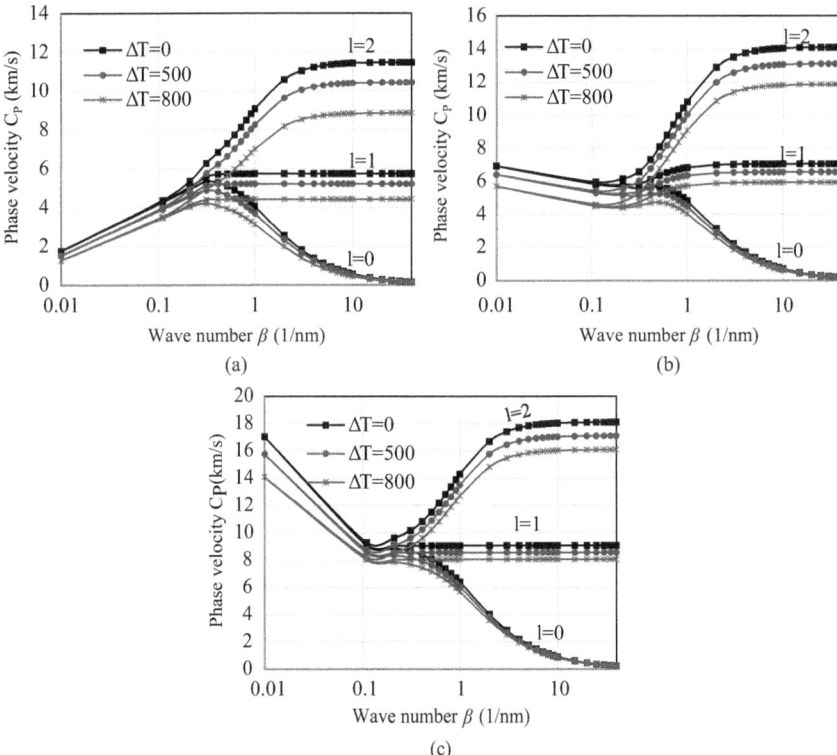

FIGURE 3.16 The effect of nonlinear temperature and length scale parameter on the phase velocity of temperature-dependent FG nanobeam for various wave modes ($\mu = 1$, $p = 1$). (a) M_0 mode, (b) M_1 mode, and (c) M_2 mode.

temperature results in a decrease in phase velocities. When the length scale value is zero ($l = 0$), higher wave modes exhibit lower sensitivity to temperature changes.

In Figure 3.17, we see how the temperature dependence of the group velocities (cg) of the FG nanobeam varies with wave number for a fixed gradient index ($p = 1$). It is clear that the number of waves has a major impact on the dispersion of the group velocity. It is clear that the group velocity of distinct length scale parameters cannot be distinguished from one another at lower wave numbers. However, for larger wave numbers, there is a clear separation between the group speeds. At a constant length scale parameter, it can be shown that a higher temperature results in a slower group velocity. Additionally, the impact of temperature variation on group velocities is negligible for $l = 0$ and higher wave numbers. Furthermore, when the length scale parameter increases, the group velocities increase for a given wave number, thanks to the stiffness-hardening effect presented by the nonlocal strain gradient theory.

Figure 3.18 shows how the escape frequency of an FG nanobeam varies as a function of temperature, nonlocality parameter ($\mu = 1$ nm^2), and length scale parameter ($l = 1$ nm^2) for different wave modes. This is achieved by putting an

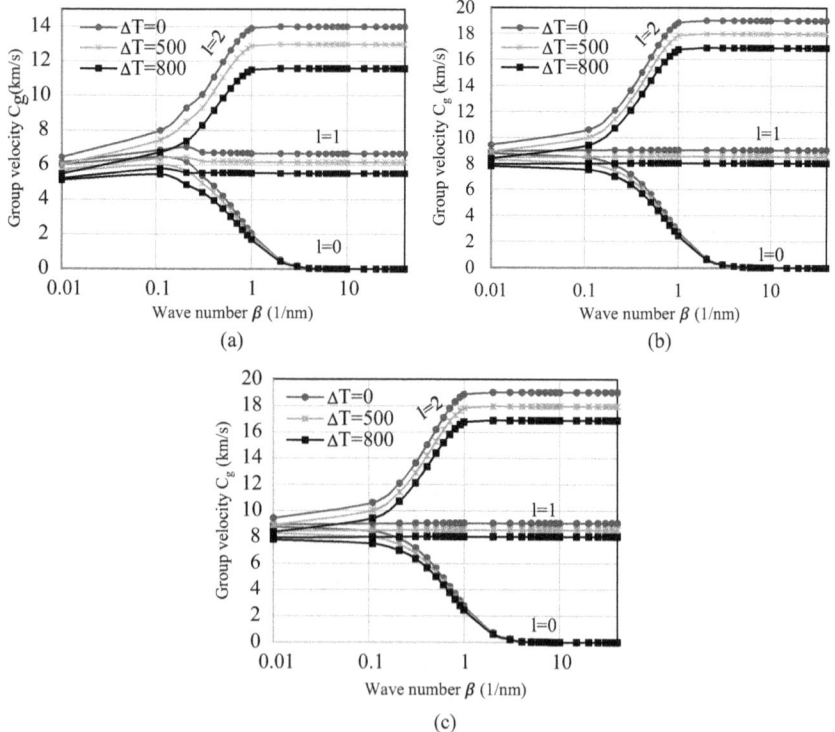

FIGURE 3.17 Effect of different temperatures on the group velocity of temperature-dependent FG nanobeam for various wave modes ($p = 1$). (a) M_0 mode, (b) M_1 mode, and (c) M_2 mode.

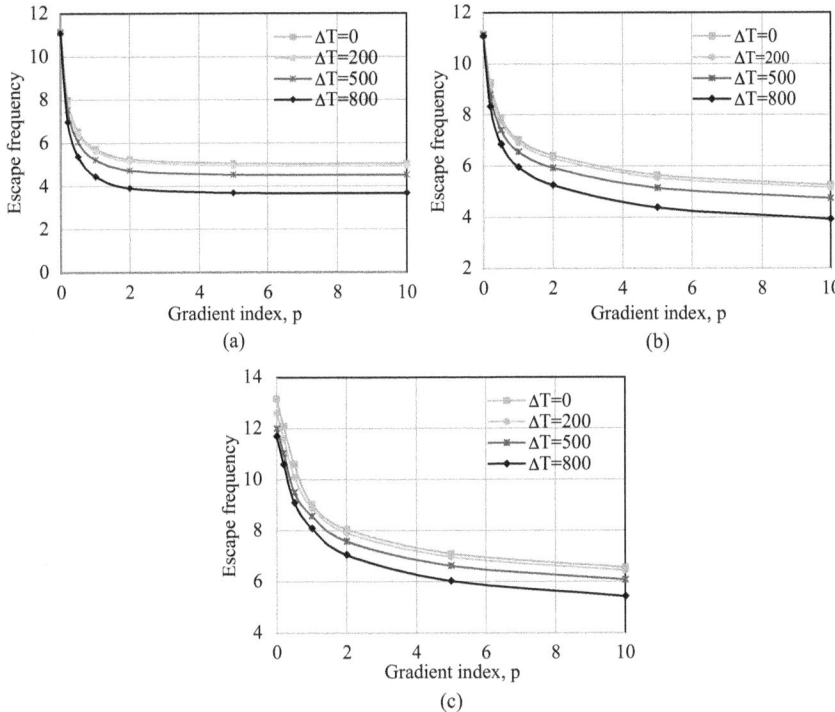

FIGURE 3.18 Effect of gradient index and nonlinear temperature changes on the escape frequency of FG nanobeam for various wave modes ($\mu = 1$ nm^2, $l = 1$ nm^2). (a) M_0 mode, (b) M_1 mode, and (c) M_2 mode.

infinite value for the wave number $\beta \to \infty$. It is evident, particularly for low gradient indices, that the escape frequency drops as the gradient index rises. At greater gradient indices, it is also clear that the differences in the escape frequencies at various temperatures are more pronounced. Also, it can be shown that the escape frequency decreases as temperature rises in a system with a constant gradient index.

3.2.3 POROSITY EFFECTS

At $p = 1$, $V = 0$, $\alpha = 0$, and $\Omega = 1$ Grad/s, Figure 3.19 considers the combined impact of porosity and electric voltage on the wave frequency of a rotating piezoelectric nanobeam made of functionally graded material. From this graphic, we learn that the frequency of an electromagnetic wave is very little affected by the applied electric voltages and that it increases as the number of tiny waves decreases. As the wave number grows, the impact weakens, and for wave numbers greater than $\beta = 0.1 \times 10^9$, a change in electric voltage cannot cause a significant shift in the wave frequency. Porosity's effects are somewhat more noticeable, meaning that adding a certain volume percentage of porosity will cause the frequency of the

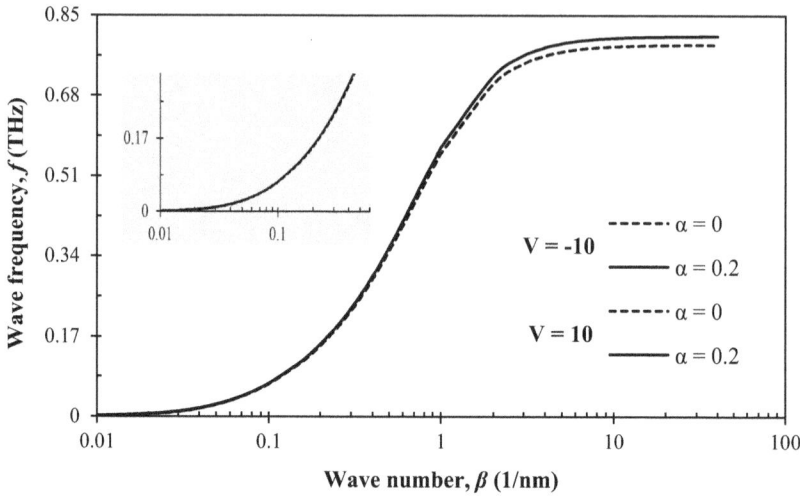

FIGURE 3.19   Variation in wave frequency of piezoelectric FG rotary nanobeam versus wave number for various applied voltages ($p = 1$, $\alpha = 0$, $\Omega = 1$ Grad/s).

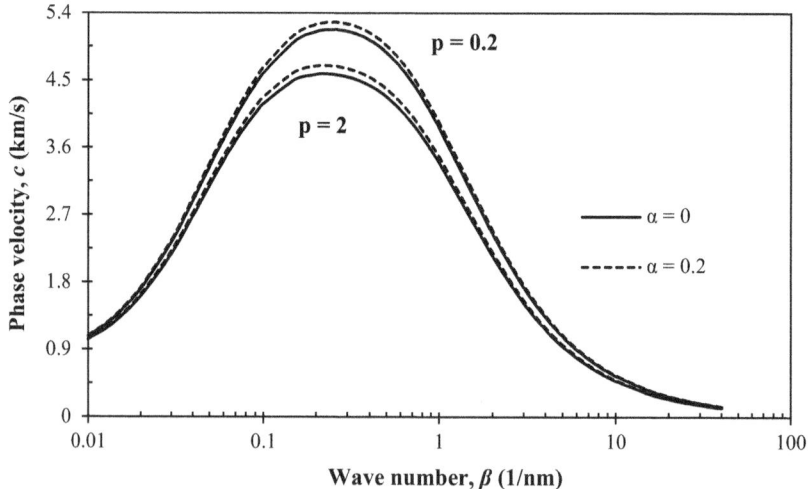

FIGURE 3.20   Variation in the phase velocity of piezoelectric FG rotary porous nanobeam versus wave number for various amounts of porosity volume fraction and gradient indices ($V = 0$, $\mu = 1$, $\Omega = 1$ Grad/s).

waves to rise. Although the effects of porosity may be seen at low wave numbers, they become more pronounced at higher wave numbers, especially those greater than $\beta = 1 \times 10^9$ [18, 61].

In addition, for different porosity values and gradient indices ($V = 0$, $\mu = 1$, and $\Omega = 1$ Grad/s), Figure 3.20 explores the relationship between the phase velocity and the wave number. It is evident that the phase velocity of FG nanobeams is negatively affected by the gradient index regardless of the volume percentage of porosity. In other words, when the gradient index rises, the phase velocity falls. The impact of porosity as a percentage of volume, however, is totally different. In fact, the volume percentage of porosity is directly related to the phase velocity of the FG porous nanobeam for any given distribution of materials. These two observed phenomena are most pronounced at the intermediate wave number range ($\beta = 0.1 \times 10^9$ and $\beta = 2 \times 10^9$).

To continue our exploration of these linked interactions, we depict the relationship between wave frequency and gradient index as a function of nonlocal parameter strength and porosity effect strength (V = 0, Ω = 1 Grad/s, and $\beta = 1 \times 10^9$) in Figure 3.21. As has been previously established, nonlocality slows the phase velocity and the wave frequency in each chosen wave number. It is also possible to see the growing impact of the porosity volume fraction. In addition, the diminishing impact of the gradient index is represented by this number. It is important to note, however, that this impact is not decisive. While gradients may have an effect on the frequency of waves, this effect is limited, and if this parameter is more than 5, the fluctuation in the frequency of waves becomes negligible. Therefore, it follows that the greatest slope of the declining trend occurs at $p = 0$ and the lowest occurs at $p > 5$.

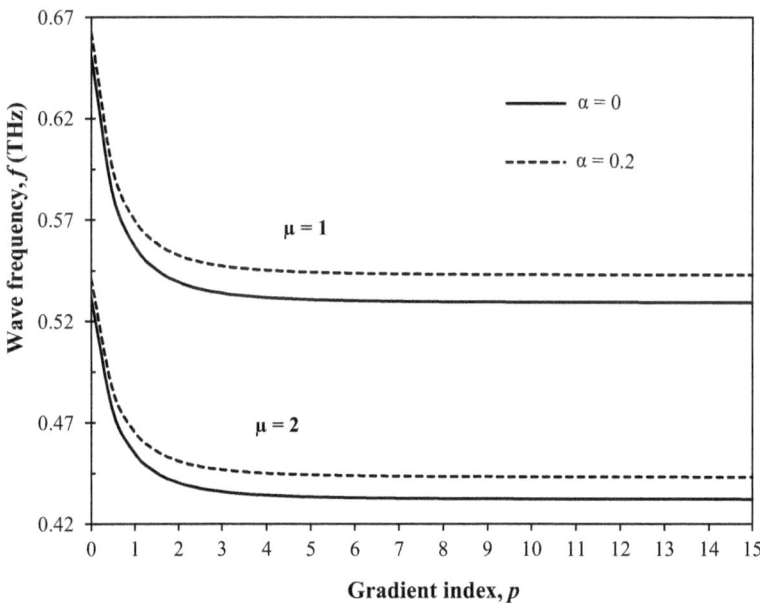

FIGURE 3.21 Variation in the wave frequency of piezoelectric FG rotary porous nanobeam versus gradient index for various amounts of porosity volume fractions and nonlocal parameters (V = 0, Ω = 1 Grad/s, $\beta = 1 \times 10^9$).

When the effects of porosity and piezoelectricity are brought to light in Figure 3.22, the changes of phase velocity versus gradient index are also investigated using the parameters: $\mu = 1$, $\Omega = 1$ Grad/s, and $\beta = 0.1 \times 10^9$. The diminishing impact of electric voltage may be seen in several ways depending on the diagram. Furthermore, an elevation in the gradient index value may promote a decrease in the phase velocity trend. The impact mentioned above causes a notable decrease in the gradient index. The impact of the volume fraction of porosity is evident. The phase velocity measured exhibits a direct proportionality to the volume fraction of porosity. The phase velocity diagram slope should be considered significant when interpreting the image, particularly for gradient indices lower than $p = 3$.

The last of the research funds will be used toward a study of escape frequency behavior. Nonlocality, material dispersion, and porosity are all examined in depth as a combined impact at $V = 0$, $\Omega = 1$ Grad/s in Figure 3.23. According to the graph, the impact of the nonlocal parameter on the escape frequency is diminishing. In addition, much as with wave frequency and phase velocity, the gradient index may be used to lower the escape frequency. Again, it's important to note that when the gradient index decreases, the impact of the material distribution parameter decreases as well. More porosity in a given volume results in a higher escape frequency, even after accounting for the attenuating effects of nonlocality and the material distribution parameter.

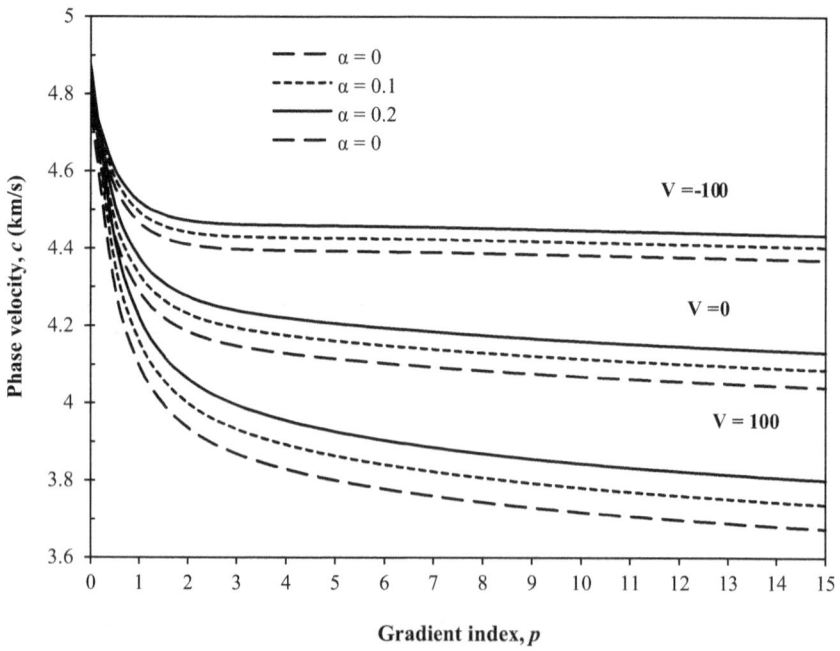

FIGURE 3.22 Variation in the phase velocity of piezoelectric FG rotary porous nanobeam versus gradient index for various amounts of porosity volume fractions and applied electric voltages ($\mu = 1$, $\Omega = 1$ Grad/s, $\beta = 0.1 \times 10^9$).

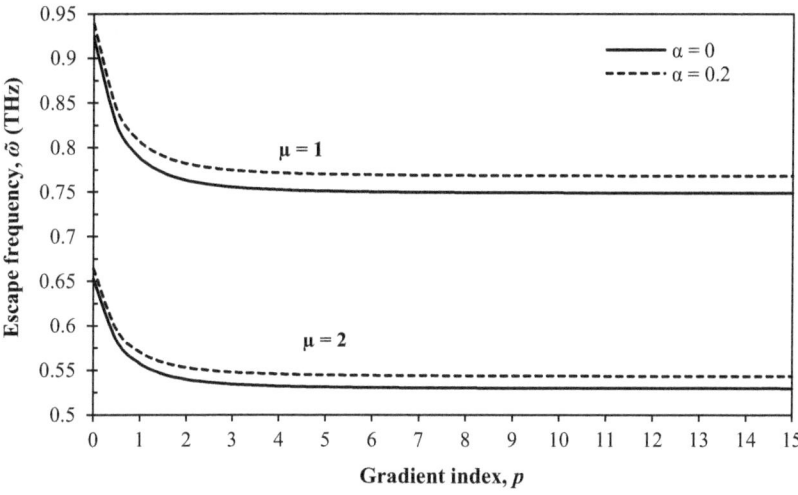

FIGURE 3.23 Variation in the escape frequency of piezoelectric FG rotary porous nanobeam versus gradient index for various amounts of porosity volume fractions and nonlocal parameters ($V = 0$, $\Omega = 1$ Grad/s).

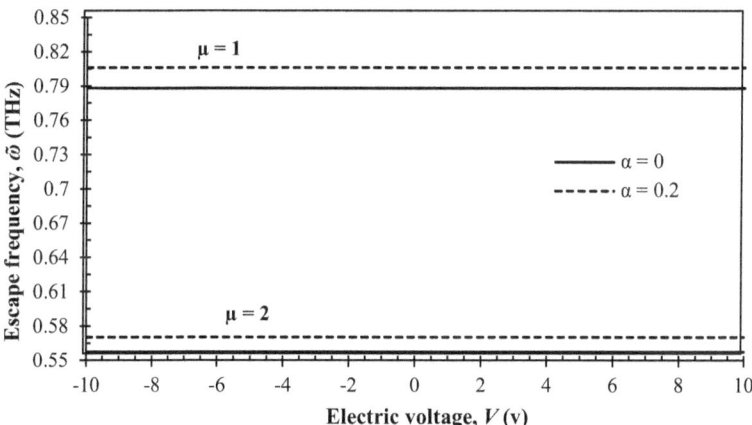

FIGURE 3.24 Variation in the escape frequency of piezoelectric FG rotary porous nanobeam versus applied electric voltage for various amounts of porosity volume fractions and nonlocal parameters ($p = 1$, $\Omega = 1$ Grad/s).

The main objective of analyzing Figure 3.24 is to demonstrate the impact of an applied electric voltage on the escape frequency of a rotating porous nanobeam that is actuated by piezoelectricity, while also taking into account the nonlocality and porosity effects at $p = 1$, $\Omega = 1$ Grad/s. The escape frequency of piezoelectric FG rotary porous nanobeams can be determined by increasing the wave number to infinity. It is noteworthy that the applied electric voltage has a minimal impact on the wave frequency at low wave numbers, and, consequently, it does not affect the escape frequency. The graphical representation demonstrates the impact of the volume fraction of porosity and the nonlocal parameter, where the former plays an expanding role while the latter plays a contracting role.

3.3 FUNCTIONALLY GRADED DOUBLE-NANOBEAM SYSTEMS: WAVE DISPERSION CHARACTERISTICS

Here, we'll use computational methods to examine the effect of different parameters on the wave dispersion properties of functionally graded double-nanobeam systems (FG-DNBSs) [19]. Plotting the differences in phase velocity versus wave number in Figure 3.25 shows the combined effects of material distribution

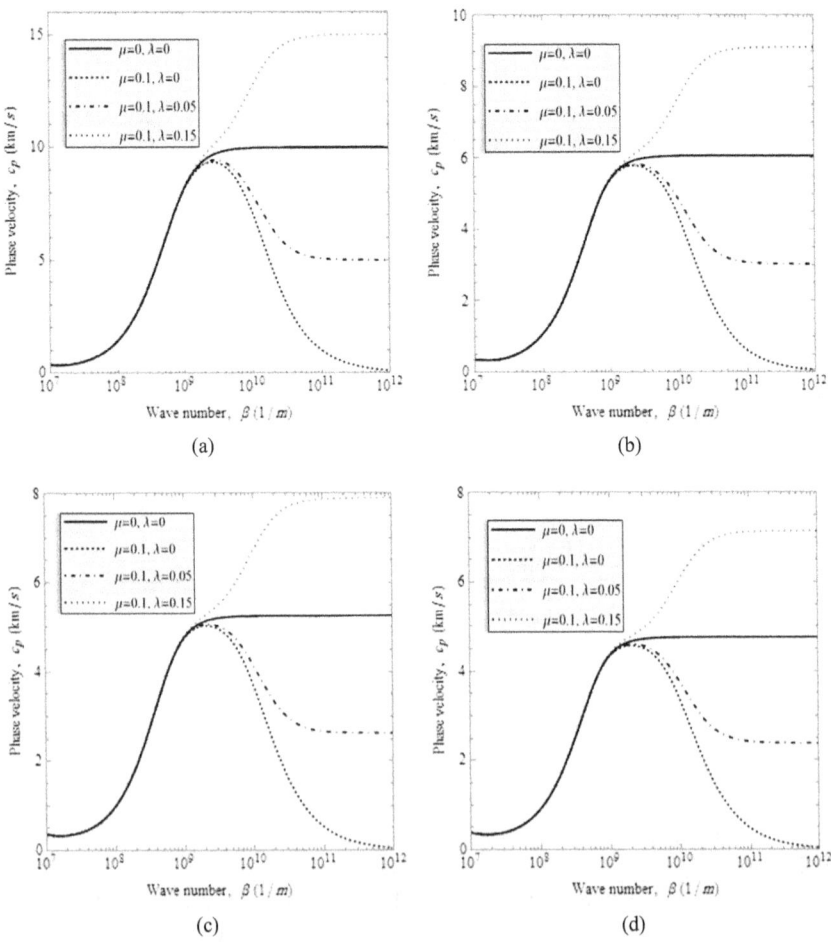

FIGURE 3.25 Variation in phase velocity versus wave number for various nonlocal and length scale parameters and different gradient indices ($h = 5\,\mathrm{nm}$, $k_w = 10^{14}$, $k_p = 1$, $K_0 = 10^{15}$). (a) $p = 0$, (b) $p = 0.5$, (c) $p = 1$, (d) $p = 5$.

parameter and scale effects. As the gradient index is expected to increase, it is clear that the responses at the phase velocity decrease. It's worth noting that the difference between the impact of altering the gradient index from $p = 0$ to $p = 0.5$ and $p = 1$ to $p = 5$ is larger at the lower value. On the other hand, there are four distinct behaviors of phase velocity that emerge from the interplay between nonlocal and length scale characteristics. Indeed, the curvature resembles a dome in the case of NE ($\lambda = 0, \mu \neq 0$). However, once the NSGT is used, there are three potential outcomes. The first rise in phase velocity with increasing wave number is shared by all three situations. Then, if the length scale parameter is lower than the nonlocal parameter ($\lambda < \mu$), the phase velocity will be weakened, and the opposite will be true if the length scale parameter is larger than the nonlocal parameter ($\lambda < \mu$). When these two scale coefficients are set to have the same value ($\lambda = \mu$), a final form is achieved, and the magnitude of the phase velocity is locked in [19].

The synchronous relative motion of FG-DNBSs is characterized by their phase velocity. The impact of nanobeam thickness and magnetic field intensity on this velocity is illustrated in Figure 3.26. This graphic should be understood as being constructed for the NE scenario. Once the strength of the magnetic field is changed, the graphic shows that the phase velocity also changes. In other words, within a limited range of wave numbers, phase velocity may be enhanced by adding magnetic field strength. In reality, increasing the strength of the magnetic field is one of the efficient methods of causing an increase in the magnitude of phase velocity in wave numbers lower than $\beta = 1 \times 10^9$. This graphic also shows how the beam's thickness may change the values of the phase velocities. Changing the nanobeam's thickness has a noticeable effect on the dome's curvature. It is obvious that when the thickness value rises, the number of waves with the same phase velocity rises as well. Changing the thickness from 5 nm to 50 nm and comparing the resulting graphs clarifies this occurrence.

In addition, Figure 3.27 is provided to show how size effects are affected by an increase in temperature for FG-DNBSs' phase velocity. The size of the temperature increase produces decreased phase velocities in a narrow band of wave numbers, close to $\beta = 0.2 \times 10^9$. At large wave numbers, it is evident that the quantity of phase velocity cannot be altered by a change in temperature. In addition, for each desired amount of nonlocality, the length scale parameter is shown to have an increasingly important bearing on the values of the phase velocities.

The impact of interlayer stiffness on the phase velocity of DNBSs is seen in Figures 3.28 and 3.29 for the out-of-phase and single-nanobeam fixed examples, respectively. Once the beams are moving in unison, it is clear that this stiffness will have no effect on the phase velocity values. In fact, as can be shown in Figure 3.28, the addition of interlayer stiffness results in an increase in phase velocity for very small wave numbers. It is important to note that in the out-of-phase situation, the phase velocity of the DNBSs might vary in comparison to the one nanobeam fixed condition if the interlayer stiffness is altered. Figure 3.29 also illustrates the FG-DNBSs' sensitivity to the interlayer stiffness at a particular wave number. Changing the value of this parameter has no effect on the system's phase velocity (in-phase motion) as a general rule. Under contrast, under out-of-phase and single-nanobeam-fixed circumstances, modifying this stiffness coefficient may be

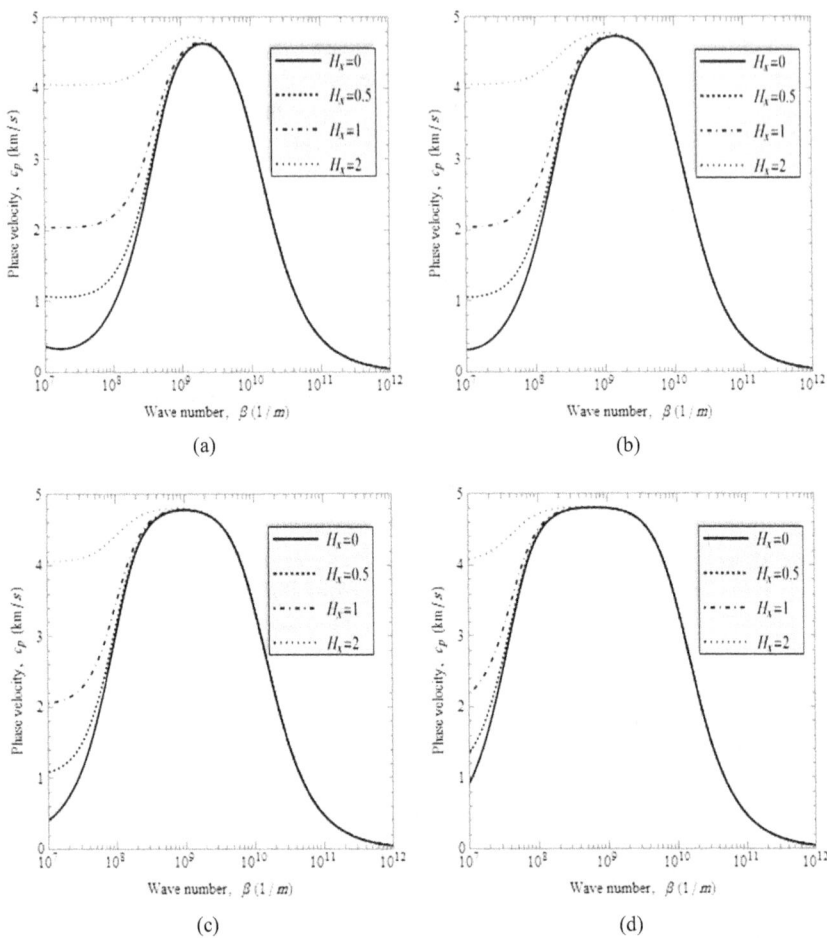

FIGURE 3.26 Variation in phase velocity versus wave number for various amounts of magnetic field intensity and different thickness values ($\mu = 0.1$, $\lambda = 0$, $p = 2$, $k_w = 10^{14}$, $k_p = 1$, $K_0 = 10^{15}$). (a) $h = 5$ nm, (b) $h = 10$ nm, (c) $h = 20$ nm, and (d) $h = 50$ nm.

used to stimulate the phase velocity. Adding the interlayer stiffness whenever an out-of-phase condition occurs changes the wave dispersion replies compared to the case of a single fixed nanobeam.

The influence of thickness and gradient index on the cut-off frequency of FG-DNBSs is next examined in Figure 3.30, which plots the fluctuations in cut-off frequency versus Winkler coefficient. It may be calculated that a higher Winkler coefficient results in a higher cut-off frequency. It's fascinating that when the thickness value rises, the cut-off frequency drops in amplitude in response. Amazingly, when the gradient index increases, the cut-off frequency also rises, making this a really remarkable result of the graphic. While the gradient index was an increasingly important factor in the previous examples, that is not the case

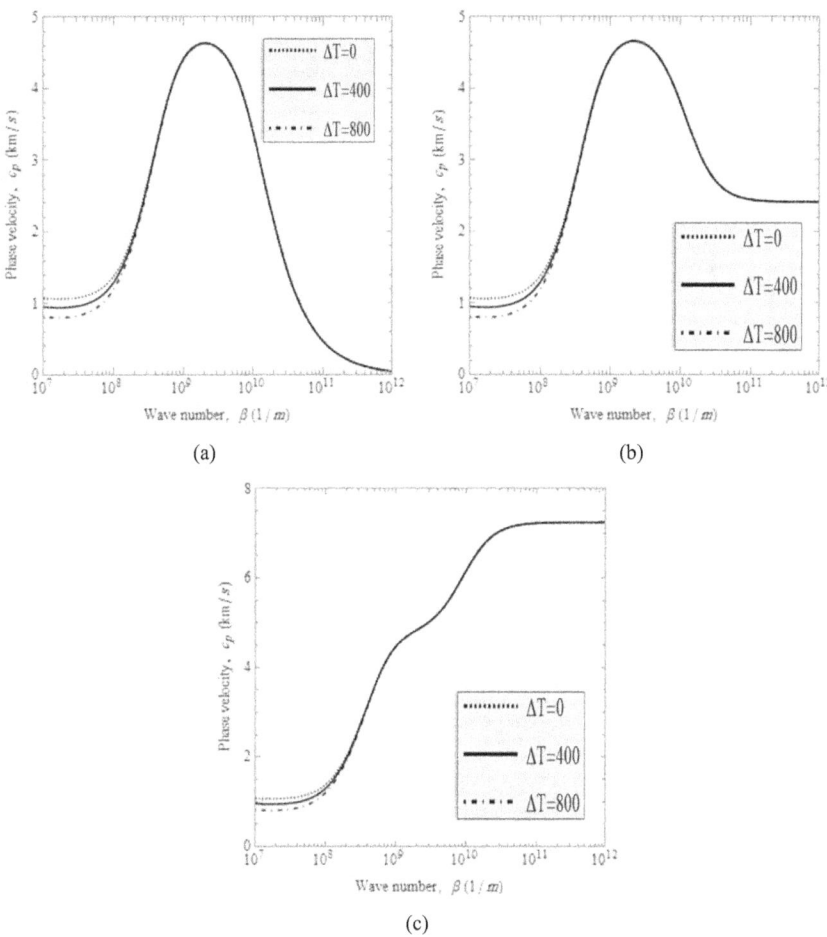

FIGURE 3.27 Variation in phase velocity versus wave number for various amounts of temperature gradient and different continuum theories (h = 5 nm, $p = 2$, $k_w = 10^{14}$, $k_p = 1$, $K_0 = 10^{15}$). (a) $\mu = 0.1$, $\lambda = 0$, (b) $\mu = 0.1$, $\lambda = 0.05$, and (c) $\mu = 0.1$, $\lambda = 0.15$.

here. The cut-off frequency is determined by the Winkler coefficient and the effective density of FG nanobeams, which are the only two independent variables. For a given Winkler coefficient, the relationship between the cut-off frequency and the density function is inverse. As we all know, a higher gradient index indicates lower densities, which results in a greater allocation to the cut-off frequency.

The next step is to record how the angular velocity of the beam affects the wave dispersion curves of DNBSs. To begin, Figure 3.31 is shown to illustrate the impact of beam rotation in the presence and absence of a thermal environment by showing the fluctuation of phase velocity versus wave number. When temperature gradients are considered, it is seen that the mechanical reaction decreases at low wave numbers. In addition, for low wave numbers, about $\beta < 0.5$ (1/nm), the

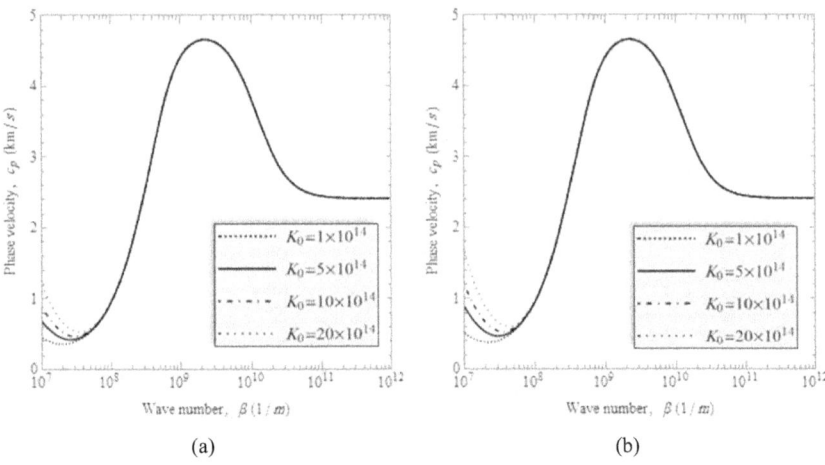

FIGURE 3.28 Influence of interlayer stiffness on the phase velocity of FG DNBS for (a) one nanobeam fixed and (b) out-of-phase motions ($h = 5\,\text{nm}$, $p = 2$, $k_w = 10^{14}$, $k_p = 1$, $\mu = 0.1$, $\lambda = 0.05$). (a) One nanobeam fixed, and (b) out-of-phase.

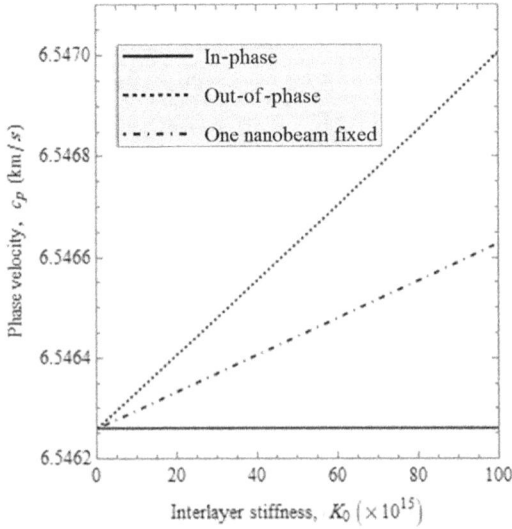

FIGURE 3.29 Influence of interlayer stiffness on the wave dispersion answers of FG DNBS for different kinds of relative motions ($h = 5\,\text{nm}$, $p = 2$, $k_w = 10^{14}$, $k_p = 1$, $\mu = \lambda = 0.1$, $\beta = 10^8$). (a) One nanobeam fixed, and (b) out-of-phase.

influence of angular velocity on the phase speed of the scattered waves becomes more important to consider. Accordingly, systems exposed to an angular velocity might display greater wave velocities.

Figure 3.32 displays the phase velocity curves of DNBSs in relation to the magnetic field strength for three selected wave numbers, assuming that the system

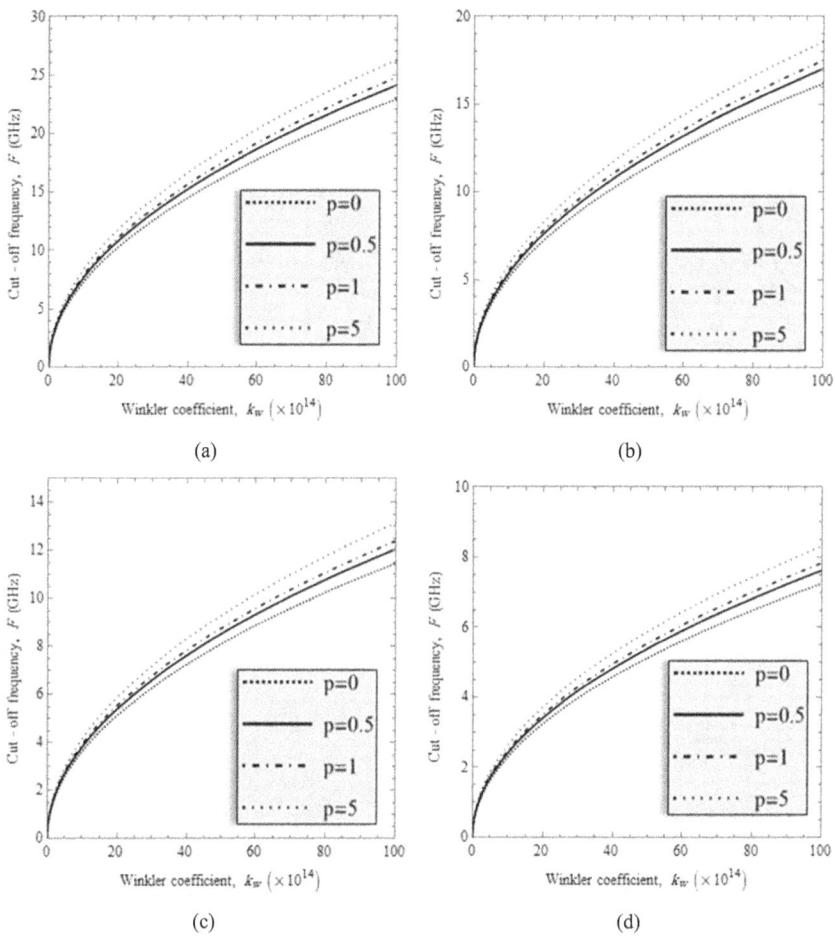

FIGURE 3.30 Variation in cut-off frequency versus Winkler coefficient for various gradient indices considering the influence of plate's thickness. (a) $h = 5$ nm, (b) $h = 10$ nm, (c) $h = 20$ nm, and (d) $h = 50$ nm.

is in a thermal environment. Increasing the strength of the magnetic field has the obvious effect of increasing the phase velocity of the waves as they spread. When we include the beams' angular velocities, we once again see this behavior. Increasing the beams' angular velocity, in other words, will increase the phase speed. It's also worth noting that when angular velocity is removed, the impact of the magnetic field's strength becomes more apparent. Finally, it's worth noting that the fluctuation in phase velocity due to the strength of the magnetic field is more pronounced for lower wave numbers.

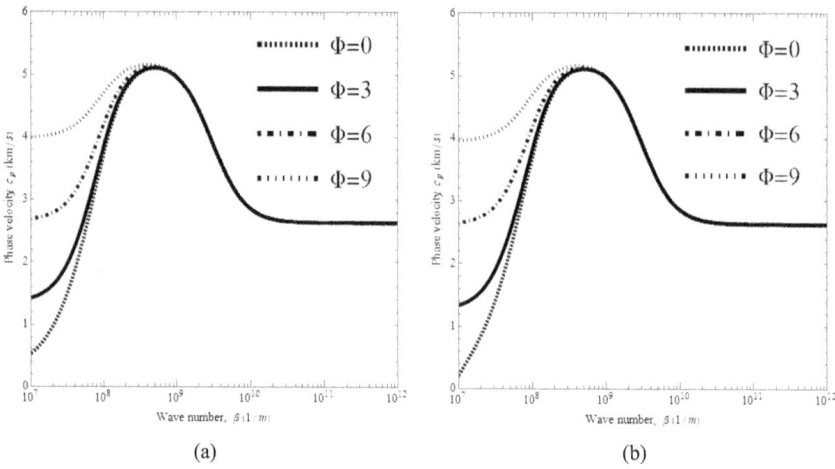

FIGURE 3.31   Variation in phase velocity versus wave number for various angular velocities of the DNBS in the out-of-phase relative motion of beams ($\mu = 0.4$, $\lambda = 0.2$, $L/h = 25$, $H_x = 0$, $k_w = 10^{14}$, $k_p = 1$, $K_0 = 10^{14}$). (a) $\Delta T = 0$ K and (b) $\Delta T = 400$ K.

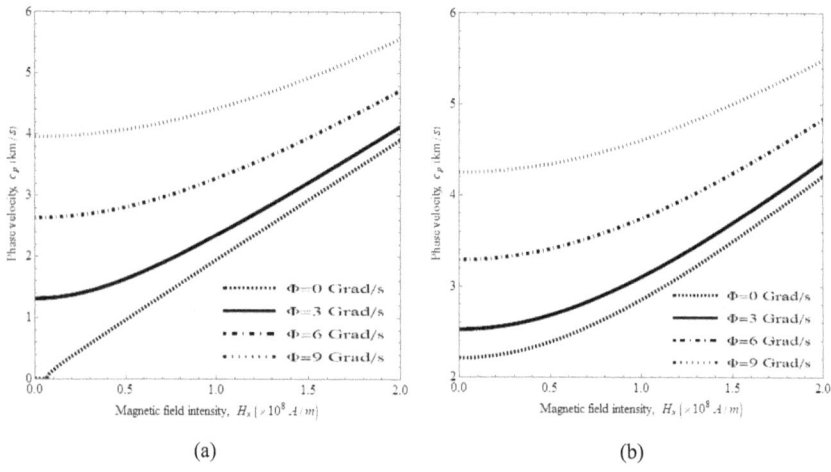

FIGURE 3.32   Variation in phase velocity versus magnetic field intensity for various angular velocities of the nanobeams in the out-of-phase motion by considering the effect of wave number ($\mu = 0.4$, $\lambda = 0.2$, $L/h = 25$, $\Delta T = 500$ K, $k_w = 10^{14}$, $k_p = 1$, $K_0 = 10^{14}$). (a) $\beta = 0.01$ (1/nm), (b) $\beta = 0.05$ (1/nm).

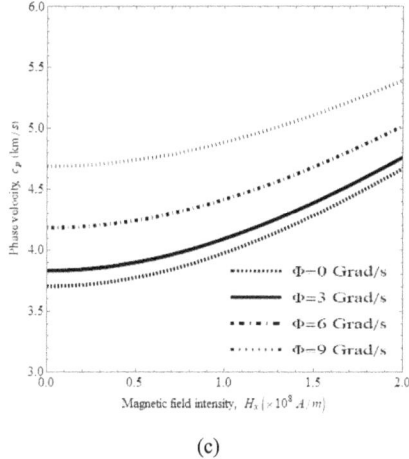

(c)

FIGURE 3.32   (Continued)   (c) $\beta = 0.01$ (1/nm).

3.4   FUNCTIONALLY GRADED PLATES: WAVE DISPERSION CHARACTERISTICS

Assumed here is a magneto-electro-elastic functionally graded spinning nano-plate with dimensions L (length) and h (thickness). BaTiO$_3$ and CoFe$_2$O$_4$ are used to create the nanoplate. Ebrahimi et al. [18, 61, 62] have employed the same material qualities in their work.

The plate is assumed to be subjected to a magnetic potential $\gamma(x, z, t)$ and electric potential $\phi(x, z, t)$. The effective material properties of an MEE-FG plate vary gradually in the thickness direction via modified power law distribution.

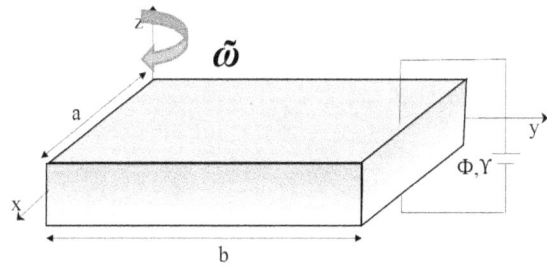

FIGURE 3.33   Geometry of an FG nanoplate under magneto-electrical field.

The displacement field of a nonlocal FG plate can be expressed using an enhanced higher-order shear deformation plate theory. This is similar to the method outlined in previous publications [57], with the only difference being the shape function.

$$f(z) = z - \frac{\sin(\xi z)}{\xi} \tag{3.73}$$

As noted in prior studies, the electric and magnetic potential fluctuations along the thickness direction are modeled using a cosine and linear variation to meet Maxwell's equation in the quasi-static approximation.

Currently, an attempt is made to locate the Euler–Lagrange equations of the MEE-FG nanoplate in light of an expanded Hamilton's principle for MEEMs. This section is identical to those of other studies [61, 62], hence it will not be rewritten here. The formal formulation of Hamilton's principle looks like this:

$$\int_0^t \delta\left(\Pi_S - \Pi_K + \Pi_W\right) dt = 0 \tag{3.74}$$

Substituting the variations of strain energy, kinetic energy, and work done by external loads into Eq. (3.70) and setting the coefficients of δu_0, δv_0, δw_b, δw_s, $\delta\phi$, and $\delta\gamma$ to zero, the Euler–Lagrange equations of MEE-FG nanoplate can be written as follows [61, 62]:

$$\frac{\partial N_x}{\partial x} + \frac{\partial N_{xy}}{\partial y} = I_0 \ddot{u}_0 - I_1 \frac{\partial \ddot{w}_b}{\partial x} - J_1 \frac{\partial \ddot{w}_s}{\partial x} \tag{3.75a}$$

$$\frac{\partial N_{xy}}{\partial x} + \frac{\partial N_y}{\partial y} = I_0 \ddot{v}_0 - I_1 \frac{\partial \ddot{w}_b}{\partial y} - J_1 \frac{\partial \ddot{w}_s}{\partial y} \tag{3.75b}$$

$$\frac{\partial^2 M_x^b}{\partial x^2} + 2\frac{\partial^2 M_{xy}^b}{\partial x \partial y} + \frac{\partial^2 M_y^b}{\partial y^2} - \left(N^E + N^H + N_{max}^R\right)\nabla^2\left(w_b + w_s\right) = I_0\left(\ddot{w}_b + \ddot{w}_s\right)$$
$$+I_1\left(\frac{\partial \ddot{u}_0}{\partial x} + \frac{\partial \ddot{v}_0}{\partial y}\right) - I_2\nabla^2\ddot{w}_b - J_2\nabla^2\ddot{w}_s$$

$$\tag{3.75c}$$

$$\frac{\partial^2 M_x^s}{\partial x^2} + 2\frac{\partial^2 M_{xy}^s}{\partial x \partial y} + \frac{\partial^2 M_y^s}{\partial y^2} + \frac{\partial Q_{xz}}{\partial x} + \frac{\partial Q_{yz}}{\partial y} - \left(N^E + N^H + N_{max}^R\right)\nabla^2\left(w_b + w_s\right)$$
$$= I_0\left(\ddot{w}_b + \ddot{w}_s\right) + J_1\left(\frac{\partial \ddot{u}_0}{\partial x} + \frac{\partial \ddot{v}_0}{\partial y}\right) - J_2\nabla^2\ddot{w}_b - K_2\nabla^2\ddot{w}_s$$

$$\tag{3.75d}$$

$$\int_A \left(\cos(\xi z) \frac{\partial D_x}{\partial x} + \cos(\xi z) \frac{\partial D_y}{\partial y} + \xi \sin(\xi z) D_z \right) dA = 0 \qquad (3.75e)$$

$$\int_A \left(\cos(\xi z) \frac{\partial B_x}{\partial x} + \cos(\xi z) \frac{\partial B_y}{\partial y} + \xi \sin(\xi z) B_z \right) dA = 0 \qquad (3.75f)$$

As per the nonlocal strain gradient theory, the stress field takes into consideration the impact of both the strain gradient stress field and nonlocal elastic stress field. This formulation is excluded in this chapter due to its similarity to formulations presented in other papers (references 61 and 62). By integrating across the plate's cross-section, the relations of stress-strain can be used to get the force-strain and moment-strain formulas of the nonlocal refined FG-MEE plate. The equations pertaining to magneto-electro-elastic solids have been extensively discussed in the references mentioned earlier. However, the rotation of the nanoplate necessitates the inclusion of additional forces and moments in these equations. Therefore, the equations concerning the rotation of the plate have been comprehensively analyzed, while the other equations remain unchanged from the aforementioned references:

$$(1 - \mu \nabla^2) \begin{Bmatrix} N_x \\ N_y \\ N_{xy} \end{Bmatrix} = (1 - \eta \nabla^2) \begin{bmatrix} A_{11} & A_{12} & 0 \\ A_{21} & A_{22} & 0 \\ 0 & 0 & A_{66} \end{bmatrix} \begin{Bmatrix} \dfrac{\partial u_0}{\partial x} \\[2mm] \dfrac{\partial v_0}{\partial y} \\[2mm] \dfrac{\partial u_0}{\partial y} + \dfrac{\partial v_0}{\partial x} \end{Bmatrix}$$

$$+ \begin{pmatrix} B_{11} & B_{12} & 0 \\ B_{21} & B_{22} & 0 \\ 0 & 0 & B_{66} \end{pmatrix} \begin{Bmatrix} -\dfrac{\partial^2 w_b}{\partial x^2} \\[2mm] -\dfrac{\partial^2 w_b}{\partial y^2} \\[2mm] -2\dfrac{\partial^2 w_b}{\partial x \partial y} \end{Bmatrix} + \begin{pmatrix} B_{11}^s & B_{12}^s & 0 \\ B_{21}^s & B_{22}^s & 0 \\ 0 & 0 & B_{66}^s \end{pmatrix} \begin{Bmatrix} -\dfrac{\partial^2 w_s}{\partial x^2} \\[2mm] -\dfrac{\partial^2 w_s}{\partial y^2} \\[2mm] -2\dfrac{\partial^2 w_s}{\partial x \partial y} \end{Bmatrix} \qquad (3.76a)$$

$$+ \begin{bmatrix} A_{31}^e \\ A_{31}^e \\ 0 \end{bmatrix} \phi + \begin{bmatrix} A_{31}^m \\ A_{31}^m \\ 0 \end{bmatrix} \gamma - \begin{bmatrix} N_x^E + N_x^H + N_{\max}^R \\ N_y^E + N_y^H + N_{\max}^R \\ 0 \end{bmatrix} \end{bmatrix}$$

$$
\left(1-\mu\nabla^2\right)\begin{Bmatrix} M_x^b \\ M_y^b \\ M_{xy}^b \end{Bmatrix} = \left(1-\eta\nabla^2\right)\left[\begin{pmatrix} B_{11} & B_{12} & 0 \\ B_{21} & B_{22} & 0 \\ 0 & 0 & B_{66} \end{pmatrix}\begin{Bmatrix} \dfrac{\partial u_0}{\partial x} \\[2mm] \dfrac{\partial v_0}{\partial y} \\[2mm] \dfrac{\partial u_0}{\partial y}+\dfrac{\partial v_0}{\partial x} \end{Bmatrix}\right.
$$

$$
+\begin{pmatrix} D_{11} & D_{12} & 0 \\ D_{21} & D_{22} & 0 \\ 0 & 0 & D_{66} \end{pmatrix}\begin{Bmatrix} -\dfrac{\partial^2 w_b}{\partial x^2} \\[2mm] -\dfrac{\partial^2 w_b}{\partial y^2} \\[2mm] -2\dfrac{\partial^2 w_b}{\partial x \partial y} \end{Bmatrix} +\begin{pmatrix} D_{11}^s & D_{12}^s & 0 \\ D_{21}^s & D_{22}^s & 0 \\ 0 & 0 & D_{66}^s \end{pmatrix}\begin{Bmatrix} -\dfrac{\partial^2 w_s}{\partial x^2} \\[2mm] -\dfrac{\partial^2 w_s}{\partial y^2} \\[2mm] -2\dfrac{\partial^2 w_s}{\partial x \partial y} \end{Bmatrix} \qquad (3.76b)
$$

$$
\left.+\begin{Bmatrix} E_{31}^e \\ E_{31}^e \\ 0 \end{Bmatrix}\phi+\begin{Bmatrix} E_{31}^m \\ E_{31}^m \\ 0 \end{Bmatrix}\gamma-\begin{Bmatrix} M_{bx}^E + M_{bx}^H + M_{bx}^R \\ M_{by}^E + M_{by}^H + M_{by}^R \\ 0 \end{Bmatrix}\right]
$$

$$
\left(1-\mu\nabla^2\right)\begin{Bmatrix} M_x^s \\ M_y^s \\ M_{xy}^s \end{Bmatrix} = \left(1-\eta\nabla^2\right)\left[\begin{pmatrix} B_{11}^s & B_{12}^s & 0 \\ B_{21}^s & B_{22}^s & 0 \\ 0 & 0 & B_{66}^s \end{pmatrix}\begin{Bmatrix} \dfrac{\partial u_0}{\partial x} \\[2mm] \dfrac{\partial v_0}{\partial y} \\[2mm] \dfrac{\partial u_0}{\partial y}+\dfrac{\partial v_0}{\partial x} \end{Bmatrix}\right.
$$

$$
+\begin{pmatrix} D_{11}^s & D_{12}^s & 0 \\ D_{21}^s & D_{22}^s & 0 \\ 0 & 0 & D_{66}^s \end{pmatrix}\begin{Bmatrix} -\dfrac{\partial^2 w_b}{\partial x^2} \\[2mm] -\dfrac{\partial^2 w_b}{\partial y^2} \\[2mm] -2\dfrac{\partial^2 w_b}{\partial x \partial y} \end{Bmatrix} +\begin{pmatrix} H_{11}^s & H_{12}^s & 0 \\ H_{21}^s & H_{22}^s & 0 \\ 0 & 0 & H_{66}^s \end{pmatrix}\begin{Bmatrix} -\dfrac{\partial^2 w_s}{\partial x^2} \\[2mm] -\dfrac{\partial^2 w_s}{\partial y^2} \\[2mm] -2\dfrac{\partial^2 w_s}{\partial x \partial y} \end{Bmatrix} \qquad (3.76c)
$$

$$
\left.+\begin{Bmatrix} F_{31}^e \\ F_{31}^e \\ 0 \end{Bmatrix}\phi+\begin{Bmatrix} F_{31}^m \\ F_{31}^m \\ 0 \end{Bmatrix}\gamma-\begin{Bmatrix} M_{sx}^E + M_{sx}^H + M_{sx}^R \\ M_{sy}^E + M_{sy}^H + M_{sy}^R \\ 0 \end{Bmatrix}\right]
$$

In addition, Eq. (3.76a–c) for the normal forces and moments created by the rotation of the plate may be derived as follows:

$$N_{\max}^{R} = \int_{0}^{L} \int_{A} \left(\rho(z) A \tilde{\omega}^2 x \right) dA dx \tag{3.77a}$$

$$M_{bx}^{R} = M_{by}^{R} = \int_{0}^{L} \int_{A} \left(\rho(z) A \tilde{\omega}^2 x \right) z dA dx \tag{3.77b}$$

$$M_{sx}^{R} = M_{sy}^{R} = \int_{0}^{L} \int_{A} \left(\rho(z) A \tilde{\omega}^2 x \right) f(z) dA dx \tag{3.77c}$$

According to previous studies [56], the variable $\tilde{\omega}$ stands for the angular velocity of the nanoplate in the aforementioned equations, and the analysis accounts for the highest axial force generated by the rotation of the plate. In order to get the nonlocal governing equations of the FG-MEE spinning nanoplate, we need to substitute the force-strain and moment-strain equations into Eqs. (3.71), which yields the following expressions in terms of displacements:

$$\left(1 - \eta \nabla^2\right)(A_{11} \frac{\partial^2 u_0}{\partial x^2} + \left(A_{12} + A_{66}\right) \frac{\partial^2 v_0}{\partial x \partial y} + A_{66} \frac{\partial^2 u_0}{\partial y^2} - B_{11} \frac{\partial^3 w_b}{\partial x^3}$$

$$- \left(B_{12} + 2B_{66}\right) \frac{\partial^3 w_b}{\partial x \partial y^2} - B_{11}^s \frac{\partial^3 w_s}{\partial x^3} - \left(B_{12}^s + 2B_{66}^s\right) \frac{\partial^3 w_s}{\partial x \partial y^2} + A_{31}^e \frac{\partial \phi}{\partial x} + A_{31}^m \frac{\partial \gamma}{\partial x}) \tag{3.78a}$$

$$+ \left(1 - \mu \nabla^2\right) \left(-I_0 \ddot{u}_0 + I_1 \frac{\partial \ddot{w}_b}{\partial x} + J_1 \frac{\partial \ddot{w}_s}{\partial x} \right) = 0$$

$$\left(1 - \eta \nabla^2\right)(A_{22} \frac{\partial^2 v_0}{\partial y^2} + \left(A_{12} + A_{66}\right) \frac{\partial^2 u_0}{\partial x \partial y} + A_{66} \frac{\partial^2 v_0}{\partial x^2} - B_{22} \frac{\partial^3 w_b}{\partial y^3}$$

$$- \left(B_{12} + 2B_{66}\right) \frac{\partial^3 w_b}{\partial x^2 \partial y} - B_{22}^s \frac{\partial^3 w_s}{\partial y^3} - \left(B_{12}^s + 2B_{66}^s\right) \frac{\partial^3 w_s}{\partial x^2 \partial y} + A_{31}^e \frac{\partial \phi}{\partial y} + A_{31}^m \frac{\partial \gamma}{\partial y}) \tag{3.78b}$$

$$+ \left(1 - \mu \nabla^2\right) \left(-I_0 \ddot{v}_0 + I_1 \frac{\partial \ddot{w}_b}{\partial y} + J_1 \frac{\partial \ddot{w}_s}{\partial y} \right) = 0$$

$$\left(1-\eta\nabla^2\right)\left(B_{11}\frac{\partial^3 u_0}{\partial x^3}+\left(B_{12}+2B_{66}\right)\frac{\partial^3 u_0}{\partial x\partial y^2}+B_{22}\frac{\partial^3 v_0}{\partial y^3}+\left(B_{12}+2B_{66}\right)\frac{\partial^3 v_0}{\partial x^2\partial y}\right.$$

$$-D_{11}\frac{\partial^4 w_b}{\partial x^4}-2\left(D_{12}+2D_{66}\right)\frac{\partial^4 w_b}{\partial x^2\partial y^2}-D_{22}\frac{\partial^4 w_b}{\partial y^4}+E_{31}^e\nabla^2\phi-D_{11}^s\frac{\partial^4 w_s}{\partial x^4}$$

$$-2\left(D_{12}^s+2D_{66}^s\right)\frac{\partial^4 w_s}{\partial x^2\partial y^2}-D_{22}^s\frac{\partial^4 w_s}{\partial y^4}+E_{31}^m\nabla^2\gamma$$

$$+\left(1-\mu\nabla^2\right)\left(-I_0\left(\ddot{w}_b+\ddot{w}_s\right)-I_1\left(\frac{\partial\ddot{u}_0}{\partial x}+\frac{\partial\ddot{v}_0}{\partial y}\right)+J_2\nabla^2\ddot{w}_b+K_2\nabla^2\ddot{w}_s\right.$$

$$\left.-\left(N^E+N^H+N_{\max}^R\right)\nabla^2\left(w_b+w_s\right)\right)=0$$

$$(3.78\text{c})$$

$$\left(1-\eta\nabla^2\right)\left(B_{11}^s\frac{\partial^3 u_0}{\partial x^3}+\left(B_{12}^s+2B_{66}^s\right)\frac{\partial^3 u_0}{\partial x\partial y^2}+B_{22}^s\frac{\partial^3 v_0}{\partial y^3}+\left(B_{12}^s+2B_{66}^s\right)\frac{\partial^3 v_0}{\partial x^2\partial y}\right.$$

$$-D_{11}^s\frac{\partial^4 w_b}{\partial x^4}-2\left(D_{12}^s+2D_{66}^s\right)\frac{\partial^4 w_b}{\partial x^2\partial y^2}-D_{22}^s\frac{\partial^4 w_b}{\partial y^4}+\left(F_{31}^e-A_{15}^e\right)\nabla^2\phi$$

$$\left.-H_{11}^s\frac{\partial^4 w_s}{\partial x^4}-2\left(H_{12}^s+2H_{66}^s\right)\frac{\partial^4 w_s}{\partial x^2\partial y^2}+F_{31}^m\nabla^2\gamma-H_{22}^s\frac{\partial^4 w_s}{\partial y^4}+A_{44}^s\nabla^2 w_s\right)$$

$$+\left(1-\mu\nabla^2\right)\left(-I_0\left(\ddot{w}_b+\ddot{w}_s\right)-J_1\left(\frac{\partial\ddot{u}_0}{\partial x}+\frac{\partial\ddot{v}_0}{\partial y}\right)+J_2\nabla^2\ddot{w}_b+K_2\nabla^2\ddot{w}_s\right.$$

$$\left.-\left(N^E+N^H+N_{\max}^R\right)\nabla^2\left(w_b+w_s\right)\right)=0$$

$$(3.78\text{d})$$

$$\left(1-\eta\nabla^2\right)\left(A_{31}^e\left(\frac{\partial u_0}{\partial x}+\frac{\partial v_0}{\partial y}\right)-E_{31}^e\nabla^2 w_b-\left(F_{31}^e-E_{15}^e\right)\nabla^2 w_s\right.$$

$$+F_{11}^e\nabla^2\phi+F_{11}^m\nabla^2\gamma-F_{33}^e\phi-F_{33}^m\gamma)=0$$

$$(3.78\text{e})$$

$$\left(1-\eta\nabla^2\right)\left(A_{31}^m\left(\frac{\partial u_0}{\partial x}+\frac{\partial v_0}{\partial y}\right)-E_{31}^m\nabla^2 w_b-\left(F_{31}^m-E_{15}^m\right)\nabla^2 w_s\right.$$

$$+F_{11}^m\nabla^2\phi+X_{11}^m\nabla^2\gamma-F_{33}^m\phi-X_{33}^m\gamma)=0$$

$$(3.78\text{f})$$

Without applying any shear force, the MEE-FG nanoplate is analyzed while being exposed to an external electric voltage, magnetic potential, and rotation on the z-axis. As a result, the electric voltage, magnetic potential, and rotation of the plate all contribute to the generation of normal in-plane forces for the purposes of the current study, with $N_{xy}^0=0$. The following equations may be used to determine the magnitude of these forces:

$$N_x^0 = N_y^0 = N^E + N^H + N_{\text{max}}^R$$

$$N^E = -\int_A \tilde{e}_{31} \frac{2V}{h} dA$$

$$N^H = -\int_A \tilde{q}_{31} \frac{2V}{h} dA \qquad (3.79)$$

$$N_{\text{max}}^R = \int_0^L \int_A \left(\rho(z) A \tilde{\omega}^2 x\right) dA dx$$

Here we solve analytically the nonlocal governing equations developed in the preceding section. Below, we shall define the displacement fields on the assumption that they are exponential:

$$\begin{Bmatrix} u_0(x,y,t) \\ v_0(x,y,t) \\ w_b(x,y,t) \\ w_s(x,y,t) \\ \phi(x,y,t) \\ \gamma(x,y,t) \end{Bmatrix} = \begin{Bmatrix} U \exp\left[i\left(k_1 x + k_2 y - \omega t\right)\right] \\ V \exp\left[i\left(k_1 x + k_2 y - \omega t\right)\right] \\ W_b \exp\left[i\left(k_1 x + k_2 y - \omega t\right)\right] \\ W_s \exp\left[i\left(k_1 x + k_2 y - \omega t\right)\right] \\ \Phi \exp\left[i\left(k_1 x + k_2 y - \omega t\right)\right] \\ Y \exp\left[i\left(k_1 x + k_2 y - \omega t\right)\right] \end{Bmatrix} \qquad (3.80)$$

where U, V, W_b, W_s, Φ, and Y are the unknown coefficients, k_1 and k_2 are the wave numbers of wave propagation along x and y directions, respectively, and finally ω is wave's angular frequency. Now, substituting Eq. (3.80) in Eqs. (3.74) reveals:

$$\left(\left[K\right] - \omega^2 \left[M\right]\right)\{\Delta\} = \{0\} \qquad (3.81)$$

Note the following about Eq. (3.81's) unknowable variables:

$$\{\Delta\} = \left\{U, V, W_b, W_s, \Phi, Y\right\}^T \qquad (3.82)$$

$$[K] = \begin{bmatrix} k_{11} & k_{12} & k_{13} & k_{14} & k_{15} & k_{16} \\ k_{21} & k_{22} & k_{23} & k_{24} & k_{25} & k_{26} \\ k_{31} & k_{32} & k_{33} & k_{34} & k_{35} & k_{36} \\ k_{41} & k_{42} & k_{43} & k_{44} & k_{45} & k_{46} \\ k_{51} & k_{52} & k_{53} & k_{54} & k_{55} & k_{56} \\ k_{61} & k_{62} & k_{63} & k_{64} & k_{65} & k_{66} \end{bmatrix}$$

$$[M] = \begin{bmatrix} m_{11} & m_{12} & m_{13} & m_{14} & m_{15} & m_{16} \\ m_{21} & m_{22} & m_{23} & m_{24} & m_{25} & m_{26} \\ m_{31} & m_{32} & m_{33} & m_{34} & m_{35} & m_{36} \\ m_{41} & m_{42} & m_{43} & m_{44} & m_{45} & m_{46} \\ m_{51} & m_{52} & m_{53} & m_{54} & m_{55} & m_{56} \\ m_{61} & m_{62} & m_{63} & m_{64} & m_{65} & m_{66} \end{bmatrix} \tag{3.83}$$

where the corresponding a_{ij} and m_{ij} are as written in Appendix B. In order to attain wave's angular frequency, the determinant of the left-hand side of Eq. (3.83) must be set to zero:

$$\left| [K] - \omega^2 [M] \right| = 0 \tag{3.84a}$$

In Eq. (3.84a) by setting $k_1 = k_2 = k$ and solving the obtained equation for ω, the wave frequency of FG-MEE nanoplate can be expressed as below:

$$\omega_0 = M_0(k), \omega_1 = M_1(k), \omega_2 = M_2(k), \omega_3 = M_3(k) \tag{3.84b}$$

Eq. (25) reveals wave's angular frequency of M_0 to M_3 modes, respectively. If these frequencies are divided by wave number, the phase velocity of each mode is obtained:

$$c_i = \frac{M_i(k)}{k}, (i = 0, 1, 2, 3) \tag{3.84c}$$

Also, the escape frequency of FG-MEE nanoplate can be derived by tending wave number to infinity:

$$\omega_i' = \lim_{k \to \infty} \frac{\omega_i}{2\pi}, i = (0, 1, 2, 3) \tag{3.84d}$$

This part of the study investigates the impact of the application of an external electric voltage and magnetic potential via NSGT on wave propagation in a $BaTiO_3$ and $CoFe_2O_4$ FG-MEE nanoplate. The influence of the length scale parameter on the wave frequency of the FG-MEE nanoplate versus wave number is depicted in Figure 3.34. The data is presented over a range of length scale parameters at $p = 1$ and $V = \Omega = 0$. When the length scale parameter is zero and the nonlocal parameter is nonzero, the wave frequency remains constant even when the wave number exceeds $k = 10 \times 10^9$. However, when the strain gradient effect is taken into account ($\eta \neq 0$), the wave frequency has a huge slope toward infinity. This behavior cannot be predicted from the results of previous studies on the wave dispersion properties of smart nanoplates using nonlocal elasticity. In addition, at low wave

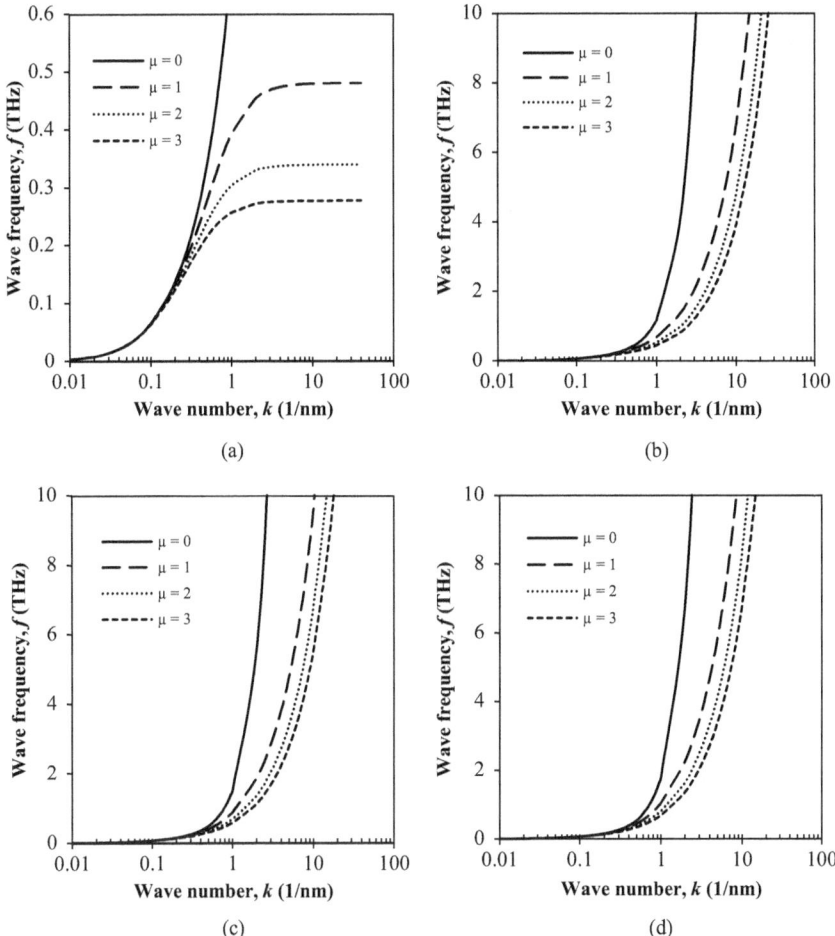

FIGURE 3.34 Variation in wave frequency of FG-MEE nanoplate versus wave number for various nonlocal and length scale parameters ($p = 1$, $V = \Omega = 0$). (a) $\eta = 0$ nm^2, (b) $\eta = 1$ nm^2, (c) $\eta = 2$, and (d) $\eta = 3$.

numbers, it is possible to report a negligible effect of nonlocal parameter on the frequency of the waves generated by the FG-MEE nanoplate. However, for wave numbers greater than $k = 0.3 \times 10^9$, this negligible influence may be transformed into a clearly detectable diminishing effect. The value of the nonlocal parameter has a tendency to drop as a consequence of the influence of nonlocal elasticity, which has a softening effect on the stiffness of the material. The wave frequency is affected in the opposite way by the length scale parameter; an increase in this parameter induces a comparable rise in the wave frequency when it is increased. The stiffness-hardening impact of the nonlocal strain gradient theory is to blame for this, and it should be noted that this aspect of the theory is separate from nonlocal elasticity. In addition, the tendency of the curve to increase becomes more pronounced as the length scale parameter increases, drawing attention to the obvious connection that exists between the length scale parameter and wave frequency. Therefore, it is possible to infer that the wave frequency of FG-MEE nanoplates is very sensitive to the values of the nonlocal and length scale parameters, particularly for rising wave numbers. This is because growing wave numbers cause the sensitivity of the wave frequency to increase. The accuracy of a wave dispersion study of smart nanoplates relies heavily on taking into account both nonlocal and length scale characteristics.

When the length scale parameter is present (when $\eta = 0.5$ nm^2) and the magneto-electric effects are absent (V = Ω = 0), Figure 3.35 illustrates the impact

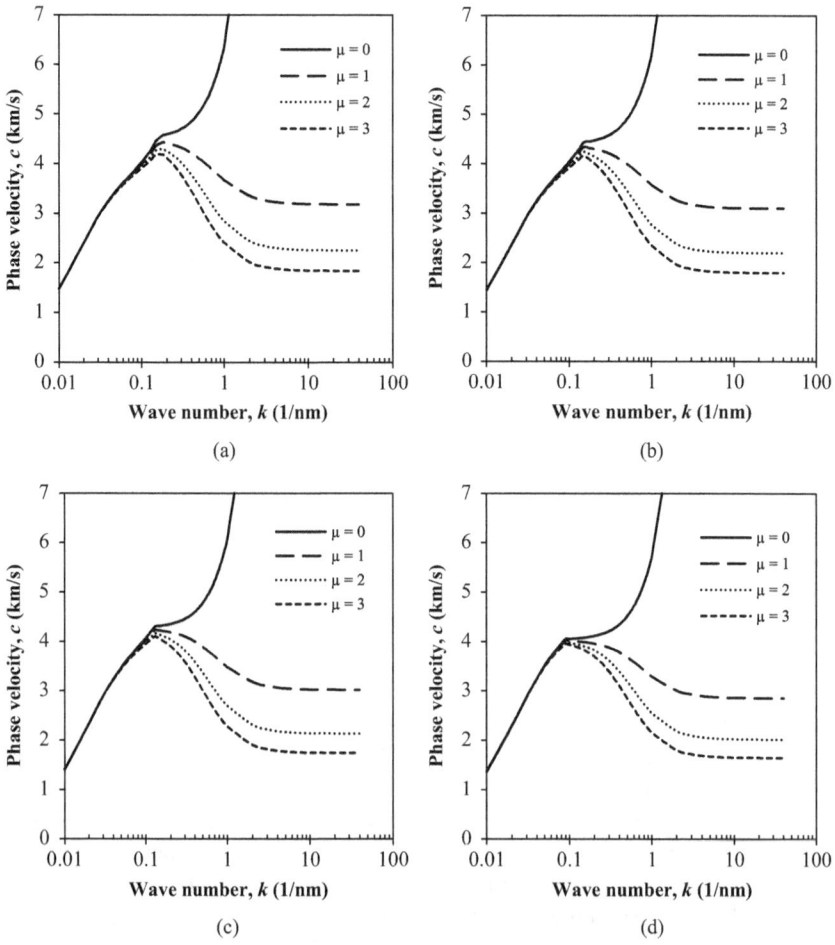

FIGURE 3.35 Variation in the phase velocity of FG-MEE nanoplate versus wave number for various nonlocal parameters and gradient indices ($\eta = 0.5$ nm^2, V = Ω = 0). (a) $p = 0.2$, (b) $p = 0.5$, (c) $p = 1$, and (d) $p = 5$.

that the gradient index has on the phase velocity of FG-MEE nanoplates. There is no variation in the general shape of the diagram regardless of the magnitude of the gradient index. When the nonlocal parameter reaches the value of 0, the resultant curvature begins to approach infinity for a given value of the gradient index. The value of the phase velocity initially increases to a maximum when the nonlocal parameter is not equal to zero; it then lowers to a minimum before remaining constant; lastly, the value of the phase velocity does not change. Once this parameter has a value that is not zero, it is necessary to take into consideration the influence that the nonlocal parameter has on the stiffness-softening of the system. This impact may be assessed by the nonzero effect that nonlocality has on the phase velocity of the FG-MEE nanosize plate.

However, it is reasonable to suppose that the maximum amount of phase velocity takes place whenever the wave number is approximately in the range of $k = 0.1 \times 10^9$ and $k = 0.2 \times 10^9$. This is because this is the range in which the phase velocity is greatest. This is due to the fact that the wave number that corresponds with the greatest magnitude of phase velocity cannot be given a particular value by modifying nonlocality for each special value of the gradient index and nonzero amounts of the nonlocal parameter. The reason for this is that it is impossible to assign a specific value to the wave number that corresponds with the highest magnitude of phase velocity. It has also been determined that the phase velocity decreases as the volume amount of the metal phase rises, which leads to another conclusion. In addition, the gradient index as well as the length scale parameter have an effect on the phase velocity of the FG-MEE size-dependent plate when $V = 0$ and 1 nm^2. These results, which are illustrated in great detail in Figure 3.36, are shown in the previous paragraph. In this figure, the impact of the gradient index's declining strength may be observed more clearly than in Figure 3.35. To elaborate, the behavior of the phase velocity varies depending on whether the length scale parameter is larger than the nonlocal parameter, lower than it, or equal to it. This is because the behavior of the phase velocity is determined by the nonlocal parameter. After a slow and even reduction, the phase velocity will suddenly pick up speed after the length scale parameter has been shown to be less than the nonlocal parameter. In the event that the conditions outlined above are met, the maximum value of phase velocity will remain unaltered. When the length scale parameter is larger than the nonlocal parameter, the phase velocity initially keeps increasing, with a wave number close to $k = 5 \times 10^9$; this occurs when the nonlocal parameter is smaller than the length scale parameter. This image also demonstrates the increasing impact of this parameter on the phase velocity of the FG-MEE nanoplate, as well as the stiffness-hardening influence created by the length scale parameter.

In addition, in Figure 3.37, we examine the effects of electric voltage applied to the FG-MEE nanoplate on its phase velocity twice: once in $\eta < \mu$, and again whenever $\eta > \mu$ at $p = 1$, $\mu = 1 \text{ nm}^2$ and $\Omega = 0$. Again, in this diagram, the course of the phase velocity varies depending on whether or not nonlocality is less than the length scale parameter. In addition, it is clear that the phase velocity increases with a larger value of the length scale parameter if the wave number is less than

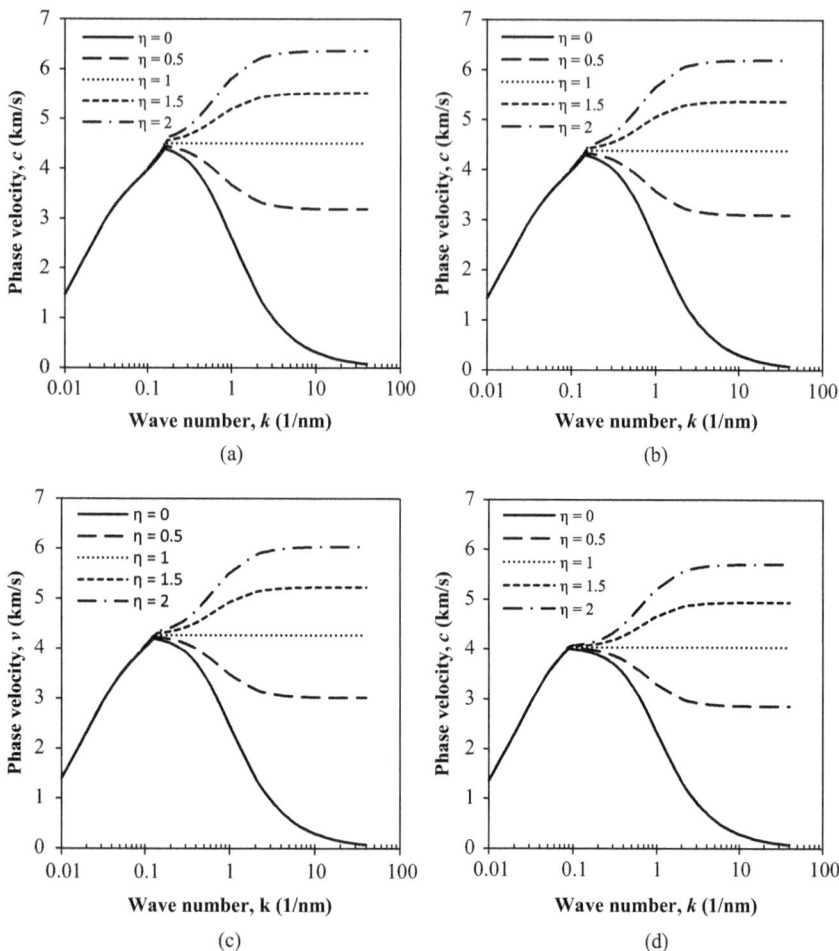

FIGURE 3.36   Variation in the phase velocity of FG-MEE nanoplate versus wave number for various length scale parameters and gradient indices ($\mu = 1$ nm², $V = \Omega = 0$). (a) $p = 0.2$, (b) $p = 0.5$, (c) $p = 1$, and (d) $p = 5$.

around $k = 0.2 \times 10^9$. The figure shows that the smart behavior of a nanoplate is restricted to low wave numbers and that once the wave number exceeds $k = 0.2 \times 10^9$, the value of the phase velocity is insensitive to changes in electric voltage. This little contribution actually decreases with increased electric voltage; therefore, a smaller phase velocity may be recorded as a result. Electric voltage's dwindling impact, on the other hand, seems to dwindle gradually with rising wave number. That is to say, the impact of electric voltage is much greater with wave numbers close to $k = 0.01 \times 10^9$ than at $k = 0.2 \times 10^9$.

However, when $p = 1$ and $\mu = 1$ nm², and with no electric voltage ($V = 0$), the effect of magnetic potential on the phase velocity of FG-MEE nanoplates is analyzed in Figure 3.38. Similar to Figure 3.37, this one bases its inquiries on two

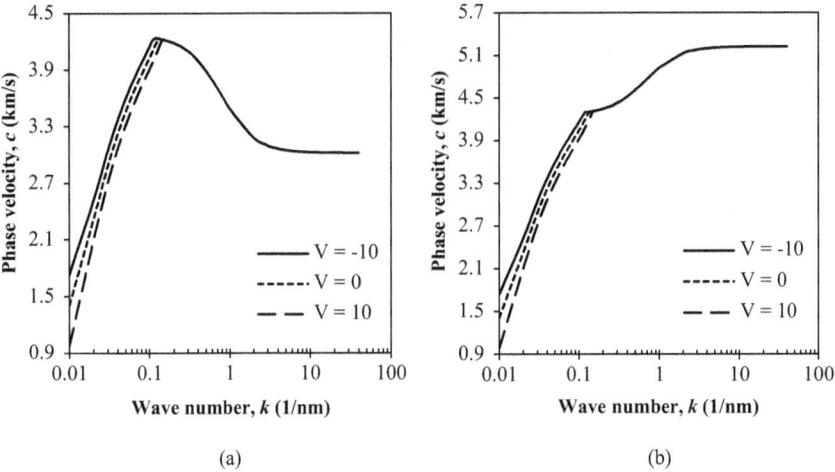

FIGURE 3.37   Variation in the phase velocity of FG-MEE nanoplate versus wave number for various electric voltages ($p = 1$, $\mu = 1$ nm^2, $\Omega = 0$). (a) $\eta = 0.5$ nm^2 and (b) $\eta = 1.5$ nm^2.

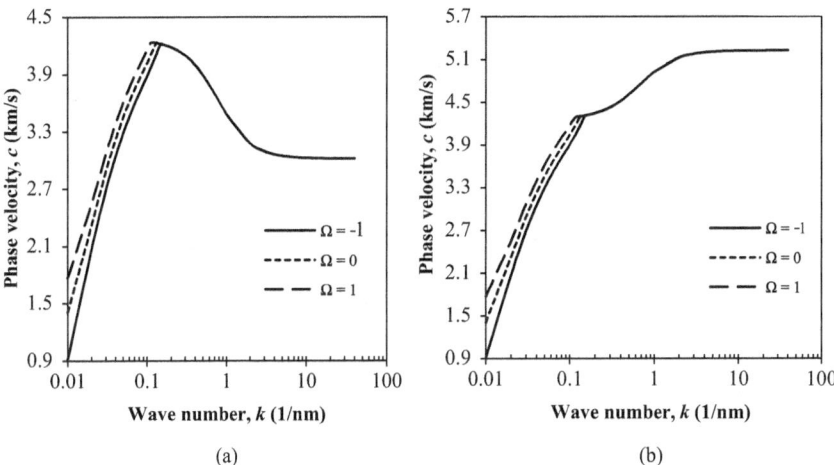

FIGURE 3.38   Variation in the phase velocity of MEE-FG nanoplate versus wave number for various magnetic potentials ($p = 1$, $\mu = 1$ nm^2, V = 0). (a) $\eta = 0.5$ nm^2 and (b) $\eta = 1.5$ nm^2.

overarching cases: ($\eta < \mu$ and $\eta > \mu$). By comparing Figures 3.37 and 3.38, we can see that the overall form is similar in both cases; yet, there is a noticeable variation between the two that helps to set them apart. The fundamental distinction is in the nature of the effect exerted on the FG-MEE nanoplate's phase velocity. The phase velocity of a nanoplate increases as the magnetic potential rises. In other words, the magnitude of phase velocity grows as magnetic potential does. Similar to the preceding illustration, the influence of magnetic potential is limited to values of k

less than $k = 0.2 \times 10^9$. As was previously indicated, this amplifying effect is most perceptible at wave numbers close to $k = 0.01 \times 10^9$.

We perform further tests beyond what is presented in Figure 3.39 to better see the effect of plate angular velocity on the phase velocity of MEE-FG nanoplates. At $p = 1$, $\mu = 1$ nm^2, $\eta = 1.5$ nm^2, $\Omega = 0$, the smart behavior of the nanoplate is again exhibited by altering the values of the applied voltage. To emphasize the effects of investigated factors, this graphic only displays phase velocity fluctuations across a narrow wave number range. Clearly, increasing the angular velocity is a simple way to boost the phase velocity. The impact of applied voltage on phase velocity is negative, as has been indicated below. Phase velocity rises based on the preceding figure; hence, it is unnecessary to produce a new diagram for this situation after magnetic potential fluctuations have been investigated. As an added bonus, the findings provided here hold true regardless of whether the length scale parameter is lower than or equal to the nonlocal value. The effectiveness of angular velocities spans a wide variety of length scales, which accounts for their independence from this parameter. In other words, when the wave number is greater than $k = 0.15 \times 10^9$, the hardening impact provided by the length scale parameter becomes obvious, but the effect of the plate's angular velocity remains visible, making the findings reported herein invariant with respect to the length scale parameter.

Figures 3.40 and 3.41 illustrate the interdependent effects of applied voltage and plate rotation and magnetic potential and plate rotation on the wave frequency of nanoplates, while the nanoplate is rotating at $p = 1$, $\mu = \eta = 1$ nm^2. The decrease in wave frequency values depicted in Figure 3.39 is attributable to an increase in

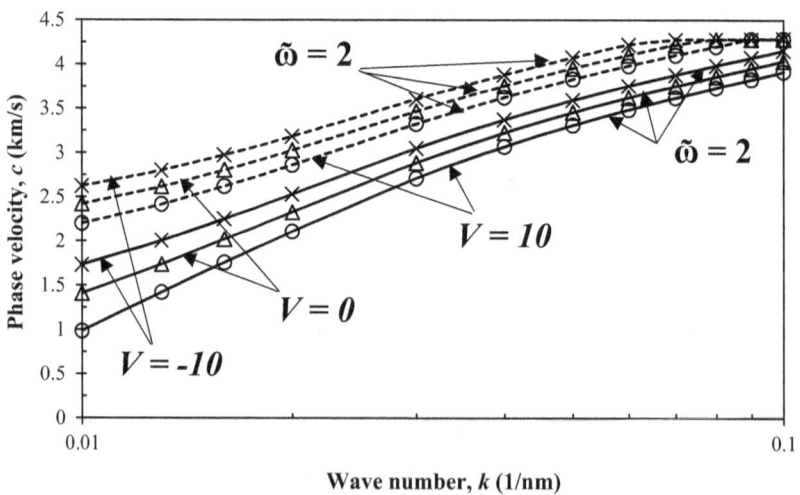

FIGURE 3.39   Variation in phase velocity of MEE-FG rotating nanoplate versus wave number for various applied electric voltage and angular velocity values ($p = 1$, $\mu = 1$ nm^2, $\eta = 1.5$ nm^2, $\Omega = 0$).

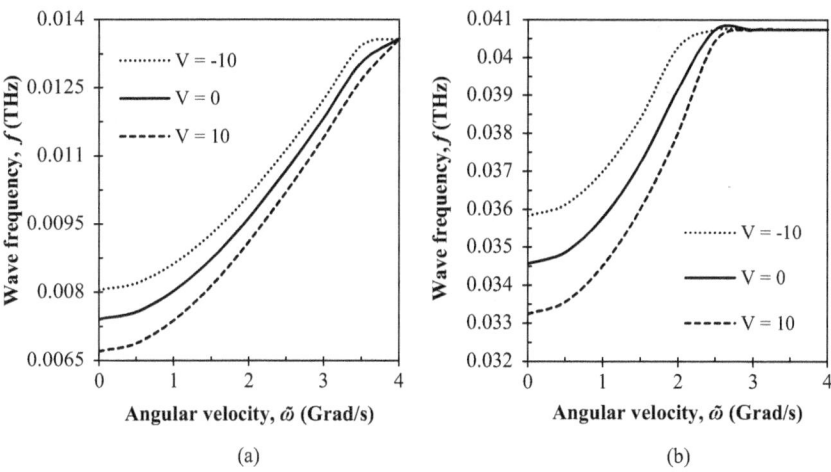

FIGURE 3.40 Variation in the wave frequency of smart FG rotating nanoplate versus angular velocity for various amounts of applied electric voltage at (a) $k = 0.02 \times 10^9$ and (b) $k = 0.06 \times 10^9$ ($p = 1, \mu = \eta = 1$ nm^2).

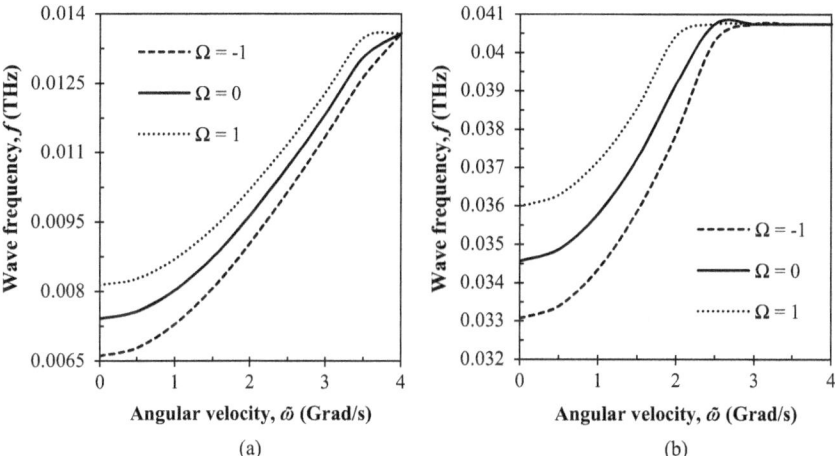

FIGURE 3.41 Variation in the wave frequency of smart FG rotating nanoplate versus angular velocity for various amounts of magnetic potentials at (a) $k = 0.02 \times 10^9$ and (b) $k = 0.06 \times 10^9$ ($p = 1, \mu = \eta = 1$ nm^2).

electric voltage. This effect is significantly more pronounced at $k = 0.02 \times 10^9$ than at $k = 0.06 \times 10^9$. Thus, as the wave number decreases, the ratio of the frequency at $\tilde{\omega} = 4$ Grad/s to the value at $\tilde{\omega} = 0$ increases. Moreover, it is crystal evident that an increase in total angular velocity corresponds qualitatively to a decrease in electric voltage. This means that if angular velocity is added, the result will be the same as if the electric voltage were decreased, namely an increase in phase velocity. It is worth noting that the frequency of waves with decreasing wave numbers, here $k = 0.02 \times 10^9$, increases when the angular velocity is adjusted from zero to $\tilde{\omega} = 4$ Grad/s. By comparing Figure 3.41 with Figure 3.38, we can see that the impact of magnetic potential differs significantly from that of applied voltage. This is not an unexpected result; earlier diagrams have shown the same thing. For each chosen magnetic potential, the angular velocity of the nanoplate in this figure increases the amount of the wave frequency.

It is fair to state that both softening and hardening effects are covered by both examples, despite the fact that in both figures nonlocal and length scale character-istics are considered to be equal. This is due to the fact that the studied wave numbers fall inside the region where all length scale parameters overlap. For wave numbers below $k = 0.15 \times 10^9$, all of the length scale parameters give the same results. The phase velocity values of MEE-FG nanoplates at $p = 1$, $\mu = \eta = 1$ nm^2 were also examined in order to establish the link between plate rotation and the intelligent properties of MEEMs. It has been observed again that increasing the electric voltage or decreasing the quantity of magnetic potential causes a decrease in the amount of wave frequency at each specific angular velocity of the nano-plate. Phase velocities are higher than they otherwise would be due to the multi-plicative effect of angular velocity. In Figures 3.40 through 3.43, it is vital to observe that the influence of wave number is taken into account. As the wave number grows, the graphs all indicate a believable rise in wave frequency or phase velocity, which is then followed by a plateau at the maximum value. Statistically, only the rotational velocity, electric voltage, and magnetic potential have any dis-cernible influence on the wave responses above a specific threshold of wave numbers.

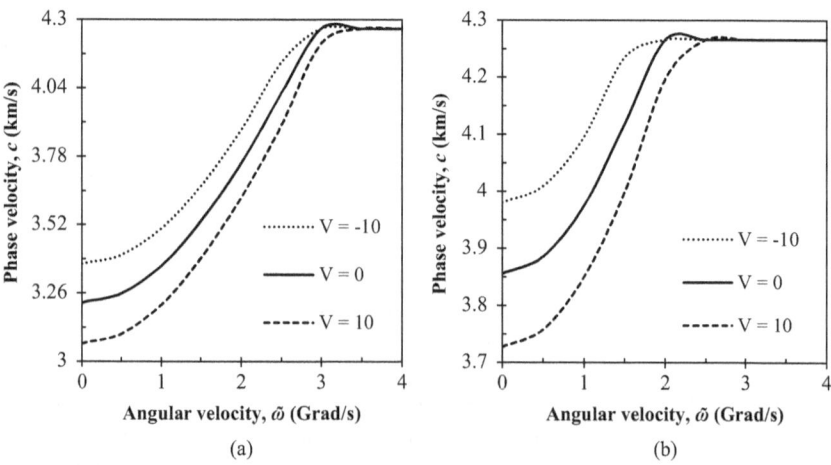

FIGURE 3.42 Variation in the phase velocity of smart FG rotating nanoplate versus angular velocity for various amounts of applied electric voltage at (a) $k = 0.04 \times 10^9$ and (b) $k = 0.08 \times 10^9$ ($p = 1$, $\mu = \eta = 1$).

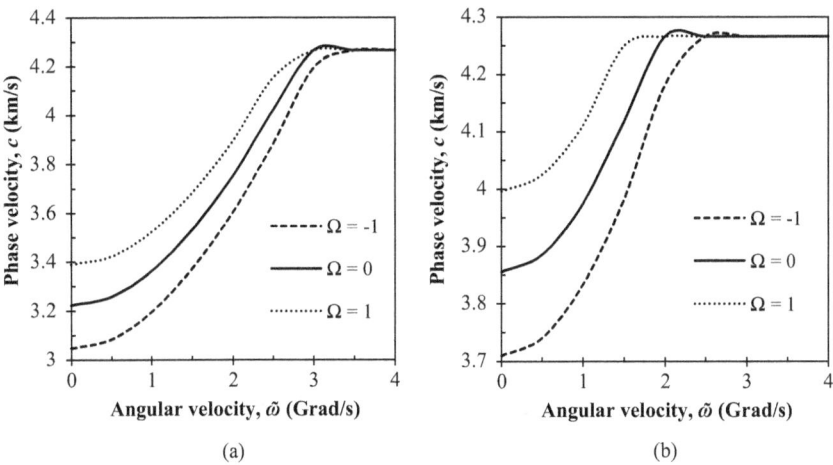

FIGURE 3.43 Variation in the phase velocity of smart FG rotating nanoplate versus angular velocity for various amounts of magnetic potentials at (a) $k = 0.04 \times 10^9$ and (b) $k = 0.08 \times 10^9$ ($p = 1, \mu = \eta = 1$).

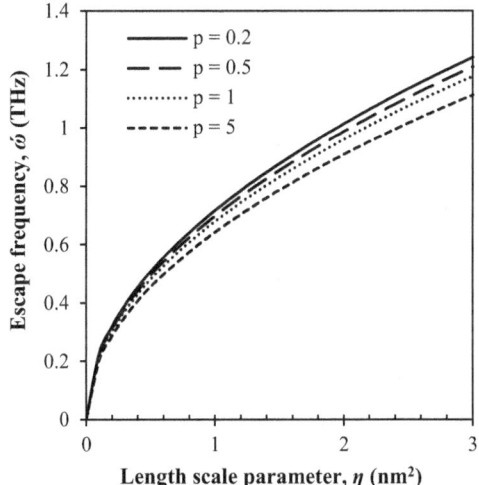

FIGURE 3.44 Variation in the escape frequency of MEE-FG nanoplate versus length scale parameter for various gradient indices ($\mu = 1$ nm^2).

As shown in Figure 3.44, when electromagnetic effects are neglected, the behavior of the escape frequency with respect to various gradient indices and length scale factors is highlighted at $\mu = 1$ nm^2. All observable influences on the FG-MEE nanoplate frequency are detailed here. Increasing the gradient index decreases the escape frequency for all values of the length scale parameter.

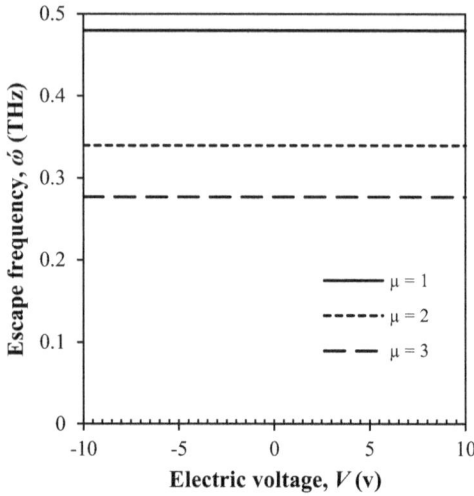

FIGURE 3.45 Variation in the escape frequency of MEE-FG nanoplate versus applied voltage for various nonlocal parameters ($\eta = 0.5$ nm^2, $p = 1$, $\Omega = 0$).

The favorable impact of the gradient index is another consideration. As a matter of fact, the escape frequency decreases as the gradient index rises. Again, the fact that a larger value for the length scale parameter results in a higher escape frequency is notable. It is important to note that the increase in escape frequency caused by a shift from 0 to 1 nm^2 in the length scale parameter is greater than the increase caused by a shift from 2 to 3 nm^2 in the same parameter.

At $\eta = 0.5$ nm^2, $p = 1$, $\Omega = 0$, Figure 3.45 investigates the effect of electric voltage on the escape frequency of an FG-MEE nanoplate for different nonzero values of nonlocal parameter. This number is consistent with previously calculated values. The overlapping of the effects of electric voltage at high wave numbers and the diminishing influence of the nonlocal parameter are both seen in this diagram. Keeping Figures 3.38 and 3.39 in mind, we can deduce that intelligent incentives are successful even in a narrow range of tiny wave numbers. According to the definition, one may easily determine the escape frequency by allowing the wave number to tend to infinity. As a consequence, the escape frequency of FG-MEE nanosize plates is insensitive to both electric and magnetic potentials. It's worth noting that the escape frequency may be decreased more quickly when going from a nonlocal parameter value of 1nm^2 to 2nm^2 than when going from a nonlocal parameter value of 2 nm^2 to 3 nm^2.

3.5 FUNCTIONALLY GRADED SHELLS: WAVE DISPERSION CHARACTERISTICS

The FGM nanoshells seen in Figure 3.46 are the focus of this analysis. Stainless steel (SUS304) and alumina (Al_2O_3), whose thermomechanical characteristics vary with temperature, are supposed to make up the FGM. The chapter makes the following assumptions, which are mentioned in reference [65]:

- Unless otherwise specified, $h = 25$ nm is the unchanging thickness of the nanoshell.
- It is expected that the nanoshell is rather thick. No research is conducted on a thick FGM nanoshell, and the ratio of radius to thickness is always set at $R/h = 50$.
- The standard temperature is set at $T_0 = 300$ K.
- Local temperature and humidity fluctuations are assumed to be linear (*i.e.*, in accordance with $\Delta T = T - T_0$ and $\Delta C = C - C_0$).
- No voids or pores are assumed to be present in the FGM's microstructure.

In this part, we collect the FGM's useful features using the well-known power law approach. This method requires that the combined volume fractions of the two phases add up to one (*i.e.*, $V_m + V_c = 1$). In addition, the volume fraction of the ceramic phase in the FGM is expressed using the following definition, which depends on the thickness [65]:

$$V_c = \left(\frac{z}{h} + \frac{1}{2} \right)^p \tag{3.85}$$

where p is the power law exponent that controls the thickness-to-volume ratio distribution. The following formula [2] may be used to easily determine any arbitrary attribute of the FGM:

$$P_{eq} = P_c V_c + P_m V_m \tag{3.86}$$

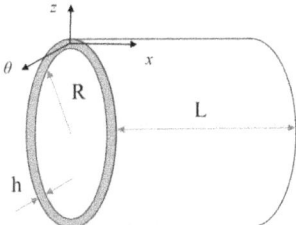

FIGURE 3.46 Schematic view of a cylindrical shell, its geometrical dimensions, and coordinate system used for modeling.

Also, P is any attribute such as Young's modulus (E), the Poisson's ratio (ν), the mass density (ρ), the coefficient of thermal expansion (α), or the coefficient of moisture concentration (β). The following definitions are used to capture the aforementioned characteristics of the component phases in order to account for the impact of local temperature on the FGM's material properties.

$$P = P_0 \left(P_{-1} T^{-1} + 1 + P_1 T + P_2 T^2 + P_3 T^3 \right) \qquad (3.87)$$

In the above definition, P_0, P_{-1}, P_1, P_2, and P_3 are considered material phase coefficients and can be found in Table 3.3 [65].

In this section, the kinematic connections between nanoshells will be discussed. The FGM nanoshell is a thin-walled continuous system; therefore, using higher-order shear deformation theory (HSDT) to describe its kinematics would be a waste of computing resources. This is why the standard FSDT for thin shells is used. Using this theory and the coordinate system shown in Figure 3.46, we can write the following expression for the nanoshell's displacement field: [65]

$$\begin{cases} u_x \left(x,\theta,z,t \right) = u\left(x,\theta,t \right) + z\varphi_x \left(x,\theta,t \right) \\ u_\theta \left(x,\theta,z,t \right) = v\left(x,\theta,t \right) + z\varphi_\theta \left(x,\theta,t \right) \\ u_z \left(x,\theta,z,t \right) = w\left(x,\theta,t \right) \end{cases} \qquad (3.88)$$

where u, v, and w denote axial, circumferential, and lateral displacements, respectively. Furthermore, φ_x and φ_θ indicate the rotations around longitudinal and circumferential directions, respectively. According to the definition of the infinitesimal strain tensor in the polar coordinate, the nonzero strains of the structure can be written as below:

TABLE 3.3
Temperature-Dependent Coefficients of Si_3N_4 and SUS304

Material	Property	P_0	P_{-1}	P_1	P_2	P_3
Si_3N_4	E [Pa]	348.43e9	0	−3.070e-4	2.160e-7	−8.946e-11
	ρ [kg/m³]	2370	0	0	0	0
	ν [−]	0.24	0	0	0	0
SUS304	E [Pa]	201.04e9	0	3.079e-4	−6.534e-7	0
	ρ [kg/m³]	8166	0	0	0	0
	ν [−]	0.3262	0	−2.002e-4	3.797e-7	0

$$
\begin{Bmatrix} \varepsilon_{xx} \\ \varepsilon_{\theta\theta} \\ \gamma_{x\theta} \\ \gamma_{xz} \\ \gamma_{\theta z} \end{Bmatrix} = \begin{Bmatrix} \dfrac{\partial u}{\partial x} + z\dfrac{\partial \varphi_x}{\partial x} \\[2mm] \dfrac{1}{R}\left(\dfrac{\partial v}{\partial \theta} + z\dfrac{\partial \varphi_\theta}{\partial \theta} + w \right) \\[2mm] \dfrac{1}{R}\dfrac{\partial u}{\partial \theta} + \dfrac{\partial v}{\partial x} + \dfrac{z}{R}\dfrac{\partial \varphi_x}{\partial \theta} + z\dfrac{\partial \varphi_\theta}{\partial x} \\[2mm] \varphi_x + \dfrac{\partial w}{\partial x} \\[2mm] \varphi_\theta + \dfrac{1}{R}\dfrac{\partial w}{\partial \theta} - \dfrac{v}{R} \end{Bmatrix} \tag{3.89}
$$

Here, we use Hamilton's energy approach to determine the motion equations. This principle states that for every given time period, the sum of the energy changes in a continuous system must equal zero [65]. In other words, it expresses:

$$
\delta \int_{t_1}^{t_2} \left(U - T + V \right) dt = 0 \tag{3.90}
$$

It uses the symbol U to represent the strain energy and the letter T to represent the kinetic energy. External loading also contributes V to the total work. Here is how you may express the virtual strain energy:

$$
\delta U = \int_{-\frac{h}{2}}^{\frac{h}{2}} \int_{0}^{2\pi} \int_{0}^{L} \sigma_{ij} \delta \varepsilon_{ij} R\,dx\,d\theta\,dz \tag{3.91}
$$

where σ_{ij} and ε_{ij} are the corresponding components of Cauchy stress and strain tensors, respectively. Furthermore, the variation in kinetic energy can be expressed by:

$$
\delta T = \int_{-\frac{h}{2}}^{\frac{h}{2}} \int_{0}^{2\pi} \int_{0}^{L} \rho(z) \left[\frac{\partial u_x}{\partial t} \frac{\partial \delta u_x}{\partial t} + \frac{\partial u_\theta}{\partial t} \frac{\partial \delta u_\theta}{\partial t} + \frac{\partial u_z}{\partial t} \frac{\partial \delta u_z}{\partial t} \right] R\,dx\,d\theta\,dz \tag{3.92}
$$

Since the FGM nanoshell is positioned in a hygrothermally stimulated environment, the variation in work done by hygrothermal forces N^T and N^H on the continuous system may be described as follows:

$$
\delta V = \int_{-\frac{h}{2}}^{\frac{h}{2}} \int_{0}^{2\pi} \int_{0}^{L} \left(N^T + N^H \right) \left(\frac{\partial^2 v}{\partial x^2} \delta v + \frac{\partial^2 w}{\partial x^2} \delta w \right) R\,dx\,d\theta\,dz \tag{3.93}
$$

where

$$N^T = \int_{\frac{-h}{2}}^{\frac{h}{2}} \frac{E(z)}{1+v(z)} \alpha_{eq} \Delta T dz, N^H = \int_{\frac{-h}{2}}^{\frac{h}{2}} \frac{E(z)}{1+v(z)} \beta_{eq} \Delta C dz \tag{3.94}$$

The data necessary to derive the motion equations has just become available. Substituting (3.91)–(3.94) into (3.90), we get the shell's motion equations. The following relations are produced if this replacement is made and a non-trivial answer of the collected identity is selected:

$$\frac{\partial N_{xx}}{\partial x} + \frac{1}{R}\frac{\partial N_{x\theta}}{\partial \theta} = I_0 \frac{\partial^2 u}{\partial t^2} + I_1 \frac{\partial^2 \varphi_x}{\partial t^2} \tag{3.95a}$$

$$\frac{\partial N_{x\theta}}{\partial x} + \frac{1}{R}\frac{\partial N_{\theta\theta}}{\partial \theta} + \frac{Q_{z\theta}}{R} + \left(N^T + N^H\right)\frac{\partial^2 v}{\partial x^2} = I_0 \frac{\partial^2 v}{\partial t^2} + I_1 \frac{\partial^2 \varphi_\theta}{\partial t^2} \tag{3.95b}$$

$$\frac{\partial Q_{xz}}{\partial x} + \frac{1}{R}\frac{\partial Q_{z\theta}}{\partial \theta} - \frac{N_{\theta\theta}}{R} + \left(N^T + N^H\right)\frac{\partial^2 w}{\partial x^2} = I_0 \frac{\partial^2 w}{\partial t^2} \tag{3.95c}$$

$$\frac{\partial M_{xx}}{\partial x} + \frac{1}{R}\frac{\partial M_{x\theta}}{\partial \theta} - Q_{xz} = I_1 \frac{\partial^2 u}{\partial t^2} + I_2 \frac{\partial^2 \varphi_x}{\partial t^2} \tag{3.95d}$$

$$\frac{\partial M_{x\theta}}{\partial x} + \frac{1}{R}\frac{\partial M_{\theta\theta}}{\partial \theta} - Q_{\theta z} = I_1 \frac{\partial^2 v}{\partial t^2} + I_2 \frac{\partial^2 \varphi_\theta}{\partial t^2} \tag{3.95e}$$

in which

$$\begin{bmatrix} N_{xx} & N_{\theta\theta} & N_{x\theta} \\ M_{xx} & M_{\theta\theta} & M_{x\theta} \end{bmatrix} = \int_{\frac{-h}{2}}^{\frac{h}{2}} \begin{bmatrix} 1 \\ z \end{bmatrix} \begin{bmatrix} \sigma_{xx} & \sigma_{\theta\theta} & \tau_{x\theta} \end{bmatrix} dz \tag{3.96a}$$

$$\begin{bmatrix} Q_{xz} & Q_{\theta z} \end{bmatrix} = \kappa_s \int_{\frac{-h}{2}}^{\frac{h}{2}} \begin{bmatrix} \tau_{xz} & \tau_{\theta z} \end{bmatrix} dz \tag{3.96b}$$

are the system's stress resultants. It is also possible to define the inertia mass moments as:

$$\left[I_0 I_1 I_2 \right] = \int\limits_{\frac{-h}{2}}^{\frac{h}{2}} \rho\left(z\right) \left[1\, z\, z^2 \right] dz \qquad (3.96c)$$

In Eq. (3.95e), κ_s is the shear correction factor that is taken to be $\dfrac{\pi^2}{12}$ in this work [65]. The system's constitutive equation is written in purely spatial terms for elastic nanostructures. This occurrence occurs because the stress and strain tensors of elastic nanostructures do not rely on time. However, stress in viscoelastic nanostructures fluctuates with time. Therefore, many constitutive models are used to examine nanoengineered systems with viscoelastic properties. Most current research on nanostructures' viscoelastic behavior has been conducted using the following terminologies [65]:

$$\left(1 - \mu^2 \nabla^2\right)\sigma_{ij} = C_{ijkl}\left(1 + \tau \frac{\partial}{\partial t}\right)\varepsilon_{kl} \qquad (3.97a)$$

while using the idea of nonlocal viscoelasticity [65]

$$\left(1 - \mu^2 \nabla^2\right)\sigma_{ij} = C_{ijkl}\left(1 - \lambda^2 \nabla^2\right)\left(1 + \tau \frac{\partial}{\partial t}\right)\varepsilon_{kl} \qquad (3.97b)$$

when applying the notion of viscoelasticity to a nonlocal strain gradient. The terms "nonlocal parameter (μ)," "length scale parameter (λ)," and "internal damping coefficient (τ)" are used throughout the preceding definitions. While this association between space and time nonlocality in nanostructures was first contested, further findings, as discussed in section 1, have validated its essential nature. When the nanostructure is exposed to waves with a wavelength similar to the characteristic wavelength of the wave medium, the resulting coupling is amplified. In fact, the aforementioned link between the preexisting nonlocalities cannot be accounted for by using Eq. (3.97a, b). The Boltzmann superposition integral and nonlocal strain gradient elasticity [37, 38] come together to form the unique theory of nonlocal strain gradient fractional time-space viscoelasticity. By applying this theory to a nanostructure subjected to hygrothermal excitations, we are able to derive a precise and universal constitutive equation for a nanoviscoelastic substance:

$$\left(1 - \mu^2 \nabla^2\right)\sigma_{ij} = C_{ijkl}\left(1 - \lambda^2 \nabla^2 + \tau \frac{\partial}{\partial t}\right)\left(\varepsilon_{kl} - \alpha_{ij}\Delta T - \beta_{ij}\Delta C\right) \qquad (3.98)$$

The nonlocal parameter μ is treated as a product of the nanostructure's thickness in the aforementioned connection, which accounts for the characteristic length of the continuous system. Taking the internal damping coefficient of the system to be reached by $\tau = \mu\sqrt{\rho_c / E_c}$ [36–38] also satisfies the connection between spatial and temporal nonlocalities. In this connection, the damping is linked to the nanostructure's characteristic length through the nonlocal parameter.

Expanding Eq. (3.98), we get the following identities for the stress-strain relationship in the nanoshell, given that the FGM nanostructure is viscoelastic.

$$\left(1-\mu^2\nabla^2\right)\sigma_{xx} = \left(1-\lambda^2\nabla^2+\tau\frac{\partial}{\partial t}\right)\left(C_{11}\varepsilon_{xx}+C_{12}\varepsilon_{\theta\theta}-C_{11}\alpha_{eq}\Delta T-C_{11}\beta_{eq}\Delta C\right) \text{ (3.99a)}$$

$$\left(1-\mu^2\nabla^2\right)\sigma_{\theta\theta} = \left(1-\lambda^2\nabla^2+\tau\frac{\partial}{\partial t}\right)\left(C_{12}\varepsilon_{xx}+C_{22}\varepsilon_{\theta\theta}-C_{22}\alpha_{eq}\Delta T-C_{22}\beta_{eq}\Delta C\right)$$

$$\text{(3.99b)}$$

$$\left(1-\mu^2\nabla^2\right)\tau_{x\theta} = \left(1-\lambda^2\nabla^2+\tau\frac{\partial}{\partial t}\right)C_{66}\gamma_{x\theta} \qquad\qquad \text{(3.99c)}$$

$$\left(1-\mu^2\nabla^2\right)\tau_{xz} = \left(1-\lambda^2\nabla^2+\tau\frac{\partial}{\partial t}\right)\kappa_s C_{44}\,\gamma_{xz} \qquad\qquad \text{(3.99d)}$$

$$\left(1-\mu^2\nabla^2\right)\tau_{\theta z} = \left(1-\lambda^2\nabla^2+\tau\frac{\partial}{\partial t}\right)\kappa_s C_{55}\gamma_{\theta z} \qquad\qquad \text{(3.99e)}$$

where

$$C_{11}=C_{22}=\frac{E(z)}{1-v(z)^2},\ C_{12}=C_{21}=\frac{v(z)E(z)}{1-v(z)^2},C_{44}=C_{55}=C_{66}=\frac{E(z)}{2(1+v(z))}$$

$$\text{(3.99f)}$$

The relations for the stress resultants shown in Eqs. (3.97a and b) are obtained by integrating Eqs. (3.99a–f) across the nanoshell thickness.

$$\left(1-\mu^2\nabla^2\right)\begin{bmatrix} N_{xx} \\ M_{xx} \\ N_{\theta\theta} \\ M_{\theta\theta} \end{bmatrix}\left(1-\lambda^2\nabla^2+\tau\frac{\partial}{\partial t}\right)$$

$$= \left(\begin{bmatrix} A_{11}\ B_{11}\ \dfrac{A_{12}}{R}\ \dfrac{B_{12}}{R} \\[2mm] B_{11}\ D_{11}\ \dfrac{B_{12}}{R}\ \dfrac{D_{12}}{R} \\[2mm] A_{12}B_{12}\ \dfrac{A_{11}}{R}\ \dfrac{B_{11}}{R} \\[2mm] B_{12}\ D_{12}\ \dfrac{B_{11}}{R}\ \dfrac{D_{11}}{R} \end{bmatrix} \begin{bmatrix} \dfrac{\partial u}{\partial x} \\[2mm] \dfrac{\partial\varphi_x}{\partial x} \\[2mm] \dfrac{\partial v}{\partial\theta}+w \\[2mm] \dfrac{\partial\varphi_\theta}{\partial\theta} \end{bmatrix} - \begin{bmatrix} N_x^T+N_x^H \\ M_x^T+M_x^H \\ N_\theta^T+N_\theta^H \\ M_\theta^T+M_\theta^H \end{bmatrix} \right) \qquad \text{(3.100a)}$$

$$\left(1-\mu^2\nabla^2\right)\begin{bmatrix} N_{x\theta} \\ M_{x\theta} \end{bmatrix} = \left(1-\lambda^2\nabla^2+\tau\frac{\partial}{\partial t}\right)\begin{bmatrix} A_{66} & B_{66} \\ B_{66} & D_{66} \end{bmatrix}\begin{bmatrix} \dfrac{1}{R}\dfrac{\partial u}{\partial\theta}+\dfrac{\partial v}{\partial x} \\ \dfrac{1}{R}\dfrac{\partial\varphi_x}{\partial\theta}+\dfrac{\partial\varphi_\theta}{\partial x} \end{bmatrix} \quad (3.100b)$$

$$\left(1-\mu^2\nabla^2\right)Q_{\theta z} = A_{55}^s\left(1-\lambda^2\nabla^2+\tau\frac{\partial}{\partial t}\right)\left(\varphi_\theta+\frac{1}{R}\frac{\partial w}{\partial\theta}-\frac{v}{R}\right) \quad (3.100c)$$

$$\left(1-\mu^2\nabla^2\right)Q_{xz} = A_{44}^s\left(1-\lambda^2\nabla^2+\tau\frac{\partial}{\partial t}\right)\left(\varphi_x+\frac{\partial w}{\partial x}\right) \quad (3.100d)$$

The following formulae may be used to determine the nanoshell's through-thickness rigidities and thermal resultants given the aforementioned identities.

$$\begin{bmatrix} A_{11} & B_{11} & D_{11} \\ A_{12} & B_{12} & D_{12} \\ A_{66} & B_{66} & D_{66} \end{bmatrix} = \int_{-\frac{h}{2}}^{\frac{h}{2}}\begin{bmatrix} C_{11} \\ C_{12} \\ C_{44} \end{bmatrix}\begin{bmatrix} 1 & z & z^2 \end{bmatrix}dz,$$

$$\begin{bmatrix} A_{44}^s \\ A_{55}^s \end{bmatrix} = \kappa_s\int_{-\frac{h}{2}}^{\frac{h}{2}}\begin{bmatrix} C_{44} \\ C_{55} \end{bmatrix}dz, \quad \begin{bmatrix} N_x^T & M_x^T \\ N_\theta^T & M_\theta^T \end{bmatrix} = \int_{-\frac{h}{2}}^{\frac{h}{2}}\begin{bmatrix} 1 & z \end{bmatrix}\begin{bmatrix} C_{11}\alpha_{eq} \\ C_{22}\alpha_{eq} \end{bmatrix}\Delta T dz, \quad (3.101)$$

$$\begin{bmatrix} N_x^H & M_x^H \\ N_\theta^H & M_\theta^H \end{bmatrix} = \int_{-\frac{h}{2}}^{\frac{h}{2}}\begin{bmatrix} 1 & z \end{bmatrix}\begin{bmatrix} C_{11}\beta_{eq} \\ C_{22}\beta_{eq} \end{bmatrix}\Delta C dz$$

This study will go forward with the development of the governing equations for the wave propagation problem in FGM nanoshells that is viscoelastically damped. The Euler–Lagrange equations of the continuum (Eqs. 3.91a–e)) must be updated by including the new constitutive relations (Eqs. 3.96a–d). Following this replacement, the following expressions will be compiled:

$$\left(1-\lambda^2\nabla^2+\tau\frac{\partial}{\partial t}\right)\left\{A_{11}\frac{\partial^2 u}{\partial x^2}+\frac{A_{66}}{R^2}\frac{\partial^2 u}{\partial\theta^2}+\frac{A_{12}+A_{66}}{R}\frac{\partial^2 v}{\partial x\partial\theta}\right.$$

$$+\frac{A_{12}}{R}\frac{\partial w}{\partial x}+B_{11}\frac{\partial^2\varphi_x}{\partial x^2}+\frac{B_{66}}{R^2}\frac{\partial^2\varphi_x}{\partial\theta^2}+\left.\frac{B_{12}+B_{66}}{R}\frac{\partial^2\varphi_\theta}{\partial x\partial\theta}\right\} \quad (3.102a)$$

$$=\left(1-\mu^2\nabla^2\right)\left\{I_0\frac{\partial^2 u}{\partial t^2}+I_1\frac{\partial^2\varphi_x}{\partial t^2}\right\}$$

$$\left(1-\lambda^2\nabla^2+\tau\frac{\partial}{\partial t}\right)\left\{\begin{array}{l}\dfrac{A_{12}+A_{66}}{R}\dfrac{\partial^2 u}{\partial x\partial\theta}+A_{66}\dfrac{\partial^2 v}{\partial x^2}+\dfrac{A_{11}}{R^2}\dfrac{\partial^2 v}{\partial\theta^2}+A_{55}^s\dfrac{v}{R^2}+\dfrac{A_{11}-A_{55}^s}{R^2}\\[4mm]\dfrac{\partial w}{\partial\theta}+\dfrac{B_{12}+B_{66}}{R}\dfrac{\partial^2\varphi_x}{\partial x\partial\theta}+B_{66}\dfrac{\partial^2\varphi_\theta}{\partial x^2}+\dfrac{B_{11}}{R^2}\dfrac{\partial^2\varphi_\theta}{\partial\theta^2}-\dfrac{A_{55}^s}{R}\varphi_\theta\end{array}\right\}$$

$$+\left(1-\mu^2\nabla^2\right)\left(N^T+N^H\right)\frac{\partial^2 v}{\partial x^2}=\left(1-\mu^2\nabla^2\right)\left\{I_0\frac{\partial^2 v}{\partial t^2}+I_1\frac{\partial^2\varphi_\theta}{\partial t^2}\right\}$$

(3.102b)

$$\left(1-\lambda^2\nabla^2+\tau\frac{\partial}{\partial t}\right)\left\{\begin{array}{l}\dfrac{A_{12}}{R}\dfrac{\partial u}{\partial x}+\dfrac{A_{11}-A_{55}^s}{R^2}\dfrac{\partial v}{\partial\theta}+\dfrac{A_{11}}{R^2}w+A_{55}^s\left(\dfrac{\partial^2 w}{\partial x^2}+\dfrac{1}{R^2}\dfrac{\partial^2 w}{\partial\theta^2}\right)\\[4mm]+\left(A_{55}^s+\dfrac{B_{12}}{R}\right)\dfrac{\partial\varphi_x}{\partial x}+\left(\dfrac{B_{11}}{R^2}+\dfrac{A_{55}^s}{R}\right)\dfrac{\partial\phi_\theta}{\partial\theta}\end{array}\right\}$$

$$+\left(1-\mu^2\nabla^2\right)\left(N^T+N^H\right)\frac{\partial^2 w}{\partial x^2}=\left(1-\mu^2\nabla^2\right)I_0\frac{\partial^2 w}{\partial t^2}$$

(3.102c)

$$\left(1-\lambda^2\nabla^2+\tau\frac{\partial}{\partial t}\right)\left\{\begin{array}{l}B_{11}\dfrac{\partial^2 u}{\partial x^2}+\dfrac{B_{66}}{R^2}\dfrac{\partial^2 u}{\partial\theta^2}+\dfrac{B_{12}+B_{66}}{R}\dfrac{\partial^2 v}{\partial x\partial\theta}+\left(\dfrac{B_{12}}{R}+A_{55}^s\right)\\[4mm]\dfrac{\partial w}{\partial x}+D_{11}\dfrac{\partial^2\varphi_x}{\partial x^2}+\dfrac{D_{66}}{R^2}\dfrac{\partial^2\varphi_x}{\partial\theta^2}+A_{55}^s\varphi_x+\dfrac{D_{12}+D_{66}}{R}\dfrac{\partial^2\varphi_\theta}{\partial x\partial\theta}\end{array}\right\}$$

$$=\left(1-\mu^2\nabla^2\right)\left\{I_1\frac{\partial^2 u}{\partial t^2}+I_2\frac{\partial^2\varphi_x}{\partial t^2}\right\}$$

(3.102d)

$$\left(1-\lambda^2\nabla^2+\tau\frac{\partial}{\partial t}\right)\left\{\begin{array}{l}\dfrac{B_{12}+B_{66}}{R}\dfrac{\partial^2 u}{\partial x\partial\theta}+B_{66}\dfrac{\partial^2 v}{\partial x^2}+\dfrac{B_{11}}{R^2}\dfrac{\partial^2 v}{\partial\theta^2}-A_{55}^s\dfrac{v}{R}+\left(\dfrac{B_{11}}{R^2}+\dfrac{A_{55}^s}{R}\right)\\[4mm]\dfrac{\partial w}{\partial\theta}+\dfrac{D_{12}+D_{66}}{R}\dfrac{\partial^2\varphi_x}{\partial x\partial\theta}+D_{66}\dfrac{\partial^2\varphi_\theta}{\partial x^2}+\dfrac{D_{11}}{R^2}\dfrac{\partial^2\varphi_\theta}{\partial\theta^2}+A_{55}^s\varphi_\theta\end{array}\right\}$$

$$=\left(1-\mu^2\nabla^2\right)\left\{I_1\frac{\partial^2 v}{\partial t^2}+I_2\frac{\partial^2\varphi_\theta}{\partial t^2}\right\}$$

(3.102e)

Researchers use a variety of techniques to determine the mechanical reactions of continuous systems at this point [65]. Here, the issue is solved analytically to get the dispersed waves' natural frequency. The FGM nanoshell has such a large variation in wavelength compared to its length that the borders of the continuous system may be thought of as being easily sustained. This analytic approach takes into account two wave numbers along both the longitudinal and circumferential axes. Not only that, but it's important to remember that the natural frequency is complex, not scalar. The method described in detail in reference [65] may be used to examine the following solution functions.

$$\begin{cases} u(x,\theta,t) = Ue^{i(\beta x + m\theta - \omega t)} \\ v(x,\theta,t) = Ve^{i(\beta x + m\theta - \omega t)} \\ w(x,\theta,t) = We^{i(\beta x + m\theta - \omega t)} \\ \phi_x(x,\theta,t) = \phi_x e^{i(\beta x + m\theta - \omega t)} \\ \phi_\theta(x,\theta,t) = \phi_\theta e^{i(\beta x + m\theta - \omega t)} \end{cases} \qquad (3.103)$$

where U, V, W, ϕ_x, and ϕ_θ are the unknown amplitudes of motion. Also, β and m represent longitudinal and circumferential wave numbers, respectively. Additionally, ω denotes the complex natural frequency of the propagated waves. Substituting the solution functions introduced in Eq. (3.103) in Eqs. (3.102a–e) and applying mathematical simplifications reveals:

$$\left\{ [\mathbf{K}] + [\mathbf{C}]\omega - \omega^2[\mathbf{M}] \right\} \{ U \quad V \quad W \quad \phi_x \quad \phi_\theta \}^T = \{\mathbf{0}\} \quad (3.104)$$

in which $\mathbf{K}$, $\mathbf{M}$, and $\mathbf{C}$ are symmetric stiffness, mass, and damping matrices, respectively. The nonzero arrays of the aforesaid matrices are as follows:

$$K_{11} = -\left(1 + \lambda^2\left(\beta^2 + \frac{m^2}{R^2}\right)\right)\left(A_{11}\beta^2 + \frac{A_{66}}{R^2}m^2\right),$$

$$K_{12} = -\left(1 + \lambda^2\left(\beta^2 + \frac{m^2}{R^2}\right)\right)\frac{A_{12} + A_{66}}{R}m\beta,$$

$$K_{13} = i\left(1 + \lambda^2\left(\beta^2 + \frac{m^2}{R^2}\right)\right)\frac{A_{12}\beta}{R},$$

$$K_{14} = -\left(1 + \lambda^2\left(\beta^2 + \frac{m^2}{R^2}\right)\right)\left(B_{11}\beta^2 + \frac{B_{66}}{R^2}m^2\right),$$

$$K_{15} = -\left(1 + \lambda^2\left(\beta^2 + \frac{m^2}{R^2}\right)\right)\frac{B_{12} + B_{66}}{R}m\beta,$$

$$K_{22} = -\left(1 + \lambda^2\left(\beta^2 + \frac{m^2}{R^2}\right)\right)\left(A_{66}\beta^2 + A_{11}\frac{m^2}{R^2} - \frac{A_{55}^s}{R^2}\right)$$

$$+ \left(1 + \mu^2\left(\beta^2 + \frac{m^2}{R^2}\right)\right)\left(N^T + N^H\right)\beta^2,$$

$$K_{23} = i \left(1 + \lambda^2 \left(\beta^2 + \frac{m^2}{R^2} \right) \right) m \frac{A_{11} - A_{55}^s}{R^2},$$

$$K_{24} = - \left(1 + \lambda^2 \left(\beta^2 + \frac{m^2}{R^2} \right) \right) \frac{B_{12} + B_{66}}{R} m \beta,$$

$$K_{25} = - \left(1 + \lambda^2 \left(\beta^2 + \frac{m^2}{R^2} \right) \right) \left(B_{66} \beta^2 + B_{11} \frac{m^2}{R^2} + \frac{A_{55}^s}{R} \right),$$

$$K_{33} = \left(1 + \lambda^2 \left(\beta^2 + \frac{m^2}{R^2} \right) \right) \left(\frac{A_{11}}{R^2} - A_{55}^s \left(\beta^2 + \frac{m^2}{R^2} \right) \right)$$

$$+ \left(1 + \mu^2 \left(\beta^2 + \frac{m^2}{R^2} \right) \right) \left(N^T + N^H \right) \beta^2,$$

$$K_{34} = i \left(1 + \lambda^2 \left(\beta^2 + \frac{m^2}{R^2} \right) \right) \beta \left(A_{55}^s + \frac{B_{12}}{R} \right),$$

$$K_{35} = i \left(1 + \lambda^2 \left(\beta^2 + \frac{m^2}{R^2} \right) \right) m \left(\frac{B_{11}}{R^2} + \frac{A_{55}^s}{R} \right),$$

$$K_{44} = - \left(1 + \lambda^2 \left(\beta^2 + \frac{m^2}{R^2} \right) \right) \left(D_{11} \beta^2 + D_{66} \frac{m^2}{R^2} - A_{55}^s \right),$$

$$K_{45} = - \left(1 + \lambda^2 \left(\beta^2 + \frac{m^2}{R^2} \right) \right) \frac{D_{12} + D_{66}}{R} m \beta,$$

$$K_{55} = - \left(1 + \lambda^2 \left(\beta^2 + \frac{m^2}{R^2} \right) \right) \left(D_{66} \beta^2 + D_{11} \frac{m^2}{R^2} - A_{55}^s \right)$$

(3.105)

$$m_{11} = - \left(1 + \mu^2 \left(\beta^2 + \frac{m^2}{R^2} \right) \right) I_0, \qquad m_{25} = - \left(1 + \mu^2 \left(\beta^2 + \frac{m^2}{R^2} \right) \right) I_1,$$

$$m_{14} = - \left(1 + \mu^2 \left(\beta^2 + \frac{m^2}{R^2} \right) \right) I_1, \qquad m_{33} = - \left(1 + \mu^2 \left(\beta^2 + \frac{m^2}{R^2} \right) \right) I_0,$$

$$m_{22} = - \left(1 + \mu^2 \left(\beta^2 + \frac{m^2}{R^2} \right) \right) I_0, \qquad m_{44} = m_{55} = - \left(1 + \mu^2 \left(\beta^2 + \frac{m^2}{R^2} \right) \right) I_2$$

(3.106)

$$C_{11} = \tau i \omega \left(A_{11} \beta^2 + \frac{A_{66}}{R^2} m^2 \right),$$

$$C_{25} = \tau i \omega \left(B_{66} \beta^2 + B_{11} \frac{m^2}{R^2} + \frac{A_{55}^s}{R^2} \right),$$

$$C_{12} = \tau i \omega \frac{A_{12} + A_{66}}{R} m \beta,$$

$$C_{33} = -\tau i \omega \left(\frac{A_{11}}{R^2} - A_{55}^s \left(\beta^2 + \frac{m^2}{R^2} \right) \right),$$

$$C_{13} = \tau \omega \beta \frac{A_{12}}{R},$$

$$C_{34} = \tau \omega m \left(A_{55}^s + \frac{B_{12}}{R} \right),$$

$$C_{14} = \tau i \omega \left(B_{11} \beta^2 + \frac{B_{66}}{R^2} m^2 \right),$$

$$C_{35} = \tau \omega m \left(\frac{B_{11}}{R^2} + \frac{A_{55}^s}{R} \right),$$

$$C_{15} = \tau i \omega \frac{B_{12} + B_{66}}{R} m \beta,$$

$$C_{44} = \tau i \omega \left(D_{11} \beta^2 + D_{66} \frac{m^2}{R^2} - A_{55}^s \right),$$

$$C_{22} = \tau i \omega \left(A_{66} \beta^2 + A_{11} \frac{m^2}{R^2} - \frac{A_{55}^s}{R^2} \right), \quad C_{45} = \tau i \omega \frac{D_{12} + D_{66}}{R} m \beta,$$

$$C_{23} = \tau \omega m \frac{A_{11} - A_{55}^s}{R^2},$$

$$C_{55} = \tau i \omega \left(D_{66} \beta^2 + D_{11} \frac{m^2}{R^2} - A_{55}^s \right)$$

$$C_{24} = \tau i \omega \frac{B_{12} + B_{66}}{R} m \beta,$$

$$(3.107)$$

The FGM nanoshell's complex frequency may be calculated by plugging its value into Eq. (3.104). The scattered waves' phase velocity in the nanoshell's axial direction (*i.e.*, $c_p = \omega/\beta$) may be calculated by dividing the measured frequency by the longitudinal wave number. To further understand how FGM nanoshells disperse hygro-thermo-viscoelastic waves, several illustrative cases will be shown below. Table 3.3 displays the input values of material attributes stated at the start of Section 2. To quantify the balance between the nanoshell's hardness and its softening effects, a non-dimensional parameter is used in this research. The equation for this variable, called the scaling factor, is $c = \lambda/\mu$.

The FGM viscoelastic nanoshell's phase velocity versus longitudinal wave number curve is shown for a range of scale factors in Figure 3.47. As the scale factor is increased from $c = 0.5$ to $c = 2.0$, the curve is seen to go upward in this example. This tendency arises because the strength of the hardening mechanism in the constitutive modeling of the nanoshell is increased when larger values are assigned to the scale factor. Therefore, the system's dynamic reactivity will increase as the through-the-thickness rigidities are accentuated. However, with low longitudinal wave numbers (*i.e.*, $\beta < 9$ [1/nm]), it is evident that phase speed does not vary. The major explanation for this phenomenon is that the stiffness of the system is modest enough in these wave numbers for the FGM nanoshell's internal dampening to be effective.

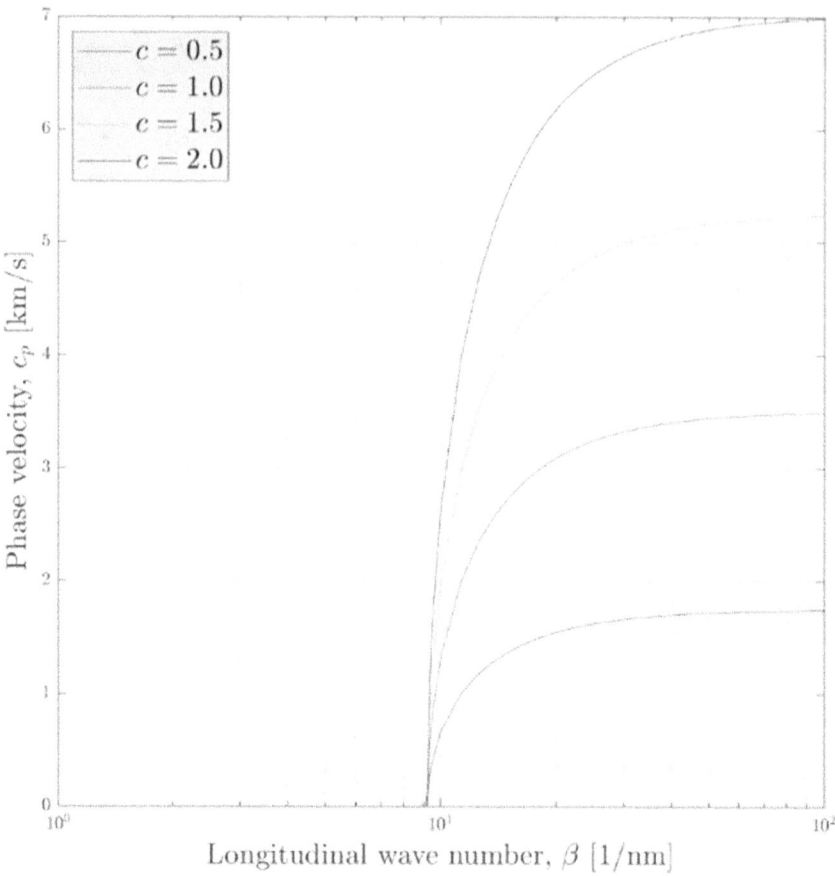

FIGURE 3.47 Variation in the phase velocity versus longitudinal wave number for various scale factors ($m = 1$, $T = 300$ K, $p = 2$, $\mu = 0.05$ h).

Figure 3.48 has a similar pattern. Here we see how the number of longitudinal waves varies as the frequency of the waves fluctuates. The frequency-wave number curve of the nanoshell exhibits a more rigid behavior with increasing scale factor. We have elaborated on the physical causes of this tendency above, so we won't go over them again.

Figure 3.49 depicts the effect of a nonlocal parameter on the dependence of the phase velocity on the circumferential wave number of viscoelastic FGM nanoshells. This graphic demonstrates that the phase speed of the FGM nanoshell slows down for any given value of the circumferential wave number as the nonlocal parameter increases. The main reason for this is that the rigidity of the nanostructure is decreased and the internal damping is increased when a nonlocal parameter is introduced. Assigning large values to the circumferential wave number also results in increased phase velocities being received from the wave medium, as expected. The dependence of the elements of the stiffness and mass matrices on the circumferential

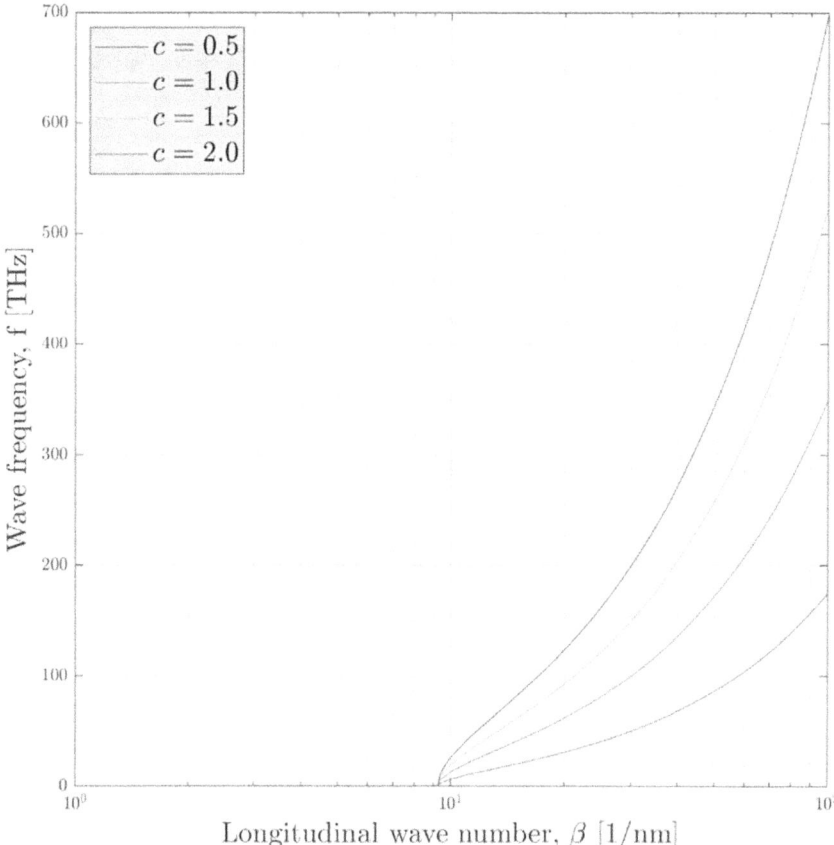

FIGURE 3.48   Variation in wave frequency versus longitudinal wave number for various scale factors ($m = 1$, $T = 300$ K, $p = 2$, $\mu = 0.05$ h).

wave number provides a straightforward justification for this tendency; in contrast, such a strong dependence is not seen in the damping matrix arrays.

Figure 3.50 also shows how the working temperature affects the fluctuation of the phase velocity of the dispersed waves in the FGM nanoshells. The dynamic responsiveness of the viscoelastic nanostructure is shown to diminish when the local temperature is raised as seen in the picture. The FGM's component stages provide a natural explanation for this observation. That is to say, both Si3N4 and SUS304 are non-negative coefficients of thermal expansion linear elastic solids. When the temperature of such substances is increased, their stiffness decreases. Adding the ambient temperature reduces the FGM's effective modulus. Because there is a direct correlation between stiffness and dynamic response, it makes sense to see a decrease in the phase velocity of the propagating waves. The reader is also encouraged to take note of the dramatic contrast in phase velocities between the examples where the scale factor is less than one and more than one, which is

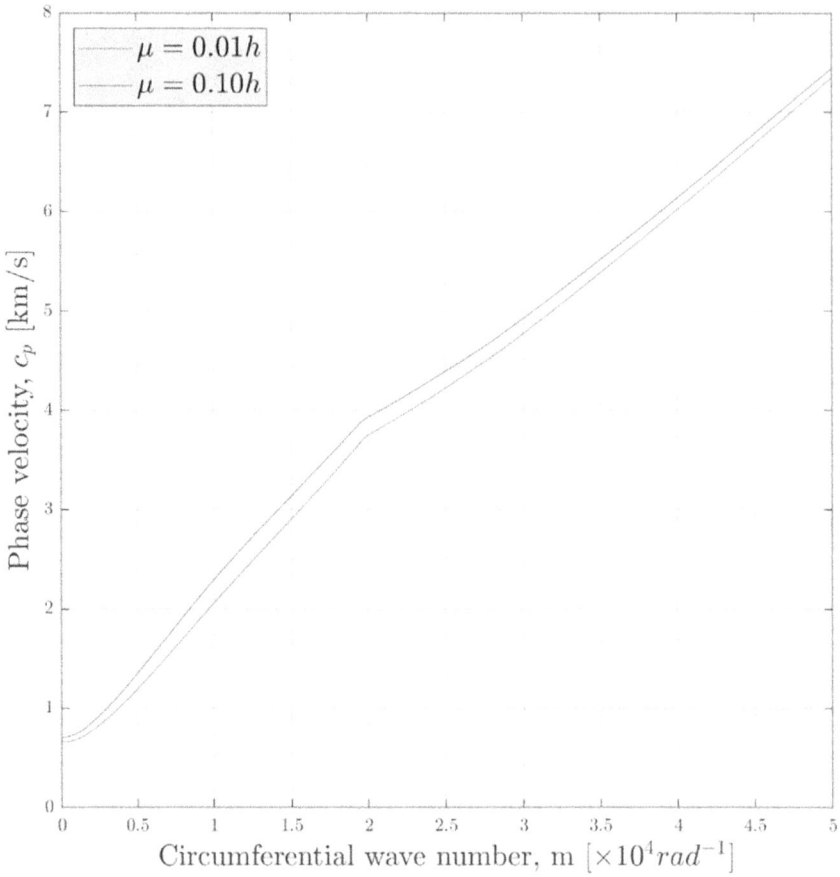

FIGURE 3.49   Variation in phase velocity versus circumferential wave number for various nonlocal parameters ($c = 0.5$, $T = 300$ K, $p = 2$).

a secondary result of this presentation. It is not surprising to see a rise in the phase velocity if a larger value is provided to the scale factor, since larger scale factors equate to stiffer nanostructures.

Figure 3.51 shows, for various gradient indices, how the phase velocity in the FGM nanoshell varies as a function of the nonlocal parameter. It is easy to determine that by increasing the gradient index, slower phase velocities are achieved. The diminishing effect of large gradient indices provides the physical basis for this trend in Young's modulus of the FGM. By selecting large gradient indices, the nanoshell softens, and its phase velocity decreases. This graph also supports the prior results of this investigation. As the level of nonlocality rises, the phase velocity naturally decreases. In addition, the stiffness-hardening mechanism provided by an increase in the length scale parameter makes it simpler to possess wave media with higher propagation speeds.

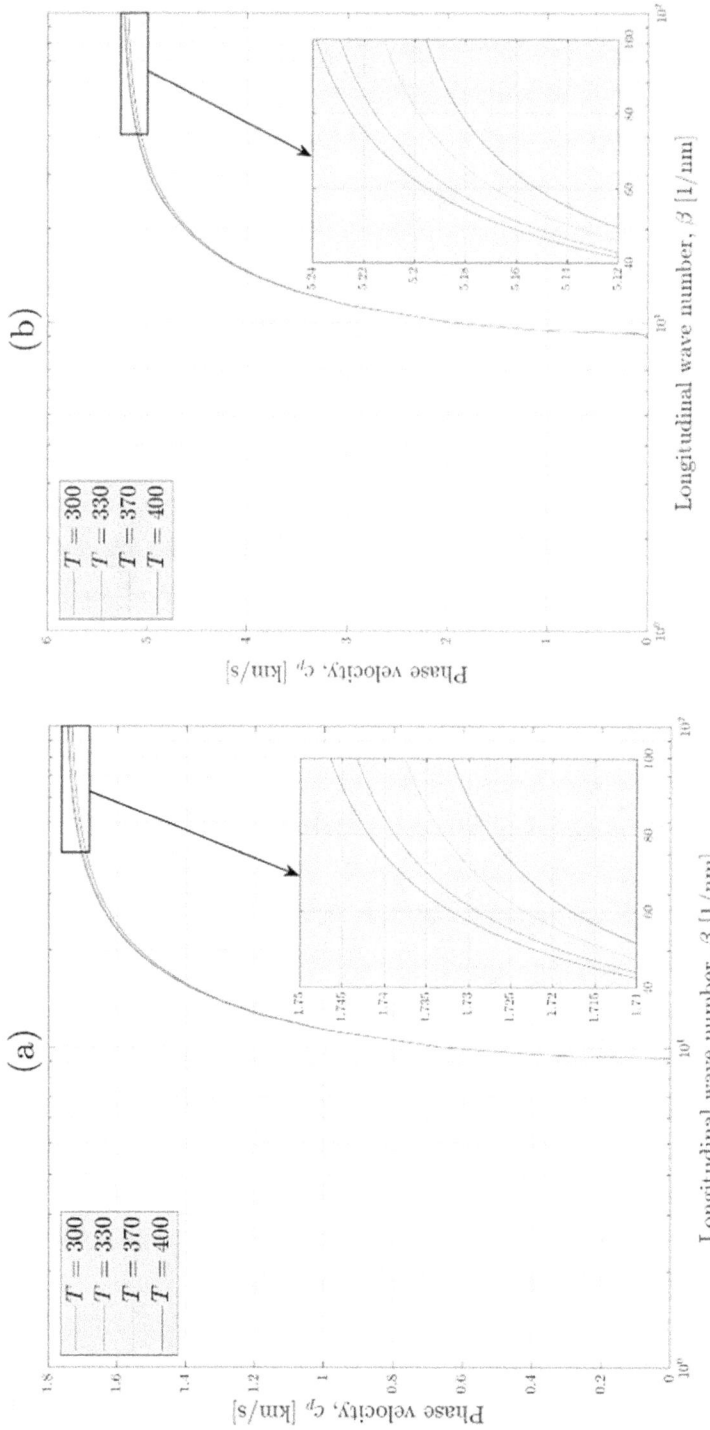

FIGURE 3.50 Variation in phase velocity versus wave number for various temperatures at (a) $c < 1$ and (b) $c > 1$ ($m = 1, p = 2, \mu = 0.05$ h).

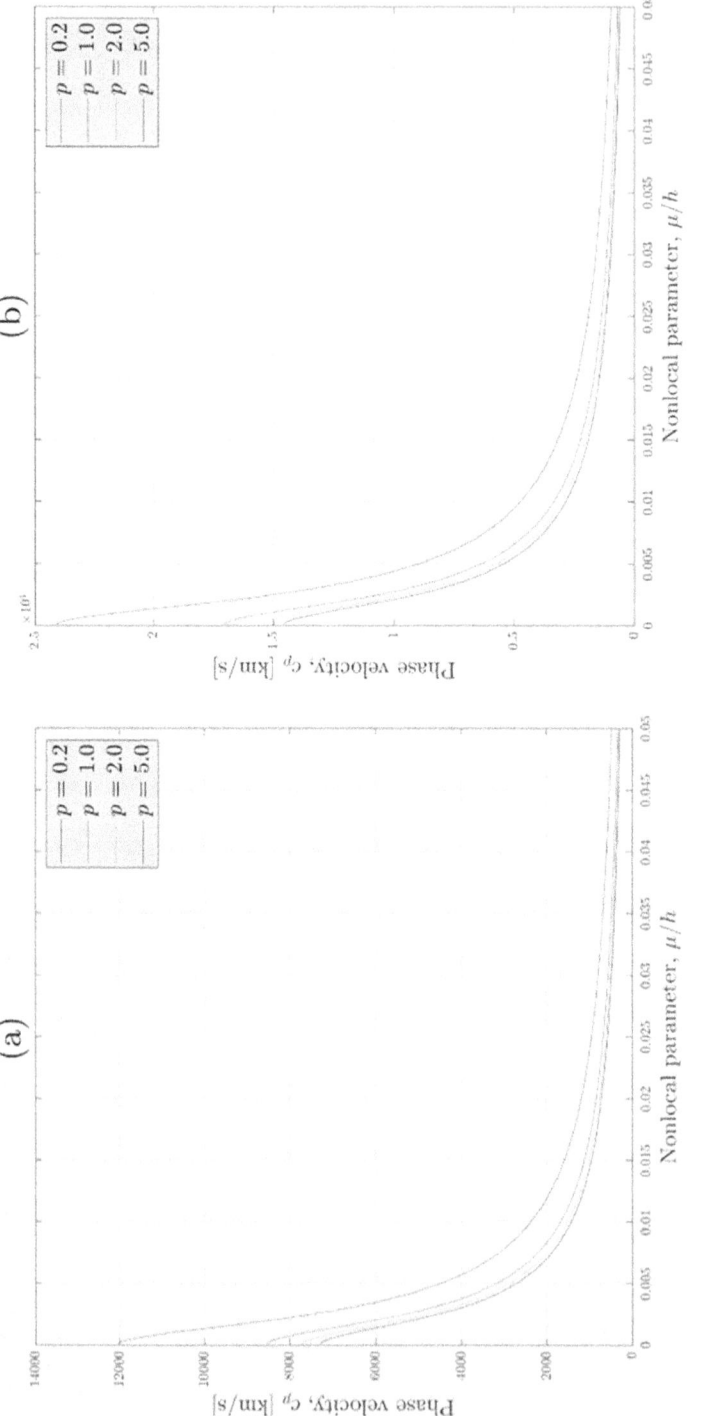

FIGURE 3.51   Variation in phase velocity versus nonlocal parameter for various gradient indices at (a) $\lambda = 5h$ and (b) $\lambda = 10h$ ($m = 1$, $T = 300$ K, $c = 1$).

3.5.1 HYGROTHERMAL EFFECTS

Figure 3.52 shows how the dynamical properties of waves propagating through FGM nanoshells are affected by a number of different size-dependent viscoelastic theorems. We consider the nanoshell to be in $\Delta T = 50$ K and $\Delta C = 50\%$ environment with hygrothermally stimulated conditions. There is a clear distinction between non-fractional decoupled conventional modeling and fractional time-space modeling. The fundamental cause of this distinction is the link in fractional modeling between the nonlocal parameter and the internal damping parameter. While this critical connection is taken into account in nonlocal strain gradient viscoelasticity theory, it is omitted in nonlocal viscoelasticity theory, which explains the viscoelastic behaviors of nanoshells. The conflict between time-space and the most fundamental types of size-dependent theories becomes increasingly apparent with small values of the nonlocal parameter. This discrepancy occurs because whereas the internal damping parameter is fixed in non-fractional simple theories, in fractional time-space theory it is understood to be a function of a nonlocal parameter. The enhanced dynamic response predicted when strain gradient effects are included is also demonstrated to be accounted for by the stiffness-hardening process caused by the length scale factor λ [65].

Figure 3.53 displays how the phase velocity versus the longitudinal wave number curve of viscoelastic FGM nanoshells has changed as a function of environmental circumstances. Here, we use the notion of nonlocal strain gradient fractional viscoelasticity to demonstrate three separate situations. A rise in both local temperature and moisture content, as shown in this case, slows the phase velocity of the propagating waves considerably. The viscoelastic FGM's thermal and moisture expansion coefficients provide sufficient physical justification for this tendency. Since the applied FGM has positive thermal and moisture expansion coefficients, its stiffness degrades at higher temperatures and higher moisture concentrations. That is to say, if either the temperature or the moisture content is increased, the FGM will act more leniently. That the phase velocity curve to slope downwards is, thus, to be expected. When considering the damping properties of viscoelastic FGM nanoshells, it is important to remember that hygrothermal damping is another mechanism that might be invoked.

Figure 3.54 displays, for $\beta = 10$ [1/nm], the hygro-thermo-viscoelastically damped phase velocity versus temperature increase curve of viscoelastic FGM nanoshells. Various size-dependent viscoelastic theories, such as those shown in Figure 3.52, are applied here. Nonlocal strain gradient fractional viscoelasticity theory predicts a dynamic curve that represents the system's maximum response. What we observed in Figure 3.52 is confirmed by this finding. Nonlocal viscoelasticity theory and nonlocal fractional viscoelasticity theory provide identical phase velocities, as seen in Figure 3.54. Figure 3.52 provides a clear explanation for this pattern. Nonlocal viscoelasticity theory and nonlocal fractional viscoelasticity theory yielded remarkably similar predictions for phase velocities in the aforementioned figure, with the nonlocal parameter held constant at $\mu = 0.01h$. The current case study takes the nonlocal parameter to be five times larger than it

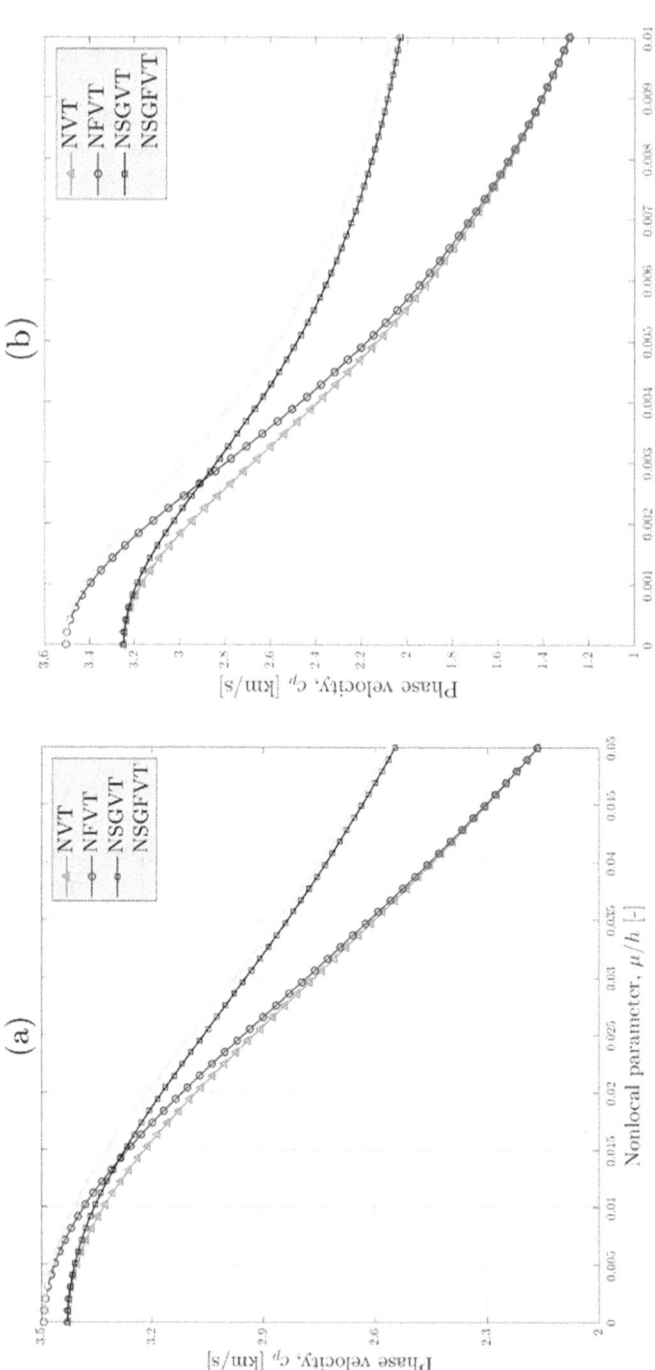

FIGURE 3.52 Phase velocity versus nonlocal parameter curve of viscoelastic FGM nanoshells in a hygrothermally affected environment for different size-dependent viscoelasticity theories at (a) $\beta = 1$ [1/nm] and (b) $\beta = 10$ [1/nm]. In this figure, it is assumed that $p = 2$, $m = 1$, $\Delta T = 50$ K, $\Delta C = 50\%$. In the cases that strain gradient effects are included, $c < 1$ is employed (NVT: nonlocal viscoelasticity theory; NFVT: nonlocal fractional viscoelasticity theory; NSGVT: nonlocal strain gradient viscoelasticity theory; NSGFVT: nonlocal strain gradient fractional viscoelasticity theory).

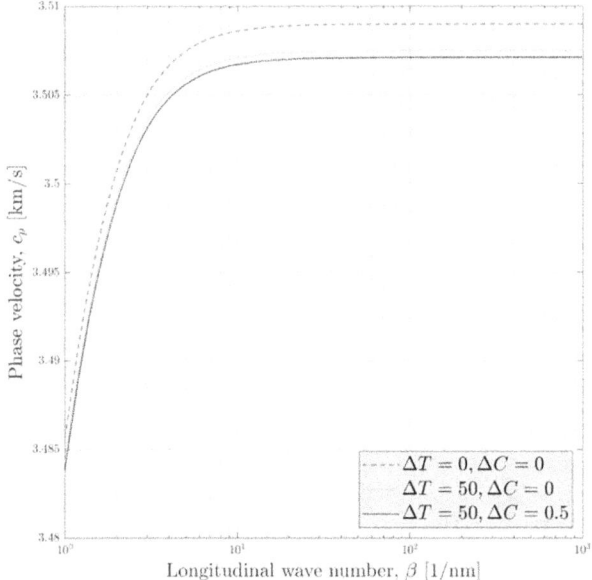

FIGURE 3.53   Variation in the velocity of viscoelastic FGM nanoshells against longitudinal wave number for various ambient circumstances ($\mu = 0.05$ h, $c = 1$, $p = 2$, $m = 1$).

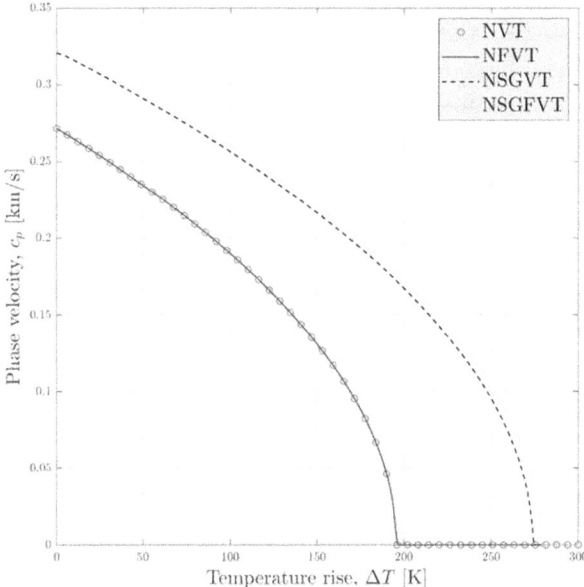

FIGURE 3.54   Variation in the phase velocity of viscoelastic FGM nanoshells versus temperature rise for different size-dependent viscoelasticity theories ($p = 2$, $m = 1$, $\Delta C = 50\%$, $c = 0.05$, $\mu = 0.05$ h, $\beta = 10$ [1/nm]).

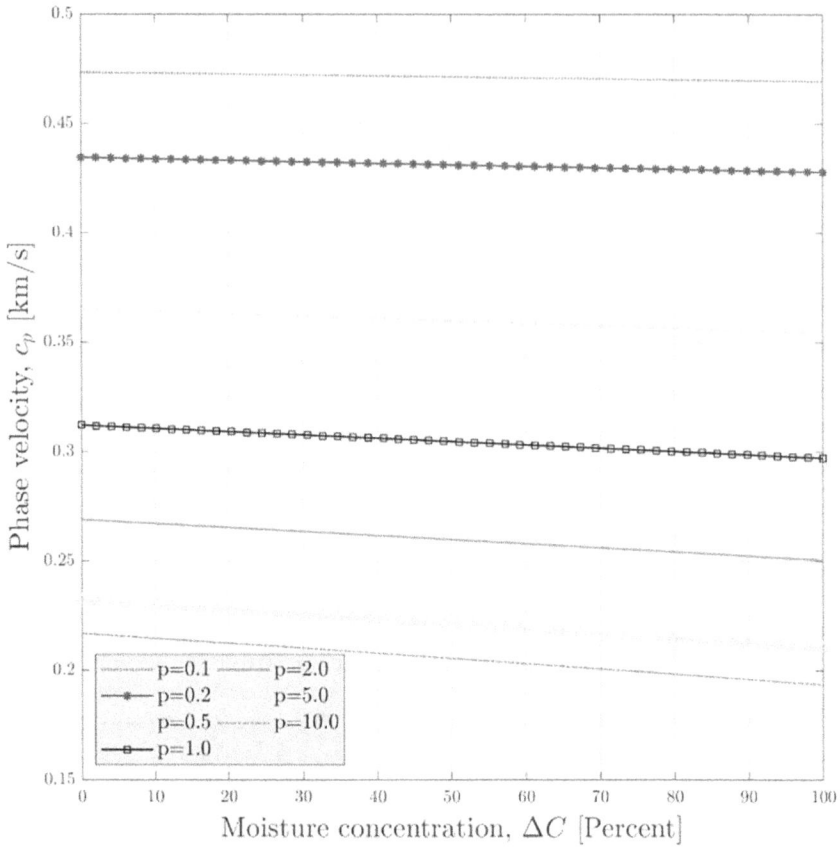

FIGURE 3.55 Variation in the phase velocity of viscoelastic FGM nanoshells versus moisture concentration for different gradient indices ($m = 1$, $\Delta T = 100$ K, $c = 0.05$, $\mu = 0.05$ h, $\beta = 10$ [1/nm]).

really is. Since both nonlocal viscoelasticity theory and nonlocal fractional viscoelasticity theory have asymptotic trends in large values of nonlocal parameters, no other trend is possible.

Finally, Figure 3.55 shows the relationship between the power law exponent p and the phase velocity variation versus the moisture content. This diagram once again demonstrates the dampening process brought on by a high concentration of moisture. The amplitude of this sort of damping, however, is obviously much lower than that of thermal damping. Furthermore, it is evident that if a larger value is supplied to the power law exponent, the phase velocity of the FGM nanoshell changes downward. This pattern has emerged as a direct result of the impact of this exponent on the suppler nature of the viscoelastic FGM. In other words, the equivalent stiffness of the FGM nanostructure decreases as a direct result of the modulus reduction process induced by the increase in the power law exponent, and this in turn reduces the dynamic responsiveness of the system.

3.6 SUMMARY

This chapter's focus is on illuminating the wave dispersion properties as they relate to the propagation of FG beams, plates, and shells. The functionally graded nanobeam has material characteristics that are spatially graded according to a Mori–Tanaka distribution. The motion equations were derived by combining the energy principles with the Euler beam model. The dependence of the attributes on the reference temperature has been incorporated in this text to offer more realistic data. The governing differential equations are obtained by using Hamilton's principle. Finally, a parametric analysis is conducted to examine the impact of several factors on the wave properties of a temperature-dependent FG nanobeam. These parameters include the temperature, the gradient index, the nonlocality parameter, and the length scale parameter.

The following is a quick summary of the findings that will be presented:

- Note that when the nonlocal and length scale parameters are changed, the wave frequency, phase velocity, and escape frequency of the FG nanobeam are all affected differently. In other words, when the length scale parameter increases, the size of the aforementioned variables increases, whereas the size of the aforementioned parameters decreases while the nonlocal parameter increases.
- The magnitude of length's effects may be broken down into three categories: When $\eta < \mu$, the phase velocity first reaches its maximum value and then begins to decay. After reaching its maximum value, phase velocity stays constant at $\eta = \mu$. Furthermore, if $\eta > \mu$, phase velocity keeps growing after reaching a certain threshold, and it stays the same after reaching its maximum value.
- When the expected volume percentage of the metal phase is increased (gradient index is added), the wave frequency, phase velocity, and escape frequency are all reduced.
- Since the FG nanobeam's rotational impact is negligible, even little alterations in this parameter may lead to noticeable changes in frequency and phase velocity for wave numbers near to $\beta = 0.2 \times 10^9$.
- It must be mentioned that once beam's angular velocity is amplified in this range, phase velocity can be enlarged. Beam phase velocity rises in proportion to its angular velocity in this area.
- Increasing the magnetic field strength leads to faster phase speeds at smaller wave numbers.
- For very small wave numbers, particularly those less than $\beta = 0.2 \times 10^9$, an increase in electric voltage may reduce phase velocity, in contrast to the effect of magnetic potential.
- The wave propagation characteristics of FG piezoelectric rotating nanobeams may be affected by fabrication-induced porosity. In fact, as the volume percentage of porosity increases, so do the wave frequencies, phase velocities, and escape velocities.
- It has been shown that when temperatures rise, the following quantities all decrease: wave frequency, phase velocity, group velocity, and escape frequency.

- In order to better distinguish between various temperature changes depending on the frequency of the waves, higher wave numbers are required.
- According to nonlocal strain gradient theory, an increase in the length scale parameter has a stiffening effect on the system, leading to an increase in the phase and group velocities.
- It's also possible that the nonlocal parameter slows down the stiffness, which in turn slows down the phase and group velocities.
- The phase and group velocities are significantly dampened by the gradient index. As the exponent in the power law increases, the stiffness of the medium decreases, and the phase velocity of the continuous system decreases. In addition, neither temperature nor gradient index significantly affects the phase or group velocities at higher wave numbers when the length scale parameter is set to zero, $l=0$.
- The phase and group velocities remain constant as the wave number increases. Therefore, the effect of a change in temperature, gradient index, nonlocal parameter, and length scale parameter on wave characteristics is very sensitive to the wave number.
- As the temperature gradient increases, the stiffness of the system decreases, and the phase velocity or wave frequency drops.
- The angular velocity of the nanobeams at low wave numbers may amplify the phase speed of the waves propagating within a DNBS. This magnifying effect disappears above a certain wave number.
- Mechanical response is greatest when nanobeams are moving out of phase with one another, next when a single nanobeam is stationary, and finally when the nanobeams are moving relative to one another but in phase.
- The phase velocity slows down more slowly than predicted by non-fractional, nonlocal theories. Since discounting this factor also means disregarding the internal damping, the nonlocal parameter's significance is large.
- It is shown that for small values of the nonlocal parameters, the viscoelastic responses derived from fractional time-space theories and non-fractional simple nonlocal theories diverge the most.
- In viscoelastic FGM nanoshells, damping due to moisture is not as significant as thermal deterioration in slowing down the phase velocity.

APPENDICES

APPENDIX A

$$k_{11} = -\left(1+\eta\beta^2\right)A_{xx}\beta^2 \qquad k_{21} = -i\left(1+\eta\beta^2\right)B_{xx}\beta^3$$

$$k_{12} = i\left(1+\eta\beta^2\right)B_{xx}\beta^3 \qquad k_{22} = -\left(1+\eta\beta^2\right)D_{xx}\beta^4$$

$$k_{13} = i\left(1+\eta\beta^2\right)B_{xx}^s\beta^3 \qquad k_{23} = -\left(1+\eta\beta^2\right)D_{xx}^s\beta^4$$

$$k_{14} = i\beta A_{31}^e\left(1+\eta\beta^2\right) \qquad k_{24} = -\left(1+\eta\beta^2\right)E_{31}^e\beta^2$$

$$k_{15} = i\beta A_{31}^m\left(1+\eta\beta^2\right) \qquad k_{25} = -\left(1+\eta\beta^2\right)E_{31}^m\beta^2$$

$$k_{31} = -i\left(1+\eta\beta^2\right)B_{xx}^s\beta^3$$

$$k_{32} = -\left(1+\eta\beta^2\right)D_{xx}^s\beta^4$$

$$k_{33} = -\left(1+\eta\beta^2\right)\left(H_{xx}^s\beta^4 + A_{xz}^s\beta^2\right)$$

$$k_{34} = -\left(1+\eta\beta^2\right)\left(F_{31}^e - E_{15}^e\right)\beta^2$$

$$k_{35} = -\left(1+\eta\beta^2\right)\left(F_{31}^m - E_{15}^m\right)\beta^2$$

$$k_{41} = i\beta A_{31}^e\left(1+\eta\beta^2\right)$$

$$k_{42} = \left(1+\eta\beta^2\right)E_{31}^e\beta^2$$

$$k_{43} = \left(1+\eta\beta^2\right)\left(F_{31}^e - E_{15}^e\right)\beta^2$$

$$k_{44} = -\left(1+\eta\beta^2\right)\left(F_{33}^e + F_{11}^e\beta^2\right)$$

$$k_{45} = -\left(1+\eta\beta^2\right)\left(F_{33}^m + F_{11}^m\beta^2\right)$$

$$k_{51} = i\beta A_{31}^m\left(1+\eta\beta^2\right)$$

$$k_{52} = \left(1+\eta\beta^2\right)E_{31}^m\beta^2$$

$$k_{53} = \left(1+\eta\beta^2\right)\left(F_{31}^m - E_{15}^m\right)\beta^2$$

$$k_{54} = -\left(1+\eta\beta^2\right)\left(F_{33}^m + F_{11}^m\beta^2\right)$$

$$k_{55} = -\left(1+\eta\beta^2\right)\left(X_{33}^m + X_{11}^m\beta^2\right)$$

and

$$m_{11} = -I_0\left(1+\mu\beta^2\right)$$

$$m_{12} = iI_1\beta\left(1+\mu\beta^2\right)$$

$$m_{13} = iJ_1\beta\left(1+\mu\beta^2\right)$$

$$m_{14} = m_{15} = 0$$

$$m_{21} = -iI_1\beta\left(1+\mu\beta^2\right)$$

$$m_{22} = -\left(I_0 + \left(\left(N_{max}^R + N^H\right.\right.\right.$$
$$\left.\left.\left. + N^E\right) + I_2\right)\beta^2\right)\left(1+\mu\beta^2\right)$$

$$m_{23} = -\left(I_0 + \left(\left(N_{max}^R + N^H\right.\right.\right.$$
$$\left.\left.\left. + N^E\right) + J_2\right)\beta^2\right)\left(1+\mu\beta^2\right)$$

$$m_{24} = m_{25} = 0$$

$$m_{31} = -iJ_1\beta\left(1+\mu\beta^2\right)$$

$$m_{32} = -\left(I_0 + \left(\left(N_{max}^R + N^H\right.\right.\right.$$
$$\left.\left.\left. + N^E\right) + J_2\right)\beta^2\right)\left(1+\mu\beta^2\right)$$

$$m_{33} = -\left(I_0 + \left(\left(N_{max}^R + N^H\right.\right.\right.$$
$$\left.\left.\left. + N^E\right) + K_2\right)\beta^2\right)\left(1+\mu\beta^2\right)$$

$$m_{34} = m_{35} = 0$$

$$m_{41} = m_{42} = m_{43} = m_{44} = m_{45} = 0$$

$$m_{51} = m_{52} = m_{53} = m_{54} = m_{55} = 0$$

Appendix B

$$k_{11} = -\left(1 + \eta\left(k_1^2 + k_2^2\right)\right)\left(A_{11}k_1^2 + A_{66}k_2^2\right)$$

$$k_{12} = -\left(1 + \eta\left(k_1^2 + k_2^2\right)\right)\left(A_{12} + A_{66}\right)k_1 k_2$$

$$k_{13} = ik_1\left(1 + \eta\left(k_1^2 + k_2^2\right)\right)\left(B_{11}k_1^2 + \left(B_{12} + 2B_{66}\right)k_2^2\right)$$

$$k_{14} = ik_1\left(1 + \eta\left(k_1^2 + k_2^2\right)\right)\left(B_{11}^s k_1^2 + \left(B_{12}^s + 2B_{66}^s\right)k_2^2\right)$$

$$k_{15} = ik_1 A_{31}^e\left(1 + \eta\left(k_1^2 + k_2^2\right)\right)$$

$$k_{16} = ik_1 A_{31}^m\left(1 + \eta\left(k_1^2 + k_2^2\right)\right)$$

$$k_{21} = -\left(1 + \eta\left(k_1^2 + k_2^2\right)\right)\left(A_{12} + A_{66}\right)k_1 k_2$$

$$k_{22} = -\left(1 + \eta\left(k_1^2 + k_2^2\right)\right)\left(A_{66}k_1^2 + A_{22}k_2^2\right)$$

$$k_{23} = ik_2\left(1 + \eta\left(k_1^2 + k_2^2\right)\right)\left(\left(B_{12} + 2B_{66}\right)k_1^2 + B_{22}k_2^2\right)$$

$$k_{24} = ik_2\left(1 + \eta\left(k_1^2 + k_2^2\right)\right)\left(\left(B_{12}^s + 2B_{66}^s\right)k_1^2 + B_{22}^s k_2^2\right)$$

$$k_{25} = ik_2 A_{31}^e\left(1 + \eta\left(k_1^2 + k_2^2\right)\right)$$

$$k_{26} = ik_2 A_{31}^m\left(1 + \eta\left(k_1^2 + k_2^2\right)\right)$$

$$k_{31} = -ik_1\left(1 + \eta\left(k_1^2 + k_2^2\right)\right)\left(B_{11}k_1^2 + \left(B_{12} + 2B_{66}\right)k_2^2\right)$$

$$k_{32} = -ik_2\left(1 + \eta\left(k_1^2 + k_2^2\right)\right)\left(\left(B_{12} + 2B_{66}\right)k_1^2 + B_{22}k_2^2\right)$$

$$k_{33} = -\left(1 + \eta\left(k_1^2 + k_2^2\right)\right)\left(D_{11}k_1^4 + 2\left(D_{12} + 2D_{66}\right)k_1^2 k_2^2 + D_{22}k_2^4\right)$$
$$+ \left(1 + \mu^2\left(k_1^2 + k_2^2\right)\right)\left(\left(N^E + N^H + N_{max}^R\right)\left(k_1^2 + k_2^2\right)\right)$$

$$k_{34} = -\left(1 + \eta\left(k_1^2 + k_2^2\right)\right)\left(D_{11}^s k_1^4 + 2\left(D_{12}^s + 2D_{66}^s\right)k_1^2 k_2^2 + D_{22}^s k_2^4\right)$$
$$+ \left(1 + \mu^2\left(k_1^2 + k_2^2\right)\right)\left(\left(N^E + N^H + N_{max}^R\right)\left(k_1^2 + k_2^2\right)\right)$$

$$k_{35} = -E_{31}^e\left(1 + \eta\left(k_1^2 + k_2^2\right)\right)\left(k_1^2 + k_2^2\right)$$

$$k_{36} = -E_{31}^m\left(1 + \eta\left(k_1^2 + k_2^2\right)\right)\left(k_1^2 + k_2^2\right)$$

$$k_{41} = -ik_1\left(1 + \eta\left(k_1^2 + k_2^2\right)\right)\left(B_{11}^s k_1^2 + \left(B_{12}^s + 2B_{66}^s\right)k_2^2\right)$$

$$k_{42} = -ik_2\left(1 + \eta\left(k_1^2 + k_2^2\right)\right)\left(\left(B_{12}^s + 2B_{66}^s\right)k_1^2 + B_{22}^s k_2^2\right)$$

$$k_{43} = -\left(1+\eta\left(k_1^2+k_2^2\right)\right)\left(D_{11}^s k_1^4 + 2\left(D_{12}^s + 2D_{66}^s\right)k_1^2 k_2^2 + D_{22}^s k_2^4\right)$$
$$+\left(1+\mu^2\left(k_1^2+k_2^2\right)\right)\left(\left(N^E + N^H + N_{max}^R\right)\left(k_1^2+k_2^2\right)\right)$$

$$k_{44} = -\left(1+\eta\left(k_1^2+k_2^2\right)\right)\left(H_{11}^s k_1^4 + 2\left(H_{12}^s + 2H_{66}^s\right)k_1^2 k_2^2 + H_{22}^s k_2^4\right)$$
$$+\left(1+\mu^2\left(k_1^2+k_2^2\right)\right)\left(\left(N^E + N^H + N_{max}^R\right)\left(k_1^2+k_2^2\right)\right) - \left(k_1^2+k_2^2\right)A_{44}^s$$

$$k_{45} = -\left(1+\eta\left(k_1^2+k_2^2\right)\right)\left(F_{31}^e - E_{15}^e\right)\left(k_1^2+k_2^2\right)$$

$$k_{46} = -F_{31}^m\left(1+\eta\left(k_1^2+k_2^2\right)\right)\left(k_1^2+k_2^2\right)$$

$$k_{51} = ik_1 A_{31}^e\left(1+\eta\left(k_1^2+k_2^2\right)\right)$$

$$k_{52} = ik_2 A_{31}^e\left(1+\eta\left(k_1^2+k_2^2\right)\right)$$

$$k_{53} = E_{31}^e\left(1+\eta\left(k_1^2+k_2^2\right)\right)\left(k_1^2+k_2^2\right)$$

$$k_{54} = \left(1+\eta\left(k_1^2+k_2^2\right)\right)\left(F_{31}^e - E_{15}^e\right)\left(k_1^2+k_2^2\right)$$

$$k_{55} = -\left(1+\eta\left(k_1^2+k_2^2\right)\right)\left(F_{31}^e + F_{11}^e\left(k_1^2+k_2^2\right)\right)$$

$$k_{56} = -\left(1+\eta\left(k_1^2+k_2^2\right)\right)\left(F_{31}^m + F_{11}^m\left(k_1^2+k_2^2\right)\right)$$

$$k_{61} = ik_1 A_{31}^m\left(1+\eta\left(k_1^2+k_2^2\right)\right)$$

$$k_{62} = ik_2 A_{31}^m\left(1+\eta\left(k_1^2+k_2^2\right)\right)$$

$$k_{63} = E_{31}^m\left(1+\eta\left(k_1^2+k_2^2\right)\right)\left(k_1^2+k_2^2\right)$$

$$k_{64} = \left(1+\eta\left(k_1^2+k_2^2\right)\right)\left(F_{31}^m - E_{15}^m\right)\left(k_1^2+k_2^2\right)$$

$$k_{65} = -\left(1+\eta\left(k_1^2+k_2^2\right)\right)\left(F_{31}^m + F_{11}^m\left(k_1^2+k_2^2\right)\right)$$

$$k_{66} = -\left(1+\eta\left(k_1^2+k_2^2\right)\right)\left(X_{31}^m + X_{11}^m\left(k_1^2+k_2^2\right)\right)$$

(B.1)

and

$$m_{11} = -I_0\left(1+\mu\left(k_1^2+k_2^2\right)\right) \qquad m_{15} = m_{16} = 0$$
$$m_{12} = 0 \qquad\qquad\qquad m_{21} = 0$$
$$m_{13} = I_1\left(1+\mu\left(k_1^2+k_2^2\right)\right) \qquad m_{22} = -I_0\left(1+\mu\left(k_1^2+k_2^2\right)\right)$$
$$m_{14} = J_1\left(1+\mu\left(k_1^2+k_2^2\right)\right) \qquad m_{23} = iI_1 k_2\left(1+\mu\left(k_1^2+k_2^2\right)\right)$$

$$m_{24} = iJ_1 k_2 \left(1 + \mu \left(k_1^2 + k_2^2\right)\right) \qquad m_{41} = -J_1 \left(1 + \mu \left(k_1^2 + k_2^2\right)\right)$$

$$m_{25} = m_{26} = 0 \qquad m_{42} = -iJ_1 k_2 \left(1 + \mu \left(k_1^2 + k_2^2\right)\right)$$

$$m_{43} = -\left(I_0 + J_2 \left(k_1^2 + k_2^2\right)\right)$$

$$m_{31} = -iI_1 k_1 \left(1 + \mu \left(k_1^2 + k_2^2\right)\right) \qquad \left(1 + \mu \left(k_1^2 + k_2^2\right)\right)$$

$$m_{44} = -\left(I_0 + K_2 \left(k_1^2 + k_2^2\right)\right)$$

$$m_{32} = -iI_1 k_2 \left(1 + \mu \left(k_1^2 + k_2^2\right)\right) \qquad \left(1 + \mu \left(k_1^2 + k_2^2\right)\right)$$

$$m_{33} = -\left(I_0 + I_2 \left(k_1^2 + k_2^2\right)\right)$$

$$\left(1 + \mu \left(k_1^2 + k_2^2\right)\right) \qquad m_{45} = m_{46} = 0$$

$$m_{34} = -\left(I_0 + J_2 \left(k_1^2 + k_2^2\right)\right)$$

$$\left(1 + \mu \left(k_1^2 + k_2^2\right)\right) \qquad m_{51} = m_{52} = m_{53} = m_{54} = m_{55} = m_{56} = 0$$

$$m_{35} = m_{36} = 0 \qquad m_{61} = m_{62} = m_{63} = m_{64} = m_{65} = m_{66} = 0$$

REFERENCES

[1] Iijima, S. (1991). Helical microtubules of graphitic carbon. *Nature*, *354*(6348), 56–58.

[2] Eringen, A. C. (1972). Linear theory of nonlocal elasticity and dispersion of plane waves. *International Journal of Engineering Science*, *10*(5), 425–435.

[3] Eringen, A. C. (1983). On differential equations of nonlocal elasticity and solutions of screw dislocation and surface waves. *Journal of Applied Physics*, *54*(9), 4703–4710.

[4] Narendar, S., & Gopalakrishnan, S. (2009). Nonlocal scale effects on wave propagation in multi-walled carbon nanotubes. *Computational Materials Science*, *47*(2), 526–538.

[5] Wang, Y. Z., Li, F. M., & Kishimoto, K. (2010). Flexural wave propagation in double-layered nanoplates with small scale effects. *Journal of Applied Physics*, *108*(6), 064519.

[6] Narendar, S., & Gopalakrishnan, S. (2012). Temperature effects on wave propagation in nanoplates. *Composites Part B: Engineering*, *43*(3), 1275–1281.

[7] Ghadiri, M., & Shafiei, N. (2015). Nonlinear bending vibration of a rotating nanobeam based on nonlocal Eringen's theory using differential quadrature method. *Microsystem Technologies*, 1–15.

[8] Ebrahimi, F., & Hosseini, S. H. S. (2016). Thermal effects on nonlinear vibration behavior of viscoelastic nanosize plates. *Journal of Thermal Stresses*, *39*(5), 606–625.

[9] Ebrahimi, F., & Shafiei, N. (2016). Influence of initial shear stress on the vibration behavior of single-layered graphene sheets embedded in an elastic medium based on Reddy's higher-order shear deformation plate theory. *Mechanics of Advanced Materials and Structures*, (Just-Accepted), 1–41.

[10] Lam, D. C. C., Yang, F., Chong, A. C. M., Wang, J., & Tong, P. (2003). Experiments and theory in strain gradient elasticity. *Journal of the Mechanics and Physics of Solids*, *51*(8), 1477–1508.

[11] Yang, F. A. C. M., Chong, A. C. M., Lam, D. C. C., & Tong, P. (2002). Couple stress based strain gradient theory for elasticity. *International Journal of Solids and Structures*, *39*(10), 2731–2743.

[12] Lim, C. W., Zhang, G., & Reddy, J. N. (2015). A higher-order nonlocal elasticity and strain gradient theory and its applications in wave propagation. *Journal of the Mechanics and Physics of Solids*, *78*, 298–313.

[13] Li, L., & Hu, Y. (2015). Buckling analysis of size-dependent nonlinear beams based on a nonlocal strain gradient theory. *International Journal of Engineering Science*, *97*, 84–94.

[14] Farajpour, A., Yazdi, M. H., Rastgoo, A., & Mohammadi, M. (2016). A higher-order nonlocal strain gradient plate model for buckling of orthotropic nanoplates in thermal environment. *Acta Mechanica*, 1–19.

[15] Li, L., Hu, Y., & Li, X. (2016). Longitudinal vibration of size-dependent rods via nonlocal strain gradient theory. *International Journal of Mechanical Sciences*, *115*, 135–144.

[16] Ramirez, F., Heyliger, P. R., & Pan, E. (2006). Free vibration response of two-dimensional magneto-electro-elastic laminated plates. *Journal of Sound and Vibration*, *292*(3), 626–644.

[17] Ebrahimi, F, & Dabbagh, A. (2019). *Wave Propagation Analysis of Smart Nanostructures* (1st ed.) Boca Raton, FL, USA: CRC Press.

[18] Ebrahimi, F., Seyfi, A., & Dabbagh, A. (2019). Dispersion of waves in FG porous nanoscale plates based on NSGT in thermal environment. *Advances in Nano Research*, *7*(5), 325–335.

[19] Ebrahimi, F., & Dabbagh, A. (2021). Magnetic field effects on thermally affected propagation of acoustical waves in rotary double-nanobeam systems. *Waves in Random and Complex Media*, *31*(1), 25–45.

[20] Ebrahimi, F., Khosravi, K., & Dabbagh, A. (2021). Wave dispersion in viscoelastic FG nanobeams via a novel spatial–temporal nonlocal strain gradient framework. *Waves in Random and Complex Media*, 1–23.

[21] Ebrahimi, F., Khosravi, K., & Dabbagh, A. (2021). A novel spatial–temporal nonlocal strain gradient theorem for wave dispersion characteristics of FGM nanoplates. *Waves in Random and Complex Media*, 1–20.

[22] Ke, L. L., Wang, Y. S., Yang, J., & Kitipornchai, S. (2014). Free vibration of size-dependent magneto-electro-elastic nanoplates based on the nonlocal theory. *Acta Mechanica Sinica*, *30*(4), 516–525.

[23] Koizumi, M., & Niino, M. (1995). Overview of FGM Research in Japan. *Mrs Bulletin*, *20*(01), 19–21.

[24] Alshorbagy, A. E., Eltaher, M. A., & Mahmoud, F. F. (2011). Free vibration characteristics of a functionally graded beam by finite element method. *Applied Mathematical Modelling*, *35*(1), 412–4425.

[25] Thai, H. T., & Vo, T. P. (2012). Bending and free vibration of functionally graded beams using various higher-order shear deformation beam theories. *International Journal of Mechanical Sciences*, *62*(1), 57–66.

[26] Thai, H. T., & Choi, D. H. (2012). A refined shear deformation theory for free vibration of functionally graded plates on elastic foundation. *Composites Part B: Engineering*, *43*(5), 2335–2347.

[27] Ebrahimi, F. (2013). Analytical investigation on vibrations and dynamic response of functionally graded plate integrated with piezoelectric layers in thermal environment. *Mechanics of Advanced Materials and Structures*, *20*(10), 854–870.

[28] Esfahani, S. E., Kiani, Y., & Eslami, M. R. (2013). Non-linear thermal stability analysis of temperature dependent FGM beams supported on non-linear hardening elastic foundations. *International Journal of Mechanical Sciences*, *69*, 10–20.

[29] Kargani, A., Kiani, Y., & Eslami, M. R. (2013). Exact solution for nonlinear stability of piezoelectric FGM Timoshenko beams under thermo-electrical loads. *Journal of Thermal Stresses*, *36*(10), 1056–1076.

[30] Ghiasian, S. E., Kiani, Y., & Eslami, M. R. (2015). Nonlinear thermal dynamic buckling of FGM beams. *European Journal of Mechanics-A/Solids*, *54*, 232–242.

[31] Ebrahimi, F., Ghasemi, F., & Salari, E. (2016). Investigating thermal effects on vibration behavior of temperature-dependent compositionally graded Euler beams with porosities. *Meccanica*, *51*(1), 223–249.

[32] Barati, M. R., Shahverdi, H., & Zenkour, A. M. (2016). Electro-mechanical vibration of smart piezoelectric FG plates with porosities according to a refined four-variable theory. *Mechanics of Advanced Materials and Structures*, (just-accepted).

[33] Barati, M. R., Sadr, M. H., & Zenkour, A. M. (2016). Buckling analysis of higher order graded smart piezoelectric plates with porosities resting on elastic foundation. *International Journal of Mechanical Sciences*, *117*, 309–320.

[34] Ebrahimi, F., & Barati, M. R. (2016). Dynamic modeling of a thermo–piezo-electrically actuated nanosize beam subjected to a magnetic field. *Applied Physics A*, *122*(4), 1–18.

[35] Ebrahimi, F., & Barati, M. R. (2016). A unified formulation for dynamic analysis of nonlocal heterogeneous nanobeams in hygro-thermal environment. *Applied Physics A*, *122*(9), 792.

[36] Ebrahimi, F., & Barati, M. R. (2016). A nonlocal higher-order refined magneto-electro-viscoelastic beam model for dynamic analysis of smart nanostructures. *International Journal of Engineering Science*, *107*, 183–196.

[37] Ebrahimi, F., Barati, M. R., & Haghi, P. (2016). Thermal effects on wave propagation characteristics of rotating strain gradient temperature-dependent functionally graded nanoscale beams. *Journal of Thermal Stresses*, 1–13.

[38] Ebrahimi, F., & Dabbagh, A. (2016). On flexural wave propagation responses of smart FG magneto-electro-elastic nanoplates via nonlocal strain gradient theory. *Composite Structures*.

[39] Mechab, B., Mechab, I., Benaissa, S., Ameri, M., & Serier, B. (2016). Probabilistic analysis of effect of the porosities in functionally graded material nanoplate resting on Winkler–Pasternak elastic foundations. *Applied Mathematical Modelling*, *40*(2), 738–749.

[40] Ebrahimi, F., & Salari, E. (2015). Thermal buckling and free vibration analysis of size dependent Timoshenko FG nanobeams in thermal environments. *Composite Structures*, *128*, 363–380.

[41] Ebrahimi, F., & Salari, E. (2015). Size-dependent free flexural vibrational behavior of functionally graded nanobeams using semi-analytical differential transform method. *Composites Part B: Engineering*, *79*, 156–169.

[42] Ebrahimi, F., & Salari, E. (2015). Size-dependent thermo-electrical buckling analysis of functionally graded piezoelectric nanobeams. *Smart Materials and Structures*, *24*(12), 125007.

[43] Ebrahimi, F., & Barati, M. R. (2016). A nonlocal higher-order shear deformation beam theory for vibration analysis of size-dependent functionally graded nano-beams. *Arabian Journal for Science and Engineering, 41*(5), 1679–1690.

[44] Ebrahimi, F., & Barati, M. R. (2016). Vibration analysis of nonlocal beams made of functionally graded material in thermal environment. *The European Physical Journal Plus, 131*(8), 279.

[45] Ebrahimi, F., & Barati, M. R. (2016). A unified formulation for dynamic analysis of nonlocal heterogeneous nanobeams in hygro-thermal environment. *Applied Physics A, 122*(9), 792.

[46] Ebrahimi, F., & Salari, E. (2016). Analytical modeling of dynamic behavior of piezo-thermo-electrically affected sigmoid and power-law graded nanoscale beams. *Applied Physics A, 122*(9), 793.

[47] Ebrahimi, F., & Barati, M. R. (2016). Vibration analysis of smart piezoelectrically actuated nanobeams subjected to magneto-electrical field in thermal environment. *Journal of Vibration and Control,* 1077546316646239.

[48] Ebrahimi, F., & Barati, M. R. (2016). A nonlocal higher-order refined magneto-electro-viscoelastic beam model for dynamic analysis of smart nanostructures. *International Journal of Engineering Science, 107,* 183–196.

[49] Ebrahimi, F., & Barati, M. R. (2017). A nonlocal strain gradient refined beam model for buckling analysis of size-dependent shear-deformable curved FG nanobeams. *Composite Structures, 159,* 174–182.

[50] Ebrahimi, F., & Barati, M. R. (2016). Buckling analysis of nonlocal third-order shear deformable functionally graded piezoelectric nanobeams embedded in elastic medium. *Journal of the Brazilian Society of Mechanical Sciences and Engineering,* 1–16.

[51] Barati, M. R., Zenkour, A. M., & Shahverdi, H. (2016). Thermo-mechanical buckling analysis of embedded nanosize FG plates in thermal environments via an inverse cotangential theory. *Composite Structures, 141,* 203–212.

[52] Ebrahimi, F., & Barati, M. R. (2016). Magnetic field effects on buckling behavior of smart size-dependent graded nanoscale beams. *The European Physical Journal Plus, 131*(7), 1–14.

[53] Ebrahimi, F., & Barati, M. R. (2016). Buckling analysis of smart size-dependent higher order magneto-electro-thermo-elastic functionally graded nanosize beams. *Journal of Mechanics,* 1–11.

[54] Ebrahimi, F., & Barati, M. R. (2017). A nonlocal strain gradient refined beam model for buckling analysis of size-dependent shear-deformable curved FG nanobeams. *Composite Structures, 159,* 174–182.

[55] Ebrahimi, F., & Barati, M. R. (2017). Hygrothermal effects on vibration characteristics of viscoelastic FG nanobeams based on nonlocal strain gradient theory. *Composite Structures, 159,* 433–444.

[56] Narendar, S., & Gopalakrishnan, S. (2011). Nonlocal wave propagation in rotating nanotube. *Results in Physics, 1*(1), 17–25.

[57] Ebrahimi, F., Barati, M. R., & Haghi, P. (2016). Thermal effects on wave propagation characteristics of rotating strain gradient temperature-dependent functionally graded nanoscale beams. *Journal of Thermal Stresses,* 1–13.

[58] Ebrahimi, F., & Dabbagh, A. (2017). Nonlocal strain gradient based wave dispersion behavior of smart rotating magneto-electro-elastic nanoplates. *Materials Research Express, 4*(2), 025003.

[59] Zhang, Y. W., Chen, J., Zeng, W., Teng, Y. Y., Fang, B., & Zang, J. (2015). Surface and thermal effects of the flexural wave propagation of piezoelectric functionally graded nanobeam using nonlocal elasticity. *Computational Materials Science, 97,* 222–226.

[60] Wu, B., Zhang, C., Chen, W., & Zhang, C. (2015). Surface effects on anti-plane shear waves propagating in magneto-electro-elastic nanoplates. *Smart Materials and Structures, 24*(9), 095017.

[61] Ebrahimi, F., & Barati, M. R. (2016). Wave propagation analysis of quasi-3D FG nanobeams in thermal environment based on nonlocal strain gradient theory. *Applied Physics A, 122*(9), 843.

[62] Ebrahimi, F., Barati, M. R., & Dabbagh, A. (2016). A nonlocal strain gradient theory for wave propagation analysis in temperature-dependent inhomogeneous nanoplates. *International Journal of Engineering Science, 107,* 169–182.

[63] Arani, A. G., Jamali, M., Ghorbanpour-Arani, A. H., Kolahchi, R., & Mosayyebi, M. (2016). Electro-magneto wave propagation analysis of viscoelastic sandwich nanoplates considering surface effects. *Proceedings of the Institution of Mechanical Engineers, Part C: Journal of Mechanical Engineering Science,* 0954406215627830.

[64] Ebrahimi, F., & Barati, M. R. (2016). Flexural wave propagation analysis of embedded S-FGM nanobeams under longitudinal magnetic field based on nonlocal strain gradient theory. *Arabian Journal for Science and Engineering,* 1–12.

[65] Ebrahimi, F., Ghazali, M., & Dabbagh, A. (2022). Hygro-thermo-viscoelastic wave propagation analysis of FGM nanoshells via nonlocal strain gradient fractional time–space theory. *Waves in Random and Complex Media,* 1–20.

[66] Ebrahimi, F., Barati, M. R., & Dabbagh, A. (2016). Wave dispersion characteristics of axially loaded magneto-electro-elastic nanobeams. *Applied Physics A, 122*(11), 949.

4 Wave Dispersion Characteristics of Fluid-Conveying Structures

4.1 BACKGROUND

Flexural wave propagation in fluid-carrying structures like cylindrical shells, carbon nanotubes, and boron nitride nanotubes is discussed in this chapter. The shell shear deformation theory (FSDT) to the first order has been used. Conventional wisdom holds that the flow of a viscous fluid is laminar, mature, Newtonian, and axially symmetric. The effect of flow velocity can be investigated using the Navier–Stokes equation. In the case of carbon nanotubes, nonlocal elasticity theory as developed by Eringen has been applied within the framework of Euler–Bernoulli beam theory. As a matter of fact, it is a well-known fact that a person's intelligence is directly proportional to his or her ability to think quickly on his or her feet. Both the modal superposition method and Newmark's direct integration method are used to obtain the time domain responses. Nonlocal parameters, heat, and the magnetic field of a harmonically moving load all play a role in the dynamic displacement of single-walled carbon nanotubes (SWCNT), and their effects are discussed. The obtained results demonstrate that the load velocity and the excitation frequency significantly affect the dynamic displacement of fluid-conveying SWCNT ratio. Furthermore, a wave propagation analysis of fluid-conducting boron nitride nanotubes (BNNTs) resting on a viscoelastic foundation subjected to multi-physical fields is presented using cylindrical shell theory, which is based on the nonlocal strain gradient theory (NSGT). Small things can have a big impact, so it's important to keep that in mind. The slip boundary wall of a nanotube and the ensuing flow are analyzed using the Knudsen number. This study's findings have been double-checked against the findings of other studies for accuracy. Wave frequencies and phase velocities are obtained by solving the governing equations analytically. We have applied Hamilton's principle to obtain the governing equations of fluid-conveying structures, and we have used an analytical method to solve these equations. Additionally, the influences of various variants have been studied and indicated within the framework of a set of figures. These variants include flow velocity, viscoelastic foundation, elastic and electric field, and geometry such as radius-to-thickness ratio.

DOI: 10.1201/9781003270263-4

4.1.1 A REVIEW ON WAVE DISPERSION ANALYSIS OF FLUID-CONVEYING STRUCTURES

Mechanical behavior is exciting to study in many ways, but one of the most intriguing is the analysis of wave propagation in different fluid-conveying structures, including beams, plates, membranes, and shells. Researchers have put forth a lot of time and effort into this question. Numerous studies have been conducted over the past decade on the buckling, bending, vibration, and wave propagation of nano/micro tubes that transmit fluid, using models such as the Euler–Bernoulli beam (EBB), the Timoshenko beam (TB), and the cylindrical shell (CS). Since the hexagonal holes on a CNT's surface are too tiny for even a molecule of fluid to flow through, simulating a CNT as a thin CS is more accurate than simulating it as a beam. Modified couple stress theory, partial nonlocal elasticity, nonlocal piezoelectricity, surface elasticity theory, and NSGT are only some of the higher-order continuum theories that have been used in mechanical modeling of these structures.

Another definition of magnetic force is the attractive or repulsive force exerted between the poles of a magnet and moving electrically charged particles. The electromagnetic force is responsible for this effect. If you're near a magnet or a conductor of electric current, you're in what's called a magnetic field. It is measured in teslas and is a vector quantity. It's no secret that the human race has come a long way from the days of cavemen and Vikings. Carbon nanotubes' magnetic characteristics and mechanical behavior in a magnetic field were the topic of their discussion. Ebrahimi and Dabbagh [2] investigated the effects of a magnetic field on the transmission of acoustic waves in a rotating double-nanobeam system when subjected to a temperature gradient. In the end, it all comes down to a simple equation: the more the force, the greater the result. By adopting a number of different boundary conditions for the nanotube, Ebrahimi and Mahmoodi [3] analyzed the impact of heat loading on the free vibration properties of carbon nanotubes (CNT) with numerous fractures. Researchers Ebrahimi and colleagues [4] employed NSGT to analyze the wave dispersion behavior of a thermally relevant, functionally graded (FG) nanobeam that had a size dependency. Using the theory of strain gradient elasticity, Wang [5] investigates the effects of inertia and strain gradients on the wave propagation of fluid-conveying single-walled carbon nanotubes. Ebrahimi and colleagues examine the wave propagation of rotating heterogeneous magneto-electro-elastic nanobeams in their paper referred to in reference [6]. Ebrahimi et al. [7] conducted research on the wave propagation of axially loaded double-layered graphene sheets (DLGSs) resting on a viscoelastic layer. They proved the influence of numerous parameters, including wave number, axial stress, Pasternak coefficient, and Winkler coefficient. Ebrahimi and Salari [8] demonstrated a nonlinear vibration analysis of electro-hygrothermally actuated embedded nanobeams with variable boundary conditions. According to the findings presented by the researchers, efficient kinds of thermal loading include both linear and uniform temperature rises in the direction of the thickness. Ebrahimi et al. [9] looked into a nonclassical beam model that was based on the Eringen nonlocal elasticity theory in order to evaluate the nonlinear vibration of a magneto-electro-hygrothermal piezoelectric structure that was resting on an elastic foundation. This was done in order to analyze the nonlinear vibration. They employed Hamilton's principle to generate the nonlinear motion equation and the boundary condition while operating inside the Euler framework of the Euler beam model with von Karman-type nonlinearity. Research shows that BNNTs may be classified as either single-, double-, or multi-walled. Single-walled boron nitride nanotubes (SWBNNTs) are one example. Using nonlocal piezoelasticity and taking into account surface stress, starting stress, and the Knudsen-dependent flow velocity impact, Arani et al. [10] calculate the electrothermal nonlocal wave propagation in fluid-conveying SWBNNTs. A study of vibration control for a rotating microshell made of multi-hybrid nanocomposite reinforcement (MHCR) and coated with a piezoelectric layer as the sensor and actuator (PLSA) is reported by Al-Furjan et al. [11]. The force exerted on their structure by the viscous fluid being conveyed was determined using an updated version of the Navier–Stokes equations. Their results demonstrate that the amplitude and vibration behavior of a spinning MHCR cylinder conveying fluid flow are significantly affected by variables such as angular velocity, fluid flow velocity, external force, PD controller, external voltage, and the MHC's weight percentage. In the end, it all comes down to the quality of the product, and the quality of the product is determined by the quality of the service provided. Honeycombs are used in their sandwich-like smart nanostructure, which also features piezoelectric face sheets that function as sensors and actuators (PFSA). Honeycombs serve as the brains of this sandwich-like smart nanostructure. Once an external voltage has been delivered to the sensor layer, a proportional-derivative controller, also known as a PD controller, is used in order to manage the sensor output. Their research demonstrates that the frequency characteristics of a cylindrical sandwich nanoshell are greatly impacted by a number of different factors, including the shape of the honeycomb core, the PD controller, the flow velocity, the length-to-radius ratio, and the applied voltage. Because of the implementation of the PD controller, there is also an increase in the critical velocity of the fluid flow inside the intelligent nanostructure. In the end, it all comes down to a matter of taste. The sensors already integrated into the planned system were also factored in as a potential tripping point. Reinforcement schemes like functionally graded distribution and uniform distribution were investigated. The governing equations and modeled cylindrical thick shell are developed from a new three-dimensional improved higher-order theory. Buckling and vibrational properties were studied in depth as a function of carbon nanotube reinforcement type and other factors. The wave propagation of a sandwich composite beam with an adjustable electro-rheological (ER) fluid core is investigated by Shariati et al. [14] The foundation layer, the ER fluid core, and the limiting layer are the three layers that come together to form the sandwich composite beam. A core of ER fluid may be discovered buried in the structure, sandwiched in between the foundational and boundary layers. Waves' dispersion qualities are investigated in relation to the electric field, the ratio of the core's thickness to that of the top layer, and the thickness of the ER core. In their investigation into the transmission of waves, Ebrahimi, F., and Sedighi [15] took into account the possibility of using a sandwich composite plate with a magneto-rheological (MR) fluid core that was capable of being adjusted. In a sandwich composite plate, the base layer, MR layer, and limiter layer make up the three layers. Within the foundational and limiting layers lies the MR layer. The findings of this study into the influence of the magnetic field highlight the importance of the magnetic field strength in determining the value of the wave dispersion characteristic. Also, as the MR fluid core is softer than the elastic layers, increasing the core-to-top layer thickness ratio reduces the wave frequency. Thus, MR fluid core functions as a damper in the same way. We conclude by discussing the impact of the remaining factors, including MR layer thickness, Winkler, Pasternak, and damping coefficients.

The wave behavior of a rectangular sandwich composite plate with an adjustable magneto-rheological fluid core was studied by Ebrahimi and Sedighi [16]. The core of magneto-rheological fluid lies between the foundation and the boundary. According to their findings, the strength of the magnetic field is the most crucial component for altering the value of the wave frequency and phase velocity, which they discovered by studying the influence of the magnetic field. Further, they demonstrated that the wave frequencies may be lowered by increasing the core-to-top layer thickness ratio, as the magneto-rheological fluid core is more compliant than elastic layers and the magneto-rheological fluid layer acts like a damper. The impact of core magneto-rheological fluid thickness on phase velocity was also examined. Ebrahimi, F., and Dabbagh [17] provide an analytical solution to the vibration issue associated with a cylindrical shell made of multi-scale hybrid nanocomposites that are used to transport fluid. The Eshelby–Mori–Tanaka homogenization approach is used to account for the agglomeration of nanofillers that arises from the high surface-to-volume ratio of nanostructures, and the Navier–Stokes equation for cylinders is extended to account for the effects of viscous flow. Each parameter's impact on the dimensionless frequency of nanocomposite shells will be shown visually. Using the longitudinal magnetic effect, Selvamani et al. [18] investigated nonlocal thermo-elastic waves in a fluid-conducting single-walled carbon nanotube supported by a polymer matrix. Based on Eringen's nonlocal elasticity theory, they were able to establish an analytical formulation. Spectral analysis was used to study the transmission of ultrasonic waves. A recent study of flexural wave propagation in anisotropic fluid-conveying cylindrical shells was conducted by Ebrahimi, F., and Seyfi [19]. Wave propagation behavior was investigated using four distinct anisotropic materials. Traditional theories of fluid dynamics assume a laminar, mature, Newtonian, and axially symmetric viscous flow. The Navier–Stokes equation was used to investigate the impact of flow velocity. Analytical solutions to their governing equations have been found. They examined how things like the speed of the flow, the ratio of the radius to the thickness, and the number of longitudinal and circumferential waves affected things. Dehghan et al. [20, 21] examine the wave propagation analysis of a fluid-conveying magneto-electro-elastic nanotube that takes into account the fluid effect. The wave frequencies and phase velocities were calculated using an analytical solution of the governing equations. Slip boundary wall of nanotube and flow are investigated using the Knudsen number. The properties of wave propagation in a fluid-carrying MEE nanotube were studied as a function of the Knudsen number, the number of modes, the length parameter, the nonlocal parameter, the fluid velocity, the fluid effect, and the slip boundary condition. Nonlocal strain gradient theory and a shell model were used by Zeighampouret et al. [22] to demonstrate wave propagation in a fluid carried by a double-walled carbon nanotube. For instance, Ebrahimi et al. [23] used FSDT to examine how waves propagate through a symmetric and asymmetric porous nanocomposite cylinder shell that was reinforced with graphene platelet. Recent research conducted by Ebrahimi and Seyfi [24] examined the wave propagation of a multi-scale hybrid nanocomposite cylinder shell based on the FSDT. This research took into consideration the influence that CNT aggregation had. A review of the relevant research shows that no studies on the study of wave propagation in cylindrical shells conveying anisotropic fluids have been published as of yet. This is something that was discovered after doing a review of the relevant research. Arani et al. [25] analyze the wave propagation in a fluid-carrying double-walled boron-nitride nanotube (DWBNNT) that is embedded in a material using the Euler–Bernoulli beam model. This is done within the framework of the strain gradient and Eringen's piezoelasticity theories. Maraghi et al. [26] presented the electro-thermo nonlinear vibration and instability of embedded DWBNNTs conveying viscous fluid within the framework of nonlocal piezoelasticity theory and the EBB model. According to the findings, the presence of an elastic medium, a change in temperature, an electric potential, and a small-scale parameter all have a substantial impact on the critical fluid velocity as well as the natural frequency. Arani et al. [27], using the nonlocal piezoelasticity CS theory, demonstrated the nonlinear vibration and instability of implanted DWBNNTs carrying viscous fluid. During the course of this investigation, the differential quadrature method (DQM) was used to determine the nonlinear frequency as well as the critical fluid velocity. Arani et al. [28] investigated the dynamic stability of DWBNNT while it was conveying a viscous fluid by making use of the nonlocal piezoelasticity theory. All of these factors—the fluid velocity, the Knudsen number, the aspect ratio, the nonlocal parameter, the fluctuations in temperature, the van der Waals forces, and the composition of the medium—were taken into consideration. Researchers from Abdollahian et al. [29] investigated the nonlocal electro-thermo-mechanical wave propagation in an implanted armchair three-walled boron nitride nanotube (TWBNNT) conveying viscous fluid while subjected to torsional strain in a sample of multi-walled boron nitride nanotubes (MWBNNTs). Arani et al. [30] apply TB theory to analyze the size-dependent wave propagation in coupled DWBNNTs carrying nanoflow systems in a recent work [30]. This research was published in the journal *Nano Letters*.

4.2 FLUID-CONVEYING CYLINDRICAL SHELLS: WAVE DISPERSION CHARACTERISTICS

In this part, we'll take a look at what happens to waves within anisotropic fluid-carrying cylindrical shells like the one seen in Figure 4.1. Anisotropic materials have been discussed, including monoclinic, triclinic, trigonal, and hexagonal materials. By using FSDT, we are able to derive the equations of motion for

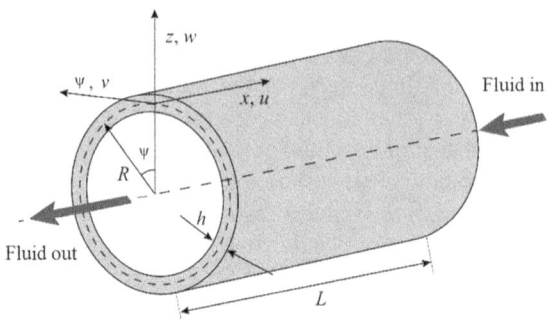

FIGURE 4.1 Schematic and coordinate of an anisotropic fluid-conveying cylindrical shell.

cylindrical shells. Newtonian, laminar, mature, and axially symmetric flow is typical for viscous fluids. Analytical solutions to the governing equations allow us to calculate the circular frequency. Finally, a set of graphics showing the impact of each parameter has been provided [19].

Hook's law is often used to refer to the behavior of a material and to connect the stresses and strains that are currently unknown. As a general rule, the stress-strain connection may be written as follows:

$$\sigma_{ij} = C_{ijkl}\varepsilon_{kl} \tag{4.1}$$

in which σ_{ij}, ε_{kl}, and C_{ijkl} denote components of Cauchy stress, strain, and elasticity tensors, respectively. Elastic properties of the monoclinic, triclinic, trigonal, and hexagonal anisotropic materials employed in this investigation are reported below. The monoclinic materials utilized have elastic properties specified in reference [19]:

$$C = \begin{bmatrix} 86.74 & -8.25 & 27.15 & -3.66 & 0 & 0 \\ -8.25 & 129.77 & -7.42 & 5.7 & 0 & 0 \\ 27.15 & -7.42 & 102.83 & 9.92 & 0 & 0 \\ -3.66 & 5.7 & 9.92 & 38.61 & 0 & 0 \\ 0 & 0 & 0 & 0 & 68.81 & 0 \\ 0 & 0 & 0 & 0 & 0 & 29.01 \end{bmatrix} \text{GPa} \tag{4.2}$$

In addition to this, the mass density of monoclinic materials is thought to be $\rho = 2649$ Kg/m³. Triclinic materials may be modeled using the matrix of elastic constants for transversely isotropic materials by performing the requisite rotations around the x_1- and x_2-axes. Following is a definition [19] of the elastic properties of triclinic materials:

$$C = \begin{bmatrix} 98.84 & 53.92 & 50.78 & -0.1 & 1.05 & 0.03 \\ 53.92 & 99.19 & 50.87 & -0.18 & 0.55 & 0.03 \\ 50.78 & 50.87 & 87.23 & -0.18 & 1.03 & 0.02 \\ -0.1 & -0.18 & -0.18 & 21.14 & 0.07 & 0.25 \\ 1.05 & 0.55 & 1.03 & 0.07 & 21.1 & -0.04 \\ 0.03 & 0.03 & 0.02 & 0.25 & -0.04 & 22.55 \end{bmatrix} \text{GPa} \quad (4.3)$$

It is generally accepted that the mass density of triclinic materials is $\rho = 7750$ Kg/m³.
The following definitions apply to the elastic properties of trigonal materials: [19]

$$C = \begin{bmatrix} 86.74 & 6.99 & 11.91 & -17.91 & 0 & 0 \\ 6.99 & 86.74 & 11.91 & 17.91 & 0 & 0 \\ 11.91 & 11.91 & 107.2 & 0 & 0 & 0 \\ -17.91 & 17.91 & 0 & 57.94 & 0 & 0 \\ 0 & 0 & 0 & 0 & 57.94 & -17.91 \\ 0 & 0 & 0 & 0 & -17.91 & 39.88 \end{bmatrix} \text{GPa} \quad (4.4)$$

The standard for trigonal materials' density of mass is $\rho = 2649$ Kg/m³.

The crystal structure of beryllium is thought to be hexagonal. In a manner analogous to the space lattice returning to the primitive point, this material behaves as if a rotation of 60 degrees around its axis of symmetry were to do so. Useful definitions for the elastic properties of hexagonal materials are as follows [19]:

$$C = \begin{bmatrix} 298.2 & 27.7 & 11 & 0 & 0 & 0 \\ 27.7 & 298.2 & 11 & 0 & 0 & 0 \\ 11 & 11 & 340.8 & 0 & 0 & 0 \\ 0 & 0 & 0 & 165.5 & 0 & 0 \\ 0 & 0 & 0 & 0 & 165.5 & 0 \\ 0 & 0 & 0 & 0 & 0 & 135.3 \end{bmatrix} \text{GPa} \quad (4.5)$$

It is generally agreed that the mass density of hexagonal materials is $\rho = 1850$ Kg/m³. Figure 4.1 depicts the coordinate system and schematic of the anisotropic cylindrical shell. Displacement fields at every location on an anisotropic cylindrical shell may be described as follows, within the context of the first-order shear deformable shell theory:

$$u_x\left(x,\varphi,z,t\right) = u\left(x,\psi,t\right) + z\theta_x\left(x,\psi,t\right) \quad (4.6)$$

$$u_\psi\left(x,\psi,z,t\right) = v\left(x,\psi,t\right) + z\theta_\psi\left(x,\psi,t\right) \quad (4.7)$$

$$u_z\left(x,\psi,z,t\right)=w\left(x,\psi,t\right) \tag{4.8}$$

where axial, circumferential, and lateral displacements as well as rotation components around axial and circumferential directions are denoted by u, v, w, θ_x, and θ_ψ, respectively. The subsequent expression for the nonzero stresses of a cylinder shell is as follows:

$$\begin{Bmatrix} \varepsilon_{xx} \\ \varepsilon_{\psi\psi} \\ \varepsilon_{x\psi} \\ \varepsilon_{xz} \\ \varepsilon_{\psi z} \end{Bmatrix} = \begin{Bmatrix} \dfrac{\partial u}{\partial x}+z\dfrac{\partial\theta_x}{\partial x} \\[2mm] \dfrac{1}{R}\left(\dfrac{\partial v}{\partial\psi}+z\dfrac{\partial\theta_\psi}{\partial\psi}+w\right) \\[2mm] \dfrac{1}{R}\dfrac{\partial u}{\partial\psi}+\dfrac{\partial v}{\partial x}+\dfrac{z}{R}\dfrac{\partial\theta_x}{\partial\psi}+z\dfrac{\partial\theta_\psi}{\partial x} \\[2mm] \theta_x+\dfrac{\partial w}{\partial x} \\[2mm] \theta_\psi+\dfrac{1}{R}\dfrac{\partial w}{\partial\psi}-\dfrac{v}{R} \end{Bmatrix} \tag{4.9}$$

Now, we use Hamilton's principle, which may be written as follows, to get at the Euler–Lagrange equations of anisotropic cylindrical shells.

$$\int_0^t \delta\left(\Pi_S-\Pi_K-\Pi_W\right)dt=0 \tag{4.10}$$

where Π_S represents strain, Π_K represents kinetic energy, and Π_W represents the work done by the external load. For a linear elastic solid, we may express the variation in strain energy as:

$$\delta\,\Pi_S=\int_{-\frac{h}{2}}^{\frac{h}{2}}\int_0^{2\pi}\int_0^L \sigma_{ij}\delta\varepsilon_{ij}Rdxd\psi\,dz \tag{4.11}$$

Additionally, the following structure may be used to represent kinetic energy variation:

$$\delta\,\Pi_K=\int_{-\frac{h}{2}}^{\frac{h}{2}}\int_0^{2\pi}\int_0^L \rho(z)\left[\left(\frac{\partial\delta u_x}{\partial t}\right)^2+\left(\frac{\partial\delta u_\psi}{\partial t}\right)^2+\left(\frac{\partial\delta u_z}{\partial t}\right)^2\right]Rdxd\psi\,dz \tag{4.12}$$

Here, we assume a fully developed, axially symmetric, Newtonian, laminar flow of a viscous fluid in anisotropic cylindrical shells [19]. That's why the Navier–Stokes equation works in this situation. Flow momentum may be described by the following equation:

$$\rho_f \frac{dV_R}{dt} = -\frac{\partial P}{\partial R} + \frac{\partial \tau_{Rx}}{\partial x} - \frac{\tau_{\psi\psi}}{R} + \frac{1}{R}\frac{\partial \tau_{R\psi}}{\partial \psi} \tag{4.13}$$

where P and ρ_f stand for pressure and density of the fluid, respectively. The following relations may be generalized as a result of the identity between the velocities and velocities at the contact sites of a fluid and a cylindrical shell:

$$V_R = \frac{dw}{dt}, \qquad \frac{d}{dt} = \frac{\partial}{\partial t} + v_x \frac{\partial}{\partial x} \tag{4.14}$$

Also, v_x denotes the mean flow velocity; shear stress (τ) and viscosity (μ_f) relations can be written in the following form:

$$\tau_{R\psi} = \frac{\mu_f}{R}\frac{\partial V_R}{\partial \psi}, \quad \tau_{\psi\psi} = 2\mu_f \frac{V_R}{R}, \quad \tau_{Rx} = \mu_f \frac{\partial V_R}{\partial \psi} \tag{4.15}$$

It is possible to express the varying amount of work done by external loadings as:

$$\delta \Pi_W = \int_{-\frac{h}{2}}^{\frac{h}{2}} \int_0^{2\pi} \int_0^L \left(\begin{array}{l} N_r + N_x \dfrac{\partial^2 w}{\partial x^2} + \dfrac{N_\psi}{R^2}\dfrac{\partial^2 w}{\partial \psi^2} \\[2mm] + \dfrac{\mu_f}{R^3}\dfrac{\partial^2 V_R}{\partial \psi^2} + \mu_f \dfrac{\partial^2 V_R}{\partial \psi^2} - \dfrac{2\mu_f}{R^2}V_R - \rho_f \dfrac{d^2 w}{dt^2} \end{array} \right) \delta w R\,dx\,d\psi\,dz \tag{4.16}$$

where the radial, axial, and circumferential loadings are denoted by the symbols N_r, N_x, and N_ψ, respectively. No consideration of loading has been made in the current investigation. After plugging in Eqs. (4.11), (4.12), and (4.16), we get the following equations, which describe the motion of a cylinder:

$$\frac{\partial N_{xx}}{\partial x} + \frac{1}{R}\frac{\partial N_{x\psi}}{\partial \psi} = I_0 \frac{\partial^2 u}{\partial t^2} + I_1 \frac{\partial^2 \theta_x}{\partial t^2} \tag{4.17a}$$

$$\frac{\partial N_{x\psi}}{\partial x} + \frac{1}{R}\frac{\partial N_{\psi\psi}}{\partial \psi} + \frac{Q_{z\psi}}{R} = I_0 \frac{\partial^2 v}{\partial t^2} + I_1 \frac{\partial^2 \theta_\psi}{\partial t^2} \tag{4.17b}$$

$$\frac{\partial Q_{xz}}{\partial x} + \frac{1}{R}\frac{\partial Q_{zy}}{\partial \psi} - \frac{N_{\psi\psi}}{R} - \rho_f h_f v_x^2 \frac{\partial^2 w}{\partial x^2} + \mu_f h_f v_x \left[\frac{\partial^3 w}{\partial x^3} + \frac{1}{R}\left(\frac{\partial^3 w}{\partial x \partial \psi^2} - 2\frac{\partial w}{\partial x} \right) \right]$$

$$= I_0 \frac{\partial^2 w}{\partial t^2} + \rho_f h_f \left[\frac{\partial^2 w}{\partial t^2} + 2v_x \frac{\partial^2 w}{\partial x \partial t} \right] + \mu_f h_f \left[\frac{\partial^3 w}{\partial t \partial x^2} + \frac{1}{R^2}\left(\frac{\partial^3 w}{\partial t \partial \psi^2} - 2\frac{\partial w}{\partial t} \right) \right]$$

$$(4.17c)$$

$$\frac{\partial M_{xx}}{\partial x} + \frac{1}{R}\frac{\partial M_{xy}}{\partial \psi} - Q_{xz} = I_1 \frac{\partial^2 u}{\partial t^2} + I_2 \frac{\partial^2 \theta_x}{\partial t^2} \qquad (4.17d)$$

$$\frac{\partial M_{xy}}{\partial x} + \frac{1}{R}\frac{\partial M_{\psi\psi}}{\partial \psi} - Q_{yz} = I_1 \frac{\partial^2 v}{\partial t^2} + I_2 \frac{\partial^2 \theta_\psi}{\partial t^2} \qquad (4.17e)$$

where

$$\begin{bmatrix} N_{xx} & N_{xy} & N_{\psi\psi} \\ M_{xx} & M_{xy} & M_{\psi\psi} \end{bmatrix} = \int_{-\frac{h}{2}}^{\frac{h}{2}} \begin{bmatrix} \sigma_{xx} & \sigma_{xy} & \sigma_{\psi\psi} \end{bmatrix} \begin{bmatrix} 1 \\ z \end{bmatrix} dz \qquad (4.17f)$$

$$\begin{bmatrix} Q_{xz} & Q_{zy} \end{bmatrix} = \kappa_s \int_{-\frac{h}{2}}^{\frac{h}{2}} \begin{bmatrix} \sigma_{xz} & \sigma_{zy} \end{bmatrix} dz \qquad (4.17g)$$

$$\begin{bmatrix} I_0 & I_1 & I_2 \end{bmatrix} = \int_{-\frac{h}{2}}^{\frac{h}{2}} \rho(z) \begin{bmatrix} 1 & z & z^2 \end{bmatrix} dz \qquad (4.17h)$$

shear correction factor is denoted by the symbol κ_s. The following relationship is obtained by integrating the previous equation across the shell's thickness:

$$\begin{bmatrix} N_{xx} \\ M_{xx} \\ N_{\psi\psi} \\ M_{\psi\psi} \end{bmatrix} = \begin{bmatrix} A_{11} & B_{11} & \dfrac{A_{12}}{R} & \dfrac{B_{12}}{R} \\ B_{11} & D_{11} & \dfrac{B_{12}}{R} & \dfrac{D_{12}}{R} \\ A_{12} & B_{12} & \dfrac{A_{11}}{R} & \dfrac{B_{11}}{R} \\ B_{12} & D_{12} & \dfrac{B_{11}}{R} & \dfrac{D_{11}}{R} \end{bmatrix} \begin{bmatrix} \dfrac{\partial u}{\partial x} \\ \dfrac{\partial \theta_x}{\partial x} \\ \dfrac{\partial v}{\partial \varphi} + w \\ \dfrac{\partial \theta_\psi}{\partial \psi} \end{bmatrix},$$

$$\begin{bmatrix} N_{x\psi} \\ M_{x\psi} \end{bmatrix} = \begin{bmatrix} A_{66} & B_{66} \\ B_{66} & D_{66} \end{bmatrix} \begin{bmatrix} \dfrac{1}{R}\dfrac{\partial u}{\partial \psi} + \dfrac{\partial v}{\partial x} \\ \dfrac{1}{R}\dfrac{\partial \theta_x}{\partial \psi} + \dfrac{\partial \theta_\psi}{\partial x} \end{bmatrix}$$

$$Q_{xz} = A_{55}^s\left(\theta_x + \frac{\partial w}{\partial x}\right), \quad Q_{\psi z} = A_{55}^s\left(\theta_\psi + \frac{1}{R}\frac{\partial w}{\partial \psi} - \frac{v}{R}\right) \qquad (4.18a)$$

in which

$$\begin{bmatrix} A_{ij} & B_{ij} & D_{ij} \end{bmatrix} = \int_{-\frac{h}{2}}^{\frac{h}{2}} C_{ij}\left[1, z, z^2\right]dz \qquad ij = 11,12,66 \qquad (4.18b)$$

$$A_{55}^s = \kappa_s \int_{-\frac{h}{2}}^{\frac{h}{2}} C_{66}dz \qquad (4.18c)$$

Anisotropic cylindrical shell governing equations may be constructed, for example, by coupling Eqs. (4.17a)–(4.17e) with Eq. (4.18).

$$A_{11}\frac{\partial^2 u}{\partial x^2} + B_{11}\frac{\partial^2 \theta_x}{\partial x^2} + \frac{A_{12}}{R}\left(\frac{\partial^2 v}{\partial x \partial \psi} + \frac{\partial w}{\partial x}\right)$$

$$+ \frac{B_{12}}{R}\frac{\partial^2 \theta_\psi}{\partial x \partial \psi} + \frac{A_{66}}{R}\left(\frac{1}{R}\frac{\partial^2 u}{\partial \psi^2} + \frac{\partial^2 v}{\partial x \partial \psi}\right) \qquad (4.19a)$$

$$+ \frac{B_{66}}{R}\left(\frac{1}{R}\frac{\partial^2 \theta_x}{\partial \psi^2} + \frac{\partial^2 \theta_\psi}{\partial x \partial \psi}\right) - I_0\frac{\partial^2 u}{\partial t^2} - I_1\frac{\partial^2 \theta_x}{\partial t^2} = 0$$

$$A_{66}\left(\frac{1}{R}\frac{\partial^2 u}{\partial x \partial \psi} + \frac{\partial^2 v}{\partial x^2}\right) + B_{66}\left(\frac{1}{R}\frac{\partial^2 \theta_x}{\partial x \partial \psi} + \frac{\partial^2 \theta_\psi}{\partial x^2}\right) + \frac{A_{12}}{R}\frac{\partial^2 u}{\partial x \partial \psi}$$

$$+ \frac{B_{12}}{R}\frac{\partial^2 \theta_x}{\partial x \partial \psi} + \frac{A_{11}}{R^2}\left(\frac{\partial^2 v}{\partial \psi^2} + \frac{\partial w}{\partial \psi}\right) \qquad (4.19b)$$

$$+ \frac{B_{11}}{R^2}\frac{\partial^2 \theta_\psi}{\partial \psi^2} + \frac{A_{55}^s}{R}\left(\theta_\psi + \frac{1}{R}\frac{\partial w}{\partial \psi} - \frac{v}{R}\right) - I_0\frac{\partial^2 v}{\partial t^2} - I_1\frac{\partial^2 \theta_\psi}{\partial t^2} = 0$$

$$A_{55}^s \left(\frac{\partial \theta_x}{\partial x} + \frac{\partial^2 w}{\partial x^2} \right) + \frac{A_{55}^s}{R} \left(\frac{\partial \theta_\psi}{\partial \psi} + \frac{1}{R} \frac{\partial^2 w}{\partial \psi^2} - \frac{1}{R} \frac{\partial v}{\partial \psi} \right) - \frac{A_{12}}{R} \frac{\partial u}{\partial x} - \frac{B_{12}}{R} \frac{\partial \theta_x}{\partial x}$$

$$- \frac{A_{11}}{R^2} \left(\frac{\partial v}{\partial \psi} + w \right) - \frac{B_{11}}{R^2} \frac{\partial \theta_\psi}{\partial \psi} - \rho_f h_f v_x^2 \frac{\partial^2 w}{\partial x^2} + \mu_f h_f v_x \left[\frac{\partial^3 w}{\partial x^3} + \frac{1}{R} \left(\frac{\partial^3 w}{\partial x \partial \psi^2} - 2 \frac{\partial w}{\partial x} \right) \right]$$

$$- I_0 \frac{\partial^2 w}{\partial t^2} - \rho_f h_f \left[\frac{\partial^2 w}{\partial t^2} + 2 v_x \frac{\partial^2 w}{\partial x \partial t} \right] - \mu_f h_f \left[\frac{\partial^3 w}{\partial t \partial x^2} + \frac{1}{R^2} \left(\frac{\partial^3 w}{\partial t \partial \psi^2} - 2 \frac{\partial w}{\partial t} \right) \right] = 0$$

(4.19c)

$$B_{11} \frac{\partial^2 u}{\partial x^2} + D_{11} \frac{\partial^2 \theta_x}{\partial x^2} + \frac{B_{12}}{R} \left(\frac{\partial^2 v}{\partial x \partial \psi} + \frac{\partial w}{\partial x} \right) + \frac{D_{12}}{R} \frac{\partial^2 \theta_\psi}{\partial x \partial \psi}$$

$$+ \frac{B_{66}}{R} \left(\frac{1}{R} \frac{\partial^2 u}{\partial \psi^2} + \frac{\partial^2 v}{\partial x \partial \psi} \right) + \frac{D_{66}}{R} \left(\frac{1}{R} \frac{\partial^2 \theta_x}{\partial \psi^2} + \frac{\partial^2 \theta_\psi}{\partial x \partial \psi} \right)$$

(4.19d)

$$- A_{55}^s \left(\theta_x + \frac{\partial w}{\partial x} \right) - I_1 \frac{\partial^2 u}{\partial t^2} - I_2 \frac{\partial^2 \theta_x}{\partial t^2} = 0$$

$$B_{66} \left(\frac{1}{R} \frac{\partial^2 u}{\partial x \partial \psi} + \frac{\partial^2 v}{\partial x^2} \right) + D_{66} \left(\frac{1}{R} \frac{\partial^2 \theta_x}{\partial x \partial \psi} + \frac{\partial^2 \theta_\psi}{\partial x^2} \right) + \frac{B_{12}}{R} \frac{\partial^2 u}{\partial x \partial \psi}$$

$$+ \frac{D_{12}}{R} \frac{\partial^2 \theta_x}{\partial x \partial \psi} + \frac{B_{11}}{R^2} \left(\frac{\partial^2 v}{\partial \psi^2} + \frac{\partial w}{\partial \psi} \right) + \frac{D_{11}}{R^2} \frac{\partial^2 \theta_\psi}{\partial \psi^2}$$

(4.19e)

$$- A_{55}^s \left(\theta_\psi + \frac{1}{R} \frac{\partial w}{\partial \psi} - \frac{v}{R} \right) - I_1 \frac{\partial^2 v}{\partial t^2} - I_2 \frac{\partial^2 \theta_\psi}{\partial t^2} = 0$$

For the purpose of solving the governing equations of anisotropic cylindrical shells, an analytical solution technique is constructed. So, the displacement fields should look like this:

$$\begin{Bmatrix} u \\ v \\ w \\ \theta_x \\ \theta_\psi \end{Bmatrix} = \begin{Bmatrix} U \exp\left[i\beta_x x + i\beta_n \psi - i\omega_n t \right] \\ V \exp\left[i\beta_x x + i\beta_n \psi - i\omega_n t \right] \\ W \exp\left[i\beta_x x + i\beta_n \psi - i\omega_n t \right] \\ \Theta_x \exp\left[i\beta_x x + i\beta_n \psi - i\omega_n t \right] \\ \Theta_\psi \exp\left[i\beta_x x + i\beta_n \psi - i\omega_n t \right] \end{Bmatrix}$$

(4.20a)

in which U, V, and W are the displacement amplitudes and Θ_x and Θ_ψ are the rotation amplitudes. Moreover, β_x and β_n stand for longitudinal and circumferential wave number, respectively, and ω_n is circular frequency. Afterward, by substituting u, v, w, θ_x, and θ_ψ from Eq. (4.20a) in Eqs. (4.19a)–(4.19e), the following equation is obtained:

$$\left(\left[K \right]_{5\times5} + i\omega_n \left[C \right]_{5\times5} - \omega^2 \left[M \right]_{5\times5} \right) \begin{bmatrix} U \\ V \\ W \\ \Theta_x \\ \Theta_\psi \end{bmatrix} = 0 \qquad (4.20b)$$

where Appendix 4.1 lists the elements of these matrices. Setting the determinant of the coefficient matrix on the left-hand side of Eq. (4.20b) to zero will solve this eigenvalue issue.

$$\left| \left[K \right]_{5\times5} + i\omega_n \left[C \right]_{5\times5} - \omega^2 \left[M \right]_{5\times5} \right| = 0 \qquad (4.20c)$$

Furthermore, by setting $\beta_x = \beta_n = k$, the phase velocity can be obtained via

$$c_p = \frac{\omega_n}{k} \qquad (4.20d)$$

Here, we provide a series of numerical figures to demonstrate the influence of these different factors on the frequency and phase velocity variations of anisotropic cylindrical shells. The thickness of the cylindrical shell is assumed to be 0.03 m in all of the diagrams, and the length and radius are assumed to be 25 times higher than the thickness of the shell.

We plot the frequency over the number of waves traveling around the circumference of an anisotropic cylinder in Figure 4.2. In any anisotropic cylinder shell, the frequency of waves rises as the circumferential wave number grows for a given longitudinal wave number. Additionally, there is a negative correlation between the value of the wave frequency and the amount of flow velocity so the wave frequency decreases as the amount of flow velocity increases. Hexagonal wave frequencies are less sensitive to changes in flow velocity compared to other wave types. The two graphs follow the same general pattern, although their wave frequencies are different. The highest wave frequency is found in the hexagonal, trigonal, monoclinic, and triclinic structures, in that order.

Anisotropic cylindrical shells' wave frequencies change as a function of the wave number around their circumference for a range of radius to shell thicknesses (R/h) as shown in Figure 4.3. The graphs show that the radius-to-thickness ratio has a diminishing effect on the variance of wave frequency values while the flow rate remains constant. This phenomenon occurs because the structure weakens with increasing radius-to-thickness ratio. Hence, greater wave frequencies occur with lower radius-to-thickness ratios. The wave frequencies are also highest and

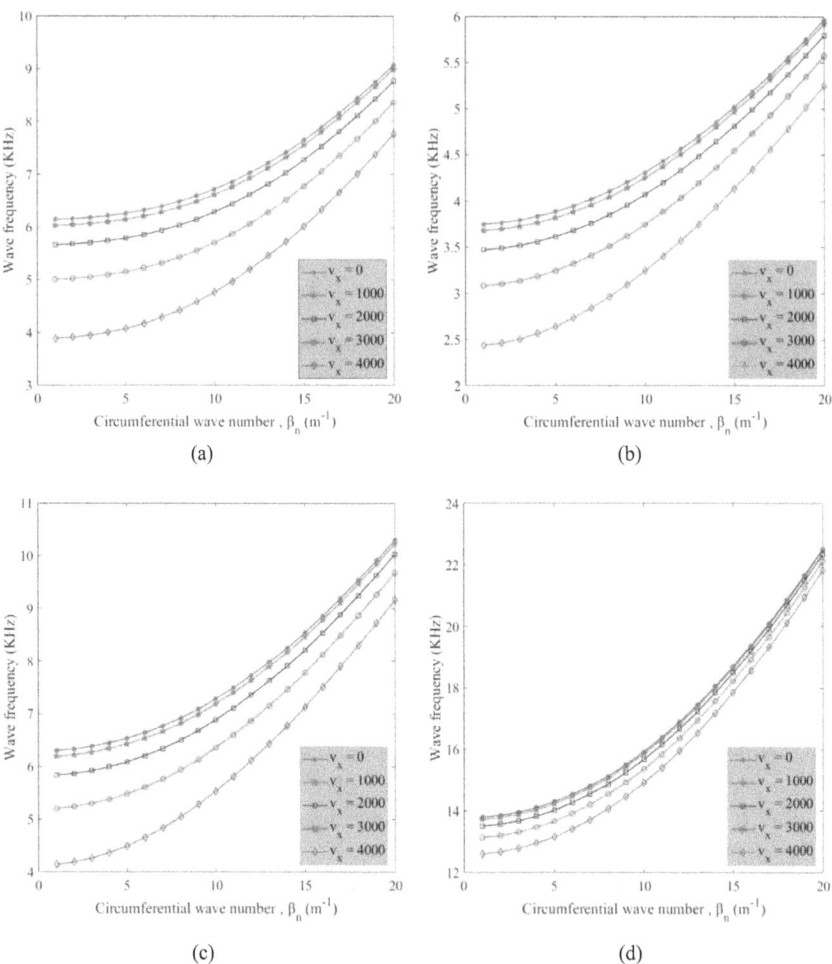

FIGURE 4.2 Variation of wave frequency against circumferential wave number for various flow velocities for different anisotropic materials. (a) Monoclinic, (b) triclinic, (c) trigonal, and (d) hexagonal.

lowest for the hexagonal and triclinic geometries, as seen in Figure 4.2. The value of the frequency of waves and the number of circumferential waves both rise [19].

Figure 4.4 depicts the relationship between the wave number and the phase velocity of an anisotropic cylindrical shell subject to a range of flow velocities. As the wave number grows, the phase velocity initially decreases to a minimal value and then rises. The critical wave number for a given flow velocity is shown by the spots where the curves intersect at their minimum. What this means is that there is a threshold wave number at which a structure becomes unstable due to the velocity of the flow. Triclinic shell has a relationship with $v_x = 0$ and $v_x = 1000$,

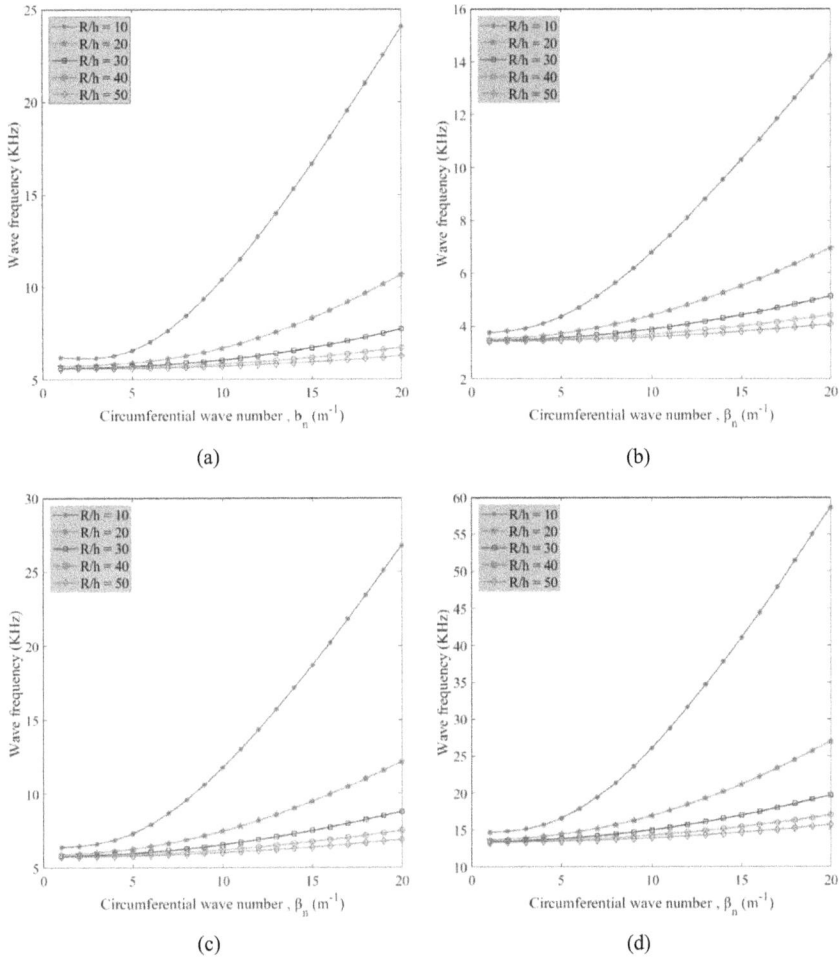

FIGURE 4.3   Variation of wave frequency against circumferential wave number for various radius-to-thickness ratios for different anisotropic materials ($v_x = 2000$). (a) Monoclinic, (b) triclinic, (c) trigonal, and (d) hexagonal.

monoclinic shell has a relationship with $v_x = 8$ and $v_x = 2000$, $k = 9$ and $v_x = 3000$, and trigonal shell has a relationship with $v_x = 9$, $v_x = 1000$, and $v_x = 2000$, and $k = 10$ and $v_x = 4000$. Additionally, flow velocity has a deleterious influence on the range of phase velocities [19].

In Figure 4.5, we see how the radius-to-thickness ratio affects the dependence of the phase velocity versus the wave number for anisotropic cylindrical shells. When the radius-to-thickness ratio is decreased, a greater phase velocity is seen. However, as was previously noted, the value of phase velocity decreases with increasing radius-to-thickness ratio and subsequently rises with increasing

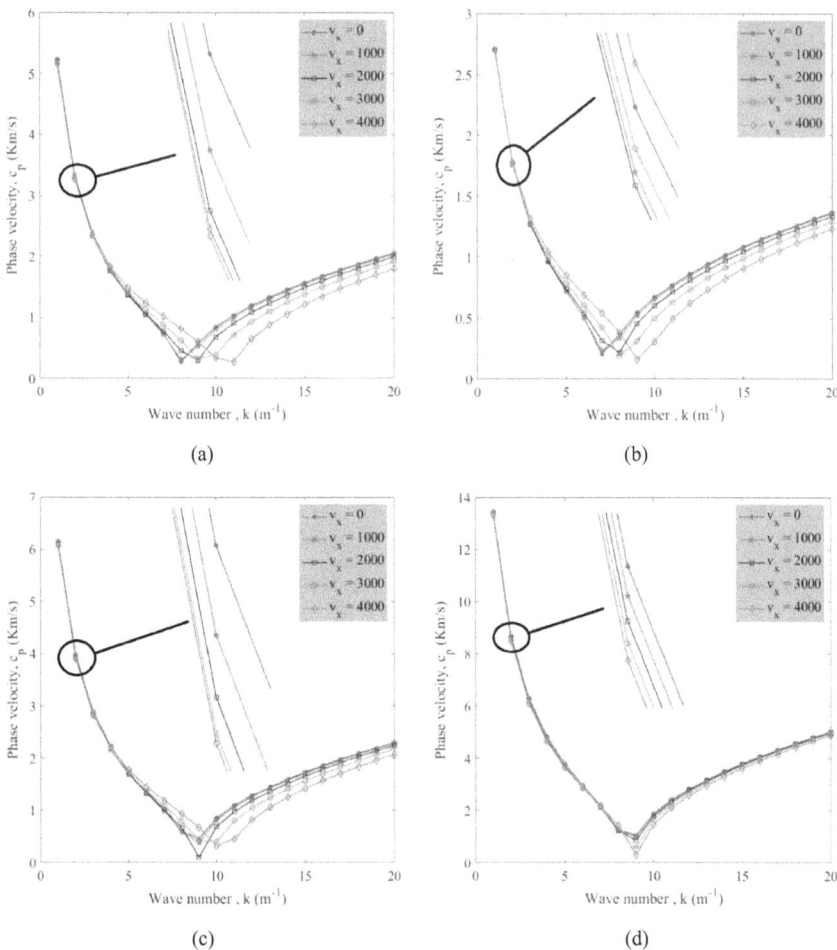

FIGURE 4.4 Variation of phase velocity against wave number for various flow veloci-ties for different anisotropic materials. (a) Monoclinic, (b) triclinic, (c) trigonal, and (d) hexagonal.

wave number. These plots are made at $v_x = 2000$, and from them, we can derive an expression for how changing the radius-to-thickness ratio affects the critical wave number at a certain flow velocity. To elaborate, we may say that the criti-cal flow velocity can occur at different wave numbers and that this is a function of the radius per unit thickness. Similar to preceding examples, the phases with the slowest velocities are triclinic, monoclinic, trigonal, and hexagonal, in that order [19].

In conclusion, Figure 4.6 illustrates the connection that exists between the flow velocity and the variation of the wave frequency of anisotropic cylindrical shells

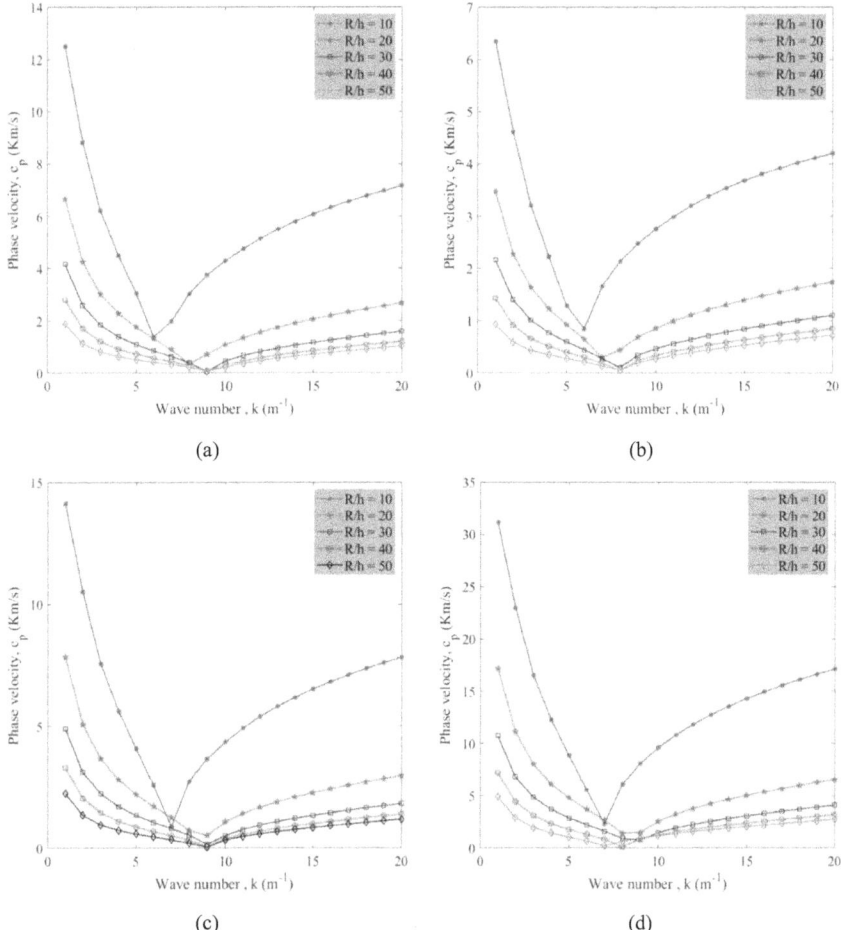

FIGURE 4.5 Variation of phase velocity against wave number for various radius-to-thickness ratios for different anisotropic materials ($v_x = 2000$). (a) Monoclinic, (b) triclinic, (c) trigonal, and (d) hexagonal.

as a function of the longitudinal wave number for a variety of radius-to-thickness ratios. It stands to reason that thinner shells with smaller radius-to-thickness ratios would be better able to withstand waves at higher frequencies than their thicker counterparts. Furthermore, at a certain ratio of radius to thickness, a greater quantity of flow velocity results in lower wave frequency values. A good option is to choose shells with a lower radius-to-thickness ratio and flow velocity. Furthermore, when the number of longitudinal waves grows, the frequency of those waves decreases to a minimum before rising again [19].

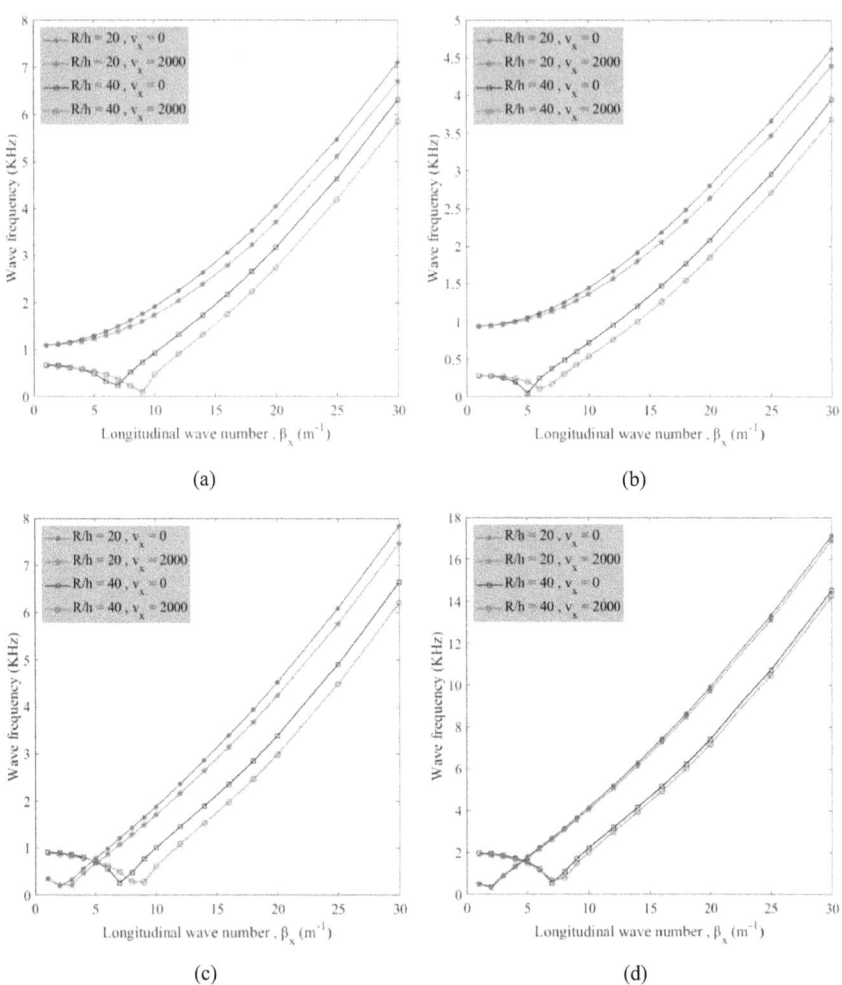

FIGURE 4.6 Variation of wave frequency against longitudinal wave number for various flow velocities and radius-to-thickness ratios for different anisotropic materials ($\beta_n = 10$). (a) Monoclinic, (b) triclinic, (c) trigonal, and (d) hexagonal.

4.3 FLUID-CONVEYING CARBON NANOTUBES: WAVE DISPERSION CHARACTERISTICS

Nonlocal elastic waves in a fluid-carrying magneto-thermo-elastic (MEE) single-walled carbon nanotube are investigated in this section under a resonant harmonic load, as seen in Figure 4.7. Nonlocal elasticity theory developed by Eringen is applied to the framework of Euler–Bernoulli beam theory. Taking into account heat and Lorenz magnetic force along with nonlocal factors, governing equations

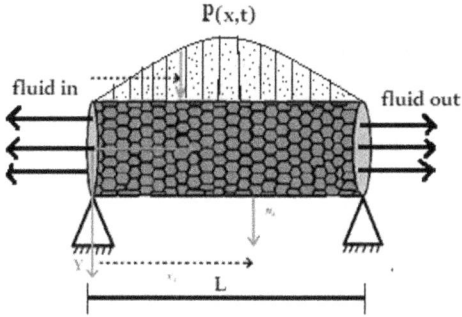

FIGURE 4.7   Geometry of the problem.

including partial differential equations for single-walled carbon nanotubes are constructed. Computed dynamic displacement is the computation of the displacement of a body of matter from its original position to its new position.

The strain across the local area is taken into account when calculating the stress state in Eringen's nonlocal theory. In the nonlocal form of elasticity, the general form of the constitutive equations includes an integral across the whole area of interest. The relative effects of the strains at different sites on the stress at a particular point are described by a nonlocal kernel function included inside the integral. For a zero-body-force linear homogeneous isotropic nonlocal elastic solid, Eringen provides the following constitutive equations [18]:

$$\Pi_{ij} + \rho\left(f_j - \ddot{u}_j\right) = 0 \tag{4.21a}$$

$$\Pi_{ij}\left(x\right) = \int_v \pi\left(\left|x - x'\right|, \alpha\right)\Pi_{ij}^c\left(x'\right)dV\left(x'\right) \tag{4.21b}$$

$$\Pi_{ij}^c = C_{ijkl}\varepsilon_{kl} \tag{4.21c}$$

$$\Pi_{ij}\left(x'\right) = \frac{1}{2}\left(\frac{\partial u_i\left(x\right)}{\partial x_j'} + \frac{\partial u_i\left(x'\right)}{\partial x_j'}\right) \tag{4.21d}$$

where Π_{ij}, ρ, f_j, u_j are the stress tensor, mass density, body force density, and displacement vector at a reference point x in the body, respectively, at the time t. Eq. (4.21c) is the classical constitutive relation, where $\Pi^c{}_{ij}(x')$ is the classical stress tensor at any point x' in the body, which is related to the linear strain tensor $e_{ij}(x')$ at the same point. Eq. (4.21d) is the classical strain-displacement relationship. The kernel function $\pi(|x - x'|, \alpha)$ is the attenuation function that incorporated the nonlocal effect in the constitutive equations. It is clear that the only difference between Eqs. (4.1)–(4.4) and the corresponding equations of classical elasticity in Eq. (4.21b) replaces the Hooke's law in Eq. (4.21c) by Eq. (4.21b). Eq. (4.21b) consists of the parameters that correspond to the nonlocal modulus, which has dimensions of (lengh)$^{-3}$ and so it depends on a characteristic length (lattice parameter, size of grain, granular distance, etc.) and "l." is an external characteristic length of the system (wavelength, crack length, size or dimensions of sample, etc.). Therefore, the nonlocal modulus can be written in the following form:

$$\pi\left(\left|x-x'\right|,\alpha\right),\tau = \frac{e_o a}{l} \tag{4.22a}$$

where $e_o a$ is a constant corresponding to the material and has to be determined for each material independently and "$|x - x'|$" is the Euclidian distance. Then, the integro-partial differential Eq. (4.21b) of nonlocal elasticity can be simplified to partial differential equation as follows:

$$\left(1-\tau^2 l^2 \nabla^2\right)\Pi_{ij}\left(x\right) = \Pi_{ij}^c\left(x\right) = C_{ijkl} e_{kl}\left(x\right) \tag{4.22b}$$

where C_{ijkl} is the elastic modulus tensor of classical isotropic elasticity and e_{ij} is the strain tensor. Where ∇^2 denotes the second-order spatial gradient applied on the stress tensor σ_{ij} and $\tau = e_o a/l$. Eringen proposed $e_o = 0.39$ by the matching of the dispersion curves via nonlocal theory for place wave and Born–Karman model of lattice dynamics at the end of the Brillouin zone ($ka = \pi$), where a is the distance between atoms, and k is the wave number in the phonon analysis.

In Figure 4.7, we see the SWCNT under a harmonic stress while it is carrying a fluid that is magneto-thermo-elastic. We considered zero gravity and incompressible internal fluid in our analytical model. When subjected to a changing harmonic load, the fluid-carrying magneto-thermo-elastic SWCNT is governed by the following partial differential equation:

$$\frac{\partial Q}{\partial X} + N_T \frac{\partial^2 Y}{\partial X^2} + q \frac{\partial^2 Y}{\partial X^2} - m_c \frac{\partial^2 \sigma_x}{\partial X^2} + F_p = p\left(x,t\right) + \rho A \frac{\partial^2 Y}{\partial t^2} \tag{4.22c}$$

where $P(x, t)$ is the distributed load in the transverse direction of x axis, ρ mass density of beam, Y is the transverse bending of the beam, and t denotes the time. The resultant shear force Q on the cross-section of the nanotube is defined in the following equilibrium equation:

$$Q = \frac{\partial M}{\partial X} \tag{4.23a}$$

The thermal term N_T represents the additional axial thermal load in terms of the temperature T and α_x is the thermal expansion of coefficient in x direction. This thermo-elastic force can be given as

$$N_T = -EA\alpha_x T \tag{4.23b}$$

The fluid force per unit length as a result of plug flow is

$$F_p = m_f \left(2v \frac{\partial^2 Y}{\partial t \partial X} + v^2 \frac{\partial^2 Y}{\partial X^2} + \frac{\partial^2 Y}{\partial t^2} \right) \tag{4.23c}$$

Applying the magnetic field $H = (H_x, 0, 0)$ in the longitudinal direction, then the Lorentz body force exerted on CNT is considered as [25]:

$$q = \eta_s A H_x^2 \frac{\partial^2 Y}{\partial X^2} \tag{4.23d}$$

in which η_s is the magnetic field permeability and H_x is the magnetic field strength. The resultant bending moment M in Eq. (4.23a) is represented in the following form:

$$M = \int_A Y \Pi_{xx} dA, \tag{4.23e}$$

where Π_{xx} is the axial stress according to nonlocal continuum theory. When modeling a one-dimensional nanotube, it is convenient to use the constitutive equation for a homogeneous isotropic elastic material, which may be written as [18]:

$$\Pi_{xx} - \left(e_0 a \right)^2 \frac{\partial^2 \Pi_{xx}}{\partial X^2} = E\varepsilon_{xx} \tag{4.24a}$$

where X is the axial coordinate, ε_{xx} is the axial strain, and $(e_0 a)$ is a nonlocal parameter that represents the impact of nonlocal scale effect on the structure. a is an internal characteristic length and E is Young's modulus. The nonlocal relations in Eq. (4.24a) can be written with temperature environment as follows:

$$\Pi_{xx} - \left(e_0 a \right)^2 \frac{\partial^2 \Pi_{xx}}{\partial X^2} = E\varepsilon_{xx} - E\alpha T \tag{4.24b}$$

In the context of Bernoulli–Euler model, the axial strain ε_{xx} for small deflection is defined as

$$\varepsilon_{xx} = -Y \frac{\partial^2 Y}{\partial X^2} \tag{4.24c}$$

where Y is the transverse coordinate in the positive direction of deflection. By using Eqs. (4.24b)–(4.24c) in Eq. (4.23e), the bending moment M can be expressed as:

$$M - \left(e_0 a\right)^2 \left[\frac{\partial^2 M}{\partial X^2}\right] = EI \frac{\partial^2 Y}{\partial X^2} \tag{4.24d}$$

where $I = \int_A Y^2 dA$ is the moment of inertia. By substituting Eqs. (4.22a)–(4.22c) into Eq. (4.24d), the nonlocal bending moment M and shear force Q can be expressed as follows:

$$M - \left(e_0 a\right)^2 \left[\left(m_c + \rho A\right)\frac{\partial^2 Y}{\partial t^2} + F_P - q + EA\alpha T + p\left(t\right)\right] = EI \frac{\partial^2 Y}{\partial X^2} \tag{4.24e}$$

and

$$Q - \left(e_0 a\right)^2 \left[\left(m_c + \rho A\right)\frac{\partial^3 Y}{\partial X^2 \partial t^2} + \frac{\partial F_P}{\partial X} - \frac{\partial^2 q}{\partial X^2} + EA\alpha T\right] = EI \frac{\partial^3 Y}{\partial X^3} \tag{4.24f}$$

Under the conditions of harmonic excitation, distributed pressure, and thermal interaction, the equation of motion for the transverse displacement is given by Eq. (4.22a):

$$
\begin{aligned}
p\left(x,t\right) &= EI \frac{\partial^4 Y\left(x,t\right)}{\partial X^4} + EA\alpha T \frac{\partial^2 y\left(x,t\right)}{\partial x^2} - \eta_s AH_x^{\,2} \frac{\partial^2 Y}{\partial X^2} + m_f v^2 \frac{\partial^2 Y\left(x,t\right)}{\partial X^2} \\
&+ 2m_f v \frac{\partial^2 Y\left(x,t\right)}{\partial t \partial X} + \left(\rho A + m_c + m_f\right)\frac{\partial^2 Y\left(x,t\right)}{\partial t^2} \\
&- \left(e_0 a\right)^2 \left(
\begin{array}{l}
EA\alpha T \dfrac{\partial^4 y\left(x,t\right)}{\partial X^4} + m_c \dfrac{\partial^2 y\left(x,t\right)}{\partial X^2} + m_f v^2 \dfrac{\partial^4 Y\left(x,t\right)}{\partial X^4} \\[2mm]
+ 2m_f v \dfrac{\partial^4 Y\left(x,t\right)}{\partial t^2 \partial X^2} + m_f \dfrac{\partial^4 Y\left(x,t\right)}{\partial t^4} - \eta_s AH_x^{\,2} \dfrac{\partial^4 Y\left(x,t\right)}{\partial X^4}
\end{array}
\right)
\end{aligned} \tag{4.25a}
$$

The SWCNT is subjected to a moving harmonic load $p(t)$, which moves in the axial direction of the nanotube with constant velocity vp as

$$p\left(x,t\right) = p\left(t\right)\delta\left(X - X_p\right) \tag{4.25b}$$

where $p(t) = p_o \sin(\Omega t)$, $\delta(X - X_p)$ is the Dirac delta function, X_p is the coordinate of the moving harmonic load, P_0 is the amplitude of the harmonic load, and Ω is the excitation frequency of the moving harmonic load. Introduction of Eq. (4.25b) into Eq. (4.25a) gives the following particle differential equation with constant coefficients in the nonlocal form:

$$
EI \frac{\partial^4 Y}{\partial X^4} - EA\alpha_x T \frac{\partial^2 Y}{\partial X^2} + m_f v^2 \frac{\partial^2 Y}{\partial X^2} + \eta_s AH_x^2 \frac{\partial^2 Y}{\partial X^2} + 2m_f v \frac{\partial^2 Y}{\partial t \partial X}
$$

$$
-\left(e_0 a\right)^2 \left(
\begin{array}{l}
\left(m_c + m_f + \rho A\right) \dfrac{\partial^4 Y}{\partial X \partial t^2} + EA\alpha_x T \dfrac{\partial^4 Y}{\partial X^4} + \rho A \dfrac{\partial^2 Y}{\partial X^2} \\[2ex]
+ m_f v^2 \dfrac{\partial^4 Y}{\partial X^4} + 2m_f v \dfrac{\partial^4 Y}{\partial t^2 \partial X^2} + m_f \dfrac{\partial^4 Y}{\partial t^4} + \eta_s AH_x^2 \dfrac{\partial^4 Y}{\partial X^4}
\end{array}
\right) \tag{4.26}
$$

$$
= P_o \sin \Omega t \, \delta\left(X - X_p\right) - \left(e_0 a\right) \frac{\partial^2 P_o \sin \Omega t}{\partial X^2}
$$

The dynamic displacement of the SWCNT in the transverse direction is represented as follows:

$$
w(x,t) = \sum_{i=1}^{\infty} \Phi_i(x) q_i(t) \tag{4.27a}
$$

$q_i(t)$ are the unknown time-dependent generalized coordinates and $\Phi_i(x)$ are the Eigen modes of an undamped simply supported beam, which can be chosen as follows:

$$
\Phi_i(x) = \sin\left(\frac{i\pi x}{l}\right), \quad i = 1, 2, 3, 4 \ldots\ldots\ldots \tag{4.27b}
$$

Using (4.27a) and (4.27b) in Eq. (4.26) produces

$$
P_o \sin \Omega t \, \delta\left(X - X_p\right) - \left(e_0 a\right) \frac{\partial^2 P_o \sin \Omega t}{\partial X^2} = EI \left(\sum_{i=1}^{\infty} \sin\left(\frac{i\pi x}{l}\right)\left(\frac{\pi}{l}\right)^4 q_i(t)\right)
$$

$$
- EA\alpha_x T \left(\sum_{i=1}^{\infty} \sin\left(\frac{i\pi x}{l}\right)\left(\frac{\pi}{l}\right)^2 q_i(t)\right) + m_f v^2 \left(\sum_{i=1}^{\infty} \sin\left(\frac{i\pi x}{l}\right)\left(\frac{\pi}{l}\right)^2 q_i(t)\right)
$$

$$
+ \eta AH_x^2 \left(\sum_{i=1}^{\infty} \sin\left(\frac{i\pi x}{l}\right)\left(\frac{\pi}{l}\right)^2 q_i(t)\right) + 2m_f v \left(\sum_{i=1}^{\infty} \sin\left(\frac{i\pi x}{l}\right)\left(\frac{\pi}{l}\right) \dot{q}_i(t)\right)
$$

$$
-\left(e_0 a\right)^2
\left(
\begin{array}{l}
\left(m_c + m_f + \rho A\right)\left(\displaystyle\sum_{i=1}^{\infty} \sin\left(\dfrac{i\pi x}{l}\right)\left(\dfrac{\pi}{l}\right)^2 \overset{**}{q_i}(t)\right) \\[4mm]
+EA\alpha T\left(\displaystyle\sum_{i=1}^{\infty} \sin\left(\dfrac{i\pi x}{l}\right)\left(\dfrac{\pi}{l}\right)^4 q_i(t)\right) + \rho A\left(\displaystyle\sum_{i=1}^{\infty} \sin\left(\dfrac{i\pi x}{l}\right)\left(\dfrac{\pi}{l}\right)^2 q_i(t)\right) \\[4mm]
+m_f v^2\left(\displaystyle\sum_{i=1}^{\infty} \sin\left(\dfrac{i\pi x}{l}\right)\left(\dfrac{\pi}{l}\right)^4 q_i(t)\right) + 2m_f v\left(\displaystyle\sum_{i=1}^{\infty} \sin\left(\dfrac{i\pi x}{l}\right)\left(\dfrac{\pi}{l}\right)^2 \overset{*}{q_i}(t)\right) \\[4mm]
+m_f\left(\displaystyle\sum_{i=1}^{\infty} \sin\left(\dfrac{i\pi x}{l}\right)\overset{**}{q_i}(t)\right) - \eta A H_x{}^2\left(\displaystyle\sum_{i=1}^{\infty} \sin\left(\dfrac{i\pi x}{l}\right)\left(\dfrac{\pi}{l}\right)^4 q_i(t)\right)
\end{array}
\right)
$$

$$(4.28a)$$

where prime denotes the derivative with respect to x, by taking the integral on both sides $\Phi_i(x)$ in the above equation can be written as

$$
\int_0^l P(x,t) - \left(e_0 a\right)^2 \frac{\partial^2 P(x,t)}{\partial X^2} - Kw\phi_i(x)\,dx = \sum_{i=1}^{\infty} qi(t)\int_0^l EI\,\overset{**}{\phi_i}\,\phi_j(x)\,dx
$$

$$
- \sum_{i=1}^{\infty} qi(t)\int_0^l EA\alpha T\,\overset{*}{\phi_i}\,\phi_j(x)\,dx + \sum_{i=1}^{\infty} qi(t)\int_0^l M_f V^2\,\overset{*}{\phi_i}\,\phi_j(x)\,dx
$$

$$
+ \sum_{i=1}^{\infty} qi(t)\int_0^l \eta A H_x{}^2\,\overset{*}{\phi_i}\,\phi_j(x)\,dx + \sum_{i=1}^{\infty} qi(t)\int_0^l 2m_f v\,\overset{*}{\phi_i}\,\phi_j(x)\,dx
$$

$$
-\left(e_0 a\right)^2
\left[
\begin{array}{l}
\displaystyle\sum_{i=1}^{\infty} \overset{*}{q_i}(t)\int_0^l (mc+mf+\rho A)\overset{*}{\phi_i}\,\phi_j(x)\,dx + \sum_{i=1}^{\infty} q_i(t)\int_0^l EA\alpha T\,\overset{**}{\phi_i}\,\phi_j(x)\,dx \\[4mm]
+ \displaystyle\sum_{i=1}^{\infty} q_i(t)\int_0^l \eta A H_x{}^2\,\overset{*}{\phi_i}\,\phi_j(x)\,dx + \sum_{i=1}^{\infty} q_i(t)\int_0^l \rho A\,\overset{*}{\phi_i}\,\phi_j(x)\,dx \\[4mm]
+ \displaystyle\sum_{i=1}^{\infty} q_i(t)\int_0^l M_f V^2\,\overset{*}{\phi_i}\,\phi_j(x)\,dx + \sum_{i=1}^{\infty} \overset{*}{q_i}(t)\int_0^l 2m_f v\,\overset{*}{\phi_i}\,\phi_j(x)\,dx \\[4mm]
+ \displaystyle\sum_{i=1}^{\infty} \overset{**}{q_i}(t)\int_0^l m_f\phi_i\phi_j(x)\,dx - \sum_{i=1}^{\infty} q_i(t)\int_0^l Kw\,\overset{*}{\phi_i}\,\phi_j(x)\,dx
\end{array}
\right]
$$

$$(4.28b)$$

Taking into account the following orthogonal requirements, the current equation may be solved.

$$\int_0^l \Phi_i \Phi_j (x) dx = \begin{cases} \dfrac{l}{2} & i = j \\ 0 & i \neq j \end{cases}$$
(4.28c)

The right-hand-side load terms were then derived by exploiting the following general property of the Dirac delta function:

$$\int_{x_1}^{x_2} \Gamma(x) \delta^{(x)} (X - X_0) dx = \begin{cases} (-1)^n g^{(n)} (x_0) & \text{if } x_1 < x_0 < x_2 = j \\ 0 & \text{otherwise} \end{cases}$$
(4.28d)

the nth derivative of the Dirac Delta function is represented as $\delta^n(X - X_p)$ and also the following relations are considered:

$$k_j = \int_0^l EI \left(\frac{i\pi}{l} \right)^4 \Phi_i \Phi_j (x) dx$$
(4.29a)

$$\alpha_j = \int_0^l \left(\eta A H_x^2 + mfv^2 + EA\alpha_x T \right) \left(\frac{i\pi}{l} \right)^2 \Phi_i \Phi_j (x) dx$$
(4.29b)

$$M_j = \int_0^l \rho A \left((e_o a)^2 - 1 + m_f + m_c \right) \left(\frac{i\pi}{l} \right)^2 \Phi_i \Phi_j (x) dx$$
(4.29c)

$$\beta_j = \int_0^l (e_o a)^2 \left(N_t + \eta A H_x^2 + m_f v^2 \right) \times \left(\frac{i\pi}{l} \right)^4 \Phi_i \Phi_j (x) dx$$
$$- (e_o a) \int_0^l (\rho A) \left(\frac{i\pi}{l} \right)^2 \Phi_i \Phi_j (x) dx$$
(4.29d)

$$D_j = (e_o a)^2 \int_0^l (N_x + 2m_f v + m_f) \left(\frac{i\pi}{l} \right)^4 \Phi_i \Phi_j (x) dx$$
$$- (e_o a) \int_0^l (\rho A) \left(\frac{i\pi}{l} \right)^2 \Phi_i \Phi_j (x) dx$$
(4.29e)

$$f_i(t) = p(x,t) \int_0^l \left[\delta(X - X_p) - (e_o a)^2 \frac{\partial^2 (X - X_p)}{\partial X^2} - Kw \right] \varphi_i d(x) \quad (4.29f)$$

This will result in the ith mode of the general deflection differential equation:

$$\ddot{q}_j(t) + \left[\omega_j^2 + \varsigma_j^2 + \psi_j^2 \right] \dot{q}_j(t) + Z_j^2 q_j(t) = \frac{1}{M_j} f_i(t) i, j = 1, 2, 3, 4 \ldots \quad (4.30a)$$

where

$$\varsigma_i = \sqrt{\frac{D_i}{M_j}}, \ \omega_j = \sqrt{\frac{k_i}{M_j}}, \ \psi_j = \sqrt{\frac{\alpha_i}{M_j}}, \ Z_i = \sqrt{\frac{\beta_i}{M_j}} \text{ and } f_i(t) = 0 \text{ for } t > 0, \text{ where}$$

$X_p(0 \leq X_P = v_p)$ is the coordinate of the following moving harmonic load. Then by considering the following equation:

$$\frac{1}{M_j} f_i(t) = S_i(t)$$

where also by employing the following equation:

$$S_i(t) = \frac{2p(t)}{\rho AL} \sin\left(\frac{i\pi x_p t}{L} \right) \quad (4.30b)$$

Eq. (4.30a) may be resolved as follows for homogeneous starting conditions:

$$q_i(t) = \frac{1}{M_j} \int_0^t S_i(\lambda) \sin m_j (t - \lambda) d\lambda \quad (4.30c)$$

The following is obtained by inserting Eq. (4.30c) into Eq. (4.30b) and integrating:

$$q_j(t) = \frac{\left((\rho A + m_f + m_c) - \rho A (e_0 a)^2 \right) p_o}{\rho AL} \sum_{i=1}^{\infty} \left[\frac{\cos\left(\Omega t - \frac{i\pi x_p(t)}{L} \right) - \cos(M_j t)}{M_j^2 - \left(\Omega t - \frac{i\pi x_p(t)}{L} \right)^2} \right.$$

$$\left. - \frac{\cos\left(\Omega t + \frac{i\pi x_p(t)}{L} \right) - \cos(M_j t)}{M_j^2 - \left(\Omega t + \frac{i\pi x_p(t)}{L} \right)^2} \right]$$

$$(4.30d)$$

Finally, the total dynamic deflection may be obtained by plugging Eq. (4.30d) into Eq. (4.27a) as shown below:

$$
w(x,t) = \frac{\left[(\rho A + m_f + m_c) - \rho A (e_0 a)^2\right] p_o}{\rho AL} \sum_{i=1}^{\infty} \left[\frac{\cos\left(\Omega t - \dfrac{i\pi x_p(t)}{L}\right) - \cos(Mt)}{M^2 - \left(\Omega t - \dfrac{i\pi x_p(t)}{L}\right)^2} \right.
$$

$$
\left. - \frac{\cos\left(\Omega t + \dfrac{i\pi x_p(t)}{L}\right) - \cos(Mt)}{M^2 - \left(\Omega t + \dfrac{i\pi x_p(t)}{L}\right)^2} \right] \sin\left(\frac{i\pi x}{L}\right)
$$

(4.30e)

Here, we look at what happens to a magneto-thermo fluid carrying a simply supported SWCNT when subjected to a moving harmonic load and a forced nonlocal wave. In the case of the elasticity waves of SWCNT, the choice of effective wall thickness and the elastic modulus E is a series of structural challenges. The numerical values of the involved parameters in the problem have been assumed to be $E = 1\ TP_a$, $\rho = 2300\ \text{kg}/\text{m}^3$, $t_p = 0.35$ nm. According to the calculation, the mass of fluid per unit length in the SWCNT is 1.52×10^{-16} kg/m^2 and the mass per unit length of the SWCNT is 2.75×10^{-15} kg/m^2. The thermal expansion coefficient in room temperature $\alpha^0 = -1.5 \times 10^{-6}$ C^{-1} and $l = 10$ nm, $d = 1$ nm. The following notation is used to express the nondimensional velocity parameter of the moving harmonic load [18]:

$$
\xi = \frac{v_p}{v_{cr}}
$$

For the SWCNT, the critical velocity is defined as:

$$
v_{cr} = \frac{\psi_1 L}{\pi}
$$

where ψ_1 is the fundamental frequency of SWCNT. The influence of the excitation frequency of the moving harmonic load Ω is represented by the frequency ratio θ, where

$$
\theta = \frac{\Omega}{\psi_1}
$$

The dimensionless t time defined by

$$t^* = \frac{X_p}{L}$$

Figures 4.8 and 4.9 show the variation of dynamic displacement of the elastic SWCNT with harmonic load velocity for the values of $N_t = 0.2$, 0.5 and $H_x = 0$ with different nonlocal constants $e_0 a = 0$, 0.5, 1.0, 1.5. From these figures, it is observed that the dynamic deflection of the SWCNT is highly influenced by the values of nonlocal parameters and has a higher magnitude at the lower values of the velocity of the harmonic loads [18].

As the values of the thermal parameters increase, a scattering effect is seen in the wave's propagation, as seen in Figure 4.9. The elastic SWCNT's dynamic displacement is compared to the velocity of a harmonic load over a range of values $N_t = 0.2$, 0.5 and $H_x = 0.5$ with different nonlocal constants $e_0 a = 0$, 0.5, 1.0, 1.5 respectively, as shown in Figures 4.10 and 4.11.

From these figures, it is clear that at the lower range of harmonic velocity, the dynamic displacement of SWCNT attains maximum value in both cases of $N_t = 0.2$ and $N_t = 0.5$, but there is deviation in elastic wave behavior when $H_x = 0.5$ in these figures. This may happen due to the effect of longitudinal magnetic field of the SWCNT [18].

Figures 4.12 and 4.13 investigate the variation dynamic displacement of the elastic SWCNT with the frequency ratio θ for the values $e_0 a = 0.2$, 0.5 and $H_x = 0$

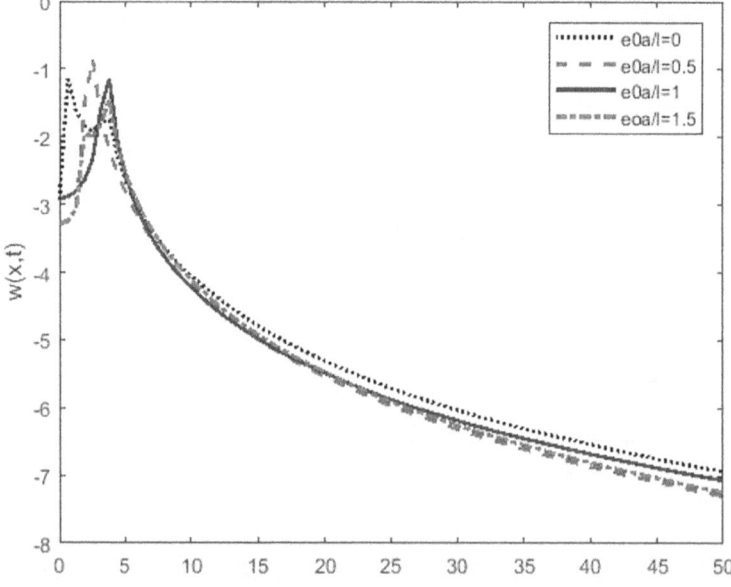

FIGURE 4.8 Variation of dynamic displacement versus harmonic load velocity with $N_T = 0.2$ $H_x = 0$.

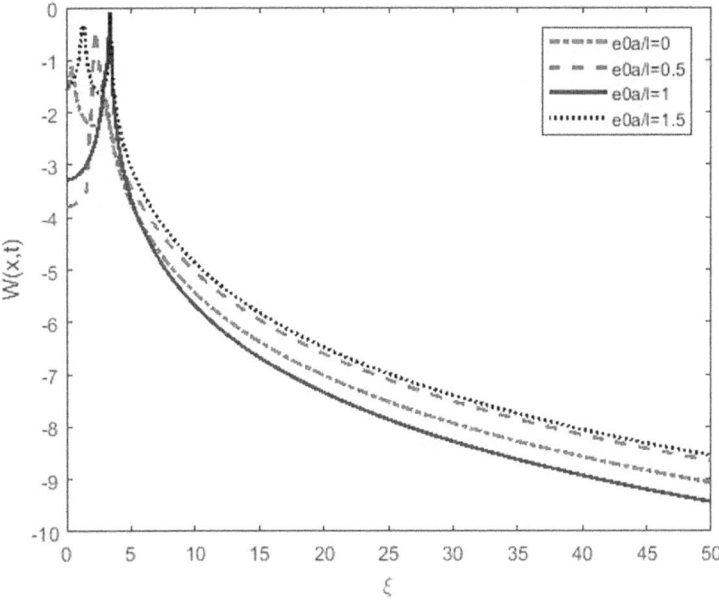

FIGURE 4.9 Variation of dynamic displacement versus harmonic load velocity with $N_T = 0.2\ H_x = 0$.

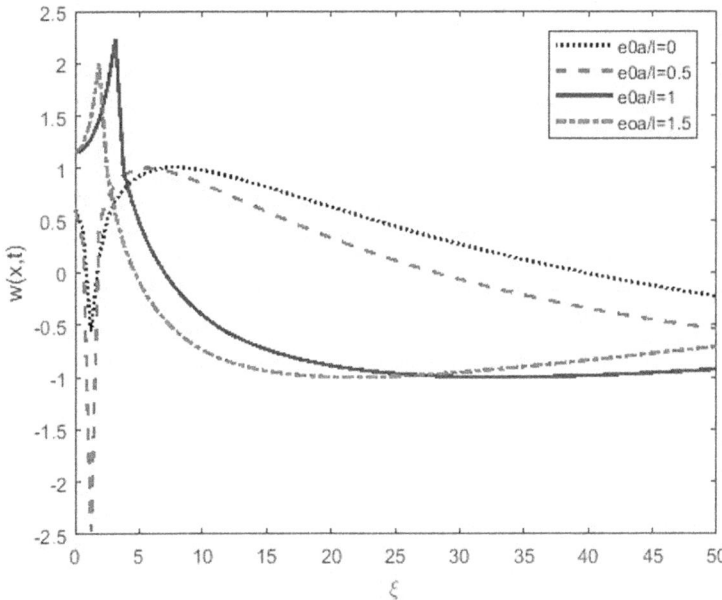

FIGURE 4.10 Variation of dynamic displacement versus harmonic load velocity with $N_T = 0.2\ H_x = 0$.

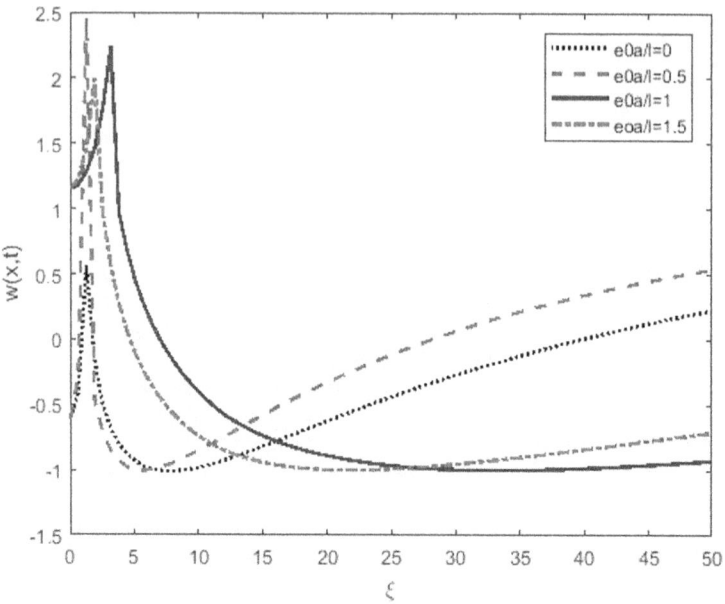

FIGURE 4.11   Variation of dynamic displacement versus harmonic load velocity with $N_T = 0.5 \, H_x = 0.5$.

with different thermal parameter values $N_t = 0.5, 1.0, 1.5, 2.0$. From these figures, it is seen that the dynamic deflection of the SWCNT is varying significantly by the varying values of the thermal parameter. Also, the dynamic displacement increases as the frequency ratio increases and starts to decrease at $\theta = 35$. Figure 4.13 presents the wave propagation nature in the trend line due to the increasing values of the nonlocal parameter and zero magnetic effect [18].

Figures 4.14 and 4.15 illustrate the comparison between the dynamic displacement of the elastic SWCNT versus the frequency ratio for the values $e_0 a = 0.2, 0.5$ and $H_x = 0.5$ with the varying thermal parameter values $N_t = 0.5, 1.0, 1.5, 2.0$, respectively. From Figure 4.14, it is clear that the dynamic displacement is getting peak values in the frequency ratio range $5 \leq \theta \leq 15$, but in Figure 4.15 it attains peak values at $\theta = 35$ due to the increase in nonlocal effect and magnetic field value of fluid-conveying SWCNT [18].

Figures 4.16 and 4.17 discuss the dispersion curves for the dynamic displacement versus time ratio with $e_0 a = 0.2, 0.5 \, N_t = 0$ of fluid-conveying elastic SWCNT for different constants of $H_x = 0.5, 1.0, 1.5, 2.0$. From these figures, it is observed that the dynamic displacement is propagating from the negative values and attains the maximum in the positive values as the time ratio increases. The crossing-over trend of the dispersion curves explains the energy exchange among the vibrational modes when the thermal parameter value is zero and increasing nonlocal parameter values [18].

Figures 4.18 and 4.19 explain the variation of the dynamic displacement over the time ratio with $e_0 a = 0.2, 0.5 \, N_t = 0.5$ of fluid-conveying elastic SWCNT for the

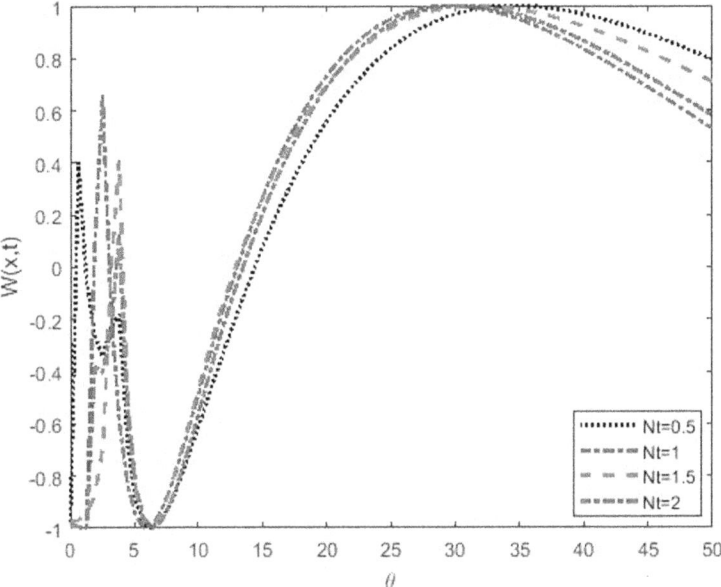

FIGURE 4.12   Variation of dynamic displacement versus frequency ratio with $e_0 a = 0.2$ $H_x = 0$.

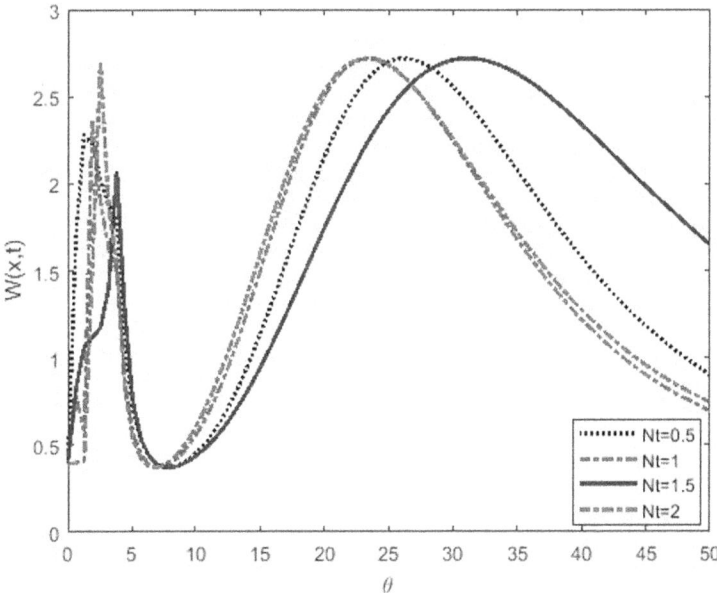

FIGURE 4.13   Variation of dynamic displacement versus frequency ratio with $e_0 a = 0.5$ $H_x = 0$.

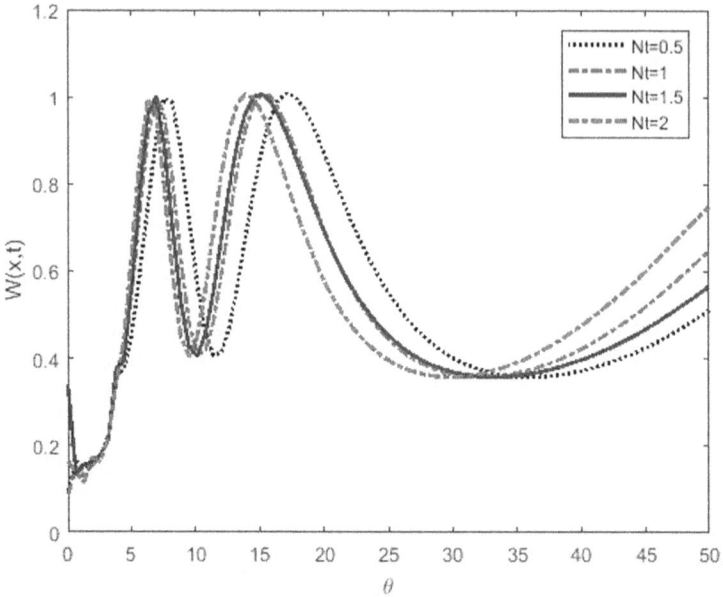

FIGURE 4.14   Variation of dynamic displacement versus frequency ratio with $e_0a = 0.2$ $H_x = 0.5$.

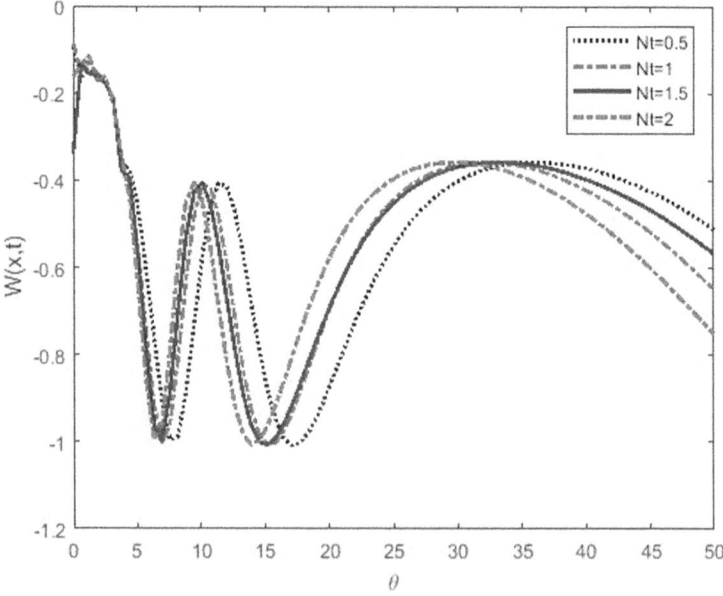

FIGURE 4.15   Variation of dynamic displacement versus frequency ratio with $e_0a = 0.5$ $H_x = 0.5$.

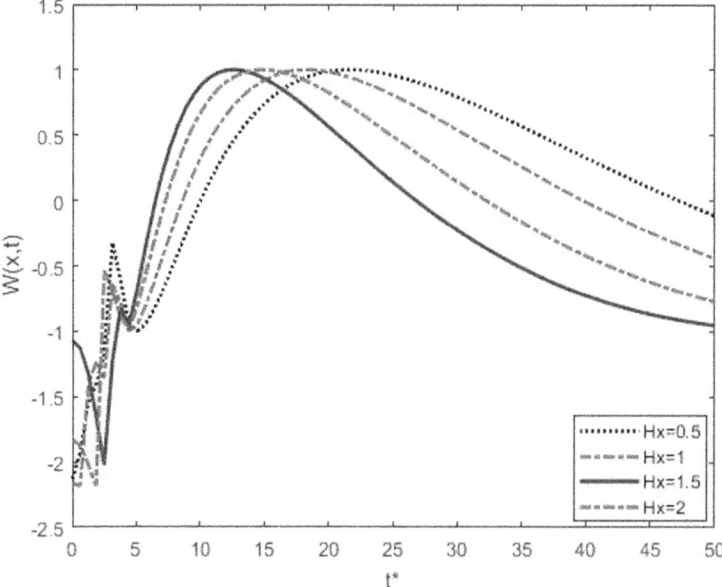

FIGURE 4.16   Variation of dynamic displacement versus frequency ratio with $e_0a = 0.2$ $N_t = 0$.

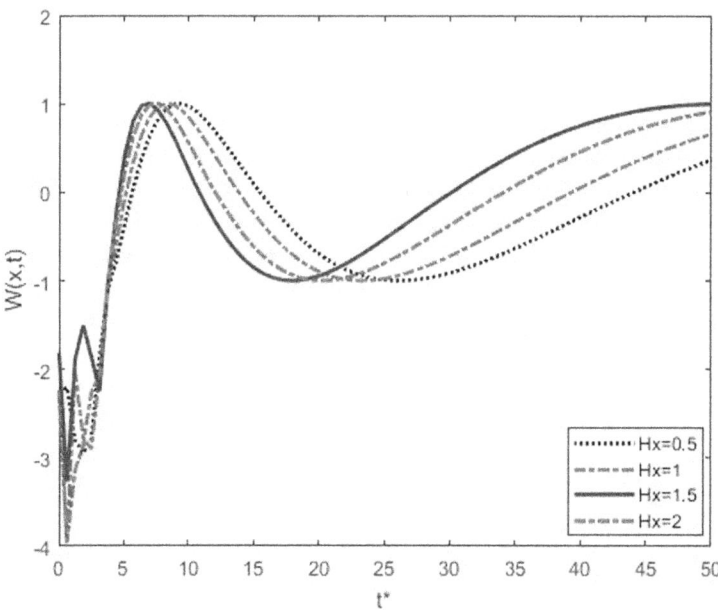

FIGURE 4.17   Variation of dynamic displacement versus frequency ratio with $e_0a = 0.5$ $N_t = 0$.

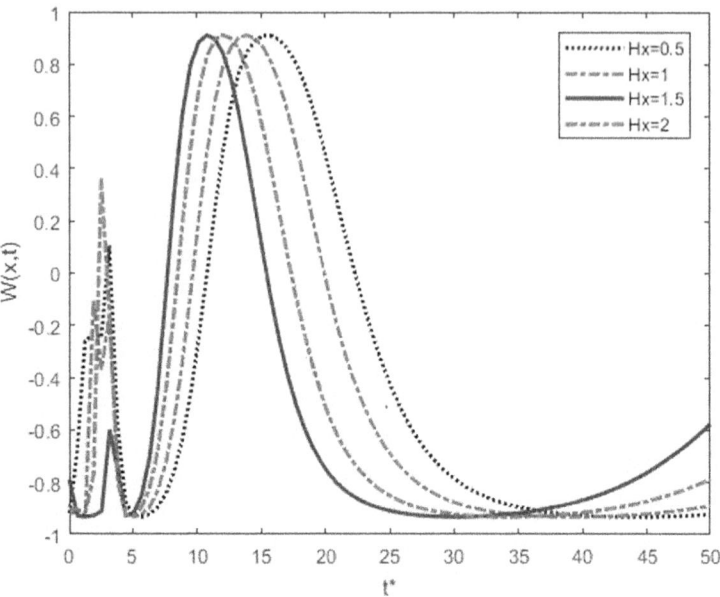

FIGURE 4.18   Variation of dynamic displacement versus frequency ratio with $e_0a = 0.2$ $N_t = 0.5$.

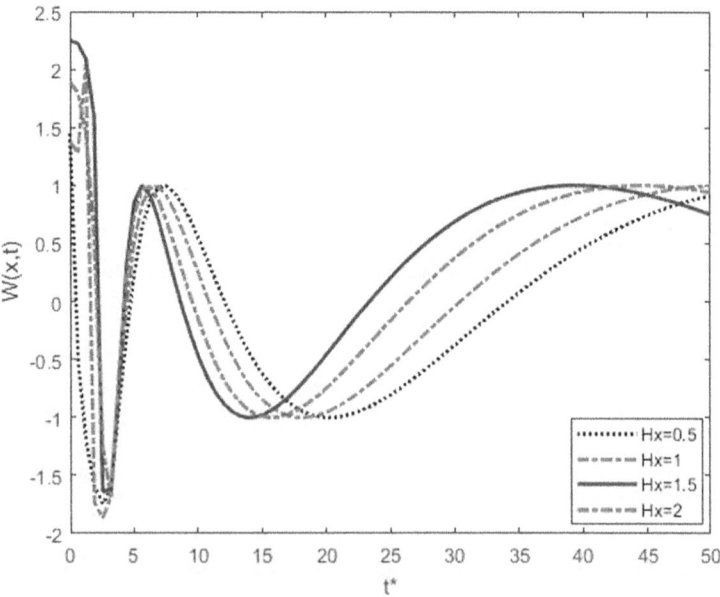

FIGURE 4.19   Variation of dynamic displacement versus frequency ratio with $e_0a = 0.5$ $N_t = 0.5$.

magnetic field values of $H_x = 0.5$, 1.0, 1.5, 2.0. According to these figures, at the lower level of time ratio, the effect of the magnetic field is found as the nonlocal parameter and thermal parameter increase. The crossing-over trend of the dispersion curves explains the energy exchange among the vibrational modes when there is an increment in thermal parameter value and nonlocal parameter values [18].

The 3D curves in Figures 4.20–4.23, clarify the variation of the dynamic displacement against the time and velocity ratio with and without fluid force for the

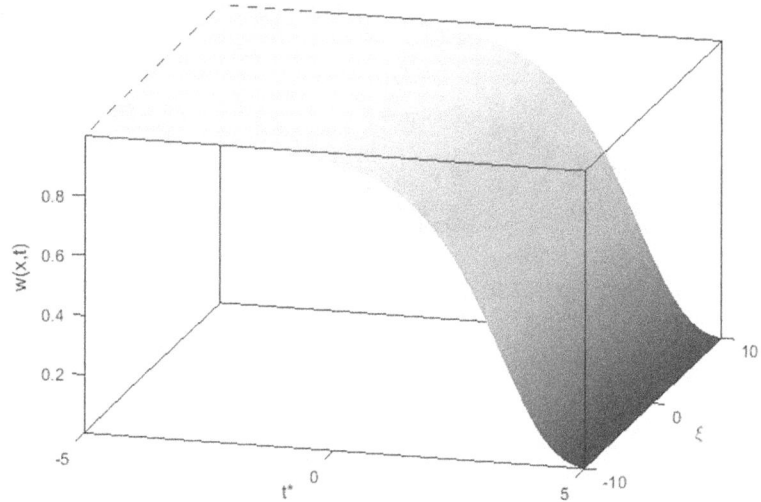

FIGURE 4.20 3D Distribution of dynamic displacement with t^* and ξ for $e_0a = 0.5$ $N_t = 0$ and $F_p = 0$.

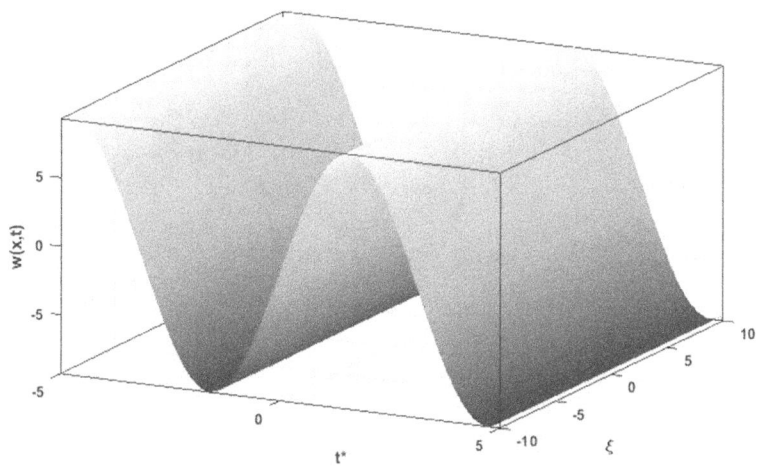

FIGURE 4.21 3D Distribution of dynamic displacement with t^* and ξ for $e_0a = 0.5$ $N_t = 0.5$ and $F_p = 0.5$.

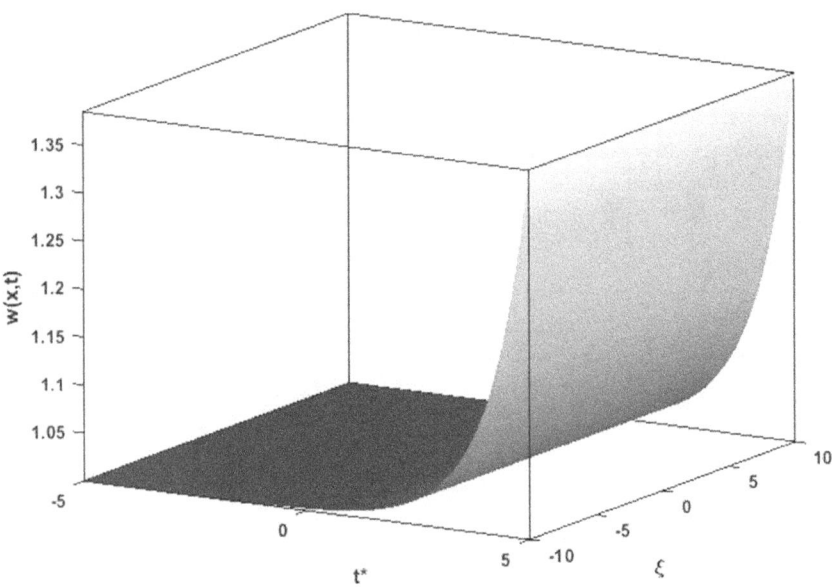

FIGURE 4.22   3D Distribution of dynamic displacement with t^* and ξ for $e_0 a = 0.5$ $H_x = 0$ and $F_p = 0$.

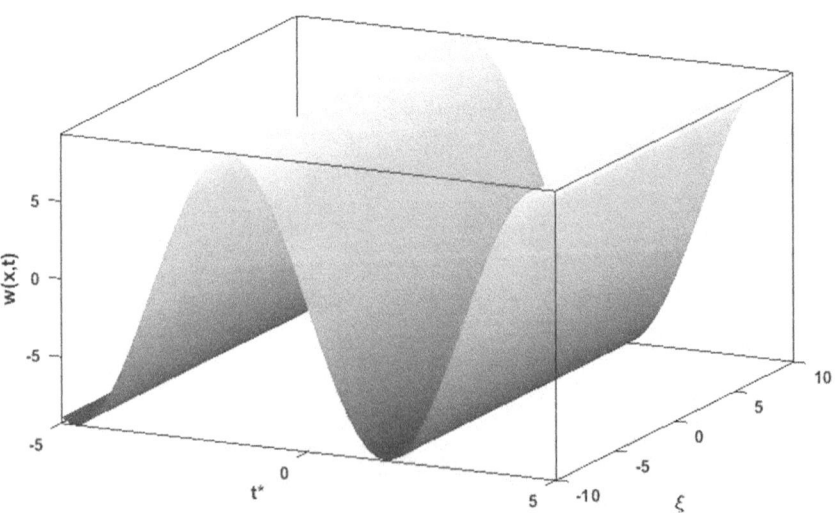

FIGURE 4.23   3D Distribution of dynamic displacement with t^* and ξ for $e_0 a = 0.5$ $H_x = 0.5$ and $F_p = 0.5$.

values of magnetic parameter, thermal constant, and nonlocal parameter. These curves explain the dependence of dynamic displacement on the different physical parameter t^*, ξ, N_t, H_x, F_p with nonlocal parameter values [18].

4.4   FLUID-CONVEYING BORON NITRIDE NANOTUBES: WAVE DISPERSION CHARACTERISTICS

This section shows the wave propagation of double- and triple-walled fluid-carrying boron nitride nanotubes (BNNTs) supported by a viscoelastic substrate in the presence of multiple physical fields according to NSGT's CS theory. Response and natural frequencies of double- and triple-walled boron nitride nanotubes carrying fluid will be modeled and discussed in terms of the effects of various parameters such as fluid velocity, viscoelastic base, Winkler and Pasternak damping coefficient, and geometry coefficient [20].

In order to account for the elastic, magnetic, and electric fields, the nonlocal elasticity theory may be extended to BNNTs. As a result, the elastic and electric fields at any given location in space-time rely not only on the elastic and electric features at that point but also on those at every other point in space time. Under the influence of electric, magnetic, and thermal fields, the following basic connection holds for boron nitride cylinder nanoshells [20]:

$$
\left(1-\left(e_0 a\right)^2 \nabla^2\right)\left\{\begin{array}{c} \sigma_{xi} \\ \sigma_{\theta i} \\ \sigma_{x\theta i} \end{array}\right\} = \left(1-l^2 \nabla^2\right)\left\{\begin{bmatrix} \tilde{c}_{11} & \tilde{c}_{12} & 0 \\ \tilde{c}_{21} & \tilde{c}_{22} & 0 \\ 0 & 0 & c_{66} \end{bmatrix}\left\{\begin{array}{c} \varepsilon_{xi} \\ \varepsilon_{\theta i} \\ \gamma_{x\theta i} \end{array}\right\}\right.
$$

$$
\left. -\begin{bmatrix} 0 & 0 & \tilde{e}_{31} \\ 0 & 0 & \tilde{e}_{32} \\ 0 & 0 & \tilde{e}_{33} \end{bmatrix}\left\{\begin{array}{c} E_{xi} \\ E_{\theta i} \\ E_{zi} \end{array}\right\}\right) - \begin{bmatrix} \tilde{\beta}_1 \\ \tilde{\beta}_2 \\ 0 \end{bmatrix} \Delta T
$$

(4.31a)

$$
\left(1-\left(e_0 a\right)^2 \nabla^2\right)\left\{\begin{array}{c} D_{xi} \\ D_{\theta i} \\ D_{Zi} \end{array}\right\} = \left(1-l^2 \nabla^2\right)\left\{\begin{bmatrix} 0 & 0 & 0 \\ 0 & 0 & 0 \\ \tilde{e}_{31} & \tilde{e}_{32} & \tilde{e}_{33} \end{bmatrix}\left\{\begin{array}{c} \varepsilon_{xi} \\ \varepsilon_{\theta i} \\ \gamma_{x\theta i} \end{array}\right\}\right.
$$

$$
\left. -\begin{bmatrix} \tilde{\in}_{11} & 0 & 0 \\ 0 & \tilde{\in}_{22} & 0 \\ 0 & 0 & \tilde{\in}_{33} \end{bmatrix}\left\{\begin{array}{c} E_{xi} \\ E_{\theta i} \\ E_{zi} \end{array}\right\}\right) + \begin{bmatrix} 0 \\ 0 \\ \tilde{p}_3 \end{bmatrix} \Delta T
$$

(4.31b)

In these equations $e_0 a$, l are nonlocal and length parameters, respectively; $\nabla^2 = \dfrac{\partial}{\partial R^2} + \dfrac{1}{R}\dfrac{\partial}{\partial R} + \dfrac{1}{R^2}\dfrac{\partial^2}{\partial \theta^2} + \dfrac{\partial^2}{\partial z^2}$. The three-dimensional stress state reduced constants for the cylindrical nanoshell are presented as:

$$
\tilde{c}_{11} = c_{11} - \frac{c_{13}^2}{c_{33}}, \; \tilde{c}_{12} = c_{12} - \frac{c_{13}^2}{c_{33}}, \; \tilde{c}_{66} = c_{66}, \; \tilde{e}_{31} = e_{31} - \frac{c_{13}e_{33}}{c_{33}}, \; \tilde{\in}_{11} = \in_{11}
$$

$$
\tilde{\in}_{33} = \in_{33} - \frac{e_{33}^2}{c_{33}}, \; \tilde{\beta}_1 = \beta_1 - \frac{c_{13}\beta_3}{c_{33}}, \; \tilde{p}_3 = p_3 + \frac{\beta_3 e_{33}}{c_{33}}
$$

(4.31c)

where in these equations:

$$\{A_{11}, A_{12}, A_{66}\} = \{\tilde{c}_{11}h, \tilde{c}_{12}h, \tilde{c}_{66}h\} \qquad \{D_{11}, D_{12}, D_{66}\} = \left\{\frac{\tilde{c}_{11}h^3}{12}, \frac{\tilde{c}_{12}h^3}{12}, \frac{\tilde{c}_{66}h^3}{12}\right\}$$

$$E_{31} = \int_{-h/2}^{h/2} \tilde{e}_{31}\beta \sin(\beta z)\, dz \qquad\qquad T_{11} = \int_{-h/2}^{h/2} \tilde{\mu}_{11} \cos^2(\beta z)\, dz$$

$$X_{11} = \int_{-h/2}^{h/2} \tilde{\in}_{11} \cos^2(\beta z)\, dz \qquad\qquad X_{22} = \int_{-h/2}^{h/2} \tilde{\in}_{11} \left[\frac{\cos(\beta z)}{R+z}\right]^2 dz \qquad\qquad (4.32)$$

$$X_{33} = \int_{-h/2}^{h/2} \tilde{\in}_{33} \left[\beta \sin(\beta z)\right]^2 dz \qquad\qquad T_{22} = \int_{-h/2}^{h/2} \tilde{\mu}_{22} \left[\frac{\cos(\beta z)}{R+z}\right]^2 dz$$

$$T_{33} = \int_{-h/2}^{h/2} \tilde{\mu}_{33} \left[\beta \sin(\beta z)\right]^2 dz$$

Table 4.1 lists the constants of the material characteristics utilized in this investigation. What's more, the electric field may be determined using these equations:

$$E_{xi} = -\frac{\partial \tilde{\Phi}}{\partial x} = \cos(\beta z)\frac{\partial \Phi_i}{\partial x} \qquad\qquad (4.33a)$$

$$E_{\theta i} = -\frac{1}{R_i + z}\frac{\partial \tilde{\Phi}}{\partial \theta} = \frac{\cos(\beta z)}{R + z}\frac{\partial \Phi_i}{\partial \theta} \qquad\qquad (4.33b)$$

$$E_{zi} = -\frac{\partial \tilde{\Phi}}{\partial z} = -\beta \sin(\beta z)\Phi_i - \frac{2\phi_0}{h} \qquad\qquad (4.33c)$$

TABLE 4.1
Properties of BNNT Material

Properties	BNNT
Elastic (Gpa)	$c_{11} = 2035, c_{22} = 2035, c_{33} = 2035$
	$c_{12} = 692, c_{13} = 692, c_{23} = 692,$
	$c_{44} = 672, c_{66} = 672$
Piezoelectric (C/m²)	$e_{31} = 0, e_{33} = 0.95, e_{15} = 0$
Dielectric	$\in_{11}/\in_0 = 1250, \in_{22}/\in_0 = 1250, \in_{33}/\in_0 = 1250$
Thermal moduli(10^5 NKm⁻²)	$\beta_1 = 4.74, \beta_3 = 4.53$
Pyroelectric (10^{-6}CN⁻¹)	$p_3 = 25$
Mass density (10^3 Kg m⁻³)	$\rho = 3.487$

Generally, electric potential is defined as:

$$\tilde{\Phi}(x,\theta,z,t) = -\cos(\beta z)\Phi(x,\theta,t) + \frac{2z\varphi_0}{h} \tag{4.34}$$

in which $\beta = \dfrac{\pi}{h}$ and $\Phi(x, \theta, t)$ is local variation of electric potential in x, θ directions. Also, φ_0 is initial value of external electric potential [33]. Schematic view BNNT containing magnetic fluid resting on viscoelastic foundation is shown in Figure 4.24. Nanotube with L length, R radius, and h thickness subjected to electric potential field $\tilde{\Phi}(x,\theta,z,t)$. Based on Love's shell theory, displacement is assumed as [21]:

$$u_z(x,\theta,z,t) = U(x,\theta,t) + z\frac{\partial W(x,\theta,t)}{\partial x} \tag{4.35a}$$

$$u_\theta(x,\theta,z,t) = U(x,\theta,t) + z\frac{\partial W(x,\theta,t)}{\partial \theta} \tag{4.35b}$$

$$u_z(x,\theta,z,t) = W(x,\theta,t) \tag{4.35c}$$

Mid-plane strains and curvatures may also be used to express strain-displacement relations derived from Love's shell theory, as shown in [20]:

$$\varepsilon_{xi} = \frac{\partial U_i}{\partial x} - z\frac{\partial^2 W_i}{\partial x^2} \tag{4.36a}$$

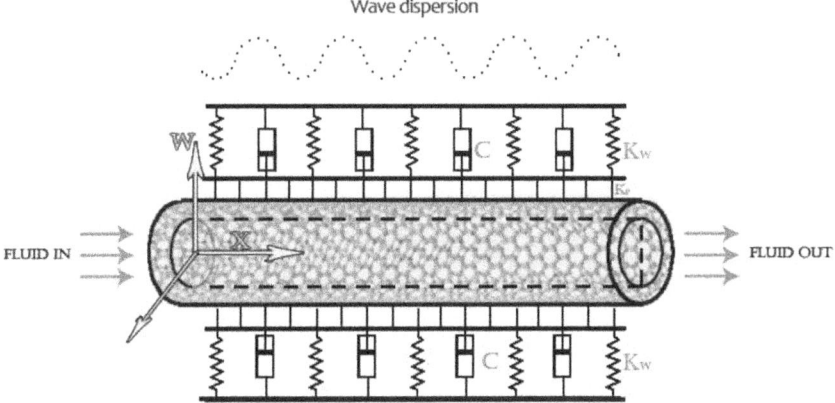

FIGURE 4.24 Schematic view of BNNT containing magnetic fluid resting on viscoelastic foundation.

$$\varepsilon_{\theta i} = \frac{1}{R_i}\left(\frac{\partial V_i}{\partial \theta} + W_i\right) - \frac{z}{R_i^2}\left(\frac{\partial^2 W_i}{\partial \theta^2} - \frac{\partial V_i}{\partial \theta}\right) \tag{4.36b}$$

$$\gamma_{x\theta i} = \frac{\partial V_i}{\partial x} + \frac{1}{R_i}\frac{\partial U_i}{\partial \theta} - \frac{z}{R_i}\left(\frac{2\partial^2 W_i}{\partial x\partial \theta} - \frac{\partial V_i}{\partial \theta}\right) \tag{4.36c}$$

Subscript i demonstrates layer number that for nanotube two layers is $i = \{1, 2\}$. Normal forces $\{N_{xi}, N_{\theta i}, N_{x\theta i}\}$ and bending moments $\{M_{xi}, M_{\theta i}, M_{x\theta i}\}$ can be defined as follows:

$$\{N_{xi}, N_{\theta i}, N_{x\theta i}\} = \int_{-h/2}^{h/2} \{\sigma_{xi}, \sigma_{\theta i}, \sigma_{x\theta i}\}\, dz \tag{4.37a}$$

$$\{M_{xi}, M_{\theta i}, M_{x\theta i}\} = \int_{-h/2}^{h/2} \{\sigma_{xi}, \sigma_{\theta i}, \sigma_{x\theta i}\}\, z\, dz \tag{4.37b}$$

The following equations are found by plugging in (4.33b–c) to (4.36c) into (4.30a,b):

$$N_{xi} - \mu^2\nabla^2 N_{xi} = A_{11}\frac{\partial U_i}{\partial x} + \frac{A_{12}}{R_i}\left(\frac{\partial V_i}{\partial \theta} + W_i\right) - l^2 A_{11}\left(\frac{1}{R_i^2}\frac{\partial^3 U_i}{\partial \theta^2 \partial x} + \frac{\partial^3 U_i}{\partial x^3}\right)$$
$$- l^2 A_{12}\left(\frac{1}{R_i^3}\frac{\partial^3 V_i}{\partial \theta^3} + \frac{1}{R_i}\frac{\partial^3 V_i}{\partial \theta \partial x^2} + \frac{1}{R_i^3}\frac{\partial^2 W_i}{\partial \theta^2} + \frac{1}{R_i}\frac{\partial^2 W_i}{\partial x^2}\right) + N_{Mxi} + N_{Exi} + N_{Txi} \tag{4.38a}$$

$$N_{\theta i} - \mu^2\nabla^2 N_{\theta i} = A_{12}\frac{\partial U_i}{\partial x} + \frac{A_{11}}{R_i}\left(\frac{\partial V_i}{\partial \theta} + W_i\right) - l^2 A_{12}\left(\frac{1}{R_i^2}\frac{\partial^3 U_i}{\partial \theta^2 \partial x} + \frac{\partial^3 U_i}{\partial x^3}\right)$$
$$- A_{11}l^2\left(\frac{1}{R_i^3}\frac{\partial^3 V_i}{\partial \theta^3} + \frac{1}{R_i}\frac{\partial^3 V_i}{\partial \theta \partial x^2} + \frac{1}{R_i^3}\frac{\partial^2 W_i}{\partial \theta^2} + \frac{1}{R_i}\frac{\partial^2 W_i}{\partial x^2}\right) \tag{4.38b}$$
$$+ N_{M\theta i} + N_{E\theta i} + N_{T\theta i}$$

$$N_{x\theta i} - e_0 a^2\nabla^2 N_{x\theta i} = A_{66}\left(\frac{\partial V_i}{\partial x} + \frac{1}{R_i}\frac{\partial U_i}{\partial \theta}\right)$$
$$- l^2 A_{66}\left(\frac{1}{R_i^2}\frac{\partial^3 V_i}{\partial x\partial \theta^2} + \frac{\partial^3 V_i}{\partial x^3} + \frac{1}{R_i^3}\frac{\partial^3 U_i}{\partial \theta^3} + \frac{1}{R_i}\frac{\partial^3 U_i}{\partial \theta \partial x^2}\right) \tag{4.38c}$$

$$M_{xi} - e_0 a^2 \nabla^2 M_{xi} = -D_{11} \frac{\partial^2 W_i}{\partial x^2} - \frac{D_{12}}{R_i^2} \left(\frac{\partial^2 W_i}{\partial \theta^2} - \frac{\partial V_i}{\partial \theta} \right) + E_{31} \Phi_i$$

$$+ l^2 D_{11} \left(\frac{1}{R_i^2} \frac{\partial^4 W_i}{\partial \theta^2 \partial x^2} + \frac{\partial^4 W_i}{\partial x^4} \right) + l^2 D_{12} \left(\frac{1}{R_i^2} \frac{\partial^4 W_i}{\partial x^2 \partial \theta^2} + \frac{1}{R_i^4} \frac{\partial^4 W_i}{\partial \theta^4} \right) \quad \text{(4.38d)}$$

$$- \frac{1}{R_i^2} \frac{\partial^3 V_i}{\partial \theta \partial x^2} - \frac{1}{R_i^4} \frac{\partial^3 V_i}{\partial \theta^3} \right) - l^2 E_{31} \left(\frac{1}{R_i^2} \frac{\partial^2 \Phi_i}{\partial \theta^2} + \frac{\partial^2 \Phi_i}{\partial x^2} \right)$$

$$M_{\theta i} - e_0 a^2 \nabla^2 M_{\theta i} = -D_{12} \frac{\partial^2 W_i}{\partial x^2} - \frac{D_{11}}{R_i^2} \left(\frac{\partial^2 W_i}{\partial \theta^2} - \frac{\partial V_i}{\partial \theta} \right) + E_{31} \Phi_i$$

$$+ l^2 D_{12} \left(\frac{1}{R_i^2} \frac{\partial^4 W_i}{\partial \theta^2 \partial x^2} + \frac{\partial^4 W_i}{\partial x^4} \right) + l^2 D_{11} \left(\frac{1}{R_i^2} \frac{\partial^4 W_i}{\partial x^2 \partial \theta^2} + \frac{1}{R_i^4} \frac{\partial^4 W_i}{\partial \theta^4} \right) \quad \text{(4.38e)}$$

$$- \frac{1}{R_i^2} \frac{\partial^3 V_i}{\partial \theta \partial x^2} - \frac{1}{R_i^4} \frac{\partial^3 V_i}{\partial \theta^3} \right) - l^2 E_{31} \left(\frac{1}{R_i^2} \frac{\partial^2 \Phi_i}{\partial \theta^2} + \frac{\partial^2 \Phi_i}{\partial x^2} \right)$$

$$M_{x\theta i} - e_0 a^2 \nabla^2 M_{x\theta i} = -\frac{D_{66}}{R_i} \left(\frac{2}{R_i} \frac{\partial^2 W_i}{\partial \theta \partial x} - \frac{\partial V_i}{\partial x} \right)$$

$$+ l^2 D_{66} \left(\frac{2}{R_i^2} \frac{\partial^4 W_i}{\partial \theta \partial x^3} + \frac{2}{R_i^4} \frac{\partial^4 W_i}{\partial \theta^3 \partial x} - \frac{1}{R_i^2} \frac{\partial^3 V_i}{\partial x^3} - \frac{1}{R_i^2} \frac{\partial^3 W_i}{\partial x \partial \theta^2} \right) \quad \text{(4.38f)}$$

$$\int_{-h/2}^{h/2} \cos(\beta z) \left[D_{xi} - e_0 a^2 \nabla^2 D_{xi} \right] dz = X_{11} \frac{\partial \Phi_i}{\partial x} - l^2 X_{11} \left(\frac{1}{R_i^2} \frac{\partial^3 \Phi_i}{\partial x \partial \theta^2} + \frac{\partial^3 \Phi_i}{\partial x^3} \right) \quad \text{(4.38g)}$$

$$\int_{-h/2}^{h/2} \frac{\cos(\beta z)}{R_i + z} \left[D_{\theta i} - e_0 a^2 \nabla^2 D_{\theta i} \right] dz = X_{22} \frac{\partial \Phi_i}{\partial \theta} - l^2 X_{22} \left(\frac{1}{R_i^2} \frac{\partial^3 \Phi_i}{\partial \theta^3} + \frac{\partial^3 \Phi_i}{\partial \theta \partial x^2} \right) \quad \text{(4.38h)}$$

$$\int_{-h/2}^{h/2} \beta \sin(\beta z) \left[D_{zi} - e_0 a^2 \nabla^2 D_{zi} \right] dz = -E_{31} \frac{\partial^2 W_i}{\partial x^2} - \frac{E_{31}}{R_i^2} \left(\frac{\partial^2 W_i}{\partial \theta^2} - \frac{\partial V_i}{\partial \theta} \right)$$

$$- X_{33} \Phi_i + l^2 E_{31} \left(\frac{\partial^4 W_i}{\partial x^4} + \frac{1}{R_i^2} \frac{\partial^4 W_i}{\partial x^2 \partial \theta^2} \right) + E_{31} \left(\frac{1}{R_i^4} \frac{\partial^4 W_i}{\partial \theta^4} + \frac{1}{R_i^2} \frac{\partial^2 W_i}{\partial x^2 \partial \theta^2} \right) \quad \text{(4.38i)}$$

$$- \frac{1}{R_i^4} \frac{\partial^3 V_i}{\partial \theta^3} - \frac{1}{R_i^2} \frac{\partial^3 V_i}{\partial x^2 \partial \theta} \right) + l^2 X_{33} \left(\frac{1}{R_i^2} \frac{\partial^2 \Phi_i}{\partial \theta^2} + \frac{\partial^2 \Phi_i}{\partial x^2} \right)$$

Hamilton's principle expresses that total variational derivation of potential and kinetic energies of device and work done by nonpotential forces in time interval $\Delta t = t_2 - t_1$ is zero [21].

$$\int_{t_1}^{t_2} \delta \left(\Pi_{ki} - \Pi_{Fi} + \Pi_{Si} \right) dt = 0 \qquad (4.39)$$

where Π_{Ki} is kinetic energy, Π_{Fi} is work done by external force, and Π_{si} is total strain energy.

Kinetic energy of nanotube is defined as:

$$\Pi_K = \frac{1}{2} \int_0^L \int_0^{2\pi} \left[I_1 \left(\frac{\partial U_i}{\partial t} \right)^2 + I_1 \left(\frac{\partial V_i}{\partial t} \right)^2 + I_1 \left(\frac{\partial W_i}{\partial t} \right)^2 \right] R_i d\theta dx \qquad (4.40)$$

in which $I_1 = \rho h$. Total strain energy is sum of elastic strain energy Π_{s1} and electric strain energy Π_{s2}.

$$\Pi_s = \Pi_{s1} + \Pi_{s2} \qquad (4.41a)$$

$$\Pi_{S1} = \frac{1}{2} \int_0^L \int_0^{2\pi} \left[N_x \frac{\partial U_i}{\partial x} + \frac{N_\theta}{R_i} \left(\frac{\partial V_i}{\partial \theta} + W \right)_i + N_{x\theta} \left(\frac{\partial V_i}{\partial x} + \frac{1}{R_i} \frac{\partial U_i}{\partial \theta} \right) \right] R_i d\theta dx$$

$$- \frac{1}{2} \int_0^L \int_0^{2\pi} \left[M_x \frac{\partial^2 W_i}{\partial x^2} + \frac{M_\theta}{R_i^2} \left(\frac{\partial^2 W_i}{\partial \theta^2} - \frac{\partial V_i}{\partial \theta} \right) + \frac{M_{x\theta}}{R_i} \left(\frac{2\partial^2 W_i}{\partial x \partial \theta} - \frac{\partial V_i}{\partial x} \right) \right] R_i d\theta dx$$

$$(4.41b)$$

$$\Pi_{S2} = -\frac{1}{2} \int_0^L \int_0^{2\pi} \int_{-h/2}^{h/2} \left[\begin{array}{c} D_{xi} \cos\left(\beta z \right) \dfrac{\partial \Phi_i}{\partial x} + D_{\theta i} \dfrac{\cos\left(\beta z \right)}{R_i + z} \dfrac{\partial \Phi_i}{\partial \theta} \\ -D_{zi} \left(\beta \sin\left(\beta z \right) \right) \Phi_i + \dfrac{2\phi_0}{h} \end{array} \right] R_i dz d\theta dx \qquad (4.41c)$$

The external force actually is the force of magnetic fluid, viscoelastic foundation, van der Waals force between two layers, and electric field on nanotube, which is calculated sum of equations Π_{F1}, Π_{F2}, Π_{F3}, Π_{F4}.

The work done by passing fluid through nanotube is presented as follows [20]:

$$\Pi_{F1} = \int_0^L \int_0^{2\pi} \left[-\frac{\rho_f R}{2} \left(\frac{\partial^2 W}{\partial t^2} + 2V_f VCF \frac{\partial^2 W}{\partial t \partial x} + \left(V_f VCF \right)^2 \frac{\partial^2 W}{\partial x^2} \right) \right] R d\theta dx \qquad (4.42)$$

where ρ_f is fluid density and *VCF* is the average velocity correction factor for fluid. Taking into account the Knudsen number for nanofluid can be obtained influences of small-scale fluid flow passing through nanostructure. *VCF* can be regarded as the average velocity correction factor for fluid by exerting nanoscale effects. The amount of correction factor in this research is defined as the ratio of fluid average velocity in slip boundary condition to fluid average velocity in no-slip boundary condition [21]:

$$VCF = \frac{V_{\text{slip}}}{V_{\text{no-slip}}} = \frac{1}{Cr(Kn)}\left(4\left(\frac{2-\sigma_v}{\sigma_v}\right)\left(\frac{Kn}{1-bKn}\right)+1\right) \qquad (4.43a)$$

$$Cr(Kn) = \frac{1}{1+\alpha Kn} \qquad (4.43b)$$

$$\alpha = \frac{2}{\pi}\alpha_0\left[\tan^{-1}\left(\alpha_1 Kn^B\right)\right] \qquad (4.43c)$$

$$\alpha_0 = 64/3\pi\left(1-(4/b)\right), \qquad (4.43d)$$

in which K_n is Knudsen number and σ_v is tangential momentum accommodation coefficient that is considered to be 0.7. In addition, $Cr(Kn)$ is the rarefaction coefficient and is defined as dynamic viscosity to total viscosity (bulk) ratio of fluid. In other words, fluid viscosity is defined as the rate of fluid passing through the nanotube in Knudsen layer, which is a function of the Knudsen number [21]. Depending on the fluid phase, Knudsen number for nano-liquids is 0.0001 to 0.001 and for nano-gases is 0.0001 to 2 [20]. Force of each layer to another is interpreted as van der Waals force between layers and is simulated in the form of spring force between two layers [20, 21]. If the force of layer i to layer $i+1$ is shown as $p_{i(i+1)}$, this force can be gained as follow:

$$p_{i(i+1)} = c\left(W_{i+1} - W_i\right) \qquad (4.44a)$$

where W_i is transverse displacement layer i. The reciprocal force of layer $i+1$ to layer i is obtained as [21]:

$$p_{(i+1)(i)} = -\frac{R_i}{R_{i+1}}p_{(i)(i+1)} \qquad (4.44b)$$

Hence, the force from the first layer to the second layer is acquired as follows:

$$p_1 = p_{12} = c\left(W_2 - W_1\right) \qquad (4.44c)$$

And the force from the second layer to the first layer can be written as follows:

$$p_{21} = -\frac{R_1}{R_2} p_{12} \tag{4.44d}$$

$$p_2 = p_{23} + p_{21} = p_{23} - \frac{R_1}{R_2} p_{12} = c(W_3 - W_2) - c\frac{R_1}{R_2}(W_2 - W_1) \tag{4.44e}$$

$$p_3 = p_{34} + p_{32} = 0 - \frac{R_2}{R_3} p_{23} = -c\frac{R_2}{R_3}(W_3 - W_2) \tag{4.44f}$$

where

$$c = \frac{320 \text{ erg} / \text{cm}^2}{0.16 d^2}, (d = 0.145 \text{ nm}, 1\text{erg} = 10^{-7} j) \tag{4.45}$$

Work done by van der Waals force between layers is calculated as:

$$\Pi_{F2}^{(i)} = \int_0^{2\pi}\int_0^L p_i R_i W_i dx d\theta, (i = 1,2,...,N) \tag{4.45}$$

Nanotube is resting on viscoelastic foundation. This foundation is simulated to springs and dampers that the nanotube is coupled with. The force from this foundation to nanotube is obtained as:

$$F_{medium} = K_w W_2 - K_G \frac{\partial^2 W_2}{\partial x^2} + C\frac{\partial W_2}{\partial t} \tag{4.46}$$

Work done by this foundation is obtained as follows [21]:

$$\Pi_{F3} = \int_0^{2\pi}\int_0^L \left(K_w W_2 - K_G \frac{\partial^2 W_2}{\partial x^2} + C\frac{\partial W_2}{\partial t} \right) R_3 dx d\theta \tag{4.47}$$

in which K_w, K_G, C are Winkler, Pasternak, and damping coefficient, respectively. Work done by electric, magnetic, and thermal fields on nanotube is computed as [20]:

$$\Pi_{F4} = \frac{1}{2}\int_0^L\int_0^{2\pi}\left[(N_{Ex} + N_{Mx} + N_{Tx})\left(\frac{\partial W_i}{\partial t}\right)^2 \right] R_i d\theta dx$$

$$+ \frac{1}{2}\int_0^L\int_0^{2\pi}\left[\frac{(N_{E\theta} + N_{M\theta} + N_{T\theta})}{R^2}\left(\frac{\partial W_i}{\partial \theta}\right)^2 \right] R_i d\theta dx \tag{4.48}$$

where $(N_{Ex}, N_{E\theta}, N_{T\theta})$, $(N_{Mx}, N_{M\theta}, N_{T\theta})$ are vertically induced forces by external electric potential ϕ_0, external magnetic potential ψ_0, and temperature changes ΔT in the x-θ direction, respectively, and is defined as follows:

$$N_{Tx} = N_{T\theta} = \tilde{\beta}_1 h \Delta T, \; N_{Ex} = N_{E\theta} = -2\tilde{e}_{31}\phi_0, \; N_{Mx} = N_{M\theta} = -2\tilde{q}_{31}\psi_0 \quad (4.49)$$

By substituting mentioned equations in Hamilton's principle, coefficient variation of each parameter is calculated as:

$$\delta U : \frac{\partial N_{xi}}{\partial x} + \frac{1}{R_i}\frac{\partial N_{x\theta i}}{\partial \theta} = I_1 \frac{\partial^2 U_i}{\partial t^2} \quad (4.50)$$

$$\delta V : \frac{\partial N_{x\theta i}}{\partial x} + \frac{1}{R_i}\frac{\partial N_{\theta i}}{\partial \theta} + \frac{1}{R_i}\frac{\partial M_{x\theta i}}{\partial x} + \frac{1}{R_i^2}\frac{\partial M_{\theta i}}{\partial \theta} = I_1 \frac{\partial^2 V_i}{\partial t^2} \quad (4.51)$$

$$\delta W : \frac{\partial M_{xi}}{\partial x^2} + \frac{2}{R_i}\frac{\partial^2 M_{x\theta i}}{\partial x \partial \theta} + \frac{1}{R_i^2}\frac{\partial^2 M_{\theta i}}{\partial \theta^2} - \frac{N_{\theta i}}{R_i} - N_{x0}\frac{\partial^2 W_i}{\partial x^2} - \frac{N_{\theta 0}}{R_i^2}\frac{\partial^2 W_i}{\partial \theta^2} - A$$

$$= I_1 \frac{\partial^2 W_i}{\partial t^2} \quad (4.52)$$

$$\delta\Phi : \int_{-h/2}^{h/2}\left[\frac{\partial D_{xi}}{\partial x}\cos(\beta z) + \frac{\cos(\beta z)}{R_i + z}\frac{\partial D_{\theta i}}{\partial \theta} + D_{zi}\beta\sin(\beta z)\right]dz = 0 \quad (4.53)$$

in these equations $N_{x0} = N_{Ex} + N_{Mx} + N_{Tx}$ and $N_{\theta 0} = N_{E\theta} + N_{M\theta} + N_{T\theta}$.
External force to nanotube has appeared as phrase A in equations:

$$A = -\frac{\rho_f R_1}{2}\left(\frac{\partial^2 W_1}{\partial t^2} + 2V_f VCF \frac{\partial^2 W_1}{\partial t \partial x} + (2V_f VCF)^2 \frac{\partial^2 W_1}{\partial x^2}\right)$$

$$+ P_i - K_w W_3 + K_g \nabla^2 W_3 - C\frac{\partial W_3}{\partial t} \quad (4.54)$$

By substituting relations for nanotube energies in Eqs. (4.50) to (4.53), the following governing equation can be obtained:

$$A_{11}\frac{\partial^2 U_i}{\partial x^2} + \frac{A_{12}}{R_i}\left(\frac{\partial^2 V_i}{\partial x \partial \theta} + \frac{\partial W}{\partial x}\right) - A_{11}\frac{l^2}{R_i^2}\frac{\partial^4 U_i}{\partial x^2 \partial \theta^2} - A_{11}l^2\frac{\partial^4 U_i}{\partial x^4}$$

$$- A_{12}\frac{l^2}{R_i^3}\frac{\partial^4 V_i}{\partial x \partial \theta^3} - A_{12}\frac{l^2}{R_i}\frac{\partial^4 V_i}{\partial x^3 \partial \theta} - A_{12}\frac{l^2}{R_i^3}\frac{\partial^3 W_i}{\partial x \partial \theta^2} - A_{12}\frac{l^2}{R_i}\frac{\partial^3 W_i}{\partial x^3}$$

$$+\frac{A_{66}}{R_i}\frac{\partial^2 V_i}{\partial x \partial\theta}+\frac{A_{66}}{R_i^2}\frac{\partial^2 U_i}{\partial\theta^2}-A_{66}\frac{l^2}{R_i^3}\frac{\partial^4 V_i}{\partial x\partial\theta^3}-A_{66}\frac{l^2}{R_i}\frac{\partial^4 V_i}{\partial x^3\partial\theta}-A_{66}\frac{l^2}{R_i^4}\frac{\partial^4 U_i}{\partial\theta^4}$$

$$-A_{66}\frac{l^2}{R_i^2}\frac{\partial^4 U_i}{\partial x^2\partial\theta^2}=\left\{1-e_0a^2\nabla^2\right\}\left(I_1\frac{\partial^2 U_i}{\partial t^2}\right) \tag{4.55}$$

$$A_{66}\frac{\partial^2 V_i}{\partial x^2}+\frac{A_{66}}{R_i}\frac{\partial^2 U_i}{\partial x\partial\theta}-A_{66}\frac{l^2}{R_i^2}\frac{\partial^4 V_i}{\partial x^2\partial\theta^2}-A_{66}l^2\frac{\partial^4 V_i}{\partial x^4}-A_{66}\frac{l^2}{R_i^3}\frac{\partial^4 U_i}{\partial x\partial\theta^3}$$

$$-A_{66}\frac{l^2}{R_i}\frac{\partial^4 U_i}{\partial x^3\partial\theta}+\frac{A_{12}}{R_i}\frac{\partial^2 U_i}{\partial x\partial\theta}+\frac{A_{11}}{R_i^2}\frac{\partial^2 V_i}{\partial\theta^2}+\frac{A_{11}}{R_i^2}\frac{\partial W_i}{\partial\theta}-A_{12}\frac{l^2}{R_i^3}\frac{\partial^4 U_i}{\partial x\partial\theta^3}$$

$$-A_{12}\frac{l^2}{R_i}\frac{\partial^4 U_i}{\partial x^3\partial\theta}-A_{11}\frac{l^2}{R_i^4}\frac{\partial^4 V_i}{\partial\theta^4}-A_{11}\frac{l^2}{R_i^2}\frac{\partial^4 V_i}{\partial x^2\partial\theta^2}-A_{11}\frac{l^2}{R_i^4}\frac{\partial^3 W_i}{\partial\theta^3}$$

$$-A_{11}\frac{l^2}{R_i^2}\frac{\partial^3 W_i}{\partial x^2\partial\theta}-\frac{D_{66}}{R_i^2}\frac{2\partial^3 W_i}{\partial x^2\partial\theta}+\frac{D_{66}}{R_i^2}\frac{\partial^2 V_i}{\partial x^2}+2D_{66}\frac{l^2}{R_i^4}\frac{\partial^5 W_i}{\partial x^2\partial\theta^3}$$

$$+2D_{66}\frac{l^2}{R_i^2}\frac{\partial^5 W_i}{\partial x^4\partial\theta}-D_{66}\frac{l^2}{R_i^4}\frac{\partial^4 V_i}{\partial x^2\partial\theta^2}-D_{66}\frac{l^2}{R_i^2}\frac{\partial^4 V_i}{\partial x^4}-\frac{D_{12}}{R_i^2}\frac{\partial^3 W_i}{\partial x^2\partial\theta} \tag{4.56}$$

$$-\frac{D_{11}}{R_i^4}\frac{\partial^3 W_i}{\partial\theta^3}+\frac{D_{11}}{R_i^4}\frac{\partial^2 V_i}{\partial\theta^2}+\frac{E_{31}}{R_i^2}\frac{\partial\Phi_i}{\partial\theta}+D_{12}\frac{l^2}{R_i^4}\frac{\partial^5 W_i}{\partial x^2\partial\theta^3}+D_{12}\frac{l^2}{R_i^2}\frac{\partial^5 W_i}{\partial x^4\partial\theta}$$

$$+D_{11}\frac{l^2}{R_i^6}\frac{\partial^5 W_i}{\partial\theta^5}+D_{11}\frac{l^2}{R_i^4}\frac{\partial^5 W_i}{\partial x^2\partial\theta^3}-D_{11}\frac{l^2}{R_i^6}\frac{\partial^4 V_i}{\partial\theta^4}-D_{11}\frac{l^2}{R_i^4}\frac{\partial^4 V_i}{\partial x^2\partial\theta^2}$$

$$-E_{31}\frac{l^2}{R_i^4}\frac{\partial^3\Phi_i}{\partial\theta^3}-E_{31}\frac{l^2}{R_i^2}\frac{\partial^3\Phi_i}{\partial x^2\partial\theta}=\left\{1-e_0a^2\nabla^2\right\}\left(I_1\frac{\partial^2 V_i}{\partial t^2}\right)$$

$$-D_{11}\frac{\partial^4 W_i}{\partial x^4}-\frac{D_{12}}{R^2}\frac{\partial^4 W_i}{\partial x^2\partial\theta^2}+\frac{D_{12}}{R^2}\frac{\partial^3 V_i}{\partial x^2\partial\theta}+E_{31}\frac{\partial^2\Phi_i}{\partial x^2}+D_{11}\frac{l^2}{R^2}\frac{\partial^6 W_i}{\partial x^4\partial\theta^2}+D_{11}l^2\frac{\partial^6 W_i}{\partial x^6}$$

$$+D_{12}\frac{l^2}{R^4}\frac{\partial^6 W_i}{\partial x^2\partial\theta^4}+D_{12}\frac{l^2}{R^2}\frac{\partial^6 W_i}{\partial x^4\partial\theta^2}-D_{12}\frac{l^2}{R^4}\frac{\partial^5 V_i}{\partial x^2\partial\theta^3}-D_{12}\frac{l^2}{R^2}\frac{\partial^5 V_i}{\partial x^4\partial\theta}$$

$$+E_{31}\frac{l^2}{R^2}\frac{\partial^4\Phi_i}{\partial x^2\partial\theta^2}+E_{31}l^2\frac{\partial^4\Phi_i}{\partial x^4}-\frac{4D_{66}}{R^2}\frac{\partial^4 W_i}{\partial x^2\partial\theta^2}+\frac{2D_{66}}{R^2}\frac{\partial^3 V_i}{\partial x^2\partial\theta}+4D_{66}\frac{l^2}{R^4}\frac{\partial^6 W_i}{\partial x^2\partial\theta^4}$$

$$+4D_{66}\frac{l^2}{R^2}\frac{\partial^6 W_i}{\partial x^4\partial\theta^2}-2D_{66}\frac{l^2}{R^4}\frac{\partial^5 V_i}{\partial x^2\partial\theta^3}-2D_{66}\frac{l^2}{R^2}\frac{\partial^5 V_i}{\partial x^4\partial\theta}-\frac{D_{12}}{R^2}\frac{\partial^4 W_i}{\partial x^2\partial\theta^2}$$

$$-\frac{D_{11}}{R^4}\frac{\partial^4 W_i}{\partial\theta^4}+\frac{D_{11}}{R^4}\frac{\partial^3 V_i}{\partial\theta^3}+\frac{E_{31}}{R^2}\frac{\partial^2\Phi_i}{\partial\theta^2}-\frac{A_{12}}{R}\frac{\partial U_i}{\partial x}-\frac{A_{11}}{R^2}\frac{\partial V_i}{\partial\theta}-\frac{A_{11}}{R^2}W_i+A_{12}\frac{l^2}{R^3}\frac{\partial^3 U_i}{\partial x\partial\theta^2}$$

$$+A_{12}\frac{l^2}{R}\frac{\partial^3 U_i}{\partial x^3}+A_{11}\frac{l^2}{R^4}\frac{\partial^3 V_i}{\partial \theta^3}+A_{11}\frac{l^2}{R^2}\frac{\partial^3 V_i}{\partial x^2 \partial \theta}+A_{11}\frac{l^2}{R^4}\frac{\partial^2 W_i}{\partial \theta^2}+A_{11}\frac{l^2}{R^2}\frac{\partial^2 W_i}{\partial x^2}$$

$$-\left\{1-e_0 a^2 \nabla^2\right\}\left(\rho_f\frac{R}{2}\left(\frac{\partial^2 W_1}{\partial t^2}+2V_{nf}VCF\frac{\partial^2 W_1}{\partial t \partial x}+\left(V_{nf}VCF\right)^2\frac{\partial^2 W_1}{\partial x^2}\right)\right.$$

$$\left.+\sigma\left(B_x^2\right)\left(\frac{\partial W_1}{\partial t}+V_{nf}VCF\frac{\partial W_1}{\partial x}\right)\right)+p_i-K_w W_3+K_G\nabla^2 W_3-C\frac{\partial W_3}{\partial t}\right)$$

$$=\left\{1-e_0 a^2 \nabla^2\right\}\left(\frac{N_{\theta 0}}{R^2}\frac{\partial^2 W_i}{\partial \theta^2}+N_{x0}\frac{\partial^2 W_i}{\partial x^2}+I_1\frac{\partial^2 W_i}{\partial t^2}\right)$$

(4.57)

$$X_{11}\frac{\partial^2 \Phi_i}{\partial x^2}-X_{11}\frac{l^2}{R^2}\frac{\partial^4 \Phi_i}{\partial x^2 \partial \theta^2}-X_{11}l^2\frac{\partial^4 \Phi_i}{\partial x^4}+X_{22}\frac{\partial^2 \Phi_i}{\partial \theta^2}-X_{22}\frac{l^2}{R^2}\frac{\partial^4 \Phi_i}{\partial \theta^4}-X_{22}l^2\frac{\partial^4 \Phi_i}{\partial x^2 \partial \theta^2}$$

$$-E_{31}\frac{\partial^2 W_i}{\partial x^2}-\frac{E_{31}}{R^2}\frac{\partial^2 W_i}{\partial \theta^2}+\frac{E_{31}}{R^2}\frac{\partial V_i}{\partial \theta}-X_{33}\Phi_i+E_{31}\frac{l^2}{R^2}\frac{\partial^4 W_i}{\partial x^2 \partial \theta^2}+E_{31}\frac{l^2}{R^4}\frac{\partial^4 W_i}{\partial \theta^4}+$$

$$E_{31}\frac{l^2}{R^2}\frac{\partial^4 W_i}{\partial x^2 \partial \theta^2}-E_{31}\frac{l^2}{R^4}\frac{\partial^3 V_i}{\partial \theta^3}-E_{31}\frac{l^2}{R^2}\frac{\partial^3 V_i}{\partial x^2 \partial \theta}+X_{33}\frac{l^2}{R^2}\frac{\partial^2 \Phi_i}{\partial \theta^2}+X_{33}l^2\frac{\partial^2 \Phi_i}{\partial x^2}=0$$

(4.58)

By applying the Fourier transform, the general wave equation, which is a differential equation of the second or fourth order, may be reduced to a series of first-order differential equations with constant coefficients. This simplification is possible because the general wave equation is a differential equation. Response of this first-order differential equation is computed as shape $A_n e^{ik_n x}$. A_n was acquired as a result of the border quantities. Response, which has been regarded exponentially for the solution of differential equations (61) to (64), has been analyzed. Therefore, in place of U, V, W, and Φ, replace them with the following phrases:

$$U_i=\overline{U_i}e^{i\left(kx+n\theta-\omega t\right)}$$

(4.59a)

$$V_i=\overline{V_i}e^{i\left(kx+n\theta-\omega t\right)}$$

(4.59b)

$$W_i=\overline{W_i}e^{i\left(kx+n\theta-\omega t\right)}$$

(4.59c)

$$\Phi_i=\overline{\Phi_i}e^{i\left(kx+n\theta-\omega t\right)}$$

(4.59d)

After simplifying the separates for each matrix layer, the following relations were found:

$$
\left[K + C\omega - M\omega^2 \right]
\begin{bmatrix}
U_1 \\
V_1 \\
W_1 \\
\Phi_1 \\
U_2 \\
V_2 \\
W_2 \\
\Phi_2 \\
U_3 \\
V_3 \\
W_3 \\
\Phi_3
\end{bmatrix} = 0
\tag{4.60}
$$

where the matrix of each parameter K, C, and M is defined as follows [20, 21]:

$$
K =
\begin{bmatrix}
K_{11}^{(1)} & K_{12}^{(1)} & K_{13}^{(1)} & K_{14}^{(1)} & K_{15}^{(1)} & K_{16}^{(1)} & K_{17}^{(1)} & K_{18}^{(1)} & K_{19}^{(1)} & K_{110}^{(1)} & K_{111}^{(1)} & K_{112}^{(1)} \\
K_{21}^{(1)} & K_{22}^{(1)} & K_{23}^{(1)} & K_{24}^{(1)} & K_{25}^{(1)} & K_{26}^{(1)} & K_{27}^{(1)} & K_{28}^{(1)} & K_{29}^{(1)} & K_{210}^{(1)} & K_{211}^{(1)} & K_{212}^{(1)} \\
K_{31}^{(1)} & K_{32}^{(1)} & K_{33}^{(1)} & K_{34}^{(1)} & K_{35}^{(1)} & K_{36}^{(1)} & K_{37}^{(1)} & K_{38}^{(1)} & K_{39}^{(1)} & K_{310}^{(1)} & K_{311}^{(1)} & K_{312}^{(1)} \\
K_{41}^{(1)} & K_{42}^{(1)} & K_{43}^{(1)} & K_{44}^{(1)} & K_{45}^{(1)} & K_{46}^{(1)} & K_{47}^{(1)} & K_{48}^{(1)} & K_{49}^{(1)} & K_{410}^{(1)} & K_{411}^{(1)} & K_{412}^{(1)} \\
K_{11}^{(2)} & K_{12}^{(2)} & K_{13}^{(2)} & K_{14}^{(2)} & K_{15}^{(2)} & K_{16}^{(2)} & K_{17}^{(2)} & K_{18}^{(2)} & K_{19}^{(2)} & K_{110}^{(2)} & K_{111}^{(2)} & K_{112}^{(2)} \\
K_{21}^{(2)} & K_{22}^{(2)} & K_{23}^{(2)} & K_{24}^{(2)} & K_{25}^{(2)} & K_{26}^{(2)} & K_{27}^{(2)} & K_{28}^{(2)} & K_{29}^{(2)} & K_{210}^{(2)} & K_{211}^{(2)} & K_{212}^{(2)} \\
K_{31}^{(2)} & K_{32}^{(2)} & K_{33}^{(2)} & K_{34}^{(2)} & K_{35}^{(2)} & K_{36}^{(2)} & K_{37}^{(2)} & K_{38}^{(2)} & K_{39}^{(2)} & K_{310}^{(2)} & K_{311}^{(2)} & K_{312}^{(2)} \\
K_{41}^{(2)} & K_{42}^{(2)} & K_{43}^{(2)} & K_{44}^{(2)} & K_{45}^{(2)} & K_{46}^{(2)} & K_{47}^{(2)} & K_{48}^{(2)} & K_{49}^{(2)} & K_{410}^{(2)} & K_{411}^{(2)} & K_{412}^{(2)} \\
K_{11}^{(3)} & K_{13}^{(3)} & K_{13}^{(3)} & K_{14}^{(3)} & K_{15}^{(3)} & K_{16}^{(3)} & K_{17}^{(3)} & K_{18}^{(3)} & K_{19}^{(3)} & K_{110}^{(3)} & K_{111}^{(3)} & K_{112}^{(3)} \\
K_{21}^{(3)} & K_{22}^{(3)} & K_{23}^{(3)} & K_{24}^{(3)} & K_{25}^{(3)} & K_{26}^{(3)} & K_{27}^{(3)} & K_{28}^{(3)} & K_{29}^{(3)} & K_{210}^{(3)} & K_{211}^{(3)} & K_{212}^{(3)} \\
K_{31}^{(3)} & K_{32}^{(3)} & K_{33}^{(3)} & K_{34}^{(3)} & K_{35}^{(3)} & K_{36}^{(3)} & K_{37}^{(3)} & K_{38}^{(3)} & K_{39}^{(3)} & K_{310}^{(3)} & K_{311}^{(3)} & K_{312}^{(3)} \\
K_{41}^{(3)} & K_{42}^{(3)} & K_{43}^{(3)} & K_{44}^{(3)} & K_{45}^{(3)} & K_{46}^{(3)} & K_{47}^{(3)} & K_{48}^{(3)} & K_{49}^{(3)} & K_{410}^{(3)} & K_{411}^{(3)} & K_{412}^{(3)}
\end{bmatrix}
$$

$$
\tag{4.61}
$$

$$
M = \begin{bmatrix}
M_{11}^{(1)} & 0 & 0 & 0 & 0 & 0 & 0 & 0 & 0 & 0 & 0 & 0 \\
0 & M_{22}^{(1)} & 0 & 0 & 0 & 0 & 0 & 0 & 0 & 0 & 0 & 0 \\
0 & 0 & M_{33}^{(1)} & 0 & 0 & 0 & 0 & 0 & 0 & 0 & 0 & 0 \\
0 & 0 & 0 & M_{44}^{(1)} & 0 & 0 & 0 & 0 & 0 & 0 & 0 & 0 \\
0 & 0 & 0 & 0 & M_{15}^{(2)} & 0 & 0 & 0 & 0 & 0 & 0 & 0 \\
0 & 0 & 0 & 0 & 0 & M_{26}^{(2)} & 0 & 0 & 0 & 0 & 0 & 0 \\
0 & 0 & 0 & 0 & 0 & 0 & M_{37}^{(2)} & 0 & 0 & 0 & 0 & 0 \\
0 & 0 & 0 & 0 & 0 & 0 & 0 & M_{48}^{(2)} & 0 & 0 & 0 & 0 \\
0 & 0 & 0 & 0 & 0 & 0 & 0 & 0 & M_{19}^{(3)} & 0 & 0 & 0 \\
0 & 0 & 0 & 0 & 0 & 0 & 0 & 0 & 0 & M_{210}^{(3)} & 0 & 0 \\
0 & 0 & 0 & 0 & 0 & 0 & 0 & 0 & 0 & 0 & M_{311}^{(3)} & 0 \\
0 & 0 & 0 & 0 & 0 & 0 & 0 & 0 & 0 & 0 & 0 & M_{412}^{(3)}
\end{bmatrix}
$$

$$
C = \begin{bmatrix}
0 & 0 & 0 & 0 & 0 & 0 & 0 & 0 & 0 & 0 & 0 & 0 \\
0 & 0 & 0 & 0 & 0 & 0 & 0 & 0 & 0 & 0 & 0 & 0 \\
0 & 0 & C_{33}^{(1)} & 0 & 0 & 0 & 0 & 0 & 0 & 0 & 0 & 0 \\
0 & 0 & 0 & 0 & 0 & 0 & 0 & 0 & 0 & 0 & 0 & 0 \\
0 & 0 & 0 & 0 & 0 & 0 & 0 & 0 & 0 & 0 & 0 & 0 \\
0 & 0 & 0 & 0 & 0 & 0 & 0 & 0 & 0 & 0 & 0 & 0 \\
0 & 0 & 0 & 0 & 0 & 0 & C_{37}^{(2)} & 0 & 0 & 0 & 0 & 0 \\
0 & 0 & 0 & 0 & 0 & 0 & 0 & 0 & 0 & 0 & 0 & 0 \\
0 & 0 & 0 & 0 & 0 & 0 & 0 & 0 & 0 & 0 & 0 & 0 \\
0 & 0 & 0 & 0 & 0 & 0 & 0 & 0 & 0 & 0 & 0 & 0 \\
0 & 0 & 0 & 0 & 0 & 0 & 0 & 0 & 0 & 0 & C_{311}^{(3)} & 0 \\
0 & 0 & 0 & 0 & 0 & 0 & 0 & 0 & 0 & 0 & 0 & 0
\end{bmatrix} \tag{4.62}
$$

Create a matrix for a three-layer nanotube by montaging two separate matrices, one for each layer. The coefficient matrix is constructed by plugging in (4.60)–(4.62) into (4.59), and the matrix members are shown in Appendix A. The link between the natural frequency and the wave speed may be determined by solving the nontrivial equation:

$$
c = \frac{\omega}{k} \tag{4.63}
$$

Using nonlocal elastic cylindrical shell theory, Hu et al. [31] studied the dispersion of transverse and torsional waves in single- and double-walled carbon nanotubes. Molecular dynamics confirmed their findings. The findings of this study are fully consistent with those of Hu's inquiry. Dimensionless frequency versus wave number is shown in Figure 4.25 for both the current investigation and the cited study [31]. The findings of local and nonlocal theories are consistent down to low wave numbers, as shown in Figure 4.25, but beyond this wave number, the quantity of dimensionless frequency for local theory with $e_0 = 0.6$ is less than the result of nonlocal theory. The obtained results by nonlocal theory are the same as obtained results by molecular dynamics. Dimensionless frequency for drawing this diagram as $\Omega = \omega L \sqrt{\dfrac{I_1}{A_{11}}}$ is regarded to be one in which Ω is dimensionless frequency. Geometry characteristics of BNNT are considered as [21]:

$$R_1 = 11.43 \text{ nm}, R_2 = 12.31 \text{ nm},$$
$$R_3 = 13.2 \text{ nm}, L/R_1 = 10,$$
$$h = 0.075 \text{ nm},$$
$$\rho = 3.4870 \text{ g/cm}^3$$

Figure 4.26 shows the phase velocity versus wave number variation of TWBNNT for various modes. Phase velocity achieves its highest value in the first mode, as predicted. Phase velocity peaks at a certain wave number and then decreases as the wave number rises. When the wave number is 0.01 (1/nm), the higher mode

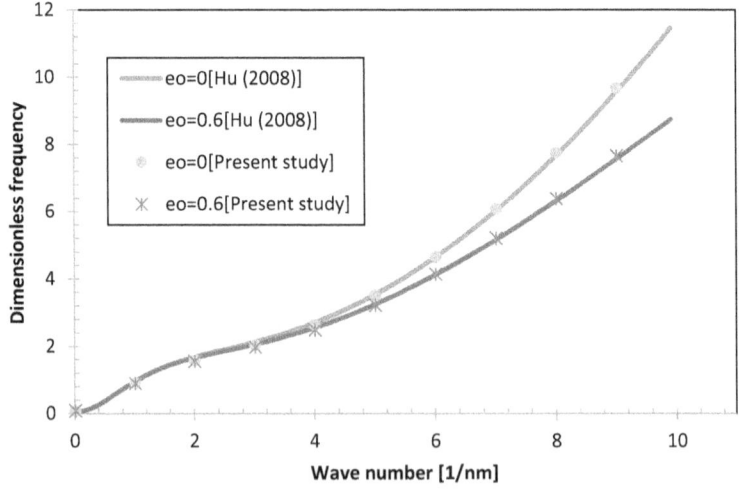

FIGURE 4.25 Diagram of dimensionless frequency versus wave number compared with reference.

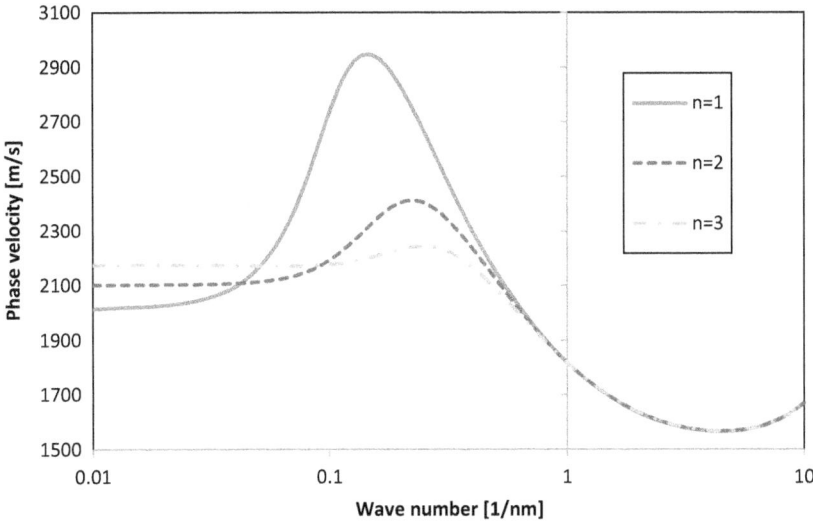

FIGURE 4.26   Diagram of phase velocity versus wave number of TWBNNT for different modes.

TABLE 4.2
Variation of Phase Velocity versus Wave Number of TWBNNT for Various Fluid Velocities

Fluid Velocity (m/s)	Wave Number (1/nm)	Phase Velocity (m/s)	Wave Frequency (THz)
100	0.1	360.307	0.0360307
100	1	848.65	0.84865
100	10	839.974	8.39974
500	0.1	1578.61	0.157861
500	1	1265.44	1.26544
500	10	1214.71	12.1471
1000	0.1	2731.4	0.27314
1000	1	1819.07	1.81907
1000	10	1668.96	16.6896

has a greater phase velocity. Charting the quantity after wave 1(1/nm) yields the same result [21].

The phase velocity versus wave number of TWBNNT varies for different fluid velocities, as shown in Table 4.2. To get the right conclusion, additional data points are needed to show how the frequency grows and the phase velocity varies with wave number in a fluid moving at a constant velocity. Additionally, at a fixed wave number, an increase in fluid velocity results in a rise in both phase velocity and frequency [21].

The influence of fluid velocity on the phase velocity versus wave number of TWBNNT is seen in Figure 4.27. More points are used to draw the diagrams, which start with an upward trend, then discover a downward trend, then level out at a constant value. As can be seen, as the fluid velocity increases, so does the phase velocity [21].

In Figure 4.28, we see how the density of the fluid affects the variations in phase velocity versus wave number of TWBNNT for different fluid velocities.

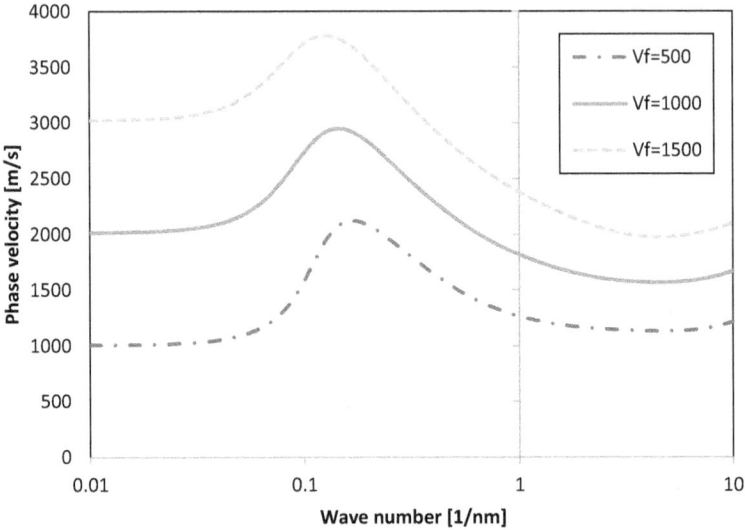

FIGURE 4.27   Diagram of phase velocity versus wave number of TWBNNT for different fluid velocities.

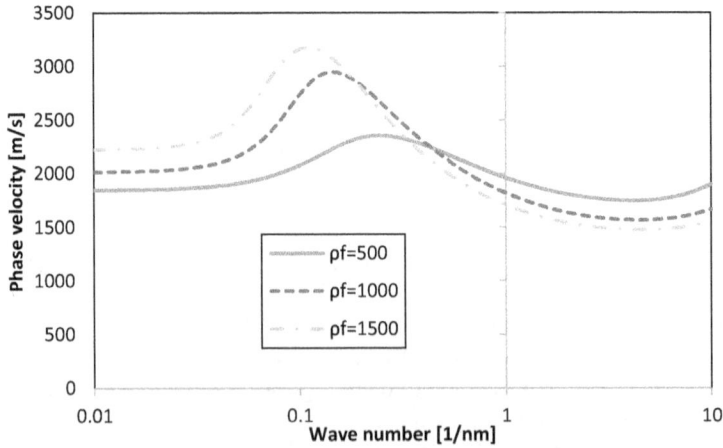

FIGURE 4.28   Diagram of phase velocity versus wave number of TWBNNT for different fluid densities.

The upward trend is clearly seen to be followed by a downward trend. As fluid density increases, a decreasing wave number gives rise to a larger peak in the figure. Phase velocity rises with increasing fluid density for wave numbers below 10^8 (1/m), but this pattern reverses beyond this wave number [21].

The relationship between the phase velocity and the wave number of TWBNNT is shown for a range of Knudsen values in Figure 4.29. Phase velocities for various Knudsen values in this figure are almost equal to one another [21], with just small deviations.

The difference becomes more noticeable with lower wave numbers, as seen in the enlarged version of Figure 4.30. Figure 4.30 shows that for every unit increase in the Knudsen number, the phase velocity rises by about 10 meters per second [21].

We shall separate out the influence of the Winkler, Pasternak, and damping coefficients on the viscoelastic basis. In Figure 4.31, we see how the Winkler coefficient affects the plot of TWBNNT phase velocity versus wave number. As the Winkler coefficient increases, the phase velocity slows down for wave numbers lower than 10^9 (1/m). This graphic starts out with a rising slope that eventually turns into a falling one. In addition, a damping condition is produced by an increase in the Winkler coefficient, which reduces the peak, ascending, and descending slope amounts. The Winkler diagram becomes more linear as the Winkler coefficient increases. Additionally, the phase velocity is insensitive to changes in the Winkler coefficient for larger wave numbers [21].

Figure 4.32 depicts the relationship between the phase velocity and wave number of TWBNNT as affected by the Pasternak coefficient. At first, the diagram shows an upward trend, but it eventually descends until it stabilizes at a value below its starting point. As the Pasternak coefficient increases, it is clear that the phase velocity grows. All graphs converge on the same final value [21].

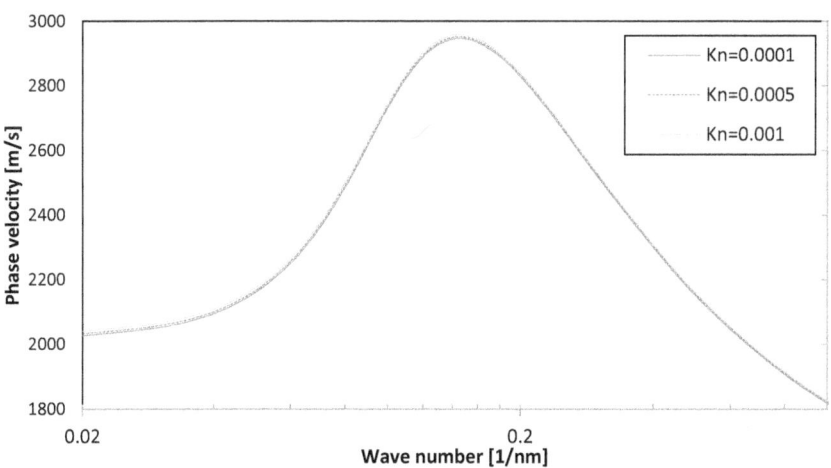

FIGURE 4.29   Diagram of phase velocity versus wave number of TWBNNT for different Knudsen numbers.

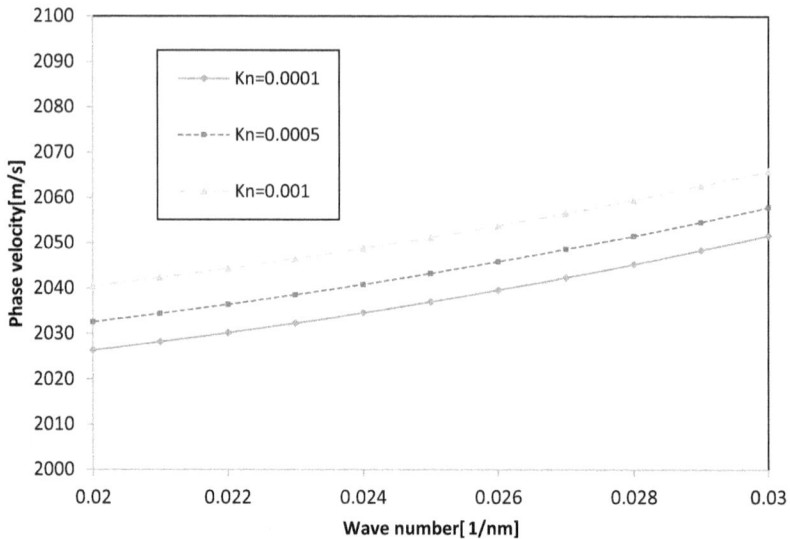

FIGURE 4.30 Magnified diagram of phase velocity versus wave number of TWBNNT for different Knudsen numbers.

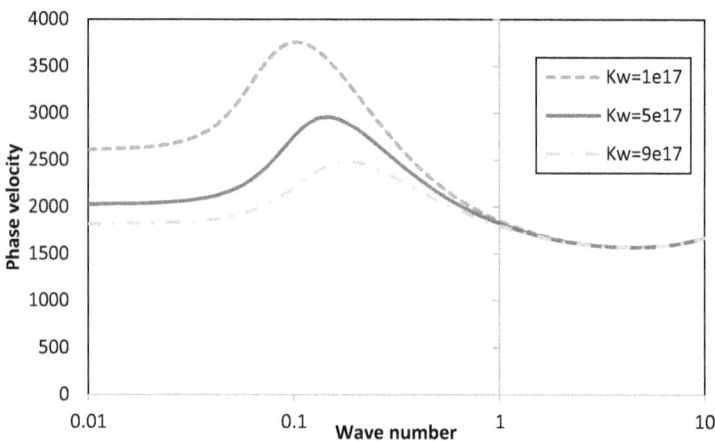

FIGURE 4.31 Diagram of phase velocity versus wave number of TWBNNT for different Winkler coefficients.

TWBNNT phase velocity versus wave number as a function of damping coefficient is shown for different values in Figure 4.33. The range of possible wave number modifications is between 0.05(1/nm) and 10(1/nm). If we split this range into three parts, the first is between 0.05 and 0.5 (1/nm), where we see a drop in phase velocity and the peak of the diagram as the damping coefficient increases. Phase velocity rises and the resulting graphical difference decreases when the

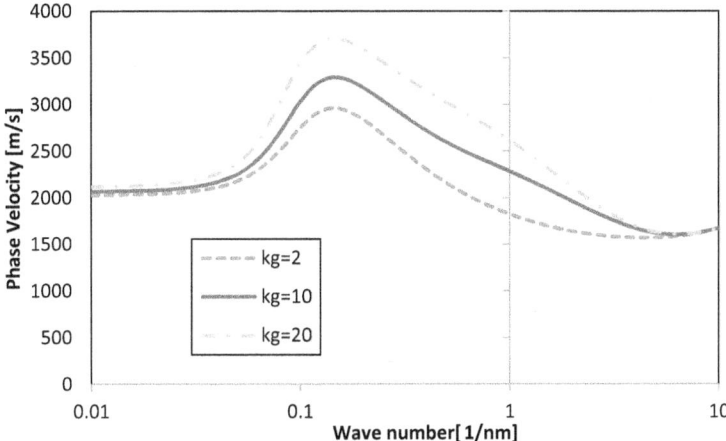

FIGURE 4.32 Diagram of phase velocity versus wave number of TWBNNT for different Pasternak coefficients.

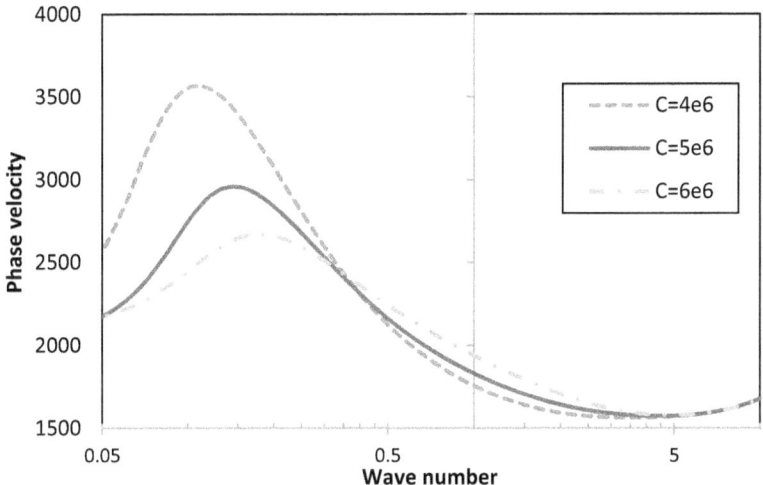

FIGURE 4.33 Diagram of phase velocity versus wave number of TWBNNT for different damping coefficients.

damping coefficient is increased across the second range of 0.5 to 5 (1/nm). Each damping coefficient's phase velocity becomes the same in the final range [21].

Figures 4.34 and 4.35 depict the effects of a magnetic field and an electric field, respectively, as physical fields on the phase velocity versus wave number of TWBNNT. In lower wave numbers, the two diagrams are similar, but in larger wave numbers, the phase velocity rises with increasing magnetic field strength (as seen in Figure 4.34). Diagrams for varying electric field strengths show just a little

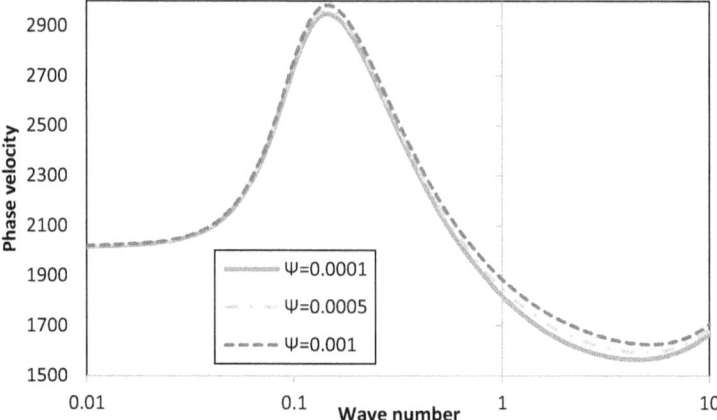

FIGURE 4.34   Diagram of phase velocity versus wave number of TWBNNT for different magnetic field intensities.

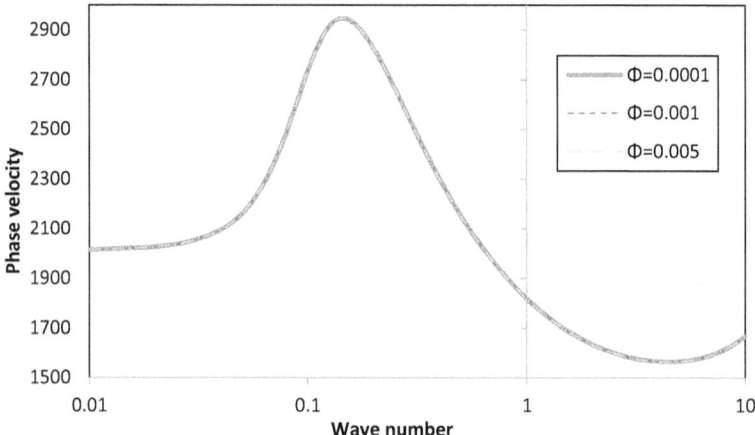

FIGURE 4.35   Diagram of phase velocity versus wave number of TWBNNT for different electric field intensities.

variation. Therefore, numerous numbers are given in Table 4.3 [21] to demonstrate this discrepancy.

Phase velocity versus wave number for TWBNNT was shown to change at various electric field strengths in Table 4.3. Increasing the strength of the electric field is seen to decrease the phase velocity. Table 4.4 displays the relationship between the phase velocity of TWBNNT and its wave number as a function of the thermal field. Phase velocity is shown to increase for a given temperature difference but then to decrease with time. As the temperature gap widens, the phase velocity slows down as well [21].

TABLE 4.3

Variation of Phase Velocity versus Wave Number of TWBNNT for Various Electric Field Intensities

Electric field intensity (v)	Wave number (1/nm)	Phase velocity (m/s)	Wave frequency (THz)
0.0001	0.01	2014.38	0.0201438
0.0001	0.1	2731.4	0.27314
0.0001	0.5	2152.7	1.07635
0.0001	1	1819.07	1.81907
0.0001	10	1668.66	16.6866
0.001	0.01	2014.37	0.0201437
0.001	0.1	2731.35	0.273135
0.001	0.5	2152.61	1.076305
0.001	1	1818.95	1.81895
0.001	10	1668.58	16.6858
0.005	0.01	2014.34	0.0201434
0.005	0.1	2731.12	0.273112
0.005	0.5	2152.18	1.07609
0.005	10	1668.26	16.6826

TABLE 4.4

Variation of Phase Velocity versus Wave Number of TWBNNT for Various Temperature Differences

Temperature difference (°C)	Wave number (1/nm)	Phase velocity (m/s)	Wave frequency (THz)
10	0.01	2014.38	0.0201438
10	0.1	2731.4	0.027314
10	0.5	2152.7	1.07635
10	1	1819.07	1.81907
10	10	1668.66	16.6866
50	0.01	2014.37	0.0201437
50	0.1	2731.32	0.0273132
50	0.5	2152.55	1.076275
50	1	1818.86	1.81886
50	10	1668.54	16.6854
100	0.01	2014.35	0.0201435
100	0.1	2731.21	0.0273121
100	0.5	2152.35	1.076175
100	1	1818.6	1.8186
100	10	1668.39	16.6839

4.5 SUMMARY

The purpose of this chapter is to illustrate the movement of waves as they travel through various fluid-carrying structures, such as cylinders, nanotubes, and nanotubes made of boron nitride. The governing differential equations are obtained by using Hamilton's principle. Finally, a parametric research is conducted to determine how variables like fluid velocity, fluid density, the Knudsen number, a viscoelastic base, and many physical fields (magnetic, electric, and thermal) affect the wave properties of fluid conveyance structures. Here, we shall recap the most salient points made at the end of this chapter:

- In anisotropic cylindrical shells, a higher flow velocity may reduce the wave frequency and phase velocity.
- Cylindrical shells may experience a critical flow velocity at different wave numbers, and this velocity can change depending on the radius-to-thickness ratio and the anisotropy of the material.
- Hexagonal, trigonal, monoclinic, and triclinic structures have the highest values of wave frequency and phase velocity.
- The resemblance between the two is only coincidental, but the resemblance is enough to make one wonder whether the two are in any way related.
- All functions are continuous, and the values of the aforementioned physical quantities are coverages to zero for small mechanical loads and other parameters.
- As the harmonic load velocity and excitation frequency values change, so does the magnitude of every dynamic displacement.
- As the nondimensional time parameter rises, the values of dynamic displacement in wave propagation are shown to grow.
- The distribution of dynamic displacement is significantly impacted by the nonlocal scale effect, as well as the magnetic and temperature field vectors.
- The dynamic displacement is affected by the fluid environment as the load velocity and nondimensional time rise.
- The phase velocity rises with the fluid velocity and the fluid density for small wave numbers, while the pattern reverses for larger wave numbers.
- For large wave numbers, the Knudsen number has no effect on the phase velocity, while for low wave numbers, the phase velocity grows with increasing Knudsen number.
- As the Winkler coefficient grows, the phase velocity decreases, and as the Pasternak coefficient increases, the phase velocity increases.
- The influence of the damping coefficient varies throughout a wide spectrum of wave numbers.
- The phase velocity rises as the strength of the magnetic field does, whereas the electric and thermal fields behave in the other way.
- Phase velocity is slowed down when electric field strength and temperature difference increase.

APPENDICES

APPENDIX A

$$
\begin{bmatrix}
K_{1,1} & K_{1,2} & K_{1,3} & K_{1,4} & K_{1,5} \\
K_{2,1} & K_{2,2} & K_{2,3} & K_{2,4} & K_{2,5} \\
K_{3,1} & K_{3,2} & K_{3,3} & K_{3,4} & K_{3,5} \\
K_{4,1} & K_{4,2} & K_{4,3} & K_{4,4} & K_{4,5} \\
K_{5,1} & K_{5,2} & K_{5,3} & K_{5,4} & K_{5,5}
\end{bmatrix}
=
$$

$$
\begin{bmatrix}
-A_{11}\beta_x^2 - \dfrac{A_{66}}{R^2}\beta_n^2 & -\beta_x\beta_n\left(\dfrac{A_{12}+A_{66}}{R}\right) & \dfrac{A_{12}}{R}\beta_x i & -\beta_x\beta_n\left(\dfrac{A_{12}+A_{66}}{R}\right) & -\beta_x\beta_n\left(\dfrac{B_{12}+B_{66}}{R}\right) \\[2em]
-\beta_x\beta_n\left(\dfrac{A_{12}+A_{66}}{R}\right) & -A_{66}\beta_x^2 - \dfrac{A_{11}}{R^2}\beta_n^2 - \dfrac{A_{55}^s}{R^2} & \beta_n\left(\dfrac{A_{11}+A_{55}^s}{R^2}\right)i & -B_{11}\beta_x^2 - \dfrac{B_{66}}{R^2}\beta_n^2 & -B_{66}\beta_x^2 - \dfrac{B_{11}}{R^2}\beta_n^2 - \dfrac{A_{55}^s}{R} \\[2em]
-\dfrac{A_{12}}{R}\beta_x i & -\beta_n\left(\dfrac{A_{11}+A_{55}^s}{R^2}\right)i & \begin{aligned}&-A_{55}^s\beta_x^2 - \dfrac{A_{11}}{R^2}+\beta_n^2\dfrac{A_{55}^s}{R^2}+\rho_f h_f v_x^2\beta_x^2\\ &-\dfrac{i\mu_f h_f v_x\beta_x}{R^2}\left(2+\beta_n^2\right)-i\mu_f h_f v_x\beta_x^2\end{aligned} & \left(A_{55}^s-\dfrac{B_{12}}{R}\right)i\beta_x & \beta_n\left(\dfrac{A_{55}^s}{R}-\dfrac{B_{11}}{R^2}\right)i \\[2em]
-B_{11}\beta_x^2 - \dfrac{B_{66}}{R^2}\beta_n^2 & -\beta_x\beta_n\left(\dfrac{B_{12}+B_{66}}{R}\right) & \left(-A_{55}^s+\dfrac{B_{12}}{R}\right)i\beta_x & -D_{11}\beta_x^2-\beta_n^2\dfrac{D_{66}}{R^2}-A_{55}^s & -D_{66}\beta_x^2-\beta_n^2\dfrac{D_{11}}{R^2}-A_{55}^s \\[2em]
-\beta_x\beta_n\left(\dfrac{B_{12}+B_{66}}{R}\right) & -B_{66}\beta_x^2 - \dfrac{B_{11}}{R^2}\beta_n^2 - \dfrac{A_{55}^s}{R} & \beta_n\left(-\dfrac{A_{55}^s}{R}+\dfrac{B_{11}}{R^2}\right)i & -\beta_x\beta_n\dfrac{D_{12}+D_{66}}{R} & -\beta_x\beta_n\dfrac{D_{12}+D_{66}}{R}
\end{bmatrix}
$$

$$
\begin{bmatrix}
C_{1,1} & C_{1,2} & C_{1,3} & C_{1,4} & C_{1,5} \\
C_{2,1} & C_{2,2} & C_{2,3} & C_{2,4} & C_{2,5} \\
C_{3,1} & C_{3,2} & C_{3,3} & C_{3,4} & C_{3,5} \\
C_{4,1} & C_{4,2} & C_{4,3} & C_{4,4} & C_{4,5} \\
C_{5,1} & C_{5,2} & C_{5,3} & C_{5,4} & C_{5,5}
\end{bmatrix}
=
\begin{bmatrix}
0 & 0 & 0 & 0 & 0 \\
0 & 0 & 0 & 0 & 0 \\
2\rho_f h_f v_x \beta_x + 2\dfrac{\mu_f h_f}{R^2}\beta_x + \mu_f h_f \beta_x^2 + \dfrac{\mu_f h_f}{R^2}\beta_n & 0 & 0 & 0 & 0 \\
0 & 0 & 0 & 0 & 0 \\
0 & 0 & 0 & 0 & 0
\end{bmatrix}
$$

$$
\begin{bmatrix}
M_{1,1} & M_{1,2} & M_{1,3} & M_{1,4} & M_{1,5} \\
M_{2,1} & M_{2,2} & M_{2,3} & M_{2,4} & M_{2,5} \\
M_{3,1} & M_{3,2} & M_{3,3} & M_{3,4} & M_{3,5} \\
M_{4,1} & M_{4,2} & M_{4,3} & M_{4,4} & M_{4,5} \\
M_{5,1} & M_{5,2} & M_{5,3} & M_{5,4} & M_{5,5}
\end{bmatrix}
=
\begin{bmatrix}
I_0 & 0 & 0 & I_1 & 0 \\
0 & I_0 & 0 & 0 & I_1 \\
0 & 0 & I_0 + \rho_f h_f & 0 & 0 \\
I_1 & 0 & 0 & I_2 & 0 \\
0 & I_1 & 0 & 0 & I_2
\end{bmatrix}
$$

Appendix B

$$K_{11}^{(1)} = -A_{11}k^2 - A_{11}\frac{l^2}{R_1^2}k^2n^2 - A_{11}l^2k^4 - \frac{A_{66}}{R_1^2}n^2 - A_{66}\frac{l^2}{R_1^4}n^4 - A_{66}\frac{l^2}{R_1^2}k^2n^2$$

$$K_{12}^{(1)} = -\frac{A_{12}}{R_1}kn - A_{12}\frac{l^2}{R_1^3}kn^3 - A_{12}\frac{l^2}{R_1}k^3n - \frac{A_{66}}{R_1}kn - A_{66}\frac{l^2}{R_1^3}kn^3 - A_{66}\frac{l^2}{R_1}k^3n$$

$$K_{13}^{(1)} = +\frac{A_{12}}{R_1}ki + A_{12}\frac{l^2}{R_1^3}kn^2i + A_{12}\frac{l^2}{R_1}k^3i$$

$$K_{14}^{(1)} = K_{15}^{(1)} = K_{16}^{(1)} = K_{17}^{(1)} = K_{18}^{(1)} = K_{19}^{(1)} = K_{110}^{(1)} = K_{111}^{(1)} = K_{112}^{(1)} = 0$$

$$K_{21}^{(1)} = \frac{A_{66}}{R_1}kn - A_{66}\frac{l^2}{R_1^3}kn^3 - A_{66}\frac{l^2}{R_1}k^3n - \frac{A_{12}}{R_1}kn - A_{12}\frac{l^2}{R_1^3}kn^3 - A_{12}\frac{l^2}{R_1}k^3n$$

$$K_{22}^{(1)} = -A_{66}k^2 - A_{66}\frac{l^2}{R_1^2}k^2n^2 - A_{66}l^2k^4 - \frac{A_{11}}{R_1^2}n^2 - A_{11}\frac{l^2}{R_1^4}n^4 - A_{11}\frac{l^2}{R_1^2}k^2n^2$$
$$-\frac{D_{66}}{R_1^2}k^2 - D_{66}\frac{l^2}{R_1^4}k^2n^2 - D_{66}\frac{l^2}{R_1^2}k^4 - \frac{D_{11}}{R_1^4}n^2 - D_{11}\frac{l^2}{R_1^6}n^4 - D_{11}\frac{l^2}{R_1^4}k^2n^2$$

$$K_{23}^{(1)} = +\frac{A_{11}}{R_1^2}ni + A_{11}\frac{l^2}{R_1^4}n^3i - A_{11}\frac{l^2}{R_1^2}k^2ni + 2\frac{D_{66}}{R_1^2}k^2ni + 2D_{66}\frac{l^2}{R_1^4}k^2n^3i$$
$$+2D_{66}\frac{l^2}{R_1^2}k^4ni + \frac{D_{12}}{R_1^2}k^2ni + D_{12}\frac{l^2}{R_1^4}k^2n^3i + D_{12}\frac{l^2}{R_1^2}k^4ni + \frac{D_{11}}{R_1^4}n^3i$$
$$+D_{11}\frac{l^2}{R_1^6}n^5i + D_{11}\frac{l^2}{R_1^4}k^2n^3$$

$$K_{24}^{(1)} = \frac{E_{31}}{R_1^2}ni + E_{31}\frac{l^2}{R_1^4}n^3i + E_{31}\frac{l^2}{R_1^2}k^2ni$$

$$K_{25}^{(1)} = 0, K_{26}^{(1)} = 0, K_{27}^{(1)} = 0, K_{28}^{(1)} = 0, K_{29}^{(1)} = 0, K_{210}^{(1)} = 0, K_{211}^{(1)} = 0,$$
$$K_{212}^{(1)} = 0$$

$$K_{31}^{(1)} = -\frac{A_{12}}{R_1}ki - A_{12}\frac{l^2}{R_1^3}kn^2i - A_{12}\frac{l^2}{R_1}k^3i$$

$$K_{32}^{(1)} = +\frac{D_{12}}{R_1^2}k^2ni + D_{12}\frac{l^2}{R_1^4}k^2n^3i + D_{12}\frac{l^2}{R_1^2}k^4ni + \frac{2D_{66}}{R_1^2}k^2ni + 2D_{66}\frac{l^2}{R_1^4}k^2n^3i$$

$$+2D_{66}\frac{l^2}{R_1^2}k^4ni - \frac{D_{11}}{R_1^4}n^3i - D_{11}\frac{l^2}{R_1^6}n^5i - D_{11}\frac{l^2}{R_1^4}n^3k^2i - \frac{A_{11}}{R_1^2}ni$$

$$-A_{11}\frac{l^2}{R_1^4}n^3i - A_{11}\frac{l^2}{R_1^2}k^2ni$$

$$K_{33}^{(1)} = -D_{11}k^4 - D_{11}\frac{l^2}{R_1^2}k^4n^2 - D_{11}l^2k^6 - 2\frac{D_{12}}{R_1^2}k^2n^2 - 2D_{12}\frac{l^2}{R_1^4}k^2n^4$$

$$-2D_{12}\frac{l^2}{R_1^2}k^4n^2 - \frac{4D_{66}}{R_1^2}k^2n^2 - 4D_{66}\frac{l^2}{R_1^4}k^2n^4 - 4D_{66}\frac{l^2}{R_1^2}k^4n^2 - \frac{D_{11}}{R_1^4}n^4$$

$$-D_{11}\frac{l^2}{R_1^6}n^6 - D_{11}\frac{l^2}{R_1^4}k^2n^4 - \frac{A_{11}}{R_1^2}W - A_{11}\frac{l^2}{R_1^4}n^2 - A_{11}\frac{l^2}{R_1^2}k^2 + \frac{N_{\theta0}}{R_1^2}n^2$$

$$+N_{x0}k^2 + \left(e_0a\right)^2\left(\frac{N_{\theta0}}{R_1^4}n^4 + \frac{N_{\theta0}}{R_1^2}k^2n^2 + N_{x0}k^4 + \frac{N_{x0}}{R_1^2}k^2n^2\right)$$

$$+\frac{\rho_f R_1}{2}\left(\left(V_f VCF\right)^2 k^2\right) + \left(e_0a\right)^2\frac{\rho_f R_1}{2}\left(\left(V_f VCF\right)^2 k^4 + \left(V_f VCF\right)^2\frac{k^2n^2}{R_1^2}\right)$$

$$-c\left(1 + e_0a^2\left(k^2 + \frac{n^2}{R_1^2}\right)\right)$$

$$K_{34}^{(1)} = -E_{31}k^2 - E_{31}\frac{l^2}{R_1^2}k^2n^2 - E_{31}l^2k^4 + \frac{E_{31}}{R_1^2}n^2 - E_{31}\frac{l^2}{R_1^2}k^2n^2 - E_{31}\frac{l^2}{R_1^4}n^4$$

$$K_{35}^{(1)} = 0, K_{36}^{(1)} = 0, K_{38}^{(1)} = 0, K_{39}^{(1)} = 0, K_{310}^{(1)} = 0, K_{311}^{(1)} = 0, K_{312}^{(1)} = 0$$

$$K_{37}^{(1)} = c\left(1 + e_0a^2\left(k^2 + \frac{n^2}{R_1^2}\right)\right)$$

$$K_{41}^{(1)} = 0$$

$$K_{42}^{(1)} = \frac{E_{31}}{R_1^2}ni + E_{31}\frac{l^2}{R_1^4}n^3i + E_{31}\frac{l^2}{R_1^2}k^2ni$$

$$K_{43}^{(1)} = E_{31}k^2 + E_{31}l^2k^4 + E_{31}\frac{l^2}{R_1^2}k^2n^2 + \frac{E_{31}}{R_1^2}n^2 + E_{31}\frac{l^2}{R_1^4}n^4 + E_{31}\frac{l^2}{R_1^2}k^2n^2$$

$$K_{44}{}^{(1)} = -X_{11}k^2 - X_{11}\frac{l^2}{R_1{}^2}k^2n^2 - X_{11}l^2n^4 - X_{22}n^2 - X_{22}\frac{l^2}{R_1{}^2}n^4 - X_{22}l^2k^2n^2$$

$$-X_{33} - X_{33}k^2 - X_{33}\frac{n^2}{R_1{}^2}$$

$$K_{45}{}^{(1)} = 0, K_{46}{}^{(1)} = 0, K_{47}{}^{(1)} = 0, K_{48}{}^{(1)} = 0, K_{49}{}^{(1)} = 0, K_{410}{}^{(1)} = 0, K_{411}{}^{(1)} = 0,$$

$$K_{412}{}^{(1)} = 0$$

$$K_{11}{}^{(2)} = 0, K_{12}{}^{(2)} = 0, K_{13}{}^{(2)} = 0, K_{14}{}^{(2)} = 0$$

$$K_{15}{}^{(2)} = -A_{11}k^2 - A_{11}\frac{l^2}{R_2{}^2}k^2n^2 - A_{11}l^2k^4 - \frac{A_{66}}{R_2{}^2}n^2 - A_{66}\frac{l^2}{R_2{}^4}n^4 - A_{66}\frac{l^2}{R_2{}^2}k^2n^2$$

$$K_{16}{}^{(2)} = -\frac{A_{12}}{R_2}kn - A_{12}\frac{l^2}{R_2{}^3}kn^3 - A_{12}\frac{l^2}{R_2}k^3n - \frac{A_{66}}{R_2}kn - A_{66}\frac{l^2}{R_2{}^3}kn^3 - A_{66}\frac{l^2}{R_2}k^3n$$

$$K_{17}{}^{(2)} = +\frac{A_{12}}{R_2}ki + A_{12}\frac{l^2}{R_2{}^3}kn^2i + A_{12}\frac{l^2}{R_2}k^3i$$

$$K_{18}{}^{(2)} = 0, K_{19}{}^{(2)} = 0, K_{110}{}^{(2)} = 0, K_{111}{}^{(2)} = 0, K_{112}{}^{(2)} = 0$$

$$K_{21}{}^{(2)} = K_{22}{}^{(2)} = K_{23}{}^{(2)} = K_{24}{}^{(2)} = 0$$

$$K_{25}{}^{(2)} = \frac{A_{66}}{R_2}kn - A_{66}\frac{l^2}{R_2{}^3}kn^3 - A_{66}\frac{l^2}{R_2}k^3n - \frac{A_{12}}{R_2}kn - A_{12}\frac{l^2}{R_2{}^3}kn^3 - A_{12}\frac{l^2}{R_2}k^3n$$

$$K_{26}{}^{(2)} = -A_{66}k^2 - A_{66}\frac{l^2}{R_2{}^2}k^2n^2 - A_{66}l^2k^4 - \frac{A_{11}}{R_2{}^2}n^2 - A_{11}\frac{l^2}{R_2{}^4}n^4 - A_{11}\frac{l^2}{R_2{}^2}k^2n^2$$

$$-\frac{D_{66}}{R_2{}^2}k^2 - D_{66}\frac{l^2}{R_2{}^4}k^2n^2 - D_{66}\frac{l^2}{R_2{}^2}k^4 - \frac{D_{11}}{R_2{}^4}n^2 - D_{11}\frac{l^2}{R_2{}^6}n^4 - D_{11}\frac{l^2}{R_2{}^4}k^2n^2$$

$$K_{27}{}^{(2)} = +\frac{A_{11}}{R_2{}^2}ni + A_{11}\frac{l^2}{R_2{}^4}n^3i - A_{11}\frac{l^2}{R_2{}^2}k^2ni + 2\frac{D_{66}}{R_2{}^2}k^2ni + 2D_{66}\frac{l^2}{R_2{}^4}k^2n^3i$$

$$+2D_{66}\frac{l^2}{R_2{}^2}k^4ni + \frac{D_{12}}{R_2{}^2}k^2ni + D_{12}\frac{l^2}{R_2{}^4}k^2n^3i + D_{12}\frac{l^2}{R_2{}^2}k^4ni + \frac{D_{11}}{R_2{}^4}n^3i + D_{11}\frac{l^2}{R_2{}^6}n^5i$$

$$+D_{11}\frac{l^2}{R_2{}^4}k^2n^3$$

$$K_{28}^{(2)} = \frac{E_{31}}{R_2^2} ni + E_{31} \frac{l^2}{R_2^4} n^3 i + E_{31} \frac{l^2}{R_2^2} k^2 ni$$

$$K_{29}^{(2)} = 0, K_{210}^{(2)} = 0, K_{211}^{(2)} = 0, K_{212}^{(2)} = 0,$$

$$K_{31}^{(2)} = 0, K_{32}^{(2)} = 0, K_{34}^{(2)} = 0$$

$$K_{33}^{(2)} = c \frac{R_1}{R_2} \left(1 + e_0 a^2 \left(k^2 + \frac{n^2}{R_2^2} \right) \right)$$

$$K_{35}^{(2)} = -\frac{A_{12}}{R_2} ki - A_{12} \frac{l^2}{R_2^3} kn^2 i - A_{12} \frac{l^2}{R_2} k^3 i$$

$$K_{36}^{(2)} = +\frac{D_{12}}{R_2^2} k^2 ni + D_{12} \frac{l^2}{R_2^4} k^2 n^3 i + D_{12} \frac{l^2}{R_2^2} k^4 ni + \frac{2D_{66}}{R_2^2} k^2 ni$$

$$+2D_{66} \frac{l^2}{R_2^4} k^2 n^3 i + 2D_{66} \frac{l^2}{R_2^2} k^4 ni - \frac{D_{11}}{R_2^4} n^3 i - D_{11} \frac{l^2}{R_2^6} n^5 i - D_{11} \frac{l^2}{R_2^4} n^3 k^2 i$$

$$-\frac{A_{11}}{R_2^2} ni - A_{11} \frac{l^2}{R_2^4} n^3 i - A_{11} \frac{l^2}{R_2^2} k^2 ni$$

$$K_{37}^{(2)} = -D_{11} k^4 - D_{11} \frac{l^2}{R_2^2} k^4 n^2 - D_{11} l^2 k^6 - 2 \frac{D_{12}}{R_2^2} k^2 n^2 - 2 D_{12} \frac{l^2}{R_2^4} k^2 n^4$$

$$-2 D_{12} \frac{l^2}{R_2^2} k^4 n^2 - \frac{4D_{66}}{R_2^2} k^2 n^2 - 4 D_{66} \frac{l^2}{R_2^4} k^2 n^4 - 4 D_{66} \frac{l^2}{R_2^2} k^4 n^2 - \frac{D_{11}}{R_2^4} n^4$$

$$-D_{11} \frac{l^2}{R_2^6} n^6 - D_{11} \frac{l^2}{R_2^4} k^2 n^4 - \frac{A_{11}}{R_2^2} W - A_{11} \frac{l^2}{R_2^4} n^2 - A_{11} \frac{l^2}{R_2^2} k^2 + \frac{N_{\theta 0}}{R_2^2} n^2$$

$$+N_{x0} k^2 + (e_0 a)^2 \left(\frac{N_{\theta 0}}{R_2^4} n^4 + \frac{N_{\theta 0}}{R_2^2} k^2 n^2 + N_{x0} k^4 + \frac{N_{x0}}{R_2^2} k^2 n^2 \right)$$

$$-c \left(1 + \frac{R_1}{R_2} \right) \left(1 + e_0 a^2 \left(k^2 + \frac{n^2}{R_2^2} \right) \right)$$

$$K_{38}^{(2)} = -E_{31} k^2 - E_{31} \frac{l^2}{R_2^2} k^2 n^2 - E_{31} l^2 k^4 + \frac{E_{31}}{R_2^2} n^2 - E_{31} \frac{l^2}{R_2^2} k^2 n^2 - E_{31} \frac{l^2}{R_2^4} n^4$$

$$K_{39}^{(2)} = 0, K_{310}^{(2)} = 0, K_{312}^{(2)} = 0$$

$$K_{311}^{(2)} = c \left(1 + e_0 a^2 \left(k^2 + \frac{n^2}{R_2^2} \right) \right)$$

$$K_{41}{}^{(2)} = 0, K_{42}{}^{(2)} = 0, K_{43}{}^{(2)} = 0, K_{44}{}^{(2)} = 0, K_{45}{}^{(2)} = 0$$

$$K_{46}{}^{(2)} = \frac{E_{31}}{R_2{}^2} ni + E_{31} \frac{l^2}{R_2{}^4} n^3 i + E_{31} \frac{l^2}{R_2{}^2} k^2 ni$$

$$K_{47}{}^{(2)} = E_{31} k^2 + E_{31} l^2 k^4 + E_{31} \frac{l^2}{R_2{}^2} k^2 n^2 + \frac{E_{31}}{R_2{}^2} n^2 + E_{31} \frac{l^2}{R_2{}^4} n^4 + E_{31} \frac{l^2}{R_2{}^2} k^2 n^2$$

$$K_{48}{}^{(2)} = -X_{11} k^2 - X_{11} \frac{l^2}{R_2{}^2} k^2 n^2 - X_{11} l^2 n^4 - X_{22} n^2 - X_{22} \frac{l^2}{R_2{}^2} n^4 - X_{22} l^2 k^2 n^2$$

$$-X_{33} - X_{33} k^2 - X_{33} \frac{n^2}{R_2{}^2}$$

$$K_{49}{}^{(2)} = 0, K_{410}{}^{(2)} = 0, K_{411}{}^{(2)} = 0, K_{412}{}^{(2)} = 0, K_{11}{}^{(3)} = 0, K_{12}{}^{(3)} = 0, K_{13}{}^{(3)} = 0,$$
$$K_{14}{}^{(3)} = 0, K_{15}{}^{(3)} = 0, K_{16}{}^{(3)} = 0, K_{17}{}^{(3)} = 0, K_{18}{}^{(3)} = 0$$

$$K_{19}{}^{(3)} = -A_{11} k^2 - A_{11} \frac{l^2}{R_3{}^2} k^2 n^2 - A_{11} l^2 k^4 - \frac{A_{66}}{R_3{}^2} n^2 - A_{66} \frac{l^2}{R_3{}^4} n^4 - A_{66} \frac{l^2}{R_3{}^2} k^2 n^2$$

$$K_{110}{}^{(3)} = -\frac{A_{12}}{R_3} kn - A_{12} \frac{l^2}{R_3{}^3} kn^3 - A_{12} \frac{l^2}{R_3} k^3 n - \frac{A_{66}}{R_3} kn - A_{66} \frac{l^2}{R_3{}^3} kn^3 - A_{66} \frac{l^2}{R_3} k^3 n$$

$$K_{111}{}^{(3)} = +\frac{A_{12}}{R_3} ki + A_{12} \frac{l^2}{R_3{}^3} kn^2 i + A_{12} \frac{l^2}{R_3} k^3 i$$

$$K_{112}{}^{(3)} = 0, K_{21}{}^{(3)} = 0, K_{22}{}^{(3)} = 0, K_{23}{}^{(3)} = 0, K_{24}{}^{(3)} = 0, K_{25}{}^{(3)} = 0, K_{26}{}^{(3)} = 0,$$
$$K_{27}{}^{(3)} = 0, K_{28}{}^{(3)} = 0$$

$$K_{29}{}^{(3)} = \frac{A_{66}}{R_3} kn - A_{66} \frac{l^2}{R_3{}^3} kn^3 - A_{66} \frac{l^2}{R_3} k^3 n - \frac{A_{12}}{R_3} kn - A_{12} \frac{l^2}{R_3{}^3} kn^3$$

$$-A_{12} \frac{l^2}{R_3} k^3 n$$

$$K_{210}{}^{(3)} = -A_{66} k^2 - A_{66} \frac{l^2}{R_3{}^2} k^2 n^2 - A_{66} l^2 k^4 - \frac{A_{11}}{R_3{}^2} n^2 - A_{11} \frac{l^2}{R_3{}^4} n^4 - A_{11} \frac{l^2}{R_3{}^2} k^2 n^2$$

$$-\frac{D_{66}}{R_3{}^2} k^2 - D_{66} \frac{l^2}{R_3{}^4} k^2 n^2 - D_{66} \frac{l^2}{R_3{}^2} k^4 - \frac{D_{11}}{R_3{}^4} n^2 - D_{11} \frac{l^2}{R_3{}^6} n^4 - D_{11} \frac{l^2}{R_3{}^4} k^2 n^2$$

$$K_{211}^{(3)} = +\frac{A_{11}}{R_3^{2}} ni + A_{11}\frac{l^2}{R_3^{4}} n^3 i - A_{11}\frac{l^2}{R_3^{2}} k^2 ni + 2\frac{D_{66}}{R_3^{2}} k^2 ni + 2D_{66}\frac{l^2}{R_3^{4}} k^2 n^3 i$$

$$+2D_{66}\frac{l^2}{R_3^{2}} k^4 ni + \frac{D_{12}}{R_3^{2}} k^2 ni + D_{12}\frac{l^2}{R_3^{4}} k^2 n^3 i + D_{12}\frac{l^2}{R_3^{2}} k^4 ni + \frac{D_{11}}{R_3^{4}} n^3 i + D_{11}\frac{l^2}{R_3^{6}} n^5 i$$

$$+D_{11}\frac{l^2}{R_3^{4}} k^2 n^3$$

$$K_{212}^{(3)} = \frac{E_{31}}{R_3^{2}} ni + E_{31}\frac{l^2}{R_3^{4}} n^3 i + E_{31}\frac{l^2}{R_3^{2}} k^2 ni$$

$$K_{31}^{(3)} = 0, K_{32}^{(3)} = 0, K_{33}^{(3)} = 0, K_{34}^{(3)} = 0, K_{35}^{(3)} = 0, K_{36}^{(3)} = 0, K_{38}^{(3)} = 0$$

$$K_{37}^{(3)} = c\frac{R_2}{R_3}\left(1 + e_0 a^2\left(k^2 + \frac{n^2}{R_3^{2}}\right)\right)$$

$$K_{39}^{(3)} = -\frac{A_{12}}{R_3} ki - A_{12}\frac{l^2}{R_3^{3}} kn^2 i - A_{12}\frac{l^2}{R_3} k^3 i$$

$$K_{310}^{(3)} = +\frac{D_{12}}{R_3^{2}} k^2 ni + D_{12}\frac{l^2}{R_3^{4}} k^2 n^3 i + D_{12}\frac{l^2}{R_3^{2}} k^4 ni + \frac{2D_{66}}{R_3^{2}} k^2 ni$$

$$+2D_{66}\frac{l^2}{R_3^{4}} k^2 n^3 i + 2D_{66}\frac{l^2}{R_3^{2}} k^4 ni - \frac{D_{11}}{R_3^{4}} n^3 i - D_{11}\frac{l^2}{R_3^{6}} n^5 i - D_{11}\frac{l^2}{R_3^{4}} n^3 k^2 i$$

$$-\frac{A_{11}}{R_3^{2}} ni - A_{11}\frac{l^2}{R_3^{4}} n^3 i - A_{11}\frac{l^2}{R_3^{2}} k^2 ni$$

$$K_{311}^{(2)} = -D_{11}k^4 - D_{11}\frac{l^2}{R_3^{2}} k^4 n^2 - D_{11}l^2 k^6 - 2\frac{D_{12}}{R_3^{2}} k^2 n^2 - 2D_{12}\frac{l^2}{R_3^{4}} k^2 n^4$$

$$-2D_{12}\frac{l^2}{R_3^{2}} k^4 n^2 - \frac{4D_{66}}{R_3^{2}} k^2 n^2 - 4D_{66}\frac{l^2}{R_3^{4}} k^2 n^4 - 4D_{66}\frac{l^2}{R_3^{2}} k^4 n^2 - \frac{D_{11}}{R_3^{4}} n^4$$

$$-D_{11}\frac{l^2}{R_3^{6}} n^6 - D_{11}\frac{l^2}{R_3^{4}} k^2 n^4 - \frac{A_{11}}{R_3^{2}} W - A_{11}\frac{l^2}{R_3^{4}} n^2 - A_{11}\frac{l^2}{R_3^{2}} k^2 + \frac{N_{\theta 0}}{R_3^{2}} n^2$$

$$+N_{x0}k^2 + (e_0 a)^2\left(\frac{N_{\theta 0}}{R_3^{4}} n^4 + \frac{N_{\theta 0}}{R_3^{2}} k^2 n^2 + N_{x0}k^4 + \frac{N_{x0}}{R_3^{2}} k^2 n^2\right)$$

$$-c\frac{R_2}{R_3}\left(1 + e_0 a^2\left(k^2 + \frac{n^2}{R_3^{2}}\right)\right) - K_w - K_G\left(k^2 + \frac{n^2}{R_3^{2}}\right)$$

$$K_{312}^{(3)} = -E_{31}k^2 - E_{31}\frac{l^2}{R_3^2}k^2n^2 - E_{31}l^2k^4 + \frac{E_{31}}{R_3^2}n^2 - E_{31}\frac{l^2}{R_3^2}k^2n^2 - E_{31}\frac{l^2}{R_3^4}n^4$$

$$K_{41}^{(3)} = 0, K_{42}^{(3)} = 0, K_{43}^{(3)} = 0, K_{44}^{(3)} = 0, K_{45}^{(3)} = 0, K_{46}^{(3)} = 0, K_{47}^{(3)} = 0,$$
$$K_{48}^{(3)} = 0, K_{49}^{(3)} = 0$$

$$K_{410}^{(3)} = \frac{E_{31}}{R_3^2}ni + E_{31}\frac{l^2}{R_3^4}n^3i + E_{31}\frac{l^2}{R_3^2}k^2ni$$

$$K_{411}^{(3)} = E_{31}k^2 + E_{31}l^2k^4 + E_{31}\frac{l^2}{R_3^2}k^2n^2 + \frac{E_{31}}{R_3^2}n^2 + E_{31}\frac{l^2}{R_3^4}n^4 + E_{31}\frac{l^2}{R_3^2}k^2n^2$$

$$K_{412}^{(3)} = -X_{11}k^2 - X_{11}\frac{l^2}{R_3^2}k^2n^2 - X_{11}l^2n^4 - X_{22}n^2 - X_{22}\frac{l^2}{R_3^2}n^4 - X_{22}l^2k^2n^2$$
$$-X_{33} - X_{33}k^2 - X_{33}\frac{n^2}{R_3^2}$$

$$M_{11}^{(1)} = M_{22}^{(1)} = -I_1 - I_1\left(e_0a\right)^2\left(k^2 + \frac{n^2}{R_1^2}\right)$$

$$M_{15}^{(2)} = M_{26}^{(2)} = M_{37}^{(2)} = -I_1 - I_1\left(e_0a\right)^2\left(k^2 + \frac{n^2}{R_2^2}\right)$$

$$M_{19}^{(3)} = M_{210}^{(3)} = M_{311}^{(3)} = -I_1 - I_1\left(e_0a\right)^2\left(k^2 + \frac{n^2}{R_3^2}\right)$$

$$M_{33}^{(1)} = -I_1 - I_1\left(e_0a\right)^2\left(k^2 + \frac{n^2}{R_1^2}\right) + \frac{\rho_f R_1}{2} + \left(e_0a\right)^2\frac{\rho_f R_1}{2}\left(k^4 + \frac{k^2n^2}{R_1^2}\right)$$

$$M_{44}^{(1)} = 0, M_{48}^{(2)} = 0, M_{412}^{(3)} = 0$$

$$C_{33}^{(1)} = -2\frac{\rho_f R_1}{2}\left(V_f VCF\right)ki - 2\left(e_0a\right)^2\frac{\rho_f R_1}{2}\left(k^3 + \frac{kn^2}{R_1^2}\right)i$$

$$C_{37}^{(2)} = 0, C_{311}^{(3)} = Ci$$

REFERENCES

[1] Ebrahimi, F., Dehghan, M., & Seyfi. A. (2019). Eringen nonlocal elasticity theory for wave propagation analysis of magneto-electro-elastic nanotubes. *Advances in Nano Research, 7*(1), 1–11.

[2] Ebirahimi, F., & Dabbagh. A. (2018). Magnetic field effects on thermally affected propagation of acoustical waves in rotary double-nanobeam system. *Wave Random Complex*, 1–21.

[3] Ebrahimi, F., & Mahmoodi, F. (2018). Vibration analysis of carbon nanotubes with multiple cracks in thermal environment. *Advances in Nano Research, 6*(1), 2287–2388.

[4] Ebrahimi, F., & Barati M. (2016). Wave propagation analysis of quasi-3D FG nanobeams in thermal environment based on nonlocal strain gradient theory. *Applied Physics A, 122*(9), 843.

[5] Wang, L. (2010). Wave propagation of fluid-conveying single-walled carbon nanotubes via gradient elasticity theory. *Computational Materials Science, 49*(4), 761–766.

[6] Ebrahimi, F., & Ali Dabbagh. (2018). Wave dispersion characteristics of rotating heterogeneous magneto-electro-elastic nanobeams based on nonlocal strain gradient elasticity theory. *Journal of Electromagnetic Waves and Applications, 32*(2), 138–169.

[7] Ebrahimi, F., & Dabbagh, A. (2018). Viscoelastic wave propagation analysis of axially motivated double-layered graphene sheets via nonlocal strain gradient theory. *Waves in Random and Complex Media*, 1–20.

[8] Ebrahimi, F. and Salari, E. (2015). Thermal buckling and free vibration analysis of size dependent Timoshenko FG Nano beams in thermal environments. *Composite Structures, 128*, 363–380.

[9] Ebrahimi, F., Boreiry, M., & Shaghaghi, G.R. (2018). Nonlinear vibration analysis of electro – hygro - thermally actuated embedded nanobeams with various with various boundary conditions. *Micro Systems Technologies, 24*(12), 5037–5054.

[10] Arani, A. G., & Roudbari, M. A. (2014). Surface stress, initial stress and Knudsen-dependent flow velocity effects on the electro-thermo nonlocal wave propagation of SWBNNTs. *Physica B: Condensed Matter, 452*, 159–165.

[11] Al-Furjan, M. S. H., Bolandi, S. Y., Habibi, M., Ebrahimi, F., Chen, G., & Safarpour, H. (2022). Enhancing vibration performance of a spinning smart nanocomposite reinforced microstructure conveying fluid flow. *Engineering with Computers, 38*(5), 4097–4112.

[12] Zhang, Y., Wang, Z., Tazeddinova, D., Ebrahimi, F., Habibi, M., & Safarpour, H. (2021). Enhancing active vibration control performances in a smart rotary sandwich thick nanostructure conveying viscous fluid flow by a PD controller. *Waves in Random and Complex Media*, 1–24.

[13] Ebrahimi, F., Hajilak, Z. E., Habibi, M., & Safarpour, H. (2019). Buckling and vibration characteristics of a carbon nanotube-reinforced spinning cantilever cylindrical 3D shell conveying viscous fluid flow and carrying spring-mass systems under various temperature distributions. *Proceedings of the Institution of Mechanical Engineers, Part C: Journal of Mechanical Engineering Science, 233*(13), 4590–4605.

[14] Shariati, A., Bayrami, S. S., Ebrahimi, F., & Toghroli, A. (2022). Wave propagation analysis of electro-rheological fluid-filled sandwich composite beam. *Mechanics Based Design of Structures and Machines, 50*(5), 1481–1490.

[15] Ebrahimi, F., & Sedighi, S. B. (2022). Wave dispersion characteristics of a rectangular sandwich composite plate with tunable magneto-rheological fluid core rested on a visco-Pasternak foundation. *Mechanics Based Design of Structures and Machines, 50*(1), 170–183.

[16] Ebrahimi, F., & Sedighi, S. B. (2021). Wave propagation analysis of a rectangular sandwich composite plate with tunable magneto-rheological fluid core. *Journal of Vibration and Control, 27*(11–12), 1231–1239.

[17] Ebrahimi, F., & Dabbagh, A. (2021). Vibration analysis of fluid-conveying multi-scale hybrid nanocomposite shells with respect to agglomeration of nanofillers. *Defence Technology, 17*(1), 212–225.

[18] Selvamani, R., Jayan, M. M. S., & Ebrahimi, F. (2021). thermomagnetic field effects on stability analysis of a single-walled fluid-conveying carbon nanotube rested on polymer matrix. *Nanoscience and Technology: An International Journal, 12*(2).

[19] Ebrahimi, F., & Seyfi, A. (2020). Propagation of flexural waves in anisotropic fluid-conveying cylindrical shells. *Symmetry, 12*(6), 901.

[20] Dehghan, M., Ebrahimi, F., & Vinyas, M. (2020). Wave dispersion characteristics of fluid-conveying magneto-electro-elastic nanotubes. *Engineering with Computers, 36*(4), 1687–1703.

[21] Dehghan, M., Ebrahimi, F., & Vinyas, M. (2019). Wave dispersion analysis of magnetic-electrically affected fluid-conveying nanotubes in thermal environment. *Proceedings of the Institution of Mechanical Engineers, Part C: Journal of Mechanical Engineering Science, 233*(19–20), 7116–7131.

[22] Zeighampour, H., Beni, Y. T., & Karimipour, I. (2017). Wave propagation in double-walled carbon nanotube conveying fluid considering slip boundary condition and shell model based on nonlocal strain gradient theory. *Microfluidics and Nanofluidics, 21*(5), 85.

[23] Ebrahimi, F., Seyfi, A., Dabbagh, A., & Tornabene, F. (2019). Wave dispersion characteristics of porous graphene platelet-reinforced composite shells. Structural Engineering and Mechanics, 71(1), 99–107.

[24] Ebrahimi, F., & Seyfi, A. (2019). Wave propagation response of multi-scale hybrid nanocomposite shell by considering aggregation effect of CNTs. *Mechanics Based Design of Structures and Machines*, 1–22.

[25] Arani, A. Ghorbanpour, R. Kolahchi, & H. Vossough. (2012). Nonlocal wave propagation in an embedded DWBNNT conveying fluid via strain gradient theory. *Physica B: Condensed Matter, 407*(21), 4281–4286.

[26] Maraghi, Z. K., et al. (2013). Nonlocal vibration and instability of embedded DWBNNT conveying viscose fluid. *Composites Part B: Engineering, 45*(1), 423–432.

[27] Arani, A. Ghorbanpour, R. K., & Z. Khoddami Maraghi. (2013). "Nonlinear vibration and instability of embedded double-walled boron nitride nanotubes based on nonlocal cylindrical shell theory." Applied Mathematical Modelling, 37(14–15), 7685–7707.

[28] Arani, G., Ali, M. H., & Kolahchi, R. (2015). Nonlocal Timoshenko beam model for dynamic stability of double-walled boron nitride nanotubes conveying nano-flow. *Proceedings of the Institution of Mechanical Engineers, Part N: Journal of Nanoengineering and Nanosystems, 229*(1), 2–16.

[29] Abdollahian, M. et al. (2013). Non-local wave propagation in embedded armchair TWBNNTs conveying viscous fluid using DQM. *Physica B: Condensed Matter, 418*, 1–15.

[30] Ghorbanpour-Arani, A. H., et al. (2017). Wave propagation of coupled double-DWBNNTs conveying fluid-systems using different nonlocal surface piezoelasticity theories. *Mechanics of Advanced Materials and Structures, 24*(14), 1159–1179.

[31] Hu, Y. G., Liew, K. M., Wang, Q., He, X. Q., & Yakobson, B. I. (2008). Nonlocal shell model for elastic wave propagation in single-and double-walled carbon nanotubes. *Journal of the Mechanics and Physics of Solids, 56*(12), 3475–3485.

5 Wave Dispersion Characteristics of Graphene-based Structures

5.1 BACKGROUND

In this chapter, we will attempt to take a comprehensive look at the wave dispersion characteristics of a variety of graphene-based structures, such as single- and double-layered graphene sheets, graphene oxide powder–reinforced nanocomposite plates, graphene platelet–reinforced structures, and graphene foam structures. This chapter will focus on the wave dispersion properties of single- and double-layered graphene sheets. All of the components, including beams, plates, and shells, are taken into consideration. Solving the governing equations of the issues, which are determined by utilizing Hamilton's principle, enables one to compute the wave frequency, phase velocity, and escape frequency of graphene-based structures. The wave number, nonlocal parameter, length scale parameter, structural damping coefficient, Winkler coefficient, Pasternak coefficient, damping coefficient, and axial stress are all parameters that are examined in detail here as being ones that impact the wave propagation behaviors of graphene-based structures. It can be shown visually that wave propagation in structures based on graphene is very sensitive to the features in question. Tables and figures have been used to display the results of the study. The material included in this chapter may be used for your own benefit but only under the following terms and conditions: The material that has been given is just for educational purposes and is not meant to be taken as professional legal advice, and it would be informative for design and fabrication of the microactuators and microsensors.

5.1.1 A REVIEW ON WAVE DISPERSION ANALYSIS OF GRAPHENE-BASED STRUCTURES

Applications for nanoscale shells, beams, and plates in nanoelectromechanical systems (NEMSs) have proliferated in recent years [1]. As a result, there has been a sharp uptick in the number of studies using these tiny components to analyze the mechanical reactions of nanostructures. There is no denying that the mechanical reactions of small structures are distinct from those of large ones. Researching the static and dynamic responses of nanoscale components necessitates the use of size-dependent continuum models. Nonlocal elasticity theory

 DOI: 10.1201/9781003270263-5

(NET) was initially presented by Eringen [2, 3] to account for the size dependence of nanostructures. According to Eringen, there are two fundamental variables that determine the stress state at any given location in a solid body: (1) the strain at that site and (2) the strain at all of the neighboring points. This hypothesis is widely used in numerous publications that investigate the mechanical behavior of nanoscale objects. For this reason, an analysis of previous efforts has been compiled. Ebrahimi and Salari [4] have attempted to map the vibrational characteristics of carbon nanotubes (CNTs) using NET. When investigating the vibration and buckling responses of nanotubes, Ebrahimi et al. [5] also take into account the combined impacts of temperature conditions and surface elasticity. Mechanical reactions of size-dependent beams and plates under different loads were studied by Ebrahimi et al. [6–19] using the NET. Within the context of the nonlocal strain gradient theory (NSGT), Ebrahimi et al. [20–23] conduct an in-depth study of the wave dispersion characteristics of nanobeams and nanoplates. In order to emphasize the impact of different factors on the vibrational responses of compositionally graded nanobeams, the NSGT was also used by Ebrahimi and Barati [24, 25]. When resting on a visco-Pasternak substrate, Ebrahimi and Barati [26] provided an NSG-based theory for vibration analysis of viscoelastic nanoplates. Recently, Ebrahimi and Barati [27] created an NSGT to study the nanobeam vibration issue in a functionally graded axial direction. Ebrahimi and Dabbagh [28] have also provided a thorough analysis of the modeling of nanocomposite materials and structures. Controlled distortions in single-layered graphene sheets (SLGSs) have the potential to build a diverse array of carbon-based structures, such as CNTs, carbon nanocones (CNCs), and nanorings [42]. Graphene sheets also have better elastic potential [43] and thermal conductivity [44] than other tiny structures built from several different kinds of materials. That's why it's crucial to learn everything you can about the mechanical responses of such structures under different loads. "Modeling and Control of a Smart Single-Layer Graphene Sheet" was presented by Arani and Ebrahimi [32]. The buckling study of graphene sheets in humid and temperature-controlled settings was given by Ebrahimi and Barati [33]. Two scale factors associated with nonlocal and strain gradient effects are included in their suggested theory. They used a two-variable shear deformation plate theory to simulate the graphene sheet, eliminating the necessity for shear correction factors. The governing equations were solved under various boundary conditions using Galerkin's technique. The buckling properties of graphene sheets were investigated as a function of a variety of parameters, including moisture content, temperature, nonlocal parameter, length scale parameter, elastic foundation, and geometrical parameters. The wave propagation issue in a thermally loaded double-layered graphene sheet (DLGS) was investigated by Ebrahimi and Dabbagh [34]. The kinematic relations of each layer of DLGS were derived using a traditional plate theory, and both uniform and linear temperature distributions were included to compare their relative effects. "Wave Propagation Responses of Double-Layered Graphene Sheets in Hygrothermal Environment" was also given in their paper [35]. Ebrahimi and Dabbagh [36] used nonlocal strain gradient theory to analyze the propagation of viscoelastic waves in axially motivated double-layered graphene sheets. Nonlocal strain gradient single-layered graphene sheets supported by an elastic medium were also investigated for their thermo-mechanical wave dispersion properties [37]. A magnetic field and its influence on the wave dispersion properties of single- and double-layered graphene sheets were also studied [38, 39]. Ebrahimi and Shafiei [40], in their investigation of the vibrational properties of SLGSs supported by Winkler–Pasternak foundations, consider the impact of initial shear stress.

Ebrahimi et al. [41] recently carried out research using a parametric approach to explore how the free vibration response of graphene oxide powder (GOP)-reinforced nanocomposites was impacted by a magnetic field that was not uniform in its distribution. In a subsequent investigation, Ebrahimi and colleagues [42] studied the free vibration behavior of GOP-reinforced nanocomposite plates implanted on a Winkler–Pasternak substrate and exposed to a variety of thermal loadings. They did this by applying the loadings to the plates in a controlled manner. Ebrahimi et al. [43] presented an alternative method for modeling the wave propagation of GPL-reinforced nanocomposite shells in a thermal environment. This method took into consideration the effect that porosity had on wave propagation. No prior study, as far as the authors are aware, has focused on modeling the wave dispersion of embedded GOP-reinforced nanocomposite plates that were exposed to various heat loadings. An analytical approach for studying the wave propagation in GOP-reinforced nanocomposite plates that have been exposed to heat stress has been developed by Ebrahimi and colleagues [44]. The plate was implanted on a two-parameter elastic substrate, and the structure was exposed to thermal loadings at varied temperature rises. This was done so that the research could be more realistically represented. We projected the effective material properties of the structure in which GOPs were distributed in a polymeric matrix across the thickness of the plate in a variety of functionally graded (FG) patterns using the Halpin–Tsai micromechanical method. This allowed us to determine how the GOPs would affect the material's performance. In addition, we spoke about the possibility that modifying some factors may change the way waves interact with a structure.

Ebrahimi et al. [45–47] presented a variety of experiments for the free vibration analysis of graphene platelet–reinforced (GPLR) composites beam and plates. In this research project, the vibrational behavior of GPLR composite structures, as well as the influence of temperature and viscoelastic basis, were investigated. Additionally, presented [48, 50] were assessments of buckling, stability, and dynamics for multi-layered GPLR composite structures, which were supported by a two-parameter viscoelastic basis. Shokrgozar et al. [51] investigated the stability of a microshell in the shape of a cylinder that was strengthened by graphene nanoplatelets. They did this by applying a uniform axial stress and using a viscoelastic basis. The governing equations were derived via the use of the energy technique, and they were then solved through the use of the generalized differential quadrature method (GDQM). According to the findings of their research, the stability of a graphene platelets reinforced composite (GPLRC) cylinder microshell is critically dependent on a number of different factors. These factors include the underlying viscoelastic material, the pattern of GPL distribution, the nonlocal parameter, the length scale parameter, the number of layers, the boundary condition, and the GPL weight function. Ignore the negative opinions of those around you and follow the lead of those who have faith in your capacity to improvise. Selvamani and Ebrahimi [53] established a unique analytical approach for generalized thermoelastic waves in a doubly connected polygonal ring reinforced by graphene platelets. This was done under the exact heat conduction equation with delay. The Eshelby–Mori–Tanaka notion of homogenizing was used for four different distribution patterns in this study. The triangular, square, pentagonal, and hexagonal GPLRC rings each had a numerical value that was given to them in order to rank them in order of quality. The consequences of the physical variations such as stress, strain, and mechanical displacement, weight % of GPL, dimensionless frequency, and temperature rise were explored, and their results were shown in tabular and graphical formats respectively. The wave propagation study of the porous graphene platelet–reinforced GPLR nanocomposite shell was investigated by Ebrahimi et al. [54, 55]. Extending the Halpin–Tsai relations for the porous nanocomposite was done with the intention of achieving uniformity in the material that was employed. Throughout the course of their investigation, they were sure to take into consideration both symmetric and asymmetric porosity distributions. The results of their simulations indicate that the stiffness reduces when the impacts of porosity are taken into consideration. It is essential to take into consideration the effect that the longitudinal wave number has on the wave dispersion curves produced by the nanocomposite. Ebrahimi et al. [56] investigated the wave propagation properties of cylindrical nanoshells composed of graphene nanoplatelet–reinforced composite (GNPRC). These nanoshells were spun to varied diameters. The nonlocal strain gradient theory, often known as NSGT, was used in this study to investigate the effect of small scale. We examined the wave propagation porosity behavior of a GNPRC cylinder nanoshell by using NSGT. The impacts of porosity, angular velocity, wave number, and graphene platelet dispersion patterns on phase velocity were studied with the use of continuum mechanics. In order to solve the problem of vibration that occurs with porous metal foam plates and shells supported by a viscoelastic substrate, Ebrahimi et al. [57, 58] developed an analytical method to take into consideration the impacts of various porosity distributions. Their research demonstrates that a significant reduction in the frequency of the porous metal foam structure is possible by increasing the structure's porosity coefficient. In addition, Selvamani et al. [59] published the results of a finite element model and analysis of a piezoelectric nanoporous metal foam nanobeam that was exposed to a hygro and nonlinear thermal field. In addition to this [60–62], they used to update higher-order shear deformable beam theory in order to investigate the propagation of flexural waves in porous metal foam beams, plates, and shells that were supported by a Winkler–Pasternak foundation. In addition, studies are conducted to determine if the porosity is distributed in a uniform, symmetric, or asymmetric manner along the thickness axis. Frequency, escape frequency, and phase velocity fluctuation under the influence of variables such as porosity distribution, porosity coefficient, slenderness ratio, wave number, and elastic foundation coefficients were addressed in depth. Frequency was found to fluctuate most under the influence of slenderness ratio. In addition, Ebrahimi and Seyfi [63, 64] investigated the propagation of elastic waves in graphene foam (GF) plates in a hygrothermal environment using an enhanced version of the higher-order shear deformation plate theory. Three distinct designs were used to conduct an investigation of the porosity distribution inside the plate. An analytical method that is based on an exponential function was also used in order to solve the obtained governing equations of GF plates that are supported by an elastic medium. Among the factors that were investigated for their impacts on the GF plate wave frequency and phase velocity variation were the aspect ratio, wave number, porosity coefficient, porosity distribution, temperature and moisture fluctuations, and porosity distribution. In the end, everything comes down to a question of taste, but if you ask me, the best method to decide whether or not a given meal is excellent is to compare it to other dishes that are comparable to it. In other words, flavor is the most important factor in determining whether or not anything is good. In GOP-reinforced nanocomposite plates, the visual results indicate that the components have a considerable influence on wave propagation.

5.2 WAVE DISPERSION CHARACTERISTICS OF SINGLE/ DOUBLE-LAYERED GRAPHENE SHEETS

5.2.1 Background

In this chapter, we will look at how waves behave while traveling through a single- and double-layered graphene sheet (SLGS, DLGS), respectively. In reality, a thorough size-dependent analysis is carried out, including not only the amplifying but also the dampening effects. In addition, the stiffness-softening effect is taken into account, and NSGT is used to provide more accurate findings. The study does not need a shear correction factor since the kinematic connections are developed within the context of an improved trigonometric two-variable plate theory. The Euler–Lagrange equations are generated using Hamilton's principle, and the governing equations are constructed in terms of displacement fields by merging these equations with the nonlocal constitutive relations of the NSGT. Wave frequency and phase velocity should be calculated once the final Eigenvalue issue is solved. The last part is dedicated to investigating the impact of a wide range of variables on the wave propagation behaviors of SLGSs and DLGSs. These variables include the wave number, nonlocal parameter, length scale parameter, structural damping coefficient, Winkler coefficient, Pasternak coefficient, damping coefficient, and axial stress. The Lorentz force produced by solving Maxwell's equation along the x-axis is determined. The impact of each parameter is enhanced by drawing several types of diagrams for them, as we do here with the hygrothermal, bending force, foundation, and magnetic field impacts.

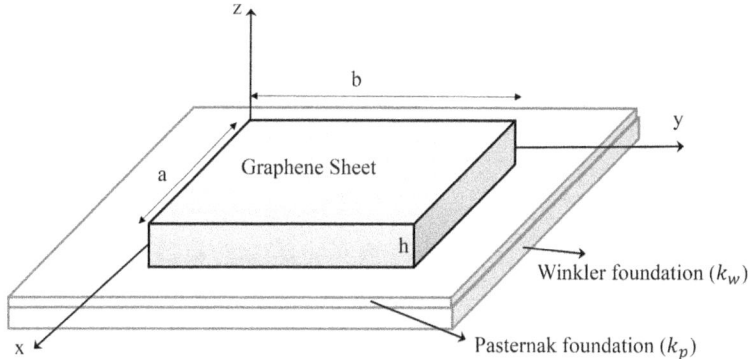

FIGURE 5.1   Geometry of a single-layered graphene sheet rested on Winkler–Pasternak foundation.

5.2.2   SINGLE-LAYERED GRAPHENE SHEETS: WAVE DISPERSION CHARACTERISTICS

The focus here is on the kinematic behaviors of graphene sheets, and how they are described. Figure 5.1 depicts a schematic of a typical embedded SLGS. An improved higher-order two-variable plate theory is used to account for shear deformation effects. This allows us to express the displacement fields as [33]:

$$U(x,y,z,t) = -z\frac{\partial w_b}{\partial x} - f(z)\frac{\partial w_s}{\partial x} \tag{5.1a}$$

$$V(x,y,z,t) = -z\frac{\partial w_b}{\partial y} - f(z)\frac{\partial w_s}{\partial y} \tag{5.1b}$$

$$W(x,y,z,t) = w_b(x,y,t) + w_s(x,y,t) \tag{5.1c}$$

where w_b and w_s are bending and shear deflections in the thickness direction, respectively.

Also, $f(z)$ is a shape function that estimates shear stress and shear strain. In the present theory, a trigonometric function is used as follows:

$$f(z) = z - \frac{h}{\pi}\sin\left(\frac{\pi z}{h}\right) \tag{5.2}$$

in Eq. (5.2), h is plate's thickness. Now, the nonzero strains can be stated as follows:

$$\begin{Bmatrix} \varepsilon_x \\ \varepsilon_y \\ \gamma_{xy} \end{Bmatrix} = z \begin{Bmatrix} -\dfrac{\partial^2 w_b}{\partial x^2} \\[2mm] -\dfrac{\partial^2 w_b}{\partial y^2} \\[2mm] -2\dfrac{\partial^2 w_b}{\partial x \partial y} \end{Bmatrix} + f(z) \begin{Bmatrix} -\dfrac{\partial^2 w_s}{\partial x^2} \\[2mm] -\dfrac{\partial^2 w_s}{\partial y^2} \\[2mm] -2\dfrac{\partial^2 w_s}{\partial x \partial y} \end{Bmatrix}, \quad \begin{Bmatrix} \gamma_{yz} \\ \gamma_{xz} \end{Bmatrix} = g(z) \begin{Bmatrix} \dfrac{\partial w_s}{\partial y} \\[2mm] \dfrac{\partial w_s}{\partial x} \end{Bmatrix} \qquad (5.3a)$$

in Eq. (5.3a), $g(z)$ can be stated as:

$$g(z) = 1 - \frac{df(z)}{dz} \qquad (5.3b)$$

The principle of Hamilton, moreover, may be stated as:

$$\int_0^t \delta(U - T + V)\,dt = 0 \qquad (5.4a)$$

in which U is strain energy, T is kinetic energy, and V is work done by external loads. The variation of strain energy can be calculated as:

$$\delta U = \int_V \sigma_{ij}\delta\varepsilon_{ij}dV = \int_V \left(\sigma_x\delta\varepsilon_x + \sigma_y\delta\varepsilon_y + \sigma_{xy}\delta\gamma_{xy} + \sigma_{yz}\delta\gamma_{yz} + \sigma_{xz}\delta\gamma_{xz}\right)dV \quad (5.4b)$$

Substituting Eq. (5.3a) in Eq. (5.4b) reveals:

$$\delta U = \int_0^a \int_0^b \left(\begin{array}{l} -M_x^b \dfrac{\partial^2 \delta w_b}{\partial x^2} - M_x^s \dfrac{\partial^2 \delta w_s}{\partial x^2} - M_y^b \dfrac{\partial^2 \delta w_b}{\partial y^2} - M_y^s \dfrac{\partial^2 \delta w_s}{\partial y^2} \\[3mm] -2M_{xy}^b \dfrac{\partial^2 \delta w_b}{\partial x \partial y} - 2M_{xy}^s \dfrac{\partial^2 \delta w_s}{\partial x \partial y} + Q_{xz}\dfrac{\partial \delta w_s}{\partial x} + Q_{yz}\dfrac{\partial \delta w_s}{\partial y} \end{array} \right) dydx \quad (5.4c)$$

The parameters in Eq. (5.4c) may be written as follows:

$$\left(M_i^b, M_i^s\right) = \int_{-h/2}^{h/2} (z, f)\sigma_i dz, \quad i = (x, y, xy)$$

$$Q_i = \int_{-h/2}^{h/2} g\sigma_i dz, \quad i = (xz, yz) \qquad (5.4d)$$

It is also possible to separate the external work done by forces into two catego-ries: the external work created by the elastic medium and the external work of the Lorentz force induced by the magnetic field. A longitudinally constant magnetic field of strength H_0 is thought to be acting upon the SLGS. Therefore, the formula for the applied body force resulting from this field is as follows [33]:

$$f_{Lz} = \eta \left[\underbrace{\nabla \times \left(\nabla \times \left(u \times H_0 \right) \right)}_{h} \right] \times H_0$$

$$\underbrace{\hspace{6cm}}_{J}$$

(5.5a)

in which η, ∇, h, and J are the magnetic permeability of SLGS, gradient operator, small disturbance of applied magnetic field and current density vector, respec-tively. Here, the magnetic field can be expressed as follows:

$$H_0 = H_x \delta_x \hat{i}$$

(5.5b)

where, δ is the Kronecker delta tensor. The applied Lorentz forces per unit volume are given by the following equation:

$$f_{Lz} = \eta H_x^2 \frac{\partial^2 \left(w_b + w_s \right)}{\partial x^2}$$

(5.5c)

In addition, the ensuing Lorentz's forces may be expressed as:

$$F_{Lz} = \int_{-h/2}^{h/2} f_{Lz} dz = \eta h H_x^2 \frac{\partial^2 \left(w_b + w_s \right)}{\partial x^2}$$

(5.5d)

Thus, the work done by external forces can be expressed as:

$$\delta V = \int_0^a \int_0^b \left(\begin{array}{l} -k_w \delta \left(w_b + w_s \right) + k_p \left(\begin{array}{l} + \dfrac{\partial \left(w_b + w_s \right)}{\partial x} \dfrac{\partial \delta \left(w_b + w_s \right)}{\partial x} \\[3mm] + \dfrac{\partial \left(w_b + w_s \right)}{\partial y} \dfrac{\partial \delta \left(w_b + w_s \right)}{\partial y} \end{array} \right) \\[8mm] -\eta h H_x^2 \dfrac{\partial^2 \left(w_b + w_s \right)}{\partial x^2} \end{array} \right) dy dx$$

(5.6a)

where k_w and k_p are Winkler and Pasternak coefficients. The variation of the kinetic energy should be written as:

$$\delta K = \int_0^a \int_0^b \left| \begin{array}{l} I_0 \left(\dfrac{\partial \left(w_b + w_s\right)}{\partial t} \dfrac{\partial \delta \left(w_b + w_s\right)}{\partial t} \right) + I_2 \left(\dfrac{\partial w_b}{\partial x \partial t} \dfrac{\partial \delta w_b}{\partial x \partial t} + \dfrac{\partial w_b}{\partial y \partial t} \dfrac{\partial \delta w_b}{\partial y \partial t} \right) \\[3mm] + K_2 \left(\dfrac{\partial w_s}{\partial x \partial t} \dfrac{\partial \delta w_s}{\partial x \partial t} + \dfrac{\partial w_s}{\partial y \partial t} \dfrac{\partial \delta w_s}{\partial y \partial t} \right) + \\[3mm] J_2 \left(\dfrac{\partial w_b}{\partial x \partial t} \dfrac{\partial \delta w_s}{\partial x \partial t} + \dfrac{\partial w_s}{\partial x \partial t} \dfrac{\partial \delta w_b}{\partial x \partial t} + \dfrac{\partial w_b}{\partial y \partial t} \dfrac{\partial \delta w_s}{\partial y \partial t} + \dfrac{\partial w_s}{\partial y \partial t} \dfrac{\partial \delta w_b}{\partial y \partial t} \right) \end{array} \right| dy dx$$

$$(5.6b)$$

in which

$$\left(I_0, I_2, J_2, K_2 \right) = \int_{-h/2}^{h/2} \left(1, z^2, zf, f^2 \right) \rho dz \tag{5.6c}$$

Inserting Eqs. (5.4c), (5.6a), and (5.6b) in Eq. (5.4a) and setting the coefficients of δw_b and δw_s to zero, the Euler–Lagrange equations of GSs can be rewritten as:

$$\frac{\partial^2 M_x^b}{\partial x^2} + 2 \frac{\partial^2 M_{xy}^b}{\partial x \partial y} + \frac{\partial^2 M_y^b}{\partial y^2} + k_p \nabla^2 \left(w_b + w_s \right) - k_w \left(w_b + w_s \right)$$

$$= I_0 \frac{\partial^2 \left(w_b + w_s \right)}{\partial t^2} - I_2 \nabla^2 \left(\frac{\partial^2 w_b}{\partial t^2} \right) - J_2 \nabla^2 \left(\frac{\partial^2 w_s}{\partial t^2} \right) - \eta h H_x^2 \frac{\partial^2 \left(w_b + w_s \right)}{\partial x^2} \tag{5.7a}$$

$$\frac{\partial^2 M_x^s}{\partial x^2} + 2 \frac{\partial^2 M_{xy}^s}{\partial x \partial y} + \frac{\partial^2 M_y^s}{\partial y^2} + \frac{\partial Q_{xz}}{\partial x} + \frac{\partial Q_{yz}}{\partial y} + k_p \nabla^2 \left(w_b + w_s \right) - k_w \left(w_b + w_s \right)$$

$$= I_0 \frac{\partial^2 \left(w_b + w_s \right)}{\partial t^2} - J_2 \nabla^2 \left(\frac{\partial^2 w_b}{\partial t^2} \right) - K_2 \nabla^2 \left(\frac{\partial^2 w_s}{\partial t^2} \right) - \eta h H_x^2 \frac{\partial^2 \left(w_b + w_s \right)}{\partial x^2} \tag{5.7b}$$

According to the nonlocal strain gradient theory, the stress field accounts for the effects of both the strain gradient stress field and the nonlocal elastic stress field. As a result, the theory may be expressed for elastic solids as follows:

$$\sigma_{ij} = \sigma_{ij}^{(0)} - \frac{d \sigma_{ij}^{(1)}}{dx} \tag{5.8a}$$

in Eq. (5.8a), the stresses $\sigma_{xx}^{(0)}$ (classical stress) and $\sigma_{xx}^{(1)}$ (higher-order stress) are corresponding to strain ε_{xx} and strain gradient $\varepsilon_{xx,\,x}$, respectively, as follows:

$$
\begin{cases}
\sigma_{ij}^{(0)} = \displaystyle\int_0^L C_{ijkl}\alpha_0\left(x,x',e_0 a\right)\varepsilon'_{kl}\left(x'\right)dx' \\[4mm]
\sigma_{ij}^{(1)} = l^2 \displaystyle\int_0^L C_{ijkl}\alpha_1\left(x,x',e_1 a\right)\varepsilon'_{kl,x}\left(x'\right)dx'
\end{cases}
\tag{5.8b}
$$

in which C_{ijkl} is the elastic coefficient; $e_0 a$ and $e_1 a$ are introduced to account for the nonlocality effects. Also, l captures the strain gradient effects. Once the nonlocal kernel functions $\alpha_0(x,\,x',\,e_0 a)$ and $\alpha_1(x,\,x',\,e_1 a)$ satisfy the developed conditions, the constitutive relation of nonlocal strain gradient theory can be expressed as below:

$$
\begin{aligned}
&\left(1-\left(e_1 a\right)^2\nabla^2\right)\left(1-\left(e_0 a\right)^2\nabla^2\right)\sigma_{ij} = C_{ijkl}\left(1-\left(e_1 a\right)^2\nabla^2\right)\varepsilon_{kl} \\[2mm]
&- C_{ijkl}l^2\left(1-\left(e_0 a\right)^2\nabla^2\right)\nabla^2\varepsilon_{kl}
\end{aligned}
\tag{5.8c}
$$

in which ∇^2 denotes the Laplacian operator. Considering $e_1 = e_0 = e$, the general constitutive relation in Eq. (5.8c) becomes:

$$
\left(1-\left(ea\right)^2\nabla^2\right)\sigma_{ij} = C_{ijkl}\left(1-l^2\nabla^2\right)\varepsilon_{kl}
\tag{5.8d}
$$

Finally, the reduced constitutive relation looks like this:

$$
\left(1-\mu^2\nabla^2\right)
\begin{Bmatrix}
\sigma_x \\ \sigma_y \\ \sigma_{xy} \\ \sigma_{yz} \\ \sigma_{xz}
\end{Bmatrix}
=\left(1-\lambda^2\nabla^2\right)
\begin{pmatrix}
Q_{11} & Q_{12} & 0 & 0 & 0 \\
Q_{12} & Q_{22} & 0 & 0 & 0 \\
0 & 0 & Q_{66} & 0 & 0 \\
0 & 0 & 0 & Q_{44} & 0 \\
0 & 0 & 0 & 0 & Q_{55}
\end{pmatrix}
\begin{Bmatrix}
\varepsilon_x \\ \varepsilon_y \\ \gamma_{xy} \\ \gamma_{yz} \\ \gamma_{xz}
\end{Bmatrix}
\tag{5.9a}
$$

in Eq. (5.9a)

$$
Q_{11} = Q_{22} = \frac{E}{1-v^2},\; Q_{12} = vQ_{11}, Q_{44} = Q_{55} = Q_{66} = \frac{E}{2\left(1+v\right)}
\tag{5.9b}
$$

where $\mu = e_0 a$ and $\lambda = l$. Substituting Eq. (5.4d) in Eq. (5.9a) gives:

$$\left(1-\mu^2\nabla^2\right)\left\{\begin{array}{c}M_x^b \\ M_y^b \\ M_{xy}^b\end{array}\right\} = \left(1-\lambda^2\nabla^2\right)\left(\begin{pmatrix}D_{11} & D_{12} & 0 \\ D_{12} & D_{22} & 0 \\ 0 & 0 & D_{66}\end{pmatrix}\left\{\begin{array}{c}-\dfrac{\partial^2 w_b}{\partial x^2} \\[2mm] -\dfrac{\partial^2 w_b}{\partial y^2} \\[2mm] -2\dfrac{\partial^2 w_b}{\partial x \partial y}\end{array}\right\}\right.$$

$$\left. +\begin{pmatrix}D_{11}^s & D_{12}^s & 0 \\ D_{12}^s & D_{22}^s & 0 \\ 0 & 0 & D_{66}^s\end{pmatrix}\left\{\begin{array}{c}-\dfrac{\partial^2 w_s}{\partial x^2} \\[2mm] -\dfrac{\partial^2 w_s}{\partial y^2} \\[2mm] -2\dfrac{\partial^2 w_s}{\partial x \partial y}\end{array}\right\}\right) \tag{5.9c}$$

$$\left(1-\mu^2\nabla^2\right)\left\{\begin{array}{c}M_x^s \\ M_y^s \\ M_{xy}^s\end{array}\right\} = \left(1-\lambda^2\nabla^2\right)\left(\begin{pmatrix}D_{11}^s & D_{12}^s & 0 \\ D_{12}^s & D_{22}^s & 0 \\ 0 & 0 & D_{66}^s\end{pmatrix}\left\{\begin{array}{c}-\dfrac{\partial^2 w_b}{\partial x^2} \\[2mm] -\dfrac{\partial^2 w_b}{\partial y^2} \\[2mm] -2\dfrac{\partial^2 w_b}{\partial x \partial y}\end{array}\right\}\right.$$

$$\left. +\begin{pmatrix}H_{11}^s & H_{12}^s & 0 \\ H_{12}^s & H_{22}^s & 0 \\ 0 & 0 & H_{66}^s\end{pmatrix}\left\{\begin{array}{c}-\dfrac{\partial^2 w_s}{\partial x^2} \\[2mm] -\dfrac{\partial^2 w_s}{\partial y^2} \\[2mm] -2\dfrac{\partial^2 w_s}{\partial x \partial y}\end{array}\right\}\right) \tag{5.9d}$$

$$\left(1-\mu^2\nabla^2\right)\left\{\begin{array}{c}Q_x \\ Q_y\end{array}\right\} = \left(1-\lambda^2\nabla^2\right)\left(\begin{pmatrix}A_{44}^s & 0 \\ 0 & A_{55}^s\end{pmatrix}\left\{\begin{array}{c}\dfrac{\partial w_s}{\partial x} \\[2mm] \dfrac{\partial w_s}{\partial y}\end{array}\right\}\right) \tag{5.9e}$$

Eqs. (5.9c) to (5.9e) that describe the cross-sectional rigidities may be expressed as follows.

$$
\begin{pmatrix}
D_{11} & D_{11}^s & H_{11}^s \\
D_{12} & D_{12}^s & H_{12}^s \\
D_{66} & D_{66}^s & H_{66}^s
\end{pmatrix}
= \int_{-h/2}^{h/2} Q_{11}\left(z^2 \ zf \ f^2\right)
\begin{bmatrix}
1 \\
v \\
\dfrac{1-v}{2}
\end{bmatrix} dz
\tag{5.10a}
$$

$$
A_{44}^s = A_{55}^s = \int_{-h/2}^{h/2} g^2 \frac{E}{2(1+v)}\,dz
\tag{5.10b}
$$

Directly derived in terms of displacements are the nonlocal governing equations of SLGSs, obtained by substituting Eqs. (5.9c) through (5.9e) into Eqs. (5.7a) and (5.7b).

$$
\begin{aligned}
&\left(1-\lambda^2\nabla^2\right)
\begin{pmatrix}
-D_{11}\dfrac{\partial^4 w_b}{\partial x^4} - 2\left(D_{12}+2D_{66}\right)\dfrac{\partial^4 w_b}{\partial x^2\partial y^2} - D_{22}\dfrac{\partial^4 w_b}{\partial y^4} \\[2mm]
-D_{11}^s\dfrac{\partial^4 w_s}{\partial x^4} - 2\left(D_{12}^s+2D_{66}^s\right)\dfrac{\partial^4 w_s}{\partial x^2\partial y^2} - D_{22}^s\dfrac{\partial^4 w_s}{\partial y^4}
\end{pmatrix}
+ \left(1-\mu^2\nabla^2\right) \\[4mm]
&\left(
-I_0\dfrac{\partial^2\left(w_b+w_s\right)}{\partial t^2}
+ I_2\left(\dfrac{\partial^4 w_b}{\partial x^2\partial t^2} + \dfrac{\partial^4 w_b}{\partial y^2\partial t^2}\right)
+ J_2\left(\dfrac{\partial^4 w_s}{\partial x^2\partial t^2} + \dfrac{\partial^4 w_s}{\partial y^2\partial t^2}\right)\right. \\[2mm]
&\left.
+ k_p\left(\dfrac{\partial^2\left(w_b+w_s\right)}{\partial x^2} + \dfrac{\partial^2\left(w_b+w_s\right)}{\partial y^2}\right) - k_w\left(w_b+w_s\right) + F_{Lz}
\right) = 0
\end{aligned}
\tag{5.11a}
$$

$$
\begin{aligned}
&\left(1-\eta^2\nabla^2\right)
\begin{pmatrix}
-D_{11}^s\dfrac{\partial^4 w_b}{\partial x^4} - 2\left(D_{12}^s+2D_{66}^s\right)\dfrac{\partial^4 w_b}{\partial x^2\partial y^2} - D_{22}^s\dfrac{\partial^4 w_b}{\partial y^4} \\[2mm]
-H_{11}^s\dfrac{\partial^4 w_s}{\partial x^4} - 2\left(H_{12}^s+2H_{66}^s\right)\dfrac{\partial^4 w_s}{\partial x^2\partial y^2} - H_{22}^s\dfrac{\partial^4 w_s}{\partial y^4} \\[2mm]
+ A_{44}^s\dfrac{\partial^2 w_s}{\partial x^2} + A_{55}^s\dfrac{\partial^2 w_s}{\partial y^2}
\end{pmatrix}
+ \left(1-\mu^2\nabla^2\right) \\[4mm]
&\left(
-I_0\dfrac{\partial^2\left(w_b+w_s\right)}{\partial t^2}
+ J_2\left(\dfrac{\partial^4 w_b}{\partial x^2\partial t^2} + \dfrac{\partial^4 w_b}{\partial y^2\partial t^2}\right)
+ K_2\left(\dfrac{\partial^4 w_s}{\partial x^2\partial t^2} + \dfrac{\partial^4 w_s}{\partial y^2\partial t^2}\right)\right. \\[2mm]
&\left.
+ k_p\left(\dfrac{\partial^2\left(w_b+w_s\right)}{\partial x^2} + \dfrac{\partial^2\left(w_b+w_s\right)}{\partial y^2}\right) - k_w\left(w_b+w_s\right) + F_{Lz}
\right) = 0
\end{aligned}
\tag{5.11b}
$$

In this part, we do an analytical solution of the nonlocal governing equations that were created in the section before this one. The definition of the displacement fields, which are thought of as being exponential, is as follows:

$$\begin{Bmatrix} w_b(x,y,t) \\ w_s(x,y,t) \end{Bmatrix} = \begin{Bmatrix} W_b \exp\left[i\left(\beta_1 x + \beta_2 y - \omega t\right)\right] \\ W_s \exp\left[i\left(\beta_1 x + \beta_2 y - \omega t\right)\right] \end{Bmatrix} \tag{5.12}$$

where W_b and W_s are the unknown coefficients; β_1 and β_2 are the wave numbers of wave propagation along x and y directions, respectively, and finally ω is wave's angular frequency. Now, substituting Eq. (5.12) to Eqs. (5.11a) and (5.11b) results in:

$$\left(\left[K\right]_{2\times2} - \omega^2\left[M\right]_{2\times2}\right)\{\Delta\} = \{0\} \tag{5.13a}$$

where the corresponding k_{ij}, m_{ij} are as written in Appendix 5A. The unknown parameters of Eq. (5.13) can be noted as follows:

$$\{\Delta\} = \{W_b, W_s\}^T \tag{5.13b}$$

The angular frequency of a wave may be attained by setting the determinant on the left side of Eq. (5.13) to zero.

$$\left|\left[K\right]_{2\times2} - \omega^2\left[M\right]_{2\times2}\right| = 0 \tag{5.13c}$$

In Eq. (5.13c) by setting $\beta_1 = \beta_2 = \beta$ and solving the obtained equation for ω, the wave's angular frequency of embedded SLGSs can be calculated. If the angular frequency is divided by wave number, the phase velocity can be obtained as below:

$$c_p = \frac{\omega}{\beta} \tag{5.14}$$

By expanding the wave number to infinity, we can also calculate the graphene sheet's escape frequency.

$$f_{esc} = \lim_{\beta \to \infty} \frac{\omega}{2\pi} \tag{5.15}$$

In this section, a comparison is made between the reactions of different SLGSs to the propagation of waves when various parameters are altered. The following is a list of the mechanical characteristics that graphene sheets have: $E = 1$ TPa, $\nu = 0.19$, $\rho = 2300$ kg/m³. Also, the thickness is presumed to be $h = 0.34$ nm. In the following diagrams, wave frequencies are calculated by dividing wave's angular frequency to $2\pi\left(f = \dfrac{\omega}{2\pi}\right)$. We may show the effects of both nonlocal and

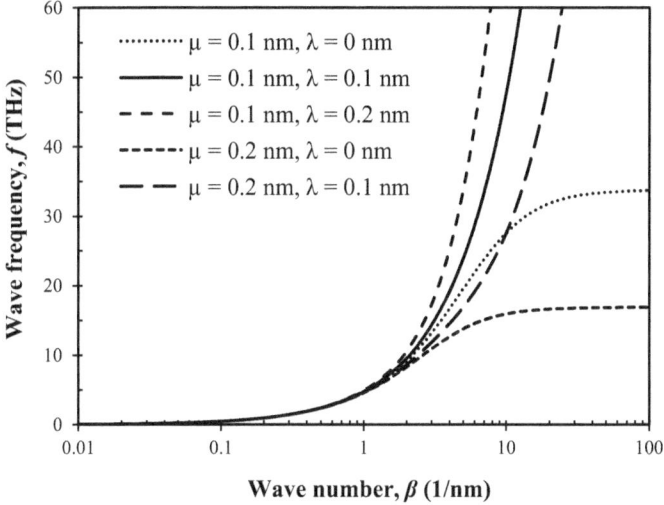

FIGURE 5.2 Variation of wave frequency versus wave number for different nonlocal and length scale parameters ($k_w = k_p = 0$, $H_x = 0$).

length scale parameters by plotting the variations of wave frequency versus wave number for different values of these parameters. The graphic reveals that a drop in wave frequency is possible when the nonlocal parameter has been raised. The frequency of waves with a β value greater than 3×10^9 is indeed reduced by this phenomenon, which is known as the stiffness-softening effect. On the other hand, the frequency of waves is affected in a different fashion by the length scale parameter. When using longer time scales, it is possible to see an increase in the frequency of the waves. This strengthening force, known as the stiffness-hardening influence, has the ability to significantly increase the wave frequency values [33].

5.2.2.1 Hygrothermal Effects

Figures 5.3 and 5.4 also provide graphs of frequency versus moisture concentration for a variety of temperature gradients and length scale parameters. To begin, it has been shown once again that an increase in the amount of moisture in the air results in a reduction in the frequency of the waves. There is a possibility that a decrease in the temperature gradient will have the same effect on wave frequency as an increase in the concentration of moisture. On the other hand, it can be shown that the strain gradient parameter has an influence. When the length scale parameter is included in the picture, two important truths become apparent: first, that it has an influence that has a stiffness-hardening effect, and, second, that it impacts the range of moisture concentrations in which the wave frequency becomes nonzero. Both of these realities are highlighted below. When the length scale parameter seems to be smaller, the wave frequency in a reduced concentration of moisture will ultimately approach zero [33].

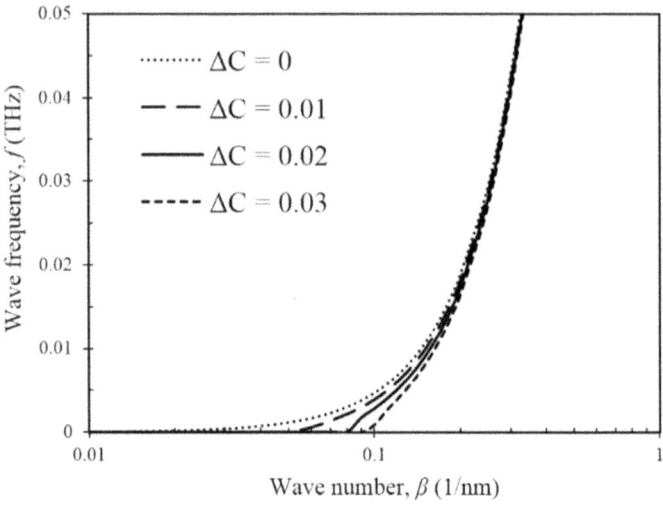

FIGURE 5.3 Effect of different moisture concentration amounts on the wave frequency of graphene sheets ($\mu = 1$ nm, $\eta = 1$ nm, $\Delta T = 0$, $k_w = k_{px} = k_{py} = 0$).

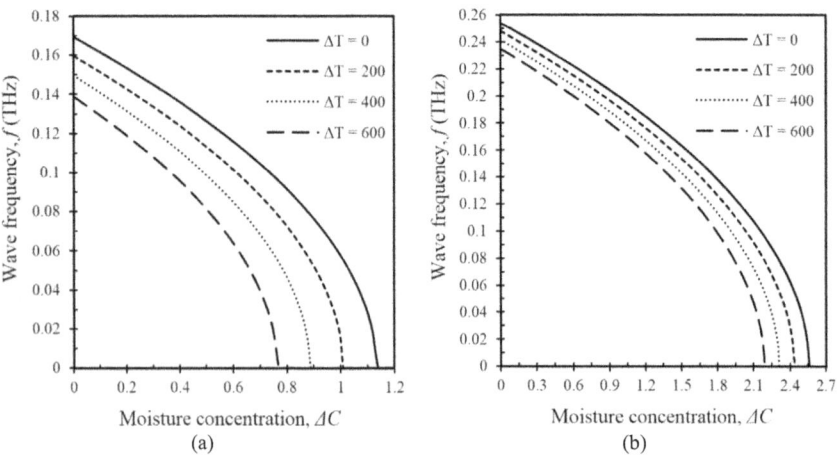

FIGURE 5.4 Variation of wave frequency versus moisture concentration for various temperature gradients at (a) $\eta = 1$ nm, (b) $\eta = 2$ nm ($\mu = 1$ nm, $\Delta T = 0$, $k_w = k_{px} = k_{py} = 0$, $\beta = 0.2 \times 10^9$).

Figure 5.5 is plotted in order to investigate the effects of changing k_{px} and k_{py} values on the behavior of wave frequency for various moisture concentration amounts. Clearly, wave frequency passes through a sinusoidal curve in each condition; however, there is a tiny difference in the two below diagrams. Actually, it is evident that the right-hand side figure has a $\dfrac{\pi}{2}$ delay in comparison with the left-hand side one. As a matter of fact, once $k_{px} < k_{py}$ wave frequency experiences

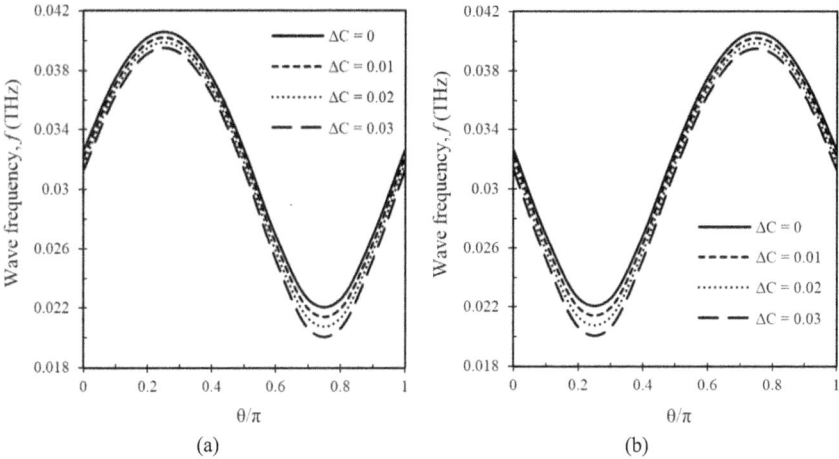

FIGURE 5.5 Variation of wave frequency versus dimensionless angle for various moisture concentrations at (a) $k_{px} > k_{py}$, (b) $k_{px} < k_{py}$ ($\mu = 1$ nm, $\eta = 1$ nm, $\Delta T = 0$, $k_w = 0$, $\beta = 0.2 \times 10^9$).

a delay compared with the condition that $k_{px} > k_{py}$. Moreover, in this figure, it can be seen that adding moisture concentration will result in a decrease in the amount of wave frequency.

In addition, the interplay between the temperature gradient and the base parameters is investigated as shown in Figure 5.6. In light of the above, it may be inferred that the growing medium coefficients have a bearing on the conclusion. The addition of a temperature gradient also clearly exposes lower wave frequencies.

5.2.2.2 Bending Force Effects

Here it is assumed that a graphene sheet is being bent in both the x and y directions by forces that vary along both axes, as indicated below [37]:

$$N_x^0 = N\left(1 - \zeta \frac{y}{b}\right), \; N_y^0 = \lambda N\left(1 - \zeta \frac{x}{a}\right) \tag{5.16}$$

in Eq. (5.16), ζ is in-plane bending load factor. Also, λ denotes biaxial load factor. Linearly changing bending forces in the plane are shown in Figure 5.7 [37]. External forces will apply work in the form of in-plane applied loads to simulate the effects of bending forces.

When $\mu = 1$ nm, Figure 5.8 illustrates the interplay between the strain gradient parameter and the linearly variable x and y bending loads on the wave frequency values. The accompanying graphics demonstrate how the NSGT-discussed stiffness-hardening process may be understood. In other words, the size dependency of graphene sheets as described by the NET is shown in Figure 5.8(a). Consistent with the NSGT's underlying assumptions, the addition of the length scale parameter amplifies the frequency of waves. Furthermore, it can be

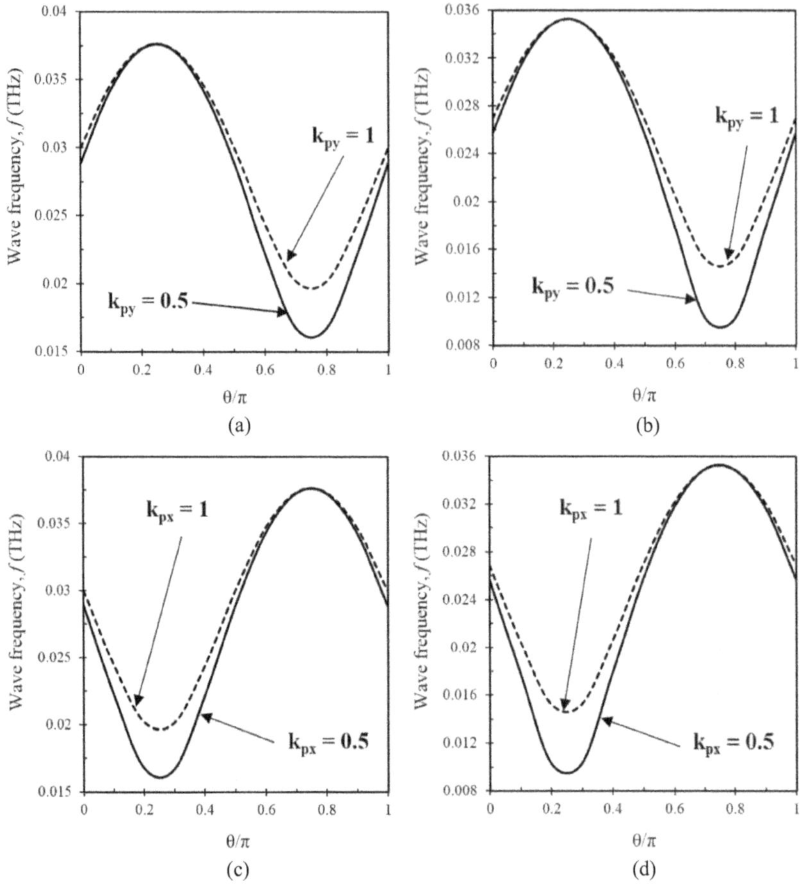

FIGURE 5.6 Effects of various foundation parameters and temperature gradients on the wave frequency of graphene sheets ($\mu = 1$ nm, $\eta = 1$ nm, $k_w = 0$, $\Delta C = 0.02$, $\beta = 0.2 \times 10^9$). (a) $k_{py} = 5$, $\Delta T = 100$, (b) $k_{py} = 5$, $\Delta T = 200$, (c) $k_{py} = 5$, $\Delta T = 100$, and (d) $k_{py} = 5$, $\Delta T = 200$.

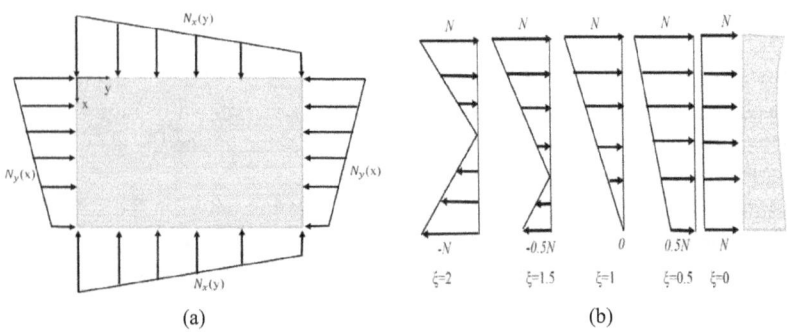

FIGURE 5.7 (a) Form of graphene sheets under in-plane varying bending forces (b) Various kinds of in-plane varying bending forces.

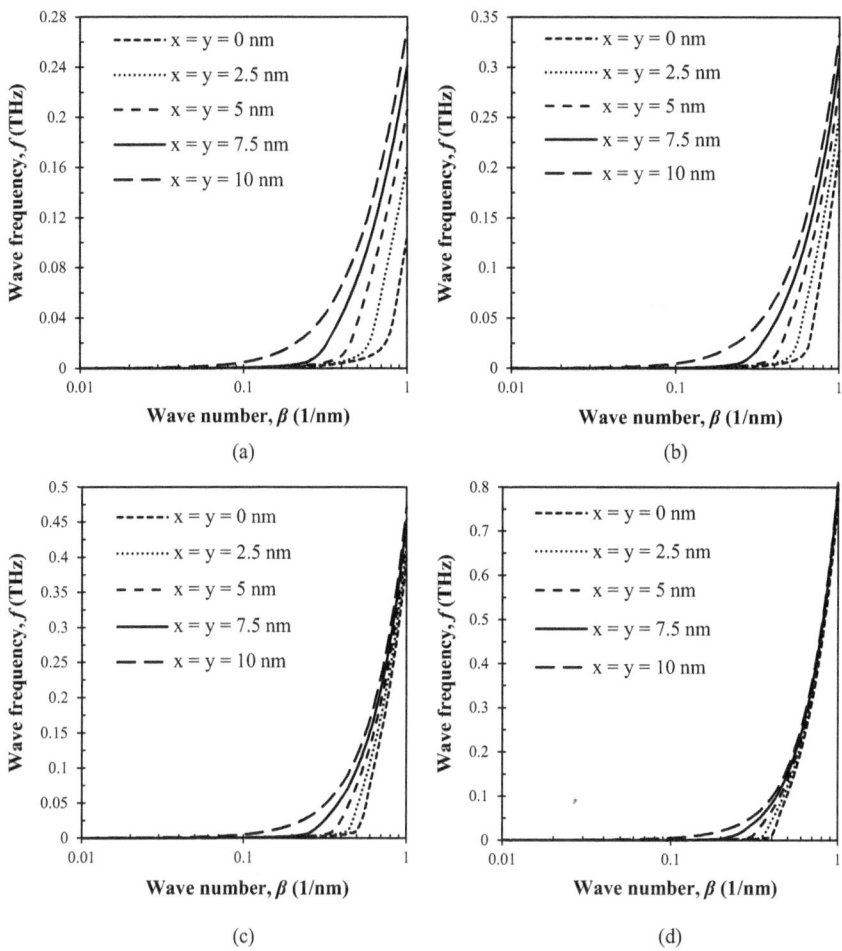

FIGURE 5.8 Variation of wave frequency versus wave number for various x and y amounts at (a) $\eta = 0$ nm (b) $\eta = 0.5$ nm (c) $\eta = 1$ nm (d) $\eta = 2$ nm ($\mu = 1$ nm, $k_w = k_p = 0$, $\zeta = \lambda = 1$, $N = 1$).

established that the larger the distance is between the loading location and the graphene sheet's origin, the higher the resulting wave frequencies. Once higher wave frequencies are the target, the loading location should be as far away from the origin as practicable.

In addition, the impacts of load factor are explored in Figure 5.9 to emphasize the implications of various bending force types, as shown in Figure 5.7. Similar to the preceding diagram, this one also accounts for impacts on a smaller scale. One of the growing coefficients that might magnify the growth in wave frequency is the load factor. Figure 5.9(c) and (d) show that the wave frequency contains non-zero quantities when the nonlocality is less than 1 and 2, respectively, but in cases

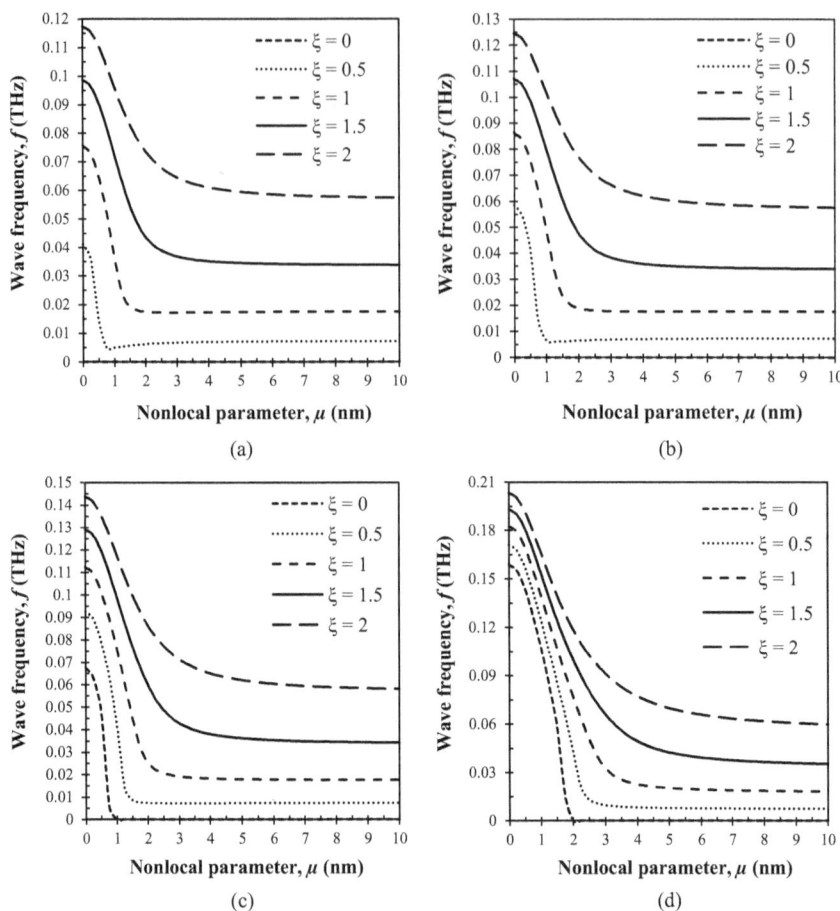

FIGURE 5.9 Variation of wave frequency versus nonlocal parameter for various load factors at (a) $\eta = 0$ nm (b) $\eta = 0.5$ nm (c) $\eta = 1$ nm (d) $\eta = 2$ nm ($k_w = k_p = 0$, $\lambda = 1$, $N = 1$).

(a) and (b), the frequency values are all zero when nonlocality fluctuates. The frequency of waves also decreases as nonlocality increases, demonstrating the stiffness-softening effect. It's important to note that the frequency of a wave never changes beyond a certain value of a nonlocal parameter ($\mu > 7$ nm).

Figure 5.10 examines the effects of the biaxial load factor. For a range of biaxial load factors, a plot of wave frequency versus nonlocal parameters is shown below. Load factor effects are clearly distinct from biaxial load factor effects. In other words, if you raise the load factor, the frequency of the waves will rise, whereas if you increase the biaxial load factor, the frequency of the waves will decrease. Another distinction is that the wave frequency is not zero for all values of the biaxial load factor, although for certain nonlocal parameters it does approach zero (at $\eta = 0$, $\eta = 0.5$). Likewise previous diagram, here, wave frequency does not vary once $\mu > 7$ nm [37].

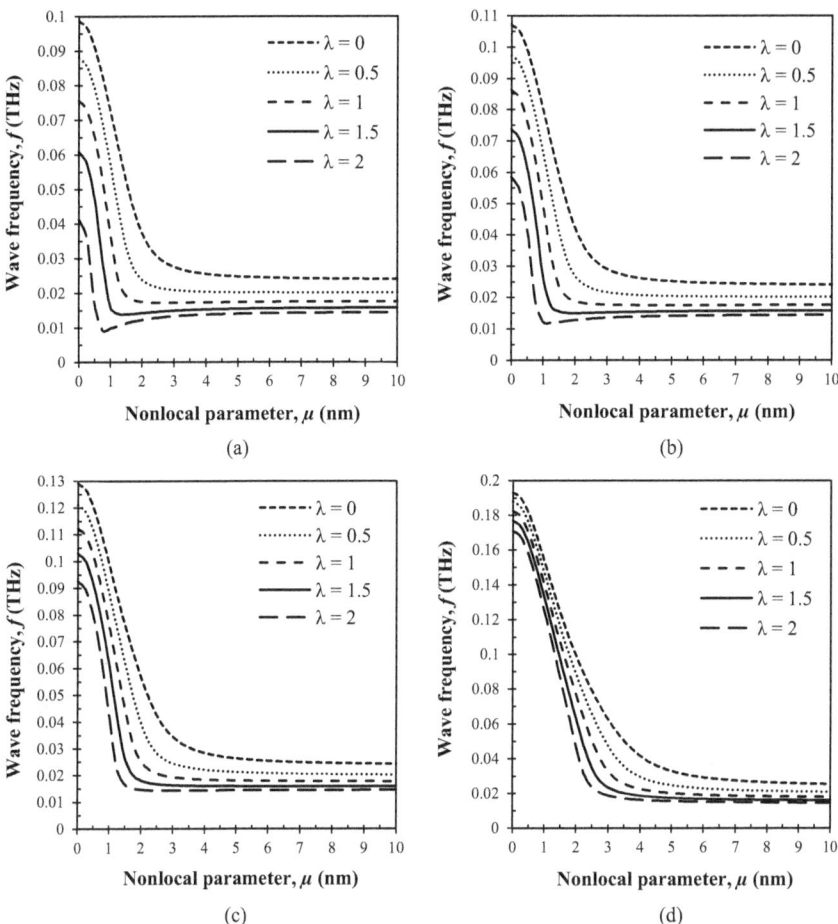

FIGURE 5.10 Variation of wave frequency versus nonlocal parameter for various biaxial load factors at (a) $\eta = 0$ nm (b) $\eta = 0.5$ nm (c) $\eta = 1$ nm (d) $\eta = 2$ nm ($k_w = k_p = 0$, $\zeta = 1$, $N = 1$).

In addition to the previously mentioned problems, the effects of the base parameters should also be investigated. The impact of the Winkler coefficient is shown in Figure 5.11. Similar to other representations, this one displays the graphene sheet's dependence on size. The essential element is that when a wave is amplified, the Winkler parameter might cause the frequency to increase. In addition, it's worth noting that the aforementioned coefficient can only affect the frequency of waves in a narrow range of wave numbers, typically below $\beta = 1 \times 10^9$ m.

Finally, Figure 5.12 is shown to elaborate on how the Pasternak coefficient might inspire numbers for wave frequencies. Here, it can be shown without a doubt that raising the Pasternak coefficient raises the value of the wave frequency. However, this graphic has a somewhat different form from the one we saw before. The most noticeable shifts occur at very low wave numbers, and they are caused

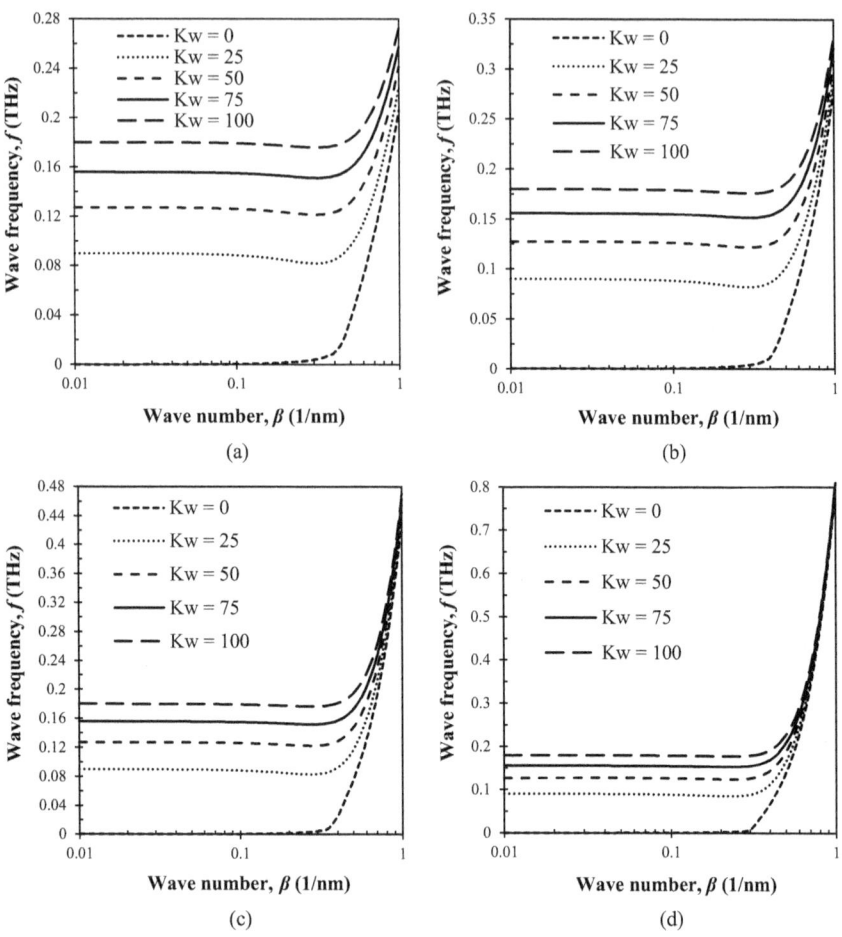

FIGURE 5.11 Variation of wave frequency versus wave number for various Winkler coefficients at (a) $\eta = 0$ nm (b) $\eta = 0.5$ nm (c) $\eta = 1$ nm (d) $\eta = 2$ nm ($\mu = 1$ nm, $k_p = 0$, $\zeta = \lambda = 1$, $N = 1$).

by a modification to the Winkler coefficient. Changes in wave numbers between $\beta = 0.1 \times 10^9$ m and $\beta = 1 \times 10^9$ m stand out most clearly in this figure.

5.2.2.3 Viscoelastic Foundation Effects

In this section, we explore the impact of a visco-Pasternak basis on the wave dispersion properties of graphene sheets, as shown in Figure 5.13. The damping effects of a viscoelastic medium are described using the Kelvin–Voigt model [39]. A visco-Pasternak substrate with a linear (Winkler) constant, a nonlinear (Pasternak) constant, and a damping (damping) constant is assumed to support the graphene sheet.

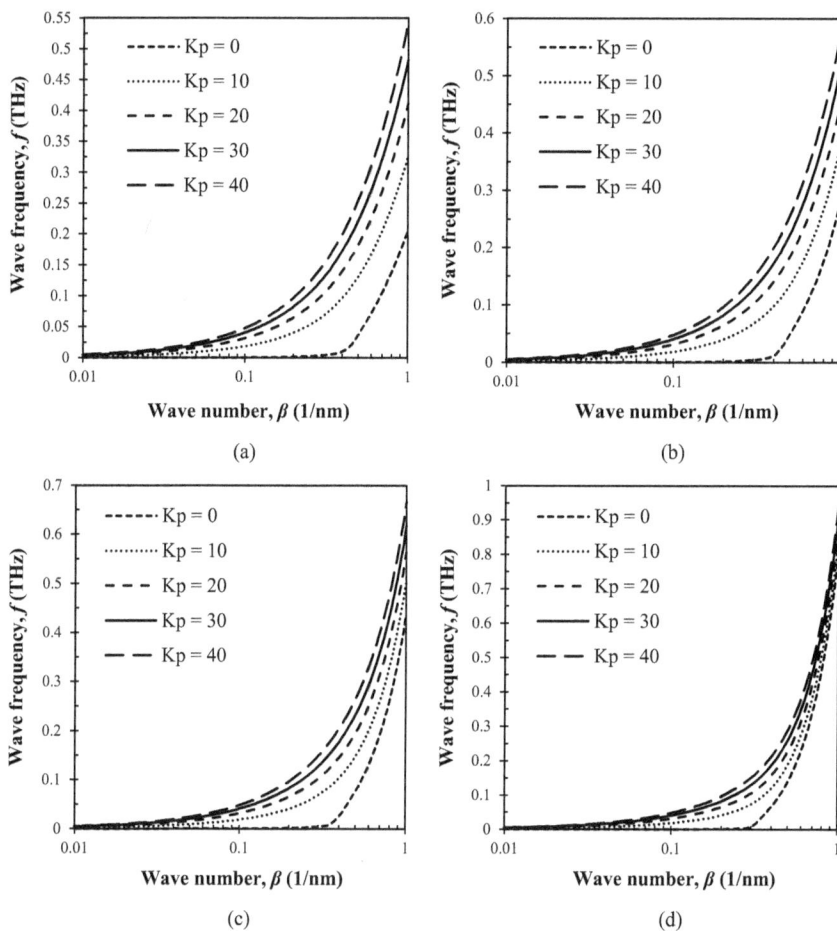

FIGURE 5.12 Variation of wave frequency versus wave number for various Pasternak coefficients at (a) $\eta = 0$ nm (b) $\eta = 0.5$ nm (c) $\eta = 1$ nm (d) $\eta = 2$ nm ($\mu = 1$ nm, $k_w = 0$, $\zeta = \lambda = 1$, $N = 1$).

The effects of the structural damping coefficient are also investigated. Here, the significance of the structural damping coefficient is shown with the aid of Figure 5.14. Once this coefficient is made 0, it is obvious that the wave frequency approaches infinity. Instead, in a damped scenario, the frequency of waves begins at zero, grows to its maximum value, and then gradually decreases back to zero as the structural damping coefficient acquires nonzero values. As the structural damping coefficient rises, the damping effect is assumed to get stronger. However, the wave number at which the greatest frequency occurs in the diagram is not the same for different amounts of the called coefficient, even if the overall form of the diagram is the same for all nonzero values of structural damping coefficient. For a given value of the structural damping coefficient [39], the diagram's peak shifts to the left.

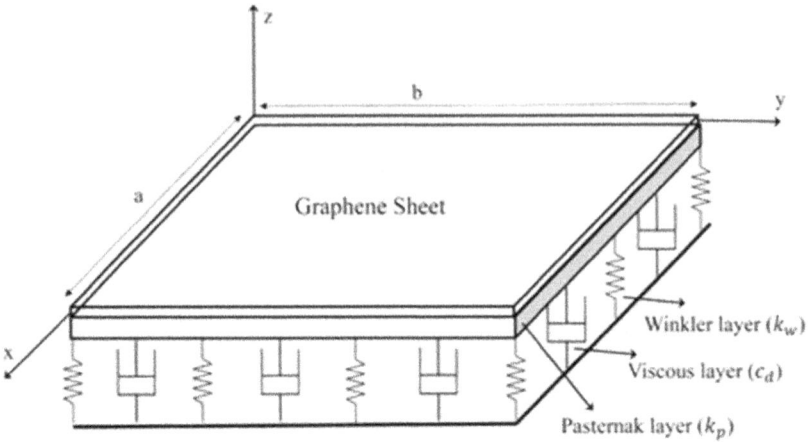

FIGURE 5.13 Geometry of a single-layered graphene sheet rested on viscoelastic medium.

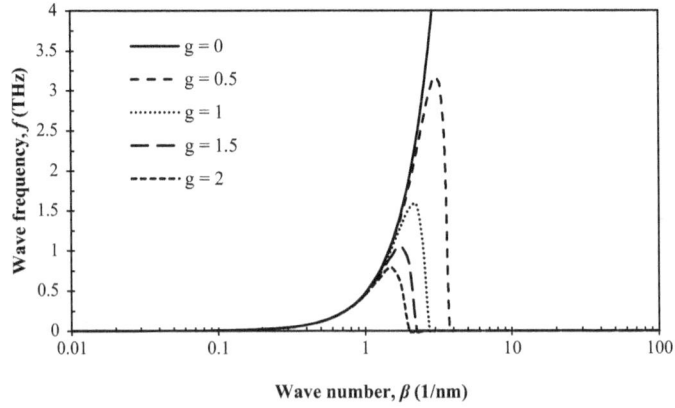

FIGURE 5.14 Effect of the structural damping coefficient on wave frequency of graphene sheets ($\mu = \eta = 1$ nm, $k_w = k_p = 0$, $C_d = 0$, $\Delta T = 0$).

   In contrast, the combined effects of small size and damping coefficient are emphasized in Figure 5.15. Consistent with earlier illustrations, there seem to be two size-dependent routes for elevating graphene sheets' wave frequencies. Applying smaller nonlocal parameters is one option, while increasing the length scale parameter is another. According to the following illustrations, wave frequency exhibits a self-suppressing pattern. When a damping coefficient is used, the frequency of the waves is often reduced. It should also be noted that at a certain threshold value of the damping coefficient, in this example about 12, the frequency of the waves becomes precisely zero [38].

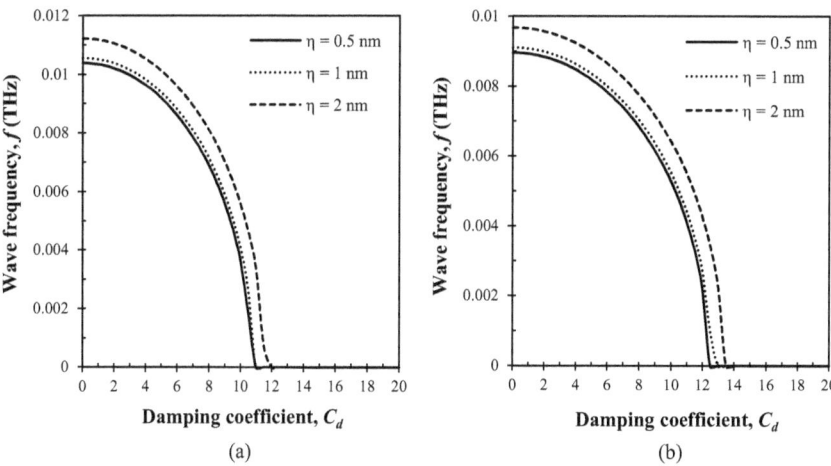

FIGURE 5.15 Variation of wave frequency versus damping coefficient for different length scale parameters at (a) $\mu = 1$ nm, (b) ($k_w = k_p = 0$, $\Delta T = 0$, $g = 0$, $\beta = 0.15 \times 10^9$ m).

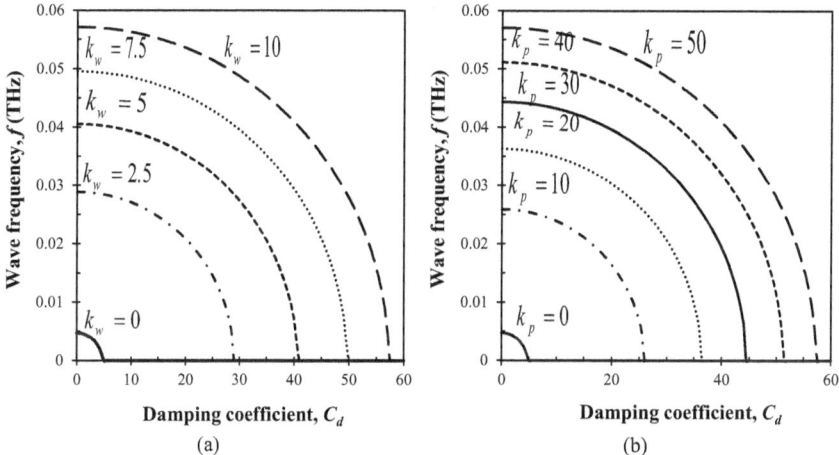

FIGURE 5.16 Variation of wave frequency versus damping coefficient for different (a) Winkler coefficients ($k_p = 0$), (b) Pasternak coefficients ($k_w = 0$). ($\mu = \eta = 1$ nm, g = 1, $\Delta T = 0$, $\beta = 0.1 \times 10^9$ m).

Now we will examine the impact of the Winkler and Pasternak coefficients by graphing the wave frequency against the damping coefficient. The possibility for amplification of wave frequencies may be shown to exist in both the Winkler (linear) and Pasternak (nonlinear) coefficients. Also, the most noticeable shift in the wave frequency responses occurs when the Winkler coefficient is shifted from $k_w = 0$ to $k_w = 2.5$. To put it another way, the most effective way to increase the frequency of waves is to modify the value of the Winkler or Pasternak coefficient from zero to the first nonzero number.

In addition, Figure 5.17 illustrates how a temperature gradient might have an effect. Pasternak coefficient's growing significance is a fact that cannot be denied. Furthermore, when the temperature gradient grows, the frequency of the waves decreases. When the Winkler coefficient is raised from 0 to 10, the frequency of the waves is clearly visible to rise, albeit only by a little amount. Whether ΔT is adjusted from 0 to 200 or 400 to 600 Kelvin, the impact of the temperature gradient is nearly the same, and the frequency of the waves increases by about the same amount in both cases.

Wave frequency versus damping coefficient for various temperature gradients is shown in Figure 5.18 to summarize the aforementioned variance. Clearly, a steeper temperature gradient results in a lower wave frequency. In addition, the damping effect may be felt by looking at the figure. When comparing Figures 5.15 and 5.18, it is clear that the frequency of waves decreases later for lower wave numbers. In reality, larger damping coefficients are required to reach zero for lower wave numbers.

5.2.2.4 Magnetic Field Effects

In addition, the effect of magnetic field strength is seen in Figure 5.19. This diagram shows the relationship between the expected change in magnetic field strength and the corresponding change in phase velocity for three key situations ($\lambda < \mu$, $\lambda = \mu$, and $\lambda > \mu$). The phase velocity of graphene sheets is affected by the strength of the magnetic field, making all three of these cases quite comparable. In reality, increasing the strength of the magnetic field has the effect of increasing the phase velocity. If $\lambda < \mu$, then the phase velocity will always decrease as the magnetic field strength increases. In addition, when the nonlocal and length scale

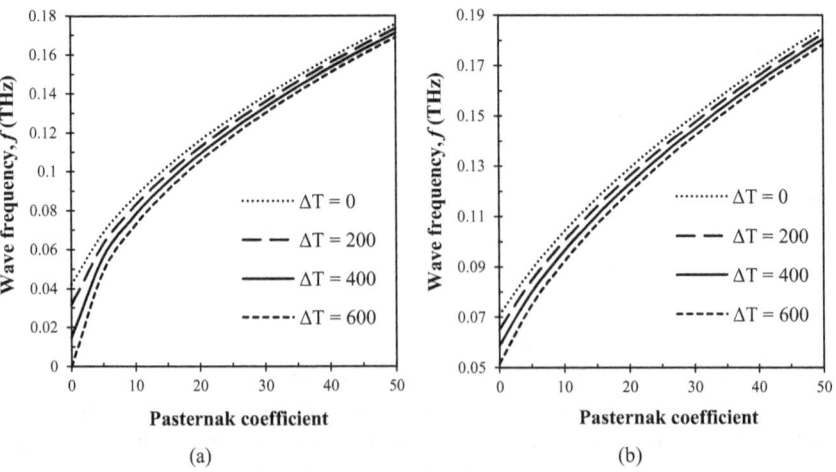

FIGURE 5.17 Variation of wave frequency versus Pasternak coefficient for different temperature gradients at (a) $k_w = 0$, (b) $k_w = 10$ ($\mu = \eta = 1$ nm, $C_d = 0$, g = 1, $\beta = 0.3 \times 10^9$ m).

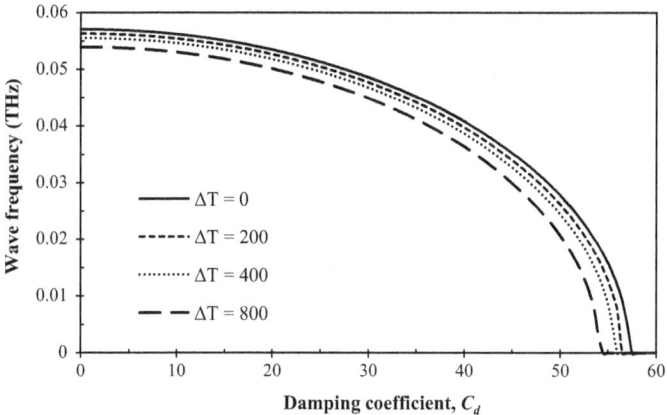

FIGURE 5.18 Mixed effects of temperature gradient and damping coefficient on wave frequency of graphene sheets ($\mu = \eta = 1$ nm, $k_w = k_p = 0$, g = 1, $\beta = 0.1 \times 10^9$ m).

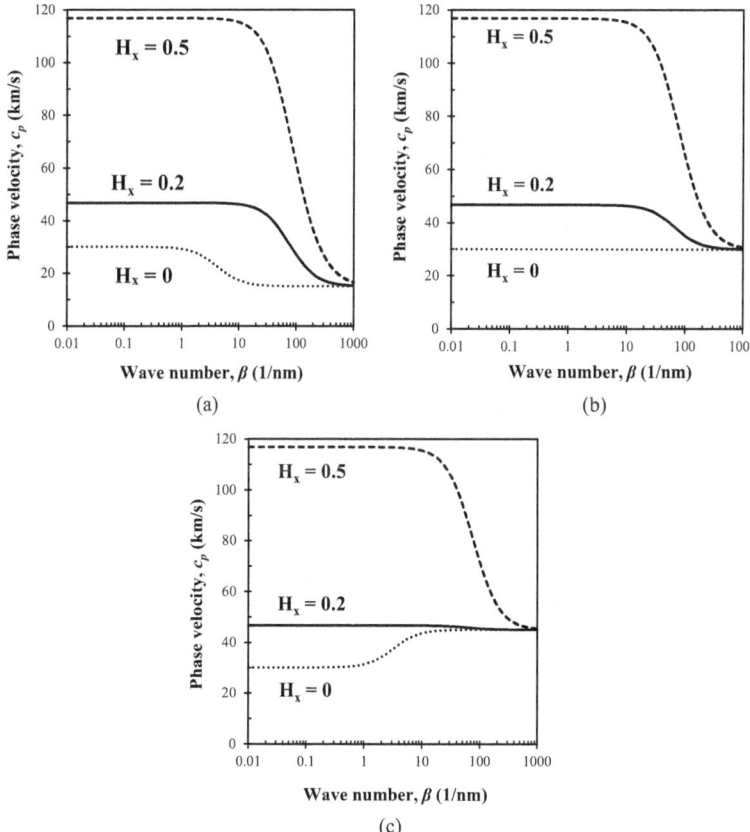

FIGURE 5.19 Variation of phase velocity versus wave number for various values of magnetic field intensity at (a) $\lambda < \mu$, (b) $\lambda = \mu$ and (c) $\lambda > \mu$ ($k_w = k_p = 0$).

parameters ($\lambda = \mu$) are both equal to one another, the phase velocity is constant for all wave numbers when the magnetic field intensity is zero, but it decreases for nonzero values of the magnetic field intensity. Furthermore, when $\lambda > \mu$, the phase velocity is fixed initially but progressively rises as the magnetic field strength is reduced to zero [38].

In addition, Figure 5.20 is depicted in order to magnify the influences of Winkler coefficient on the phase velocity of SLGSs for three major conditions ($\lambda < \mu$, $\lambda = \mu$, and $\lambda > \mu$). It is clear that Winkler coefficient affects phase velocity amounts once wave number is assumed to be smaller than $\beta = 0.04 \times 10^9$. Obviously, this parameter is powerful enough to motivate phase velocity values in

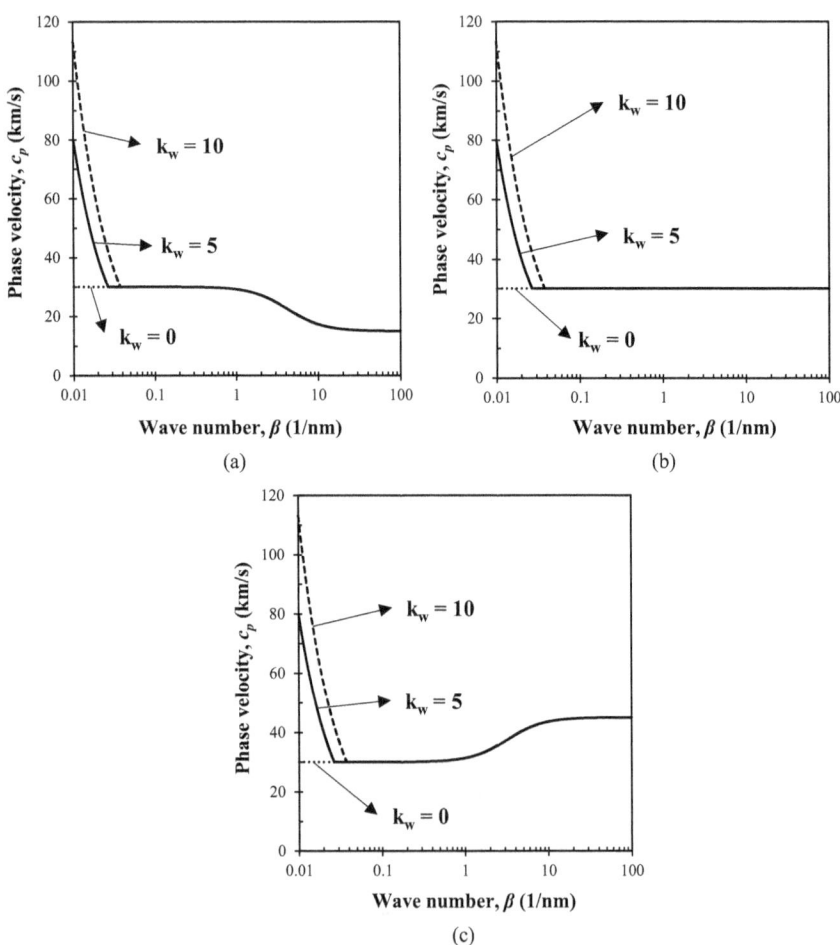

FIGURE 5.20 Variation of phase velocity versus wave number for various values of Winkler coefficient at (a) $\lambda < \mu$, (b) $\lambda = \mu$ and (c) $\lambda > \mu$ ($k_p = 0$, $H_x = 0$).

the mentioned range of wave numbers. On the other hand, it is of significance to advert that phase velocity's behavior is a bit different in higher wave numbers once nonlocal and length scale parameters are supposed to obtain various amounts compared to each other. Precisely, phase velocity experiences a decreasing, unchangeable, and increasing trend in the high wave numbers in the cases of $\lambda < \mu$, $\lambda = \mu$ and $\lambda > \mu$, respectively [38].

Now, it is turn to highlight the effects of nonlinear foundation parameter on the phase velocity of SLGSs for the same major conditions ($\lambda < \mu$, $\lambda = \mu$ and $\lambda > \mu$). Based on Figure 5.21, it shall be regarded that the effects of the Pasternak coefficient can be well observed in a wide range of wave numbers. Clearly, higher phase

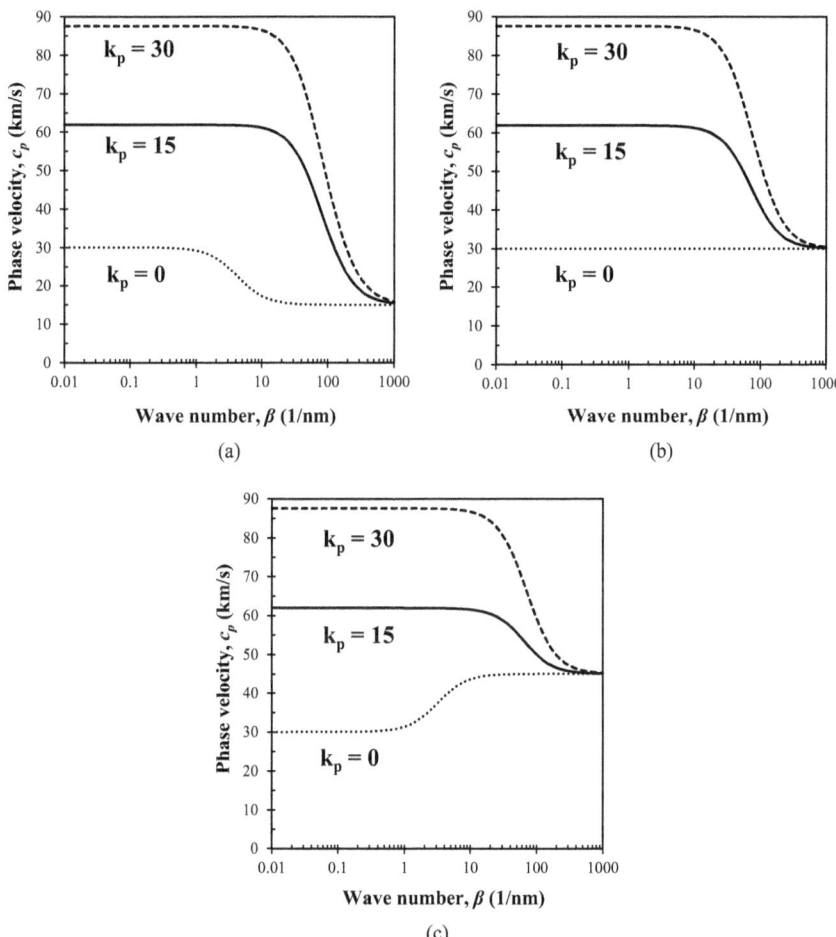

FIGURE 5.21 Variation of phase velocity versus wave number for various values of Pasternak coefficient at (a) $\lambda < \mu$, (b) $\lambda = \mu$, and (c) $\lambda > \mu$ ($k_w = 0$, $H_x = 0$).

velocity amounts are achieved if greater Pasternak coefficients are employed. Moreover, it is clear to everybody that the size-dependent behavior of SLGSs happens in this diagram in the same form as occurred in the previous figure. In other words, once Pasternak coefficient is set to zero ($k_p = 0$), wave responses of SLGSs in high wave numbers are hugely influenced by the ratio of nonlocal to length scale parameters. It is remarkably worth that phase velocity shows a similar response in each of the three main cases once the Pasternak parameter obtained a nonzero value ($k_p \neq 0$). In fact, in this condition ($k_p \neq 0$), phase velocity achieves the same amount in the beginning and then starts a continuous decrease to its final value.

Finally, Figure 5.22 plots the fluctuation of escape frequency versus nonlocal parameter for different length scale parameters, providing coverage of the combined impacts of nonlocal and length scale factors. Parameters for the other players in this diagram are not included in the figure since they have no effect on the escape frequency ($\beta \to \infty$). Here, prior findings are reaffirmed. Actually, there is a distinct downward trend in escape frequency as nonlocality grows. However, by increasing the values for the length scale parameters, the escape frequency may be increased [38].

5.2.3   Double-layered Graphene Sheets: Wave Dispersion Characteristics

The focus here is on the kinematic behaviors of graphene sheets, and how they are described. A diagrammatic representation of embedded double-layered graphene sheet DLGS is shown in Figure 5.23.

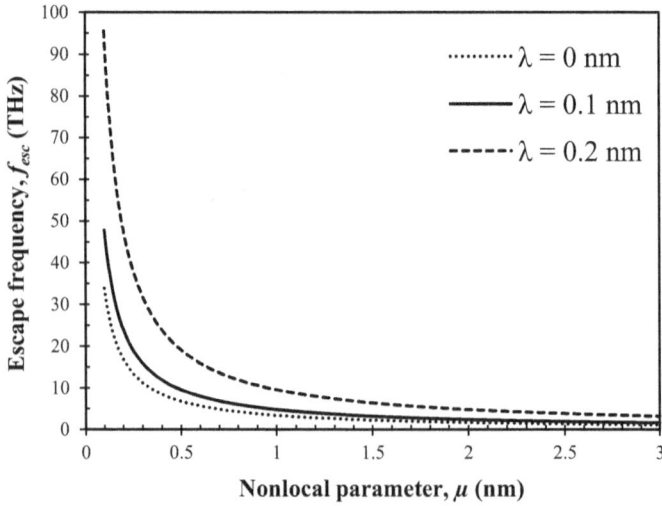

FIGURE 5.22   Variation of escape frequency versus nonlocal parameter for various values of length scale parameter ($k_w = k_p = 0$, $H_x = 0$).

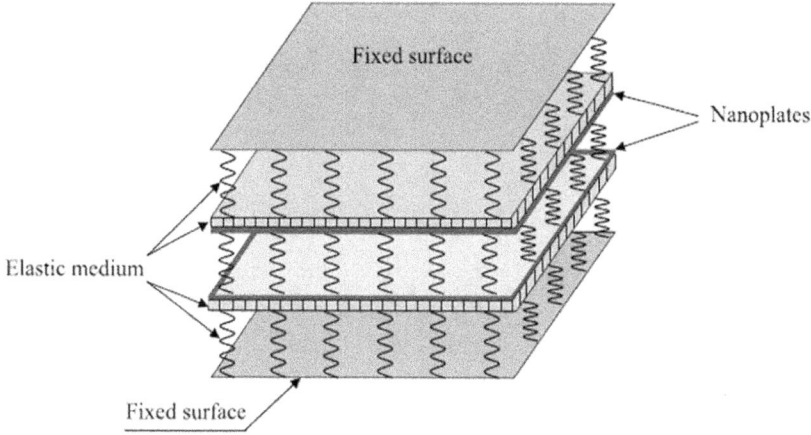

FIGURE 5.23 Geometry of a double-layered graphene sheet rested on Winkler–Pasternak foundation.

Here, we may express the displacement fields as:

$$\begin{cases} U_i(x,y,z) = -z\dfrac{\partial w_i}{\partial x} \\[2mm] V_i(x,y,z) = -z\dfrac{\partial w_i}{\partial y} \quad , i = (1,2) \\[2mm] W_i(x,y,z) = w_i(x,y) \end{cases} \tag{5.17}$$

where w_i is bending deflection of the i-th plate in the thickness direction, respectively. Now, we can express the nonzero stresses in each graphene layer as:

$$\begin{Bmatrix} \varepsilon_{xx,i} \\ \varepsilon_{yy,i} \\ \gamma_{xy,i} \end{Bmatrix} = z \begin{Bmatrix} -\dfrac{\partial^2 w_i}{\partial x^2} \\[2mm] -\dfrac{\partial^2 w_i}{\partial y^2} \\[2mm] -2\dfrac{\partial^2 w_i}{\partial x \partial y} \end{Bmatrix}, i = (1,2) \tag{5.18}$$

The principle of Hamilton, moreover, may be stated as:

$$\int_0^t \delta(U - T + V)\,dt = 0 \tag{5.19}$$

in which U is strain energy, T is kinetic energy, and V is work done by external loads. The variation of strain energy for each plate can be calculated as:

$$\delta U = \int_V \sigma_{mn,i} \delta \varepsilon_{mn,i} dV = \int_V \left(\sigma_{xx,i} \delta \varepsilon_{xx,i} + \sigma_{yy,i} \delta \varepsilon_{yy,i} + \sigma_{xy,i} \delta \gamma_{xy,i} \right) dV, i = (1,2) \quad (5.20)$$

Putting Eq. (5.18) into Eq. (5.20) leads to the following:

$$\delta U_i = \int_0^a \int_0^b \left(-M_{xx,i} \frac{\partial^2 \delta w_i}{\partial x^2} - 2M_{xy,i} \frac{\partial^2 \delta w_i}{\partial x \partial y} - M_{yy,i} \frac{\partial^2 \delta w_i}{\partial y^2} \right) dy dx, i = (1,2) \quad (5.21)$$

The parameters in Eq. (5.21) may be written as follows.

$$M_{j,i} = \int_{-h/2}^{h/2} z \sigma_{j,i} dz, j = (xx, yy, xy), i = (1,2) \quad (5.22)$$

In addition, the following demonstrates how the output of external forces varies:

$$\delta V_i = \int_0^a \int_0^b \left(\begin{array}{c} N_x^0 \dfrac{\partial w_i}{\partial x} \dfrac{\partial \delta w_i}{\partial x} + N_y^0 \dfrac{\partial w_i}{\partial y} \dfrac{\partial \delta w_i}{\partial y} \\[2mm] -k_w \delta w_i + k_p \left(\dfrac{\partial w_i}{\partial x} \dfrac{\partial \delta w_i}{\partial x} + \dfrac{\partial w_i}{\partial y} \dfrac{\partial \delta w_i}{\partial y} \right) - C_d \delta \dfrac{\partial w_i}{\partial t} \end{array} \right) dy dx, i = (1,2) \quad (5.23)$$

where N_x^0, N_y^0 are in-plane applied loads; k_w, k_p, and C_d are Winkler, Pasternak, and damping coefficients, respectively. The variation of the kinetic energy will be written as:

$$\delta K = \int_0^a \int_0^b \left(I_0 \left(\frac{\partial w_i}{\partial t} \frac{\partial \delta w_i}{\partial t} \right) + I_2 \left(\frac{\partial w_i}{\partial x \partial t} \frac{\partial \delta w_i}{\partial x \partial t} + \frac{\partial w_i}{\partial y \partial t} \frac{\partial \delta w_i}{\partial y \partial t} \right) \right) dy dx \quad (5.24)$$

in which

$$(I_0, I_2) = \int_{-h/2}^{h/2} \left(1, z^2 \right) \rho dz \quad (5.25)$$

The Euler–Lagrange equations for each graphene sheet may be recast by inserting Eqs. (5.21), (5.23), and (5.24) into Eq. (5.19) and zeroing out the coefficients of δw_i.

$$\frac{\partial^2 M_{xx,i}}{\partial x^2} + 2\frac{\partial^2 M_{xy,i}}{\partial x \partial y} + \frac{\partial^2 M_{yy,i}}{\partial y^2} + N_x^0 \frac{\partial^2 w_i}{\partial x^2} + N_y^0 \frac{\partial^2 w_i}{\partial y^2}$$

$$- k_w w_i - C_d \frac{\partial w_i}{\partial t} + k_p \nabla^2 w_i = I_0 \frac{\partial^2 w_i}{\partial t^2} - I_2 \nabla^2 \left(\frac{\partial^2 w_i}{\partial t^2}\right), i = (1,2) \quad (5.26)$$

where $N_x^0 = N_y^0 = N$, where N is the axial load that is being applied. According to the nonlocal strain gradient theory, Eqs. 5.8a–d for the stress field take into consideration not only the effects of the strain gradient stress field but also those of the nonlocal elastic stress field. The theory may be presented in this manner with regard to elastic solids:

$$\sigma_{ij} = \sigma_{ij}^{(0)} - \frac{d\sigma_{ij}^{(1)}}{dx} \quad (5.27)$$

in Eq. (5.27), the stresses $\sigma_{xx}^{(0)}$ (classical stress) and $\sigma_{xx}^{(1)}$ (higher-order stress) are corresponding to strain ε_{xx} and strain gradient $\varepsilon_{xx,x}$, respectively, as follows:

$$\begin{cases} \sigma_{ij}^{(0)} = \int_0^L C_{ijkl}\alpha_0\left(x,x',e_0a\right)\varepsilon'_{kl}\left(x'\right)dx' \\ \\ \sigma_{ij}^{(1)} = l^2 \int_0^L C_{ijkl}\alpha_1\left(x,x',e_1a\right)\varepsilon'_{kl,x}\left(x'\right)dx' \end{cases} \quad (5.28)$$

in which C_{ijkl} is the elastic coefficient; e_0a and e_1a are introduced to account for the nonlocality effects. Also, l captures the strain gradient effects. Once the nonlocal kernel functions $\alpha_0(x, x', e_0a)$ and $\alpha_1(x, x', e_1a)$ satisfy the developed conditions, the constitutive relation of nonlocal strain gradient theory can be expressed as below:

$$\left(1-\left(e_1a\right)^2 \nabla^2\right)\left(1-\left(e_0a\right)^2 \nabla^2\right)\sigma_{ij} = C_{ijkl}\left(1-\left(e_1a\right)^2 \nabla^2\right)\varepsilon_{kl}$$

$$- C_{ijkl}l^2\left(1-\left(e_0a\right)^2 \nabla^2\right)\nabla^2\varepsilon_{kl} \quad (5.29)$$

in which ∇^2 denotes the Laplacian operator. Considering $e_1 = e_0 = e$, the general constitutive relation in Eq. (5.31) becomes:

$$\left(1-\left(ea\right)^2 \nabla^2\right)\sigma_{ij} = C_{ijkl}\left(1-l^2\nabla^2\right)\varepsilon_{kl} \quad (5.30)$$

Finally, the reduced constitutive relation looks like this:

$$\left(1-\mu^2\nabla^2\right)\begin{Bmatrix}\sigma_{xx}\\\sigma_{yy}\\\sigma_{xy}\end{Bmatrix}=\left(1-\eta^2\nabla^2\right)\begin{pmatrix}Q_{11}&Q_{12}&0\\Q_{12}&Q_{22}&0\\0&0&Q_{66}\end{pmatrix}\begin{Bmatrix}\varepsilon_{xx}\\\varepsilon_{yy}\\\varepsilon_{xy}\end{Bmatrix}\tag{5.31}$$

in Eq. (5.31)

$$Q_{11}=Q_{22}=\frac{E}{1-v^2},Q_{12}=vQ_{11},Q_{66}=\frac{E}{2(1+v)}\tag{5.32}$$

where $\mu=e_0 a$ and $\eta=l$. Now, employing the Kelvin–Voigt model for viscoelastic nanoplates and inserting Eq. (5.22) in Eq. (5.31) gives:

$$\left(1-\mu^2\nabla^2\right)\begin{Bmatrix}M_{xx}\\M_{yy}\\M_{xy}\end{Bmatrix}=\left(1-\eta^2\nabla^2\right)\left(1+g\frac{\partial}{\partial t}\right)\begin{pmatrix}D_{11}&D_{12}&0\\D_{12}&D_{22}&0\\0&0&D_{66}\end{pmatrix}\begin{Bmatrix}-\dfrac{\partial^2 w_i}{\partial x^2}\\-\dfrac{\partial^2 w_i}{\partial y^2}\\-2\dfrac{\partial^2 w_i}{\partial x\partial y}\end{Bmatrix}\tag{5.33}$$

The structural damping coefficient, g, may be written as follows for the cross-sectional rigidities in Eq. (5.33):

$$\begin{Bmatrix}D_{11}\\D_{12}\\D_{66}\end{Bmatrix}=\int_{-h/2}^{h/2}Q_{11}z^2\begin{Bmatrix}1\\v\\\dfrac{1-v}{2}\end{Bmatrix}dz\tag{5.34}$$

It is possible to immediately construct the nonlocal governing equation in terms of displacements for each layer of DLGSs by putting Eq. (5.33) into Eq. (5.26).

$$\left(1-\eta^2\nabla^2\right)\left(1+g\frac{\partial}{\partial t}\right)\left(D_{11}\frac{\partial^4 w_i}{\partial x^4}+2\left(D_{12}+2D_{66}\right)\frac{\partial^4 w_i}{\partial x^2\partial y^2}+D_{22}\frac{\partial^4 w_i}{\partial y^4}\right)$$
$$+\left(1-\mu^2\nabla^2\right)\begin{pmatrix}I_0\dfrac{\partial^2 w_i}{\partial t^2}-I_2\left(\dfrac{\partial^4 w_i}{\partial x^2\partial t^2}+\dfrac{\partial^4 w_i}{\partial y^2\partial t^2}\right)+\\k_w w_i+C_d\dfrac{\partial w_i}{\partial t}-\left(k_p-N\right)\left(\dfrac{\partial^2 w_i}{\partial x^2}+\dfrac{\partial^2 w_i}{\partial y^2}\right)\end{pmatrix}=0,i=(1,2)\tag{5.35}$$

Without considering interactions between layers, Eq. (5.35) represents the nonlocal governing equations of each layer. To explain this observation, we use the van der Waals (vdW) model.

$$\left(1-\eta^2\nabla^2\right)\left(1+g\frac{\partial}{\partial t}\right)\left(D_{11}\frac{\partial^4 w_1}{\partial x^4}+2\left(D_{12}+2D_{66}\right)\frac{\partial^4 w_1}{\partial x^2 \partial y^2}+D_{22}\frac{\partial^4 w_1}{\partial y^4}\right)$$

$$+\left(1-\mu^2\nabla^2\right)\begin{pmatrix} I_0\dfrac{\partial^2 w_1}{\partial t^2}-I_2\left(\dfrac{\partial^4 w_1}{\partial x^2 \partial t^2}+\dfrac{\partial^4 w_1}{\partial y^2 \partial t^2}\right)+k_w w_1 + C_d \dfrac{\partial w_1}{\partial t} \\[2ex] -\left(k_p-N\right)\left(\dfrac{\partial^2 w_1}{\partial x^2}+\dfrac{\partial^2 w_1}{\partial y^2}\right)+C\left(w_1-w_2\right) \end{pmatrix} = 0 \qquad (5.36)$$

$$\left(1-\eta^2\nabla^2\right)\left(1+g\frac{\partial}{\partial t}\right)\left(D_{11}\frac{\partial^4 w_2}{\partial x^4}+2\left(D_{12}+2D_{66}\right)\frac{\partial^4 w_2}{\partial x^2 \partial y^2}+D_{22}\frac{\partial^4 w_2}{\partial y^4}\right)$$

$$+\left(1-\mu^2\nabla^2\right)\begin{pmatrix} I_0\dfrac{\partial^2 w_2}{\partial t^2}-I_2\left(\dfrac{\partial^4 w_2}{\partial x^2 \partial t^2}+\dfrac{\partial^4 w_2}{\partial y^2 \partial t^2}\right)+k_w w_2 + C_d \dfrac{\partial w_2}{\partial t} \\[2ex] -\left(k_p-N\right)\left(\dfrac{\partial^2 w_2}{\partial x^2}+\dfrac{\partial^2 w_2}{\partial y^2}\right)+C\left(w_2-w_1\right) \end{pmatrix} = 0 \qquad (5.37)$$

The vdW interaction coefficient is denoted by C in the foregoing equations. Following their derivation in the preceding section, the nonlocal governing equations will now be solved analytically. The following is a definition of the displacement fields, which are considered to be exponential:

$$\begin{Bmatrix} w_1(x,y,t) \\ w_2(x,y,t) \end{Bmatrix} = \begin{Bmatrix} W_1 \exp\left[i\left(\beta_1 x + \beta_2 y - \omega t\right)\right] \\ W_2 \exp\left[i\left(\beta_1 x + \beta_2 y - \omega t\right)\right] \end{Bmatrix} \qquad (5.38)$$

where W_1 and W_2 are the unknown coefficients; β_1 and β_2 are the wave numbers of wave propagation along x and y directions, respectively, and finally ω is wave's angular frequency. Now, substituting Eq. (5.38) into Eqs. (5.36) and (5.37) results in:

$$\left(\left[K\right]_{2\times2}-\omega^2\left[M\right]_{2\times2}\right)\{\Delta\}=\{0\} \qquad (5.39)$$

where the corresponding k_{ij}, m_{ij} are as written in Appendix 5B. The unknown parameters of Eq. (5.39) can be noted as follows:

$$\{\Delta\}=\{W_1,W_2\}^T \qquad (5.40)$$

The determinant on the left-hand side of Eq. (5.39) must be zero in order to get the angular frequency of the wave:

$$\left|\left[K\right]_{2\times2} - \omega^2\left[M\right]_{2\times2}\right| = 0 \tag{5.41}$$

In Eq. (5.41) by setting $\beta_1 = \beta_2 = \beta$ and solving the obtained equation for ω, the wave's angular frequency of embedded DLGSs can be calculated. Also, the escape frequency of DLGSs can be derived by tending wave number to infinity:

$$f_{esc} = \lim_{\beta\to\infty} \frac{\omega}{2\pi} \tag{5.42}$$

Here, we evaluate how different DLGSs react to changes in key factors that affect wave propagation. Graphene sheets are characterized by their material qualities as: $E = 1$ TPa, $\nu = 0.19$, $\rho = 2300$ kg/m³. Also, the thickness is presumed to be $h = 0.34$ nm. In addition, the vdW interaction coefficient can be supposed to be $C = -108$ GPa/nm [43]. Calculating wave frequencies in the following diagrams is accomplished by using the formula $2\pi\left(f = \dfrac{\omega}{2\pi}\right)$ as a division factor to the wave's angular frequency. The major purpose of Figure 5.24 is to illustrate the influence of a nonlocal parameter on the wave frequency behavior of DLGSs. The decrease in wave frequency associated with growing nonlocality is graphically shown here. The purpose of this illustration is to explain how Eringen's theory may be used to anticipate the characteristics of DLGSs despite their tiny size.

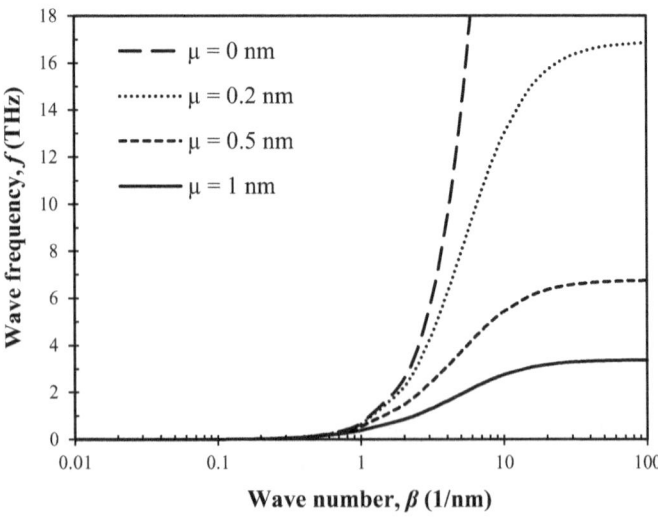

FIGURE 5.24   Influences of different nonlocal parameters on wave frequency of DLGSs ($\eta = 0$ nm, $k_w = k_p = 0$).

Nonlocality has a reasonable impact that reduces the frequency of waves. The mechanical responses of the structures are constrained to have lower values due to this phenomenon, which is referred to in different studies as a stiffness-softening effect. Also, remember that if nonlocality is ignored, the frequency of a wave approaches infinity ($\mu = 0$).

The values of phase velocities of DLGSs are also investigated in this manner. Once the length scale parameter is disregarded, the focus shifts to Figure 5.25, which is devoted to the study of the effects caused by the nonlocal parameter on the phase velocity. Clearly, this graphic demonstrates the same impact as the one before it. In this diagram, however, the phase velocity changes in two distinct ways depending on the values of the nonlocal parameters. As the wave number increases in the local condition ($\mu = 0$), the phase velocity first increases to a maximum and then stays constant. The other aesthetic qualifier ($\mu \neq 0$) is associated with the need that localized effects be taken into account. In the second, the phase velocity first grows to a maximum before shrinking down to zero.

Two earlier representations attempted to do the same thing by showing how changing a nonlocal parameter affected the DLGSs' wave frequency and phase velocity when the strain gradient stress field was disregarded. Here, we'll look at two diagrams—one for wave frequency and another for phase velocity—to examine the effects of strain gradients. Therefore, Figure 5.26 is drawn to emphasize the impact of the length scale parameter on the DLGSs' wave frequency. Nonlocal elasticity ($\eta = 0$) makes it abundantly clear that the frequency of waves increases gradually to a maximum value and then remains constant. When the strain gradient is eliminated ($\eta \neq 0$), the frequency of the waves dramatically rises and

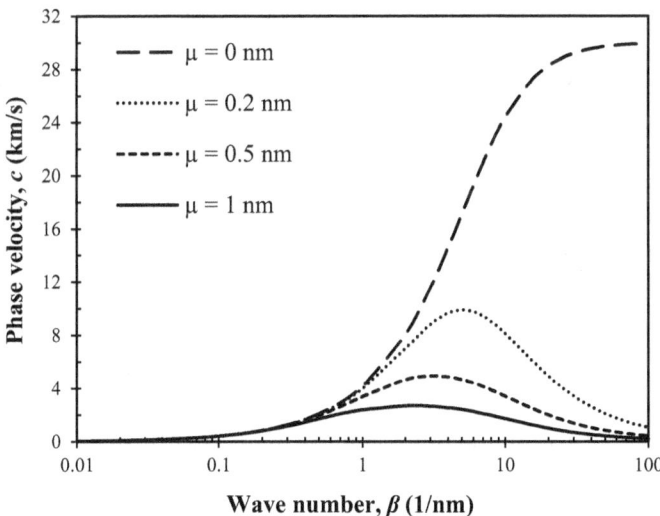

FIGURE 5.25 Influences of different nonlocal parameters on phase velocity of DLGSs ($\eta = 0$ nm, $k_w = k_p = 0$).

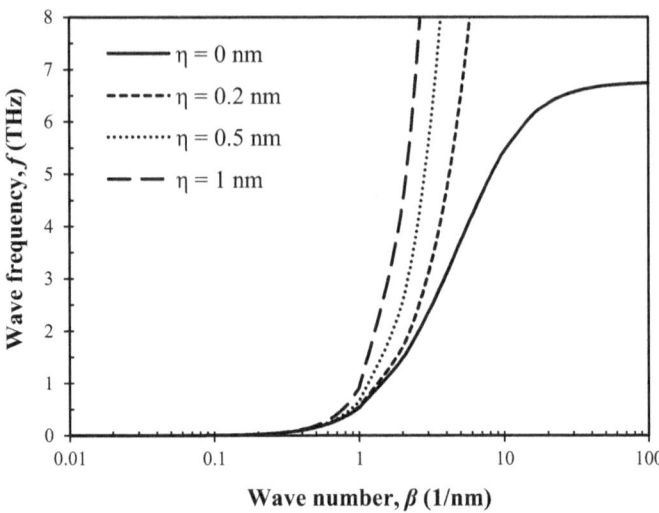

FIGURE 5.26 Influences of different length scale parameters on wave frequency of DLGSs ($\mu = 0.5$ nm, $k_w = k_p = 0$).

approaches infinity. It's important to remember that different values of the length scale parameter have different effects on the tendency of the speed of frequency to infinity. In other words, the amplitude of the wave frequency grows as the value of the length scale parameter grows. When the wave number is believed to be greater than $\beta = 10 \times 10^9$, dramatic shifts are possible.

In addition, the purpose of Figure 5.27 is to demonstrate how the length scale parameter influences the phase velocity of DLGSs. Once NET is administered ($\eta = 0$), the only noticeable impact is the lessening of stiffness. If $\eta \neq 0$, it is self-evident that the phase velocity first increases to its maximum value and then stabilizes. Similar to previous findings, however, this one suggests that for bigger values of the length scale parameters, the phase velocity may be higher. Certainly, findings in Figure 5.27 are a direct reflection of those of earlier figures.

The effect of nonlocal and length scale characteristics on the wave frequency of DLGSs relative to changes in the wave number is also shown in Figure 5.28. For the NE ($\eta = 0$), it is evident that as the nonlocal parameter increases, the wave frequency decreases. With each successive addition of the wave number, the frequency of the waves eventually reaches its maximum value. Once strain gradient elasticity ($\eta \neq 0$) is taken into account, the frequency of waves approaches infinity as the wave number increases. Additionally, the length scale parameter here functions to boost the wave frequency. In other words, if the nonlocal parameter is assumed to be fixed, then a larger value for the length scale parameter might lead to a higher frequency of waves.

The dependence of the phase velocity on the wave number over a range of nonlocal and length scale characteristics is shown in Figure 5.29. When the NET is used, the phase velocity first climbs to a maximum and then gradually decreases

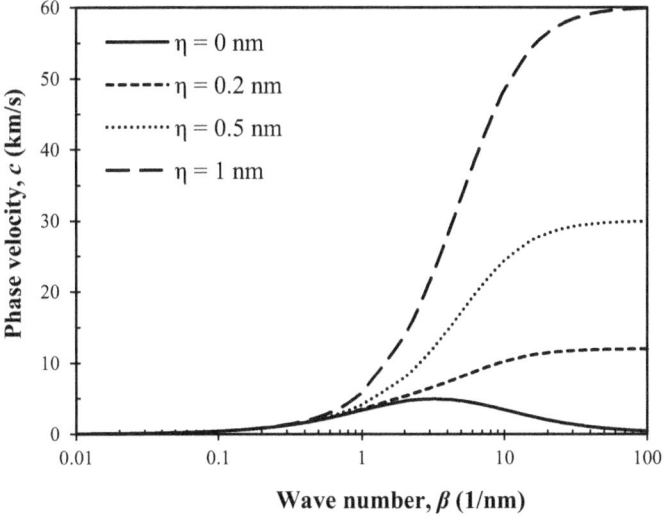

FIGURE 5.27 Influences of different length scale parameters on phase velocity of DLGSs ($\mu = 0.5$ nm, $k_w = k_p = 0$).

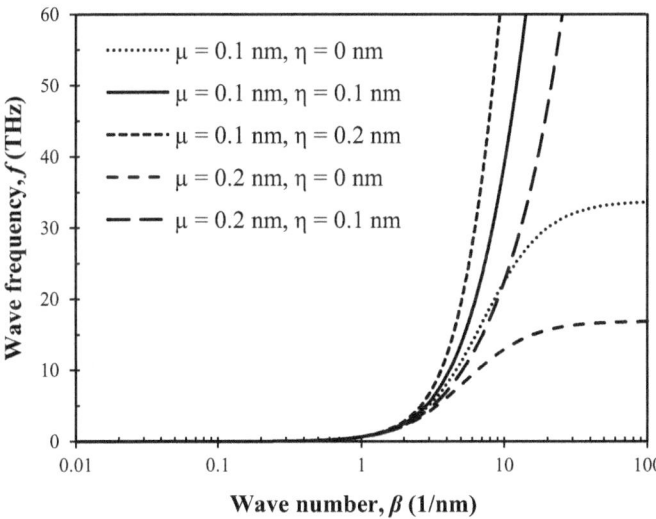

FIGURE 5.28 Coupled effect of nonlocal and length scale parameters on wave frequency of DLGSs ($k_w = k_p = 0$, $C_d = 0$, $\Delta T = \Delta C = 0$).

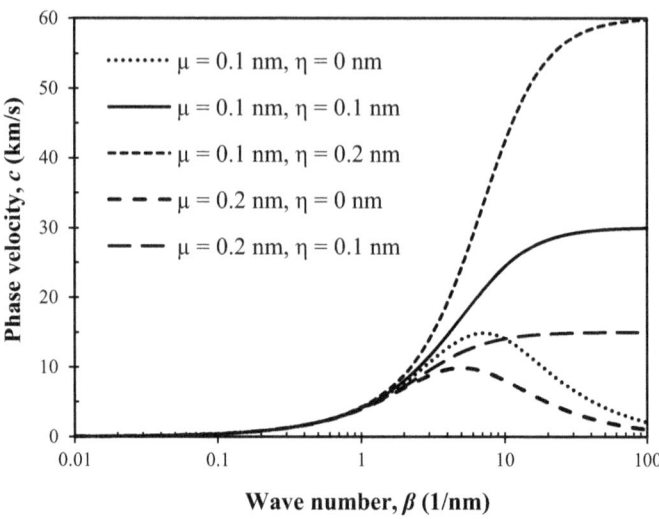

FIGURE 5.29 Coupled effect of nonlocal and length scale parameters on phase velocity of DLGSs ($k_w = k_p = 0$, $C_d = 0$, $\Delta T = \Delta C = 0$).

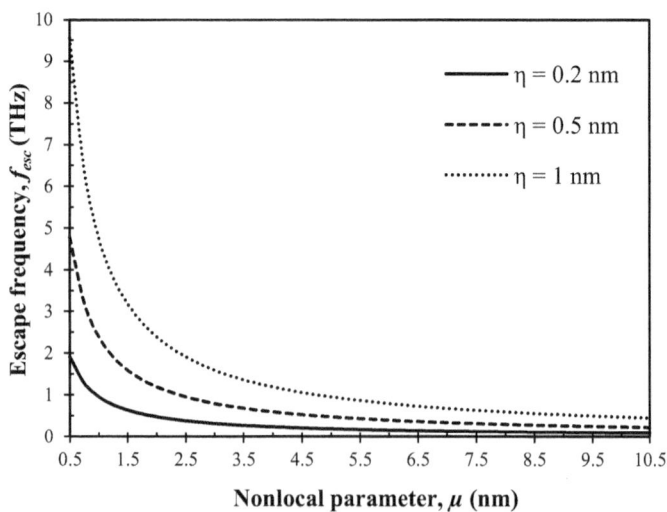

FIGURE 5.30 Influences of different length scale parameters on escape frequency of DLGSs ($k_w = k_p = 0$).

with increasing wave number ($\eta = 0$). It is also important to note that the nonlocal parameter slows down the DLGSs' phase velocity in the same way as it slows down their wave frequency. Choosing a larger nonlocal value does allow for straightforward subtraction of phase velocity. In addition, if NSGT is used ($\eta \neq 0$), the phase velocity becomes larger as the wave number increases, and once it reaches its maximum value, it stays constant. It should be noted that by increasing the length scale parameter, a bigger value of phase velocity may be attained.

Finally, in the absence of underlying factors, variations in escape frequency as a function of the nonlocal parameter are presented for three distinct length scale values. The major reason foundation effects are ignored is because the calculation of wave frequency requires tending the wave number to infinity, which is beyond the efficient range of these parameters. It has been noticed that when the nonlocal parameter is raised, the escape frequency decreases. Furthermore, once a value of the nonlocal parameter is established, it is evident that the escape frequency does not change. This variation, however, is not uniform across all factors of scale. That is, for nonlocal parameters greater than $\mu = 4$ nm, $\mu = 7$ nm, and $\mu = 10$ nm for length scale parameters $\eta = 0.2$ nm, $\eta = 0.5$ nm and $\eta = 1$ nm, respectively, the escape frequency is fixed.

5.2.3.1 Foundation Effects

An additional vital aspect is, without a doubt, the impacts of base parameters on DLGSs' wave propagation solutions. In Figure 5.31, we display the changes in phase velocity versus wave number to examine the impact of the Winkler coefficient on the phase velocity of DLGSs. It can be seen from the figure that the range of wave numbers within which the Winkler coefficient has an effect on the phase velocity values is limited to somewhere around $\beta = 0.05 \times 10^9$. In addition, Figure 6 shows that a larger value for the aforementioned coefficient has a noticeable impact on the manner that phase velocity increases. The addition of the linear component of the foundation allows for larger phase velocities within a certain range of wave numbers, as a consequence [35].

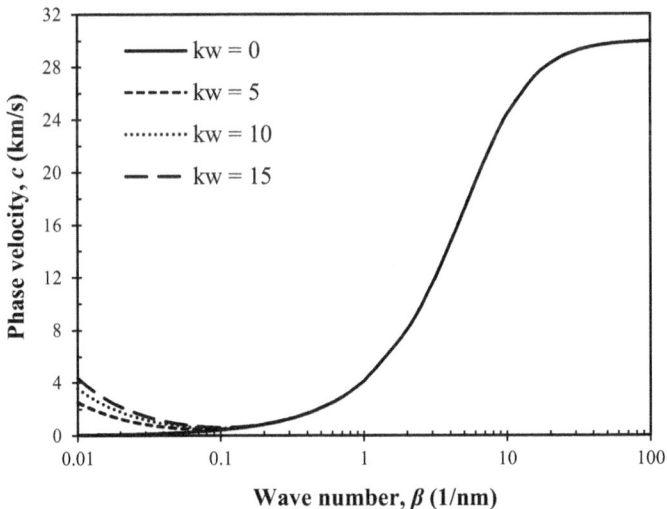

FIGURE 5.31 Influences of different Winkler coefficients on phase velocity of DLGSs ($\mu = \eta = 0.2$ nm, $k_p = 0$).

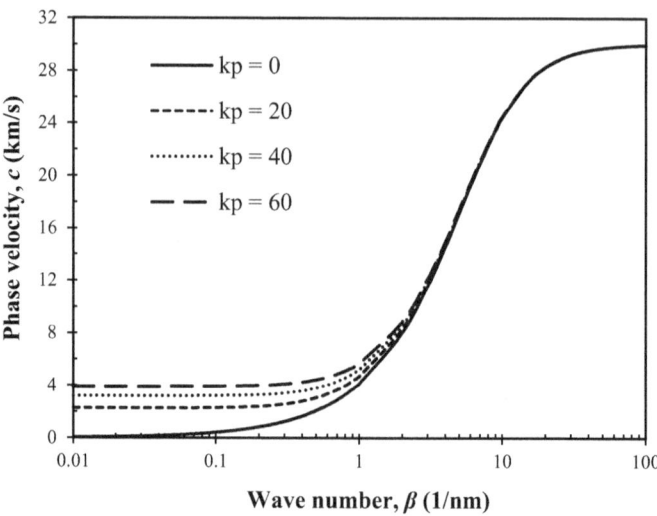

FIGURE 5.32   Influences of different Pasternak coefficients on phase velocity of DLGSs ($\mu = \eta = 0.2$ nm, $k_w = 0$).

Furthermore, the primary goal in constructing Figure 5.32 is to emphasize the influence of the Pasternak coefficient on the phase velocity of DLGSs. Once again, the constraint on the effective range of any linear or nonlinear base parameter may be seen as a point of similarity in this figure. However, the effective range of the Pasternak coefficient is larger, and it is calculated to be the wave numbers smaller than $\beta = 2 \times 10^9$. Adding a Pasternak coefficient is like adding a Winkler coefficient in that it increases the phase velocity. There is just a little difference between the linear and nonlinear coefficients of the elastic media in terms of their ability to increase phase velocity. The efficiency ranges they cover are one distinction; the form of the resulting diagram upon changing any of the aforementioned factors is another. Adding the Pasternak coefficient causes a steady rise in the phase velocity, whereas modifying the Winkler coefficient causes a fall in the phase velocity initially, followed by a steady increase.

In Figure 5.33, we can see how changing the Winkler and Pasternak coefficients affects the DLGSs' wave frequency. It is evident that larger wave frequencies may be attained in a constant quantity of each of these parameters by selecting a higher value for another coefficient. The Winkler coefficient may have a larger effect on the frequency of waves in the low-frequency range than the Pasternak coefficient. The impact of the Pasternak coefficient is, however, more readily apparent with higher wave numbers than $\beta = 0.2 \times 10^9$. It follows that larger values for linear or nonlinear medium properties allow for greater wave frequencies.

Figure 5.34 illustrates the impact of viscosity on the dispersion properties of propagating waves by plotting wave frequency versus wave number for various values of the structural damping coefficient. This helps to highlight how viscosity influences these features. It is clear from the illustration that DLGSs may have a damping effect, reducing the frequency of waves after they had first increased.

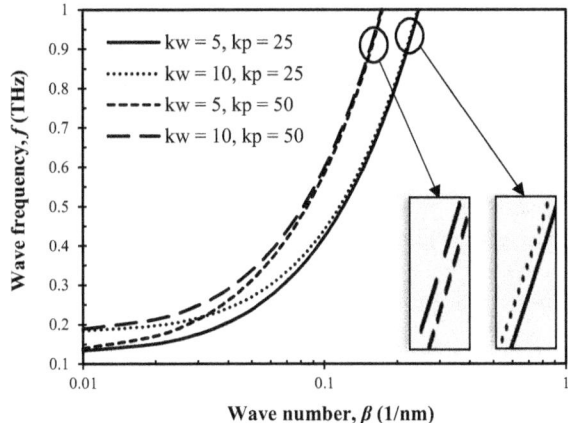

FIGURE 5.33 Coupled effect of Winkler and Pasternak coefficients on wave frequency of DLGSs ($\mu = \eta = 0.1$ nm, $C_d = 0$, $\Delta T = \Delta C = 0$).

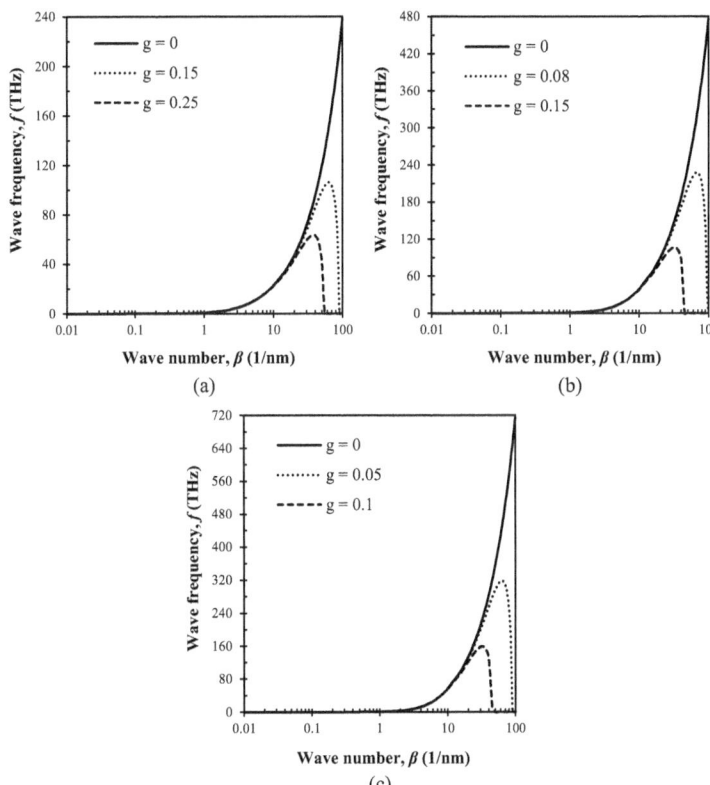

FIGURE 5.34 Variation of wave frequency versus wave number for different structural damping coefficient values at (a) $\eta < \mu$, (b) $\eta = \mu$, and (c) $\eta > \mu$ ($\mu = 0.2$ nm, $k_w = k_p = 0$, $C_d = 0$, $N = 0$).

As the number of waves increases, the frequency increases as well, as long as the structural damping coefficient is disregarded. The small-scale effects, in addition to the graphene sheets' damping effect, are discussed. Although the overall form of the figure is the same for all three cases ($\eta < \mu$, $\eta = \mu$ and $\eta > \mu$), the value of the wave frequency differs. In other words, with fixed values of the nonlocal parameter, larger wave numbers than $\beta = 7 \times 10^9$ see an increase in wave frequency as the length scale parameter increases. The effectiveness of an NSG-based theory in assessing the stiffness-hardening effects of nanoscale structures is shown in this image.

Also, the size-dependent damping behavior of DLGSs is amplified by studying another vital aspect of the propagating waves. Figure 5.35 depicts the relationship between phase velocity and wave number as a function of structural damping

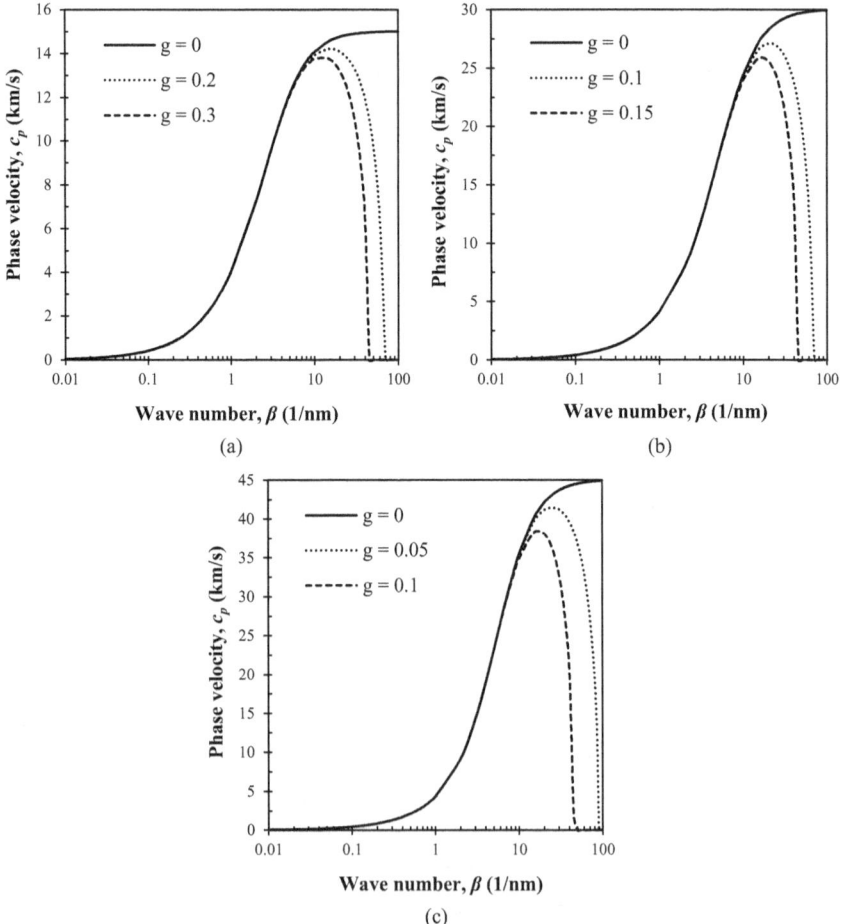

FIGURE 5.35   Variation of phase velocity versus wave number for different structural damping coefficient values at (a) $\eta < \mu$, (b) $\eta = \mu$, and (c) $\eta > \mu$ ($\mu = 0.2$ nm, $k_w = k_p = 0$, $C_d = 0$, $N = 0$).

coefficient. Similar to Figure 5.34, the phase velocity of DLGSs may be slowed by setting the structural damping coefficient ($g \neq 0$) to a nonzero value. However, when this coefficient is made null ($g = 0$), the phase velocity initially rises to its maximum value and then stabilizes. The initial structural damping coefficient has the capacity to entirely damp the wave frequency values or phase velocity amounts; therefore, it's important to keep an eye on it as well. since the length scale parameter grows, the minimum coefficient decreases, since it is the first effective nonzero value of the structural damping coefficient. The NSGT's new stiffness-hardening action is also clearly shown in Figure 3, which is otherwise identical to the earlier picture.

In Figure 5.36, we examine the effects of the linear basis parameter by displaying the shifts in wave frequency versus the shifts in wave number. The addition of

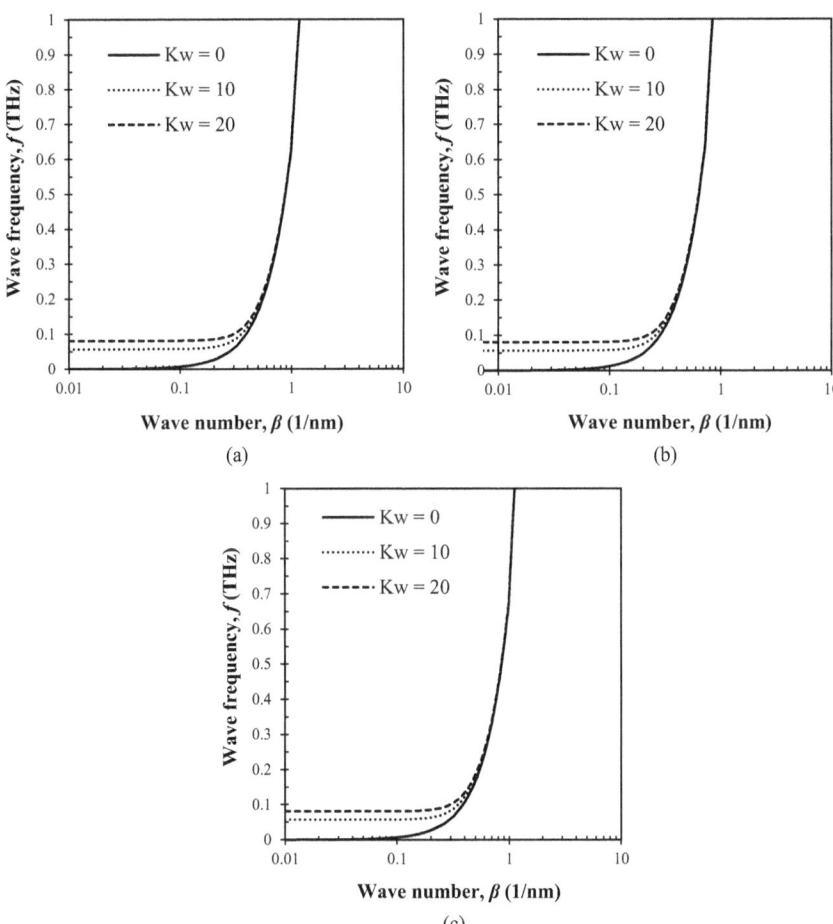

FIGURE 5.36 Variation of wave frequency versus wave number for different Winkler coefficients at (a) $\eta < \mu$, (b) $\eta = \mu$, and (c) $\eta > \mu$ ($\mu = 0.2$ nm, $k_p = 0$, $C_d = 0$, $N = 0$).

the Winkler coefficient makes it possible to increase the frequency of a wave to achieve a certain wave number. However, the effectiveness of this impact varies along the axis of wave number. Whenever the wave number is expected to be greater than $\beta = 0.5 \times 10^9$, adjusting the Winkler coefficient has no effect on the wave frequency. This graphic also shows that adjusting the length scale parameter is not an effective technique to increase the frequency of the waves at low wave numbers.

The effects of the nonlinear constant of the foundation on the wave frequency of DLGSs are emphasized in Figure 5.37. One of the more effective methods of increasing wave frequency, especially in wave numbers lower than $\beta = 1 \times 10^9$, is to raise the Pasternak coefficient. Once the wave number is within the given range,

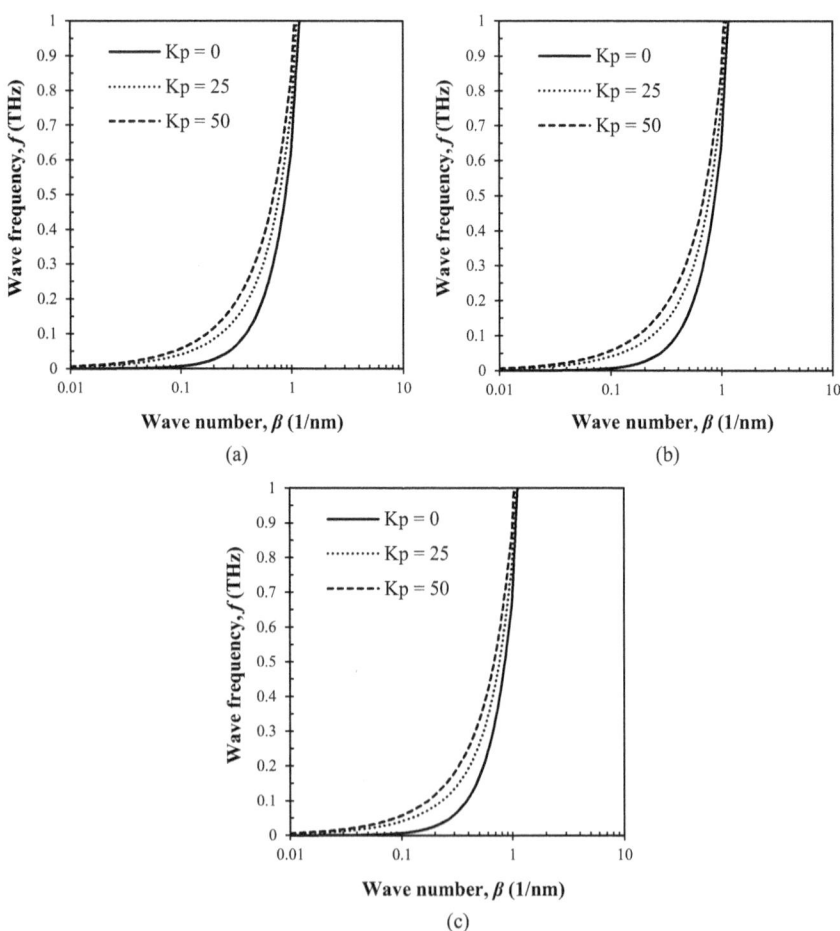

FIGURE 5.37 Variation of wave frequency versus wave number for different Pasternak coefficients at (a) $\eta < \mu$, (b) $\eta = \mu$, and (c) $\eta > \mu$ ($\mu = 0.2$ nm, $k_w = 0$, $C_d = 0$, $N = 0$).

it is obvious that increasing the length scale parameter has no influence on the overall frequency of the waves.

The damping qualities of a viscoelastic substrate are studied within the scope of Figure 5.38, after the study of linear and nonlinear foundation parameters. This graphic shows the relationship between the damping coefficient and the required wave number over a range of length scales. The graphic shows that the frequency of a wave gradually drops to zero as the damping coefficient increases, but it may be enhanced by the addition of the length scale parameter. Clearly, Figure 5.38(a) is plotted for $\beta = 3 \times 10^9$, and Figure 5.38(b) is plotted for $\beta = 6 \times 10^9$, two distinct wave numbers. It is clear that larger damping coefficients are needed to control higher wave frequencies as the wave number increases. To damp the wave

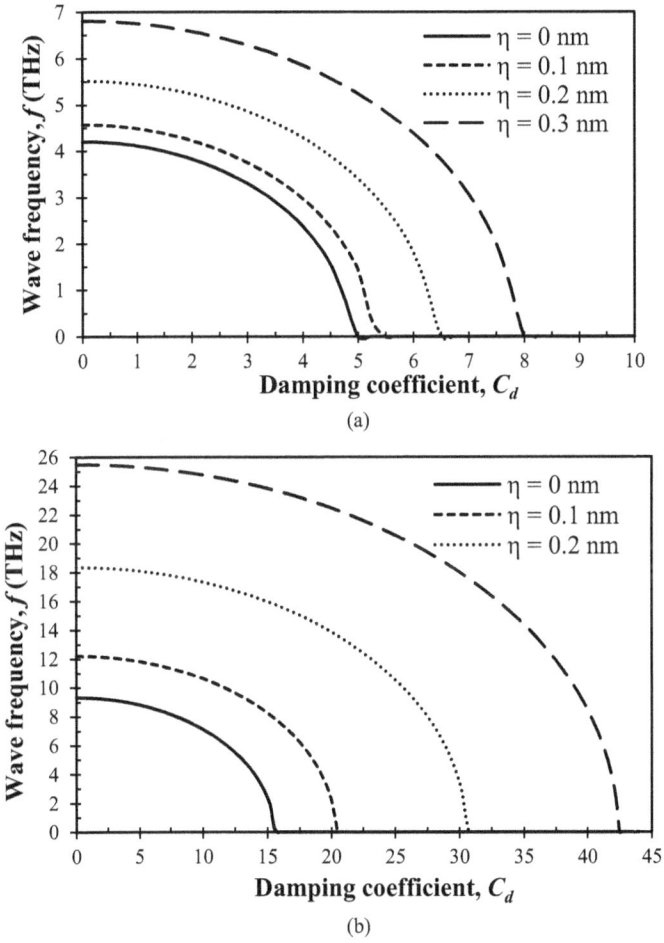

FIGURE 5.38 Variation of wave frequency versus damping coefficient for different length scale parameters at (a) $\beta = 3 \times 10^9$ and (b) $\beta = 6 \times 10^9$ ($\mu = 0.2$ nm, $k_w = k_p = 0$, $N = 0$).

frequency of DLGSs, a little increase in the quantity of wave number may need the use of greater values of the damping coefficient.

The influence of axial load on the wave dispersion properties of DLGSs is shown graphically in Figure 5.39, which shows the fluctuation of wave frequency versus wave number. Depending on the plot, a little increase in axial stress might cause a drop in wave frequency readings. However, there is a narrow window of wave numbers around $\beta = 1 \times 10^9$, where this phenomenon may be effective. Raising the axial load is not a viable option to raising the wave number for waves with magnitudes greater than the amount specified. In the $\beta = 1 \times 10^9$ range of wave numbers, for example, increasing the wave frequency by increasing the length scale parameter is not possible.

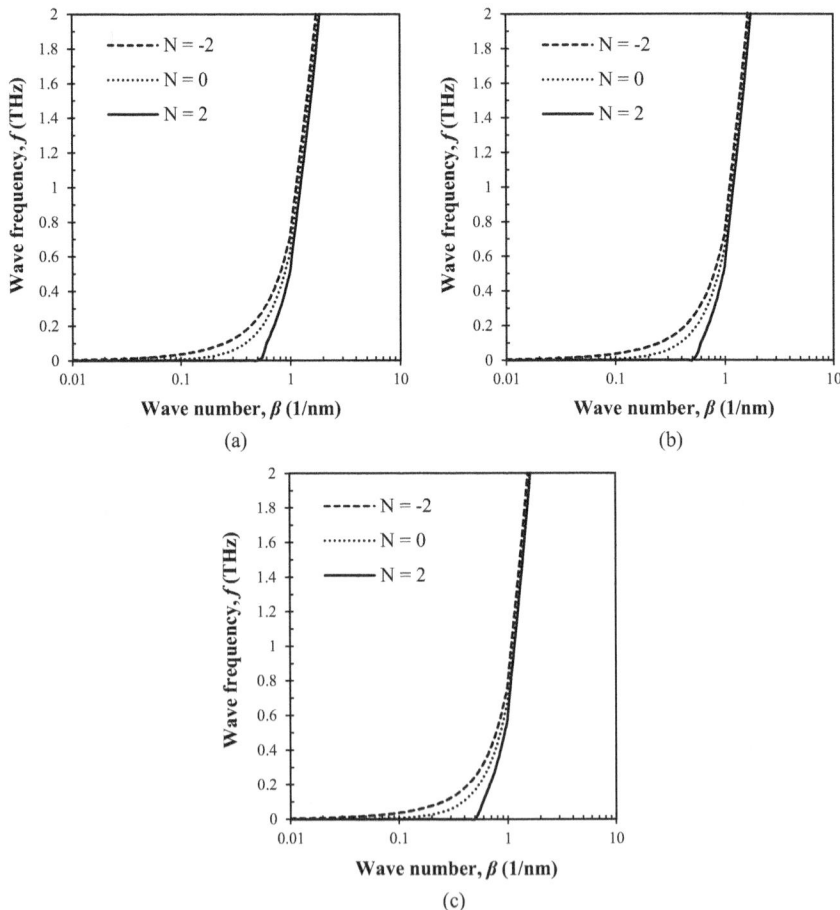

FIGURE 5.39   Variation of wave frequency versus wave number for different axial forces at (a) $\eta < \mu$, (b) $\eta = \mu$, and (c) $\eta > \mu$ ($\mu = 0.2$ nm, $k_w = k_p = 0$, $C_d = 0$).

The same topic is shown in Figure 5.40, which plots the shift in phase velocity versus the wave number. In fact, it is possible to draw the conclusion that the magnitude of phase velocity will grow once the value of axial load has been increased. Furthermore, once the axial load is meant to be modified, only a certain range of wave numbers engage in the operation of altering phase velocity. This graph displays the key result of the NSGT, which is an increase in the amount of phase velocity whenever the length scale parameter is assumed to be increased. For wave numbers bigger than $\beta = 5 \times 10^9$, Figure 5.40 suggests that a change in phase velocity may serve as a motivating factor.

Additionally, in Figure 5.41, the variation of phase velocity with relation to axial load is displayed in a desired wave number for a variety of amounts of nonlocal parameter and in the absence of graphene's viscous qualities. Once one

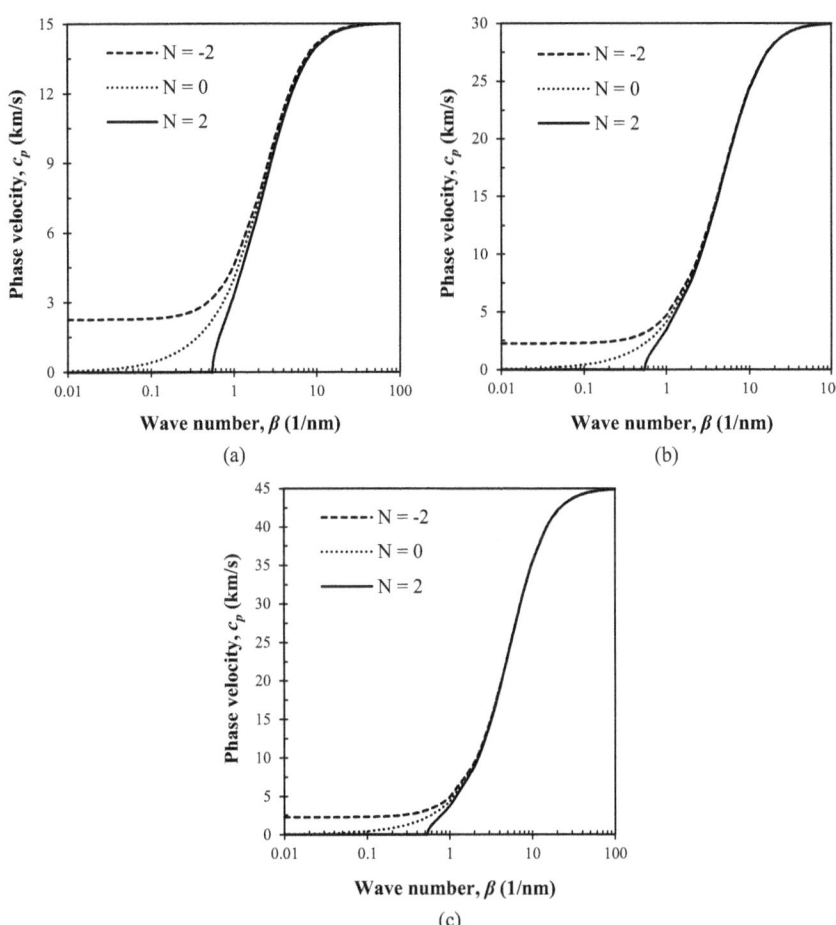

FIGURE 5.40 Variation of phase velocity versus wave number for different axial forces at (a) $\eta < \mu$, (b) $\eta = \mu$, and (c) $\eta > \mu$ ($\mu = 0.2$ nm, $k_w = k_p = 0$, $C_d = 0$).

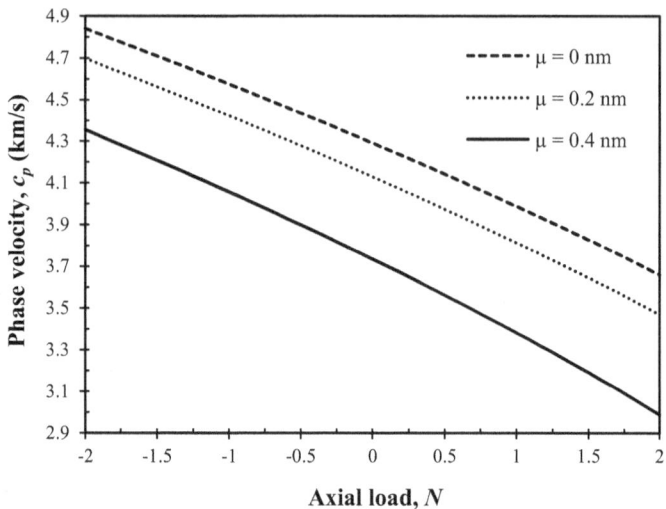

FIGURE 5.41 Variation of phase velocity versus axial load for different amounts of nonlocal parameter ($\eta = 0.2$ nm, $k_w = k_p = 0$, $C_d = 0$, $g = 0$, $\beta = 1 \times 10^9$).

of the nonlocal factors has been chosen, the major objective will be to investigate the effect that axial load, the graphic shows that the phase velocity may increase under compression relative to tension. Using larger nonlocal parameters naturally yields slower phase velocities as well. Thus, Eringen et al.'s [1, 2] stiffness-softening effect may be included into the current model. Since both the axial load and the nonlocal parameter reduce the phase velocity when their values are raised, it follows that both are categorized as decreasing parameters.

Consequently, the effect of size dependence and axial load on phase velocity values will be examined, as shown in Figure 5.42. When an axial tension load ($N = 2$) is taken into account, it is clear from the diagram that the phase velocity begins at its maximum and gradually decreases to zero. When the axial force is disregarded, however ($N = 0$), the phase velocity begins at its maximum and decreases to a nonzero value. However, when the nonlocal parameter increases, the phase velocity falls under any of the aforementioned situations (with or without axial stress). This figure depicts a growing influence of length scale parameter, which can be understood as an increase in the value of phase velocity after length scale parameter is applied in addition to the softening effect of nonlocal parameter. This rise in value occurs after length scale parameter is applied.

Finally, Figure 5.43 summarizes the intertwined influences of nonlocal and length scale characteristics on the escape frequency of DLGSs. Since the effectiveness of the preceding extra impacts is limited by a fixed value of wave numbers, this graphic ignores them in favor of the path that leads to escape frequency: increasing the wave frequency to infinity. Both the softening and hardening impacts on stiffness are clearly seen in this diagram. Actually, the escape frequency may be increased by either decreasing the nonlocal parameters or increasing the length scale parameters.

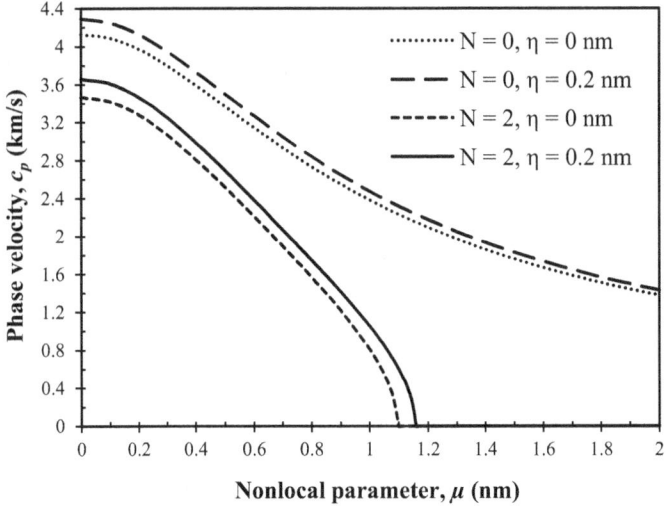

FIGURE 5.42 Variation of phase velocity versus nonlocal parameter for different amounts of length scale parameter and axial load ($k_w = k_p = 0$, $C_d = 0$, $g = 0$, $\beta = 1 \times 10^9$).

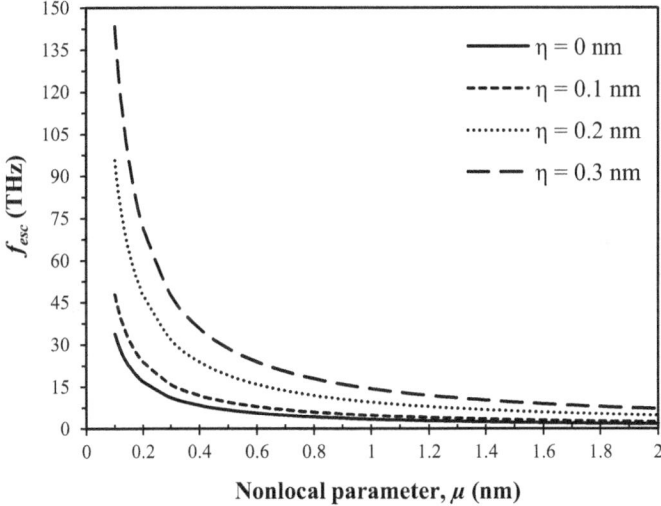

FIGURE 5.43 Variation of escape frequency versus nonlocal parameter for different amounts of length scale parameter ($k_w = k_p = 0$, $C_d = 0$, $N = 0$, $g = 0$).

5.2.3.2 Hygrothermal Effects

The relationship between wave frequency and wave number as a function of temperature gradient and moisture content is shown in Figure 5.44. Changing the temperature gradient or the content of moisture may clearly impact the frequency of the waves. As was previously expected, a steeper temperature gradient causes

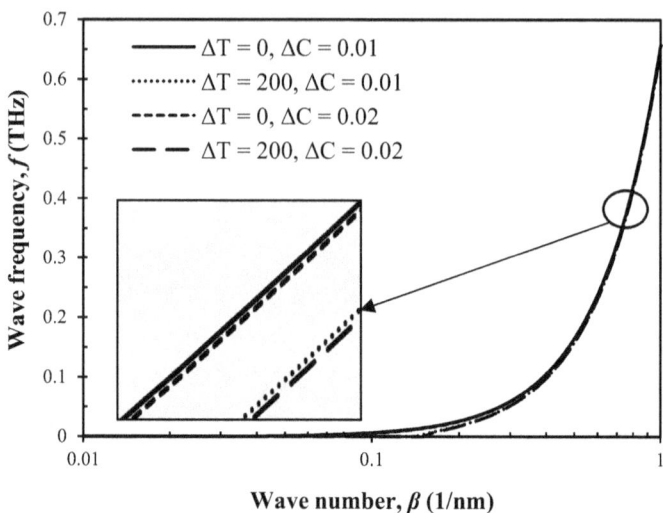

FIGURE 5.44 Coupled effect of temperature gradient and moisture concentration on wave frequency of DLGSs ($\mu = \eta = 0.1$ nm, $k_w = k_p = 0$, $C_d = 0$).

a lower frequency of waves. In addition, varying the amount of moisture present produces the same kind of response. That is to say, after an increase in moisture content is established, lower wave frequencies are attained. When the wave number is increased to infinity, changes in the frequency of the waves no longer register. Therefore, using a hygrothermal environment rather than a thermal one is one technique to produce lower wave frequencies [36].

Additionally, for various amounts of moisture, the phase velocity variation versus the temperature gradient is presented in Figure 5.45. When this number is increased, phase velocities may be slowed by either the temperature gradient or the moisture content. The phase velocity in each chosen moisture concentration gradually decreases from its greatest value to zero. If there is a greater concentration of moisture, this effect occurs across a narrower temperature difference.

In addition, the relationship between phase velocity and damping coefficient in dry and humid circumstances is examined in Figure 5.46. It is clear that, as compared to a Winkler–Pasternak base, the wave dispersion response of graphene sheets is attenuated when they are supported by a viscoelastic substrate. It should be noted that DLGSs show similar reactions to wave propagation in both dry and humid conditions. It's possible, nevertheless, that the phase velocity value will be somewhat lower under hygrothermal conditions than in thermal ones. Selecting a hygrothermal environment, thus, might be regarded as an option for getting lower phase velocities, despite the fact that phase velocity is insensitive to the thermal or hygrothermal condition of the environment.

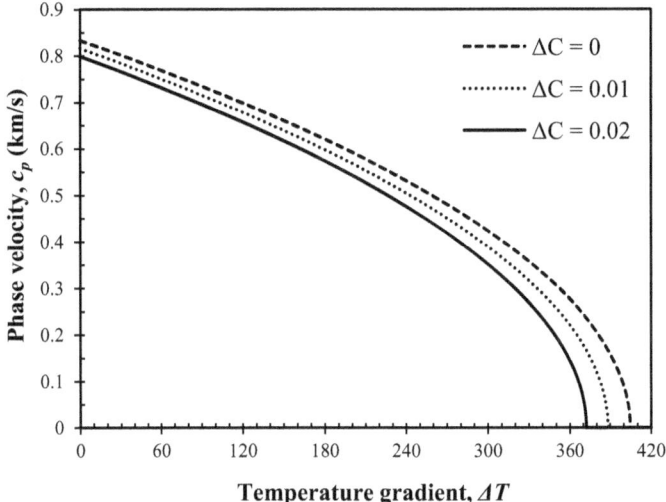

FIGURE 5.45   Variation of phase velocity of a DLGS versus temperature gradient for different moisture concentrations ($\mu = \eta = 0.1$ nm, $k_w = k_p = 0$, $C_d = 0$, $\beta = 0.2 \times 10^9$).

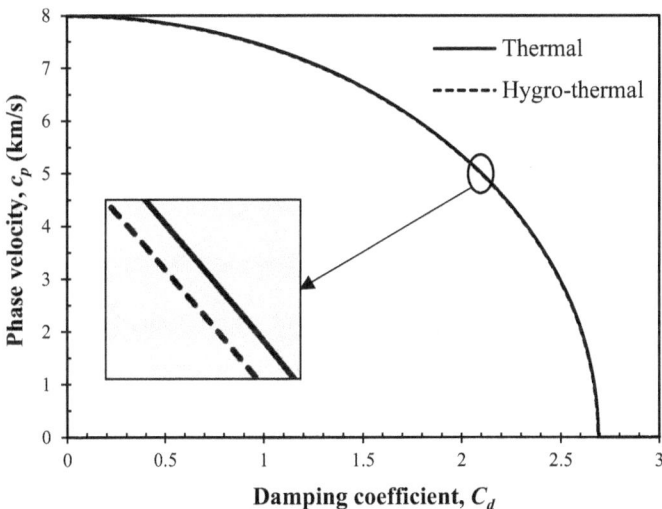

FIGURE 5.46   Variation of phase velocity of a DLGS versus damping coefficient for both thermal and hygrothermal conditions ($\mu = \eta = 0.1$ nm, $k_w = k_p = 0$, $\beta = 2 \times 10^9$).

Finally, the effects of the viscoelastic medium on the wave frequency of DLGSs in both dry and humid environments are shown graphically in Figures 5.47 and 5.48. When the fluctuation in wave frequency is plotted against the damping coefficient of the visco-Pasternak basis, it is clear that the frequency is influenced by

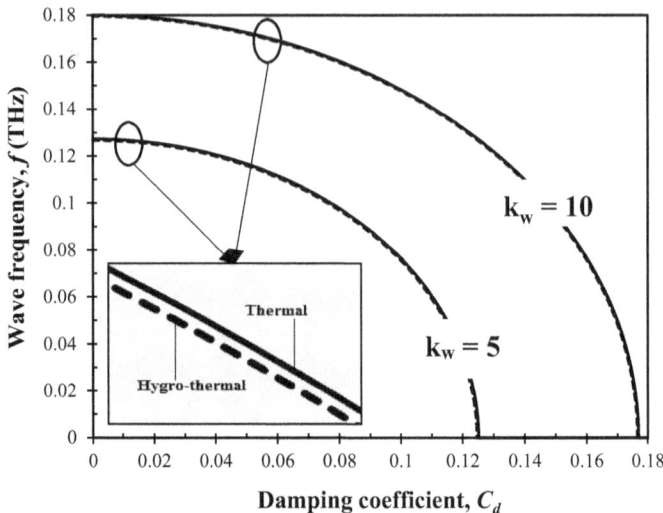

FIGURE 5.47 Variation of wave frequency of a DLGS versus damping coefficient for various Winkler coefficients and both thermal and hygrothermal conditions ($\mu = \eta = 0.1$ nm, $k_p = 0$, $\beta = 0.2 \times 10^9$).

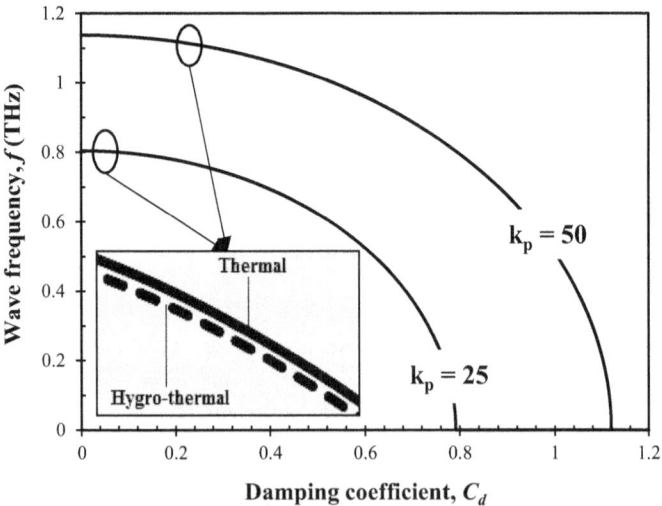

FIGURE 5.48 Variation of wave frequency of a DLGS versus damping coefficient for various Pasternak coefficients and both thermal and hygrothermal conditions ($\mu = \eta = 0.1$ nm, $k_w = 0$, $\beta = 0.2 \times 10^9$).

the damping. Once a hygrothermal environment is taken into account, the frequency of the waves will decrease, as was previously anticipated. In reality, the quantity of wave frequency may be increased by either the linear (Winkler) or nonlinear (Pasternak) factors. A lesser damping coefficient may be the difference between success and failure when it comes to a Winkler. It's also worth noting that unlike the Winkler coefficient, the Pasternak coefficient needs larger values to cause a greater increase in the quantity of wave frequency. Since the Pasternak coefficient may significantly amplify wave frequency relative to the Winkler coefficient for a given set of linear and nonlinear foundation parameters, it is the obvious choice if a greater wave frequency is desired.

5.3   WAVE DISPERSION CHARACTERISTICS OF GRAPHENE OXIDE-BASED STRUCTURES

5.3.1   BACKGROUND

This section was prepared so that additional information may be gained about the behavior of GOP-reinforced nanocomposite plates when they are exposed to heat stress. The research is improved by subjecting the plates to thermal loadings such as sinusoidal temperature increase, linear temperature rise, and uniform temperature rise, as well as by embedding them on an elastic substrate. This makes the study more representative of real-world conditions. Wave frequency and phase velocity of the GOP-reinforced nanocomposite plate may be obtained by analytically solving the governing differential equations derived by using Hamilton's principle and updated higher-order plate theory. This can be done thus in order to acquire these values. When distributing GOPs in a polymeric matrix over the thickness of a plate, several functionally graded FG patterns are taken into consideration and may be used. In order to make an accurate prediction of the effective material features of the structure, the Halpin–Tsai micromechanical approach is also used. Finally, the accuracy of the proposed model is validated by comparing it to previous studies, and the ramifications of various parameters on the wave propagation behavior of the GOP-reinforced nanocomposite plates are discussed in depth [65].

5.3.2   GRAPHENE OXIDE-BASED PLATES: WAVE DISPERSION CHARACTERISTICS

One well-known class of composites, FG composites [65], have unique qualities because their material properties change throughout the material's thickness. Here, we assume a matrix polymer and a set of GOPs as reinforcement, and we look at how these GOPs may be dispersed over the plate's thickness using a variety of FG patterns. Figure 5.49 depicts a schematic of a nanocomposite plate reinforced with GOPs and supported by a Winkler–Pasternak base, with a variety of FG patterns of GOPs' distribution taken into account.

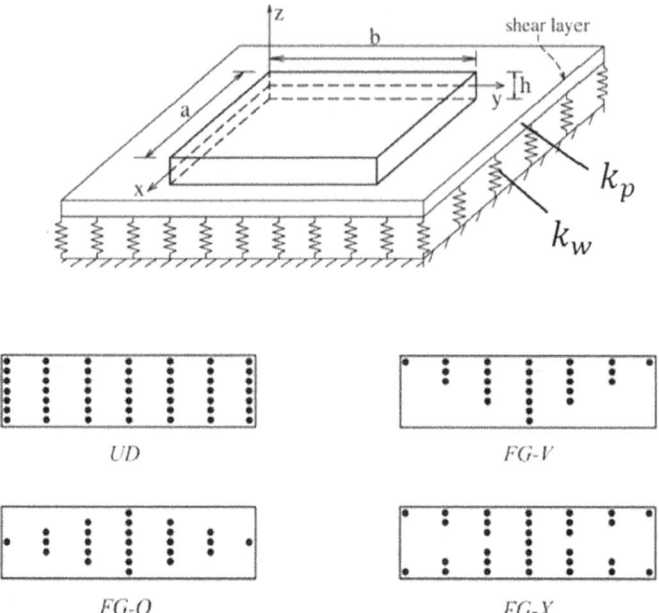

FIGURE 5.49 Configuration of the GOP-reinforced nanocomposite plate rested on elastic foundation considering different types of GOPs' distribution pattern. Simple modeling [41, 42] may be used to determine the optimal placement for the nanofillers to create the desired patterns.

$$
\begin{cases}
V_{GOP} = V_{GOP}^* & \text{GOPR-U} \\[2mm]
V_{GOP} = \left(2 - 4\dfrac{|z|}{h}\right)V_{GOP}^* & \text{GOPR-O} \\[2mm]
V_{GOP} = 4\dfrac{|z|}{h}V_{GOP}^* & \text{GOPR-X} \\[2mm]
V_{GOP} = \left(1 + 2\dfrac{z}{h}\right)V_{GOP}^* & \text{GOPR-V}
\end{cases}
\tag{5.43}
$$

in which V_{GOP}^* is the total GOPs' volume fraction and can be defined as follows:

$$
V_{GOP}^* = \frac{W_{GOP}}{W_{GOP} + \left(\rho_{GOP} / \rho_{M}\right)\left(1 - W_{GOP}\right)}
\tag{5.44}
$$

where ρ and W represent mass density and weight fraction, respectively, and, also, GOP and M subscripts denote GOP reinforcements and the matrix, respectively. Furthermore, in order to achieve equivalent mechanical properties of GOP-reinforced plate, the Halpin–Tsai homogenization technique is implemented [65]. Now, the effective Young's modulus can be computed in the following form:

$$E_{eq} = 0.49E_l + 0.51E_t \tag{5.45}$$

in which E_l and E_t are longitudinal and transverse Young's modulus of the composite plate, respectively. Both these variants can be calculated as [65]:

$$E_l = \frac{1 + \xi_l \eta_l V_{GOP}}{1 - \eta_l V_{GOP}} \times E_M, \quad E_t = \frac{1 + \xi_t \eta_t V_{GOP}}{1 - \eta_t V_{GOP}} \times E_M \tag{5.46}$$

where

$$\eta_l = \frac{(E_{GOP}/E_M) - 1}{(E_{GOP}/E_M) + \xi_l}, \quad \eta_t = \frac{(E_{GOP}/E_M) - 1}{(E_{GOP}/E_M) + \xi_t} \tag{5.47}$$

where, ξ_l and ξ_t are the geometry factors that can be written as [41]:

$$\xi_l = \xi_t = \frac{2d_{GOP}}{h_{GOP}} \tag{5.48}$$

in which d_{GOP} and h_{GOP} are the diameter and thickness of the GOPs, respectively. Afterward, the equivalent Poisson's ratio and mass density of the composite can be achieved in the framework of the rule of mixture, which can be written as:

$$v_{eq} = v_{GOP} V_{GOP} + v_M V_M \tag{5.49}$$

$$\rho_{eq} = \rho_{GOP} V_{GOP} + \rho_M V_M \tag{5.50}$$

where V stands for the volume fraction. Plus, the volume fractions are related to each other as:

$$V_{GOP} + V_M = 1 \tag{5.51}$$

Furthermore, the graphene oxide powders-reinforced (GOPR) nanocomposite's coefficient of thermal expansion (CTE) may be written as [42]:

$$\alpha_{eq} = \alpha_M + \frac{\alpha_M + \alpha_{GOP}}{\frac{1}{K_M} + \frac{1}{K_{GOP}}} \left[\frac{1}{K_{eq}} + \frac{1}{K_M} \right] \tag{5.52}$$

in which the bulk moduli (K) and the CTE (α) are shown. Because of these oversimplifications, the outcome of the traditional plate hypothesis is suspect. Therefore, in order to anticipate the shear stress and strain of the plate, the scientists built higher-order shear deformable plate models and theoretically specified a shape function. In order to determine the plate's kinematic relations, this research uses a more developed version of the sinusoidal plate theory. According to this theory, the following equation describes a plate's displacement field:

$$u_1\left(x,y,z\right) = u\left(x,y\right) - z\frac{\partial w_b}{\partial x} - f\left(z\right)\frac{\partial w_s}{\partial x} \tag{5.53}$$

$$u_2\left(x,y,z\right) = v\left(x,y\right) - z\frac{\partial w_b}{\partial y} - f\left(z\right)\frac{\partial w_s}{\partial y} \tag{5.54}$$

$$u_3\left(x,y,z\right) = w_b\left(x,y\right) + w_s\left(x,y\right) \tag{5.55}$$

where, u, w_b, and w_s denote longitudinal displacement bending and shear deflections, respectively. The corresponding shape function is $f\left(z\right) = z - \dfrac{h}{\pi}\sin\left(\dfrac{\pi z}{h}\right)$. The nonzero strains of the plate can be expressed by the following equations:

$$\begin{Bmatrix} \varepsilon_x \\ \varepsilon_y \\ \gamma_{xy} \end{Bmatrix} = \begin{Bmatrix} \varepsilon_x^0 \\ \varepsilon_y^0 \\ \gamma_{xy}^0 \end{Bmatrix} + z\begin{Bmatrix} k_x^b \\ k_y^b \\ k_{xy}^b \end{Bmatrix} + f\left(z\right)\begin{Bmatrix} k_x^s \\ k_y^s \\ k_{xy}^s \end{Bmatrix}, \begin{Bmatrix} \gamma_{yz} \\ \gamma_{xz} \end{Bmatrix} = g\begin{Bmatrix} \gamma_{yz}^s \\ \gamma_{xz}^s \end{Bmatrix} \tag{5.56}$$

where $g\left(z\right) = 1 - \dfrac{df\left(z\right)}{dz}$

$$\begin{Bmatrix} \varepsilon_x^0 \\ \varepsilon_y^0 \\ \gamma_{xy}^0 \end{Bmatrix} = \begin{Bmatrix} \dfrac{\partial u}{\partial x} \\ \dfrac{\partial v}{\partial y} \\ \dfrac{\partial u}{\partial y} + \dfrac{\partial v}{\partial x} \end{Bmatrix}, \begin{Bmatrix} k_x^b \\ k_y^b \\ k_{xy}^b \end{Bmatrix} = \begin{Bmatrix} -\dfrac{\partial^2 w_b}{\partial x^2} \\ -\dfrac{\partial^2 w_b}{\partial y^2} \\ -2\dfrac{\partial^2 w_b}{\partial x\partial y} \end{Bmatrix},$$

$$\begin{Bmatrix} k_x^s \\ k_y^s \\ k_{xy}^s \end{Bmatrix} = \begin{Bmatrix} -\dfrac{\partial^2 w_s}{\partial x^2} \\ -\dfrac{\partial^2 w_s}{\partial y^2} \\ -2\dfrac{\partial^2 w_s}{\partial x\partial y} \end{Bmatrix}, \begin{Bmatrix} \gamma_{yz}^s \\ \gamma_{xz}^s \end{Bmatrix} = \begin{Bmatrix} \dfrac{\partial w_s}{\partial y} \\ \dfrac{\partial w_s}{\partial x} \end{Bmatrix}, \tag{5.57}$$

The Euler–Lagrange equation is now derived using the Hamiltonian method, and it may be written as:

$$\int_{t_0}^{t_1}\delta\left(\Pi_s - \Pi_k + \Pi_w\right)dt = 0 \tag{5.58}$$

Here, Π_s, Π_k, and Π_w stand for the strain energy, kinetic energy, and work done by external forces, respectively. The variation of strain energy is written as:

$$\delta\Pi_s = \int_V \sigma_{ij}\delta\varepsilon_{ij}dV = \int_V \left(\sigma_x\delta\varepsilon_x + \sigma_y\delta\varepsilon_y + \sigma_{xy}\delta\gamma_{xy} + \sigma_{yz}\delta\gamma_{yz} + \sigma_{xz}\delta\gamma_{xz}\right)dV \quad (5.59)$$

Putting Eqs. (5.53) through (5.57) into Eq. (5.59) produces:

$$\delta\Pi_s = \int_0^L \left(\begin{aligned} &N_x\frac{\partial\delta u}{\partial x} - M_x^b\frac{\partial^2\delta w_b}{\partial x^2} - M_x^s\frac{\partial^2\delta w_s}{\partial x^2} + Q\frac{\partial w_s}{\partial x} + N_y\frac{\partial\delta v}{\partial y}\\ &-M_y^b\frac{\partial^2\delta w_b}{\partial y^2} - M_y^s\frac{\partial^2\delta w_s}{\partial y^2} + N_{xy}\left(\frac{\partial\delta u}{\partial y}+\frac{\partial\delta v}{\partial x}\right) - 2M_{xy}^b\frac{\partial^2\delta w_b}{\partial x\partial y}\\ &-2M_{xy}^s\frac{\partial^2\delta w_s}{\partial x\partial y} + Q_{yz}\frac{\partial\delta w_s}{\partial y} + Q_{xz}\frac{\partial\delta w_s}{\partial x} \end{aligned} \right) dx \quad (5.60)$$

The following are the definitions of the variables used in the preceding equation:

$$\left[N_i, M_i^b, M_i^s\right] = \int_A \left[1, z, f(z)\right]\sigma_i dA, \quad i=(x,y,xy) \quad (5.61)$$

$$Q_i = \int_A g(z)\sigma_i dA, \quad i=(xz,yz) \quad (5.62)$$

Kinetic energy may be represented as a function of:

$$\delta\Pi_k = \int_0^a\int_0^b \left(\begin{aligned} &I_0\left(\frac{\partial u}{\partial t}\frac{\partial\delta u}{\partial t} + \frac{\partial v}{\partial t}\frac{\partial\delta v}{\partial t} + \frac{\partial(w_b+w_s)}{\partial t}\frac{\partial\delta(w_b+w_s)}{\partial t}\right)\\ &-I_1\left(\frac{\partial u}{\partial t}\frac{\partial\delta w_b}{\partial x\partial t} + \frac{\partial w_b}{\partial x\partial t}\frac{\partial\delta u}{\partial t} + \frac{\partial v}{\partial t}\frac{\partial\delta w_b}{\partial y\partial t} + \frac{\partial w_b}{\partial y\partial t}\frac{\partial\delta v}{\partial t}\right)\\ &+I_2\left(\frac{\partial w_b}{\partial x\partial t}\frac{\partial\delta w_b}{\partial x\partial t} + \frac{\partial w_b}{\partial y\partial t}\frac{\partial\delta w_b}{\partial y\partial t}\right) + K_2\left(\frac{\partial w_s}{\partial x\partial t}\frac{\partial\delta w_s}{\partial x\partial t} + \frac{\partial w_s}{\partial y\partial t}\frac{\partial\delta w_s}{\partial y\partial t}\right)\\ &-J_1\left(\frac{\partial u}{\partial t}\frac{\partial\delta w_s}{\partial x\partial t} + \frac{\partial w_b}{\partial x\partial t}\frac{\partial\delta u}{\partial t} + \frac{\partial v}{\partial t}\frac{\partial\delta w_s}{\partial y\partial t} + \frac{\partial w_s}{\partial y\partial t}\frac{\partial\delta v}{\partial t}\right)\\ &J_2\left(\frac{\partial w_b}{\partial x\partial t}\frac{\partial\delta w_s}{\partial x\partial t} + \frac{\partial w_s}{\partial x\partial t}\frac{\partial\delta w_b}{\partial x\partial t} + \frac{\partial w_b}{\partial y\partial t}\frac{\partial\delta w_s}{\partial y\partial t} + \frac{\partial w_s}{\partial y\partial t}\frac{\partial\delta w_b}{\partial y\partial t}\right) \end{aligned} \right) dxdy$$

$$(5.63)$$

$$\left(I_0, I_1, J_1, I_2, J_2, K_2\right) = \int_{-h/2}^{h/2} \left(1, z, f, z^2, zf\left(z\right), f^2\left(z\right)\right) \rho\left(z\right) dz \qquad (5.64)$$

The first variation expression for the work done by external forces is as follows:

$$\delta\Pi_w = \int_0^a \int_0^b \begin{vmatrix} N_x^T \dfrac{\partial\delta\left(w_b + w_s\right)}{\partial x} \dfrac{\partial\left(w_b + w_s\right)}{\partial x} + N_y^T \dfrac{\partial\delta\left(w_b + w_s\right)}{\partial y} \dfrac{\partial\left(w_b + w_s\right)}{\partial y} + \\[2mm] 2\delta N_{xy}^0 \dfrac{\partial\left(w_b + w_s\right)}{\partial x} \dfrac{\partial\left(w_b + w_s\right)}{\partial y} - k_w\left(w_b + w_s\right)\delta\left(w_b + w_s\right) + \\[2mm] k_p\left(\dfrac{\partial\left(w_b + w_s\right)}{\partial x} \dfrac{\partial\delta\left(w_b + w_s\right)}{\partial x} + \dfrac{\partial\left(w_b + w_s\right)}{\partial y} \dfrac{\partial\delta\left(w_b + w_s\right)}{\partial y}\right) \end{vmatrix} dxdy$$

$$(5.65)$$

In Eq. (5.65), k_w and k_p are Winkler and Pasternak coefficient and also it is assumed that the nanocomposite plate is subjected to a biaxial thermal loading ($N_x^T = N_y^T = N^T$); also, the shear loading is ignored ($N_{xy}^0 = 0$). The thermal loading (N^T) can be computed in the following form:

$$N^T = \int_{-\frac{h}{2}}^{\frac{h}{2}} \left(\frac{E_{eq}}{1 - v_{eq}} \alpha_{eq}\Delta T\right) dz \qquad (5.66)$$

This study examines the impact of three types of thermal loading, namely uniform temperature rise (UTR), linear temperature rise (LTR), and sinusoidal temperature rise (STR), on the wave propagation behavior of a structure. The research draws upon existing literature that has explored thermal environments [65]. The initial temperature is assumed to be $T_0 = 300$ K. The resulting temperature (T) can be expressed for each type of temperature increase.

$$\begin{cases} T = T_0 + \Delta T & \text{UTR} \\[3mm] T = T_0 + \Delta T\left(\dfrac{1}{2} + \dfrac{z}{h}\right) & \text{LTR} \\[3mm] T = T_0 + \Delta T\left(1 - \cos\left(\dfrac{\pi}{4} + \dfrac{\pi z}{2h}\right)\right) & \text{STR} \end{cases} \qquad (5.67)$$

where $\Delta T = T - T_0$ is the temperature change. By substituting Eqs. (5.60), (5.63), and (5.65) into Eq. (5.58) and setting the coefficients of δu, δv, δw_b, and δw_s to zero, the following Euler–Lagrange equation can be obtained as:

$$\frac{\partial N_x}{\partial x} + \frac{\partial N_{xy}}{\partial y} = I_0 \frac{\partial^2 u}{\partial t^2} - I_1 \frac{\partial^3 w_b}{\partial x \partial t^2} - J_1 \frac{\partial^3 w_s}{\partial x \partial t^2} \tag{5.68}$$

$$\frac{\partial N_{xy}}{\partial x} + \frac{\partial N_y}{\partial y} = I_0 \frac{\partial^2 v}{\partial t^2} - I_1 \frac{\partial^3 w_b}{\partial y \partial t^2} - J_1 \frac{\partial^3 w_s}{\partial y \partial t^2} \tag{5.69}$$

$$\frac{\partial^2 M_x^b}{\partial x^2} + 2 \frac{\partial^2 M_{xy}^b}{\partial x \partial y} + \frac{\partial^2 M_y^b}{\partial y^2} + \frac{\partial Q}{\partial x} - N^T \nabla^2 (w_b + w_s) - k_w (w_b + w_s)$$

$$+ k_p \nabla^2 (w_b + w_s) = I_0 \frac{\partial^2 (w_b + w_s)}{\partial t^2} + I_1 \left(\frac{\partial^3 u}{\partial x \partial t^2} + \frac{\partial^3 v}{\partial y \partial t^2} \right) \tag{5.70}$$

$$- I_2 \nabla^2 \left(\frac{\partial^2 w_b}{\partial t^2} \right) - J_2 \nabla^2 \left(\frac{\partial^2 w_s}{\partial t^2} \right)$$

$$\frac{\partial^2 M_x^s}{\partial x^2} + 2 \frac{\partial^2 M_{xy}^s}{\partial x \partial y} + \frac{\partial^2 M_y^s}{\partial y^2} + \frac{\partial Q_{xy}}{\partial x} + \frac{\partial Q_{yz}}{\partial y} - N^T \nabla^2 (w_b + w_s) - k_w (w_b + w_s)$$

$$- k_p \nabla^2 (w_b + w_s) = I_0 \frac{\partial^2 (w_b + w_s)}{\partial t^2} + J_1 \left(\frac{\partial^3 u}{\partial x \partial t^2} + \frac{\partial^3 v}{\partial y \partial t^2} \right) \tag{5.71}$$

$$- J_2 \nabla^2 \left(\frac{\partial^2 w_b}{\partial t^2} \right) - K_2 \nabla^2 \left(\frac{\partial^2 w_s}{\partial t^2} \right)$$

The Laplacian operator is denoted by ∇^2. The basic elastic equations of isotropic solids are derived from their elastic stress-strain relations. Beam constitutive equations may be given in the form of:

$$\sigma_{ij} = C_{ijkl} \varepsilon_{kl} \tag{5.72}$$

In Eq. (5.72), σ_{ij}, ε_{kl}, and C_{ijkl} are related to componentass of Cauchy stress, strain, and elasticity tensors, respectively. Whenever extending the aforementioned equation for a shear deformable plate, the following relations can be reached:

$$\begin{bmatrix} \sigma_{xx} \\ \sigma_{yy} \\ \sigma_{yz} \\ \sigma_{xz} \\ \sigma_{xy} \end{bmatrix} = \begin{bmatrix} Q_{11} & Q_{12} & 0 & 0 & 0 \\ Q_{12} & Q_{22} & 0 & 0 & 0 \\ 0 & 0 & Q_{44} & 0 & 0 \\ 0 & 0 & 0 & Q_{55} & 0 \\ 0 & 0 & 0 & 0 & Q_{66} \end{bmatrix} \begin{bmatrix} \varepsilon_{xx} \\ \varepsilon_{yy} \\ \varepsilon_{yz} \\ \varepsilon_{xz} \\ \varepsilon_{xy} \end{bmatrix} \tag{5.73}$$

where

$$Q_{11} = Q_{22} = \frac{E_{eq}}{1 - V_{eq}^2}, \ Q_{12} = \frac{E_{eq}V_{eq}}{1 - V_{eq}^2}, \ Q_{44} = Q_{55} = Q_{66} = G_{eq} \qquad (5.74)$$

in which E_{eq} and G_{eq} are the Young and shear moduli of the nanocomposite, respectively. Integrating from Eqs. (5.68) to (5.71) over the cross-section area of the beam, the following equations can be written for the stress resultants:

$$
\begin{bmatrix} N_x \\ N_y \\ N_{xy} \end{bmatrix} = \begin{bmatrix} A_{11} & A_{12} & 0 \\ A_{21} & A_{22} & 0 \\ 0 & 0 & A_{66} \end{bmatrix} \begin{bmatrix} \dfrac{\partial u}{\partial x} \\ \dfrac{\partial v}{\partial y} \\ \dfrac{\partial u}{\partial y} + \dfrac{\partial v}{\partial x} \end{bmatrix} - \begin{bmatrix} B_{11} & B_{12} & 0 \\ B_{21} & B_{22} & 0 \\ 0 & 0 & B_{66} \end{bmatrix} \begin{bmatrix} \dfrac{\partial^2 w_b}{\partial u^2} \\ \dfrac{\partial^2 w_b}{\partial y^2} \\ 2\dfrac{\partial^2 w_b}{\partial x \partial y} \end{bmatrix}
$$

$$
- \begin{bmatrix} B_{11}^s & B_{12}^s & 0 \\ B_{21}^s & B_{22}^s & 0 \\ 0 & 0 & B_{66}^s \end{bmatrix} \begin{bmatrix} \dfrac{\partial^2 w_s}{\partial u^2} \\ \dfrac{\partial^2 w_s}{\partial y^2} \\ 2\dfrac{\partial^2 w_s}{\partial x \partial y} \end{bmatrix} \qquad (5.75)
$$

$$
\begin{bmatrix} M_x^b \\ M_y^b \\ M_{xy}^b \end{bmatrix} = \begin{bmatrix} B_{11} & B_{12} & 0 \\ B_{21} & B_{22} & 0 \\ 0 & 0 & B_{66} \end{bmatrix} \begin{bmatrix} \dfrac{\partial u}{\partial x} \\ \dfrac{\partial v}{\partial y} \\ \dfrac{\partial u}{\partial y} + \dfrac{\partial v}{\partial x} \end{bmatrix} - \begin{bmatrix} D_{11} & D_{12} & 0 \\ D_{21} & D_{22} & 0 \\ 0 & 0 & D_{66} \end{bmatrix} \begin{bmatrix} \dfrac{\partial^2 w_b}{\partial u^2} \\ \dfrac{\partial^2 w_b}{\partial y^2} \\ 2\dfrac{\partial^2 w_b}{\partial x \partial y} \end{bmatrix}
$$

$$
- \begin{bmatrix} D_{11}^s & D_{12}^s & 0 \\ D_{21}^s & D_{22}^s & 0 \\ 0 & 0 & D_{66}^s \end{bmatrix} \begin{bmatrix} \dfrac{\partial^2 w_s}{\partial u^2} \\ \dfrac{\partial^2 w_s}{\partial y^2} \\ 2\dfrac{\partial^2 w_s}{\partial x \partial y} \end{bmatrix} \qquad (5.76)
$$

$$
\begin{bmatrix} M_x^s \\ M_y^s \\ M_{xy}^s \end{bmatrix} = \begin{bmatrix} B_{11}^s & B_{12}^s & 0 \\ B_{21}^s & B_{22}^s & 0 \\ 0 & 0 & B_{66}^s \end{bmatrix} \begin{bmatrix} \dfrac{\partial u}{\partial x} \\ \dfrac{\partial v}{\partial y} \\ \dfrac{\partial u}{\partial y} + \dfrac{\partial v}{\partial x} \end{bmatrix} - \begin{bmatrix} D_{11}^s & D_{12}^s & 0 \\ D_{21}^s & D_{22}^s & 0 \\ 0 & 0 & D_{66}^s \end{bmatrix} \begin{bmatrix} \dfrac{\partial^2 w_b}{\partial u^2} \\ \dfrac{\partial^2 w_b}{\partial y^2} \\ 2\dfrac{\partial^2 w_b}{\partial x \partial y} \end{bmatrix}
$$

$$
- \begin{bmatrix} H_{11}^s & H_{12}^s & 0 \\ H_{21}^s & H_{22}^s & 0 \\ 0 & 0 & H_{66}^s \end{bmatrix} \begin{bmatrix} \dfrac{\partial^2 w_s}{\partial u^2} \\ \dfrac{\partial^2 w_s}{\partial y^2} \\ 2\dfrac{\partial^2 w_s}{\partial x \partial y} \end{bmatrix}
\tag{5.77}
$$

$$
\begin{bmatrix} Q_x \\ Q_y \end{bmatrix} = \begin{bmatrix} A_{44}^s & 0 \\ 0 & A_{55}^s \end{bmatrix} \begin{bmatrix} \dfrac{\partial w_s}{\partial x} \\ \dfrac{\partial w_s}{\partial y} \end{bmatrix}
\tag{5.78}
$$

The relations for the cross-sectional rigidities are provided by Eqs. (5.75)–(5.78).

$$
\begin{bmatrix} A_{11} & B_{11} & D_{11} \\ A_{12} & B_{12} & D_{12} \\ A_{66} & B_{66} & D_{66} \end{bmatrix} = \int_{-\frac{h}{2}}^{\frac{h}{2}} Q_{11}\left(1, z, z^2\right) \begin{bmatrix} 1 \\ V_{eq} \\ \dfrac{1 - V_{eq}}{2} \end{bmatrix} dz
$$

$$
\begin{bmatrix} B_{11}^s & D_{11}^s & H_{11}^s \\ B_{12}^s & D_{12}^s & H_{12}^s \\ B_{66}^s & D_{66}^s & H_{66}^s \end{bmatrix} = \int_{-\frac{h}{2}}^{\frac{h}{2}} Q_{11}\left(f(z), zf(z), f^2(z)\right) \begin{bmatrix} 1 \\ V_{eq} \\ \dfrac{1 - V_{eq}}{2} \end{bmatrix} dz
\tag{5.79}
$$

$$
\left(A_{22}, B_{22}, D_{22}, B_{22}^s, D_{22}^s, H_{22}^s\right) = \left(A_{11}, B_{11}, D_{11}, B_{11}^s, D_{11}^s, H_{11}^s\right)
\tag{5.80}
$$

$$A_{44}^s = A_{55}^s = \int_{-\frac{h}{2}}^{\frac{h}{2}} \frac{E_{eq}}{2(1+v_{eq})} g^2(z) dz \tag{5.81}$$

The governing equations of a GOP-reinforced plate may be found by just plugging the corresponding values from Eqs. (5.75) to (5.78) into Eqs. (5.68) to (5.71).

$$A_{11}\frac{\partial^2 u}{\partial x^2} + (A_{12} + A_{66})\frac{\partial^2 v}{\partial x \partial y} + A_{66}\frac{\partial^2 u}{\partial y^2} - B_{11}\frac{\partial^3 w_b}{\partial x^3} - (B_{12} + 2B_{66})\frac{\partial^3 w_b}{\partial x \partial y^2} - B_s^{11}\frac{\partial^3 w_s}{\partial x^3}$$

$$- (B_{12}^s + 2B_{66}^s)\frac{\partial^3 w_s}{\partial x \partial y^2} - I_0\frac{\partial^2 u}{\partial t^2} + I_1\frac{\partial^2 w_b}{\partial x \partial t^2} + J_1\frac{\partial^2 w_s}{\partial x \partial t^2} = 0$$

$$\tag{5.82}$$

$$A_{22}\frac{\partial^2 v}{\partial y^2} + (A_{12} + A_{66})\frac{\partial^2 u}{\partial x \partial y} + A_{66}\frac{\partial^2 v}{\partial x^2} - B_{11}\frac{\partial^3 w_b}{\partial x^3} - B_{22}\frac{\partial^3 w_b}{\partial y^3}$$

$$- (B_{12} + 2B_{66})\frac{\partial^3 w_b}{\partial x^2 \partial y} - B_s^{22}\frac{\partial^3 w_s}{\partial y^3} - (B_{12}^s + 2B_{66}^s)\frac{\partial^3 w_s}{\partial x^2 \partial y} - I_0\frac{\partial^2 v}{\partial t^2} \tag{5.83}$$

$$+ I_1\frac{\partial^2 w_b}{\partial y \partial t^2} + J_1\frac{\partial^2 w_s}{\partial y \partial t^2} = 0$$

$$B_{11}\frac{\partial^3 u}{\partial x^3} + (B_{12} + 2B_{66})\frac{\partial^3 u}{\partial x \partial y^2} + B_{22}\frac{\partial^3 v}{\partial y^3} - (B_{12} + 2B_{66})\frac{\partial^3 v}{\partial x^2 \partial y} - D_{11}\frac{\partial^4 w_b}{\partial x^4}$$

$$-2(D_{12} + 2D_{66})\frac{\partial^4 w_b}{\partial x^2 \partial y^2} - D_{22}\frac{\partial^4 w_b}{\partial y^4} - D_{11}^s\frac{\partial^4 w_s}{\partial x^4} - 2(D_{12}^s + 2D_{66}^s)\frac{\partial^4 w_s}{\partial x^2 \partial y^2}$$

$$-D_{22}^s\frac{\partial^4 w_s}{\partial y^4} - I_0\frac{\partial^2 (w_b + w_s)}{\partial t^2} - I_1\left(\frac{\partial^2 u}{\partial x \partial t^2} + \frac{\partial^2 v}{\partial y \partial t^2}\right) + I_2\nabla^2\frac{\partial^2 w_b}{\partial t^2} \tag{5.84}$$

$$+ J_2\nabla^2\frac{\partial^2 w_s}{\partial t^2} + (k_p - N^T)\nabla^2(w_b + w_s) - k_w(w_b + w_s) = 0$$

$$B_{11}^s\frac{\partial^3 u}{\partial x^3} + (B_{12}^s + 2B_{66}^s)\frac{\partial^3 u}{\partial x \partial y^2} + B_{22}^s\frac{\partial^3 v}{\partial y^3} - (B_{12}^s + 2B_{66}^s)\frac{\partial^3 v}{\partial x^2 \partial y} - D_{11}^s\frac{\partial^4 w_b}{\partial x^4}$$

$$-2(D_{12}^s + 2D_{66}^s)\frac{\partial^4 w_b}{\partial x^2 \partial y^2} - D_{22}^s\frac{\partial^4 w_b}{\partial y^4} - H_{11}^s\frac{\partial^4 w_s}{\partial x^4} - 2(H_{12}^s + 2H_{66}^s)\frac{\partial^4 w_s}{\partial x^2 \partial y^2}$$

$$- H_{22}^s\frac{\partial^4 w_s}{\partial y^4} + A_{44}^s\nabla^2 w_s - I_0\frac{\partial^2 (w_b + w_s)}{\partial t^2} - J_1\left(\frac{\partial^2 u}{\partial x \partial t^2} + \frac{\partial^2 v}{\partial y \partial t^2}\right) \tag{5.85}$$

$$+ J_2\nabla^2\frac{\partial^2 w_b}{\partial t^2} + K_2\nabla^2\frac{\partial^2 w_s}{\partial t^2} + (k_p - N^T)\nabla^2(w_b + w_s) - k_w(w_b + w_s) = 0$$

In order to get answers to the governing equations of a GOP-reinforced nano-composite plate, an analytical solution approach is used. The following equations are assumed to describe the displacement field's constituent parts:

$$u = U \exp\left[i\left(\beta_1 x + \beta_2 y - \omega_n t\right)\right] \tag{5.86a}$$

$$v = V \exp\left[i\left(\beta_1 x + \beta_2 y - \omega_n t\right)\right] \tag{5.86b}$$

$$w_b = W_b \exp\left[i\left(\beta_1 x + \beta_2 y - \omega_n t\right)\right] \tag{5.86c}$$

$$w_s = W_s \exp\left[i\left(\beta_1 x + \beta_2 y - \omega_n t\right)\right] \tag{5.86d}$$

Where U, V, W_b, and W_s represent wave amplitudes, respectively. Also, β_1 and β_2 are longitudinal and transverse wave numbers, respectively, and ω_n is circular frequency dispersed waves. By substituting u, v, w_b, and w_s from Eqs. (5.86a)–(5.86d) into Eqs. (5.82)–(53), the following equation is gained:

$$\left[\left[K\right]_{4\times4} - \omega_n^2 \left[M\right]_{4\times4}\right] \begin{bmatrix} U \\ V \\ W_b \\ W_s \end{bmatrix} = 0 \tag{5.87}$$

in which K and M are the stiffness and mass matrices, respectively. Besides, all components of these matrices are presented in Appendix 5C.

Eq. (5.87) states that the determinant of the coefficient matrix must be zero in order to solve the eigenvalue issue.

$$\left| \left[K\right]_{4\times4} - \omega_n^2 \left[M\right]_{4\times4} \right| = 0 \tag{5.88}$$

Additionally, the phase velocity may be calculated by dividing the wave frequency by the wave number when $\beta_1 = \beta_2 = \beta$. Here, we present the phase velocity and wave frequency responses of the structure under various conditions, including three types of thermal loadings, four patterns of GOP distribution, Winkler–Pasternak foundation parameters, aspect ratio, and GOP weight fraction, using various graphical illustrations. The parameters of the examined structure's component materials, a polymeric matrix, and GOPs as nanofillers are summarized in Table 5.1. The thickness of the nanocomposite plate is assumed to be $h = 5$ cm. For convenience, we will now write the Winkler–Pasternak base parameters in their non-dimensional forms [41, 42]:

$$K_w = k_w \frac{a^4}{DD}, \quad K_p = k_p \frac{a^2}{DD}, \quad DD = \frac{E_M h^3}{12\left(1 - v_M^2\right)}$$

TABLE 5.1

Material Properties of the FG-GOP Nanocomposite Plates

Polymeric Matrix	Graphene Oxide Powder (GOP)
$E_m = 3$ GPa	$E_{GOP} = 444.8$ GPa
$\rho_m = 1.2$ g/cm^3	$\rho_{GOP} = 1.2$ g/cm^3
$\nu_m = 0.34$	$\nu_{GOP} = 0.165$
$\alpha_m = 7 \times 10^{-5}$ 1/°C	$\alpha_{GOP} = -5 \times 10^{-5}$ 1/°C

Figure 5.50 is an analysis of how the wave frequency response of the nanocomposite plate varies depending on the FG-reinforcing pattern of the GOP and the thermal loadings. This analysis is presented in the form of a diagram. The historical pattern of the frequency's change with respect to the wave number is the first thing that will be investigated. As can be observed in each of the photographs that follow, the frequency of the structure's waves increases as the number of waves in the structure rises. This trend of growing heights is maintained by higher wave numbers, notably after wave number 20, and it reaches its highest point. It is clear that the structure that is subjected to thermal loading with STR has a higher wave frequency than the other types of thermal loadings, and it is also clear that by ignoring the thermal effect in Figure 2d, wave frequency will be led to the highest range. Subfigures of this figure are provided to show the effect of the thermal loadings with various temperature rises. The surprising finding that the X-distribution pattern of GOP has a broader range of wave frequencies (from wave number 0 to

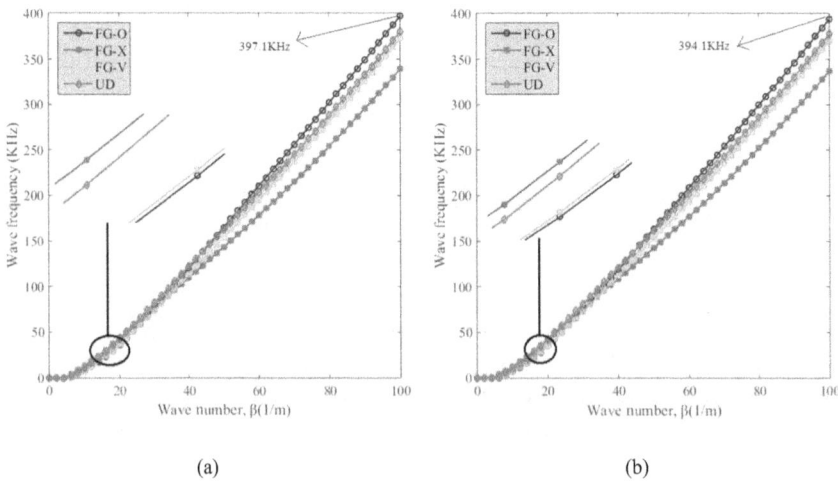

(a)   (b)

FIGURE 5.50 Effect of the various distribution patterns of the GOPs on the wave frequency of the plate under thermal loading with (a) STR temperature rise, (b) LTR temperature rise.

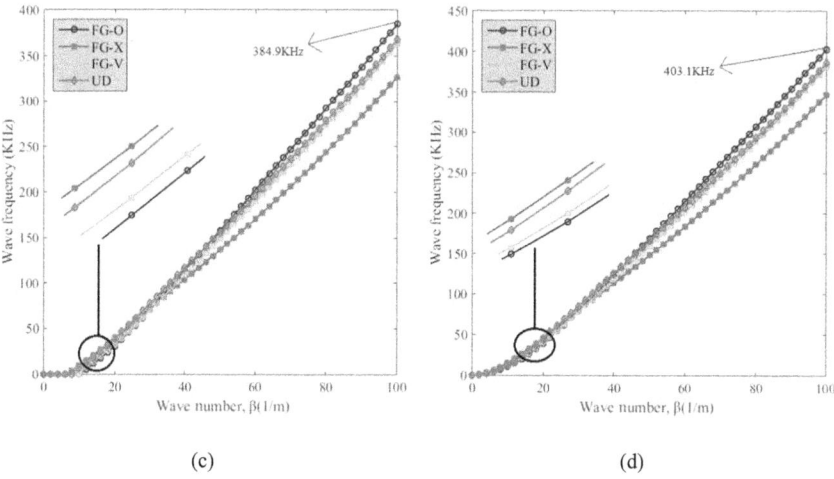

(c)   (d)

FIGURE 5.50 (Continued) (c) UTR temperature rise, and also (d) without considering thermal effect ($\Delta T = 400$, $W_{GOP} = 5\%$, $K_w = 500$, $K_p = 100$).

wave number 20) compared to the UD-, V-, and O-patterns, as shown in this figure. As a consequence, this pattern is currently the best choice for reinforcing the plate with GOP nanofillers. After this threshold, however, the frequency range of the nanocomposite plates flips, with the O-pattern resulting in the highest frequency range of the other types and the X-type resulting in the lowest frequency range. This is because the O-pattern has the biggest frequency range of all the other kinds.

In Figure 5.51, we illustrate the variations in the phase velocity response of the structure versus temperature rise (ΔT) to better understand how the GOP-reinforced nanocomposite plates disperse waves under three distinct thermal loadings. Each of the four distribution patterns of FG nanocomposite plates is given its own subplot, so their effects may be studied separately. Figure 5.51 shows the phase velocity changes versus temperature gradient for different forms of FG reinforcing, allowing one to better understand the impact that temperature gradient may have on the phase velocity of the GOP-reinforced plate. Clearly, when the temperature gradient rises, the y-axis of every subplot begins to decline. The phase velocity decreases because the structure's rigidity relaxes after being subjected to thermal loadings. To mitigate this loss of stiffness, heat loadings should be applied differently. This diagram takes into account three distinct thermal loadings for the aforementioned objective. The previous figure's findings that STR thermal loading has less of an effect on phase velocity values may be readily seen once again. At a fixed wave number ($\beta_x = \beta_y = 10$), the plate is reinforced with four distinct patterns of GOP distribution, with the most effective pattern being shown in the picture. Based on the phase velocity numbers shown in the individual graphs. It is conceivable that the X-pattern plate provides the most desirable phase velocity responses. Note that at temperatures above 350 degrees Celsius, the phase velocity in Figure 5.51d for a structure reinforced with an O-distribution pattern and

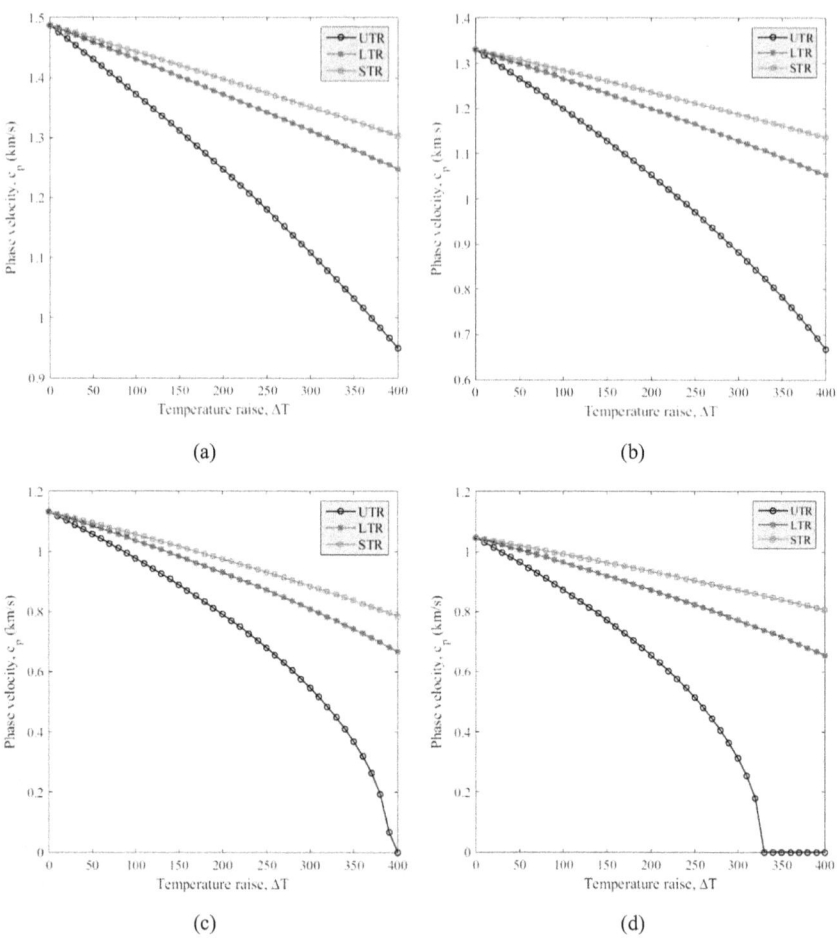

FIGURE 5.51 Variation of the phase velocity versus temperature rise for (a) X, (b) U, (c) V, and (d) O-distribution patterns of the GOPs through the thickness considering various types of thermal loading ($W_{GOP} = 5$ %, $K_w = 500$, $K_p = 100$, $\beta_x = \beta_y = 10$).

subjected to a UTR type of thermal loading becomes zero, indicating that the structure could not have a phase velocity response under these conditions.

It has been shown that the wave propagation in a GOP-reinforced nanocomposite plate is very sensitive to the weight fraction of the GOPs, as can be seen in Figure 5.52. The wave frequency responses of the plate alter at different wave numbers as the fraction of GOPs added to the plate varies from 0% to 10%. The frequency of waves increases with the quantity of GOPs present, albeit this rise is not very noticeable at low wave counts. When the wave numbers are large enough, it is clear that the values of wave frequency may be significantly increased by adding the GOPs' weight fraction.

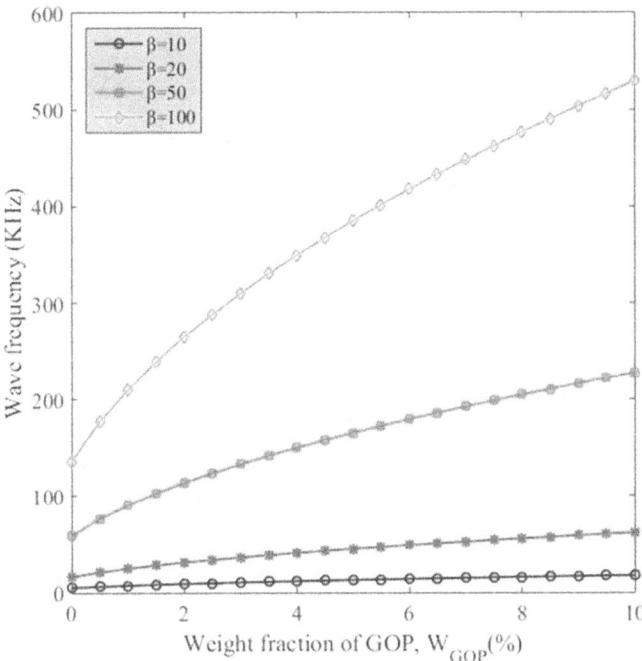

FIGURE 5.52 Variation of the wave frequency of the GOP-reinforced nanocomposite plate versus nanofillers' weight fraction with respect to different wave numbers ($\Delta T = 100(STR)$, $K_w = 500$, $K_p = 100$, UD).

In Figure 5.53, the phase velocity of the GOP-reinforced nanocomposite plate is evaluated in the same circumstances as in Figure 5.50; however, there is one significant difference. This figure's objective is to illustrate that previous results may be generalized to the phase velocity of the GOP-reinforced nanocomposite plate, and are not, as a result, restricted to the investigation of wave frequency response alone. Generally speaking, the vast majority of the time it's important to note that a plate with UD of GOPs has the most consistent behavior among the different kinds, ranking second in terms of phase velocity values over the whole range of wave numbers from $\beta = 0$ *to* $\beta = 100$. In addition, Figure 5.54 provides a comprehensive overview, in terms of thermal loadings with varying degrees of temperature increase, of the effects of the plate's aspect ratio on the phase velocity of the GOP-reinforced nanocomposite plate. This graphic demonstrates that until the plate's length is equal to twice its breadth, the phase velocity decreases as the aspect ratio increases. After this point, the phase velocity is generally unaffected by additional changes in the plate's aspect ratio, despite the fact that this value has been reached. At the wave number $\beta = 10$ (under 20), it is clear that strengthening the structure with an O-pattern of GOPs' distribution results in the lowest range of phase velocity, while the X-pattern results in the highest range of phase velocity. This is because the O-pattern strengthens the structure more than the X-pattern

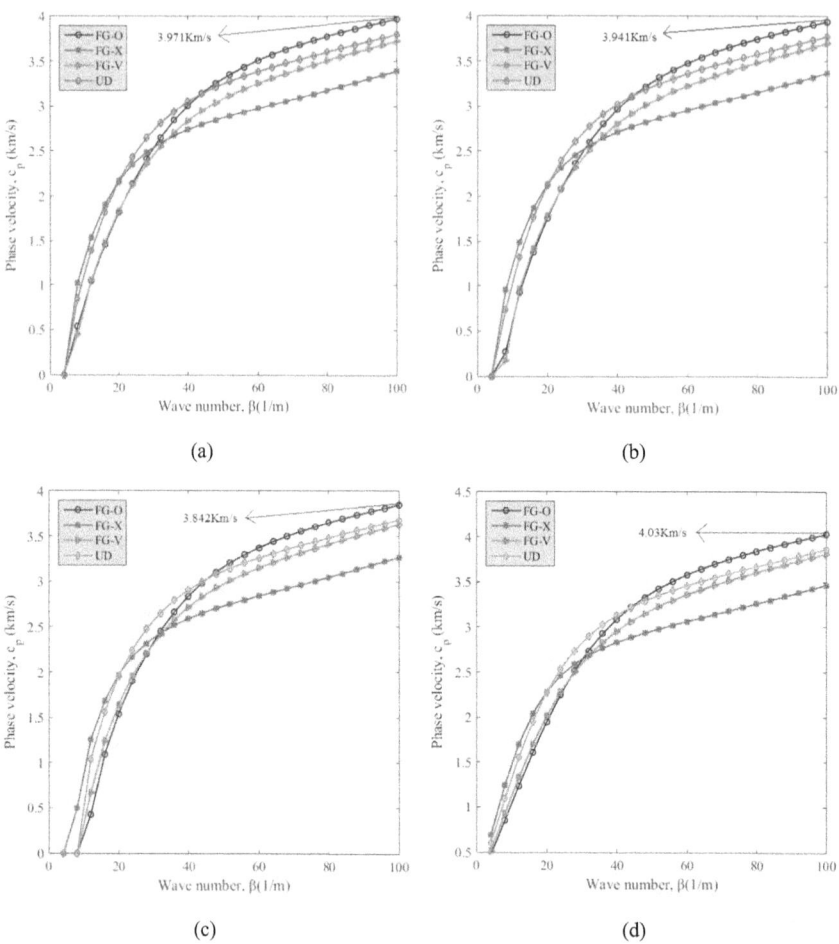

FIGURE 5.53   Effect of the various distribution patterns of the GOPs on the phase velocity of the plate under thermal loading with (a) STR temperature rise, (b) LTR temperature rise, (c) UTR temperature rise, and also (d) without considering thermal effect ($\Delta T = 400$, $W_{GOP} = 5$ %, $K_w = 500$, $K_p = 100$, $\beta_x = \beta_y$).

does. Even if there isn't much of a difference in the phase velocity values under STR and LTR thermal loadings, STR is still the preferable sort of thermal loading for this architecture because of how it distributes the heat.

In addition, Figure 5.55 provides a summary of the investigation into the simultaneous effects of the foundation parameters and various types of temperature rise on the wave frequency of the nanocomposite plate by graphing its variations against Pasternak and Winkler coefficients. This investigation was carried out in order to determine how the wave frequency was affected. This figure demonstrates that the Pasternak layer has a more significant level of influence on the structure than does the Winkler layer. This makes perfect sense when you consider that the

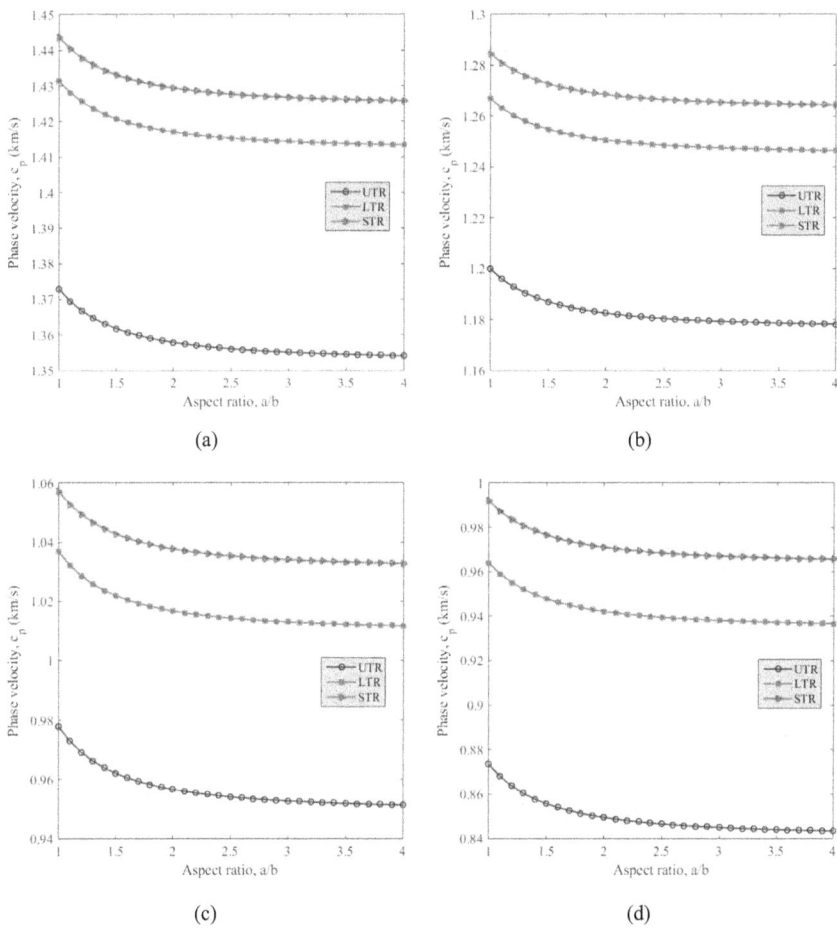

FIGURE 5.54   Variation of the phase velocity against plate's aspect ratio considering different types of thermal loadings for various patterns of GOPs' distribution ($\Delta T = 100$, $W_{GOP} = 5\%$, $K_w = 100$, $K_p = 500$, $\beta_x = \beta_y = 10$). (a) FG-X, (b) UD, (c) FG-V, and (d) FG-O.

Pasternak layer interacts in a continuous manner with the plate, but the Winkler layer interacts in a discontinuous manner with the nanocomposite plate. The inclusion of the thermal effect in this picture demonstrates that the stiffness improvement brought on by embedding the structure on elastic foundations is sufficient to overcome the stiffness hardening brought on by the thermal loadings, as seen by the increasing trend in all plots. Without thinking about temperature impacts, the nanocomposite plate's wave frequency behavior is certain to improve.

Figure 5.56 examines the impact of the length-to-thickness ratio and the use of an elastic medium on the wave frequency of a structure. The graph displays the changes in wave frequency as a function of the plate's length-to-thickness ratio. The experiment was conducted using various foundation coefficients and four

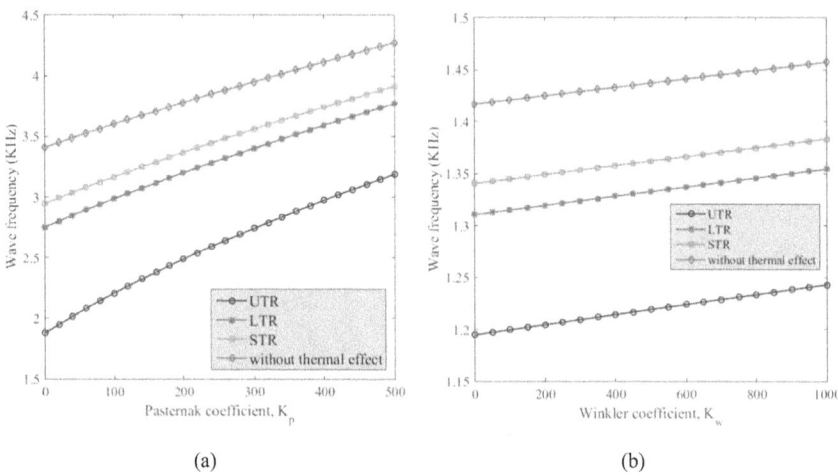

(a)   (b)

FIGURE 5.55 Variation of the plate's wave frequency versus (a) Pasternak and (b) Winkler coefficient of the elastic substrate by considering various types of thermal loadings ($W_{GOP(UD)}$ = 5%). (a) ΔT = 100, K_w = 100, $\beta_x = \beta_y$ = 5 and (b) ΔT = 20, KP = 100, $\beta_x = \beta_y$ = 3.

types of GOP distribution for plate reinforcement. The diagram indicates that, for small values of the elastic foundation coefficients, there is no significant alteration in the wave frequency as the length-to-thickness ratio increases. In contrast, when the length-to-thickness ratio of the plate increases, a declining tendency will be seen in the upper range of foundation coefficients. The more you think about it,

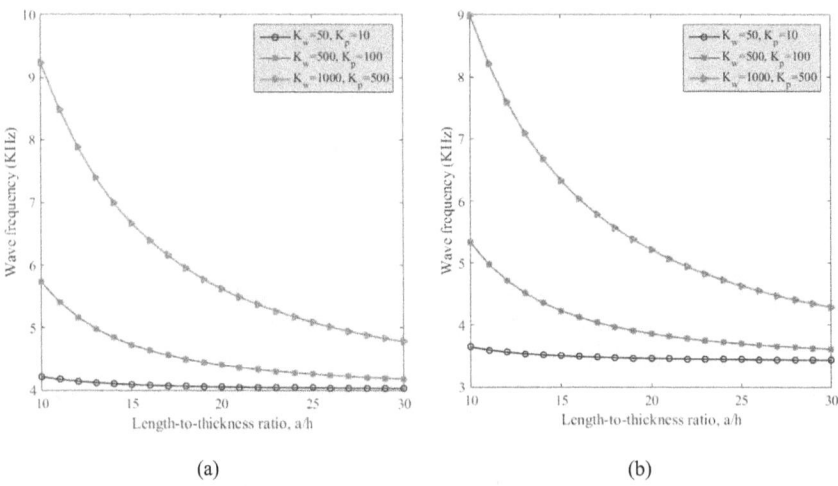

(a)   (b)

FIGURE 5.56 Illustration of the coupled effects of plate's length-to-thickness ratio and elastic foundation on the wave frequency response of GOP-reinforced nanocomposite plate once GOPs are dispersed through the thickness of the plate with (a) FG-X pattern, (b) UD.

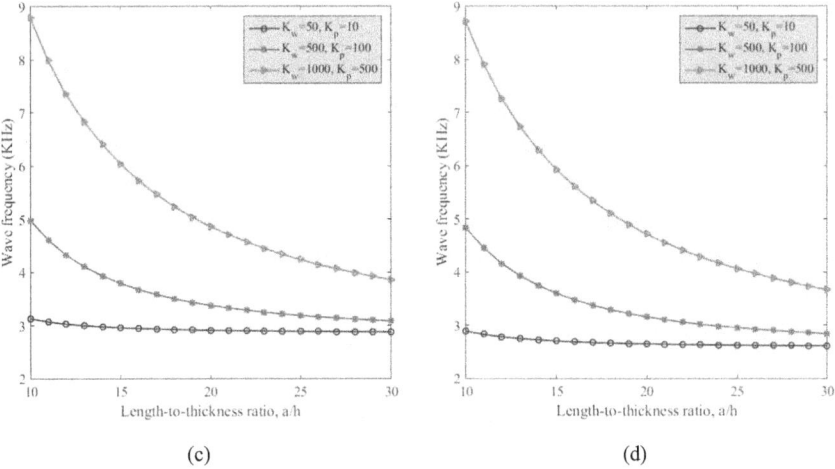

(c)   (d)

FIGURE 5.56 (Continued)   (c) FG-V pattern, and (d) FG-O pattern ($\Delta T = 0$, $W_{GOP} = 5\,\%$, $\beta_x = \beta_y = 5$).

the more you'll realize that the more elastic foundation coefficients grow, the more you'll have to sacrifice in order to keep up with the pace.

5.4 WAVE DISPERSION CHARACTERISTICS OF GRAPHENE PLATELET–REINFORCED STRUCTURES

5.4.1 BACKGROUND

This chapter is devoted mostly to discussing the wave propagation characteristics of various graphene platelet–reinforced (GPLR) structures subjected to thermo-magnetic external loadings, such as beams, rings, plates, shells, and nanoshells. The equivalent modulus of elasticity for composite structures is determined within the context of the Halpin–Tsai model, and several GPL distribution patterns are taken into account. Both symmetric and asymmetric porosity distributions are taken into account in this study, and the effective values of density and Poisson's ratio are attained by using relations of the mixing rule. The impact of multiple graphene platelet (GPL) releases is presented in order to deliver more accurate information. The shear deformation effects in the equations of motion are also accounted for by combining Hamilton's principle with a more detailed theory. In order to illustrate the effects of different factors on the phase speed of propagated waves, the most important findings are provided in the context of a set of diagrams. Furthermore, various graphics depicting the impact of all contributor factors are shown. When the effects of porosity are taken into account, the simulations show a reduction in stiffness. The influence of the longitudinal wave number on the nanocomposite structure's wave dispersion curves [52–56] should also be taken into account.

5.4.2 GRAPHENE PLATELET–REINFORCED BEAMS: WAVE DISPERSION CHARACTERISTICS

To characterize the effective Young's modulus of the GPLR beam, a Halpin–Tsai model is constructed here. It is assumed in this study that the GPLs exist in the polymer matrix as rectangular solid fillers of three different kinds (GPLR-O, GPLR-U, and GPLR-X), as shown in Figure 5.57.

Below [52] is a formulation of the effective modulus of elasticity based on the Voigt–Reuss model.

$$E_e = \frac{3}{8}E_\Lambda + \frac{5}{8}E_\Theta \qquad (5.89)$$

where E_Λ and E_Θ are the longitudinal and transverse Young's modulus, which can be denoted based on the Halpin–Tsai model in the following form:

$$E_\Lambda = \frac{1 + \xi_L \eta_L V_{GPL}}{1 - \eta_L V_{GPL}} \times E_M, \quad E_\Theta = \frac{1 + \xi_W \eta_W V_{GPL}}{1 - \eta_W V_{GPL}} \times E_M \qquad (5.90)$$

where

$$\eta_L = \frac{\left(E_{GPL} / E_M\right) - 1}{\left(E_{GPL} / E_M\right) + \xi_L}, \quad \eta_W = \frac{\left(E_{GPL} / E_M\right) - 1}{\left(E_{GPL} / E_M\right) + \xi_W} \qquad (5.91)$$

in which E_M and E_{GPL} stand for Young's modulus of the polymer matrix and GPL, respectively. In addition, V_{GPL} is the GPL's volume fraction; ξ_L and ξ_W are the parameters specifying the geometry and size of the nanofillers, which can be formulated as:

$$\xi_L = \frac{2l_{GPL}}{h_{GPL}}, \xi_W = \frac{2w_{GPL}}{h_{GPL}} \qquad (5.92)$$

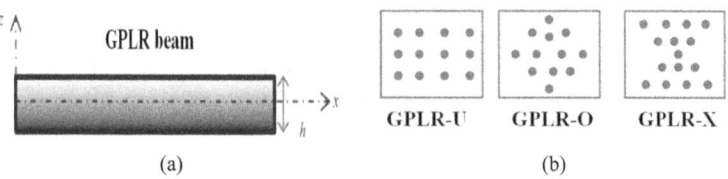

(a)   (b)

FIGURE 5.57 (a) Configuration of a GPLR beam, (b) cross-section of different patterns of reinforcements.

in Eq. 4, l_{GPL}, w_{GPL}, and h_{GPL} are the average length, width, and thickness of the GPL, respectively. Moreover, the effective mass density, Poisson's ratio, and thermal expansion coefficient of the GPLR beam can be defined as follows based on the mixture's rule:

$$\rho_e = \rho_{GPL} V_{GPL} + \rho_M V_M \tag{5.93}$$

$$v_e = v_{GPL} V_{GPL} + v_M V_M \tag{5.94}$$

$$\alpha_e = \alpha_{GPL} V_{GPL} + \alpha_M V_M \tag{5.95}$$

where V_M denotes the volume fraction of the polymer matrix and can be obtained by $V_M = 1 - V_{GPL}$. Besides, subscript "e" stands for the GPL/polymer composite. It shall be considered that the volume fraction of the GPL can be calculated based on the following equation:

$$V_{GPL}^* = \frac{g_{GPL}}{g_{GPL} + \left(\rho_{GPL} / \rho_M\right)\left(1 - g_{GPL}\right)} \tag{5.96}$$

in which, g_{GPL} is the weight fraction of the GPLs in the composite beam. Also, the equivalent volume fraction can be separately modified for different distributions as:

$$
\begin{cases}
V_{GPL} = V_{GPL}^* & \text{GPLR–U} \\[2mm]
V_{GPL} = \left(2 - 4\dfrac{|z|}{h}\right) V_{GPL}^* & \text{GPLR–O} \\[2mm]
V_{GPL} = 4\dfrac{|z|}{h} V_{GPL}^* & \text{GPLR–X}
\end{cases} \tag{5.97}
$$

The GPLR beam's displacement field may be expressed using a revised version of the theory of beams subjected to higher-order shear deformation.

$$u_x\left(x,z,t\right) = u\left(x,t\right) - z\frac{\partial w_b}{\partial x} - f\left(z\right)\frac{\partial w_s}{\partial x} \tag{5.98}$$

$$u_z\left(x,z,t\right) = w_b\left(x,t\right) + w_s\left(x,t\right) \tag{5.99}$$

In Eqs. (5.98) and (5.99), u is longitudinal displacement and w_b and w_s are bending and shear deflections, respectively. Also, $f(z)$ is a shape function that estimates the distribution of shear stress or strain through the beam thickness.

Due to the applied shape function, presented model can be utilized needless of any shear correction factor. In this research, this function can be expressed as:

$$f(z) = z\left(1 - \frac{1}{2}\left(\frac{h^2}{4} - \frac{z^2}{3}\right)\right) \qquad (5.100)$$

Eqs. (5.101) and (5.104) may be used to represent the beam's nonzero strains:

$$\varepsilon_{xx} = \frac{\partial u}{\partial x} - z\frac{\partial^2 w_b}{\partial x^2} - f(z)\frac{\partial^2 w_s}{\partial x^2}, \gamma_{xz} = g(z)\frac{\partial w_s}{\partial x} \qquad (5.101)$$

where

$$g(z) = 1 - \frac{df(z)}{dz} \qquad (5.102)$$

Now, Hamilton's principle can be defined as:

$$\int_0^t \delta(U - T + V)dt = 0 \qquad (5.103)$$

where U, T, and V account for strain energy, kinetic energy, and work done by external forces, respectively. The variation of strain energy is written as:

$$\delta U = \int_V \left(\sigma_{xx}\delta\varepsilon_{xx} + \sigma_{xz}\delta\gamma_{xz}\right)dV \qquad (5.104)$$

Substituting Eqs. (5.98)–(5.102) in Eq. (5.104) yields:

$$\delta U = \int_0^L \left(N\frac{\partial \delta u}{\partial x} - M_b\frac{\partial^2 w_b}{\partial x^2} - M_s\frac{\partial^2 w_s}{\partial x^2} + Q\frac{\partial w_s}{\partial x}\right)dx \qquad (5.105)$$

in which the stress resultants N, M_b, M_s, and Q are expressed as:

$$[N, M_b, M_s] = \int_A [1, z, f]\sigma_{xx}dA, \qquad (5.106)$$

$$Q = \int_A g(z)\sigma_{xz}dA \qquad (5.107)$$

However, when the beam is meant to be exposed to a magnetic field, the Lorentz force, created by the magnetic field, must be tolerated. In this case, we use Maxwell's equations to deduce the mathematical form of the Lorentz force that has been induced. In this scenario, it is assumed that a longitudinally stable magnetic field of strength H is applied to the beam. Therefore, the following expression may be used to describe the body force exerted by this field:

$$f_{Lz} = \eta \left[\nabla \times \underbrace{\left(\overbrace{\nabla \times (u \times H)}^{h} \right)}_{J} \right] \times H \qquad (5.108)$$

in which η, ∇, h, and J are the magnetic permeability of beam, gradient operator, small disturbance of applied magnetic field, and current density vector, respectively. Here, the magnetic field can be expressed as follows:

$$H = (H_x, 0, 0) \qquad (5.109)$$

The Lorentz forces exerted per unit volume may be calculated by plugging the results of Eqs. (5.98) and (5.102) into Eq. (5.108):

$$f = \eta H_x^2 \frac{\partial^2 w}{\partial x^2} \qquad (5.110)$$

The answer comes from using Eq. (5.110)'s integral across the beam's cross-section region.

$$f_{Lz} = \eta A H_x^2 \frac{\partial^2 w}{\partial x^2} \qquad (5.111)$$

The external work done by varying forces may now be expressed as follows:

$$\delta V = -\int_0^L \left(N^T \frac{\partial^2 \delta w}{\partial x^2} + \eta A H_x^2 \frac{\partial^2 w}{\partial x^2} \right) dx \qquad (5.112)$$

where N^T represents the force produced by the increase in temperature, which may be written as:

$$N^T = \int_A E_e \alpha_e \Delta T dA \qquad (5.113)$$

The first variation for a change in kinetic energy is:

$$\delta T = \int_V \left(\dot{u}_x \delta \dot{u}_x + \dot{u}_z \delta \dot{u}_z \right) \rho(z) dV$$

$$= \int_0^L \begin{pmatrix} I_0 \left(\dfrac{\partial u}{\partial t} \dfrac{\partial \delta u}{\partial t} + \dfrac{\partial \left(w_b + w_s \right)}{\partial t} \dfrac{\partial \delta \left(w_b + w_s \right)}{\partial t} \right) - I_1 \left(\dfrac{\partial u}{\partial t} \dfrac{\partial^2 \delta w_b}{\partial x \partial t} + \dfrac{\partial^2 w_b}{\partial x \partial t} \dfrac{\partial \delta u}{\partial t} \right) \\[4mm] - J_1 \left(\dfrac{\partial u}{\partial t} \dfrac{\partial^2 \delta w_s}{\partial x \partial t} + \dfrac{\partial^2 w_s}{\partial x \partial t} \dfrac{\partial \delta u}{\partial t} \right) + I_2 \dfrac{\partial^2 w_b}{\partial x \partial t} \dfrac{\partial^2 \delta w_b}{\partial x \partial t} + K_2 \dfrac{\partial^2 w_s}{\partial x \partial t} \dfrac{\partial^2 \delta w_s}{\partial x \partial t} + \\[4mm] J_2 \left(\dfrac{\partial^2 w_b}{\partial x \partial t} \dfrac{\partial^2 \delta w_s}{\partial x \partial t} + \dfrac{\partial^2 w_s}{\partial x \partial t} \dfrac{\partial^2 \delta w_b}{\partial x \partial t} \right) \end{pmatrix} dx$$

$$(5.114)$$

All time differentiation is indicated by the dot-superscript in the preceding equations, and the mass inertias are presented as follows:

$$\left[I_0, I_1, J_1, I_2, J_2, K_2 \right] = \int_A \left[1, z, f(z), z^2, zf(z), f^2(z) \right] \rho_e dA \qquad (5.115)$$

By substituting Eqs. (5.105), (5.112), and (5.114) into equation (5.103) and setting the coefficients of δu, δw_b, and δw_s to zero, the Euler–Lagrange equations of GPLR beam can be written as:

$$\frac{\partial N}{\partial x} = I_0 \frac{\partial^2 u}{\partial t^2} - I_1 \frac{\partial^3 w_b}{\partial x \partial t^2} - J_1 \frac{\partial^3 w_s}{\partial x \partial t^2}, \qquad (5.116)$$

$$\frac{\partial^2 M_b}{\partial x^2} + f_{Lz} - N^T \frac{\partial^2 \left(w_b + w_s \right)}{\partial x^2} = I_0 \frac{\partial^2 \left(w_b + w_s \right)}{\partial t^2} + I_1 \frac{\partial^3 u}{\partial x \partial t^2} - I_2 \frac{\partial^4 w_b}{\partial t^2 \partial x^2} - J_2 \frac{\partial^4 w_s}{\partial t^2 \partial x^2}, \qquad (5.117)$$

$$\frac{\partial^2 M_s}{\partial x^2} + \frac{\partial Q}{\partial x} + f_{Lz} - N^T \frac{\partial^2 \left(w_b + w_s \right)}{\partial x^2} = I_0 \frac{\partial^2 \left(w_b + w_s \right)}{\partial t^2} + J_1 \frac{\partial^3 u}{\partial x \partial t^2}$$
$$- J_2 \frac{\partial^4 w_b}{\partial t^2 \partial x^2} - K_2 \frac{\partial^4 w_s}{\partial t^2 \partial x^2} \qquad (5.118)$$

Now, the thermoelastic stress-strain relations of isotropic materials are reviewed for the purpose of deriving the fundamental elastic equations of solids. Here, following constitutive equations can be expressed once thermal effects are regarded:

$$\sigma_{ij} = C_{ijkl} \left(\varepsilon_{kl} - \alpha_{ij} T \right) \qquad (5.119)$$

in which α_{ij} denotes thermal expansion coefficient and T stands for temperature in terms of Kelvin degrees. Therefore, these relations can be modified as follows for beams:

$$\sigma_{xx} = E_e\left(\varepsilon_{xx} - \alpha_e \Delta T\right), \tag{5.120a}$$

$$\sigma_{xz} = G_e \gamma_{xz} \tag{5.120b}$$

In addition, the following relations may be derived after the integration from Eq. (5.120) has been performed across the cross-section area of the beam:

$$N = A\frac{\partial u}{\partial x} - B\frac{\partial^2 w_b}{\partial x^2} - B_s\frac{\partial^2 w_s}{\partial x^2} - N^T, \tag{5.121}$$

$$M_b = B\frac{\partial u}{\partial x} - D\frac{\partial^2 w_b}{\partial x^2} - D_s\frac{\partial^2 w_s}{\partial x^2} - M_b^T, \tag{5.122}$$

$$M_s = B_s\frac{\partial u}{\partial x} - D_s\frac{\partial^2 w_b}{\partial x^2} - H_s\frac{\partial^2 w_s}{\partial x^2} - M_s^T, \tag{5.123}$$

$$Q = A_s\frac{\partial w_s}{\partial x} \tag{5.124}$$

the expression of cross-sectional rigidities may be done as follows:

$$\left[A, B, D, B_s, D_s, H_s\right] = \int_A \left[1, z, z^2, f(z), zf(z), f^2(z)\right]E_e dA, \tag{5.125}$$

$$A_s = \int_A g^2(z)G_e dA \tag{5.126}$$

and

$$\left[N^T, M_b^T, M_s^T\right] = \int_A \left[1, z, f(z)\right]E_e\alpha_e\Delta T dA \tag{5.127}$$

Now, the governing equations of GPLR beams may be stated in the following manner by putting Eqs. (5.121)–(5.124) in Eqs. (5.116)–(5.118).

$$\left(A\frac{\partial^2 u}{\partial x^2} - B\frac{\partial^3 w_b}{\partial x^3} - B_s\frac{\partial^3 w_s}{\partial x^3}\right) + \left(-I_0\frac{\partial^2 u}{\partial t^2} + I_1\frac{\partial^3 w_b}{\partial x\partial t^2} + J_1\frac{\partial^3 w_s}{\partial x\partial t^2}\right) = 0, \tag{5.128}$$

$$\left(B\frac{\partial^3 u}{\partial x^3} - D\frac{\partial^4 w_b}{\partial x^4} - D_s\frac{\partial^4 w_s}{\partial x^4}\right) + \left(\begin{array}{c} -I_0\dfrac{\partial^2\left(w_b + w_s\right)}{\partial t^2} - I_1\dfrac{\partial^3 u}{\partial x\partial t^2} + I_2\dfrac{\partial^4 w_b}{\partial t^2\partial x^2} + \\[4mm] J_2\dfrac{\partial^4 w_s}{\partial t^2\partial x^2} + \left(\eta AH_x^2 - N^T\right)\dfrac{\partial^2\left(w_b + w_s\right)}{\partial x^2} \end{array}\right) = 0,$$

$$(5.129)$$

$$\left(B_s\frac{\partial^3 u}{\partial x^3} - D_s\frac{\partial^4 w_b}{\partial x^4} - H_s\frac{\partial^4 w_s}{\partial x^4} + A_s\frac{\partial^2 w_s}{\partial x^2}\right)$$
$$+ \left(\begin{array}{c} -I_0\dfrac{\partial^2\left(w_b + w_s\right)}{\partial t^2} - J_1\dfrac{\partial^3 u}{\partial x\partial t^2} + J_2\dfrac{\partial^4 w_b}{\partial t^2\partial x^2} + \\[4mm] K_2\dfrac{\partial^4 w_s}{\partial t^2\partial x^2} + \left(\eta AH_x^2 - N^T\right)\dfrac{\partial^2\left(w_b + w_s\right)}{\partial x^2} \end{array}\right) = 0$$

$$(5.130)$$

An analytical solution of the nonlocal governing equations that were generated in the stage before this one is the last step that has to be taken in this process. It is assumed that the displacement fields are exponential, and the definition of these fields may be seen below [52]:

$$\begin{bmatrix} u(x,t) \\ w_b(x,t) \\ w_s(x,t) \end{bmatrix} = \begin{bmatrix} U\exp\left[i\left(\beta x - \omega t\right)\right] \\ W_b\exp\left[i\left(\beta x - \omega t\right)\right] \\ W_s\exp\left[i\left(\beta x - \omega t\right)\right] \end{bmatrix}$$

$$(5.131)$$

where U, W_b, and W_s are the unknown coefficients and β is the wave numbers of wave propagation in x direction, and finally ω is wave's angular frequency. Now, substituting Eq. (5.130) to Eqs. (5.128)–(5.129) reveals:

$$\left(\left[K\right]_{3\times3} - \omega^2\left[M\right]_{3\times3}\right)\left[\Delta\right] = \left[0\right]$$

$$(5.132)$$

The following is a list of the unknown parameters of Eq. (5.131), which may be noted as follows:

$$\left[\Delta\right] = \left[U, W_b, W_s\right]^T$$

$$(5.133)$$

In order to obtain wave's angular frequency, the determinant of the left-hand side of Eq. (5.131) must be set to zero:

$$\left|\left[K\right]_{3\times3} - \omega^2\left[M\right]_{3\times3}\right| = 0$$

$$(5.134)$$

By working through the derived equation to find, it is possible to determine the angular frequency of the wave in the GPLR beam. After dividing these angular frequencies by the total number of waves, the phase velocity may be calculated as follows:

$$c_p = \frac{\omega}{\beta} \tag{5.135}$$

In this part, a number of images are shown in order to provide clarity on the impact that a variety of factors have on the wave speed of GPLR beams. Epoxy and GPL fibers are expected to be used in the construction of the beam, which is the basic structure of the beam. It is possible to describe the following material qualities of reinforcing GPLs and epoxy [52]:

$$l_{GPL} = 2.5\,\mu\text{m}, w_{GPL} = 1.5\,\mu\text{m},$$
$$h_{GPL} = 1.5\,\text{nm}, \rho_{GPL} = 1062.5\,\text{kg} / \text{m}^3,$$
$$v_{GPL} = 0.186, \alpha_{GPL} = 60 \times 10^{-6}\,1/\text{K},$$
$$\text{and } E_{GPL} = 1.01\,\text{TPa}. E_M = 3\,\text{GPa},$$
$$\rho_M = 1200\,\text{kg} / \text{m}^3, v_M = 0.34, \text{and}$$
$$\alpha_M = 5 \times 10^{-6}\,1/\text{K}.$$

On the basis of this information, Figure 5.58 was created in order to illustrate how the phase velocity of propagating waves is affected by the weight percentage of GPLs as well as the effects of various distributions of GPLs. When the distribution of GPLs is changed, it is simple to see that this leads to a wide variety of possible replies. In point of fact, when a GPLR-X beam is used, the wave speed may be seen to be at its absolute maximum, but when a GPLR-O beam is used, the response can be reduced to its bare minimum. The use of a GPLR-U beam makes it abundantly clear that a phase velocity in the middle range may be acquired. It is important to note that high wave numbers are associated with a more pronounced manifestation of this phenomenon. In addition to this, another influence may be detected as a result of the fluctuation in the weight percentage of GPLs. To put it another way, the weight fraction of GPLs has the potential to alter the speed of the wave in an increasing manner. If I'm being really straightforward with you, an increase in the weight fraction of GPLs may lead to a reciprocal increase in the magnitude of phase velocity.

In Figure 5.59, the primary focus is on bringing attention to the influence that the strength of the magnetic field has on the phase velocity of the GPLR beams. In light of the image, it is possible to grasp the concept that phase speed may be enhanced even in tiny wave numbers, hence leading to an increase in the magnetic field strength. However, this impact cannot be reported in high wave number regions because, in such an area, all of the curves converge to a limited value and

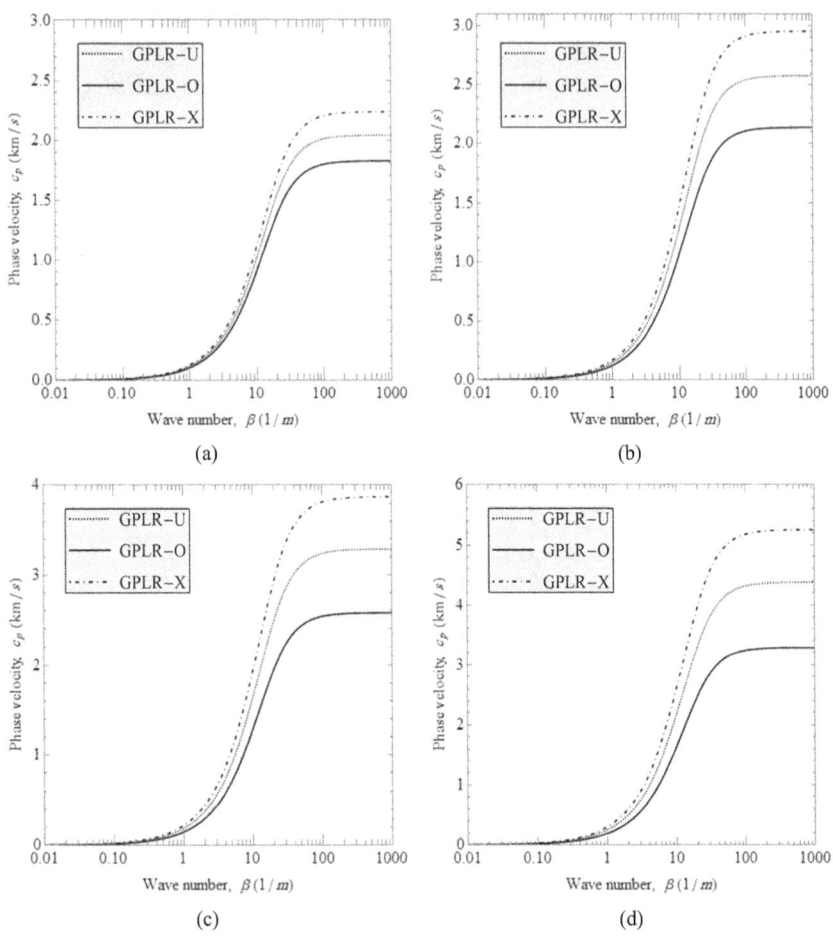

FIGURE 5.58 Variation of phase velocity versus wave number for various distributions and weight fractions of GPLs. (a) $g_{GPL} = 0.2\%$, (b) $g_{GPL} = 0.5\%$, (c) $g_{GPL} = 1\%$, and (d) $g_{GPL} = 2\%$.

magneto-elastic interactions are too weak to create a change in the value of wave dispersion replies. As a result, high wave number regions are not subject to this influence. If the wave number is more than $\beta = 30$, then it is unlikely that a significant difference would be seen regardless of how the magnetic field strength changes. Additionally, the effects of the different methods of distribution may be seen once again. In point of fact, one has to pay attention to the fact that the following assortment is really important for the wave velocity in the designs: GPLR-X>GPLR-U>GPLR-O [52].

In addition, the purpose of Figure 5.60 is to investigate the thermo-magnetic effects on the phase velocity of GPLR beams, taking into account the wave number and the distribution type. It should come as no surprise that the phase velocity

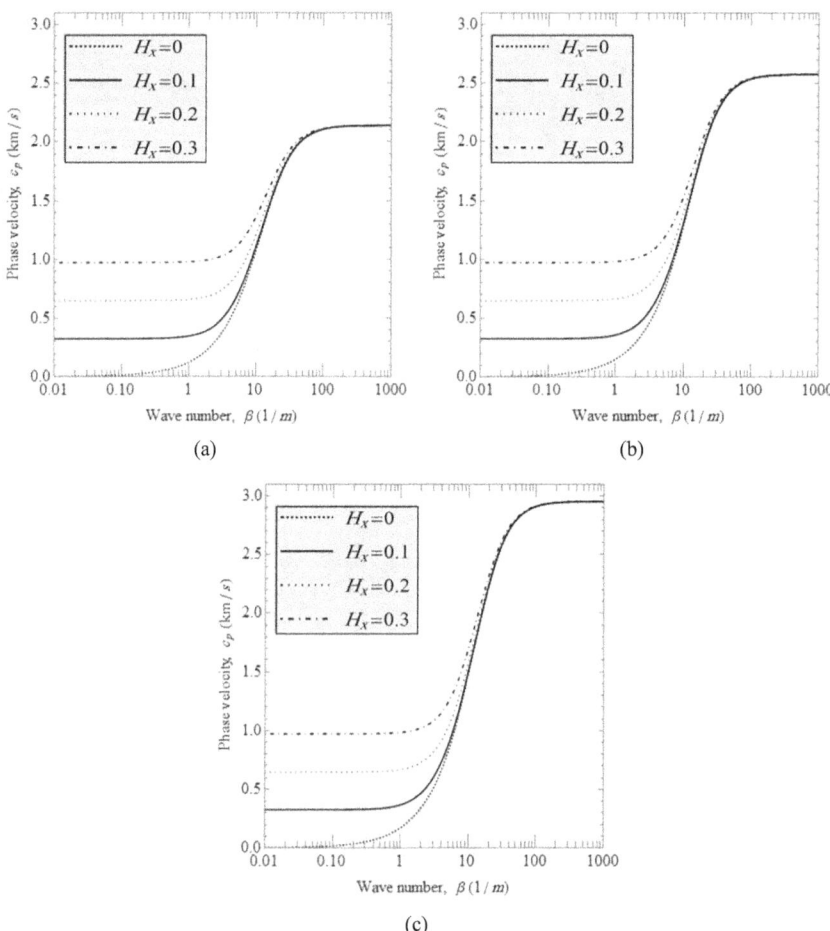

FIGURE 5.59 Coupled effects of magnetic field intensity and GPLs' distribution on the phase velocity of GPLR beams. (a) GPLR-O, (b) GPLR-U, and (c) GPLR-X.

vanishes when the temperature increase is severe enough. It is possible, but not guaranteed, that this decreasing tendency will culminate in a full dampening of the wave response; nonetheless, this is the typical outcome. In point of fact, there are several situations in which an increase in the temperature rise might cause a marginal slowing down of the wave speed. The occurrence of this phenomenon is more easily noticed in waves with bigger numbers. In point of fact, once a damping index has been created for the efficacy of temperature increase, which is responsible for measuring the speed difference percentage, it is possible to report that this index is larger in circumstances in which tiny wave numbers are employed as opposed to cases in which high wave numbers are utilized. This is because small wave numbers are responsible for measuring the speed difference percentage. Once the quantity of wave speed is measured, the results show that GPLR-X

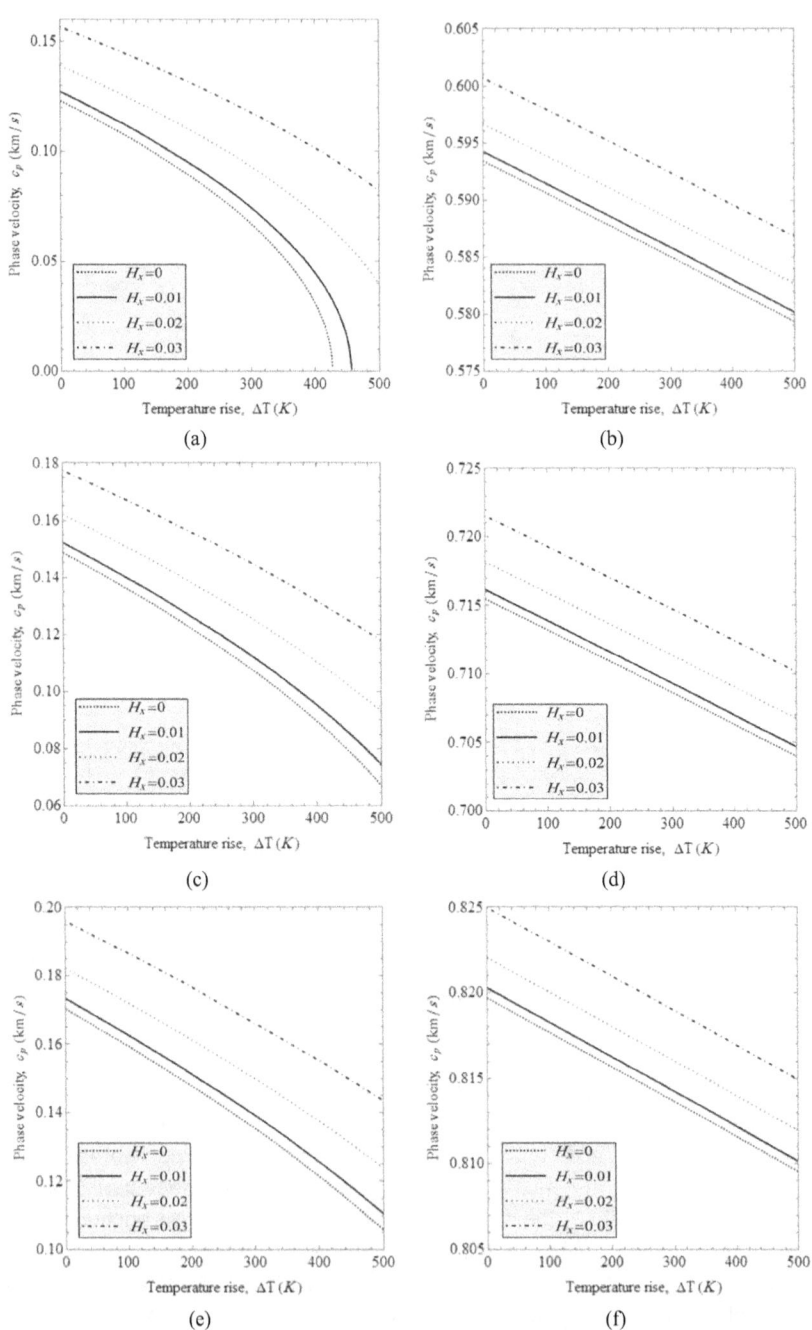

FIGURE 5.60　Thermo-magnetical effects on phase velocity of GPLR beams with respect to wave number and various distributions of GPLs. (a) GPLR-O, $\beta = 1$, (b) GPLR-O, $\beta = 5$, (c) GPLR-U, $\beta = 1$, (d) GPLR-U, $\beta = 5$, (e) GPLR-X, $\beta = 1$, and (f) GPLR-X, $\beta = 5$.

has the highest response, followed by GPLR-U and GPLR-O, which correlate with the following spots. This was previously anticipated. One other thing that has been shown to be true is that an increase in the magnetic field's influence leads to a rise in the wave dispersion responses at any temperature rise value that is required.

At the very end, the impact of the length-to-width ratio of GPLs is discussed in Figure 5.61. This figure plots the changes in phase velocity versus temperature rise for various distributions of GPLs. It is plain to observe that the phase velocity follows a path that is inexorably descending, while the temperature rise is steadily climbing. The highest value of phase velocity in each temperature increase

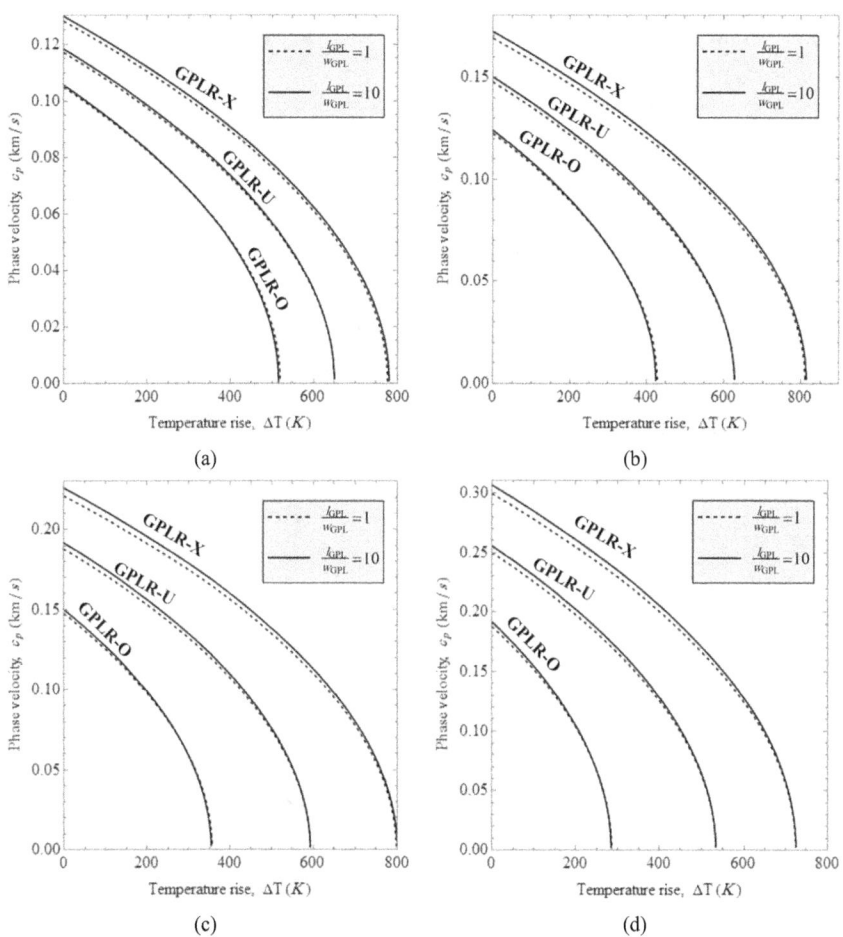

FIGURE 5.61 Effect of GPLs' weight fraction and length-to-width ratio on variation of phase velocity in GPLR beams considering different types of GPLs' distribution. (a) g_{GPL} = 0.2%, (b) g_{GPL} = 0.5%, (c) g_{GPL} = 1%, and (d) g_{GPL} = 2%.

emerges after the GPLR-X beam has been selected, just as it did in the earlier figures. The most important conclusion that can be drawn from this image is that the length-to-width ratio of GPLs has an increasing impact on the wave velocity. To put it another way, if you choose a reinforcing element that is thinner, you will be able to achieve a larger phase velocity at temperatures that have only slightly increased. It is important to point out that this worsening influence disappears when the temperature becomes really high. In point of fact, when the temperature rises to high levels, a change in the length-to-breadth ratio will have no effect at all on the phase velocity quantities. Additionally, it is observable that greater responses are attainable by using a larger weight fraction for the reinforcing GPLs, since this can be shown to be the case.

5.4.3 Graphene Platelet–reinforced Plates: Wave Dispersion Characteristics

As was said in the section before this one [52], the effective Young's modulus of the Graphene platelet-reinforced (GPLR) plates will be described using a model established by Halpin and Tsai, which will be presented in this section. For the sake of this study, the GPLs are envisioned as being in the shape of rectangular solid fillers that have been dispersed throughout the polymer matrix in an even fashion.

On the basis of the Voigt–Reuss model, which was described in Section 5.4.2, the effective modulus of elasticity may be derived as follows: As a result, in accordance with the Kirchhoff plate theory, the plate's displacement field may be expressed as follows [52]:

$$u_x\left(x,y,z,t\right) = u\left(x,y,t\right) - z\frac{\partial w\left(x,y,t\right)}{\partial x} \qquad (5.136)$$

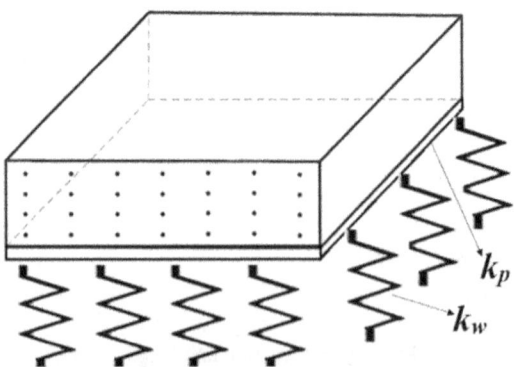

FIGURE 5.62   Schematic of a GPL-reinforced plate.

$$u_y(x,y,z,t) = v(x,y,t) - z\frac{\partial w(x,y,t)}{\partial y} \tag{5.137}$$

$$u_z(x,y,z,t) = w(x,y,t) \tag{5.138}$$

in Eqs. (5.13)–(5.138), u, v, and w are mid-plane displacement components of the plate. Now, the nonzero strains of the plate can be expressed by the following equations:

$$\begin{Bmatrix} \varepsilon_{xx} \\ \varepsilon_{yy} \\ \varepsilon_{xy} \end{Bmatrix} = \begin{Bmatrix} \varepsilon_{xx}^0 \\ \varepsilon_{yy}^0 \\ \varepsilon_{xy}^0 \end{Bmatrix} + z \begin{Bmatrix} \kappa_{xx} \\ \kappa_{yy} \\ \kappa_{xy} \end{Bmatrix} \tag{5.139}$$

where

$$\begin{Bmatrix} \varepsilon_{xx}^0 \\ \varepsilon_{yy}^0 \\ \varepsilon_{xy}^0 \end{Bmatrix} = \begin{Bmatrix} \dfrac{\partial u}{\partial x} \\[2mm] \dfrac{\partial v}{\partial y} \\[2mm] \dfrac{\partial u}{\partial y} + \dfrac{\partial v}{\partial x} \end{Bmatrix}, \quad \begin{Bmatrix} \kappa_{xx} \\ \kappa_{yy} \\ \kappa_{xy} \end{Bmatrix} = \begin{Bmatrix} -\dfrac{\partial^2 w}{\partial x^2} \\[2mm] -\dfrac{\partial^2 w}{\partial y^2} \\[2mm] -2\dfrac{\partial^2 w}{\partial x \partial y} \end{Bmatrix} \tag{5.140}$$

Now, in order to explain Hamilton's principle, the following form may be used:

$$\int_0^t \delta(U - T + V)\,dt = 0 \tag{5.141}$$

where U, T, and V account for strain energy, kinetic energy, and work done by external forces, respectively. Now, the variation of strain energy for core and face sheets can be formulated as:

$$\delta U = \int_V \left(\sigma_{xx}\delta\varepsilon_{xx} + \sigma_{yy}\delta\varepsilon_{yy} + \sigma_{xy}\delta\varepsilon_{xy} \right) dV \tag{5.142}$$

The following is what we get when we plug Eq. (5.139) into Eq. (5.142):

$$\delta U = \int_0^L \left(N_{xx}\frac{\partial \delta u}{\partial x} - M_{xx}\frac{\partial^2 w}{\partial x^2} + N_{yy}\frac{\partial \delta u}{\partial y} - M_{yy}\frac{\partial^2 w}{\partial y^2} \right.$$
$$\left. + N_{xy}\left(\frac{\partial u}{\partial y} + \frac{\partial v}{\partial x} \right) - 2M_{xy}\frac{\partial^2 w}{\partial x \partial y} \right) dx \tag{5.143}$$

in which the stress resultants N and M are expressed as:

$$[N_i, M_i] = \int_A [1, z] \sigma_i dA, \quad i = (xx, yy, xy) \tag{5.144}$$

The first variation in the amount of work done by forces applied from the outside may be described using the following form:

$$\delta V = \int_0^L \int_A \left(N_x^0 \frac{\partial w}{\partial x} \frac{\partial \delta w}{\partial x} + N_y^0 \frac{\partial w}{\partial y} \frac{\partial \delta w}{\partial y} + 2N_{xy}^0 \frac{\partial w}{\partial x} \frac{\partial w}{\partial y} \right) dA dx \tag{5.145}$$

where N_x^0, N_y^0, and N_{xy}^0 are in-plane loads applied to the plate. The first variation of kinetic energy can be expressed as:

$$\delta T = \int_0^L \left(\begin{array}{l} I_0 \left(\dfrac{\partial u}{\partial t} \dfrac{\partial \delta u}{\partial t} + \dfrac{\partial v}{\partial t} \dfrac{\partial \delta v}{\partial t} + \dfrac{\partial w}{\partial t} \dfrac{\partial \delta w}{\partial t} \right) \\[2ex] -I_1 \left(\dfrac{\partial u}{\partial t} \dfrac{\partial^2 \delta w}{\partial x \partial t} + \dfrac{\partial^2 w}{\partial x \partial t} \dfrac{\partial \delta u}{\partial t} + \dfrac{\partial v}{\partial t} \dfrac{\partial^2 \delta w}{\partial y \partial t} + \dfrac{\partial^2 w}{\partial y \partial t} \dfrac{\partial \delta v}{\partial t} \right) \\[2ex] +I_2 \left(\dfrac{\partial^2 w}{\partial x \partial t} \dfrac{\partial^2 \delta w}{\partial x \partial t} + \dfrac{\partial^2 w}{\partial y \partial t} \dfrac{\partial^2 \delta w}{\partial y \partial t} \right) \end{array} \right) dx \tag{5.146}$$

The differentiation with respect to time is denoted by the dot-superscript in all of the equations, and the mass inertias that are employed in the equations that are presented earlier are provided in the following form:

$$[I_0, I_1, I_2] = \int_A [1, z, z^2] \rho_e dA \tag{5.147}$$

By substituting Eqs. (5.143), (5.144), and (5.146) into Eq. (5.141) and setting the coefficients of δu, δv, and δw to zero, the Euler–Lagrange equations of graphene platelet-reinforced plates (GPLRP) can be written as:

$$\frac{\partial N_{xx}}{\partial x} + \frac{\partial N_{xy}}{\partial y} = I_0 \frac{\partial^2 u}{\partial t^2} - I_1 \frac{\partial^3 w}{\partial x \partial t^2} \tag{5.148}$$

$$\frac{\partial N_{xy}}{\partial x} + \frac{\partial N_{yy}}{\partial y} = I_0 \frac{\partial^2 v}{\partial t^2} - I_1 \frac{\partial^3 w}{\partial y \partial t^2} \tag{5.149}$$

$$\frac{\partial^2 M_{xx}}{\partial x^2} + 2\frac{\partial^2 M_{xy}}{\partial x \partial y} + \frac{\partial^2 M_{yy}}{\partial y^2} - k_w w + k_p \nabla^2 w$$

$$= I_0 \frac{\partial^2 w}{\partial t^2} + I_1 \left(\frac{\partial^3 u}{\partial x \partial t^2} + \frac{\partial^3 v}{\partial y \partial t^2} \right) - I_2 \nabla^2 \frac{\partial^2 w}{\partial t^2}$$

(5.150)

In the NE, the stress field is said to take into account the impacts of the nonlocal elastic stress field. On the basis of this theory, the stress condition at a particular position inside a solid body may be expressed as [52]:

$$\sigma_{ij} = \int \alpha \left(\left| x - x' \right|, \tau \right) C_{ijkl} \varepsilon_{kl} \left(x' \right) dV \left(x' \right)$$

(5.151)

where σ_{ij}, ε_{kl}, and C_{ijkl} are components of stress, strain, and elastic stiffness, respectively. On the other hand, $\alpha(|x - x'|, \tau)$ is a nonlocal kernel that is responsible for the effect of strain at point x' on the stress at point x. Also, $(|x - x'|)$ is Euclidian distance between x and x'. Likewise, τ is a dimensionless constant value, which can be calculated as follows:

$$\tau = \frac{e_0 a}{l}$$

(5.152)

where e_0 has a particular value for each material; also, a and l are internal and external characteristic lengths, respectively. Once the nonlocal kernel function $\alpha(|x - x'|, \tau)$ satisfies the developed conditions, the constitutive relation of this theory can be expressed as below:

$$\left(1 - \left(e_0 a \right)^2 \nabla^2 \right) \sigma_{ij} = C_{ijkl} \varepsilon_{kl}$$

(5.153)

in which ∇^2 denotes the Laplacian operator. Finally, the simplified constitutive relation can be written as follows:

$$\left(1 - \mu^2 \nabla^2 \right) \begin{bmatrix} \sigma_{xx} \\ \sigma_{yy} \\ \sigma_{xy} \end{bmatrix} = \begin{bmatrix} C_{11} & C_{12} & 0 \\ C_{21} & C_{22} & 0 \\ 0 & 0 & C_{66} \end{bmatrix} \begin{bmatrix} \varepsilon_{xx} \\ \varepsilon_{yy} \\ \varepsilon_{xy} \end{bmatrix}$$

(5.154)

in above equation

$$C_{11} = C_{22} = \frac{E_e}{1 - v_e^2}, C_{12} = v_e C_{11}, C_{66} = \frac{E_e}{2(1 + v_e)}$$

(5.155)

where $\mu = e_0 a$. Now, inserting Eq. (5.144) in Eq. (5.154) reveals:

$$
\left(1 - \mu^2 \nabla^2\right)\begin{bmatrix} N_{xx} \\ N_{yy} \\ N_{xy} \end{bmatrix} = \left(\begin{bmatrix} A_{11} & A_{12} & 0 \\ A_{21} & A_{22} & 0 \\ 0 & 0 & A_{66} \end{bmatrix} \begin{bmatrix} \dfrac{\partial u}{\partial x} \\ \dfrac{\partial v}{\partial y} \\ \dfrac{\partial u}{\partial y} + \dfrac{\partial v}{\partial x} \end{bmatrix} \right.
$$

$$
\left. + \begin{bmatrix} B_{11} & B_{12} & 0 \\ B_{21} & B_{22} & 0 \\ 0 & 0 & B_{66} \end{bmatrix} \begin{bmatrix} -\dfrac{\partial^2 w}{\partial x^2} \\ -\dfrac{\partial^2 w}{\partial y^2} \\ -2\dfrac{\partial^2 w}{\partial x \partial y} \end{bmatrix} \right),
$$

(5.156)

$$
\left(1 - \mu^2 \nabla^2\right)\begin{bmatrix} M_{xx} \\ M_{yy} \\ M_{xy} \end{bmatrix} = \left(\begin{bmatrix} B_{11} & B_{12} & 0 \\ B_{21} & B_{22} & 0 \\ 0 & 0 & B_{66} \end{bmatrix} \begin{bmatrix} \dfrac{\partial u}{\partial x} \\ \dfrac{\partial v}{\partial y} \\ \dfrac{\partial u}{\partial y} + \dfrac{\partial v}{\partial x} \end{bmatrix} \right.
$$

$$
\left. + \begin{bmatrix} D_{11} & D_{12} & 0 \\ D_{21} & D_{22} & 0 \\ 0 & 0 & D_{66} \end{bmatrix} \begin{bmatrix} -\dfrac{\partial^2 w}{\partial x^2} \\ -\dfrac{\partial^2 w}{\partial y^2} \\ -2\dfrac{\partial^2 w}{\partial x \partial y} \end{bmatrix} \right)
$$

(5.157)

The cross-sectional rigidities may be stated as follows using Eqs. (5.156) and (5.157), which are as follows:

$$
\begin{bmatrix} A_{11} & B_{11} & D_{11} \\ A_{12} & B_{12} & D_{12} \\ A_{66} & B_{66} & D_{66} \end{bmatrix} = \int_A \begin{bmatrix} 1 & z & z^2 \end{bmatrix} \begin{bmatrix} C_{11} \\ C_{12} \\ C_{66} \end{bmatrix} dA
$$

(5.158)

It is possible to directly construct, in terms of displacements, the nonlocal governing equation of GPLR-MPs as follows. This may be done by entering Eqs. (5.156) and (5.157) into Eqs. (5.148)–(5.150), respectively:

$$\left(A_{11}\frac{\partial^2 u}{\partial x^2} + \left(A_{12} + A_{66} \right)\frac{\partial^2 v}{\partial x \partial y} + A_{66}\frac{\partial^2 u}{\partial y^2} - B_{11}\frac{\partial^3 w}{\partial x^3} - \left(B_{12} + 2B_{66} \right)\frac{\partial^3 w}{\partial x \partial y^2} \right)$$
$$+ \left(1 - \mu^2 \nabla^2 \right)\left(-I_0 \frac{\partial^2 u}{\partial t^2} + I_1 \frac{\partial^3 w}{\partial x \partial t^2} \right) = 0 \tag{5.159}$$

$$\left(A_{22}\frac{\partial^2 v}{\partial y^2} + \left(A_{12} + A_{66} \right)\frac{\partial^2 u}{\partial x \partial y} + A_{66}\frac{\partial^2 v}{\partial x^2} - B_{22}\frac{\partial^3 w}{\partial y^3} - \left(B_{12} + 2B_{66} \right)\frac{\partial^3 w}{\partial x^2 \partial y} \right)$$
$$+ \left(1 - \mu^2 \nabla^2 \right)\left(-I_0 \frac{\partial^2 v}{\partial t^2} + I_1 \frac{\partial^3 w}{\partial y \partial t^2} \right) = 0 \tag{5.160}$$

$$\left(\begin{array}{c} B_{11}\dfrac{\partial^3 u}{\partial x^3} + \left(B_{12} + 2B_{66} \right)\dfrac{\partial^3 u}{\partial x \partial y^2} + B_{22}\dfrac{\partial^3 v}{\partial y^3} + \left(B_{12} + 2B_{66} \right)\dfrac{\partial^3 v}{\partial x^2 \partial y} \\[2mm] -D_{11}\dfrac{\partial^4 w}{\partial x^4} - 2\left(D_{12} + 2D_{66} \right)\dfrac{\partial^4 w}{\partial x^2 \partial y^2} - D_{22}\dfrac{\partial^4 w}{\partial y^4} \end{array} \right)$$
$$+ \left(1 - \mu^2 \nabla^2 \right)\left(\begin{array}{c} -I_0 \dfrac{\partial^2 w}{\partial t^2} - I_1 \left(\dfrac{\partial^3 u}{\partial x \partial t^2} + \dfrac{\partial^3 v}{\partial y \partial t^2} \right) + I_2 \nabla^2 \dfrac{\partial^2 w}{\partial t^2} \\[2mm] -k_w w + k_p \nabla^2 w \end{array} \right) = 0 \tag{5.161}$$

Once again, the governing equations that were generated in the previous part will have an analytical solution found for them. It is going to be assumed that the displacement fields are exponential, and they will be defined as follows:

$$\begin{bmatrix} u(x,y,t) \\ v(x,y,t) \\ w(x,y,t) \end{bmatrix} = \begin{bmatrix} U\exp\left[i\left(\beta_1 x + \beta_2 y - \omega t \right) \right] \\ V\exp\left[i\left(\beta_1 x + \beta_2 y - \omega t \right) \right] \\ W\exp\left[i\left(\beta_1 x + \beta_2 y - \omega t \right) \right] \end{bmatrix} \tag{5.162}$$

where U, V, and W are the unknown coefficients, β_1 and β_2 are the wave numbers of wave propagation along x and y directions, respectively; finally, ω is wave's angular frequency. Now, substituting Eq. (5.162) in Eqs. (5.159)–(5.61) reveals:

$$\left([K]_{3\times 3} - \omega^2 [M]_{3\times 3} \right)\{\Delta\} = \{0\} \tag{5.163}$$

The following is a list of the unknown parameters of Eq. (5.163), which may be found:

$$\{\Delta\} = \{U, V, W\}^T \tag{5.164}$$

where the corresponding k_{ij} and m_{ij} are as written in Appendix 5E. In order to attain wave's angular frequency, the determinant of the left-hand side of Eq. (5.163) must be set to zero:

$$\left| \left[K \right]_{3\times3} - \omega^2 \left[M \right]_{3\times3} \right| = 0 \qquad (5.165)$$

After solving the obtained equation, the angular frequency of the wave that is produced by GPLR-MPs may be stated as follows:

$$\omega = M\left(\beta \right) \qquad (5.166)$$

By calculating Eq. (5.166), one may determine the frequency of the angular waves produced by dispersed waves. In this section, we will provide a few numerical examples with the objective of gaining an understanding of the effects that all of the included parameters have on the wave propagation answers of the GPL-reinforced microplate. The graphene platelet (GPL) includes the following important features: $l_{GPL} = 2.5$ μm, $w_{GPL} = 1.5$ μm, $h_{GPL} = 1.5$ nm, $\rho_{GPL} = 1062.5$ kg/m³, $\nu_{GPL} = 0.186$, and $E_{GPL} = 1.01$ TPa. Besides, the matrix's properties are: $E_M = 3$ GPa, $\rho_M = 1200$ kg/m³, $\nu_M = 0.34$. For the sake of keeping things as simple as possible, the dimensionless version of the foundation parameters may be written as follows [52].

$$K_w = \frac{k_w L^4}{D_c}, K_p = \frac{k_p L^2}{D_c} \qquad (5.167)$$

in which

$$D_c = \frac{E_c h^3}{12\left(1 - v_c^2 \right)} \qquad (5.178)$$

On display in Figure 5.63 is a representation meant to demonstrate the mixed influence that nonlocal parameter and weight fraction of GPLs have on the wave frequency of waves that have propagated. When the wave number is considered to be more than $\beta = 1 \times 10^9$, it is self-evident that the greater the nonlocality, the lower the wave frequency; this indicates that any increase in the value of the nonlocal parameter exposes a reduction in the size of the wave replies. On the other hand, it is not hard to establish that greater wave frequencies are achieved anytime larger GPLs' weight fractions are applied. This is a well-known fact. It is possible that it is noteworthy that this phenomenon is more noticeable in the wave numbers that are more than $\beta = 5 \times 10^9$. As a result, the wave frequency may be lowered by either including a nonlocal parameter or reducing the weight fraction of the GPLs [52]. This brings us to a quick conclusion.

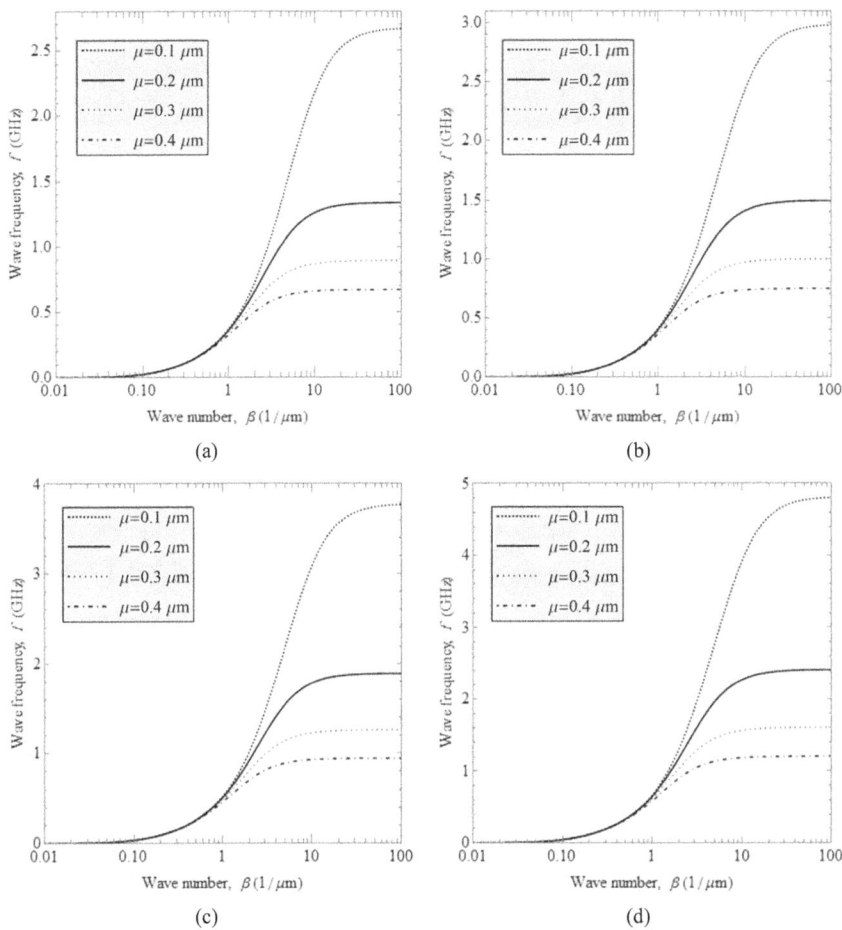

FIGURE 5.63 Effect of weight fraction of GPL on wave frequency of GPLR-MPs for various amounts of nonlocal parameter (l_{GPL} = 2.5 μm, w_{GPL} = 1.5 μm, h_{GPL} = 1.5 nm, K_w = K_p = 0). (a) g_{GPL} = 0.1%, (b) g_{GPL} = 0.2%, (c) g_{GPL} = 0.5%, and (d) g_{GPL} = 1%.

Figure 5.64 illustrates the influence that foundation settings have on the wave dispersion characteristics of GPL-reinforced microplates, constructing a graph that shows the relationship between phase velocity and wave number. Greater nonlocal parameters may be used in this figure, just as in the one that came before it, in order to accomplish the aim of achieving lower wave speeds. Altering the values of the spring constants in the foundation may obviously have an effect on the wave responses. Despite this, the usefulness of this strategy is restricted to a narrow range of tiny wave numbers. To put it another way, the phase velocity of the element may be enhanced anytime either linear or nonlinear coefficients of the elastic media are applied [52].

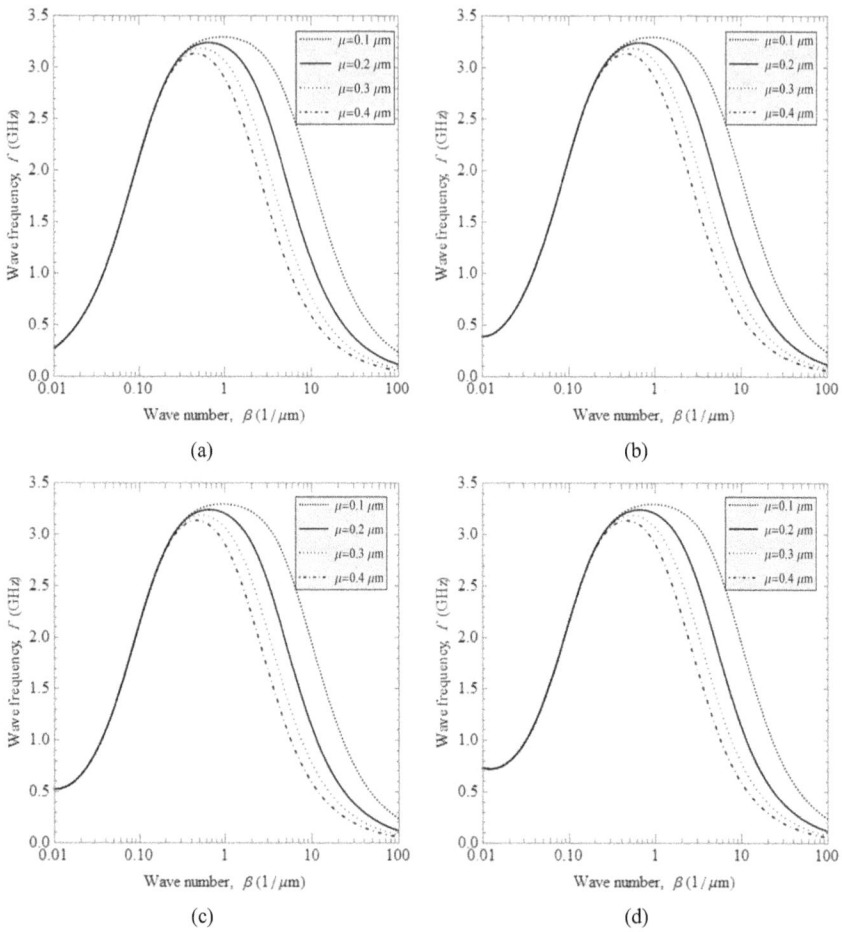

FIGURE 5.64   Effect of foundation parameters on phase velocity of GPLR-MPs for various amounts of nonlocal parameter (l_{GPL} = 2.5 μm, w_{GPL} = 1.5 μm, h_{GPL} = 1.5 nm, L/h = 10). (a) K_w = 0, K_p = 0, (b) K_w = 50, K_p = 0, (c) K_w = 50, K_p = 10, and (d) K_w = 100, K_p = 25.

Figure 5.65 illustrates the effect that a number of different elements have on the phase velocity of GPL-reinforced microplates. These factors include wave number, slenderness ratio, and foundation parameters, among others. Once again, the figure makes it clear that an increase in the total amount of either the Winkler or the Pasternak coefficients may result in an increase in the speed of waves that are being propagated. This is the case regardless of whatever coefficient is increased, in addition, at a certain amount of wave number, it is possible to lower the phase velocity by picking plates that are narrower and have bigger slenderness ratios. This is because narrower plates have a higher slenderness ratio. When picking plates with a certain value of wave number, it is feasible to do this. When compared to the situation in which the wave number is believed to be $\beta = 0.02 \times 10^9$, it is crucial to note that once the wave number is supposed to be $\beta = 0.01 \times 10^9$,

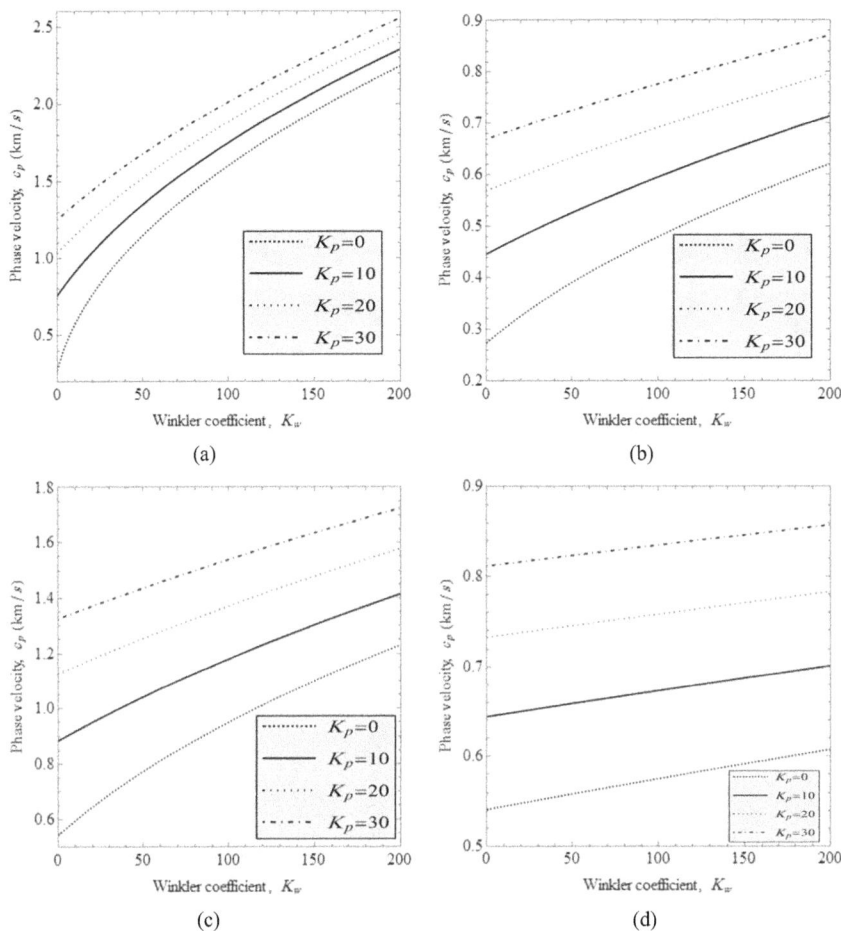

FIGURE 5.65 Effect of foundation parameters on phase velocity of GPLR-MPs for various wave numbers and slenderness ratios ($l_{GPL} = 2.5$ μm, $w_{GPL} = 1.5$ μm, $h_{GPL} = 1.5$ nm, $\mu = 0.2$ μm). (a) $\beta = 0.01 \times 10^9$, $L/h = 5$, (b) $\beta = 0.01 \times 10^9$, $L/h = 10$, (c) $\beta = 0.02 \times 10^9$, $L/h = 5$, and (d) $\beta = 0.02 \times 10^9$, $L/h = 10$.

more perceptible fluctuations in the phase velocity are able to be detected when the Winkler coefficient rises. This is important to keep in mind since the scenario in which the wave number is thought to be $\beta = 0.02 \times 10^9$ is described in the following way [52].

In addition, a different figure (Figure 5.66) is dedicated to analyzing the influence of the length-to-width ratio of the reinforcing element while charting the change in phase velocity versus a nonlocal parameter. In point of fact, the influence of nonlocality on the stiffness and softness of a system may be plainly seen; to put it another way, once nonlocality is increased, phase velocity travels along a route that steadily decreases. On the other hand, one may draw the conclusion that with any desired quantity of nonlocal parameter, it is possible to attain larger

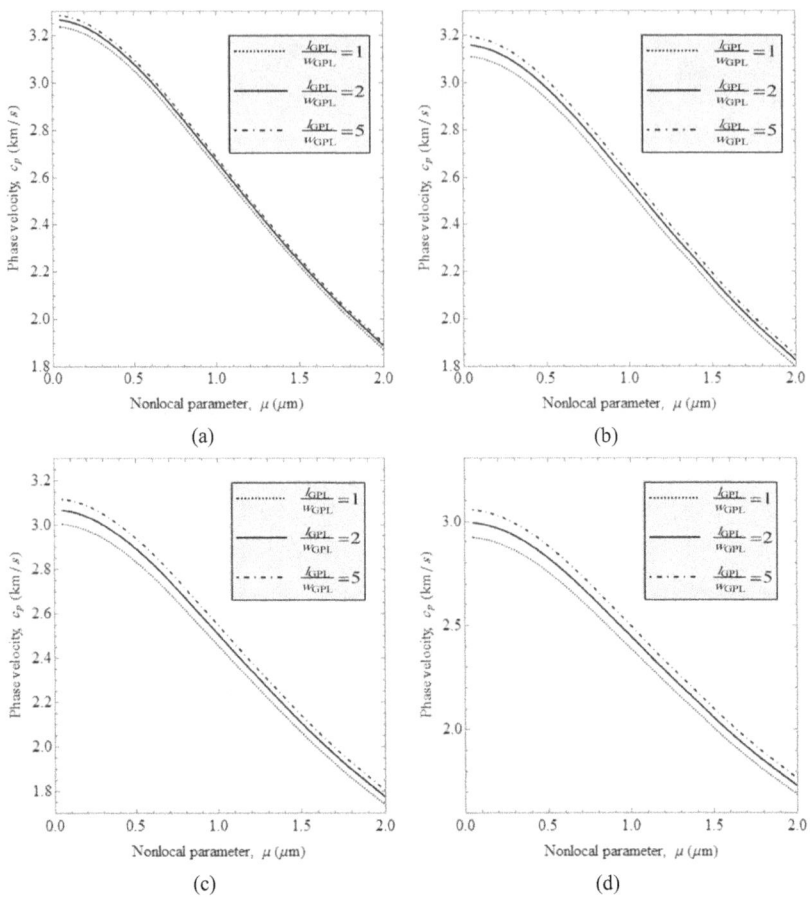

FIGURE 5.66 Variation of phase velocity versus nonlocal parameter for various GPLs' length-to-width ratios and GPLs' thickness ($\beta = 0.5 \times 10^9$, $w_{GPL} = 1.5$ μm, $g_{GPL} = 0.5\%$). (a) $h_{GPL} = 1.5$ nm, (b) $h_{GPL} = 3$ nm, (c) $h_{GPL} = 4.5$ nm, and (d) $h_{GPL} = 6$ nm.

phase velocities if the length-to-width ratio of GPLs is amplified. This is the case regardless of the amount of nonlocal parameter. On the other hand, once greater thicknesses of GPLs are applied, the pace of the waves may be slowed down. As a result, taking into account the information shown in Figure 5.66, it is possible to improve phase velocity by using reinforcing components that have better l/w ratios but thinner thicknesses [52].

In conclusion, Figure 5.67 illustrates the variation of cut-off frequency versus Winkler coefficient for different values of the slenderness ratio and weight percentage of GPLs. There is no way to get around the fact that the cut-off frequency rises along with an increase in the Winkler coefficient. Additionally, it is possible to calculate that an increase in the amount of GPLs' weight fraction may directly lead to a rise in the magnitude of cut-off frequency; yet, a corresponding change in the value of slenderness ratio affects the wave dispersion response in a

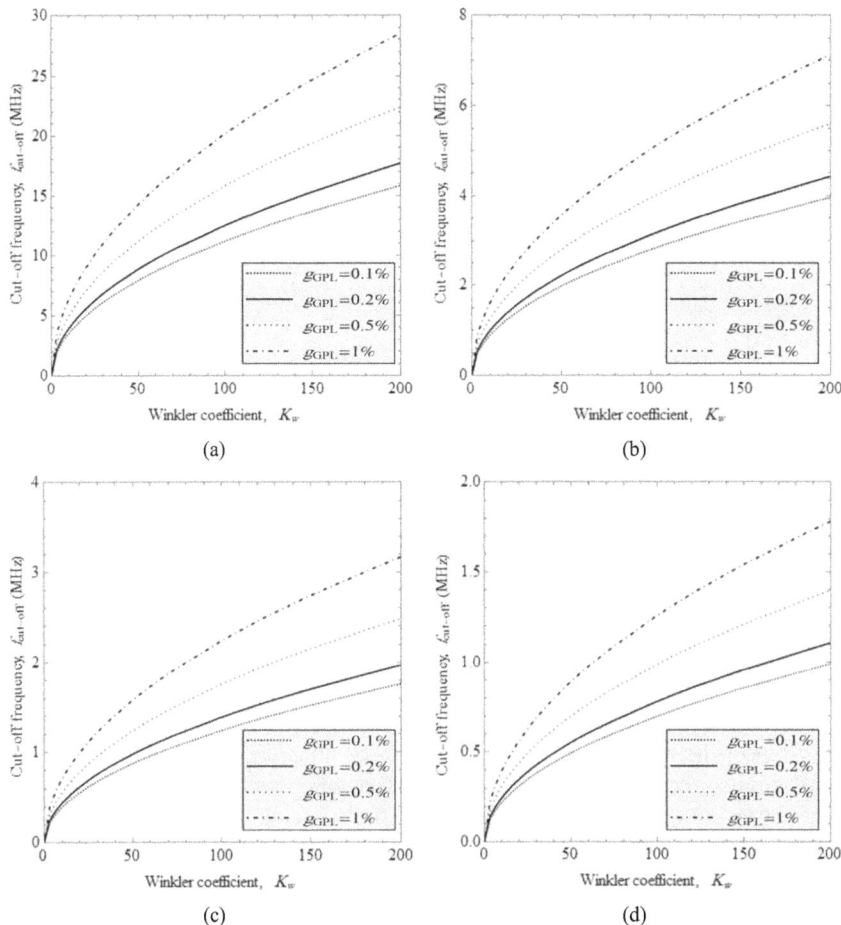

FIGURE 5.67 Variation of cut-off frequency versus Winkler coefficient for different GPL weight fractions with respect to influence of various slenderness ratios ($l_{GPL} = 2.5\ \mu m$, $w_{GPL} = 1.5\ \mu m$, $h_{GPL} = 1.5$ nm). (a) $L/h = 5$, (b) $L/h = 10$, (c) $L/h = 15$, and (d) $L/h = 20$.

completely different manner. This is due to the fact that the slenderness ratio is responsible for determining the wave dispersion response. When the slenderness ratio is increased, the amount of cut-off frequencies that are obtained is reduced as a direct consequence of this change [52].

5.4.4 GRAPHENE PLATELET–REINFORCED SHELLS: WAVE DISPERSION CHARACTERISTICS

Take, for example, the cylinder made of porous GPLR nanocomposite that is seen in Figure 5.68 [54].

Once again, the Halpin–Tsai model is used in order to characterize the mechanical characteristics that are identical to those of the GPLR beams. Porosities may

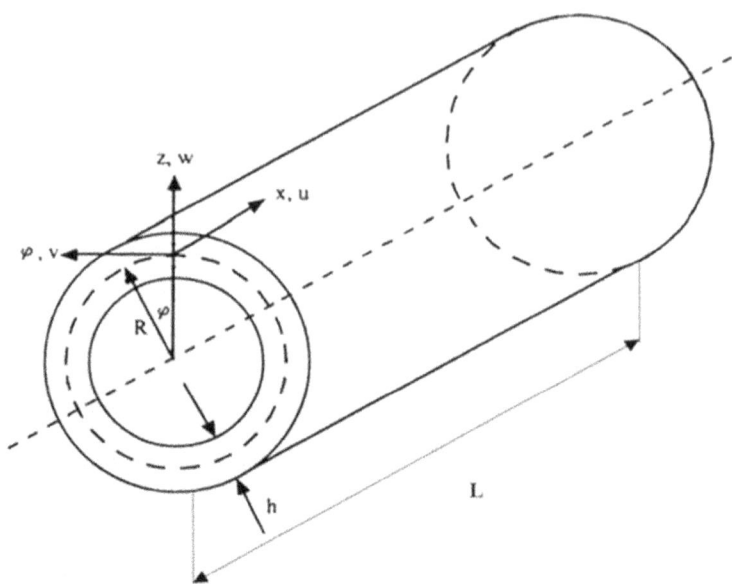

FIGURE 5.68   Geometry and coordinate system of a cylindrical shell.

show up in the material in either a uniform or nonuniform distribution pattern depending on the circumstances. The homogenization will be examined based on symmetric and asymmetric distributions in the subsequent formulations that are shown below. For the purpose of computing the mechanical characteristics of a porous nanocomposite shell, the following form [54] may be used:

$$E(z) = E_1\left(1 - e_0\lambda(z)\right),$$
$$G(z) = \frac{E(z)}{2\left(1 + v(z)\right)},$$
$$\rho(z) = \rho_1\left(1 - e_m\lambda(z)\right)$$

(5.169)

where $\lambda(z)$ is the function that defines how the structure's porosity is distributed along its thickness. For symmetric and asymmetric distribution patterns for the porous nanocomposite, this function may be stated in the following form:

$$\lambda(z) = \cos\left(\frac{\pi z}{h}\right) \qquad \text{for symmetric distribution}$$
$$\lambda(z) = \cos\left(\frac{\pi z}{2h} + \frac{\pi}{4}\right) \qquad \text{for asymmetric distribution}$$

(5.170)

Also, E_1, G_1, and ρ_1 are the maximum amounts of elasticity modulus, shear modulus, and mass density in the nanocomposite, respectively. Besides, e_0 and e_m are the porosity and mass density coefficients, respectively. These variants can be calculated as:

$$e_0 = 1 - \frac{E_2}{E_1} = 1 - \frac{G_2}{G_1},$$

$$e_m = \frac{1.121\left[1 - \sqrt[2.3]{1 - e_0\lambda(z)}\right]}{\lambda(z)} \tag{5.171}$$

The following is an example of how the Poisson's ratio of a porous GPLR nanocomposite material may be expressed mathematically:

$$v(z) = 0.221\tilde{p} + v_1\left(0.342\tilde{p}^2 - 1.21\tilde{p} + 1\right) \tag{5.172}$$

where

$$\tilde{p} = 1.121\left(1 - \sqrt[2.3]{1 - e_0\lambda(z)}\right) \tag{5.173}$$

The maximal elasticity modulus of the nanocomposite may now be expressed using the following form:

$$E_1 = \frac{3}{8}E_\Lambda + \frac{5}{8}E_\Theta \tag{5.174}$$

where E_Λ and E_Θ are the elastic moduli in the longitudinal and transverse directions that can be obtained as:

$$E_\Lambda = \frac{1 + \xi_L\eta_L V_{GPL}}{1 - \eta_L V_{GPL}} \times E_M, \quad E_\Theta = \frac{1 + \xi_W\eta_W V_{GPL}}{1 - \eta_W V_{GPL}} \times E_M \tag{5.175}$$

where

$$\eta_L = \frac{\left(E_{GPL}/E_M\right) - 1}{\left(E_{GPL}/E_M\right) + \xi_L}, \quad \eta_W = \frac{\left(E_{GPL}/E_M\right) - 1}{\left(E_{GPL}/E_M\right) + \xi_W} \tag{5.176}$$

in which E_M and E_{GPL} are related to matrix's elastic moduli and GPLs' elastic moduli, respectively. In addition, V_{GPL} is the GPLs' volume fraction; ξ_L and ξ_W are geometrical coefficients which can be formulated as:

$$\xi_L = \frac{2l_{GPL}}{h_{GPL}}, \quad \xi_W = \frac{2w_{GPL}}{h_{GPL}} \tag{5.177}$$

in Eq., l_{GPL}, w_{GPL}, and h_{GPL} are the average length, width, and thickness of the GPL, respectively. Moreover, the maximum mass density and Poisson's ratio coefficient of the GPLR nanocomposite can be defined as follows:

$$\rho_1 = \rho_{GPL}V_{GPL} + \rho_M V_M \tag{5.178}$$

$$v_1 = v_{GPL}V_{GPL} + v_M V_M \tag{5.179}$$

where V_M stands for the volume fraction of the polymer matrix and can be calculated by $V_M = 1 - V_{GPL}$. Moreover, subscript "e" stands for the equivalent material properties of the inhomogeneous material. Also, the volume fraction of the GPL can be calculated as:

$$V^*_{GPL} = \frac{g_{GPL}}{g_{GPL} + \left(\rho_{GPL} / \rho_M\right)\left(1 - g_{GPL}\right)} \tag{5.180}$$

in which, g_{GPL} is the weight fraction of the GPLs. Also, the equivalent volume fraction can be separately modified for different porosity distributions as:

$$V_{GPL} = V^*_{GPL}\left[1 - \lambda(z)\right] \tag{5.181}$$

Following is an expression that may be used to describe the displacement fields of a shell, which is based on the first-order shear deformable shell theory:

$$\begin{aligned}
u_x\left(x,\varphi,z,t\right) &= u\left(x,\varphi,t\right) + z\theta_x\left(x,\varphi,t\right), \\
u_\varphi\left(x,\varphi,z,t\right) &= v\left(x,\varphi,t\right) + z\theta_\varphi\left(x,\varphi,t\right), \\
u_z\left(x,\varphi,z,t\right) &= w\left(x,\varphi,t\right)
\end{aligned} \tag{5.182}$$

in which u, v, and w are axial, circumferential, and lateral displacements, respectively. Furthermore, θ_x and θ_φ are the rotation components about axial and circumferential directions, respectively. Henceforward, the nonzero strains of a shell-type element can be written in the following form:

$$\varepsilon_{xx} = \frac{\partial u}{\partial x} + z\frac{\partial \theta_x}{\partial x}, \; \varepsilon_{\varphi\varphi} = \frac{1}{R}\left(\frac{\partial v}{\partial \varphi} + z\frac{\partial \theta_\varphi}{\partial \varphi} + w\right), \; \varepsilon_{x\varphi} = \frac{1}{R}\frac{\partial u}{\partial \varphi} + \frac{\partial v}{\partial x} + \frac{z}{R}\frac{\partial \theta_x}{\partial \varphi} + z\frac{\partial \theta_\varphi}{\partial x},$$

$$\varepsilon_{xz} = \theta_x + \frac{\partial w}{\partial x}, \; \varepsilon_{\varphi z} = \theta_\varphi + \frac{1}{R}\frac{\partial w}{\partial \varphi} - \frac{v}{R}$$

$$(5.183)$$

This part will expand the dynamic form of the principle of virtual work, commonly known as Hamilton's principle, for cylindrical shells in order to get at the Euler–Lagrange equations for a nanocomposite shell. This will be done in order to arrive at the Euler–Lagrange equations for a nanocomposite shell. The definition of Hamilton's principle may be stated in a number of different ways, including the one that is provided below:

$$\int_0^t \delta(U - K)\,dt = 0 \qquad (5.184)$$

where U and K are strain energy and kinetic energy, respectively. The variation of strain energy for a linear elastic solid can be expressed as:

$$\delta U = \int_{-\frac{h}{2}}^{\frac{h}{2}}\int_0^{2\pi}\int_0^L \sigma_{ij}\delta\varepsilon_{ij}R\,dx\,d\varphi\,dz \qquad (5.185)$$

In addition, the variation in kinetic energy may be represented by the following equation:

$$\delta K = \int_{-\frac{h}{2}}^{\frac{h}{2}}\int_0^{2\pi}\int_0^L \rho(z)\left[\left(\frac{\partial \delta u_x}{\partial t}\right)^2 + \left(\frac{\partial \delta u_\varphi}{\partial t}\right)^2 + \left(\frac{\partial \delta u_z}{\partial t}\right)^2\right]R\,dx\,d\varphi\,dz \qquad (5.186)$$

When Eqs. (5.185) and (5.186) are added to Eq. (5.184), the motion equations for cylindrical shells may be stated as follows:

$$\frac{\partial N_{xx}}{\partial x} + \frac{1}{R}\frac{\partial N_{x\varphi}}{\partial \varphi} = I_0\frac{\partial^2 u}{\partial t^2} + I_1\frac{\partial^2 \theta_x}{\partial t^2}, \qquad (5.187)$$

$$\frac{\partial N_{x\varphi}}{\partial x} + \frac{1}{R}\frac{\partial N_{\varphi\varphi}}{\partial \varphi} + \frac{Q_{z\varphi}}{R} = I_0\frac{\partial^2 v}{\partial t^2} + I_1\frac{\partial^2 \theta_\varphi}{\partial t^2}, \qquad (5.188)$$

$$\frac{\partial Q_{xz}}{\partial x} + \frac{1}{R}\frac{\partial Q_{z\varphi}}{\partial \varphi} - \frac{N_{\varphi\varphi}}{R} = I_0 \frac{\partial^2 w}{\partial t^2}, \tag{5.189}$$

$$\frac{\partial M_{xx}}{\partial x} + \frac{1}{R}\frac{\partial M_{x\varphi}}{\partial \varphi} - Q_{xz} = I_1 \frac{\partial^2 u}{\partial t^2} + I_2 \frac{\partial^2 \theta_x}{\partial t^2}, \tag{5.190}$$

$$\frac{\partial M_{x\varphi}}{\partial x} + \frac{1}{R}\frac{\partial M_{\varphi\varphi}}{\partial \varphi} - Q_{\varphi z} = I_1 \frac{\partial^2 v}{\partial t^2} + I_2 \frac{\partial^2 \theta_\varphi}{\partial t^2} \tag{5.191}$$

where

$$\left[N_{xx}, N_{\varphi\varphi}, N_{x\varphi}\right] = \int_{-\frac{h}{2}}^{\frac{h}{2}} \left[\sigma_{xx}, \sigma_{\varphi\varphi}, \sigma_{x\varphi}\right] dz,$$

$$\left[M_{xx}, M_{\varphi\varphi}, M_{x\varphi}\right] = \int_{-\frac{h}{2}}^{\frac{h}{2}} \left[\sigma_{xx}, \sigma_{\varphi\varphi}, \sigma_{x\varphi}\right] z dz, \tag{5.192}$$

$$\left[Q_{xz}, Q_{z\varphi}\right] = \kappa_s \int_{-\frac{h}{2}}^{\frac{h}{2}} \left[\sigma_{xz}, \sigma_{z\varphi}\right] dz$$

and

$$\left[I_0, I_1, I_2\right] = \int_{-\frac{h}{2}}^{\frac{h}{2}} \left[1, z, z^2\right] \rho(z) dz \tag{5.193}$$

in which κ_s is shear correction factor. The stress-strain relationship of a multi-scale hybrid nanocomposite can be expressed in the following form:

$$\sigma_{ij} = C_{ijkl}\varepsilon_{kl} \tag{5.194}$$

where σ_{ij}, ε_{kl}, and C_{ijkl} are components of Cauchy stress, strain, and elasticity tensors, respectively. Integrating from Eq. (5.194) over the shell's thickness, the following relation can be achieved:

$$\begin{bmatrix} N_{xx} \\ M_{xx} \\ N_{\varphi\varphi} \\ M_{\varphi\varphi} \end{bmatrix} = \begin{bmatrix} A_{11} & B_{11} & \dfrac{A_{12}}{R} & \dfrac{B_{12}}{R} \\ B_{11} & D_{11} & \dfrac{B_{12}}{R} & \dfrac{D_{12}}{R} \\ A_{12} & B_{12} & \dfrac{A_{11}}{R} & \dfrac{B_{11}}{R} \\ B_{12} & D_{12} & \dfrac{B_{11}}{R} & \dfrac{D_{11}}{R} \end{bmatrix} \begin{bmatrix} \dfrac{\partial u}{\partial x} \\ \dfrac{\partial \theta_x}{\partial x} \\ \dfrac{\partial v}{\partial \varphi} + w \\ \dfrac{\partial \theta_\varphi}{\partial \varphi} \end{bmatrix},$$

$$\begin{bmatrix} N_{x\varphi} \\ M_{x\varphi} \end{bmatrix} = \begin{bmatrix} A_{66} & B_{66} \\ B_{66} & D_{66} \end{bmatrix} \begin{bmatrix} \dfrac{1}{R}\dfrac{\partial u}{\partial \varphi} + \dfrac{\partial v}{\partial x} \\ \dfrac{1}{R}\dfrac{\partial \theta_x}{\partial \varphi} + \dfrac{\partial \theta_\varphi}{\partial x} \end{bmatrix}, \qquad (5.195)$$

$$Q_{xz} = A_{55}^s\left(\theta_x + \frac{\partial w}{\partial x}\right), \quad Q_{\varphi z} = A_{55}^s\left(\theta_\varphi + \frac{1}{R}\frac{\partial w}{\partial \varphi} - \frac{v}{R}\right)$$

where

$$\left[A_{11}, B_{11}, D_{11}\right] = \int_{-\frac{h}{2}}^{\frac{h}{2}} \left[1, z, z^2\right]\frac{E}{1-v^2}\, dz, \left[A_{12}, B_{12}, D_{12}\right] = \int_{-\frac{h}{2}}^{\frac{h}{2}} \left[1, z, z^2\right]\frac{vE}{1-v^2}\, dz,$$

$$(5.196)$$

$$\left[A_{66}, B_{66}, D_{66}\right] = \int_{-\frac{h}{2}}^{\frac{h}{2}} \left[1, z, z^2\right]\frac{E}{2(1+v)}\, dz, \quad A_{55}^s = \kappa_s \int_{-\frac{h}{2}}^{\frac{h}{2}} \frac{E}{2(1+v)}\, dz$$

The coupled partial differential governing equations of a multi-scale hybrid nanocomposite shell are able to be defined in the following manner if they are given enough information:

$$A_{11}\frac{\partial^2 u}{\partial x^2} + B_{11}\frac{\partial^2 \theta_x}{\partial x^2} + \frac{A_{12}}{R}\left(\frac{\partial^2 v}{\partial x \partial \varphi} + \frac{\partial w}{\partial x}\right) + \frac{B_{12}}{R}\frac{\partial^2 \theta_\varphi}{\partial x \partial \varphi}$$

$$+ \frac{A_{66}}{R}\left(\frac{1}{R}\frac{\partial^2 u}{\partial \varphi^2} + \frac{\partial^2 v}{\partial x \partial \varphi}\right) + \frac{B_{66}}{R}\left(\frac{1}{R}\frac{\partial^2 \theta_x}{\partial \varphi^2} + \frac{\partial^2 \theta_\varphi}{\partial x \partial \varphi}\right) \qquad (5.197)$$

$$- I_0\frac{\partial^2 u}{\partial t^2} - I_1\frac{\partial^2 \theta_x}{\partial t^2} = 0,$$

$$A_{66}\left(\frac{1}{R}\frac{\partial^2 u}{\partial x \partial \varphi}+\frac{\partial^2 v}{\partial x^2}\right)+B_{66}\left(\frac{1}{R}\frac{\partial^2 \theta_x}{\partial x \partial \varphi}+\frac{\partial^2 \theta_\varphi}{\partial x^2}\right)$$

$$+\frac{A_{12}}{R}\frac{\partial^2 u}{\partial x \partial \varphi}+\frac{B_{12}}{R}\frac{\partial^2 \theta_x}{\partial x \partial \varphi}+\frac{A_{11}}{R^2}\left(\frac{\partial^2 v}{\partial \varphi^2}+\frac{\partial w}{\partial \varphi}\right)+\frac{B_{11}}{R^2}\frac{\partial^2 \theta_\varphi}{\partial \varphi^2} \qquad (5.198)$$

$$+\frac{A_{55}^s}{R}\left(\theta_\varphi+\frac{1}{R}\frac{\partial w}{\partial \varphi}-\frac{v}{R}\right)-I_0\frac{\partial^2 v}{\partial t^2}-I_1\frac{\partial^2 \theta_\varphi}{\partial t^2}=0,$$

$$A_{55}^s\left(\frac{\partial \theta_x}{\partial x}+\frac{\partial^2 w}{\partial x^2}\right)+\frac{A_{55}^s}{R}\left(\frac{\partial \theta_\varphi}{\partial \varphi}+\frac{1}{R}\frac{\partial^2 w}{\partial \varphi^2}-\frac{1}{R}\frac{\partial v}{\partial \varphi}\right)-\frac{A_{12}}{R}\frac{\partial u}{\partial x}$$

$$-\frac{B_{12}}{R}\frac{\partial \theta_x}{\partial x}-\frac{A_{11}}{R^2}\left(\frac{\partial v}{\partial \varphi}+w\right)-\frac{B_{11}}{R^2}\frac{\partial \theta_\varphi}{\partial \varphi}-I_0\frac{\partial^2 w}{\partial t^2}=0, \qquad (5.199)$$

$$B_{11}\frac{\partial^2 u}{\partial x^2}+D_{11}\frac{\partial^2 \theta_x}{\partial x^2}+\frac{B_{12}}{R}\left(\frac{\partial^2 v}{\partial x \partial \varphi}+\frac{\partial w}{\partial x}\right)+\frac{D_{12}}{R}\frac{\partial^2 \theta_\varphi}{\partial x \partial \varphi}$$

$$+\frac{B_{66}}{R}\left(\frac{1}{R}\frac{\partial^2 u}{\partial \varphi^2}+\frac{\partial^2 v}{\partial x \partial \varphi}\right)+\frac{D_{66}}{R}\left(\frac{1}{R}\frac{\partial^2 \theta_x}{\partial \varphi^2}+\frac{\partial^2 \theta_\varphi}{\partial x \partial \varphi}\right) \qquad (5.200)$$

$$-A_{55}^s\left(\theta_x+\frac{\partial w}{\partial x}\right)-I_1\frac{\partial^2 u}{\partial t^2}-I_2\frac{\partial^2 \theta_x}{\partial t^2}=0,$$

$$B_{66}\left(\frac{1}{R}\frac{\partial^2 u}{\partial x \partial \varphi}+\frac{\partial^2 v}{\partial x^2}\right)+D_{66}\left(\frac{1}{R}\frac{\partial^2 \theta_x}{\partial x \partial \varphi}+\frac{\partial^2 \theta_\varphi}{\partial x^2}\right)$$

$$+\frac{B_{12}}{R}\frac{\partial^2 u}{\partial x \partial \varphi}+\frac{D_{12}}{R}\frac{\partial^2 \theta_x}{\partial x \partial \varphi}+\frac{B_{11}}{R^2}\left(\frac{\partial^2 v}{\partial \varphi^2}+\frac{\partial w}{\partial \varphi}\right)+\frac{D_{11}}{R^2}\frac{\partial^2 \theta_\varphi}{\partial \varphi^2} \qquad (5.201)$$

$$-A_{55}^s\left(\theta_\varphi+\frac{1}{R}\frac{\partial w}{\partial \varphi}-\frac{v}{R}\right)-I_1\frac{\partial^2 v}{\partial t^2}-I_2\frac{\partial^2 \theta_\varphi}{\partial t^2}=0$$

The goal of the current section is to solve the partial differential governing equations of the issue that have been obtained. In order to accomplish this goal, the components of the displacement field are assumed to contain exponential solutions in the following manner:

$$u(x,\varphi,t)=U\exp\left(i\left[\beta x+n\varphi-\omega t\right]\right),$$

$$v(x,\varphi,t)=V\exp\left(i\left[\beta x+n\varphi-\omega t\right]\right),$$

$$w(x,\varphi,t)=W\exp\left(i\left[\beta x+n\varphi-\omega t\right]\right), \qquad (5.202)$$

$$\theta_x(x,\varphi,t)=\Theta_x\exp\left(i\left[\beta x+n\varphi-\omega t\right]\right),$$

$$\theta_\varphi(x,\varphi,t)=\Theta_\varphi\exp\left(i\left[\beta x+n\varphi-\omega t\right]\right)$$

where β and n are axial and circumferential wave numbers, respectively, and ω is the circular frequency. Once Eq. (5.202) is inserted in Eqs. (5.197)–(5.201), the following eigenvalue equation is obtained:

$$\left(\mathbf{K} - \omega^2 \mathbf{M}\right)\{\Delta\} = \mathbf{0} \tag{5.203}$$

where $\mathbf{K}$ and $\mathbf{M}$ are stiffness and mass matrices, respectively, and Δ denotes a column vector containing the wave amplitudes (U, V, W, Θ_x, and Θ_φ). The components of stiffness and mass matrices are presented in Appendix 5F. The frequency of the dispersed waves can be easily computed by solving Eq. (5.203) for ω. This process can be performed by setting the determinant of the left-hand-side of Eq. (5.203) equal to zero:

$$\left|\mathbf{K} - \omega^2 \mathbf{M}\right| = 0 \tag{5.204}$$

Afterward, the velocity of the propagated waves can be calculated via:

$$c_p = \frac{\omega}{\beta} \tag{5.205}$$

Now, a set of numerical results will be created in order to explain the influence of each parameter on the wave dispersion behaviors of GPLR nanocomposite shells. These results will be made in order to show the relationship between the parameters and the wave dispersion behaviors. This will be done in order to ensure that the results may be comprehended by a wider audience. The dimensionless form of the wave frequency, which can be found by using reference [54], will be provided in the subsequent numerical drawings. These may be found below.

$$\Omega = 100\omega h \sqrt{\frac{\rho_M}{E_M}} \tag{5.206}$$

The relationship between the dimensionless wave frequency and the radius-to-thickness ratio for various longitudinal wave numbers is shown in Figure 5.69 for the case when $n = 2$, $W_{GPL} = 1\%$, $e_0 = 0.4$. and $\beta = 1(1/m)$. This relationship is valid for two distinct kinds of porosity distribution in GPL-reinforced shells. In $\beta = 1(1/m)$, the frequency of dimensionless waves is greater for asymmetric distributions than they are for symmetric distributions. Both symmetric and asymmetric distributions provide the same results, which are dimensionless wave frequencies ($\beta = 10(1/m)$). In Figure 5.69c, initially, the asymmetric distribution and the symmetric distribution have the same dimensionless wave frequency. However, as shown, the dimensionless wave frequencies of the asymmetric distribution steadily decrease until they are lower than those of the symmetric distribution. To put it another way, as the number of longitudinal waves increases, the difference in magnitude between symmetric and asymmetric frequencies decreases. Additionally,

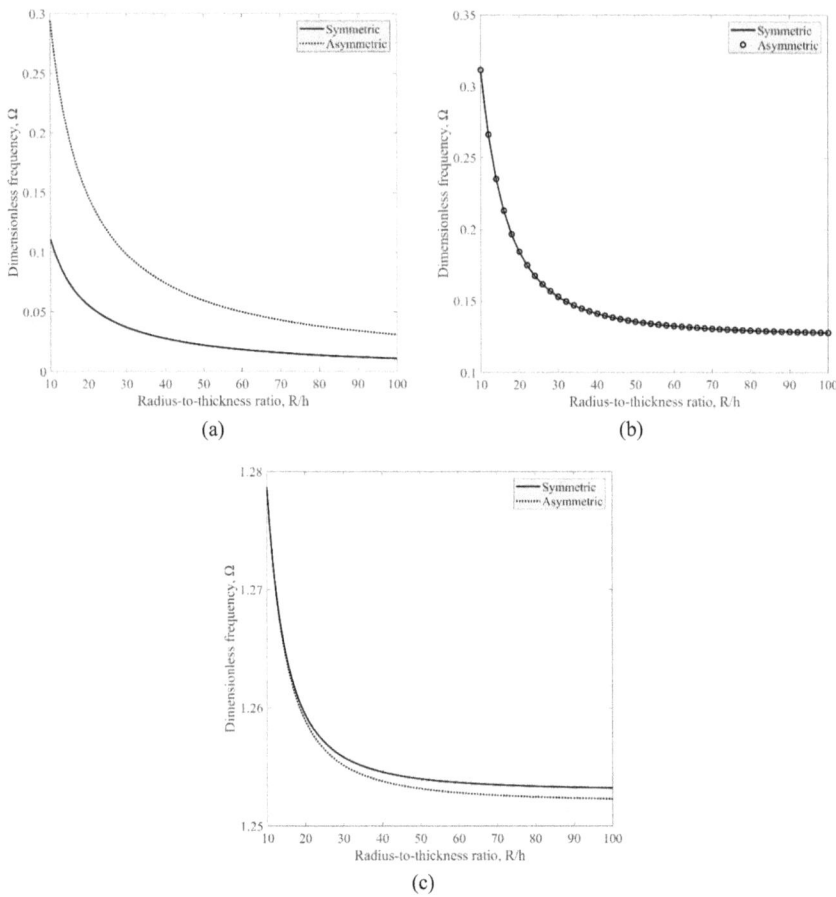

FIGURE 5.69 Illustration of coupled effects of longitudinal wave number and porosity distribution on the variations of dimensionless wave frequency of GPL-reinforced shells versus radius-to-thickness ratio ($n = 2$, $W_{GPL} = 1\%$, $e_0 = 0.4$). (a) $\beta = 1$ (1/m), (b) $\beta = 10$ (1/m), and (c) $\beta = 100$ (1/m).

the dimensionless wave frequencies will decrease as the radius-to-thickness ratio of the object increases. Because of their increased flexibility, nanocomposite shells with a larger radius-to-thickness ratio indicate such a behavior [54].

The effects of the weight percentage of GPLs on variations in dimensionless wave frequency versus radius-to-thickness ratio are shown in Figure 5.70. For this illustration, symmetric porosity distribution ($n = 2$, $\beta = 1$, $e_0 = 0.4$) was used. It should come as no surprise that nanocomposite shells that include a greater proportion of GPLs by weight also have higher dimensionless wave frequencies. It indicates that the effective stiffness of the porous shell may be improved by dispersing a little quantity of GPL into the matrix of the material. Also, it is possible to work out that the structure will grow softer as the radius-to-thickness ratio rises and that this tendency would continuously lower the system's frequency [54].

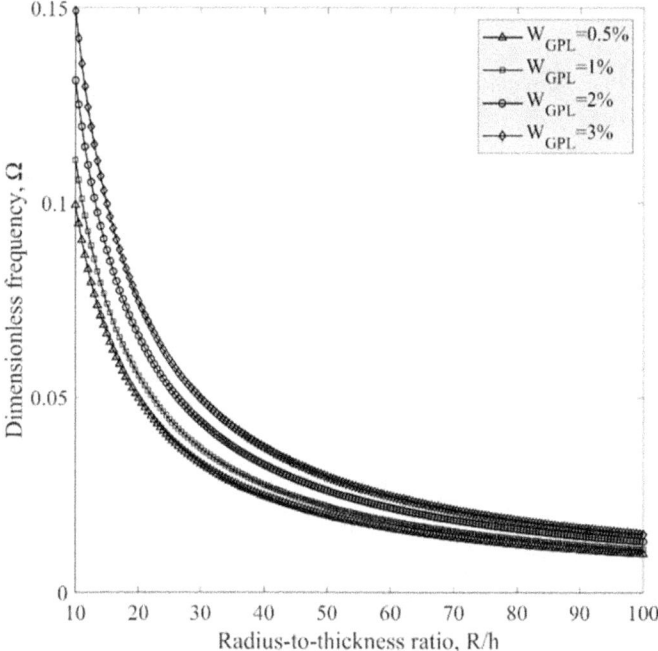

FIGURE 5.70 Variation of dimensionless wave frequency versus radius-to-thickness ratio for various weight fractions of GPLs by considering symmetric porosity distribution ($n = 2, \beta = 1, e_0 = 0.4$).

When assuming symmetric porosity distribution ($R/h = 20, e_0 = 0.4$), Figure 5.71 presents the mixed effects of longitudinal wave number and the weight percentage of GPLs on the dimensionless wave frequency of porous shells versus circumferential wave number. When the value of the circumferential wave number is added, it is possible to see that the dimensionless frequency increases. This is something that can be noticed. Again, elevating the circumferential wave number results in an increase in the dimensionless wave frequency in each weight fraction. Additionally, nanocomposite shells with a greater weight percentage of GPLs have been shown to exhibit higher dimensionless wave frequencies. To put it another way, a rise in the weight fraction of GPLs results in a higher dimensionless wave frequency. In addition, the dimensionless wave frequency may be amplified if a greater value is given to the longitudinal wave number [54]. This can happen anytime the value of the longitudinal wave number is increased.

In addition, Figure 5.72 illustrates the impacts of various porosity coefficients on the wave frequency curves of nanocomposite shells while assuming symmetric porosity distribution ($R/h = 20, n = 2, \beta = 1$). Porous structures, on the other hand, have porosity coefficients that aren't equal to zero, hence it stands to reason that perfect structures ($e_0 = 0$) have a greater capacity for dimensionless wave frequency. To put it another way, when the coefficient of porosity is increased, one

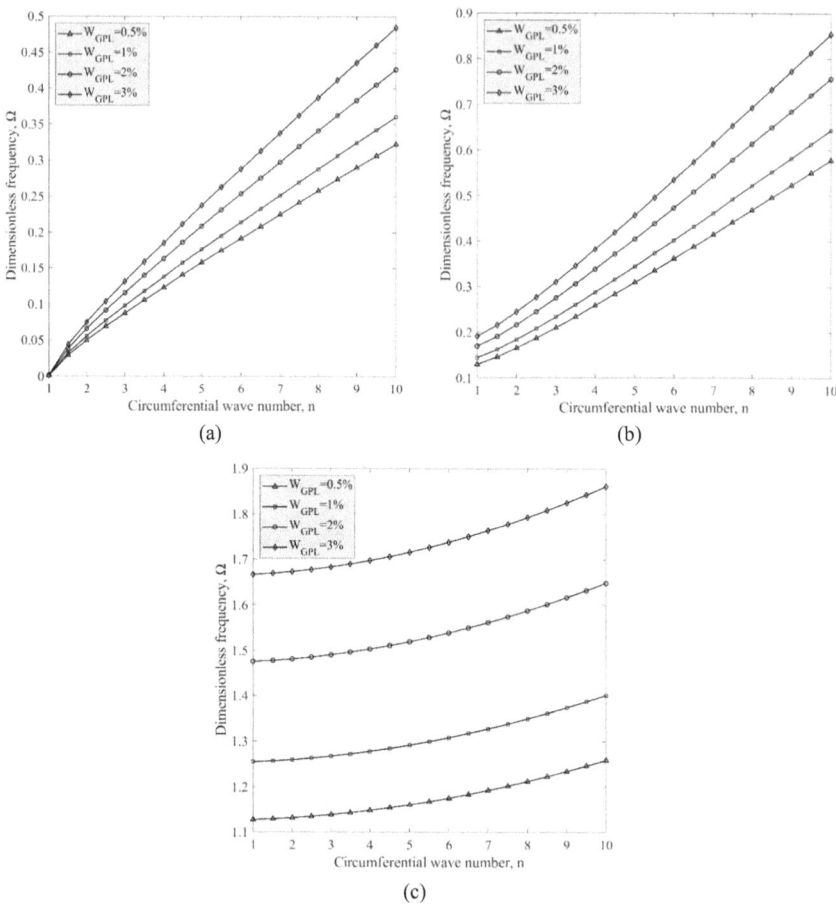

FIGURE 5.71 Illustration of coupled effects of both circumferential and longitudinal wave numbers and GPLs' weight fraction on the dimensionless wave frequency of porous shells using symmetric distribution for porosity ($R/h = 20$, $e_0 = 0.4$). (a) $\beta = 1$ (1/m), (b) $\beta = 10$ (1/m), and (c) $\beta = 100$ (1/m).

should expect to find dimensionless wave frequencies that are lower. Additionally, it has been shown that the mechanical frequency increases when a larger weight fraction is used for the GPLs. This finding is consistent with the previous one.

In conclusion, the change in phase velocity with weight fraction of GPLs is shown in Figure 5.73 for GPLR porous nanocomposites with symmetric porosity distribution and various porosity coefficients. On the basis of this figure, one may draw the conclusion that the phase velocity of a porous shell will rise if the weight percentage of GPLs increases, regardless of the sort of porosity coefficient being considered. It has been shown that keeping the weight percentage of GPLs at a fixed amount results in a reduction in phase velocity when the porosity coefficient is allowed to increase. In point of fact, nanocomposite shells that are flawless have the greatest value of phase velocity [54].

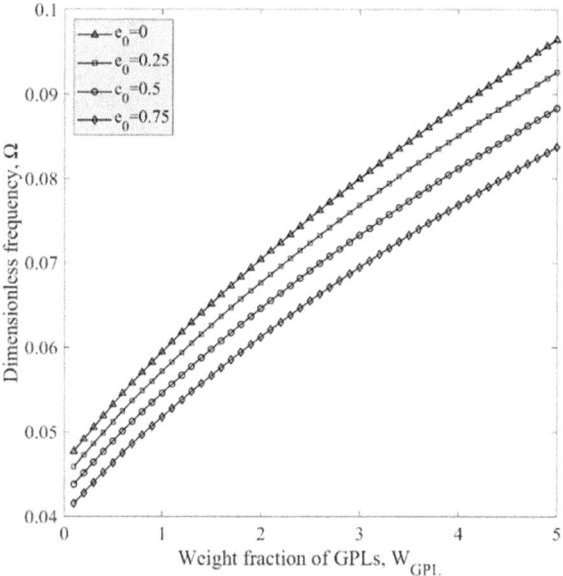

FIGURE 5.72   Variation of dimensionless wave frequency versus GPLs' weight fraction for different porosity coefficients by considering symmetric porosity distribution ($R/h = 20$, $n = 2, \beta = 1$).

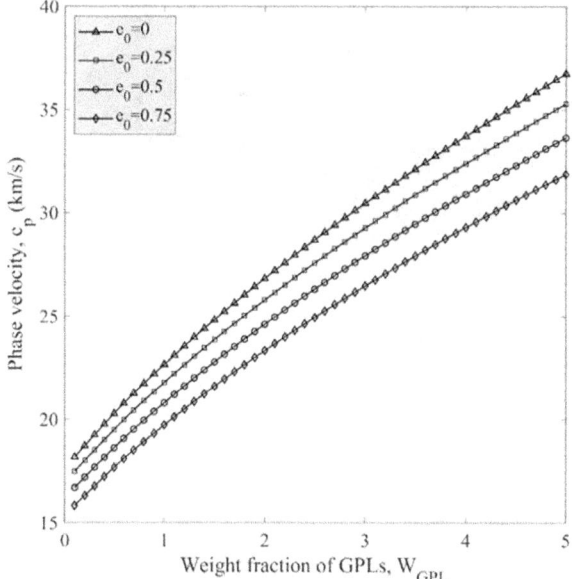

FIGURE 5.73   Variation of phase velocity versus GPLs' weight fraction for different porosity coefficients by considering symmetric porosity distribution ($R/h = 20, n = 2, \beta = 1$).

5.4.5   GRAPHENE PLATELET–REINFORCED NANOSHELLS: WAVE DISPERSION CHARACTERISTICS

A GPLRC cylindrical nanoshell in its thermal environment is seen in Figure 5.74, where h and R, which stand for the nanoshell's thickness and the middle surface radius, respectively, describe the structure. The composite material known as graphene nanoplatelet (GNP) forms the nanoshell in the shape of a cylinder [56].

In addition, Figure 5.75 depicts three unique nonuniform patterns for the porous material. Following is an example of how the general constitutive equation might be built [56], which is predicated on the idea of a nonlocal strain gradient:

$$\left(1 - \mu^2 \nabla^2\right) t_{ij} = C_{ijck} \left(1 - l^2 \nabla^2\right) \varepsilon_{ck} \qquad (5.207)$$

In which, $\nabla^2 = \partial^2/\partial x^2 + \partial^2/\partial(\theta R)^2$; t_{ij}, ε_{ck}, and C_{ijck} are the components of nonlocal strain gradient stress tensor, strain tensor, and elasticity tensor, respectively.

The following form provides an explanation of the tensor of the nonlocal strain gradient stress:

$$t_{ij} = \sigma_{ij} - \nabla \sigma_{ij}^{(1)} \qquad (5.208)$$

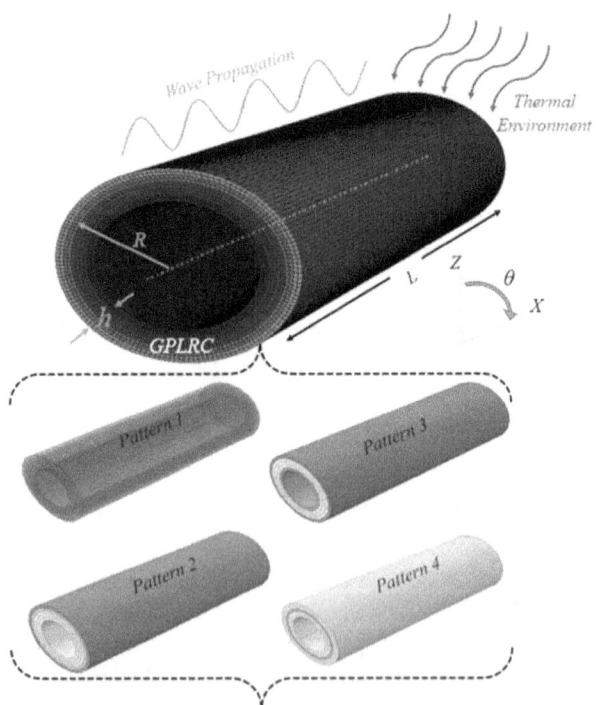

FIGURE 5.74   Geometry of GNPRC cylindrical shell in thermal environment.

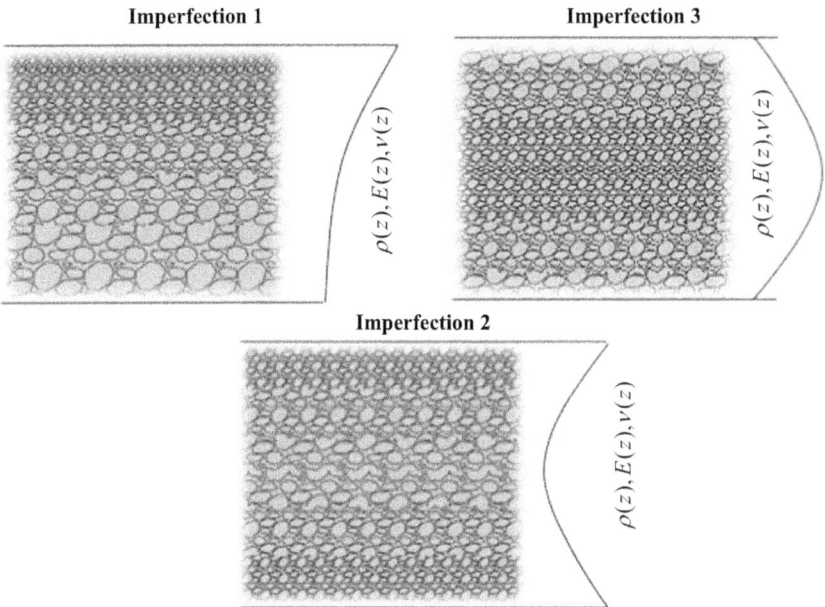

Imperfection 1   **Imperfection 3**

Imperfection 2

FIGURE 5.75 Schematic representation of the porous cylindrical nanoshell with different porosity dispersion patterns.

where σ_{ij} and $\sigma_{ij}^{(1)}$ are classical and size-dependent stresses, respectively. The μ and l parameters denote the influence of the non-invariant stress field and higher-order strain gradient stress field. Experimentation is used to ascertain the calibrated values of the size-dependent factors indicated earlier in the sentence. In the context of the proposed model, these parameters are regarded as being both stable and constant. The strain tensor is represented by the notation [56]:

$$\varepsilon_{ij} = \frac{1}{2}\left(u_{i,j} + u_{j,i}\right) \tag{5.209}$$

where u_i represents the individual components of the vector of motion. Nanostructure matrices are made using GNPRC materials. Reference [56] is a representation of the volume fraction functions for these materials.

$$Pattern1 : V_{GPL}\left(k\right) = V_{GPL}^* \tag{5.210}$$

$$Pattern2 : V_{GPL}\left(k\right) = 2V_{GPL}^* \left|2k - N_L - 1\right| / N_L \tag{5.211}$$

$$Pattern3 : V_{GPL}^*(k) = 2V_{GPL}^* \left[1 - \left(|2k - N_L - 1| / N_L \right) \right] \qquad (5.212)$$

$$Pattern4 : V_{GPL}^*(k) = 2V_{GPL}^* (2k - 1) / N_L \qquad (5.213)$$

where k is the number of layers of the microstructure, N_L is the total number of layers, and V_{GPL}^* is the total volume fraction of GNPs. The relation between V_{GPL}^* and their weight fraction g_{GPL} can be expressed by:

$$V_{GPL}^* = \frac{g_{GPL}}{g_{GPL} + \left(\rho_{GPL} / \rho_m \right) \left(1 - g_{GPL} \right)} \qquad (5.214)$$

in which ρ_{GPL} and ρ_m are the mass densities of GNP and the polymer matrix. Based on Halpin–Tsai model, the elastic modulus of composites reinforced with randomly GNP approximated by reference [56]:

$$E = \frac{3}{8} E_L + \frac{5}{8} E_T,$$

$$E_L = \frac{1 + \xi_L n_L V_{GPL}}{1 - n_L V_{GPL}} E_m, \quad E_T = \frac{1 + \xi_T n_T V_{GPL}}{1 - n_T V_{GPL}} E_m \qquad (5.215)$$

where E is effective modulus of composites reinforced with GNPs, and E_L and E_T are the longitudinal and transverse moduli for a unidirectional lamina. In Eq. (5.215), the GNP geometry factors (ξ_L and ξ_T) and other parameters are given by reference [56]:

$$\xi_L = 2 \left(\mathbb{Z}_{GPL} / h_{GPL} \right), \quad \xi_t = 2 \left(b_{GPL} / h_{GPL} \right),$$

$$n_L = \frac{\left(E_{GPL} / E_m \right) - 1}{\left(E_{GPL} / E_m \right) + \xi_L}, \quad n_T = \frac{\left(E_{GPL} / E_m \right) - 1}{\left(E_{GPL} / E_m \right) + \xi_T} \qquad (5.216)$$

where $\mathbb{Z}_{GPL}, h_{GPL}, b_{GPL}$ are the average length, thickness, and width of the GPLs. By using rule of mixture, mass density ρ_c and Poisson's ratio ν_c of the GPL/polymer nanocomposite are expressed as:

$$\bar{E} = E_{GPL} V_{GPL} + E_M V_M,$$
$$\bar{\rho} = \rho_{GPL} V_{GPL} + \rho_M V_M,$$
$$\bar{\nu} = \nu_{GPL} V_{GPL} + \nu_M V_M, \qquad (5.217)$$
$$\bar{\alpha} = \alpha_{GPL} V_{GPL} + \alpha_M V_M.$$

As shown in reference [56], it is possible to acquire the mechanical characteristics of the porous GPLR cylinder shell with varying porosity distributions.

$$E_{eff} = \bar{E}\left[1-\Gamma_p S(z)\right],$$
$$\rho_{eff} = \bar{\rho}\left[1-\Gamma_m S(z)\right],$$
$$\alpha_{eff} = \bar{\alpha}\left[1-\Gamma_p S(z)\right], \tag{5.218}$$
$$V_{eff} = 0.221\left(1-\frac{\rho_{eff}}{\bar{\rho}}\right)+\bar{v}\left[1+0.342\left(1-\frac{\rho_{eff}}{\bar{\rho}}\right)^2-1.21\left(1-\frac{\rho_{eff}}{\bar{\rho}}\right)\right]$$

where

$$\vartheta(z)=\begin{cases}\cos\left(\dfrac{\pi z}{2h}+\dfrac{\pi}{4}\right) & \text{for imperfection 1}\\[2mm]\cos\left(\dfrac{\pi z}{h}\right) & \text{for imperfection 2}\\[2mm]1-\cos\left(\dfrac{\pi z}{h}\right) & \text{for imperfection 3}\end{cases} \tag{5.219}$$

$$\Gamma_m = 1.121\left[1-\left(1-\lambda\vartheta(z)\right)^{1/2.3}\right]/\vartheta(z)$$

In Eqs. (5.218) and (5.219) λ and Γ_m are porosity and density coefficient, respectively.

The displacement field of the cylinder shell is as follows, according to the first shear deformation theory:

$$U(x,\theta,z,t)=u(x,\theta,z)+z\psi_x(x,\theta,z)$$
$$V(x,\theta,z,t)=v(x,\theta,z)+z\psi_\theta(x,\theta,z) \tag{5.220}$$
$$W(x,\theta,z,t)=w(x,\theta,z,t)$$

Here, $u(x, \theta, z)$, $v(x, \theta, z)$, and $w(x, \theta, z, t)$ indicate displacements of the neutral surface along x, θ, and z directions, respectively, and ψ_x, ψ_θ show rotations of a cross-section around θ and x- directions. Substituting Eq. (5.220) into Eq. (5.209), the components of the strain tensor are extracted as follows:

$$\varepsilon_{xx}=\frac{\partial u}{\partial x}+z\frac{\partial\psi_x}{\partial x}, \quad \varepsilon_{\theta\theta}=\frac{1}{R}\frac{\partial v}{\partial\theta}+\frac{z}{R}\frac{\partial\psi_\theta}{\partial\theta}+\frac{w}{R}$$

$$\varepsilon_{xz}=\frac{1}{2}\left(\psi_x+\frac{\partial w}{\partial x}\right)$$

$$\varepsilon_{x\theta}=\frac{1}{2}\left(\frac{1}{R}\frac{\partial u}{\partial\theta}+\frac{\partial v}{\partial x}\right)+\frac{z}{2}\left(\frac{1}{R}\frac{\partial\psi_x}{\partial\theta}+\frac{\partial\psi_\theta}{\partial x}\right) \tag{5.221}$$

$$\varepsilon_{\theta z}=\frac{1}{2}\left(\psi_\theta+\frac{1}{R}\frac{\partial w}{\partial\theta}-\frac{v}{R}\right)$$

The stress-strain relationship in three dimensions may be written as:

$$
\begin{bmatrix} \sigma_{xx} \\ \sigma_{\theta\theta} \\ \sigma_{zz} \\ \sigma_{x\theta} \\ \sigma_{xz} \\ \sigma_{\theta z} \end{bmatrix} = \begin{bmatrix} \overline{Q}_{11} & \overline{Q}_{12} & \overline{Q}_{13} & 0 & 0 & 0 \\ \overline{Q}_{12} & \overline{Q}_{22} & \overline{Q}_{23} & 0 & 0 & 0 \\ \overline{Q}_{13} & \overline{Q}_{23} & \overline{Q}_{33} & 0 & 0 & 0 \\ 0 & 0 & 0 & \overline{Q}_{44} & 0 & 0 \\ 0 & 0 & 0 & 0 & \overline{Q}_{55} & 0 \\ 0 & 0 & 0 & 0 & 0 & \overline{Q}_{66} \end{bmatrix} \begin{bmatrix} \varepsilon_{xx} - \alpha_1 \Delta T \\ \varepsilon_{\theta\theta} - \alpha_2 \Delta T \\ \varepsilon_{zz} - \alpha_3 \Delta T \\ \varepsilon_{x\theta} \\ \varepsilon_{xz} \\ \varepsilon_{\theta z} \end{bmatrix} \tag{5.222}
$$

In Eq. (5.222) α_i and ΔT are thermal expansion (in x, θ, and z directions) and temperature changes, respectively. In addition, the stiffness coefficients are obtained in reference [56]. The Hamilton principle may be used to derive the governing equations and boundary conditions of the cylindrical shell from the nonlocal strain gradient theory and first-order shear deformation theory (FSDT), as shown below [56]:

$$
\int_{t_1}^{t_2} \left(\delta K - \delta \Pi_s + \delta \Pi_w \right) dt = 0 \tag{5.223}
$$

where K is the kinetic energy, Π_s is strain energy, and Π_w is the work done corresponding to external applied forces. It is anticipated that the high-temperature environment may be evenly spread over the thickness of a typical cylinder-shaped nanoshell. In accordance with the task done, the temperature may be increased to the desired level. Hence, the work done depending on the temperature change can be obtained as [56]:

$$
W = \frac{1}{2} \iint_A \left[\left(N_1^T \right) \left(\frac{\partial w_0}{\partial x} \right)^2 + \left(N_2^T \right) \left(\frac{\partial v_0}{\partial x} \right)^2 \right] R \, dx \, d\theta \tag{5.224}
$$

where N_1^T and N_2^T are the thermal resultants that can be obtained as follows:

$$
N_1^T = \int_{-h/2}^{h/2} \left(C_{11} + C_{12} \right) \alpha \left(T - T_0 \right) dz,
$$
$$
N_2^T = \int_{-h/2}^{h/2} \left(C_{21} + C_{22} \right) \alpha \left(T - T_0 \right) dz. \tag{5.225}
$$

The formula for the thermal expansion coefficient is:

$$
\alpha = \begin{bmatrix} \alpha_1 & \alpha_2 & 0 & 0 & 0 \end{bmatrix}^T \tag{5.226}
$$

It is assumed that the temperature varies linearly along the thickness from T_m at the outer surface to T_c at the inner surface.

$$\delta K = \iiint_{Z \, A} \rho \left\{ \begin{array}{l} \left(\dfrac{\partial u}{\partial t} + z \dfrac{\partial \psi_x}{\partial t} \right) \left(\dfrac{\partial}{\partial t} \delta u + z \dfrac{\partial}{\partial t} \delta \psi_x \right) \\[2mm] + \left(\dfrac{\partial v}{\partial t} + z \dfrac{\partial \psi_\theta}{\partial t} \right) \left(\dfrac{\partial}{\partial t} \delta v + z \dfrac{\partial}{\partial t} \delta \psi_\theta \right) + \left(\dfrac{\partial w}{\partial t} \right) \dfrac{\partial}{\partial t} \delta w \\[2mm] +\Omega \left[\begin{array}{l} w \left(\dfrac{\partial}{\partial t} \delta v + z \dfrac{\partial}{\partial t} \delta \psi_\theta \right) - \left(v + z \psi_\theta \right) \left(\dfrac{\partial}{\partial t} \delta w \right) + \\[2mm] \delta w \left(\dfrac{\partial v}{\partial t} + z \dfrac{\partial \psi_\theta}{\partial t} \right) - \left(\delta v + z \delta \psi_\theta \right) \left(\dfrac{\partial}{\partial t} w \right) \end{array} \right] \\[2mm] +\Omega^2 \left[\left(v + z \psi_\theta \right) \left(\delta v + z \delta \psi_\theta \right) + w \delta w \right] \end{array} \right\} R \, dz \, dx \, d\theta$$

(5.227)

Following is a definition of the strain energy of a system in terms of NSGT:

$$\Pi_s = \frac{1}{2} \iiint_V \left(\sigma_{ij} \varepsilon_{ij} + \sigma_{ij}^{(1)} \nabla \varepsilon_{ij} \right) dV \ \Rightarrow \ \delta \Pi_s = \iiint_S t_{ij} \delta \varepsilon_{ij} dV + \iint_A \sigma_{ij}^{(1)} \delta \varepsilon_{ij} \Big|_0^L \ dS \quad (5.228)$$

The strain energy variations may be seen as follows when Eq. (5.222) is substituted into Eq. (5.228),

$$\delta \Pi_s = \iint_A \left\{ \begin{array}{l} N_{xx} \dfrac{\partial}{\partial x} \delta u + M_{xx} \dfrac{\partial}{\partial x} \delta \psi_x + N_{\theta\theta} \left(\dfrac{1}{R} \dfrac{\partial}{\partial \theta} \delta v + \dfrac{1}{R} \delta w \right) \\[2mm] + \dfrac{1}{R} M_{\theta\theta} \dfrac{\partial}{\partial \theta} \delta \psi_\theta + Q_{xz} \left(\delta \psi_x + \dfrac{\partial}{\partial x} \delta w \right) \\[2mm] + N_{x\theta} \left(\dfrac{1}{R} \dfrac{\partial}{\partial \theta} \delta u + \dfrac{\partial}{\partial x} \delta v \right) + M_{x\theta} \left(\dfrac{1}{R} \dfrac{\partial}{\partial \theta} \delta \psi_x + \dfrac{\partial}{\partial x} \delta \psi_\theta \right) \\[2mm] + Q_{z\theta} \left(\delta \psi_\theta + \dfrac{1}{R} \dfrac{\partial}{\partial \theta} \delta w - \dfrac{1}{R} \delta v \right) \end{array} \right\} R \, dx \, d\theta$$

$$+ \int \left\{ \begin{array}{l} N_{xx}^{(1)} \dfrac{\partial}{\partial x} \delta u + M_{xx}^{(1)} \dfrac{\partial}{\partial x} \delta \psi_x + N_{\theta\theta}^{(1)} \left(\dfrac{1}{R} \dfrac{\partial}{\partial \theta} \delta v + \dfrac{1}{R} \delta w \right) \\[2mm] + \dfrac{1}{R} M_{\theta\theta}^{(1)} \dfrac{\partial}{\partial \theta} \delta \psi_\theta + Q_{xz}^{(1)} \left(\delta \psi_x + \dfrac{\partial}{\partial x} \delta w \right) + N_{x\theta}^{(1)} \left(\dfrac{1}{R} \dfrac{\partial}{\partial \theta} \delta u + \dfrac{\partial}{\partial x} \delta v \right) \\[2mm] + M_{x\theta}^{(1)} \left(\dfrac{1}{R} \dfrac{\partial}{\partial \theta} \delta \psi_x + \dfrac{\partial}{\partial x} \delta \psi_\theta \right) + Q_{\theta z}^{(1)} \left(\delta \psi_\theta + \dfrac{1}{R} \dfrac{\partial}{\partial \theta} \delta w - \dfrac{1}{R} \delta v \right) \end{array} \right\} \Big|_0^L \ R \, d\theta$$

(5.229)

where the resulting forces and velocities are:

$$\begin{Bmatrix} N_{xx}, N_{\theta\theta}, N_{x\theta} \\ N_{xx}^{(1)}, N_{\theta\theta}^{(1)}, N_{x\theta}^{(1)} \end{Bmatrix} = \int_{-h/2}^{h/2} \begin{Bmatrix} t_{xx}, t_{\theta\theta}, t_{x\theta} \\ \sigma_{xx}^{(1)}, \sigma_{\theta\theta}^{(1)}, \sigma_{x\theta}^{(1)} \end{Bmatrix} dz$$

$$\begin{Bmatrix} M_{xx}, M_{\theta\theta}, M_{x\theta} \\ M_{xx}^{(1)}, M_{\theta\theta}^{(1)}, M_{x\theta}^{(1)} \end{Bmatrix} = \int_{-h/2}^{h/2} \begin{Bmatrix} t_{xx}, t_{\theta\theta}, t_{x\theta} \\ \sigma_{xx}^{(1)}, \sigma_{\theta\theta}^{(1)}, \sigma_{x\theta}^{(1)} \end{Bmatrix} z\,dz \qquad (5.230)$$

$$\begin{Bmatrix} Q_{xz}, Q_{\theta z} \\ Q_{xz}^{(1)}, Q_{\theta z}^{(1)} \end{Bmatrix} = \int_{-h/2}^{h/2} K_s \begin{Bmatrix} t_{xz}, t_{\theta z} \\ \sigma_{xz}^{(1)}, \sigma_{\theta z}^{(1)} \end{Bmatrix} dz$$

The governing equations for a cylindrical shell in the FSDT and the nonlocal strain gradient theory may be obtained by substituting Eqs. (5.224), (5.227), and (5.228) into Eq. (5.223), and then integrating by parts. These equations can be found in Appendix 5G [56]. Now the displacement fields for investigation of the wave propagation analysis of the structure is defined as follows [56]:

$$\begin{Bmatrix} U(x,\theta,z,t) \\ V(x,\theta,z,t) \\ W(x,\theta,z,t) \\ \psi_x(x,\theta,z,t) \\ \psi_\theta(x,\theta,z,t) \end{Bmatrix} = \begin{Bmatrix} u_0 \exp(sx + n\theta - \omega t)i \\ v_0 \exp(sx + n\theta - \omega t)i \\ w_0 \exp(sx + n\theta - \omega t)i \\ \psi_{x_0} \exp(sx + n\theta - \omega t)i \\ \psi_{\theta_0} \exp(sx + n\theta - \omega t)i \end{Bmatrix} \qquad (5.231)$$

Where u_0, v_0, w_0, ψ_{x_0}, and ψ_{θ_0} are wave amplitude parameters and s and n are wave numbers along the directions of x and θ, respectively; also, ω is called frequency. By replacing Eq. (5.231) in governing equations, we Eq. (56):

$$\left([K] - \omega^2 [M] \right) \{d\} = \{0\} \qquad (5.232)$$

where

$$\{d\} = \{u_0 \quad v_0 \quad w_0 \quad \psi_{x_0} \quad \psi_{\theta_0}\} \qquad (5.233)$$

In addition, Eq. (5.234) may be used to determine the wave's phase velocity of dispersion:

$$c = \frac{\omega}{s} \qquad (5.234)$$

In Eq. (5.234), c_p and s are called phase velocity and wave number of a laminated nanocomposite cylindrical shell. Particle speeds as they go through a laminated nanocomposite cylinder shell are described by these characteristics. Classical continuum theory is used to calculate the phase velocity considering $\mu = 0$. These parameters are propagation speeds of the particles in a laminated nanocomposite cylindrical shell. Considering $\mu = 0$, the phase velocity of classical continuum theory is computed. In the section under "Results," it is explained how the phase velocity of the porous GPLRC cylinder changes according to the temperature, the GPL distribution pattern, the porosity distribution pattern, and the GPL weight function. In Table 5.2 [56], you will find a table that provides an overview of the material properties of the GPLRC nanoshell.

Temperature-dependent GPLRC material is poly, referred to as polymethyl methacrylate (PMMA), is selected for the matrix, and the material properties of which are assumed to be $\alpha_m = 45(1 + 0.0005\Delta T) \times 10^{-6}$/K and $E = (3.52–0.0034T)$ GPa, in which $T = T_0 + \Delta T$. Table 5.3 displays the results of a parametric analysis into the impacts of various GNP designs, radii, porosity coefficients, and thicknesses on the nanostructure's phase velocity. Higher levels of phase velocity in a cylinder-shaped nanoshell are seen for all GNP-reinforcing configurations when compared to pure epoxy. This is because the nanostructure becomes more stable after being reinforced by GNP. The phase velocity is greatest for the ideal nanostructure and lowest for the imperfect one. The phase velocity of nanostructures with GPL distribution Pattern 3 is greater than that of nanostructures with other patterns across the board for all values of radius, porosity coefficient, and thickness. The third GPL distribution pattern indicates an improvement in the stability of nanostructures. In other words, the phase velocity is greater for Pattern 3 than it is for any of the other patterns. The effects of the porosity coefficient and the thickness on the phase velocity may be shown to be both inverse and direct [56].

The porosity coefficient and nanoshell thickness are shown to vary as a function of $\Delta T=280$ K, $g_{GPL} = 1\%$, $\mu = 0.55$ nm, and $l = 0.35$ nm in Figures 5.76–5.79. For a given amount of thickness and porosity coefficient, these graphs reveal, for example, that a rise in phase velocity is seen initially when the wave number increases. The phase velocity drops from its highest value as the wave number increases, eventually reaching its lowest value. As the wave number is increased from the minimal value, the phase velocity improves. An incredible conclusion may be drawn from these numbers. As the figures' thickness grows, the phase velocity's lowest and maximum values move to the left. When the thickness is increased, the maximum values of the phase velocity are seen at the minimum

TABLE 5.2
Material Properties of the Epoxy and GPL

Material Properties	Epoxy	GPL
Young's modulus (*GPa*)	3	1010
Density (kg m^{-3})	1200	1062.5
Poisson's ratio	0.34	0.186

TABLE 5.3

The Effect of Different Radius, Thickness, Pattern of GNP, and Porosity on Phase Velocity (km/s) of GNPRC Nanoshell with $L/R = 10$, $R/h = 10$, $g_{GPL} = 1\%$, and $R_1 = 1$ nm

		$R = 1$ nm		$R = 2$ nm		$R = 3$ nm	
		$h = R_1/10$	$h = R_1/15$	$h = R_1/10$	$h = R_1/15$	$h = R_1/10$	$h = R_1/15$
Pure epoxy	λ						
	0	0.495497	0.498817	0.482761	0.481870	0.383836	0.382537
	0.1	0.478968	0.478375	0.466550	0.465699	0.371060	0.369822
	0.2	0.461674	0.461106	0.449592	0.448780	0.357674	0.356497
	0.3	0.443524	0.442979	0.431795	0.431025	0.343607	0.342492
Pattern 1	λ						
	0	1.134734	1.133326	1.105545	1.103509	0.879111	0.876143
	0.1	1.096850	1.095494	1.068391	1.066445	0.849816	0.846987
	0.2	1.057217	1.055915	1.029525	1.027670	0.819127	0.816437
	0.3	1.015623	1.014377	0.988743	0.986981	0.786880	0.784332
Pattern 2	λ						
	0	1.134070	1.133255	1.104375	1.103204	0.877242	0.875497
	0.1	1.096203	1.095419	1.067265	1.066148	0.848027	0.846365
	0.2	1.056588	1.055836	1.028446	1.027383	0.817420	0.815841
	0.3	1.015014	1.014295	0.987712	0.986703	0.785257	0.783763
Pattern 3	λ						
	0	1.136214	1.1342168	1.107497	1.104602	0.881652	0.877472
	0.1	1.098269	1.0963454	1.070255	1.067486	0.852236	0.848250
	0.2	1.058574	1.0567263	1.031300	1.028660	0.821426	0.817635
	0.3	1.016916	1.0151473	0.990427	0.987918	0.789057	0.785464
Pattern 4	λ						
	0	1.136123	1.134533	1.103621	1.102568	0.876721	0.874952
	0.1	1.098182	1.096621	1.066568	1.065556	0.847552	0.845860
	0.2	1.058490	1.057021	1.027807	1.026835	0.816991	0.815377
	0.3	1.016837	1.015430	0.987132	0.986200	0.784876	0.783339

values of the wave number, providing a more holistic picture. These numbers suggest that, for small values of the wave number, the phase velocity may be increased by making the nanostructure thicker. Additionally, at larger wave number values, an unexpected influence of porosity on the phase velocity is shown when the thickness is increased. The influence of the porosity coefficient on the phase velocity is shown to diminish with increasing thickness. Increasing the porosity coefficient and the thickness from 1 nm to 1.5 nm results in graphs that are quite near to each other, as seen in the figures. These data show that as the thickness increases, the phase velocity decreases for wave numbers between 2(1/nm) and 6(1/nm). Out of these wave numbers, raising the porosity coefficient results in a faster phase velocity. Finally, the findings imply that when the thickness increases from 1 to 2 nm, the influence of porosity on the phase velocity shifts.

The effects of the GNP weight function and radius on the phase velocity of a GNPRC nanoshell with $\Delta T = 280k$, $\Gamma_P = 0.1$, $\mu = 0.055$ nm, $l = 0.035$ nm, and $h = R/10$ are shown in Figures 5.80–5.83. All of these graphs illustrate that the

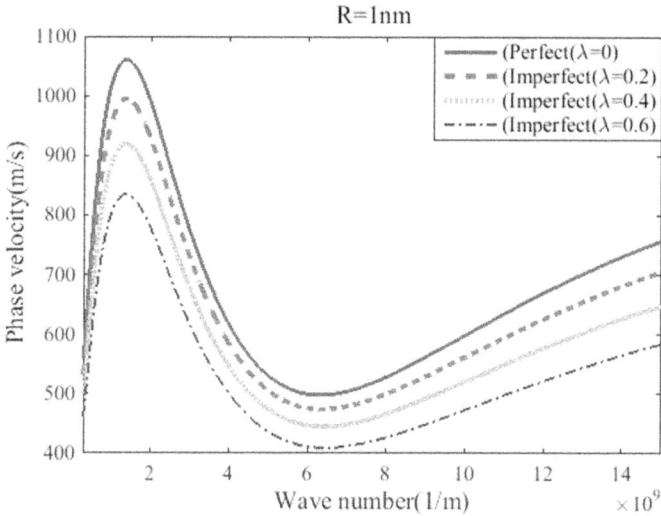

FIGURE 5.76   Effect of porosity coefficient and wave number on the phase velocity of the GPLRC nanoshell considering $R = 1$ nm.

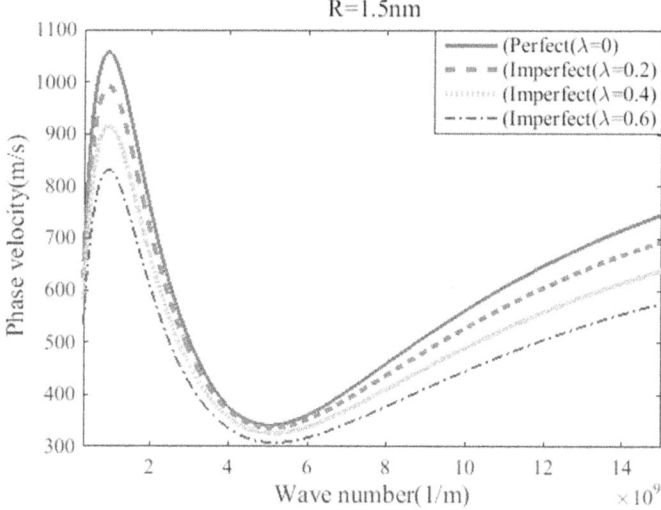

FIGURE 5.77   Effect of porosity coefficient and wave number on the phase velocity of the GPLRC nanoshell considering $R = 1.5$ nm.

phase velocity rises as the GNP weight function rises; put another way, the GNP weight function positively affects the phase velocity. These numbers also suggest that the phase velocity drops across the board as the radius of the GNPRC nanoshell becomes larger. These figures also reveal an astonishing conclusion: as the radius grows, the influence of the GNP weight function on the phase velocity

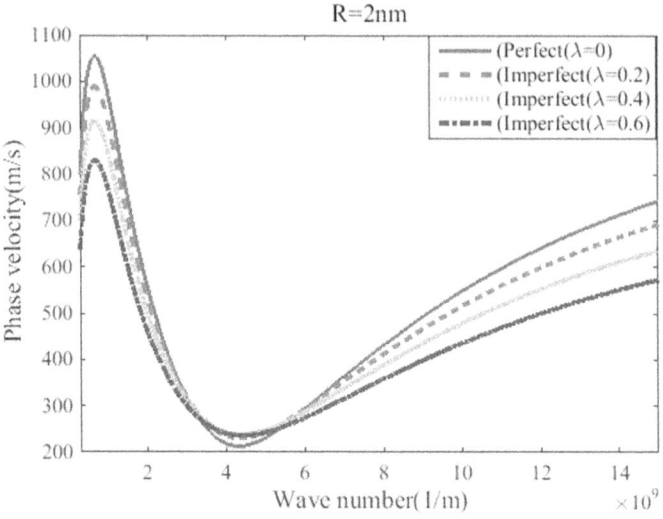

FIGURE 5.78 Effect of porosity coefficient and wave number on the phase velocity of the GPLRC nanoshell considering $R = 2$ nm.

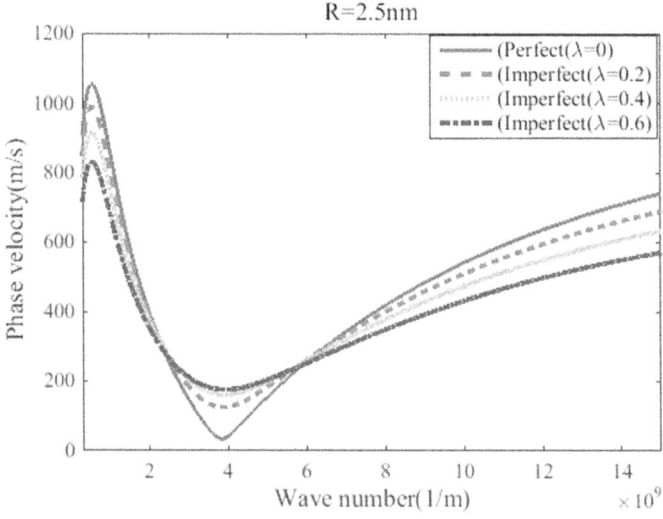

FIGURE 5.79 Effect of porosity coefficient and wave number on the phase velocity of the GPLRC nanoshell considering $R = 2.5$ nm.

weakens. In other words, as the radius grows, the graphs in these figures become more similar to one another. Therefore, the influence of radius on phase velocity is counter to that of the GNP weight function. This is due to the fact that the stability of the GNPRC nanoshell degrades with an increase in radius, but the stability of the GNPRC nanoshell increases with an increase in the GNP weight function.

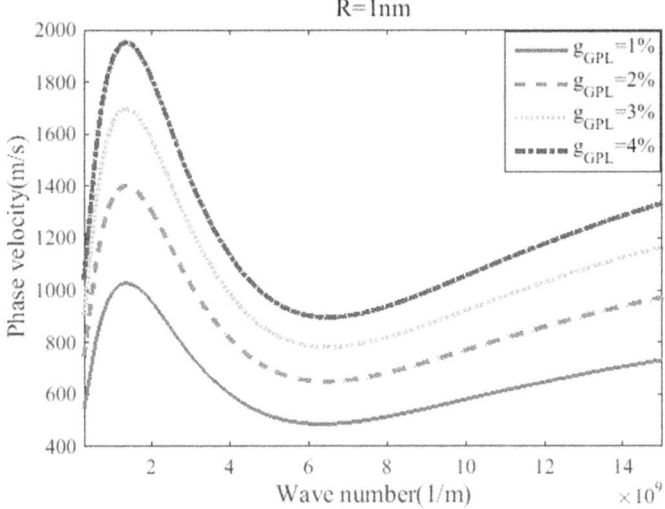

FIGURE 5.80   Effect of GPL weight function and wave number on the phase velocity of the GPLRC nanoshell considering $R = 1$ nm.

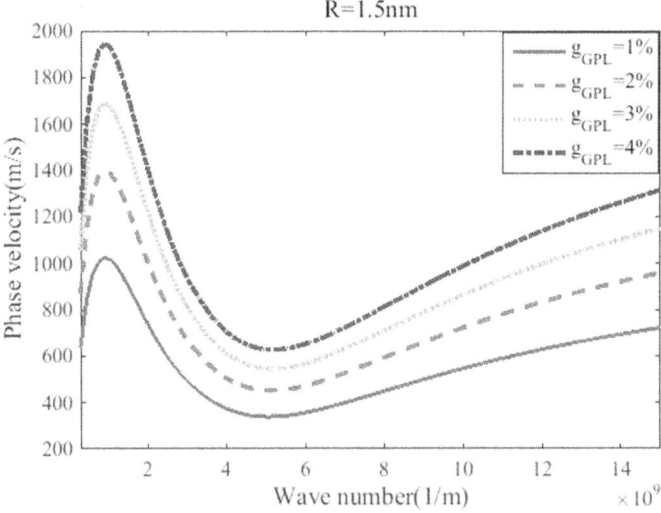

FIGURE 5.81   Effect of GPL weight function and wave number on the phase velocity of the GPLRC nanoshell considering $R = 1.5$ nm.

One can see, for a given radius and GNP weight function, that the phase velocity increases with increasing wave number. The phase velocity drops from its highest value as the wave number increases, eventually reaching its lowest value. As the wave number is increased from the minimal value, the phase velocity improves. The smoothness with which the graph transitions between states is notable.

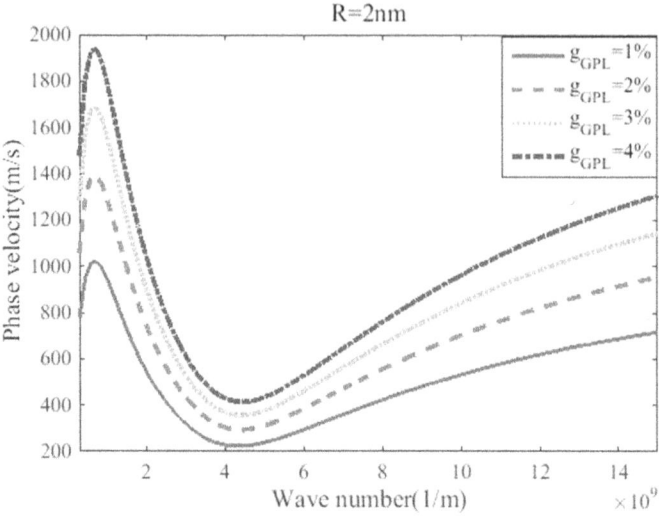

FIGURE 5.82   Effect of GPL weight function and wave number on the phase velocity of the GPLRC nanoshell considering $R = 3$ nm.

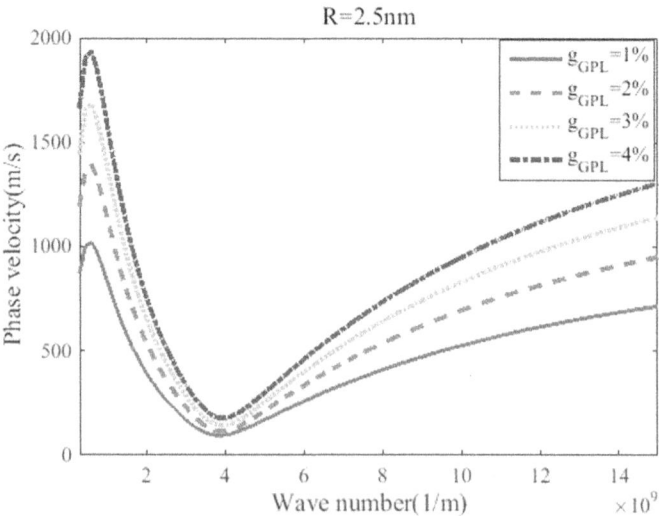

FIGURE 5.83   Effect of GPL weight function and wave number on the phase velocity of the GPLRC nanoshell considering $R = 2.5$ nm.

This research is the first-ever report of an increase in the difference between the lowest and highest values of the phase velocity as a function of the radius. These data provide additional evidence that the phase velocity's lowest and maximum values shift to the left as the radius increases. As the radius grows, the maximum values of phase velocity are seen at the minimum values of the wave number, providing a more holistic picture. This study offers the practical recommendation [56] that while developing a GPLRC nanostructure, the GNP weight function and radius should be considered concurrently.

The effects of temperature variations, the GNP weight function, and porosity on the phase velocity of a GNPRC nanoshell with $\mu = 0.055$ nm, $l = 0.035$ nm, and $h = R/10$ are shown in Figures 5.84–5.87. The phase velocity is affected both directly and indirectly by the GNP weight function and porosity, as has been discussed before. These numbers also show that the phase velocity decreases with an increase in temperature changes; this is in contrast to the effect that the GNP weight function has on the phase velocity, which is the same as the effect that porosity has on the phase velocity. One striking feature of these figures (for a given value of porosity coefficient and GNP weight function) is that when the temperature is increased from low to high (0–600 K), the phase velocity declines at a constant rate. The phase velocity is noticeably impacted by temperature fluctuations in the 600–700-degree range. As a corollary explanation, raising this parameter causes a drastic slowing of the phase velocity for large values of temperature change. This research reports for the first time that the impact of temperature variations on the phase velocity increases with increasing porosity effect [56].

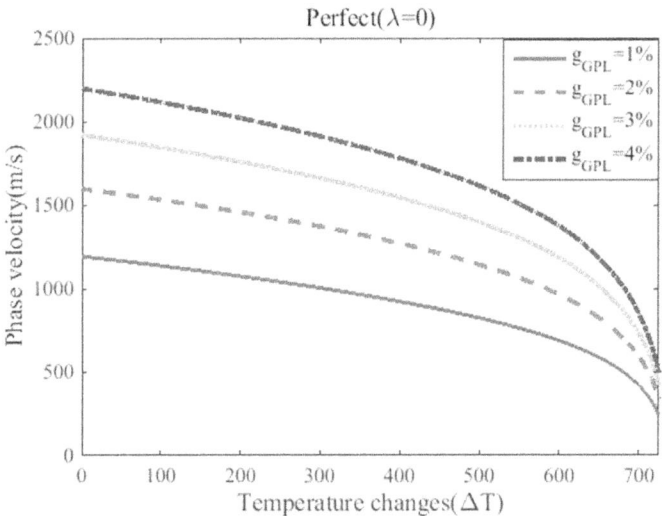

FIGURE 5.84 Effect of GPL weight function and temperature change on the phase velocity of the GPLRC nanoshell considering $\lambda = 0$.

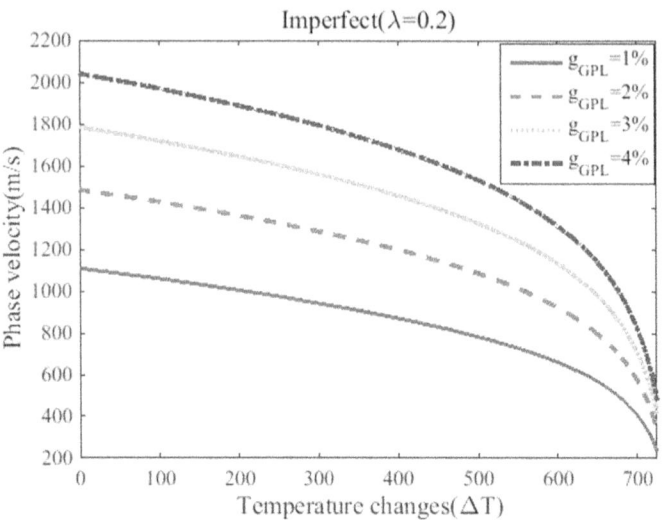

FIGURE 5.85 Effect of GPL weight function and temperature change on the phase velocity of the GPLRC nanoshell considering $\lambda = 0.2$.

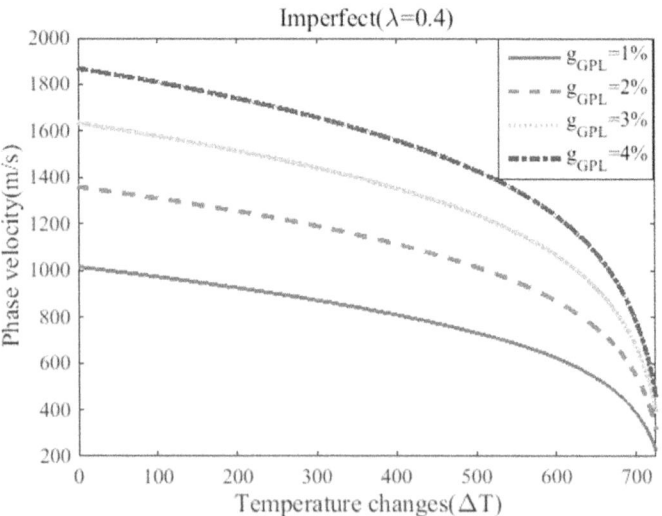

FIGURE 5.86 Effect of GPL weight function and temperature change on the phase velocity of the GPLRC nanoshell considering $\lambda = 0.4$.

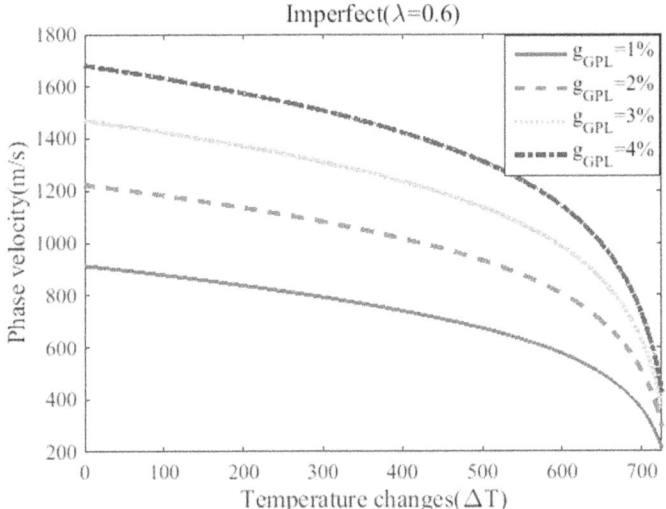

FIGURE 5.87 Effect of GPL weight function and temperature change on the phase velocity of the GPLRC nanoshell considering $\lambda = 0.6$.

5.5   WAVE DISPERSION CHARACTERISTICS OF GRAPHENE FOAM STRUCTURES

5.5.1   BACKGROUND

The premise of this chapter is based on the premise that the reader has a basic understanding of the concepts of probability and statistics, and is familiar with the basic concepts of probability and statistics. While the pores are dispersed equally and symmetrically along the thickness direction, the impacts of the hygrothermal environment are also considered. Incorporating the kinetic relation from revised higher-order shear deformation theories, the Hamiltonian technique is used to determine the motion equations of porous graphene foam (PGF) beams and plates. Within the framework of a set of figures, the influence of a number of parameters such as the wave number, porosity coefficient, porosity distribution, Winkler–Pasternak coefficients, temperature, moisture, and aspect ratio, on the variation of wave frequency and phase velocity of PGF structures, is examined and presented. Some tables and figures [63, 64] demonstrate the findings achieved.

5.5.2   GRAPHENE FOAM BEAM WAVE DISPERSION CHARACTERISTICS

Here, we assume that L is the length of the PGF beam and h is the thickness of the PGF beam, both of which are located in Figure 5.88 and rest on an elastic medium [63].

Along the thickness axis, there are several potential patterns for the distribution of porosity. These patterns include a uniform porosity distribution, symmetric

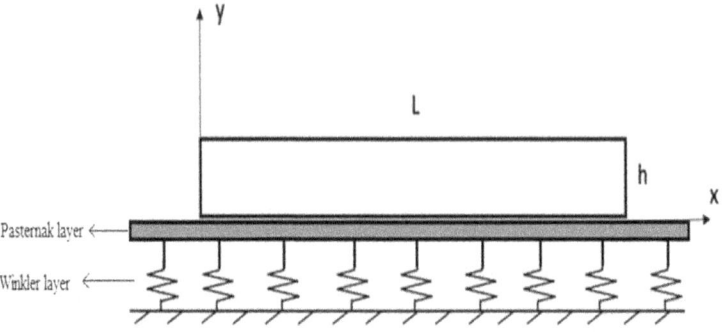

FIGURE 5.88   The configuration of graphene foam beam resting on an elastic medium.

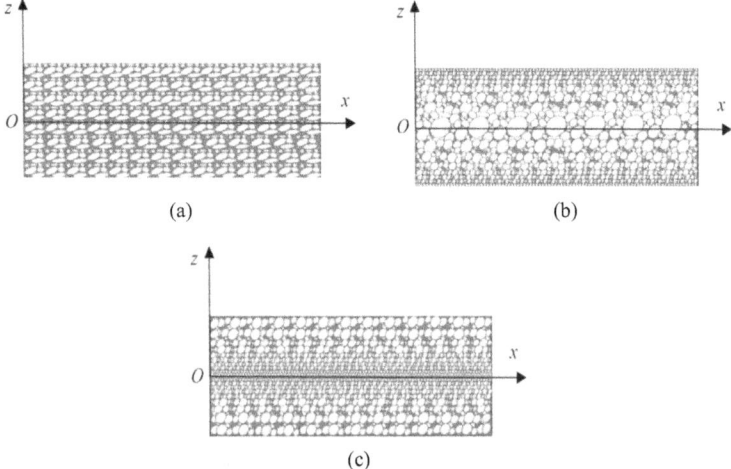

FIGURE 5.89   (a) Schematic of various types of porosity distribution through the beam's thickness for (b) symmetric porosity distribution I, and (c) symmetric porosity distribution II.

porosity distribution I, and symmetric porosity distribution II. Figure 5.89 shows a schematic showing the porosity over the beam's thickness. To begin, the formula [63] may be used to determine the effective mechanical characteristics of a material with a uniform porosity distribution.

$$E(z) = E_1 e_2 \qquad\qquad (5.235a)$$

$$G(z) = G_1 e_2 \qquad\qquad (5.235b)$$

$$\rho(z) = \rho_1 e_{m2} \qquad\qquad (5.235c)$$

$$\alpha(z) = \alpha_1 e_2 \tag{5.235d}$$

$$\beta(z) = \beta_1 e_2 \tag{5.235e}$$

$$\upsilon(z) = \upsilon_1 \tag{5.235f}$$

In addition, a type I beam with a symmetric porosity distribution has effective mechanical characteristics that may be expressed as [63]:

$$E(z) = E_1 \left[1 - e_0 \cos\left(\frac{\pi z}{h} \right) \right] \tag{5.236a}$$

$$G(z) = G_1 \left[1 - e_0 \cos\left(\frac{\pi z}{h} \right) \right] \tag{5.236b}$$

$$\rho(z) = \rho_1 \left[1 - e_{m0} \cos\left(\frac{\pi z}{h} \right) \right] \tag{5.236c}$$

$$\alpha(z) = \alpha_1 \left[1 - e_0 \cos\left(\frac{\pi z}{h} \right) \right] \tag{5.236d}$$

$$\beta(z) = \beta_1 \left[1 - e_0 \cos\left(\frac{\pi z}{h} \right) \right] \tag{5.236e}$$

$$\upsilon(z) = \upsilon_1 \tag{5.236f}$$

Mechanical characteristics of a type II beam with symmetric porosity distribution may be expressed as:

$$E(z) = E_1 \left[1 - e_1 \left(1 - \cos\left(\frac{\pi z}{h} \right) \right) \right] \tag{5.237a}$$

$$G(z) = G_1 \left[1 - e_1 \left(1 - \cos\left(\frac{\pi z}{h} \right) \right) \right] \tag{5.237b}$$

$$\rho(z) = \rho_1 \left[1 - e_{m1} \left(1 - \cos\left(\frac{\pi z}{h} \right) \right) \right] \tag{5.237c}$$

$$\alpha(z) = \alpha_1 \left[1 - e_1 \left(1 - \cos\left(\frac{\pi z}{h} \right) \right) \right]$$ (5.237d)

$$\beta(z) = \beta_1 \left[1 - e_1 \left(1 - \cos\left(\frac{\pi z}{h} \right) \right) \right]$$ (5.237e)

$$\upsilon(z) = \upsilon_1$$ (5.237f)

where E_1, G_1, ρ_1, α_1 β_1, and υ_1 are Young's modulus, shear modulus, mass density, thermal expansion coefficient, moisture expansion coefficient, and Poisson's ratio of the solid pure graphene, respectively. In addition, e_0, e_1, e_2 stand for porosity coefficient corresponding to symmetric porosity distribution I; symmetric porosity distribution II, and uniform porosity distribution, respectively. Also, e_{m0}, e_{m1}, e_{m2} mass density coefficient corresponding to symmetric porosity distribution I, symmetric porosity distribution II, and uniform porosity distribution, respectively. In addition, we may express the connection between the elastic modulus and the mass density as [63]:

$$\frac{E(z)}{E_1} = \left[\frac{\rho(z)}{\rho_1} \right]^{2.73}$$ (5.238)

Using Eq. (5.238), the following form may be used to express the relationship between the elastic modulus coefficient and the mass density coefficient over all possible porosity distributions:

$$e_{m2} = \left[e_2 \right]^{\frac{1}{2.73}}$$ (5.239a)

$$1 - e_{m0} \cos\left(\frac{\pi z}{h} \right) = \left[1 - e_0 \cos\left(\frac{\pi z}{h} \right) \right]^{\frac{1}{2.73}}$$ (5.239b)

$$1 - e_{m1} \left(1 - \cos\left(\frac{\pi z}{h} \right) \right) = \left[1 - e_1 \left(1 - \cos\left(\frac{\pi z}{h} \right) \right) \right]^{\frac{1}{2.73}}$$ (5.239c)

It is also possible to calculate e_1 and e_2, given e_0 using the following equation, which describes the relationship between the porosity coefficients of all porosity distributions.

$$\int_0^{\frac{h}{2}} \left[1 - e_0 \cos\left(\frac{\pi z}{h} \right) \right]^{\frac{1}{2.73}} dz = \int_0^{\frac{h}{2}} \left[e_2 \right]^{\frac{1}{2.73}} dz$$ (5.240a)

$$\int_0^{\frac{h}{2}} \left[1 - e_0 \cos\left(\frac{\pi z}{h}\right)\right]^{\frac{1}{2.73}} dz = \int_0^{\frac{h}{2}} \left[1 - e_1\left(1 - \cos\left(\frac{\pi z}{h}\right)\right)\right]^{\frac{1}{2.73}} dz \quad (5.240b)$$

Using revised higher-order shear deformation beam theory, the displacement field of beams may be formulated as follows to realize the kinetic relation of PGF beam.

$$u_x(x,z,t) = u(x,t) - z\frac{\partial w_b(x,t)}{\partial x} - f(z)\frac{\partial w_s(x,t)}{\partial x} \quad (5.241)$$

$$u_z(x,z,t) = w_b(x,t) + w_s(x,t) \quad (5.242)$$

where, u, w_b, and w_s represent longitudinal displacement and bending and shear deflections, respectively. Shear stress or strain along the thickness axis may be estimated using the form function $f(z)$. The provided model may be used without a shear correction component because of the exerted shape function. The form function for the current study may be expressed as:

$$f(z) = -\frac{z}{4} - \frac{5z^3}{3h^2} \quad (5.243)$$

Nonzero beam stresses include the following:

$$\varepsilon_{xx} = \frac{\partial u}{\partial x} - z\frac{\partial^2 w_b}{\partial x^2} - f(z)\frac{\partial^2 w_s}{\partial x^2}, \gamma_{xz} = g(z)\frac{\partial w_s}{\partial x} \quad (5.244)$$

where $g(z) = 1 - \frac{df}{dz}$. Within the context of Hamilton's principle, the equations of motion for a PGF beam may be expressed as:

$$\int_{t_0}^{t_1} \delta(U - T + V)dt = 0 \quad (5.245)$$

in which U, T, and V denote strain energy, kinetic energy, and work done by external forces. The variation of strain energy can be defined in the following form:

$$\delta U = \int_V \left(\sigma_{xx}\delta\varepsilon_{xx} + \sigma_{xz}\delta\gamma_{xz}\right)dV \quad (5.246)$$

The following equation, Eq. (5.246), may be rewritten by inserting Eqs. (5.241)–(5.244):

$$\delta U = \int_0^L \left(N \frac{\partial \delta u}{\partial x} - M_b \frac{\partial^2 \delta w_b}{\partial x^2} - M_s \frac{\partial^2 \delta w_s}{\partial x^2} + Q \frac{\partial \delta w_s}{\partial x} \right) dx \qquad (5.247)$$

in which the stress resultants N, M_b, M_s, and Q can be calculated as follows:

$$\begin{bmatrix} N & M_b & M_s \end{bmatrix} = \int_A \begin{bmatrix} 1 & z & f(z) \end{bmatrix} \sigma_{xx} dA, \quad Q = \int_A g(z) \sigma_{xz} dA \qquad (5.248)$$

As an example of the first variation in kinetic energy, consider the following:

$$\delta T = \int_V \left(\dot{u}_x \delta \dot{u}_x + \dot{u}_z \delta \dot{u}_z \right) \rho(z) dV$$

$$= \int_0^L \left\{ \begin{array}{l} I_0 \left(\dfrac{\partial u}{\partial t} \dfrac{\partial \delta u}{\partial t} + \dfrac{\partial (w_b + w_s)}{\partial t} \dfrac{\partial \delta (w_b + w_s)}{\partial t} \right) \\[2ex] - I_1 \left(\dfrac{\partial u}{\partial t} \dfrac{\partial^2 \delta w_b}{\partial x \partial t} + \dfrac{\partial^2 w_b}{\partial x \partial t} \dfrac{\partial \delta u}{\partial t} \right) \\[2ex] + I_2 \dfrac{\partial^2 w_b}{\partial x \partial t} \dfrac{\partial^2 \delta w_b}{\partial x \partial t} - J_1 \left(\dfrac{\partial u}{\partial t} \dfrac{\partial^2 \delta w_s}{\partial x \partial t} + \dfrac{\partial^2 w_s}{\partial x \partial t} \dfrac{\partial \delta u}{\partial t} \right) \\[2ex] + J_2 \left(\dfrac{\partial^2 w_b}{\partial x \partial t} \dfrac{\partial^2 \delta w_s}{\partial x \partial t} + \dfrac{\partial^2 w_s}{\partial x \partial t} \dfrac{\partial^2 \delta w_b}{\partial x \partial t} \right) + K_2 \dfrac{\partial^2 w_s}{\partial x \partial t} \dfrac{\partial^2 \delta w_s}{\partial x \partial t} \end{array} \right\} dx \qquad (5.249)$$

Differentiation with respect to time is in all equations, and the mass inertias needed for the aforementioned equation may be calculated as:

$$\begin{bmatrix} I_0 & I_1 & I_2 \\ J_1 & J_2 & K_2 \end{bmatrix} = \int_A \begin{bmatrix} 1 & z & z^2 \\ f(z) & zf(z) & f^2(z) \end{bmatrix} \rho(z) dA \qquad (5.250)$$

Afterward, the variation of the external work done by external forces can be expressed in the following form:

$$\delta V = \int_0^L \left[\left(k_p - N^T - N^H \right) \frac{\partial^2 \delta (w_b + w_s)}{\partial x^2} - k_w \delta (w_b + w_s) \right] dx \qquad (5.251)$$

in which k_p, k_w are Pasternak and Winkler coefficients, respectively. Moreover, N^T and N^H represent hygrothermal loading, which can be defined as [63]:

$$N^H = \int_{-\frac{h}{2}}^{\frac{h}{2}} \frac{E(z)}{1-\upsilon} \beta(z) \Delta C dz \tag{5.252}$$

$$N^T = \int_{-\frac{h}{2}}^{\frac{h}{2}} \frac{E(z)}{1-\upsilon} \alpha(z) \Delta T dz \tag{5.253}$$

here, ΔC is moisture concentration and ΔT is temperature change. In addition, the following equations may be used to calculate the uniform temperature change (UTC), linear temperature change (LTC), and sinusoidal temperature change (STC): [63]

$$\begin{cases} T = T_0 + \Delta T & \text{UTC} \\ T = T_0 + \Delta T \left(\dfrac{1}{2} + \dfrac{z}{h} \right) & \text{LTC} \\ T = T_0 + \Delta T \left(1 - \cos\left(\dfrac{\pi}{4} + \dfrac{\pi z}{2h} \right) \right) & \text{STC} \end{cases} \tag{5.254}$$

in which T_0 is considered to be room temperature ($T_0 = 300$ K).

By substituting Eqs. (5.247), (5.249), and (5.251) into Eq. (5.245) and setting the coefficients of δu, δw_b, and δw_s to zero, the Euler–Lagrange equations of PGF beams can be obtained as follows:

$$\frac{\partial N}{\partial x} = I_0 \frac{\partial^2 u}{\partial t^2} - I_1 \frac{\partial^3 w_b}{\partial x \partial t^2} - J_1 \frac{\partial^3 w_s}{\partial x \partial t^2} \tag{5.255}$$

$$\frac{\partial^2 M_b}{\partial x^2} = I_0 \frac{\partial^2 (w_b + w_s)}{\partial t^2} + I_1 \frac{\partial^3 u}{\partial x \partial t^2} - I_2 \frac{\partial^4 w_b}{\partial t^2 \partial x^2} - J_2 \frac{\partial^4 w_s}{\partial t^2 \partial x^2}$$
$$+ k_w (w_b + w_s) - (k_p - N^T - N^H) \frac{\partial^2 (w_b + w_s)}{\partial x^2} \tag{5.256}$$

$$\frac{\partial^2 M_s}{\partial x^2} + \frac{\partial Q}{\partial x} = I_0 \frac{\partial^2 (w_b + w_s)}{\partial t^2} + J_1 \frac{\partial^3 u}{\partial x \partial t^2} - J_2 \frac{\partial^4 w_b}{\partial t^2 \partial x^2}$$
$$- K_2 \frac{\partial^4 w_s}{\partial t^2 \partial x^2} + k_w (w_b + w_s) - (k_p - N^T - N^H) \frac{\partial^2 (w_b + w_s)}{\partial x^2} \tag{5.257}$$

In addition, the elastic stress-strain relations of PGF beam are investigated with the intention of establishing the basic elastic equations of such materials. In this case, the constitutive equation may be expressed as follows:

$$\sigma_{ij} = C_{ijkl}\varepsilon_{kl} \tag{5.258}$$

where σ_{ij}, ε_{kl}, and C_{ijkl} represent stress, strain, and elastic stiffness tensors, respectively. Therefore, these relations can be modified for beams as:

$$\sigma_{xx} = E_{11}\varepsilon_{xx} \tag{5.259}$$

$$\sigma_{xz} = G_{12}\gamma_{xz} \tag{5.260}$$

The following relations are obtained by integrating Eqs. (5.259) and (5.260) across the beam's cross-section area:

$$N = A\frac{\partial u}{\partial x} - B\frac{\partial^2 w_b}{\partial x^2} - B_s\frac{\partial^2 w_s}{\partial x^2} \tag{5.261}$$

$$M_b = B\frac{\partial u}{\partial x} - D\frac{\partial^2 w_b}{\partial x^2} - D_s\frac{\partial^2 w_s}{\partial x^2} \tag{5.262}$$

$$M_s = B_s\frac{\partial u}{\partial x} - D_s\frac{\partial^2 w_b}{\partial x^2} - H_s\frac{\partial^2 w_s}{\partial x^2} \tag{5.263}$$

$$Q = A_s\frac{\partial w_s}{\partial x} \tag{5.264}$$

in which

$$\begin{bmatrix} A & B & B_s \\ D & D_s & H_s \end{bmatrix} = \int_A \begin{bmatrix} 1 & z & f(z) \\ z^2 & zf(z) & f^2(z) \end{bmatrix} E_{11}dA \tag{5.265}$$

$$A_s = \int_A g^2(z)G_{12}dA \tag{5.266}$$

By substituting Eqs. (5.261)–(5.264) in Eqs. (5.255)–(5.257), the governing equations of PGF beam can be obtained as:

$$A\frac{\partial^2 u}{\partial x^2} - B\frac{\partial^3 w_b}{\partial x^3} - B_s\frac{\partial^3 w_s}{\partial x^3} - I_0\frac{\partial^2 u}{\partial t^2} + I_1\frac{\partial^3 w_b}{\partial x\partial t^2} + J_1\frac{\partial^3 w_s}{\partial x\partial t^2} = 0 \tag{5.267}$$

$$B\frac{\partial^3 u}{\partial x^3} - D\frac{\partial^4 w_b}{\partial x^4} - D_s\frac{\partial^4 w_s}{\partial x^4} - I_0\frac{\partial^2\left(w_b+w_s\right)}{\partial t^2} - I_1\frac{\partial^3 u}{\partial x\partial t^2}$$

$$+I_2\frac{\partial^4 w_b}{\partial t^2\partial x^2} + J_2\frac{\partial^4 w_s}{\partial t^2\partial x^2} + k_w\left(w_b+w_s\right) - \left(k_p - N^T - N^H\right)\frac{\partial^2\left(w_b+w_s\right)}{\partial x^2} = 0 \tag{5.268}$$

$$B_s\frac{\partial^3 u}{\partial x^3} - D_s\frac{\partial^4 w_b}{\partial x^4} - H_s\frac{\partial^4 w_s}{\partial x^4} + A_s\frac{\partial^2 w_s}{\partial x^2} - I_0\frac{\partial^2\left(w_b+w_s\right)}{\partial t^2} - J_1\frac{\partial^3 u}{\partial x\partial t^2}$$

$$+J_2\frac{\partial^4 w_b}{\partial t^2\partial x^2} + K_2\frac{\partial^4 w_s}{\partial t^2\partial x^2} + k_w\left(w_b+w_s\right) - \left(k_p - N^T - N^H\right)\frac{\partial^2\left(w_b+w_s\right)}{\partial x^2} = 0 \tag{5.269}$$

The PGF beam's governing equations are solved analytically. It is also expected that the constituents of the displacement field are [63]:

$$\begin{Bmatrix} u \\ w_b \\ w_s \end{Bmatrix} = \begin{Bmatrix} U \\ W_b \\ W_s \end{Bmatrix} e^{i\left(\beta x - \omega t\right)} \tag{5.270}$$

in which U, W_b, and W_s denote wave amplitudes, respectively. β and ω are wave number and circular frequency dispersed waves, respectively. Replacing for u, w_b, and w_s from Eq. (5.269) in Eqs. (5.266)–(5.268), the following equation is attained:

$$\left(\left[K - \omega^2 M\right]_{3\times 3}\right)\begin{bmatrix} U \\ W_b \\ W_s \end{bmatrix} = 0 \tag{5.271}$$

where K and M denote stiffness and mass matrices, respectively, and components of these matrices are presented in Appendix 5H. The eigenvalue issue stated above is resolved for ω, allowing for the determination of the wave frequency. Zero must be entered for the following determinant.

$$\left|K - \omega^2 M\right|_{3\times 3} = 0 \tag{5.272}$$

In addition, if you divide the wave frequency by the wave number, you'll get the phase velocity. Table 5.4 lists the mechanical parameters of the graphene foam used as the beam's constitutive material [63], and Table 5.5 lists the foam coefficients for different porosity distributions [63]. The impacts of these variables on the wave frequency and phase velocity change of PGF beam are also discussed [63]. In addition, $h = 5$ cm is used as the estimated PGF beam thickness in all

TABLE 5.4
Mechanical Properties of Graphene Foam

E (Gpa)	ρ (Kg/m³)	ν	α (1/K)	β
1020	2300	0.19	1.6×10^{-6}	2.6×10^{-3}

TABLE 5.5
Coefficients of Foam for Various Porosity Distribution

e_0	e_1	e_2
0.05	0.0872	0.9681
0.1	0.1734	0.9360
0.15	0.2586	0.9038
0.2	0.3426	0.8713
0.25	0.4253	0.8387
0.3	0.5065	0.8058
0.35	0.5861	0.7726
0.4	0.6637	0.7391
0.45	0.7389	0.7053
0.5	0.8112	0.6711

illustrations. Here, for the purpose of brevity, we'll present Winkler–Pasternak foundation parameters [63]:

$$K_W = \frac{k_W L^4}{X^*}, \quad K_P = \frac{k_P L^2}{X^*}, \quad X^* = \frac{E_1 h^3}{12\left(1-v^2\right)}$$

It can be shown in Figure 5.90 that the phase velocity of PGF measured against the porosity coefficient changes depending on the porosity distribution. It is clear that the porosity coefficient has a declining impact on the value of the phase velocity of PGF, and it is also clear that the curve of symmetric porosity distribution II is the most tolerant of this reducing effect. This is because the curve of symmetric porosity distribution II has the highest porosity overall. In other words, the symmetric porosity distribution II curve flattens down more steeply as the porosity coefficient increases than do the other curves. When compared to the values of phase velocity for the other porosity distributions, the value of phase velocity for the symmetric porosity distribution I is lowered by the least amount. This pattern, in which the phase velocity value lowers with increasing quantities of porosity coefficient, may be explained by the fact that porosity, as a defect, diminishes the structure's stiffness [63]. Specifically, the value of the phase velocity falls as the porosity coefficient increases. Figure 5.91 depicts the relationship between the wave number and the variation in the phase velocity of a PGF beam for a range of

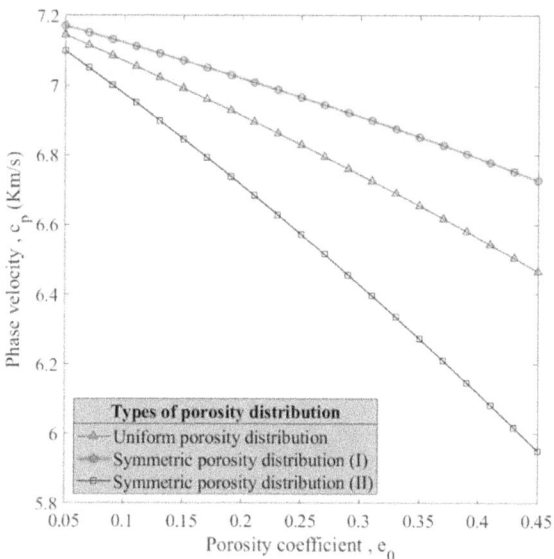

FIGURE 5.90   Variation of phase velocity of PGF beam versus porosity coefficient for different types of porosity distribution ($\Delta T = 60$ (UTC), $\Delta C = 0.05$, $\dfrac{L}{h} = 20$, $\beta = 30$, $K_W = 30$, $K_p = 10$).

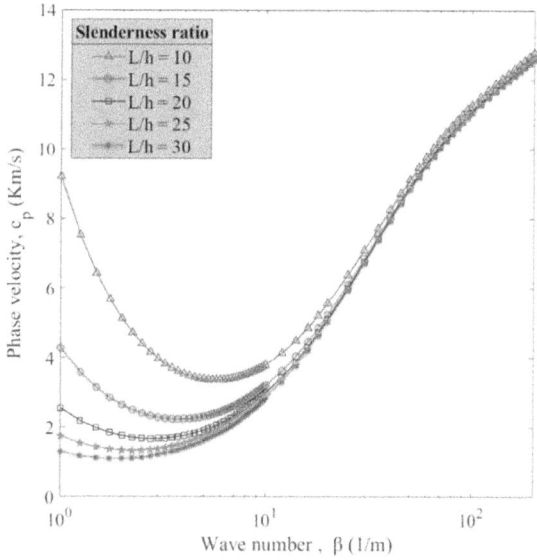

FIGURE 5.91   Variation of phase velocity of PGF beam versus wave number for different slenderness ratios ($e_0 = 0.2$ (sym 2), $\Delta T = 50$ (STC), $\Delta C = 0.05$, $K_P = 20$, $K_W = 60$).

slenderness ratios. This figure shows that a greater value of phase velocity may be achieved by selecting a smaller slenderness ratio. On the other side, when the aspect ratio thins, phase velocity loses some of its significance. In addition, the lower the slenderness ratio, the more the curves tend to decrease from 1 to 10, after which they tend to rise. Phase velocity variation of PGF beam as a function of wave number is shown for (a) uniform porosity distribution, (b) symmetric porosity distribution I, and (c) symmetric porosity distribution II in Figure 5.92. The arithmetic mean of the three is the same as the mean of the other two, and the mean of the three is the mean of the other two. In addition, it has been shown that

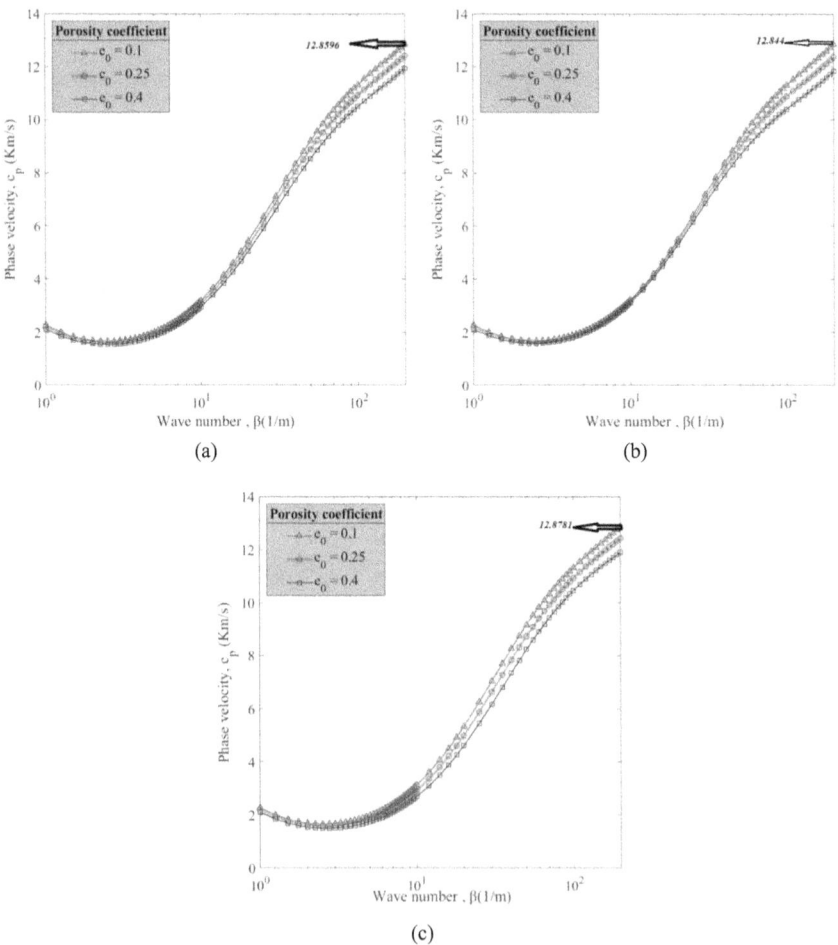

(a)

(b)

(c)

FIGURE 5.92 Variation of phase velocity of PGF beam versus wave number for various porosity coefficients for (a) uniform porosity distribution (b) symmetric porosity distribution I, and (c) symmetric porosity distribution II ($\Delta T = 50$ (LTC), $\dfrac{L}{h} = 20$, $\Delta C = 0.03$, $K_W = 40$, $K_p = 20$).

a smaller value for the porosity coefficient results in a higher value for the phase velocity. The value of the phase velocity at $\beta = 200$ and $e_0 = 0.1$ is provided in each picture so that the impact of varying porosity distribution can be seen. The value is lowest for symmetric porosity distribution I and greatest for symmetric porosity distribution II [63].

Take a look at Figure 5.93, which depicts the Pasternak coefficient against the phase velocity variation of a PGF beam as a function of temperature. Figure 5.93 demonstrates the growing significance of the Pasternak coefficient in an elastic medium's shear layer. When a larger value is selected for the Pasternak coefficient, the value of phase velocity increases. In addition, a magnification is utilized to see the effects of various temperature changes more clearly. Magnifying curves make it easy to deduce that (a) the effect of temperature change or thermal loading on the variation of phase velocity of PGF beam is smaller as the temperature increases and (b) the effect of STC, LTC, and UTC is smaller as the temperature increases [63].

The effects of the Winkler coefficient and the porosity coefficients on the frequency shift of the PGF beam are illustrated in Figure 5.94. Winkler coefficient,

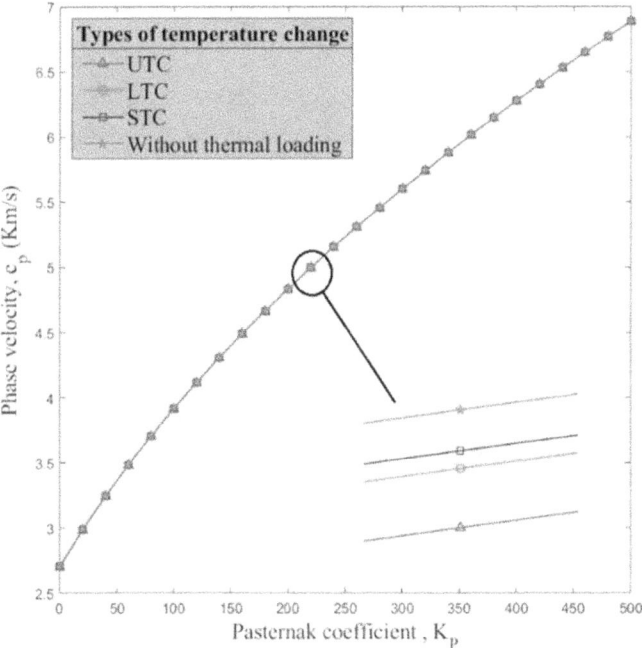

FIGURE 5.93 Illustration of variation of phase velocity of PGF beam versus Pasternak coefficient for different types of temperature change ($e_0 = 0.2$ (sym2), $\Delta T = 50$, $\Delta C = 0.05$, $\dfrac{L}{h} = 20, \beta = 10, K_W = 60$).

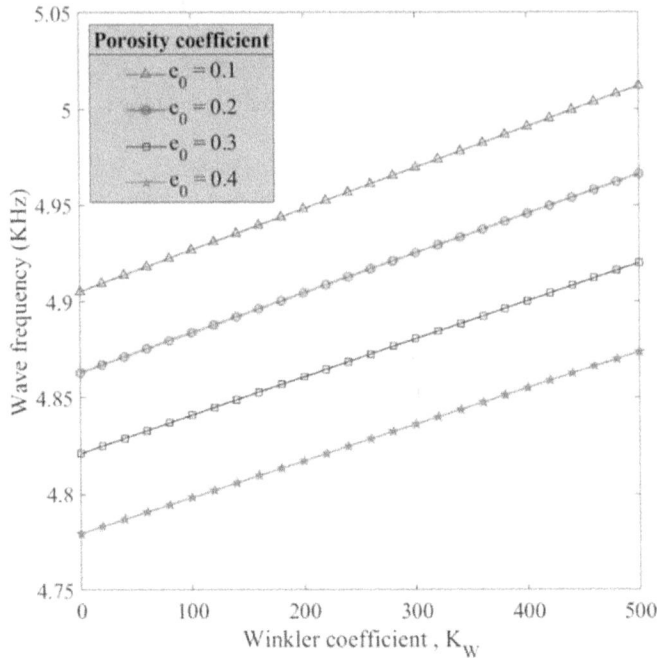

FIGURE 5.94   Variation of wave frequency of PGF beam versus Winkler coefficient for various porosity coefficients ($\Delta T = 20$ (STC), $\Delta C = 0.05, \dfrac{L}{h} = 20, \beta = 10, K_P = 10$).

as linear layer of elastic material, performs growing role, same as Pasternak coefficient. The fact that the structure is more rigid and can withstand greater natural frequencies is due to the presence of an elastic medium. Furthermore, porosity is a flaw and has a negligible effect on the frequency of the PGF beam's waves, as was mentioned before [63].

In Figure 5.95, we see how temperature shifts affect the relationship between the PGF beam's wave frequency and its moisture content. The value of the wave frequency decreases as the quantity of moisture concentration increases because, as with a change in temperature, the stiffness of the structure decreases with an increase in moisture concentration. In addition, the value of the frequency of waves is reduced by a growing amount of temperature change under conditions of constant moisture concentration [63].

The influence of the porosity coefficient on the frequency shift of a PGF beam is shown in Figure 5.96 for three different types of porosity distributions: (a) a uniform porosity distribution; (b) a symmetric porosity distribution (I); and (c) an asymmetric porosity distribution (II). It is abundantly evident that as the number of waves increases, the value of the frequency of a wave also increases. On the other hand, there is an increasing correlation between the value of wave frequency and the wave number. In addition to this, the frequency of the waves will decrease

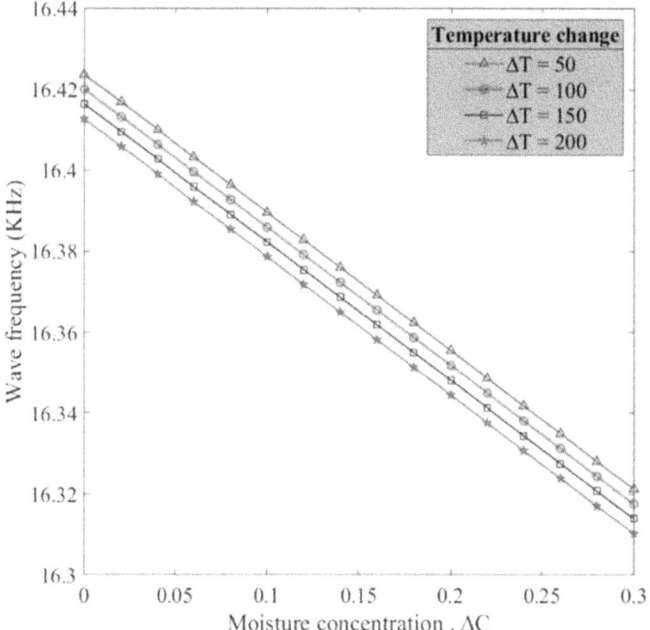

FIGURE 5.95 Variation of wave frequency of PGF beam versus moisture concentration for various amounts of temperature change ($e_0 = 0.2\,(sym2)$, $\dfrac{L}{h} = 20$, $\beta = 20$, $K_w = 50$, $K_p = 25$).

as the porosity coefficient rises. The wave frequency is found to be highest for the second symmetric porosity distribution, followed by the uniform porosity distribution, and finally the first symmetric porosity distribution [63]. This was discovered by comparing three different figures.

5.5.3 GRAPHENE FOAM PLATE WAVE DISPERSION CHARACTERISTICS

Uniform porosity distribution (UPD), symmetric porosity distribution I (SPD-1), and symmetric porosity distribution II (SPD-2) are all taken into consideration in this section for porosity distribution along the plate thickness direction. To put it simply, the more you know about the subject, the better off you'll be. To begin, the SPD-1 plate's effective mechanical characteristics may be calculated as [64]:

$$P(z) = P_1 \left[1 - e_0 \cos\left(\frac{\pi z}{h} \right) \right] \tag{5.273}$$

where $P(z)$ stands for mechanical properties including elastic modulus (E), mass density (ρ), Poisson ratio (ν), thermal expansion coefficient (α), and moisture expansion coefficient (β^H). Also, P_1 is material properties of the solid pristine graphene.

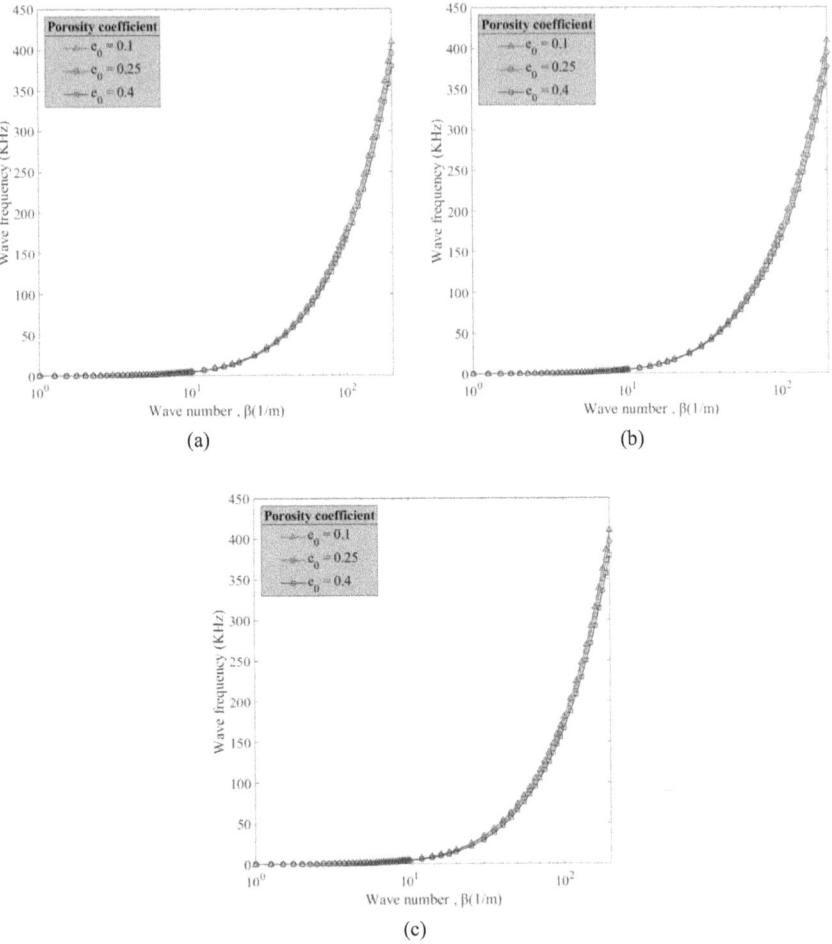

FIGURE 5.96 Variation of wave frequency of PGF beam versus wave number for different porosity coefficients for (a) uniform porosity distribution (b) symmetric porosity distribution I, and (c) symmetric porosity distribution II.

Furthermore, the following equations may be used to determine the SPD-2 plate's effective mechanical properties:

$$P(z) = P_1\left[1 - e_1\left(1 - \cos\left(\frac{\pi z}{h}\right)\right)\right] \tag{5.274}$$

The following equations may be used to calculate the effective mechanical characteristics of a UPD plate:

$$P(z) = P_1 e_2 \tag{5.275}$$

In above relations, e_0, e_1, and e_2 are porosity coefficient corresponding to SPD-1, SPD-2, and UPD, respectively. Also, e_{m0}, e_{m1}, e_{m2} mass density coefficients stand for SPD-1, SPD-2, and UPD, respectively. Additionally, the elastic modulus is related to mass density in the following way [64]:

$$\frac{E(z)}{E_1} = \left[\frac{\rho(z)}{\rho_1}\right]^{2.73} \tag{5.276}$$

For all possible porosity distributions, the relationship between the elastic modulus coefficient and the mass density coefficient may be written as:

$$e_{m2} = \left[e_2\right]^{\frac{1}{2.73}} \tag{5.277}$$

$$1 - e_{m0}\cos\left(\frac{\pi z}{h}\right) = \left[1 - e_0\cos\left(\frac{\pi z}{h}\right)\right]^{\frac{1}{2.73}} \tag{5.278}$$

$$1 - e_{m1}\left(1 - \cos\left(\frac{\pi z}{h}\right)\right) = \left[1 - e_1\left(1 - \cos\left(\frac{\pi z}{h}\right)\right)\right]^{\frac{1}{2.73}} \tag{5.279}$$

In addition, given an e_0, one can easily determine the values of e_1 and e_2 by solving the following equations [64]. This holds true for all porosity distributions' porosity coefficients

$$\int_0^{\frac{h}{2}}\left[1 - e_0\cos\left(\frac{\pi z}{h}\right)\right]^{\frac{1}{2.73}}dz = \int_0^{\frac{h}{2}}\left[e_2\right]^{\frac{1}{2.73}}dz \tag{5.280}$$

$$\int_0^{\frac{h}{2}}\left[1 - e_0\cos\left(\frac{\pi z}{h}\right)\right]^{\frac{1}{2.73}}dz = \int_0^{\frac{h}{2}}\left[1 - e_1\left(1 - \cos\left(\frac{\pi z}{h}\right)\right)\right]^{\frac{1}{2.73}}dz \tag{5.281}$$

Table 5.5 [64] lists the foam coefficients for a number of different porosity distributions. The GF plate kinetic and kinematic relations are attained by using a revised high-speed dynamic testing (HSDT) of plate with a new form function. In reference [64], we get an expression for the GF plate displacement field that works as follows:

$$u_x(x,y,z,t) = u(x,y,t) - z\frac{\partial w_b(x,y,t)}{\partial x} - f(z)\frac{\partial w_s(x,y,t)}{\partial x} \tag{5.282}$$

$$u_y(x,y,z,t) = u(x,y,t) - z\frac{\partial w_b(x,y,t)}{\partial y} - f(z)\frac{\partial w_s(x,y,t)}{\partial y} \tag{5.283}$$

$$u_z(x,y,z,t) = w_b(x,y,t) + w_s(x,y,t) \tag{5.284}$$

where u, w_b, and w_s each stand for the longitudinal displacement, bending, and shear deflections that are a function of in-plane coordinates and time, respectively. In addition, the shear stress or strain distribution along the thickness axis may be estimated using the shape function denoted by $f(z)$. The form function used allows the introduced model to be applied without the requirement for a shear correction factor. The form function for this study may be expressed as [64]:

$$f(z) = \frac{zh^2}{5h^2 + 4z^2} - \frac{4z^3}{27h^2} \tag{5.285}$$

The plate's nonzero stresses are as follows:

$$\begin{Bmatrix} \varepsilon_{xx} \\ \varepsilon_{yy} \\ \gamma_{xy} \end{Bmatrix} = \begin{Bmatrix} \varepsilon_{xx}^0 \\ \varepsilon_{yy}^0 \\ \gamma_{xy}^0 \end{Bmatrix} + z \begin{Bmatrix} \kappa_{xx}^b \\ \kappa_{yy}^b \\ \kappa_{xy}^b \end{Bmatrix} + f(z) \begin{Bmatrix} \kappa_{xx}^s \\ \kappa_{yy}^s \\ \kappa_{xy}^s \end{Bmatrix}, \quad \begin{Bmatrix} \gamma_{xz} \\ \gamma_{yz} \end{Bmatrix} = g(z) \begin{Bmatrix} \gamma_{xz}^0 \\ \gamma_{yz}^0 \end{Bmatrix} \tag{5.286}$$

$$\begin{Bmatrix} \varepsilon_{xx}^0 \\ \varepsilon_{yy}^0 \\ \gamma_{xy}^0 \end{Bmatrix} = \begin{Bmatrix} \dfrac{\partial u}{\partial x} \\[2mm] \dfrac{\partial v}{\partial y} \\[2mm] \dfrac{\partial u}{\partial y} + \dfrac{\partial v}{\partial x} \end{Bmatrix}, \quad \begin{Bmatrix} \kappa_{xx}^b \\ \kappa_{yy}^b \\ \kappa_{xy}^b \end{Bmatrix} = \begin{Bmatrix} -\dfrac{\partial^2 w_b}{\partial x^2} \\[2mm] -\dfrac{\partial^2 w_b}{\partial y^2} \\[2mm] -2\dfrac{\partial^2 w_b}{\partial x \partial y} \end{Bmatrix},$$

$$\begin{Bmatrix} \kappa_{xx}^s \\ \kappa_{yy}^s \\ \kappa_{xy}^s \end{Bmatrix} = \begin{Bmatrix} -\dfrac{\partial^2 w_s}{\partial x^2} \\[2mm] -\dfrac{\partial^2 w_s}{\partial y^2} \\[2mm] -2\dfrac{\partial^2 w_s}{\partial x \partial y} \end{Bmatrix}, \quad \begin{Bmatrix} \gamma_{xz}^0 \\ \gamma_{yz}^0 \end{Bmatrix} = \begin{Bmatrix} \dfrac{\partial w_s}{\partial x} \\[2mm] \dfrac{\partial w_s}{\partial y} \end{Bmatrix} \tag{5.287}$$

in which $g(z) = 1 - \dfrac{df(z)}{dz}$. The motion equations of GF plate will be developed on the basis of Hamilton's principle, which can be expressed as:

$$\int_{t_0}^{t_1} \delta(U - T + W)\,dt = 0 \tag{5.288}$$

in which U, T, and W are strain energy, kinetic energy, and work done by external forces. Here is a definition of strain energy fluctuation [64]:

$$\delta U = \int_V \sigma_{ij}\delta\varepsilon_{ij}dV = \int_V \left(\sigma_{xx}\delta\varepsilon_{xx} + \sigma_{yy}\delta\varepsilon_{yy} + \sigma_{xy}\delta\gamma_{xy} + \sigma_{yz}\delta\gamma_{yz} + \sigma_{xz}\delta\gamma_{xz}\right)dV \quad (5.289)$$

The following equation is obtained by putting Eqs. (5.297) through (5.302) into Eq. (5.304):

$$\delta U = \int_0^a \left(\begin{array}{l} N_{xx}\dfrac{\partial\delta u}{\partial x} - M_{xx}^b\dfrac{\partial^2\delta w_b}{\partial x^2} - M_{xx}^s\dfrac{\partial^2\delta w_s}{\partial x^2} + N_{yy}\dfrac{\partial\delta v}{\partial y} \\[2mm] -M_{yy}^b\dfrac{\partial^2\delta w_b}{\partial y^2} - M_{yy}^s\dfrac{\partial^2\delta w_s}{\partial y^2} + N_{xy}\left(\dfrac{\partial\delta u}{\partial y} + \dfrac{\partial\delta v}{\partial x}\right) - 2M_{xy}^b\dfrac{\partial^2\delta w_b}{\partial x\partial y} \\[2mm] -2M_{xy}^s\dfrac{\partial^2\delta w_s}{\partial x\partial y} + Q_{xz}\dfrac{\partial\delta w_s}{\partial x} + Q_{yz}\dfrac{\partial\delta w_s}{\partial y} \end{array}\right)dx \quad (5.290)$$

where the values of the variables in the preceding equation may be found by solving the following:

$$\left(N_{ij} \quad M_{ij}^b \quad M_{ij}^s\right) = \int_A \sigma_{ij}\left(1 \quad z \quad f(z)\right)dA, \qquad ij = \left(xx \quad yy \quad xy\right) \quad (5.291)$$

$$Q_{kz} = \int_A \sigma_{kz}g(z)dA, \qquad k = \left(x \quad y\right) \quad (5.292)$$

Here are several ways to depict the variation in kinetic energy:

$$\delta T = \int_0^a\int_0^b \left(\begin{array}{l} I_0\left(\dfrac{\partial u}{\partial t}\dfrac{\partial\delta u}{\partial t} + \dfrac{\partial v}{\partial t}\dfrac{\partial\delta v}{\partial t} + \dfrac{\partial\left(w_b+w_s\right)}{\partial t}\dfrac{\partial\delta\left(w_b+w_s\right)}{\partial t}\right) \\[3mm] -I_1\left(\dfrac{\partial u}{\partial t}\dfrac{\partial\delta w_b}{\partial x\partial t} + \dfrac{\partial w_b}{\partial x\partial t}\dfrac{\partial\delta u}{\partial t} + \dfrac{\partial v}{\partial t}\dfrac{\partial\delta w_b}{\partial y\partial t} + \dfrac{\partial w_b}{\partial y\partial t}\dfrac{\partial\delta v}{\partial t}\right) \\[3mm] +I_2\left(\dfrac{\partial w_b}{\partial x\partial t}\dfrac{\partial\delta w_b}{\partial x\partial t} + \dfrac{\partial w_b}{\partial y\partial t}\dfrac{\partial\delta w_b}{\partial y\partial t}\right) + K_2\left(\dfrac{\partial w_s}{\partial x\partial t}\dfrac{\partial\delta w_s}{\partial x\partial t} + \dfrac{\partial w_s}{\partial y\partial t}\dfrac{\partial\delta w_s}{\partial y\partial t}\right) \\[3mm] -J_1\left(\dfrac{\partial u}{\partial t}\dfrac{\partial\delta w_s}{\partial x\partial t} + \dfrac{\partial w_b}{\partial x\partial t}\dfrac{\partial\delta u}{\partial t} + \dfrac{\partial v}{\partial t}\dfrac{\partial\delta w_s}{\partial y\partial t} + \dfrac{\partial w_s}{\partial y\partial t}\dfrac{\partial\delta v}{\partial t}\right) \\[3mm] J_2\left(\dfrac{\partial w_b}{\partial x\partial t}\dfrac{\partial\delta w_s}{\partial x\partial t} + \dfrac{\partial w_s}{\partial x\partial t}\dfrac{\partial\delta w_b}{\partial x\partial t} + \dfrac{\partial w_b}{\partial y\partial t}\dfrac{\partial\delta w_s}{\partial y\partial t} + \dfrac{\partial w_s}{\partial y\partial t}\dfrac{\partial\delta w_b}{\partial y\partial t}\right) \end{array}\right)dxdy$$

$$(5.293)$$

The mass inertias needed for the aforementioned equation may be calculated as where t is the time variable.

$$\begin{pmatrix} I_0 & I_1 & I_2 \\ J_1 & J_2 & K_2 \end{pmatrix} = \int_A \rho(z) \begin{pmatrix} 1 & z & z^2 \\ f(z) & zf(z) & f^2(z) \end{pmatrix} dA \qquad (5.294)$$

The subsequent form for expressing the varying effort done by external forces is as follows:

$$\delta W = \int_0^a \int_0^b \left(\begin{array}{l} \left(N_{xx}^T + N_{xx}^H\right) \dfrac{\partial \delta\left(w_b + w_s\right)}{\partial x} \dfrac{\partial\left(w_b + w_s\right)}{\partial x} \\[3mm] +\left(N_{yy}^T + N_{yy}^H\right) \dfrac{\partial \delta\left(w_b + w_s\right)}{\partial y} \dfrac{\partial\left(w_b + w_s\right)}{\partial y} \\[3mm] +2\delta N_{xy}^0 \dfrac{\partial\left(w_b + w_s\right)}{\partial x} \dfrac{\partial\left(w_b + w_s\right)}{\partial y} \end{array} \right) dxdy \qquad (5.295)$$

in which N^T and N^H represent hygrothermal loading, which can be calculated as follows:

$$N^H = \int_{-\frac{h}{2}}^{\frac{h}{2}} \frac{E(z)}{1-\upsilon(z)} \beta^H(z)\Delta C dz \qquad (5.296)$$

$$N^T = \int_{-\frac{h}{2}}^{\frac{h}{2}} \frac{E(z)}{1-\upsilon(z)} \alpha(z)\Delta T dz \qquad (5.297)$$

In this case, ΔC is the relative humidity and ΔT is the difference in temperature. Furthermore, different patterns for increases in temperature and humidity are taken into account. Uniform temperature and humidity increase (UTR and UMR), linear temperature and humidity rise (LTR and LMR), and sinusoidal temperature and humidity rise (STR and SMR) may be expressed as follows [64]:

$$\begin{cases} T = T_0 + \Delta T & \text{UTR} \\[3mm] T = T_0 + \Delta T\left(\dfrac{1}{2} + \dfrac{z}{h}\right) & \text{LTR} \\[3mm] T = T_0 + \Delta T\left(1 - \cos\left(\dfrac{\pi}{4} + \dfrac{\pi z}{2h}\right)\right) & \text{STR} \end{cases} \qquad (5.298)$$

$$\begin{cases} C = C_0 + \Delta C & \text{UMR} \\ C = C_0 + \Delta C\left(\dfrac{1}{2} + \dfrac{z}{h}\right) & \text{LMR} \\ C = C_0 + \Delta C\left(1 - \cos\left(\dfrac{\pi}{4} + \dfrac{\pi z}{2h}\right)\right) & \text{SMR} \end{cases} \qquad (5.299)$$

in which T_0 is considered to be room temperature ($T_0 = 300°\text{K}$). By inserting Eqs. (5.290), (5.293), and (5.295) into Eq. (5.288) and setting the coefficients of δu, δw_b, and δw_s to zero, the Euler–Lagrange equations of GF plate can be obtained as follows:

$$\frac{\partial N_{xx}}{\partial x} + \frac{\partial N_{xy}}{\partial y} = I_0\frac{\partial^2 u}{\partial t^2} - I_1\frac{\partial^3 w_b}{\partial x \partial t^2} - J_1\frac{\partial^3 w_s}{\partial x \partial t^2} \qquad (5.300)$$

$$\frac{\partial N_{xy}}{\partial x} + \frac{\partial N_{yy}}{\partial y} = I_0\frac{\partial^2 v}{\partial t^2} - I_1\frac{\partial^3 w_b}{\partial y \partial t^2} - J_1\frac{\partial^3 w_s}{\partial y \partial t^2} \qquad (5.301)$$

$$\begin{aligned}
&\frac{\partial^2 M_{xx}^b}{\partial x^2} + 2\frac{\partial^2 M_{xy}^b}{\partial x \partial y} + \frac{\partial^2 M_{yy}^b}{\partial y^2} - \left(N^T + N^H\right)\nabla^2\left(w_b + w_s\right) \\
&= I_0\frac{\partial^2\left(w_b + w_s\right)}{\partial t^2} + I_1\left(\frac{\partial^3 u}{\partial x \partial t^2} + \frac{\partial^3 v}{\partial y \partial t^2}\right) \\
&-I_2\nabla^2\left(\frac{\partial^2 w_b}{\partial t^2}\right) - J_2\nabla^2\left(\frac{\partial^2 w_s}{\partial t^2}\right)
\end{aligned} \qquad (5.302)$$

$$\begin{aligned}
&\frac{\partial^2 M_{xx}^s}{\partial x^2} + 2\frac{\partial^2 M_{xy}^s}{\partial x \partial y} + \frac{\partial^2 M_{yy}^s}{\partial y^2} + \frac{\partial Q_{xz}}{\partial x} + \frac{\partial Q_{yz}}{\partial y} \\
&-\left(N^T + N^H\right)\nabla^2\left(w_b + w_s\right) = I_0\frac{\partial^2\left(w_b + w_s\right)}{\partial t^2} \\
&+J_1\left(\frac{\partial^3 u}{\partial x \partial t^2} + \frac{\partial^3 v}{\partial y \partial t^2}\right) - J_2\nabla^2\left(\frac{\partial^2 w_b}{\partial t^2}\right) - K_2\nabla^2\left(\frac{\partial^2 w_s}{\partial t^2}\right)
\end{aligned} \qquad (5.303)$$

Laplacian operator is represented by ∇^2. The elastic stress-strain relationship of GF plates is also analyzed in order to get the basic elastic equations applicable to such materials. The following equation is the constitutive one in this case:

$$\sigma_{ij} = C_{ijkl}\varepsilon_{kl} \qquad (5.304)$$

where σ_{ij}, ε_{kl}, and C_{ijkl} stand for stress, strain, and elastic stiffness tensors, respectively. Therefore, these relations can be modified for plates in the following form:

$$
\begin{bmatrix} \sigma_{xx} \\ \sigma_{yy} \\ \sigma_{yz} \\ \sigma_{xz} \\ \sigma_{xy} \end{bmatrix} = \begin{bmatrix} Q_{11} & Q_{12} & 0 & 0 & 0 \\ Q_{12} & Q_{22} & 0 & 0 & 0 \\ 0 & 0 & Q_{44} & 0 & 0 \\ 0 & 0 & 0 & Q_{55} & 0 \\ 0 & 0 & 0 & 0 & Q_{66} \end{bmatrix} \begin{bmatrix} \varepsilon_{xx} \\ \varepsilon_{yy} \\ \varepsilon_{yz} \\ \varepsilon_{xz} \\ \varepsilon_{xy} \end{bmatrix} \qquad (5.305)
$$

$$
Q_{11} = Q_{22} = \frac{E(z)}{1 - \upsilon^2(z)}, \; Q_{12} = \frac{E(z)\upsilon(z)}{1 - \upsilon^2(z)}, \; Q_{44} = Q_{55} = Q_{66} = G(z) \quad (5.306)
$$

The relations between the cross-sectional rigidities and the corresponding equations are as follows [64]:

$$
\begin{bmatrix} N_{xx} \\ N_{yy} \\ N_{xy} \end{bmatrix} = \begin{bmatrix} A_{11} & A_{12} & 0 \\ A_{21} & A_{22} & 0 \\ 0 & 0 & A_{66} \end{bmatrix} \begin{bmatrix} \dfrac{\partial u}{\partial x} \\ \dfrac{\partial v}{\partial y} \\ \dfrac{\partial u}{\partial y} + \dfrac{\partial v}{\partial x} \end{bmatrix} - \begin{bmatrix} B_{11} & B_{12} & 0 \\ B_{21} & B_{22} & 0 \\ 0 & 0 & B_{66} \end{bmatrix} \begin{bmatrix} \dfrac{\partial^2 w_b}{\partial x^2} \\ \dfrac{\partial^2 w_b}{\partial y^2} \\ 2\dfrac{\partial^2 w_b}{\partial x \partial y} \end{bmatrix}
$$

$$
- \begin{bmatrix} B_{11}^s & B_{12}^s & 0 \\ B_{21}^s & B_{22}^s & 0 \\ 0 & 0 & B_{66}^s \end{bmatrix} \begin{bmatrix} \dfrac{\partial^2 w_s}{\partial x^2} \\ \dfrac{\partial^2 w_s}{\partial y^2} \\ 2\dfrac{\partial^2 w_s}{\partial x \partial y} \end{bmatrix} \qquad (5.307)
$$

$$
\begin{bmatrix} M_{xx}^b \\ M_{yy}^b \\ M_{xy}^b \end{bmatrix} = \begin{bmatrix} B_{11} & B_{12} & 0 \\ B_{21} & B_{22} & 0 \\ 0 & 0 & B_{66} \end{bmatrix} \begin{bmatrix} \dfrac{\partial u}{\partial x} \\ \dfrac{\partial v}{\partial y} \\ \dfrac{\partial u}{\partial y} + \dfrac{\partial v}{\partial x} \end{bmatrix} - \begin{bmatrix} D_{11} & D_{12} & 0 \\ D_{21} & D_{22} & 0 \\ 0 & 0 & D_{66} \end{bmatrix} \begin{bmatrix} \dfrac{\partial^2 w_b}{\partial x^2} \\ \dfrac{\partial^2 w_b}{\partial y^2} \\ 2\dfrac{\partial^2 w_b}{\partial x \partial y} \end{bmatrix}
$$

$$
\begin{bmatrix} D_{11}^s & D_{12}^s & 0 \\ D_{21}^s & D_{22}^s & 0 \\ 0 & 0 & D_{66}^s \end{bmatrix} \begin{bmatrix} \dfrac{\partial^2 w_s}{\partial x^2} \\ \dfrac{\partial^2 w_s}{\partial y^2} \\ 2\dfrac{\partial^2 w_s}{\partial x \partial y} \end{bmatrix} \qquad (5.308)
$$

$$\begin{bmatrix} M_{xx}^s \\ M_{yy}^s \\ M_{xy}^s \end{bmatrix} = \begin{bmatrix} B_{11}^s & B_{12}^s & 0 \\ B_{21}^s & B_{22}^s & 0 \\ 0 & 0 & B_{66}^s \end{bmatrix} \begin{bmatrix} \dfrac{\partial u}{\partial x} \\ \dfrac{\partial v}{\partial y} \\ \dfrac{\partial u}{\partial y} + \dfrac{\partial v}{\partial x} \end{bmatrix} - \begin{bmatrix} D_{11}^s & D_{12}^s & 0 \\ D_{21}^s & D_{22}^s & 0 \\ 0 & 0 & D_{66}^s \end{bmatrix} \begin{bmatrix} \dfrac{\partial^2 w_b}{\partial x^2} \\ \dfrac{\partial^2 w_b}{\partial y^2} \\ 2\dfrac{\partial^2 w_b}{\partial x \partial y} \end{bmatrix}$$

$$- \begin{bmatrix} H_{11}^s & H_{12}^s & 0 \\ H_{21}^s & H_{22}^s & 0 \\ 0 & 0 & H_{66}^s \end{bmatrix} \begin{bmatrix} \dfrac{\partial^2 w_s}{\partial x^2} \\ \dfrac{\partial^2 w_s}{\partial y^2} \\ 2\dfrac{\partial^2 w_s}{\partial x \partial y} \end{bmatrix}$$

(5.309)

$$\begin{bmatrix} Q_{xz} \\ Q_{yz} \end{bmatrix} = \begin{bmatrix} A_{44}^s & 0 \\ 0 & A_{55}^s \end{bmatrix} \begin{bmatrix} \dfrac{\partial w_s}{\partial x} \\ \dfrac{\partial w_s}{\partial y} \end{bmatrix}$$

(5.310)

The GF plate governing equations are found by plugging in Eqs. (5.307)–(5.310) into Eqs.

$$\left(A_{11} \quad A_{12} \quad A_{66} \right) = \int_{-\frac{h}{2}}^{\frac{h}{2}} Q_{11} \left(1 \quad v(z) \quad \frac{1-v(z)}{2} \right) dz \qquad (5.311)$$

$$\left(B_{11} \quad B_{12} \quad B_{66} \right) = \int_{-\frac{h}{2}}^{\frac{h}{2}} Q_{11} z \left(1 \quad v(z) \quad \frac{1-v(z)}{2} \right) dz \qquad (5.312)$$

$$\left(D_{11} \quad D_{12} \quad D_{66} \right) = \int_{-\frac{h}{2}}^{\frac{h}{2}} Q_{11} z^2 \left(1 \quad v(z) \quad \frac{1-v(z)}{2} \right) dz \qquad (5.313)$$

$$\left(B_{11}^s \quad B_{12}^s \quad B_{66}^s \right) = \int_{-\frac{h}{2}}^{\frac{h}{2}} Q_{11} f(z) \left(1 \quad v(z) \quad \frac{1-v(z)}{2} \right) dz \qquad (5.314)$$

$$\left(D_{11}^{s} \quad D_{12}^{s} \quad D_{66}^{s} \right) = \int_{-\frac{h}{2}}^{\frac{h}{2}} Q_{11} z f\left(z \right) \left(1 \quad v\left(z \right) \quad \frac{1-v\left(z \right)}{2} \right) dz \qquad (5.315)$$

$$\left(H_{11}^{s} \quad H_{12}^{s} \quad H_{66}^{s} \right) = \int_{-\frac{h}{2}}^{\frac{h}{2}} Q_{11} f^{2}\left(z \right) \left(1 \quad v\left(z \right) \quad \frac{1-v\left(z \right)}{2} \right) dz \qquad (5.316)$$

$$\begin{aligned} \left(A_{22} \quad B_{22} \quad D_{22} \quad B_{22}^{s} \quad D_{22}^{s} \quad H_{22}^{s} \right) \\ = \left(A_{11} \quad B_{11} \quad D_{11} \quad B_{11}^{s} \quad D_{11}^{s} \quad H_{11}^{s} \right) \end{aligned} \qquad (5.317)$$

$$A_{44}^{s} = A_{55}^{s} = \int_{-\frac{h}{2}}^{\frac{h}{2}} G\left(z \right) g^{2}\left(z \right) dz \qquad (5.318)$$

By substituting Eqs. (5.307)–(5.310) in Eqs. (5.300)–(5.301), the governing equations of GF plate can be obtained as:

$$\begin{aligned} & A_{11} \frac{\partial^{2} u}{\partial x^{2}} + \left(A_{12} + A_{66} \right) \frac{\partial^{2} v}{\partial x \partial y} + A_{66} \frac{\partial^{2} u}{\partial y^{2}} - B_{11} \frac{\partial^{3} w_{b}}{\partial x^{3}} - \left(B_{12} + 2B_{66} \right) \frac{\partial^{3} w_{b}}{\partial x \partial y^{2}} - B_{s}^{11} \frac{\partial^{3} w_{s}}{\partial x^{3}} \\ & - \left(B_{12}^{s} + 2B_{66}^{s} \right) \frac{\partial^{3} w_{s}}{\partial x \partial y^{2}} - I_{0} \frac{\partial^{2} u}{\partial t^{2}} + I_{1} \frac{\partial^{2} w_{b}}{\partial x \partial t^{2}} + J_{1} \frac{\partial^{2} w_{s}}{\partial x \partial t^{2}} = 0 \end{aligned}$$

$$(5.319)$$

$$\begin{aligned} & A_{22} \frac{\partial^{2} v}{\partial y^{2}} + \left(A_{12} + A_{66} \right) \frac{\partial^{2} u}{\partial x \partial y} + A_{66} \frac{\partial^{2} v}{\partial x^{2}} - B_{11} \frac{\partial^{3} w_{b}}{\partial x^{3}} - B_{22} \frac{\partial^{3} w_{b}}{\partial y^{3}} \\ & - \left(B_{12} + 2B_{66} \right) \frac{\partial^{2} w_{b}}{\partial x^{2} \partial y} - \left(B_{12}^{s} + 2B_{66}^{s} \right) \frac{\partial^{3} w_{s}}{\partial x^{2} \partial y} - I_{0} \frac{\partial^{2} v}{\partial t^{2}} \\ & + I_{1} \frac{\partial^{2} w_{b}}{\partial y \partial t^{2}} + J_{1} \frac{\partial^{2} w_{s}}{\partial y \partial t^{2}} = 0 \end{aligned} \qquad (5.320)$$

$$\begin{aligned} & B_{11} \frac{\partial^{3} u}{\partial x^{3}} + \left(B_{12} + 2B_{66} \right) \frac{\partial^{3} u}{\partial x \partial y^{2}} + B_{22} \frac{\partial^{3} v}{\partial y^{3}} - \left(B_{12} + 2B_{66} \right) \frac{\partial^{3} v}{\partial x^{2} \partial y} - D_{11} \frac{\partial^{4} w_{b}}{\partial x^{4}} \\ & - 2\left(D_{12} + 2D_{66} \right) \frac{\partial^{4} w_{b}}{\partial x^{2} \partial y^{2}} - D_{22} \frac{\partial^{4} w_{b}}{\partial y^{4}} - D_{11}^{s} \frac{\partial^{4} w_{s}}{\partial x^{4}} - 2\left(D_{12}^{s} + 2D_{66}^{s} \right) \frac{\partial^{4} w_{s}}{\partial x^{2} \partial y^{2}} \end{aligned}$$

$$-D_{22}^s \frac{\partial^4 w_s}{\partial y^4} - I_0 \frac{\partial^2 (w_b + w_s)}{\partial t^2} - I_1 \left(\frac{\partial^2 u}{\partial x \partial t^2} + \frac{\partial^2 v}{\partial y \partial t^2} \right) + I_2 \nabla^2 \frac{\partial^2 w_b}{\partial t^2} + J_2 \nabla^2 \frac{\partial^2 w_s}{\partial t^2}$$

$$-\left(N^T + N^H\right)\nabla^2 \left(w_b + w_s\right) = 0$$

(5.321)

$$B_{11}^s \frac{\partial^3 u}{\partial x^3} + \left(B_{12}^s + 2B_{66}^s\right)\frac{\partial^3 u}{\partial x \partial y^2} + B_{22}^s \frac{\partial^3 v}{\partial y^3} - \left(B_{12}^s + 2B_{66}^s\right)\frac{\partial^3 v}{\partial x^2 \partial y} - D_{11}^s \frac{\partial^4 w_b}{\partial x^4}$$

$$-2\left(D_{12}^s + 2D_{66}^s\right)\frac{\partial^4 w_b}{\partial x^2 \partial y^2} - D_{22}^s \frac{\partial^4 w_b}{\partial y^4} - H_{11}^s \frac{\partial^4 w_s}{\partial x^4} - 2\left(H_{12}^s + 2H_{66}^s\right)\frac{\partial^4 w_s}{\partial x^2 \partial y^2}$$

$$-H_{22}^s \frac{\partial^4 w_s}{\partial y^4} + A_{44}^s \nabla^2 w_s - I_0 \frac{\partial^2 (w_b + w_s)}{\partial t^2} - J_1 \left(\frac{\partial^2 u}{\partial x \partial t^2} + \frac{\partial^2 v}{\partial y \partial t^2} \right) + J_2 \nabla^2 \frac{\partial^2 w_b}{\partial t^2}$$

$$+K_2 \nabla^2 \frac{\partial^2 w_s}{\partial t^2} - \left(N^T + N^H\right)\nabla^2 \left(w_b + w_s\right) = 0$$

(5.322)

The resulting GF plate governing equations in the hygrothermal environment are solved using an analytical technique. More so, the constituents of the displacement field are expected to be:

$$\left\{ \begin{matrix} u \\ v \\ w_b \\ w_s \end{matrix} \right\} = \left\{ \begin{matrix} U \\ V \\ W_b \\ W_s \end{matrix} \right\} e^{i(\beta_x x + \beta_y y - \omega_n t)}$$

(5.323)

in which U, V, W_b, and W_s stand for wave amplitudes, respectively. β_x and β_y are wave numbers through x and y directions, and ω_n is angular frequency dispersed waves. Substituting for u, v, w_b, and w_s from Eq. (5.323) in Eqs. (5.277)–(5.322) the following equation is attained [64]:

$$\left(\left[K - \omega_n^2 M\right]_{4\times 4}\right)\left[\begin{matrix} U \\ V \\ W_b \\ W_s \end{matrix} \right] = 0$$

(5.324)

where K and M are stiffness and mass matrices, respectively, and components of these matrices are given in Appendix 5I. Above eigenvalue problem is solved for ω_n and wave frequency can be attained. The following determinant must be set to zero:

$$\left\| \left[K - \omega_n^2 M \right]_{4\times4} \right\| = 0 \qquad\qquad (5.325)$$

Table 5.4 lists the mechanical parameters of GF, the constitutive material of the plate used in this study. Wave frequency and phase velocity variations of GF plates, as well as the effects of other factors, are also discussed. In addition, $h =$ 0.05 [m] is used as the GF plate thickness in all illustrations [64]. Changes in (a) wave frequency and (b) phase velocity versus wave number for a range of porosity coefficients are shown in Figure 5.97. $\Delta T = 50$ (UTR), $\Delta C = 0.05$ (UMR), UPD, $a/h = 25$, and $a/b = 1$ are the parameters used to create these graphs. As can be observed, the frequency and phase velocity of waves are attenuated as the porosity coefficient increases. This demonstrates the drawback of having porous structures. These graphs also reveal that the dissimilarity between curves is less for smaller wave numbers than it is at larger ones. The porosity impact may be deduced to be more noticeable with larger wave numbers. Furthermore, as the number of waves grows, so does their frequency and phase velocity [64]. In Figure 5.98 ($\Delta T = 50$ (UTR), $\Delta C = 0.05$ (UMR), $e_0 = 0.2$, $a/h = 25$, $a/b = 1$), we see how the frequency and phase velocity of waves change as a function of wave number over a range of porosity distributions. Once again, the rising influence of wave number is seen graphically here. $\beta < 200$ and $\beta > 200$ should be the dividing lines in both diagrams. In the first section, SPD-1 has the greatest wave frequency and phase velocity, while in the second section, SPD-2 has the higher values. This observation demonstrates that the impact of porosity distribution over the thickness varies with wave number [64].

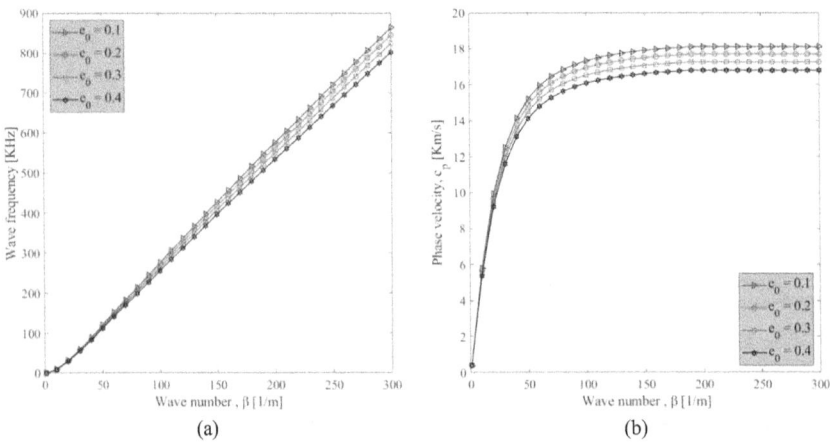

FIGURE 5.97 Variation of (a) wave frequency and (b) phase velocity versus wave number for different porosity coefficients ($\Delta T = 50$ (UTR), $\Delta C = 0.05$ (UMR), UPD, $a/h = 25$, $a/b = 1$).

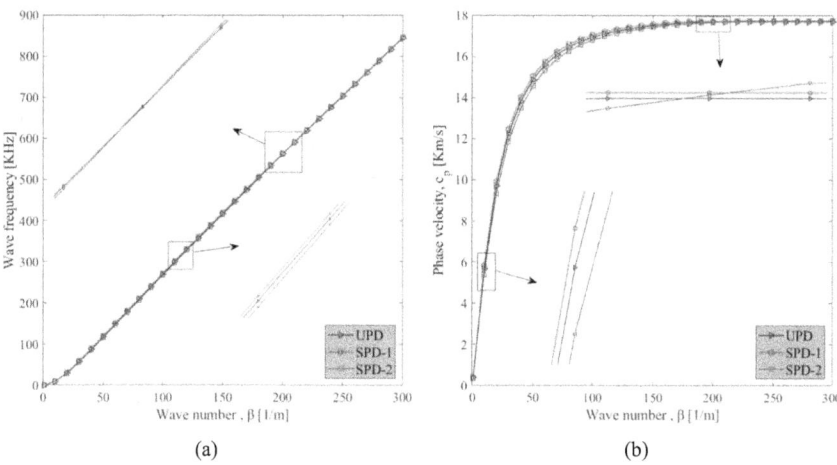

(a)   (b)

FIGURE 5.98 Variation of (a) wave frequency and (b) phase velocity versus wave number for various porosity distributions ($\Delta T = 50$ (UTR), $\Delta C = 0.05$ (UMR), $e_0 = 0.2$, $a/h = 25$, $a/b = 1$).

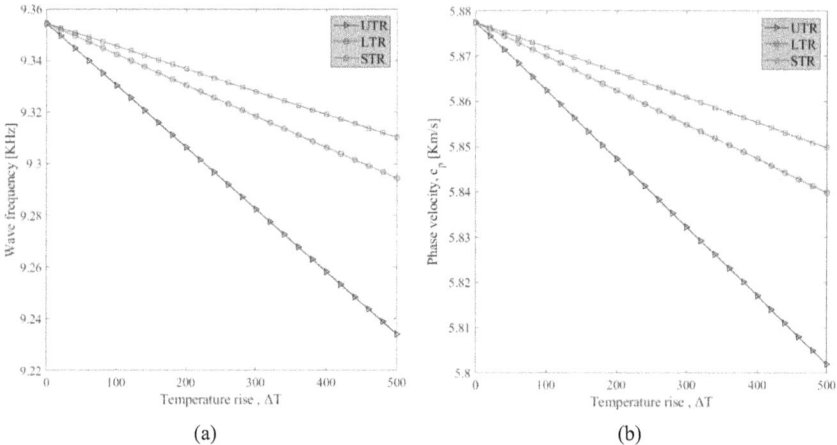

(a)   (b)

FIGURE 5.99 Variation of (a) wave frequency and (b) phase velocity versus temperature rise for different patterns of temperature rise (SPD-1, $\Delta C = 0.05$ (UMR), $e_0 = 0.1$, $\beta = 10$, $a/h = 25$, $a/b = 1$).

For various patterns of temperature increase (SPD-1, $\Delta C = 0.05$ (UMR), $e_0 = 0.1$, $\beta = 10$, $a/h = 25$, $a/b = 1$), Figure 5.99 depicts the correlation between (a) wave frequency and (b) phase velocity and (c) temperature. These numbers suggest that as temperatures increase, there is less of an effect on wave frequency and phase velocity fluctuation due to temperature change. The warming impact of this fact is the reason behind it. The frequency and phase velocity of waves are less affected by a decrease in temperature from a sinusoidal increase than they are from a linear climb or a uniform rise.

In Figure 5.100 (SPD-1, $\Delta T = 100$ (STR), $e_0 = 0.1$, $\beta = 10$, $a/h = 25$, $a/b = 1$), the examples of change of (a) wave frequency and (b) phase velocity against moisture increase for different forms of moisture rise have been shown. The pattern of the diagrams follows the same pattern as the diagrams in Figure 5.100. It is possible to draw conclusions from this graph that wave frequency and phase velocity are negatively related to the amount of moisture that is present. To put it another way, when the amount of moisture in the air increases, the wave frequency and phase velocity both drop. The wave frequency and phase velocity are at their lowest levels along the curve of UMR, which is followed by the curves of LTR and STR [64].

In Figure 5.101, we see the change of (a) wave frequency and (b) phase velocity versus porosity coefficient for various porosity distributions ($\Delta T = 100$ (STR), $\Delta C = 0.1$ (SMC), $\beta = 15$, $a/h = 25$, $a/b = 1$). The wave frequency and phase velocity both drop as the porosity coefficient is increased. The reason for this behavior is that as the porosity coefficient increases, the structure gets weaker. To put it another way, the structure's stiffness decreases, and, as a result, lower wave frequency and phase velocity are seen. The reason for this behavior is that the porosity coefficient increases. Additionally, it is possible to specify that the quantity of SPD-2 degradation for larger porosity coefficients is bigger than that of other ones [64].

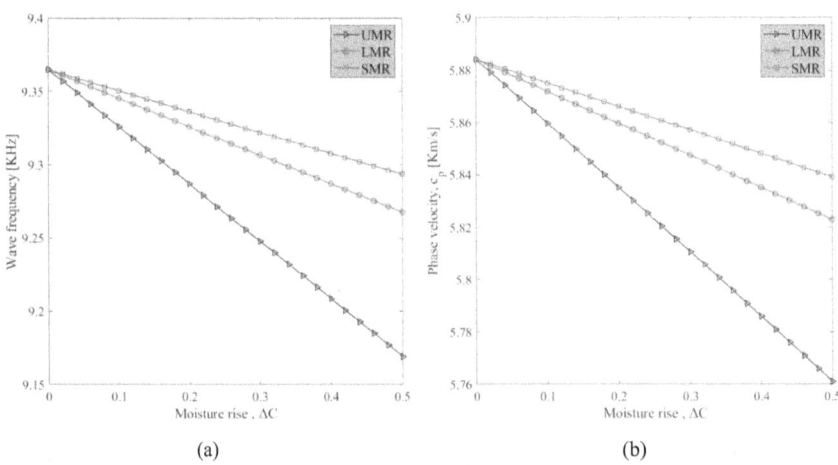

(a)   (b)

FIGURE 5.100   Variation of (a) wave frequency and (b) phase velocity versus moisture rise for different patterns of moisture rise (SPD-1, $\Delta T = 100$ (STR), $e_0 = 0.1$, $\beta = 10$, $a/h = 25$, $a/b = 1$).

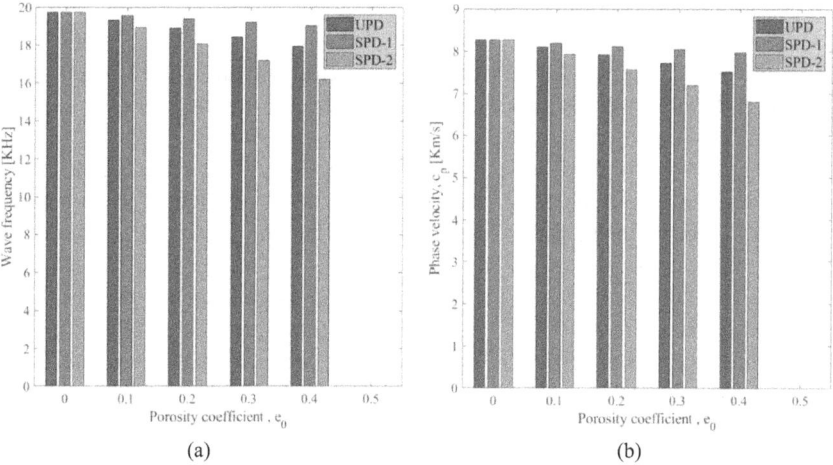

FIGURE 5.101 Variation of (a) wave frequency and (b) phase velocity versus porosity coefficient for various porosity distributions ($\Delta T = 100$ (STR), $\Delta C = 0.1$ (SMC), $\beta = 15$, $a/h = 25$, $a/b = 1$).

5.6 SUMMARY

The purpose of the present chapter was to conduct a review of the wave dispersion properties of graphene-based structures. These structures include single- and double-layered graphene sheets, graphene oxide powder–reinforced nanocomposite plates, graphene platelet–reinforced structures, and graphene foam structures. Following is a rundown of the most significant takeaways from this chapter, in no particular order:

- If either the length scale parameter or the nonlocal parameter is elevated, the wave dispersion responses of graphene sheets have the potential to grow in size.
- An increase in the wave frequency of graphene sheets may be accomplished in a few different ways. Two of these approaches include increasing the length scale parameter of graphene sheets and decreasing the nonlocal parameter.
- It is possible to see a decrease in the wave frequencies of graphene sheets if there is a temperature gradient or an increase in the content of moisture.
- The behavior of wave frequency does not stay constant across the board for all of the fundamental parameters for graphene sheets. This is one of the limitations of graphene.
- If the qualities of the base of the graphene sheets are enhanced, then the graphene sheets may be able to attain greater wave frequencies.
- Once the temperature gradient is magnified, numerical simulations predict that there will be a decline in the wave frequency values for graphene

sheets. This will be the case since graphene sheets have a very low dielectric constant.

- The wave frequency of the graphene sheets will ultimately reach zero, and the phenomenon that causes this will become apparent in waves with smaller numbers at a later time.
- The wave dispersion responses of SLGSs may be enhanced by either decreasing the nonlocal parameters or increasing the length scale parameters. Both of these options are available. Both of these choices are open to consideration.

    The only situation in which the Winkler parameter has the ability to induce an increase in the value of the phase velocity of SLGSs is one in which there is a very small range of wave numbers. This range of wave numbers is roughly within.

- The Pasternak coefficient, in the same vein as the Winkler parameter, has adequate ability to magnify the phase velocity values of SLGSs. This is the same thing that can be said about the Winkler parameter. Do not make the mistake of assuming that it only applies to high wave numbers since the impact of this parameter may be observed in low wave numbers as well as high wave numbers; thus, do not think that it only applies to low wave numbers.
- When nonzero values are given to the magnetic field intensity, it is possible to attain bigger phase velocity quantities of SLGSs as compared to the situation in which this parameter is set to zero. This is in contrast to the scenario in which it is set to zero. In contrast to this, the scenario in which this parameter is not changed at all is described below.
- The behavior of the phase velocity of SLGSs changes in predictable ways based on the length scale parameters that are under investigation. In the event of high wave numbers, the phase velocity will either follow a path that is falling, remaining constant, or increasing. This behavior will be determined by the length scale parameter that is being studied.
- As the nonlocality of the system increases, the number of times that SLGSs are able to successfully escape the system gradually diminishes. In addition, the use of length scale parameters of a greater magnitude may allow for the achievement of higher escape frequencies.
- In graphene sheets, an increase in the total amount of wave frequency or phase velocity may be created for graphene sheets by the linear or nonlinear foundation parameters, depending on the kind of foundation they have. Graphene sheets have a nonlinear foundation. On the other hand, the wave number has to be somewhat low in order to see this effect.
- The structural damping coefficient has the potential to generate a damping effect on the wave frequency and phase velocity of DLVGSs.
- In addition to this, the damping coefficient of the foundation has the potential to lower the wave propagation responses of DLGSs.
- According to the results, one of the other ways to produce a decrease in the size of the wave responses of DLGSs is to apply bigger axial loads.

This was one of the options that was tested. Several distinct approaches are possible for achieving this goal.

- Either raising the length scale parameter or reducing the nonlocal parameter is an easy way to increase the wave frequency and phase velocity of the DLGSs. This may be done in a straightforward manner.

- After the DLGSs have been amplified, linear and nonlinear medium factors have the potential to either enhance the wave frequency of the DLGSs or increase their phase velocity.

- The DLGSs' wave frequency or phase velocity may finally be dampened in each needed wave number whenever the damping coefficient is recorded. This is possible whenever the phrase "whenever the damping coefficient is recorded."

- When the hygrothermal condition is employed instead of the thermal condition, the wave dispersion responses of DLGSs are dramatically decreased when compared to those seen under the thermal condition.

- An increase in the temperature gradient or a higher concentration of moisture may be used to achieve the goal of reducing the wave frequency and phase velocity quantities of DLGSs in a way that is consistent with the actual world.

- The optimal FG-pattern for strengthening the structure with GOPs was shown to be fully reliant on the wave number in such a way that if the wave number is less than 20, the best pattern is FG-X, and if the wave number is more than 20, the best pattern is FG-O. This was discovered via research. After it was shown that the wave number is an essential factor in determining the optimal FG-pattern for enhancing the structure with GOPs, it was possible to make this breakthrough finding.

- The wave frequency of the structure can be raised by adding amounts of the GOPs' weight fraction; nevertheless, this rising tendency will be enhanced in bigger wave numbers.

- The structure can be raised by adding quantities of the GOPs' weight fraction.

- The structure's wave frequency may be enhanced by increasing the quantity of the GOPs' weight fraction that is present.

  After conducting research on the wave propagation behavior of GOP-reinforced nanocomposite plates, it was discovered that embedding the structure on an elastic foundation would improve the plates' behavior. This was discovered as a result of the research. It is essential to take into consideration the fact that the influence of the Pasternak parameter is noticeably greater than that of the Winkler parameter.

- The demonstration of the stiffness-softening effect was a decrease in the phase velocity of GOP-reinforced nanocomposite plates when the temperature gradient of the thermal environment was increased. This proved that the stiffness-softening effect occurred. This decrease became apparent when the temperature of the thermal environment had been raised to a greater level.

- If the elastic coefficients of the GOP-reinforced nanocomposite plates are high enough, utilizing a length-to-thickness ratio that is bigger will result in a narrower range of the phase velocity and wave frequency of the plate. This can only happen if the elastic coefficients of the GOP-reinforced nanocomposite plates are high enough.

- It was found that the thermal loading with STR may have a lesser influence on the GOP-reinforced nanocomposite plate compared to the other types of thermal loadings. This was observed. Consequently, the STR type of thermal loading is most efficient, followed by the LTR type, and subsequently by the UTR type.

- It is conceivable for wave propagation responses to be boosted after the weight fraction of GPLs has been introduced into GPLR composite beams. This is achievable after the weight fraction of GPLs has been incorporated.

- When the intensity of the magnetic field is increased in GPLR composite beams, another influence that gradually becomes more noticeable may be seen. This is due to the fact that it has the effect of making it more apparent.

- The GPLR-X beams allow for the greatest amount of phase velocity to be acquired; on the other hand, the GPLR-O beams allow for the least amount of phase velocity to be obtained. This phenomenon is particularly noticeable in low wave numbers for GPLR composite beams. Once there is an increase in temperature, there is a steady slowing down of the wave's speed. In addition, a response falling anywhere in the middle of the range may be recorded for the GPLR-U beams.

- It is conceivable to discern a little bit of an increase in the amplitude of the phase velocity that is present in GPLR composite beams when utilizing bigger length-to-width ratios of GPLs. This is something that may be seen.

- It is thought that the normal stress shown by a GPLRC poly ring is tensile at the lower radial distance and approaches zero as the value of the radial distance rises. This is because the GPLRC poly ring has more radial connections.

- The mechanical displacement achieves a bigger amplitude in GPLRC poly rings with lower aspect ratios, and it propagates toward a negative value in GPLRC poly rings with higher aspect ratios. In contrast, the mechanical displacement reaches a smaller amplitude in GPLRC poly rings with lower aspect ratios.

- When the aspect ratio is raised, changing the weight fraction of the GPL may result in a lower normal strain. This is because changing the weight fraction of the GPL has the effect of increasing the aspect ratio.

- When the temperature of the GPLRC poly ring changes, the frequency of the ring initially decreases to a level where it is no longer detectable and then gradually rises until it reaches its highest possible level.

- When the proportion of GPL in a polymer's total weight goes up, the wave frequency of all four types of poly rings goes up as well.

- The rising weight fraction and aspect ratio of the GPL have a major influence on the wave frequency, and this can be seen in both the GL and GN-III models. This result can be shown to be true.

The plate that has a hexagonal cross-section obtains higher values for all of its physical quantities across the board. Convergence may be achieved more quickly using the Fourier expansion collocation technique for surfaces that include irregularities.

- It is possible to decrease the wave dispersion responses of the GPL-reinforced composite plates by increasing the value of the nonlocal parameters.
- The foundation characteristics of the GPL-reinforced composite plates have the ability to accelerate the wave speed even when there are only a limited number of waves.
- The cut-off frequency of the GPL-reinforced composite plates is a function of the Winkler parameter, and it is possible to raise this coefficient in order to get a higher value for the frequency. Because the cut-off frequency is a function of the Winkler parameter, this is feasible.
- One can arrive at more accurate results when utilizing a bigger weight percentage of GPL in composite plates for GPLs.
- Choosing length-to-breadth ratios that are bigger for GPL-reinforced composite plates results in stronger responses. This is along the same lines as the previous point.
- It is feasible to figure out the answers to the issues by using higher slenderness ratios for the GPL-reinforced composite plates.
- According to the results, increasing the thickness of the GNPRC cylindrical nanoshell causes extremum values of phase velocity to occur in a range of wave numbers that is lower than the range in which the initial values were found.
- If the thickness of the cylindrical nanoshell generated by GNPRC is increased, it may be possible to reduce the impact that the porosity coefficient has on the phase velocity.
- Extremum values of phase velocity are presented in the lower values of the wave number when the radius of the GNPRC cylindrical nanoshell is raised. This is the case when the lower values of the wave number are increased.
- When the value of the radius of the GNPRC cylindrical nanoshell is made bigger, the difference in phase velocity between the lowest possible value and the highest possible value is increased. This occurs because the lowest possible value represents the lowest possible value.
- An elastic medium with a Winkler layer functioning as the linear layer and a Pasternak layer functioning as the shear layer has a growing impact on the fluctuation of wave frequency and phase velocity of the PGF beam.
- Additionally, a decrease in wave frequency and phase velocity structures may be attributed to an increase in temperature change as well as an increase in the quantity of moisture concentration.
- Porosity, a kind of defect, has a dampening impact on the wave frequency and phase velocity of the PGF beam. This is because porosity is an imperfection.
- The porosity of the GF plate structure, which is regarded as a defect, has a reducing influence on the wave frequency and phase velocity of the GF plate. This effect is caused by the fact that the porosity of the structure exists.
- An increase in the wave number brings about an increase in the wave frequency as well as an increase in the phase velocity for the GF plate structure.

- The fluctuation in wave frequency and phase velocity of the GF plate structure is adversely affected by temperature and moisture changes, which have a detrimental impact overall. This occurs when both the temperature and the amount of moisture in the air increase.

  Both the wave frequency and the phase velocity of the GF plate structure are reduced when placed in an environment characterized by high levels of humidity and temperature.
- The value of wave frequency and phase velocity are both reduced when a GF plate structure is present in a hygrothermal environment with a uniform rise, linear increase, or sinusoidal rise, respectively. This is because the GF plate structure dampens the rise in temperature. This is true for all three different kinds of increase.

APPENDICES

APPENDIX A

$$
\left\{
\begin{aligned}
k_{11} =\ & -\left(1+\eta^2\left(\beta_1^2+\beta_2^2\right)\right)\left(D_{11}\beta_1^4+2\left(D_{12}+2D_{66}\right)\beta_1^2\beta_2^2+D_{22}\beta_2^4\right) \\
& +\left(1+\mu^2\left(\beta_1^2+\beta_2^2\right)\right)\left(-k_p\left(\beta_1^2+\beta_2^2\right)-k_w-\eta hH_x^2\beta_1^2\right) \\
k_{12} =\ & -\left(1+\eta^2\left(\beta_1^2+\beta_2^2\right)\right)\left(D_{11}^s\beta_1^4+2\left(D_{12}^s+2D_{66}^s\right)\beta_1^2\beta_2^2+D_{22}^s\beta_2^4\right) \\
& +\left(1+\mu^2\left(\beta_1^2+\beta_2^2\right)\right)\left(-k_p\left(\beta_1^2+\beta_2^2\right)-k_w-\eta hH_x^2\beta_1^2\right) \\
k_{21} =\ & -\left(1+\eta^2\left(\beta_1^2+\beta_2^2\right)\right)\left(D_{11}^s\beta_1^4+2\left(D_{12}^s+2D_{66}^s\right)\beta_1^2\beta_2^2+D_{22}^s\beta_2^4\right) \\
& +\left(1+\mu^2\left(\beta_1^2+\beta_2^2\right)\right)\left(-k_p\left(\beta_1^2+\beta_2^2\right)-k_w-\eta hH_x^2\beta_1^2\right) \\
k_{22} =\ & -\left(1+\eta^2\left(\beta_1^2+\beta_2^2\right)\right)\left(\begin{array}{c}H_{11}^s\beta_1^4+2\left(H_{12}^s+2H_{66}^s\right)\beta_1^2\beta_2^2+H_{22}^s\beta_2^4 \\ +A_{55}^s\beta_1^2+A_{44}^s\beta_2^2\end{array}\right) \\
& +\left(1+\mu^2\left(\beta_1^2+\beta_2^2\right)\right)\left(-k_p\left(\beta_1^2+\beta_2^2\right)-k_w-\eta hH_x^2\beta_1^2\right)
\end{aligned}
\right.
\tag{A.1}
$$

$$
\left\{
\begin{aligned}
m_{11} &= -\left(1+\mu^2\left(\beta_1^2+\beta_2^2\right)\right)\left(I_0+I_2\left(\beta_1^2+\beta_2^2\right)\right) \\
m_{12} &= -\left(1+\mu^2\left(\beta_1^2+\beta_2^2\right)\right)\left(I_0+J_2\left(\beta_1^2+\beta_2^2\right)\right) \\
m_{21} &= -\left(1+\mu^2\left(\beta_1^2+\beta_2^2\right)\right)\left(I_0+J_2\left(\beta_1^2+\beta_2^2\right)\right) \\
m_{22} &= -\left(1+\mu^2\left(\beta_1^2+\beta_2^2\right)\right)\left(I_0+K_2\left(\beta_1^2+\beta_2^2\right)\right)
\end{aligned}
\right.
\tag{A.2}
$$

APPENDIX B

$$k_{11} = k_{22} = \left(1 + \eta^2 \left(\beta_1^2 + \beta_2^2\right)\right)\left(1 + g\frac{\partial}{\partial t}\right)\left(D_{11}\beta_1^4 + 2\left(D_{12} + 2D_{66}\right)\beta_1^2\beta_2^2 + D_{22}\beta_2^4\right)$$

$$+ \left(1 + \mu^2 \left(\beta_1^2 + \beta_2^2\right)\right)\left(k_w + \left(k_p - N\right)\left(\beta_1^2 + \beta_2^2\right) - i\omega C_d + C\right)$$

$$k_{12} = k_{21} = -\left(1 + \mu^2 \left(\beta_1^2 + \beta_2^2\right)\right)C$$

(B.1)

$$m_{11} = m_{22} = \left(1 + \mu^2 \left(\beta_1^2 + \beta_2^2\right)\right)\left(I_0 + I_2\left(\beta_1^2 + \beta_2^2\right)\right)$$

$$m_{12} = m_{21} = 0$$

(B.2)

APPENDIX C

$$k_{11} = -A_{11}\beta_1^2 - A_{66}\beta_2^2$$

$$k_{12} = \left(A_{12} + A_{66}\right)\beta_1\beta_2$$

$$k_{13} = iB_{11}\beta_1^3 + i\left(B_{12} + 2B_{66}\right)\beta_2^2\beta_1$$

$$k_{14} = iB_{11}^s\beta_1^3 + i\left(B_{12}^s + 2B_{66}^s\right)\beta_2^2\beta_1$$

$$k_{22} = -A_{22}\beta_1^2 - A_{66}\beta_2^2$$

$$k_{23} = iB_{22}\beta_2^3 + i\left(B_{12} + 2B_{66}\right)\beta_1^2\beta_2$$

$$k_{24} = B_{22}^s\beta_2^3 + i\left(B_{12}^s + 2B_{66}^s\right)\beta_1^2\beta_2$$

$$k_{33} = -D_{11}\beta_1^4 - 2\times\left(D_{12} + 2D_{66}\right)\beta_1^2\beta_2^2$$

$$\quad - D_{22}\beta_2^4 - k_w - \left(k_p - N^T\right)\left(\beta_1^2 + \beta_2^2\right)$$

$$k_{34} = -D_{11}^s\beta_1^4 - 2\times\left(D_{12}^s + 2D_{66}^s\right)\beta_1^2\beta_2^2$$

$$\quad - D_{22}^s\beta_2^4 - k_w - \left(k_p - N^T\right)\left(\beta_1^2 + \beta_2^2\right)$$

$$k_{44} = -H_{11}^s\beta_1^4 - 2\times\left(H_{12}^s + 2H_{66}^s\right)\beta_1^2\beta_2^2$$

$$\quad - H_{22}^s\beta_2^4 - k_w - \left(k_p - N^T\right)\left(\beta_1^2 + \beta_2^2\right) - A_{55}^s\beta_1^2 - A_{44}^s\beta_2^2$$

$$m_{11} = m_{22} = I_0$$

$$m_{12} = 0$$

$$m_{13} = -I_1\beta_1 i$$

$$m_{14} = -J_1\beta_1 i$$

$$m_{23} = -I_1\beta_2 i$$

$$m_{24} = -J_1\beta_2 i$$

$$m_{33} = I_0 - I_2\left(\beta_1^2 + \beta_2^2\right)$$
$$m_{34} = I_0 - J_2\left(\beta_1^2 + \beta_2^2\right)$$
$$m_{44} = I_0 - K_2\left(\beta_1^2 + \beta_2^2\right)$$

APPENDIX D

$$a_n^j = 2\left\{p(p-1)J_p(\alpha_i ax) + (\alpha_i ax)J_{p+1}(\alpha_i ax)\right\}\cos 2(\theta - \gamma_i)\cos m\theta$$
$$- x^2\left\{(\alpha_i a)^2 + \left[\bar{\lambda} + 2\cos^2(\theta - \gamma_i)\right] + a_i\right\}J_p(\alpha_i ax)\cos m\theta \qquad \text{(D1)}$$
$$+ 2p\left\{(p-1)J_p(\alpha_i ax) - (\alpha_i ax)J_{p+1}(\alpha_i ax)\right\}\sin m\theta \sin 2(\theta - \gamma_i), \quad j = 1,2$$

$$a_n^3 = 2\left\{p(p-1)J_p(\alpha_3 ax) - (\alpha_3 ax)J_{p+1}(\alpha_3 ax)\right\}\cos m\theta \cos 2(\theta - \gamma_i)$$
$$+ 2\left\{\left[p(p-1) - (\alpha_3 ax)^2\right]J_p(\alpha_3 ax) + (\alpha_3 ax)J_{p+1}(\alpha_3 ax)\right\}\sin m\theta \sin 2(\theta - \gamma_i) \qquad \text{(D2)}$$

$$a_n^4 = 2\left\{p(p-1)Y_p(\alpha_4 ax) - (\alpha_4 ax)Y_{p+1}(\alpha_4 ax)\right\}\cos m\theta \cos 2(\theta - \gamma_i)$$
$$+ 2\left\{\left[p(p-1) - (\alpha_4 ax)^2\right]Y_p(\alpha_4 ax) + (\alpha_4 ax)Y_{p+1}(\alpha_4 ax)\right\}\sin m\theta \sin 2(\theta - \gamma_i) \qquad \text{(D3)}$$

$$a_n^j = 2\left\{p(p-1)Y_p(\alpha_i ax) + (\alpha_i ax)Y_{p+1}(\alpha_i ax)\right\}\cos 2(\theta - \gamma_i)\cos m\theta$$
$$- x^2\left\{(\alpha_i a)^2 + \left[\bar{\lambda} + 2\cos^2(\theta - \gamma_i)\right] + a_i\right\}Y_p(\alpha_i ax)\cos m\theta \qquad \text{(D4)}$$
$$+ 2p\left\{(p-1)Y_p(\alpha_i ax) - (\alpha_i ax)Y_{p+1}(\alpha_i ax)\right\}\sin m\theta \sin 2(\theta - \gamma_i), \quad j = 5,6$$

$$b_n^j = 2\left\{\left[p(p-1) - (\alpha_i ax)^2\right]J_p(\alpha_i ax) + (\alpha_i ax)J_{p+1}(\alpha_i ax)\right\}\cos m\theta \sin 2(\theta - \gamma_i)$$
$$+ 2p\left\{(\alpha_i ax)J_{p+1}(\alpha_i ax) - (p-1)J_p(\alpha_i ax)\right\}\sin m\theta \cos 2(\theta - \gamma_i), j = 1,2 \qquad \text{(D5)}$$

$$b_n^3 = \left\{(p-1)J_p(\alpha_3 ax) - (\alpha_3 ax)J_{p+1}(\alpha_3 ax)\right\}\cos m\theta \sin 2(\theta - \gamma_i)$$
$$- \left\{2(\alpha_3 ax)J_{p+1}(\alpha_3 ax) - \left[(\alpha_3 ax)^2 - 2p(p-1)\right]J_p(\alpha_3 ax)\right\} \qquad \text{(D6)}$$
$$2p\sin m\theta \cos 2(\theta - \gamma_i)$$

$$b_n^4 = \left\{ (p-1)Y_p(\alpha_4 ax) - (\alpha_4 ax)Y_{p+1}(\alpha_4 ax) \right\} \cos m\theta \sin 2(\theta - \gamma_i)$$
$$- \left\{ 2(\alpha_4 ax)Y_{p+1}(\alpha_4 ax) - \left[(\alpha_4 ax)^2 - 2p(p-1) \right] Y_p(\alpha_4 ax) \right\} 2p \sin m\theta \cos 2(\theta - \gamma_i)$$

$$\text{(D7)}$$

$$b_n^j = 2 \left\{ \left[p(p-1) - (\alpha_i ax)^2 \right] Y_p(\alpha_i ax) + (\alpha_i ax)Y_{p+1}(\alpha_i ax) \right\} \cos m\theta \sin 2(\theta - \gamma_i)$$
$$+ 2p \left\{ (\alpha_i ax)Y_{p+1}(\alpha_i ax) - (p-1)Y_p(\alpha_i ax) \right\} \sin m\theta \times \cos 2(\theta - \gamma_i), j = 5,6$$

$$\text{(D8)}$$

$$c_n^j = d_i \left\{ p \cos \left(\overline{n-1}\theta + \gamma_i \right) J_p(\alpha_i ax) - (\alpha_i ax)J_{p+1}(\alpha_i ax) \cos(\theta - \gamma_i) \cos m\theta \right\},$$
$$j = 1,2$$

$$\text{(D9)}$$

$$c_n^3 = 0.0, \ c_n^4 = 0.0 \qquad \text{(D10)}$$

$$c_n^j = d_i \left\{ p \cos \left(\overline{n-1}\theta + \gamma_i \right) Y_p(\alpha_i ax) - (\alpha_i ax)Y_{p+1}(\alpha_i ax) \cos(\theta - \gamma_i) \cos m\theta \right\},$$
$$j = 5,6$$

$$\text{(D11)}$$

The expressions $\bar{a}_n^j, \bar{b}_n^j$ & $\bar{c}_n^j$ are obtained by interchanging $\cos n\theta$ and $\sin n\theta$ in Eqs. (D1)–(D11).

Appendix E

$$k_{11} = -\left(A_{11}\beta_1^2 + A_{66}\beta_2^2 \right),$$
$$k_{12} = -\left(A_{12} + A_{66} \right)\beta_1\beta_2,$$
$$k_{13} = i\beta_1 \left(B_{11}\beta_1^2 + (B_{12} + 2B_{66})\beta_2^2 \right),$$
$$k_{21} = \left(A_{12} + A_{66} \right)\beta_1\beta_2,$$
$$k_{22} = -\left(A_{66}\beta_1^2 + A_{22}\beta_2^2 \right),$$
$$k_{23} = i\beta_2 \left((B_{12} + 2B_{66})\beta_1^2 + B_{22}\beta_2^2 \right),$$
$$k_{31} = -i\beta_1 \left(B_{11}\beta_1^2 + (B_{12} + 2B_{66})\beta_2^2 \right),$$
$$k_{32} = -i\beta_2 \left((B_{12} + 2B_{66})\beta_1^2 + B_{22}\beta_2^2 \right),$$
$$k_{33} = -\left(D_{11}\beta_1^4 + 2(D_{12} + 2D_{66})\beta_1^2\beta_2^2 + D_{22}\beta_2^4 \right)$$

$$\text{(E.1)}$$

$$m_{11} = -I_0\left(1+\mu^2\left(\beta_1^2+\beta_2^2\right)\right),$$

$$m_{12} = 0,$$

$$m_{13} = -iI_1\beta_1\left(1+\mu^2\left(\beta_1^2+\beta_2^2\right)\right),$$

$$m_{21} = 0,$$

$$m_{22} = -I_0\left(1+\mu^2\left(\beta_1^2+\beta_2^2\right)\right),$$

(E.2)

$$m_{23} = -iI_1\beta_2\left(1+\mu^2\left(\beta_1^2+\beta_2^2\right)\right),$$

$$m_{31} = iI_1\beta_1\left(1+\mu^2\left(\beta_1^2+\beta_2^2\right)\right),$$

$$m_{32} = -iI_1\beta_2\left(1+\mu^2\left(\beta_1^2+\beta_2^2\right)\right),$$

$$m_{33} = -\left(I_0-I_2\left(\beta_1^2+\beta_2^2\right)\right)\left(1+\mu^2\left(\beta_1^2+\beta_2^2\right)\right)$$

Appendix F

$$K_{11} = -A_{11}\beta^2 - \frac{\left(A_{66}n^2\right)}{R^2}, \quad K_{12} = -\frac{A_{12}n\beta}{R} - \frac{A_{66}n\beta}{R},$$

$$K_{13} = -\frac{A_{12}\beta i}{R}, \quad K_{14} = -B_{11}\beta^2 - \frac{B_{66}n^2}{R^2}, \quad K_{15} = -\frac{B_{12}n\beta}{R} - \frac{B_{66}n\beta}{R},$$

$$K_{22} = -A_{66}\beta^2 - \frac{B_{12}n^2}{R^2} - \frac{\tilde{A}_{66}}{R^2}, \quad K_{23} = -\frac{A_{11}ni}{R^2} + \frac{\tilde{A}_{66}\,ni}{R^2},$$

$$K_{24} = -\frac{B_{66}n\beta}{R} - \frac{B_{12}n\beta}{R}, \quad K_{25} = -B_{66}\beta^2 - \frac{B_{11}n^2}{R^2} - \frac{\tilde{A}_{66}}{R^2},$$

(F.1)

$$K_{33} = -\tilde{A}_{66}\beta^2 - \frac{A_{11}}{R^2} - \frac{\tilde{A}_{66}\,n^2}{R^2} + k_W + k_P\beta^2 + k_P\frac{n^2}{R^2},$$

$$K_{34} = -\tilde{A}_{66}\beta i - \frac{B_{12}\beta i}{R}, \quad K_{35} = -\frac{B_{11}ni}{R^2} - \frac{\tilde{A}_{66}\,ni}{R},$$

$$K_{44} = -D_{11}\beta^2 - \frac{D_{66}n^2}{R^2} - \tilde{A}_{66}, \quad K_{45} = -\frac{D_{66}n\beta}{R} - \frac{D_{12}n\beta}{R},$$

$$K_{55} = -D_{66}\beta^2 - \frac{D_{11}n^2}{R^2} - \tilde{A}_{66}$$

$$M_{11} = I_0, \quad M_{14} = I_1, \quad M_{22} = I_0, \quad M_{25} = I_1$$

$$M_{33} = I_0, \quad M_{44} = I_2, \quad M_{55} = I_2$$

(F.2)

APPENDIX G

$$\delta u : A_{11}\left(\frac{\partial^2 u}{\partial x^2}+l^2\frac{\partial^4 u}{\partial x^4}-\frac{l^2}{R^2}\frac{\partial^4 u}{\partial x^2\partial\theta^2}\right)+B_{11}\left(\frac{\partial^2\psi_x}{\partial x^2}+l^2\frac{\partial^4\psi_x}{\partial x^4}-\frac{l^2}{R^2}\frac{\partial^4\psi_x}{\partial x^2\partial\theta^2}\right)$$

$$+A_{12}\left(\frac{1}{R}\frac{\partial^2 v}{\partial x\partial\theta}+\frac{1}{R}\frac{\partial w}{\partial x}-\frac{l^2}{R^2}\frac{\partial^4 v}{\partial x^3\partial\theta}-\frac{l^2}{R^2}\frac{\partial^3 w}{\partial x^3}-\frac{l^2}{R^3}\frac{\partial^4 v}{\partial x\partial\theta^3}-\frac{l^2}{R^3}\frac{\partial^3 w}{\partial x\partial\theta^2}\right)$$

$$+B_{12}\left(\frac{1}{R}\frac{\partial^2\psi_\theta}{\partial x\partial\theta}-\frac{l^2}{R}\frac{\partial^4\psi_\theta}{\partial x^3\partial\theta}-\frac{l^2}{R^3}\frac{\partial^4\psi_\theta}{\partial x\partial\theta^3}\right)-N_h\left(\frac{1}{R}\frac{\partial^2 v}{\partial x\partial\theta}-\frac{1}{R^2}\frac{\partial^2 u}{\partial\theta^2}\right)$$

$$+\frac{A_{66}}{R}\left(\frac{1}{R}\frac{\partial^2 u}{\partial\theta^2}+\frac{\partial^2 v}{\partial x\partial\theta}-\frac{l^2}{R}\frac{\partial^4 u}{\partial x^2\partial\theta^2}-l^2\frac{\partial^4 v}{\partial x^3\partial\theta}-\frac{l^2}{R^3}\frac{\partial^4 u}{\partial\theta^4}-\frac{l^2}{R^2}\frac{\partial^4 v}{\partial x\partial\theta^3}\right)$$

$$+\frac{B_{66}}{R}\left(\frac{1}{R}\frac{\partial^2\psi_x}{\partial\theta^2}+\frac{\partial^2\psi_\theta}{\partial x\partial\theta}-\frac{l^2}{R}\frac{\partial^4\psi_x}{\partial x^2\partial\theta^2}-l^2\frac{\partial^4\psi_\theta}{\partial x^3\partial\theta}-\frac{l^2}{R^3}\frac{\partial^4\psi_x}{\partial\theta^4}-\frac{l^2}{R^2}\frac{\partial^4\psi_\theta}{\partial x\partial\theta^3}\right)$$

$$=\left(1-\mu^2\nabla^2\right)\left(I_0\frac{\partial^2 u}{\partial t^2}+I_1\frac{\partial^2\psi_x}{\partial t^2}\right)$$

(G.1)

$$\delta v : \frac{A_{12}}{R}\left(\frac{\partial^2 u}{\partial x\partial\theta}+l^2\frac{\partial^4 u}{\partial x^3\partial\theta}-\frac{l^2}{R^2}\frac{\partial^4 u}{\partial x\partial\theta^3}\right)+\frac{B_{12}}{R}\left(\frac{\partial^2\psi_x}{\partial x\partial\theta}+l^2\frac{\partial^4\psi_x}{\partial x^3\partial\theta}-\frac{l^2}{R^2}\frac{\partial^4\psi_x}{\partial x\partial\theta^3}\right)$$

$$+\frac{A_{22}}{R}\left(\frac{1}{R}\frac{\partial^2 v}{\partial\theta^2}+\frac{1}{R}\frac{\partial w}{\partial\theta}-\frac{l^2}{R^2}\frac{\partial^4 v}{\partial x^2\partial\theta^2}-\frac{l^2}{R^2}\frac{\partial^3 w}{\partial x^2\partial\theta}-\frac{l^2}{R^3}\frac{\partial^4 v}{\partial\theta^4}-\frac{l^2}{R^3}\frac{\partial^3 w}{\partial\theta^3}\right)$$

$$+\frac{B_{22}}{R}\left(\frac{1}{R}\frac{\partial^2\psi_\theta}{\partial\theta^2}+\frac{l^2}{R}\frac{\partial^4\psi_\theta}{\partial x^2\partial\theta^2}-\frac{l^2}{R^3}\frac{\partial^4\psi_\theta}{\partial\theta^4}\right)-N_2^T\left(\frac{\partial^2 v}{\partial x^2}\right)$$

$$+A_{66}\left(\frac{1}{R}\frac{\partial^2 u}{\partial x\partial\theta}+\frac{\partial^2 v}{\partial x^2}-\frac{l^2}{R}\frac{\partial^4 u}{\partial x^3\partial\theta}-l^2\frac{\partial^4 v}{\partial x^4}-\frac{l^2}{R^3}\frac{\partial^4 u}{\partial x\partial\theta^3}-\frac{l^2}{R^2}\frac{\partial^4 v}{\partial x^2\partial\theta^2}\right)$$

$$+B_{66}\left(\frac{1}{R}\frac{\partial^2\psi_x}{\partial x\partial\theta}+\frac{\partial^2\psi_\theta}{\partial x^2}-\frac{l^2}{R}\frac{\partial^4\psi_x}{\partial x^3\partial\theta}-l^2\frac{\partial^4\psi_\theta}{\partial x^4}-\frac{l^2}{R^3}\frac{\partial^4\psi_x}{\partial x\partial\theta^3}-\frac{l^2}{R^2}\frac{\partial^4\psi_\theta}{\partial x^2\partial\theta^2}\right)$$

$$+\frac{K_s A_{44}}{R}\left(\psi_\theta+\frac{1}{R}\frac{\partial w}{\partial\theta}-\frac{v}{R}-l^2\frac{\partial^2\psi_\theta}{\partial x^2}-\frac{l^2}{R^2}\frac{\partial^3 w}{\partial x^2\partial\theta}+\frac{l^2}{R^2}\frac{\partial^2 v}{\partial x^2}\right.$$

$$\left.-\frac{l^2}{R^2}\frac{\partial^2\psi_\theta}{\partial\theta^2}-\frac{l^2}{R^3}\frac{\partial^3 w}{\partial\theta^3}+\frac{l^2}{R^3}\frac{\partial^2 v}{\partial\theta^2}\right)=\left(1-\mu^2\nabla^2\right)\left(I_0\left[\frac{\partial^2 v}{\partial t^2}\right]+I_1\left[\frac{\partial^2\psi_\theta}{\partial t^2}\right]\right)$$

(G.2)

$$\delta w : \frac{A_{12}}{R}\left(\frac{\partial u}{\partial x}+l^2\frac{\partial^3 u}{\partial x^3}+\frac{l^2}{R^2}\frac{\partial^3 u}{\partial x\partial\theta^2}\right)+\frac{B_{12}}{R}\left(\frac{\partial\psi_x}{\partial x}+l^2\frac{\partial^3\psi_x}{\partial x^3}+\frac{l^2}{R^2}\frac{\partial^3\psi_x}{\partial x\partial\theta^2}\right)$$

$$+\frac{A_{22}}{R}\left(-\frac{1}{R}\frac{\partial v}{\partial\theta}-\frac{w}{R}+\frac{l^2}{R^2}\frac{\partial^3 v}{\partial x^2\partial\theta}+\frac{l^2}{R^2}\frac{\partial^2 w}{\partial x^2}+\frac{l^2}{R^3}\frac{\partial^3 v}{\partial\theta^3}+\frac{l^2}{R^3}\frac{\partial^2 w}{\partial\theta^2}\right)$$

$$+\frac{B_{22}}{R}\left(-\frac{1}{R}\frac{\partial\psi_\theta}{\partial\theta}+\frac{l^2}{R^2}\frac{\partial^3\psi_\theta}{\partial x^2\partial\theta}+\frac{l^2}{R^3}\frac{\partial^3\psi_\theta}{\partial\theta^3}\right)-N_1^T\left(\frac{\partial^2 w}{\partial x^2}\right)$$

$$+K_sA_{55}\left(\frac{\partial\psi_x}{\partial x}-\frac{\partial^2 w}{\partial x^2}-l^2\frac{\partial^3\psi_x}{\partial x^3}-l^2\frac{\partial^4 w}{\partial x^4}-\frac{l^2}{R^2}\frac{\partial^3\psi_x}{\partial x\partial\theta^2}-\frac{l^2}{R^2}\frac{\partial^4 w}{\partial x^2\partial\theta^2}\right)$$

$$+\frac{K_sA_{44}}{R}\left(\frac{\partial\psi_\theta}{\partial\theta}-\frac{1}{R}\frac{\partial^2 w}{\partial\theta^2}-\frac{1}{R}\frac{\partial v}{\partial\theta}-l^2\frac{\partial^3\psi_\theta}{\partial x^2\partial\theta}-\frac{l^2}{R}\frac{\partial^4 w}{\partial x^2\partial\theta^2}+\frac{l^2}{R}\frac{\partial^3 v}{\partial x^2\partial\theta}\right.$$

$$\left.-\frac{l^2}{R^2}\frac{\partial^3\psi_\theta}{\partial\theta^3}-\frac{l^2}{R^3}\frac{\partial^4 w}{\partial\theta^4}+\frac{l^2}{R^3}\frac{\partial^3 v}{\partial\theta^3}\right)=\left(1-\mu^2\nabla^2\right)\left(I_0\left(\frac{\partial^2 w}{\partial t^2}\right)\right)$$

$$\tag{G.3}$$

$$\delta\psi_x : B_{11}\left(\frac{\partial^2 u}{\partial x^2}-l^2\frac{\partial^4 u}{\partial x^4}-\frac{l^2}{R^2}\frac{\partial^4 u}{\partial x^2\partial\theta^2}\right)+D_{11}\left(\frac{\partial^2\psi_x}{\partial x^2}-l^2\frac{\partial^4\psi_x}{\partial x^4}-\frac{l^2}{R^2}\frac{\partial^4\psi_x}{\partial x^2\partial\theta^2}\right)$$

$$+B_{12}\left(\frac{1}{R}\frac{\partial^2 v}{\partial x\partial\theta}+\frac{1}{R}\frac{\partial w}{\partial x}-\frac{l^2}{R^2}\frac{\partial^4 v}{\partial x^3\partial\theta}-\frac{l^2}{R^2}\frac{\partial^3 w}{\partial x^3}-\frac{l^2}{R^3}\frac{\partial^4 v}{\partial x\partial\theta^3}-\frac{l^2}{R^3}\frac{\partial^3 w}{\partial x\partial\theta^2}\right)$$

$$+D_{12}\left(\frac{1}{R}\frac{\partial^2\psi_\theta}{\partial x\partial\theta}+\frac{l^2}{R^2}\frac{\partial^4\psi_\theta}{\partial x^3\partial\theta}-\frac{l^2}{R^3}\frac{\partial^4\psi_\theta}{\partial x\partial\theta^3}\right)$$

$$+\frac{B_{66}}{R}\left(\frac{1}{R}\frac{\partial^2 u}{\partial\theta^2}+\frac{\partial^2 v}{\partial x\partial\theta}-\frac{l^2}{R}\frac{\partial^4 u}{\partial x^2\partial\theta^2}-l^2\frac{\partial^4 v}{\partial x^3\partial\theta}-\frac{l^2}{R^3}\frac{\partial^4 u}{\partial\theta^4}-\frac{l^2}{R^2}\frac{\partial^4 v}{\partial x\partial\theta^3}\right)$$

$$+\frac{D_{66}}{R}\left(\frac{1}{R}\frac{\partial^2\psi_x}{\partial\theta^2}+\frac{\partial^2\psi_\theta}{\partial x\partial\theta}-\frac{l^2}{R}\frac{\partial^4\psi_x}{\partial x^2\partial\theta^2}-l^2\frac{\partial^4\psi_\theta}{\partial x^3\partial\theta}-\frac{l^2}{R^3}\frac{\partial^4\psi_x}{\partial\theta^4}-\frac{l^2}{R^2}\frac{\partial^4\psi_\theta}{\partial x\partial\theta^3}\right)$$

$$-K_sA_{55}\left(\psi_x+\frac{\partial w}{\partial x}-l^2\frac{\partial^2\psi_x}{\partial x^2}-l^2\frac{\partial^3 w}{\partial x^3}-\frac{l^2}{R^2}\frac{\partial^2\psi_x}{\partial\theta^2}-\frac{l^2}{R^2}\frac{\partial^3 w}{\partial x\partial\theta^2}\right)$$

$$=\left(1-\mu^2\nabla^2\right)\left(I_1\frac{\partial^2 u}{\partial t^2}+I_2\frac{\partial^2\psi_x}{\partial t^2}\right)$$

$$\tag{G.4}$$

$$\delta\psi_\theta : \frac{B_{12}}{R}\left(\frac{\partial^2 u}{\partial x \partial\theta} - l^2\frac{\partial^4 u}{\partial x^3\partial\theta} - \frac{l^2}{R^2}\frac{\partial^4 u}{\partial x\partial\theta^3}\right) + \frac{D_{12}}{R}\left(\frac{\partial^2\psi_x}{\partial x\partial\theta} - l^2\frac{\partial^4\psi_x}{\partial x^3\partial\theta} - \frac{l^2}{R^2}\frac{\partial^4\psi_x}{\partial x\partial\theta^3}\right)$$

$$+\frac{B_{22}}{R}\left(\frac{1}{R}\frac{\partial^2 v}{\partial\theta^2} - \frac{1}{R}\frac{\partial w}{\partial\theta} - \frac{l^2}{R^2}\frac{\partial^4 v}{\partial x^2\partial\theta^2} - \frac{l^2}{R^2}\frac{\partial^3 w}{\partial x^2\partial\theta} - \frac{l^2}{R^3}\frac{\partial^4 v}{\partial\theta^4} - \frac{l^2}{R^3}\frac{\partial^3 w}{\partial\theta^3}\right)$$

$$+\frac{D_{22}}{R}\left(\frac{1}{R}\frac{\partial^2\psi_\theta}{\partial\theta^2} - \frac{l^2}{R}\frac{\partial^4\psi_\theta}{\partial x^2\partial\theta^2} - \frac{l^2}{R^3}\frac{\partial^4\psi_\theta}{\partial\theta^4}\right)$$

$$+B_{66}\left(\frac{1}{R}\frac{\partial^2 u}{\partial x\partial\theta} - \frac{\partial^2 v}{\partial x^2} - \frac{l^2}{R}\frac{\partial^4 u}{\partial x^3\partial\theta} - l^2\frac{\partial^4 v}{\partial x^4} - \frac{l^2}{R^3}\frac{\partial^4 u}{\partial x\partial\theta^3} - \frac{l^2}{R^2}\frac{\partial^4 v}{\partial x^2\partial\theta^2}\right)$$

$$+D_{66}\left(\frac{1}{R}\frac{\partial^2\psi_x}{\partial x\partial\theta} - \frac{\partial^2\psi_\theta}{\partial x^2} - \frac{l^2}{R}\frac{\partial^4\psi_x}{\partial x^3\partial\theta} - l^2\frac{\partial^4\psi_\theta}{\partial x^4} - \frac{l^2}{R^3}\frac{\partial^4\psi_x}{\partial x\partial\theta^3} - \frac{l^2}{R^2}\frac{\partial^4\psi_\theta}{\partial x^2\partial\theta^2}\right)$$

$$-K_s A_{44}\left(\psi_\theta + \frac{1}{R}\frac{\partial w}{\partial\theta} - \frac{v}{R} - l^2\frac{\partial^2\psi_\theta}{\partial x^2} - \frac{l^2}{R}\frac{\partial^3 w}{\partial x^2\partial\theta} + \frac{l^2}{R}\frac{\partial^2 v}{\partial x^2} - \frac{l^2}{R^2}\frac{\partial^2\psi_\theta}{\partial\theta^2} - \frac{l^2}{R^3}\frac{\partial^3 w}{\partial\theta^3}\right)$$

$$+\frac{l^2}{R^3}\frac{\partial^2 v}{\partial\theta^2}\right) = \left(1-\mu^2\nabla^2\right)\left[I_1\left(\frac{\partial^2 v}{\partial t^2}\right) + I_2\left(\frac{\partial^2\psi_\theta}{\partial t^2}\right)\right]$$

$$(G.5)$$

where the defined parameters in Eqs. (G.1)–(G.5) are described as:

$$\left\{A_{11}\ A_{12}\ A_{22}\ A_{66}\ A_{44}\ A_{55}\right\} = \int_{-h/2}^{h/2}\left\{C_{11}\quad C_{12}\quad C_{22}\quad C_{66}\quad C_{44}\quad C_{55}\right\}dz$$

$$\left\{B_{11}\quad B_{12}\quad B_{22}\quad B_{66}\right\} = \int_{-h/2}^{h/2}\left\{C_{11}\quad C_{12}\quad C_{22}\quad C_{66}\right\}zdz$$

$$\left\{D_{11}\quad D_{12}\quad D_{22}\quad D_{66}\right\} = \int_{-h/2}^{h/2}\left\{C_{11}\quad C_{12}\quad C_{22}\quad C_{66}\right\}z^2dz$$

$$\left\{I_0\quad I_1\quad I_2\right\} = \int_{-h/2}^{h/2}\rho(z,T)\left\{1\quad z\quad z^2\right\}zdz$$

$$(G.6)$$

APPENDIX H

$$k_{11} = -A\beta^2$$

$$k_{12} = Bi\beta^3$$

$$k_{13} = B_s i\beta^3$$

$$k_{22} = -D\beta^4 - k_W - \left(k_P - N^T - N^H\right)\beta^2$$

$$k_{23} = -D_s\beta^4 - k_W - \left(k_P - N^T - N^H\right)\beta^2$$

$$k_{33} = -H_s\beta^4 - A_s\beta^2 - k_W - \left(k_P - N^T - N^H\right)\beta^2$$

$$m_{11} = I_0$$

$$m_{12} = -I_1\beta i$$

$$m_{13} = -J_1\beta i$$

$$m_{22} = I_0 + I_2\beta^2$$

$$m_{23} = I_0 + J_2\beta^2$$

$$m_{33} = I_0 + K_2\beta^2$$

APPENDIX I

$$k_{11} = -A_{11}\beta_x^2 - A_{66}\beta_y^2$$

$$k_{12} = \left(A_{12} + A_{66}\right)\beta_x\beta_y$$

$$k_{13} = iB_{11}\beta_x^3 + i\left(B_{12} + 2B_{66}\right)\beta_y^2\beta_x$$

$$k_{14} = iB_{11}^s\beta_x^3 + i\left(B_{12}^s + 2B_{66}^s\right)\beta_y^2\beta_x$$

$$k_{22} = -A_{22}\beta_x^2 - A_{66}\beta_y^2$$

$$k_{23} = iB_{22}\beta_y^3 + i\left(B_{12} + 2B_{66}\right)\beta_x^2\beta_y$$

$$k_{24} = B_{22}^s\beta_y^3 + i\left(B_{12}^s + 2B_{66}^s\right)\beta_x^2\beta_y$$

$$k_{33} = -D_{11}\beta_x^4 - 2\times\left(D_{12} + 2D_{66}\right)\beta_x^2\beta_y^2$$
$$- D_{22}\beta_y^2 + \left(N^T + N^H\right)\left(\beta_x^2 + \beta_y^2\right)$$

$$k_{34} = -D_{11}^s\beta_x^4 - 2\times\left(D_{12}^s + 2D_{66}^s\right)\beta_x^2\beta_y^2$$
$$- D_{22}^s\beta_y^2 + \left(N^T + N^H\right)\left(\beta_x^2 + \beta_y^2\right)$$

$$k_{44} = -H_{11}^s\beta_x^4 - 2\times\left(H_{12}^s + 2H_{66}^s\right)\beta_x^2\beta_y^2$$
$$- H_{22}^s\beta_y^2 + \left(N^T + N^H\right)\left(\beta_x^2 + \beta_y^2\right) - A_{55}^s\beta_x^2 - A_{44}^s\beta_y^2$$

$$m_{11} = m_{22} = I_0$$

$$m_{12} = 0$$

$$m_{13} = -I_1 \beta_x i$$

$$m_{14} = -J_1 \beta_x i$$

$$m_{23} = -I_1 \beta_y i$$

$$m_{24} = -J_1 \beta_y i$$

$$m_{33} = I_0 - I_2 \left(\beta_x^2 + \beta_y^2 \right)$$

$$m_{34} = I_0 - J_2 \left(\beta_x^2 + \beta_y^2 \right)$$

$$m_{44} = I_0 - K_2 \left(\beta_x^2 + \beta_y^2 \right)$$

REFERENCES

[1] Ebrahimi, F., & Salari, E. (2015). Thermal buckling and free vibration analysis of size dependent Timoshenko FG nanobeams in thermal environments. *Composite Structures*, *128*, 363–380.

[2] Eringen, A. C. (1972). Linear theory of nonlocal elasticity and dispersion of plane waves. *International Journal of Engineering Science*, *10*(5), 425–435.

[3] Eringen, A. C. (1983). On differential equations of nonlocal elasticity and solutions of screw dislocation and surface waves. *Journal of Applied Physics*, *54*(9), 4703–4710.

[4] Ebrahimi, F., & Salari, E. (2015). Thermo-mechanical vibration analysis of a single-walled carbon nanotube embedded in an elastic medium based on higher-order shear deformation beam theory. *Journal of Mechanical Science and Technology*, *29*(9), 3797–3803.

[5] Ebrahimi, F., Shaghaghi, G. R., & Boreiry, M. (2016). An investigation into the influence of thermal loading and surface effects on mechanical characteristics of nanotubes. *Structural Engineering and Mechanics*, *57*(1), 179–200.

[6] Ebrahimi, F., & Barati, M. R. (2016). A nonlocal higher-order shear deformation beam theory for vibration analysis of size-dependent functionally graded nanobeams. *Arabian Journal for Science and Engineering*, *41*(5), 1679–1690.

[7] Ebrahimi, F., & Barati, M. R. (2016). Vibration analysis of nonlocal beams made of functionally graded material in thermal environment. *The European Physical Journal Plus*, *131*(8), 279.

[8] Ebrahimi, F., & Barati, M. R. (2016). A unified formulation for dynamic analysis of nonlocal heterogeneous nanobeams in hygro-thermal environment. *Applied Physics A*, *122*(9), 792.

[9] Ebrahimi, F., Barati, M. R., & Haghi, P. (2016). Nonlocal thermo-elastic wave propagation in temperature-dependent embedded small-scaled nonhomogeneous beams. *The European Physical Journal Plus*, *131*(11), 383.

[10] Ebrahimi, F., Barati, M. R., & Dabbagh, A. (2016). Wave dispersion characteristics of axially loaded magneto-electro-elastic nanobeams. *Applied Physics A*, *122*(11), 949.

[11] Ebrahimi, F., Dabbagh, A., & Barati, M. R. (2016). Wave propagation analysis of a size-dependent magneto-electro-elastic heterogeneous nanoplate. *The European Physical Journal Plus*, *131*(12), 433.

[12] Ebrahimi, F., & Barati, M. R. (2016). Static stability analysis of smart magneto-electro-elastic heterogeneous nanoplates embedded in an elastic medium based on a four-variable refined plate theory. *Smart Materials and Structures*, *25*(10), 105014.

[13] Ebrahimi, F., & Barati, M. R. (2016). Thermal buckling analysis of size-dependent FG nanobeams based on the third-order shear deformation beam theory. *Acta Mechanica Solida Sinica*, *29*(5), 547–554.

[14] Ebrahimi, F., & Barati, M. R. (2016). Magneto-electro-elastic buckling analysis of nonlocal curved nanobeams. *The European Physical Journal Plus*, *131*(9), 346.

[15] Ebrahimi, F., & Hosseini, S. H. S. (2016). Thermal effects on nonlinear vibration behavior of viscoelastic nanosize plates. *Journal of Thermal Stresses*, *39*(5), 606–625.

[16] Ebrahimy, F., & Hosseini, S. H. S. (2016). Nonlinear electroelastic vibration analysis of NEMS consisting of double-viscoelastic nanoplates. *Applied Physics A*, *122*(10), 922.

[17] Ebrahimi, F., & Dabbagh, A. (2017). Wave propagation analysis of smart rotating porous heterogeneous piezo-electric nanobeams. *The European Physical Journal Plus*, *132*, 1–15.

[18] Ebrahimi, F., Barati, M. R., & Haghi, P. (2017). Wave propagation analysis of size-dependent rotating inhomogeneous nanobeams based on nonlocal elasticity theory. *Journal of Vibration and Control*, 1077546317711537.

[19] Ebrahimi, F., & Barati, M. R. (2017). Vibration analysis of viscoelastic inhomogeneous nanobeams incorporating surface and thermal effects. *Applied Physics A*, *123*(1), 5.

[20] Ebrahimi, F., Barati, M. R., & Haghi, P. (2016). Thermal effects on wave propagation characteristics of rotating strain gradient temperature-dependent functionally graded nanoscale beams. *Journal of Thermal Stresses*, 1–13.

[21] Ebrahimi, F., Barati, M. R., & Dabbagh, A. (2016). A nonlocal strain gradient theory for wave propagation analysis in temperature-dependent inhomogeneous nanoplates. *International Journal of Engineering Science*, *107*, 169–182.

[22] Ebrahimi, F., & Dabbagh, A. (2017). On flexural wave propagation responses of smart FG magneto-electro-elastic nanoplates via nonlocal strain gradient theory. *Composite Structures*, *162*, 281–293.

[23] Ebrahimi, F., & Dabbagh, A. (2017). Nonlocal strain gradient based wave dispersion behavior of smart rotating magneto-electro-elastic nanoplates. *Materials Research Express*, *4*(2), 025003.

[24] Ebrahimi, F., & Barati, M. R. (2017). Vibration analysis of piezoelectrically actuated curved nanosize FG beams via a nonlocal strain-electric field gradient theory. *Mechanics of Advanced Materials and Structures*, 1–10.

[25] Ebrahimi, F., & Barati, M. R. (2017). Through-the-length temperature distribution effects on thermal vibration analysis of nonlocal strain-gradient axially graded nanobeams subjected to nonuniform magnetic field. *Journal of Thermal Stresses*, *40*(5), 548–563.

[26] Ebrahimi, F., & Barati, M. R. (2017). Damping vibration analysis of smart piezoelectric polymeric nanoplates on viscoelastic substrate based on nonlocal strain gradient theory. *Smart Materials and Structures*, *26*(6), 065018.

[27] Ebrahimi, F., & Barati, M. R. (2017). Buckling analysis of nonlocal strain gradient axially functionally graded nanobeams resting on variable elastic medium. *Proceedings of the Institution of Mechanical Engineers, Part C: Journal of Mechanical Engineering Science*, 0954406217713518.

[28] Ebrahimi, F., & Dabbagh, A. (2019). A comprehensive review on modeling of nanocomposite materials and structures. *Journal of Computational Applied Mechanics*, *50*(1), 197–209.

[29] Arani, A. G., & Jalaei, M. H. (2016). Nonlocal dynamic response of embedded single-layered graphene sheet via analytical approach. *Journal of Engineering Mathematics*, *98*(1), 129–144.

[30] Lee, C., Wei, X., Kysar, J. W., & Hone, J. (2008). Measurement of the elastic properties and intrinsic strength of monolayer graphene. *Science*, *321*(5887), 385–388.

[31] Seol, J. H., Jo, I., Moore, A. L., Lindsay, L., Aitken, Z. H., Pettes, M. T., … & Mingo, N. (2010). Two-dimensional phonon transport in supported graphene. *Science*, *328*(5975), 213–216.

[32] Arani, A. G., & Ebrahimi, F. (2015). Modeling and Control of a Smart Single-Layer Graphene Sheet. In Graphene-New Trends and Developments. IntechOpen.

[33] Ebrahimi, F., & Barati, M. R. (2019). Hygrothermal effects on static stability of embedded single-layer graphene sheets based on nonlocal strain gradient elasticity theory. *Journal of Thermal Stresses*, *42*(12), 1535–1550.

[34] Ebrahimi, F., & Dabbagh, A. (2018). On wave dispersion characteristics of double-layered graphene sheets in thermal environments. *Journal of Electromagnetic Waves and Applications*, *32*(15), 1869–1888.

[35] Ebrahimi, F., & Dabbagh, A. (2019). Wave Propagation Responses of Double-Layered Graphene Sheets in Hygrothermal Environment. Handbook of Graphene, Volume 8: Technology and Innovations, 289.

[36] Ebrahimi, F., & Dabbagh, A. (2020). Viscoelastic wave propagation analysis of axially motivated double-layered graphene sheets via nonlocal strain gradient theory. *Waves in Random and Complex Media*, *30*(1), 157–176.

[37] Ebrahimi, F., & Dabbagh, A. (2019). Thermo-mechanical wave dispersion analysis of nonlocal strain gradient single-layered graphene sheet rested on elastic medium. *Microsystem Technologies*, *25*, 587–597.

[38] Fardshad, R. E., Mohammadi, Y., & Ebrahimi, F. (2019). Modeling wave propagation in graphene sheets influenced by magnetic field via a refined trigonometric two-variable plate theory. *Structural Engineering and Mechanics, An Int'l Journal*, *72*(3), 329–338.

[39] Ebrahimi, F., & Dabbagh, A. (2018). Effect of humid-thermal environment on wave dispersion characteristics of single-layered graphene sheets. *Applied Physics A*, *124*, 1–11.

[40] Ebrahimi, F., & Shafiei, N. (2017). Influence of initial shear stress on the vibration behavior of single-layered graphene sheets embedded in an elastic medium based on Reddy's higher-order shear deformation plate theory. *Mechanics of Advanced Materials and Structures*, *24*(9), 761–772.

[41] Ebrahimi, F., Dabbagh, A., & Civalek, Ö. (2019). Vibration analysis of magnetically affected graphene oxide-reinforced nanocomposite beams. *Journal of Vibration and Control*, 1077546319861002.

[42] Ebrahimi, F., Nouraei, M., & Dabbagh, A. (2019). Thermal vibration analysis of embedded graphene oxide powder-reinforced nanocomposite plates. *Engineering with Computers*, 1–17.

[43] Ebrahimi, F., Habibi, M., & Safarpour, H. (2018). On modeling of wave propagation in a thermally affected GNP-reinforced imperfect nanocomposite shell. *Engineering with Computers*, 1–15.

[44] Ebrahimi, F., Nouraei, M., & Dabbagh, A. (2019). Modeling vibration behavior of embedded graphene-oxide powder-reinforced nanocomposite plates in thermal environment. *Mechanics Based Design of Structures and Machines*, 1–24.

[45] Qaderi, S., Ebrahimi, F., & Mahesh, V. (2019). Free vibration analysis of graphene platelets–reinforced composites plates in thermal environment based on higher-order shear deformation plate theory. *International Journal of Aeronautical and Space Sciences*, *20*, 902–912.

[46] Qaderi, S., & Ebrahimi, F. (2022). Vibration analysis of polymer composite plates reinforced with graphene platelets resting on two-parameter viscoelastic foundation. *Engineering with Computers*, 1–17.

[47] Qaderi, S., Ebrahimi, F., & Seyfi, A. (2019). An investigation of the vibration of multi-layer composite beams reinforced by graphene platelets resting on two parameter viscoelastic foundation. *SN Applied Sciences*, *1*, 1–10.

[48] Ebrahimi, F., & Qaderi, S. (2019). Stability analysis of embedded graphene platelets reinforced composite plates in thermal environment. *The European Physical Journal Plus*, *134*(7), 349.

[49] Qaderi, S., Ebrahimi, F., & Vinyas, M. (2019). Dynamic analysis of multi-layered composite beams reinforced with graphene platelets resting on two-parameter viscoelastic foundation. *The European Physical Journal Plus*, *134*, 1–11.

[50] Shariati, A., Qaderi, S., Ebrahimi, F., & Toghroli, A. (2022). On buckling characteristics of polymer composite plates reinforced with graphene platelets. *Engineering with Computers*, 1–12.

[51] Shokrgozar, A., Ghabussi, A., Ebrahimi, F., Habibi, M., & Safarpour, H. (2022). Viscoelastic dynamics and static responses of a graphene nanoplatelets-reinforced composite cylindrical microshell. *Mechanics Based Design of Structures and Machines*, *50*(2), 509–536.

[52] Ebrahimi, F., & Dabbagh, A. (2018). Wave dispersion characteristics of embedded graphene platelets-reinforced composite microplates. *The European Physical Journal Plus*, *133*, 1–13.

[53] Selvamani, R., & Ebrahimi, F. (2021). Wave propagation characteristics of thermoelastic graphene platelet reinforced polygonal ring with phase lags. *Computational Mathematics and Modeling*, *32*(4), 453–477.

[54] Ebrahimi, F., Seyfi, A., Dabbagh, A., & Tornabene, F. (2019). Wave dispersion characteristics of porous graphene platelet-reinforced composite shells. *Structural Engineering and Mechanics, An Int'l Journal*, *71*(1), 99–107.

[55] Ebrahimi, F., Mohammadi, K., Barouti, M. M., & Habibi, M. (2021). Wave propagation analysis of a spinning porous graphene nanoplatelet-reinforced nanoshell. *Waves in Random and Complex Media*, *31*(6), 1655–1681.

[56] Ebrahimi, F., Habibi, M., & Safarpour, H. (2019). On modeling of wave propagation in a thermally affected GNP-reinforced imperfect nanocomposite shell. *Engineering with Computers*, *35*, 1375–1389.

[57] Ebrahimi, F., Dabbagh, A., & Taheri, M. (2021). Vibration analysis of porous metal foam plates rested on viscoelastic substrate. *Engineering with Computers*, *37*, 3727–3739.

[58] Ebrahimi, F., Dabbagh, A., & Rastgoo, A. (2019). Vibration analysis of porous metal foam shells rested on an elastic substrate. *The Journal of Strain Analysis for Engineering Design, 54*(3), 199–208.

[59] Selvamani, R., Rexy, J. B., & Ebrahimi, F. (2022). Finite element modeling and analysis of piezoelectric nanoporous metal foam nanobeam under hygro and nonlinear thermal field. *Acta Mechanica, 233*(8), 3113–3132.

[60] Ebrahimi, F., & Seyfi, A. (2021). A wave propagation study for porous metal foam beams resting on an elastic foundation. *Waves in Random and Complex Media*, 1–15.

[61] Ebrahimi, F., & Seyfi, A. (2022). Studying propagation of wave of metal foam rectangular plates with graded porosities resting on Kerr substrate in thermal environment via analytical method. *Waves in Random and Complex Media, 32*(2), 832–855.

[62] Ebrahimi, F., & Seyfi, A. (2022). Studying propagation of wave in metal foam cylindrical shells with graded porosities resting on variable elastic substrate. *Engineering with Computers*, 1–17.

[63] Ebrahimi, F., & Seyfi, A. (2022). On hygrothermal wave dispersion characteristics of embedded graphene foam. *Waves in Random and Complex Media*, 1–20.

[64] Ebrahimi, F., & Seyfi, A. (2022). On wave propagation characteristics of hygrothermally excited graphene foam plates. *Waves in Random and Complex Media*, 1–20.

[65] Ebrahimi, F., Nouraei, M., & Seyfi, A. (2022). Wave dispersion characteristics of thermally excited graphene oxide powder-reinforced nanocomposite plates. *Waves in Random and Complex Media, 32*(1), 204–232.

6 Wave Dispersion Characteristics of Reinforced Nanocomposites

6.1 BACKGROUND

In this chapter, we will investigate the wave propagation of carbon nanotubes. The agglomeration of CNT is explored in relation to the development of reinforced nanocomposites and hybrid nanocomposite structures, such as beams, plates, and shells. The study aims to analyze the impact of nanoparticle agglomeration on wave dispersion in multi-scale hybrid nanocomposites. Furthermore, an updated higher-order theory is employed to calculate the kinetic relationships, eliminating the requirement for an additional component to accommodate shear deformation. The achievement was facilitated through the application of the higher-order theory. To attain this goal of homogenizing the component material, the Eshelby–Mori–Tanaka model and the rule of the mixture are used. Furthermore, the governing equations of the system can be obtained by applying Hamilton's principle. Subsequently, the governing equations are subject to analytical solution to facilitate an increase in the wave frequency. The dispersion solution is obtained by resolving the eigenvalue problem, and an analytical methodology is employed to solve the governing equations. Upon completion of the process, it is possible to acquire the wave frequency and phase velocity of the nanocomposite structure. This study examines the impact of different factors, including longitudinal and circumferential wave numbers, volume fraction of carbon fibers within the cluster, and volume fraction of carbon nanotubes within the cluster, on the changes in wave frequency and phase velocity of multi-scale hybrid nanocomposites. The findings are presented through a series of illustrations.

6.1.1 A REVIEW ON WAVE DISPERSION ANALYSIS OF NANOCOMPOSITES

The potentials and capabilities that composites possess are truly astonishing. Some examples of these potentials and capabilities include resistance to fatigue and higher operating temperatures. Composites also have high particular strengths and specific moduli. As a result of these advantages, composites have quickly established themselves as a leading candidate in the competition to find applications in a variety of industries. This class of materials may be put to use in a wide number of technological applications, such as the building of aeronautical

DOI: 10.1201/9781003270263-6

structures, lightweight optical systems, automobiles, maritime facilities, and chemical plants. You might also categorize them as composites that are based on metal, composites that are based on ceramic, or composites that are based on polymer. The mechanical characteristics of composite materials have been the subject of a significant amount of research and study, which has then resulted in scientists reporting their results. Researchers and scientists have also done this research and study. When at least one of the dimensions of an element in a composite material is on the nanoscale, the composite material is referred to as nanocomposites. Nanocomposites are a kind of composite material. It is feasible to use nanoparticles as nano-reinforcement, which will lead to an improvement in the structure's mechanical properties. This will be accomplished as a consequence of the improvement. Nanocomposites have attracted the interest of a sizeable number of scientists who investigate the static, dynamic, vibrational, and stable behavior of structures for the reason that was just indicated. Materials that are based on carbon, such as carbon nanotubes (CNT), graphene, and graphene platelets (GPL), are attractive candidates for use as reinforcement in composites. For instance, Al-Furjan and coworkers [1] looked into the wave dispersion properties of spinning cylindrical shells made of laminated nanocomposite materials. Their analysis was based on four distinct continuum mechanics models. Ebrahimi and Dabbagh [2] gave a comprehensive study on the topic of modeling nanocomposite materials and structures. The review was written in the form of an article. In a separate piece of research, the authors [3] did a review-style examination of the effects that the entanglement and waviness of nanotubes have on the mechanical characteristics of CNTR polymer nanocomposites. This analysis was carried out in a separate piece of research. Nouraei and coworkers [4] developed a model to explore the dynamic properties of a thermally impacted embedded laminated nanocomposite beam with multi-scale hybrid reinforcement. Additionally, Ebrahimi and colleagues [5] studied the impact of the curved shape of the CNTs as well as the viscoelastic features of the polymer on the dynamic response of hybrid nanocomposite beams. Dabbagh et al. [6] proposed a finite element vibration analysis of multi-scale hybrid nanocomposite beams, which was based on an improved beam theory. The approach of finite elements was used throughout the course of this research. Dabbagh and Ebrahimi [7] reported on the post-buckling analysis of meta-nanocomposite beams by taking into consideration the aggregation of CNTs in another piece of study that they conducted. In addition to this, they presented a revised theory, which they put to use in an investigation of the static stability and thermal buckling of agglomerated multi-scale hybrid nanocomposites [8, 9]. A multi-scale hybrid nanocomposite is a unique composite material that may be introduced as a structure that includes reinforcements of diverse scales. This structure can be created by introducing it as a multi-scale hybrid nanocomposite. These reinforcements are embedded into the structure's matrix in various locations. It possesses mechanical capabilities that are superior to those of other reinforced composites when compared to the features of those other composites. In light of this fact, a number of researchers have lately investigated the responses shown by multi-scale hybrid nanocomposite structures.

Safarpour et al. [10] looked at the sample in order to investigate the nonlinear dynamics of a multi-scale hybrid nanocomposite disk. In addition, Ebrahimi and Habibi [11, 12] used the finite element method (FEM) within the framework of high-speed dynamic testing (HSDT) to show that polymer–CNT–fiber multi-scale nanocomposite structures exhibit nonlinear eccentric low-velocity impact response and nonlinear dynamic features. This allowed them to demonstrate both the nonlinear eccentric low-velocity impact response and the nonlinear dynamic features of the structures. The hygrothermal effect modeling framework was taken into consideration while doing this. Ebrahimi and Dabbagh [13] did research to investigate the vibrational behavior of multi-scale hybrid (CF/CNTs/epoxy) nanocomposite beams while they were resting on an elastic foundation in a temperature environment. Specifically, the researchers were interested in how the beams behaved while they were in the temperature environment. Their results were derived from an enhanced model of the beam.

Ebrahimi and his colleagues published a number of ground-breaking works in the field of hybrid nanocomposite research, some of the most notable of which were the static stability analysis [14–16], the vibration analysis [17–21], and the nonlinear forced vibrations analysis [22]. Their most recent paper, given in reference [23], focused on the post-buckling behavior of imperfect multi-scale hybrid nanocomposite beams that were resting on a nonlinearly stiff substrate. The wave propagation characteristics of carbon nanotubes-reinforced composite (CNTRC) beams were examined by Ebrahimi and Rostami [24] when the beams were at rest on an elastic base. They were able to do this by taking into account four distinct patterns of CNT dispersion inside a polymer matrix. The new book *Mechanics of Multiscale Hybrid Nanocomposites* by Ebrahimi and Dabbagh [25] provides a practical and application-based analysis of the static and dynamic behaviors of multi-scale hybrid nanocomposites. Their book details the two-step micromechanical homogenization procedures used to forecast the mechanical behavior and material features of nanocomposites. Energy-based methods are used to include the strain-displacement relations of shear deformable beam, plate, and shell theories. Finally, the book concludes with a discussion of the implications of these findings. These techniques can be found in their book. The implications of utilizing a variety of nanofillers are broken down in depth, presenting readers with the most effective strategies for increasing the stiffness of nanocomposite materials. Methods of solving the problem were described in detail, including analytical (Navier, Galerkin, etc.) and numerical (Ritz, Rayleigh-Ritz, etc.) approaches, along with examples and procedures. Ebrahimi and Dabbagh [26] highlighted the static and dynamic behaviors of nanocomposite single- or multi-layered structures within the context of continuum mechanics–based techniques in their other innovative and comprehensive book. This book was written by Ebrahimi and Dabbagh. They evaluated the mechanical properties of polymeric matrices that had been enhanced by using a variety of nanofillers and nanoparticles such as graphene platelets, carbon nanotubes, and graphene oxides. In his second book, Ebrahimi offered new trends and advancements of nanocomposites [27], wherein this one includes equivalent characteristics of nanocomposites that may be produced by homogenization procedures that are based on micromechanics approaches. The wave dispersion characteristics of agglomerated multi-scale hybrid nanocomposite beams, plates, and shells were investigated by researchers Ebrahimi et al. [28–32]. In addition to this, we also took into account the effect that the agglomeration of nanoparticles has on the wave dispersion of multi-scale hybrid nanocomposites. In these investigations, the component material consists of both macro- and nano reinforcements, which were evenly distributed throughout the polymer matrix. These reinforcements were measured in microns. The process of homogenization was carried out in line with well-known micromechanical methodologies, such as the Eshelby–Mori–Tanaka model and the rule of the mixing. These approaches were used in order to achieve the desired results. According to the results of their inquiry, the mechanical responses of the system are proven to be lowered if the nanotubes are positioned within the clusters. This was shown to be the case whenever the clusters included the nanotubes.

6.2 CNTR NANOCOMPOSITE BEAM WAVE DISPERSION CHARACTERISTICS

As can be seen in Figure 6.1, it is assumed that the CNTR nanocomposite beam is supported by a Winkler–Pasternak elastic foundation using the coordinate system that has been defined. The nanocomposite beam is assumed to have a length L, which corresponds to the beam's length, and a thickness h, which represents the structure's thickness. It is required to mention that carbon-based reinforcements are used, namely carbon fiber and carbon nanotubes [28].

The Eshelby–Mori–Tanaka model [26] is used to predict the effective mechanical characteristics of multi-scale hybrid nanocomposites when CNT aggregation is taken into consideration. This model's goal is to estimate the effective mechanical properties of the nanocomposites. In addition to this, the rule of mixing is used so that one can account for the distribution of the cluster fractions (CFs) across the matrix. Today, we shall discuss the efficacy of carbon fiber-reinforced (CFR) composites by presenting the facts listed below: [26]

$$E_{11} = V_F E_{11}^F + V_{NCM} E^{NCM} \tag{6.1}$$

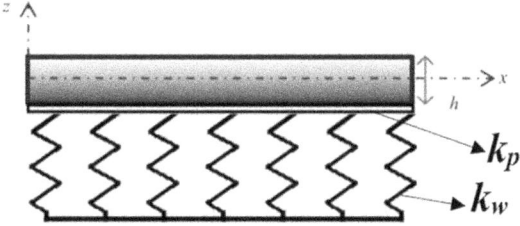

FIGURE 6.1 The shape of a multi-scale hybrid composite beam embedded on an elastic medium.

$$\frac{1}{E_{22}} = \frac{1}{E_{22}^F} + \frac{V_{NCM}}{E^{NCM}} - V_F V_{NCM} - \frac{\dfrac{\left(v^F\right)^2 E^{NCM}}{E_{22}^F} + \dfrac{\left(v^{NCM}\right)^2 E_{22}^F}{E^{NCM}} - 2v^F v^{NCM}}{V_F E_{22}^F + V_{NCM} E^{NCM}} \qquad (6.2)$$

$$\frac{1}{G_{12}} = \frac{V_F}{G_{12}^F} + \frac{V_{NCM}}{G^{NCM}} \qquad (6.3)$$

$$\rho = V_F \rho^F + V_{NCM} \rho^{NCM} \qquad (6.4)$$

$$v_{12} = V_F v^F + V_{NCM} v^{NMC} \qquad (6.5)$$

where E is Young's modulus, G is shear modulus, v is Poisson's ratio, and ρ mass density. Therewith, the superscripts NCM and F represent nanocomposite matrix and fiber, respectively. Furthermore, V_F denotes volume fractions of fiber, and V_{NCM} stands for volume fractions of nanocomposite matrix, respectively. It is clear that the noted volume fractions can be related to each other by:

$$V_F + V_{NCM} = 1 \qquad (6.6)$$

The next step is to investigate how carbon nanotubes (CNTs) influence the effective mechanical characteristics of materials. The amount of influence that CNTs have on the structure's stiffness is significantly increased when they are believed to be nanoparticles. There are instances in which they are not spread equally across the matrix, and the quantity of CNTs that may be found in any location can vary. It is important to note that the aggregation of CNTs in some locations would have a significant impact on the investigation of the nanocomposite's mechanical behavior. This is something that should be mentioned. CNTs have been broken up into two distinct categories as a result of further research. The CNTs inside the inclusions are linked to one set, while the CNTs embedded in the matrix are connected to the other. The volume of CNTs within the inclusions W_r^{in} (clusters) may be related to the volume of CNTs in the matrix by the following formula [26]:

$$W_r = W_r^{in} + W_r^M \qquad (6.7)$$

The volume of CNTs as a percentage of the whole structure may be expressed as:

$$W = W_r + W_M \qquad (6.8)$$

where W_M is the volume of the matrix the CNTs are dispersed in and W_r is the volume of the CNTs. W_M is the larger of the two volumes. It is possible to calculate the volume percentage of each component by using the formula [26], which requires one to divide these volumes by the total volume (W).

$$V_r = \frac{W_r}{W}, \quad V_M = \frac{W_M}{W} \tag{6.9}$$

In the matrix, the volume percentage of CNTs may be divided into two parts: those that are placed inside the clusters and those that are located outside of the clusters, very similarly to how the volume of CNTs (W) can be divided into two parts. As a consequence of this fact, we are going to add two additional factors in order to be able to represent this issue as:

$$\mu = \frac{W_{in}}{W}, \quad \eta = \frac{W_r^{in}}{W_r} \tag{6.10}$$

where the volume percentage of clusters is denoted by μ and the volume fraction of CNTs contained inside the clusters is denoted by. It is important to keep in mind that $\mu \leq \eta$ is one of the constraints imposed by this technique. In the condition ($\mu \leq \eta, \eta \neq 1$), part of the CNTs are contained inside the cluster, while the rest of the CNTs are dispersed across the matrix. It is possible to construct the following relations by combining the Eqs. (6.8) and (6.9), as shown as below [26]:

$$\frac{W_r^{in}}{W_{in}} = \frac{V_r \eta}{\mu} \tag{6.11}$$

$$\frac{W_r^M}{W - W_{in}} = \frac{V_r (1 - \eta)}{1 - \mu} \tag{6.12}$$

The percentage of CNTs by volume in the matrix may also be expressed as a function of the thickness z, as shown below [26]:

$$V_r(z) = \left[\frac{\rho_r}{w_r \rho_M} - \frac{\rho_r}{\rho_M} + 1 \right]^{-1} \left(\frac{z}{h} + \frac{1}{2} \right)^P \tag{6.13}$$

where ρ_r and ρ_M are mass densities of CNT and matrix, respectively. Also, w_r indicates the mass fraction of nanofillers that can be expressed as follows [26]:

$$w_r = \frac{M_r}{M_r + M_M} \tag{6.14}$$

where M_r refers to the mass of CNTs and M_M refers to the mass of the matrix, respectively. It is important to remember that V_r may be divided into two different variants when discussing the concerns associated with the agglomeration phenomena. The location of the agglomerated nanoparticles in relation to the matrix is the primary distinction between these two different forms. In this case, it is assumed that the matrix and the aggregated CNTs are positioned at the bottom and the top of the beam, respectively. It is highly recommended that you read "Wave Dispersion Characteristics of Agglomerated Multi-scale Hybrid Nanocomposite Beams" [28] in order to get additional knowledge about this topic. Following the connections established by the Eshelby–Mori–Tanaka micromechanical model [26, 27], it is now possible to arrive at effective mechanical characteristics. According to this approach, we may write down the bulk moduli and shear moduli of clusters like this [26]:

$$K_{in}(z) = K_M + \frac{V_r \eta \left(\delta_r - 3K_M \alpha_r \right)}{3 \left(\mu - V_r \eta + V_r \eta \alpha_r \right)} \tag{6.15}$$

$$G_{in}(z) = G_M + \frac{V_r \eta \left(\eta_r - 2G_M \beta_r \right)}{2 \left(\mu - V_r \eta + V_r \eta \beta_r \right)} \tag{6.16}$$

in which the bulk moduli and shear moduli of the matrix are connected to the K_M and G_M values, respectively. It is also possible to define the bulk and shear moduli of the remaining pieces as [26]:

$$K_{out}(z) = K_M + \frac{V_r (1-\eta)(\delta_r - 3K_M \alpha_r)}{3 \left(1 - \mu - V_r (1-\eta) + V_r (1-\eta) \alpha_r \right)} \tag{6.17}$$

$$G_{out}(z) = G_M + \frac{V_r (1-\eta)(\eta_r - 2G_M \beta_r)}{3 \left(1 - \mu - V_r (1-\eta) + V_r (1-\eta) \beta_r \right)} \tag{6.18}$$

In Eqs. (6.15)–(6.18), the mechanical terms α_r, β_r, δ_r, and η_r can be formulated in the following form [26, 27]:

$$\alpha_r = \frac{3(K_M + G_M) + k_r + l_r}{3(G_M + k_r)} \tag{6.19}$$

$$\beta_r = \frac{1}{5} \left(\frac{4G_M + 2k_r + l_r}{3(G_M + k_r)} + \frac{4G_M}{G_M + p_r} + \frac{2 \left(G_M \left(3K_M + G_M \right) + G_M \left(3K_M + 7G_M \right) \right)}{G_M \left(3K_M + G_M \right) + m_r \left(3K_M + 7G_M \right)} \right) \tag{6.20}$$

$$\delta_r = \frac{1}{3} \left(n_r + 2l_r + \frac{(2k_r + l_r)(3K_M + G_M - l_r)}{G_M + k_r} \right) \tag{6.21}$$

$$\eta_r = \frac{1}{5}\left(\frac{2}{3}(n_r - l_r) + \frac{8G_M p_r}{G_M + p_r} + \frac{(2k_r - l_r)(2G_M + l_r)}{3(G_M + k_r)}\right.$$

$$\left. + \frac{8m_r G_M(3K_M + 4G_M)}{3K_M(m_r + G_M) + G_M(7m_r + G_M)}\right) \tag{6.22}$$

where k_r, l_r, m_r, n_r, and p_r are elastic Hill's coefficients of CNTs, which might change among CNT types depending on the CNT's chirality. "Chirality" is a term that describes the arrangement of the carbon atoms in the CNT. In this paper, Hill's constants are applied to single-walled carbon nanotubes with a chirality of (10,10). Using the homogenization procedure, the nanocomposite beam's equivalent bulk and shear moduli may be calculated using the following formulae [25, 26]:

$$K(z) = K_{out}\left(1 + \frac{\mu\left(\frac{K_{in}}{K_{out}} - 1\right)}{1 + (1-\mu)\left(\frac{K_{in}}{K_{out}} - 1\right)\frac{1 + v_{out}}{3(1 - v_{out})}}\right) \tag{6.23}$$

$$G(z) = G_{out}\left(1 + \frac{\mu\left(\frac{G_{in}}{G_{out}} - 1\right)}{1 + (1-\mu)\left(\frac{G_{in}}{G_{out}} - 1\right)\frac{8 - 10v_{out}}{15(1 - v_{out})}}\right) \tag{6.24}$$

in which v_{out} denotes the Poisson's ratio of the matrix and can be expressed as:

$$v_{out} = \frac{3K_{out} - 2G_{out}}{6K_{out} + 2G_{out}} \tag{6.25}$$

At some point in the future, the corresponding Young's moduli and Poisson's ratio of CNTR nanocomposites will be able to be calculated as follows:

$$E(z) = \frac{9K(z) \times G(z)}{3K(z) + G(z)} \tag{6.26}$$

$$v(z) = \frac{3K(z) - 2G(z)}{6K(z) + 2G(z)} \tag{6.27}$$

Besides, the equivalent mass density of the CNTR nanocomposite can be written using the fundamentals of the mixture's rule in the following form:

$$\rho(z) = (\rho_r - \rho_M)V_r + \rho_M \tag{6.28a}$$

Furthermore, the following equation may be used to describe the thermal expansion coefficient of multi-scale hybrid composites [28].

$$\alpha_{11} = \frac{V_F E_{11}^F \alpha_{11}^F + V_{NCM} E^{NCM} \alpha^{NCM}}{V_F E_{11}^F + V_{NCM} E^{NCM}} \tag{6.28b}$$

where α^{NCM} stands for the thermal expansion coefficient of the nanocomposite matrix, which is expressed as [11, 12]:

$$\alpha^{NCM} = \frac{1}{2} \left(\begin{array}{c} \left(\dfrac{V_{CNT} E_{11}^{CNT} \alpha_{11}^{CNT} + V_M E^M \alpha^M}{V_{CNT} E_{11}^{CNT} + V_M E^M} \right) \left(1 - \nu^{NCM} \right) \\ + \left(1 + \nu^M \right) V_M \alpha^M + \left(1 + \nu^{CNT} \right) V_{CNT} \alpha^{CNT} \end{array} \right) \tag{6.28c}$$

A refined version of the parabolic beam theory will now be introduced to establish the beam's kinematic relations. An application of a shape function is made in this theory for the purpose of estimating the shear strain and stress. Following is an example of how the displacement fields of the beam might be shown here [28]:

$$u_x(x, z, t) = u(x, t) - z \frac{\partial w_b(x, t)}{\partial x} - f(z) \frac{\partial w_s(x, t)}{\partial x} \tag{6.29}$$

$$u_z(x, z, t) = w_b(x, t) + w_s(x, t) \tag{6.30}$$

where, u, w_b, and w_s are related to longitudinal displacement, bending deflection, and shear deflection through z-axis, respectively. Moreover, $f(z)$ is the shape function of the theorem. In the present study, this function is regarded to be $f(z) = -\dfrac{z}{4} + \dfrac{5z^3}{3h^2}$. The nonzero strains of the beam can be written as [28]:

$$\varepsilon_{xx} = \frac{\partial u}{\partial x} - z \frac{\partial^2 w_b}{\partial x^2} - f(z) \frac{\partial^2 w_s}{\partial x^2}, \; \gamma_{xz} = g(z) \frac{\partial w_s}{\partial x} \tag{6.31}$$

where

$$g(z) = 1 - \frac{df(z)}{dz} \tag{6.32}$$

In order to expand the motion equation of the beam, Hamilton's principle is utilized. The following is one possible formulation for defining Hamilton's principle [28]:

$$\int_0^t \delta(U - T) dt = 0 \tag{6.33}$$

In Eq. (6.33), U and T are strain energy and kinetic energy, respectively. The variation of strain energy is provided as [28]:

$$\delta U = \int_0^L \left(N \frac{\partial \delta u}{\partial x} - M_b \frac{\partial^2 \delta w_b}{\partial x^2} - M_s \frac{\partial^2 \delta w_s}{\partial x^2} + Q \frac{\partial \delta w_s}{\partial x} \right) dx \qquad (6.34)$$

When the bending moments and axial forces are given as follows:

$$[N, M_b, M_s] = \int_A \left[1, z, f(z) \right] \sigma_{xx} dA,$$

$$Q = \int_A g(z) \sigma_{xz} dA \qquad (6.35)$$

in which $g(z) = 1 - \dfrac{df(z)}{dz}$. Next, the first variation of kinetic energy can be written in the following form [28]:

$$\delta T = \int_V \left(\dot{u}_x \delta \ddot{u}_x + \dot{u}_z \delta \ddot{u}_z \right) \rho(z) dV \qquad (6.36)$$

$$= \int_0^L \begin{pmatrix} I_0 \left(\dfrac{\partial u}{\partial t} \dfrac{\partial \delta u}{\partial t} + \dfrac{\partial (w_b + w_s)}{\partial t} \dfrac{\partial \delta (w_b + w_s)}{\partial t} \right) - I_1 \left(\dfrac{\partial u}{\partial t} \dfrac{\partial^2 \delta w_b}{\partial x \partial t} + \dfrac{\partial^2 w_b}{\partial x \partial t} \dfrac{\partial \delta u}{\partial t} \right) \\ -J_1 \left(\dfrac{\partial u}{\partial t} \dfrac{\partial^2 \delta w_s}{\partial x \partial t} + \dfrac{\partial^2 w_s}{\partial x \partial t} \dfrac{\partial \delta u}{\partial t} \right) + I_2 \dfrac{\partial^2 w_b}{\partial x \partial t} \dfrac{\partial^2 \delta w_b}{\partial x \partial t} \\ +J_2 \left(\dfrac{\partial^2 w_b}{\partial x \partial t} \dfrac{\partial^2 \delta w_s}{\partial x \partial t} + \dfrac{\partial^2 w_s}{\partial x \partial t} \dfrac{\partial^2 \delta w_b}{\partial x \partial t} \right) + K_2 \dfrac{\partial^2 w_s}{\partial x \partial t} \dfrac{\partial^2 \delta w_s}{\partial x \partial t} \end{pmatrix} dx$$

The following is a formulation of the mass inertias that were utilized in the preceding equation:

$$[I_0, I_1, I_2, J_1, J_2, K_2] = \int_{-h/2}^{h/2} \left[1, z, z^2, f(z), zf(z), f^2(z) \right] \rho(z) dz \qquad (6.37)$$

The following equations may be formed when Eqs. (6.34) and (6.36) are substituted into Eq. (6.33), and the coefficients of δu, δw_b, and δw_s are set to zero [28]. These equations describe the motion of nanocomposite beams and may be constructed when the following conditions are met:

$$\frac{\partial N}{\partial x} = I_0 \frac{\partial^2 u}{\partial t^2} - I_1 \frac{\partial^3 w_b}{\partial x \partial t^2} - J_1 \frac{\partial^3 w_s}{\partial x \partial t^2} \tag{6.38}$$

$$\frac{\partial^2 M_b}{\partial x^2} = I_0 \frac{\partial^2 \left(w_b + w_s\right)}{\partial t^2} + I_1 \frac{\partial^3 u}{\partial x \partial t^2} - I_2 \frac{\partial^4 w_b}{\partial t^2 \partial x^2} - J_2 \frac{\partial^4 w_s}{\partial t^2 \partial x^2} \tag{6.39}$$

$$\frac{\partial^2 M_s}{\partial x^2} + \frac{\partial Q}{\partial x} = I_0 \frac{\partial^2 \left(w_b + w_s\right)}{\partial t^2} + J_1 \frac{\partial^3 u}{\partial x \partial t^2} - J_2 \frac{\partial^4 w_b}{\partial t^2 \partial x^2} - K_2 \frac{\partial^4 w_s}{\partial t^2 \partial x^2} \tag{6.40}$$

Elastic stress-strain relations in orthotropic composite materials are used to derive the underlying elastic equations of these materials. Depending on the details of the model, the beam might be characterized by the constitutive equations shown below:

$$\sigma_{ij} = C_{ijkl} \varepsilon_{kl} \tag{6.41}$$

In the above equation σ_{ij}, C_{ijkl}, and ε_{kl} denote stress, elastic stiffness, and strain components, respectively. Thus, the mentioned relations can be amended for beams in the following form:

$$\sigma_{xx} = E_{11} \varepsilon_{xx} \tag{6.42}$$

in which

$$\sigma_{xz} = G_{12} \gamma_{xz} \tag{6.43}$$

It is possible to establish the following relations by integrating the expressions in equations (6.42) and (6.43) across the cross-section area of the beam:

$$N = A \frac{\partial u}{\partial x} - B \frac{\partial^2 w_b}{\partial x^2} - B_s \frac{\partial^2 w_s}{\partial x^2} \tag{6.44a}$$

$$M_b = B \frac{\partial u}{\partial x} - D \frac{\partial^2 w_b}{\partial x^2} - D_s \frac{\partial^2 w_s}{\partial x^2} \tag{6.44b}$$

$$M_s = B_s \frac{\partial u}{\partial x} - D_s \frac{\partial^2 w_b}{\partial x^2} - H_s \frac{\partial^2 w_s}{\partial x^2} \tag{6.44c}$$

$$Q = A_s \frac{\partial w_s}{\partial x} \tag{6.44d}$$

where

$$\left[A,B,D,B_s,D_s,H_s\right] = \int_A \left[1,z,z^2,f(z),zf(z),f^2(z)\right]E_{11}dA$$

$$A_s = \int_A g^2(z)G_{12}dA \tag{6.44e}$$

The governing equations of multi-scale hybrid nanocomposite beams may be obtained in the following method [28] by inserting Eqs. (6.44a), (6.44d), and (6.44e) into Eqs. (6.38), (6.40), and (6.38).

$$A\frac{\partial^2 u}{\partial x^2} - B\frac{\partial^3 w_b}{\partial x^3} - B_s\frac{\partial^3 w_s}{\partial x^3} - I_0\frac{\partial^2 u}{\partial t^2} + I_1\frac{\partial^3 w_b}{\partial x \partial t^2} + J_1\frac{\partial^3 w_s}{\partial x \partial t^2} = 0, \tag{6.45a}$$

$$B\frac{\partial^3 u}{\partial x^3} - D\frac{\partial^4 w_b}{\partial x^4} - D_s\frac{\partial^4 w_s}{\partial x^4} - I_0\frac{\partial^2 (w_b + w_s)}{\partial t^2}$$

$$- I_1\frac{\partial^3 u}{\partial x \partial t^2} + I_2\frac{\partial^4 w_b}{\partial t^2 \partial x^2} + J_2\frac{\partial^4 w_s}{\partial t^2 \partial x^2} \tag{6.45b}$$

$$= 0$$

$$B_s\frac{\partial^3 u}{\partial x^3} - D_s\frac{\partial^4 w_b}{\partial x^4} - H_s\frac{\partial^4 w_s}{\partial x^4} + A_s\frac{\partial^2 w_s}{\partial x^2} - I_0\frac{\partial^2 (w_b + w_s)}{\partial t^2}$$

$$- J_1\frac{\partial^3 u}{\partial x \partial t^2} + J_2\frac{\partial^4 w_b}{\partial t^2 \partial x^2} + K_2\frac{\partial^4 w_s}{\partial t^2 \partial x^2} = 0 \tag{6.45c}$$

The analytical solution technique has been put into action so that the governing equations of the multi-scale hybrid nanocomposite beam may be solved. It is presumed that the equation [28], which has the following form:

$$\begin{Bmatrix} u \\ w_b \\ w_s \end{Bmatrix} = \begin{Bmatrix} U\exp[i(\beta x - \omega t)] \\ W_b\exp[i(\beta x - \omega t)] \\ W_s\exp[i(\beta x - \omega t)] \end{Bmatrix} \tag{6.46}$$

in which U, W_b, and W_s represent wave amplitudes, respectively. β is wave number and ω is circular frequency dispersed waves. By substituting u, w_b, and w_s from Eq. (6.46) in Eqs. (6.45a)–(6.45c) the following equation is gained:

$$\left(\left[K \right]_{3\times3} - \omega^2 \left[M \right]_{3\times3} \right) \begin{bmatrix} U \\ W_b \\ W_s \end{bmatrix} = 0 \qquad (6.47)$$

where K and M are matrices that represent stiffness and mass, respectively; their individual components are listed in Appendix 6A. The eigenvalue issue may be solved by adjusting the coefficient matrix on the left side of Eq. (6.46) such that its determinant equals zero.

$$\left| \left[K \right]_{3\times3} - \omega^2 \left[M \right]_{3\times3} \right| = 0 \qquad (6.48)$$

The following set of photos has been compiled with the intention of providing a clearer understanding of the influence that a variety of parameters have on the wave propagation analysis of multi-scale hybrid nanocomposite beam [28]. Epoxy serves as the matrix in a multi-scale hybrid nanocomposite beam, while carbon fiber and CNT are used as reinforcements in this kind of material. References [11] and [12] are the sources from which one may get the material properties of the CF and the matrix, respectively.

Figure 6.2 depicts an analysis of the relationship between the wave frequency and the volume fraction of the CF for a variety of cluster volume fractions and

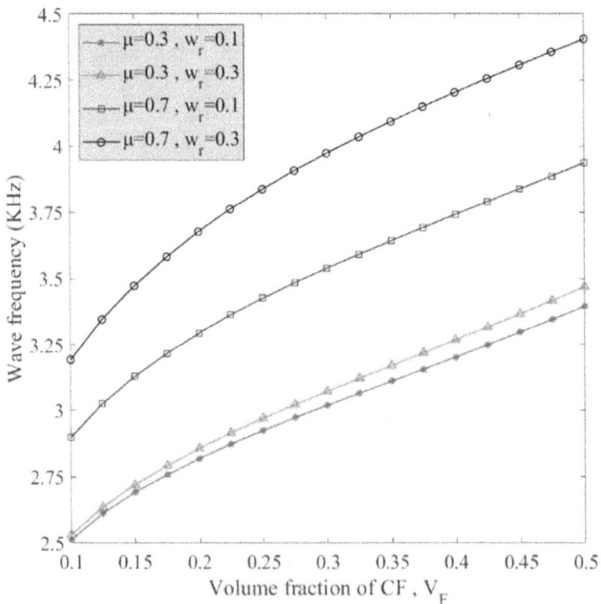

FIGURE 6.2 Illustration of coupled effects of various volume fractions of clusters and various mass fractions of the CNTs on wave frequency versus CF's volume fraction.

CNT mass fractions. It has been shown that an increase in the volume percentage of CF results in an increase in the wave frequency. The reason for this is that a higher CF volume fraction results in a more rigid structure. Furthermore, the nanocomposite beam with a higher mass percentage of CNT has a higher wave frequency compared to the nanocomposite beam with a lower mass percentage of CNT. Wave frequencies of multi-scale hybrid nanocomposite beams are shown to be more affected by the volume percentage of clusters. In other words, the cluster's wave frequency will rise as its volume fraction grows. The highest frequency is seen in the structure with the highest CNT mass percentage and the highest cluster volume fraction [28].

To be more specific, the goal of showing Figure 6.3 is to illustrate the impact that the volume percentage of the cluster has on the variance of wave frequency versus wave number. This may be found by looking at the relationship between the two variables. It can be seen that the wave frequency raises along with the number of waves that are present in the nanocomposite beam as the number of waves grows. In addition to this, there is the possibility that an increase in the wave frequency will occur if the cluster's density continues to grow. In fact, when the cluster's volume percentage becomes a more significant component, the wave frequency of multi-scale hybrid nanocomposite beams rises [28].

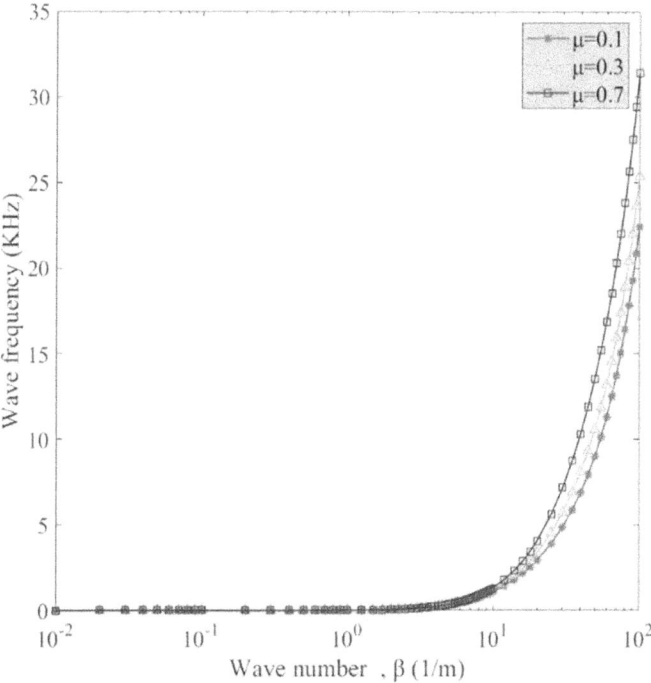

FIGURE 6.3 Variation of the wave frequency of multi-scale nanocomposite beam versus the wave number for different volume fractions of clusters.

Figure 6.4 illustrates how the relationship between wave frequency and wave number changes depending on the volume percentage of the CF. The wave frequency of the nanocomposite beam is shown to rise in a manner that is exponentially proportional to the number of waves it has, as can be seen. It is evident that raising the volume percentage of CF does not have any discernible impact; hence, the magnifier is used in order to clarify the trend of the volume fraction of each CF. It has been discovered that, for a given wave number, the volume fraction of CF that accounts for the greatest portion of the total has the highest wave frequency [28].

Figure 6.5 shows that the change in wave frequency, evaluated in terms of wave number, is affected by both the mass fraction and the volume fraction of CNTs inside the cluster. It is easy to see that the wave frequency will keep on climbing even if the total number of waves will keep on growing. This is something that can be seen. As was stated before, an increase in the mass fraction of CNT results in an increase in the wave frequency. Despite this, it is obvious that the wave frequency will decrease as the volume fraction of CNT enclosed inside the cluster increases. As a consequence of this, it is necessary to differentiate between the effects brought about by the volume fraction of the cluster and the volume fraction of the CNTs that are housed inside the clusters. It is feasible to get to the

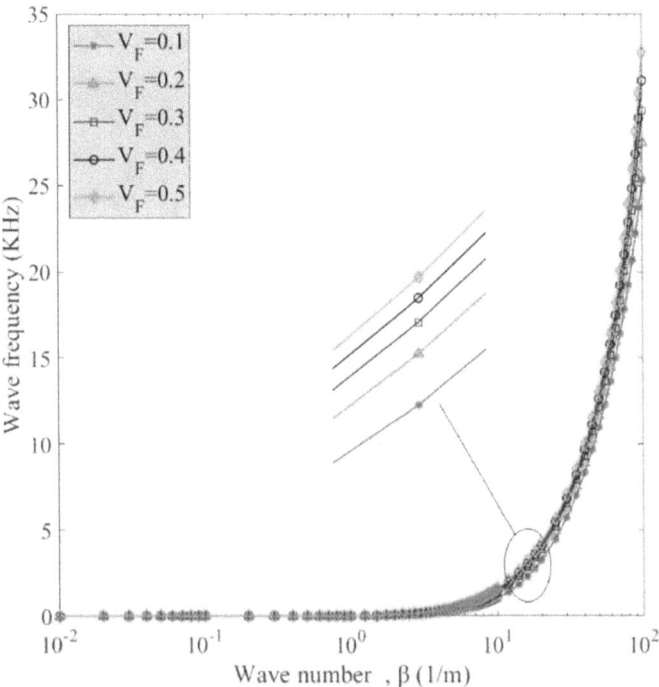

FIGURE 6.4 Variation of the wave frequency of multi-scale nanocomposite beam versus the wave number for the various volume fraction of CF.

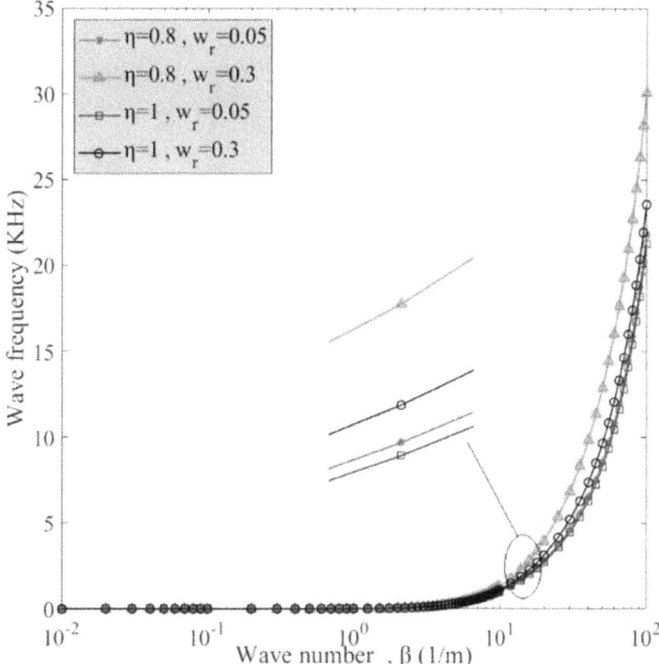

FIGURE 6.5 Variation of the wave frequency of multi-scale nanocomposite beam versus the wave number for different mass fractions of CNTs and volume fractions of CNT inside the cluster.

conclusion that the total volume percentage of CNT that is contained inside the cluster has to be lowered in order to attain a higher wave frequency [28].

Figure 6.6 shows how the frequency shift of multi-scale hybrid nanocomposite beams is affected by the gradient index and the mass fraction of CNT relative to the volume percentage of CNT inside the cluster. This is shown in contrast to the previous illustration, which depicted the volume fraction of CNT. When the relevant graphs are compared, it is possible to observe that there is a correlation between a rise in the gradient index and a drop in the wave frequency. Due to this tendency, it is recommended that just a minimal amount of gradient index be used while trying to raise the structure's wave frequency. This is because of the potential for the gradient index to have the opposite effect. In addition, each of the diagrams shows a propensity to go downhill, but the slopes of the lines are different. This data implies that reducing wave frequency may be achieved by increasing the volume percentage of CNT contained inside the cluster. Additionally, the frequency of the waves generated by the structure increases as the mass percentage of CNT in the structure increases as well. In addition, it is essential to take into account the fact that the structure becomes stiffer as the proportion of CNT in its mass increases. In addition, the reason why diagrams have a negative slope is

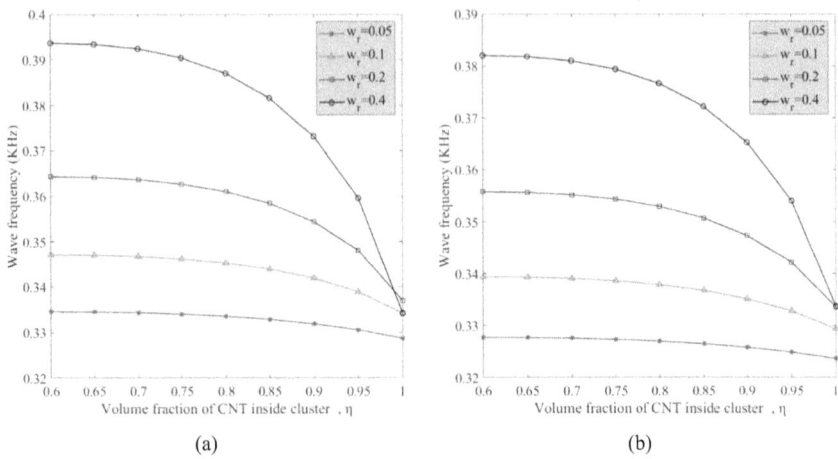

FIGURE 6.6 Change of the wave frequency of multi-scale nanocomposite beam versus the volume fraction of CNT inside the cluster for a various mass fraction of CNT for (a) $P = 1$ and (b) $P = 2$.

that the impact of the volume fraction of CNTs inside the cluster outweighs the influence of the mass fraction of CNTs contained within the cluster [28]. This is the reason why diagrams have a negative slope.

Several alternative cluster volume percentages and gradient indices are shown in Figure 6.7 to illustrate the correlation between wave frequency and the volume fraction of CNT encompassed within the cluster. When there is a greater proportion of CNT contained inside the cluster's volume, the wave frequency will be lower. This is quite easy to notice. Once $\mu = 0.6$ has been reached, the wave frequency will be at its greatest value; thereafter, it will decrease as the volume fraction of the cluster decreases. This phenomenon is due to the structure's stiffness being reduced, which is why it occurs. In addition to this, according to the figure, the wave frequency of the multi-scale nanocomposite beam will drop when the gradient index is increased [28].

In comparison to the volume percentage of CF, the phase velocity of the nanocomposite beam varies with the mass fraction of CNT and the volume percentage of the cluster, as shown in Figure 6.8. The results of this investigation are shown in the figure. The schematics are clearly moving in an upward manner, which is quite simple to notice. This demonstrates that the phase velocity of the multi-scale hybrid nanocomposite beam will grow as the volume percentage of the CFs increases, which will ultimately result in the beam having a higher degree of rigidity. In addition to this, there is a correlation between the growth of the cluster's volume percentage and a rise in the phase velocity [28].

For various cluster volume percentages, the wave number describes the fluctuation in phase velocity seen in Figure 6.9. In the beginning, each phase velocity will be equal to zero. After that, however, each figure will start to increase

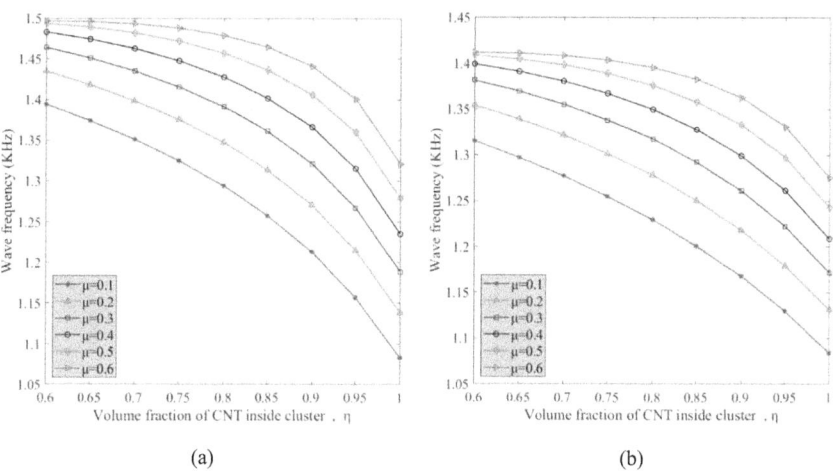

(a)   (b)

FIGURE 6.7   The effect of volume fractions of clusters on the variation of the wave frequency of multi-scale nanocomposite beam versus volume fractions of CNT inside the cluster for (a) $P = 1$ and (b) $P = 2$.

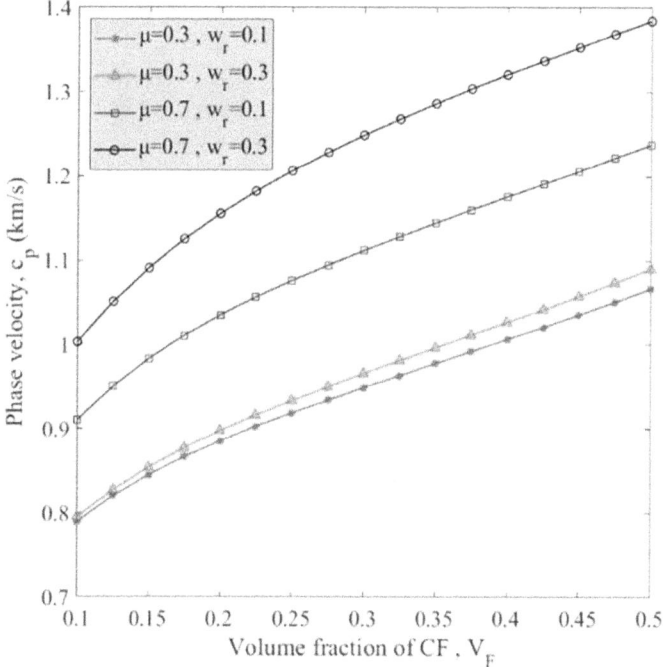

FIGURE 6.8   Variation of the phase velocity of multi-scale nanocomposite beam versus CF's volume fraction for different volume fractions of clusters and mass fractions of CNTs.

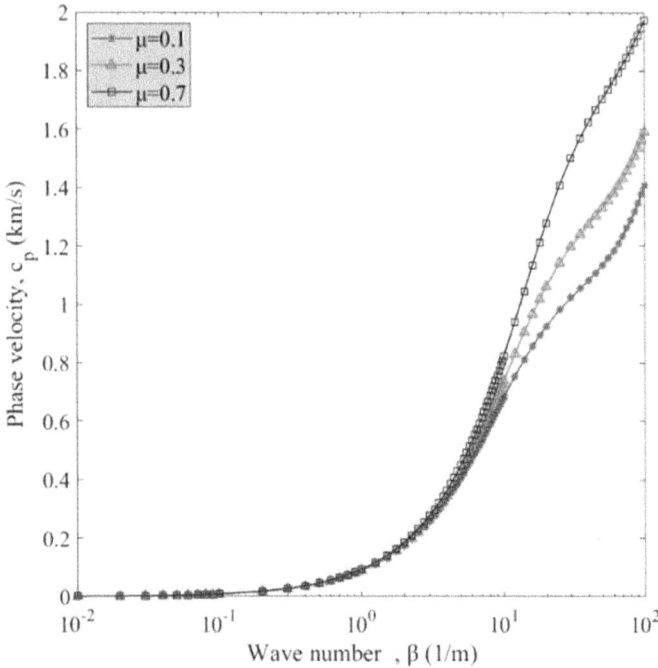

FIGURE 6.9 Illustration of the variation of phase velocity versus wave number for different volume fractions of clusters.

gradually. This is anything that is visible to the naked eye. It is conceivable to get the conclusion that an increase in the volume fraction of the cluster results in an increase in the phase velocity of the beam [28]. This is something that may be proven via experimentation.

Figure 6.10 presents an illustration of the fluctuation of phase velocity versus wave number for different volume percentages of CF. This relationship is studied. In light of the image, one may draw the conclusion that a rise in the CF volume percentage carries with it the possibility for an accompanying expansion of the phase velocity. To put it another way, if the nanocomposite beam has a larger CF volume percentage, then it is possible to get a bigger quantity of phase velocity. Similar to Figure 6.9, when the wave number is low, the phase velocity is equal to zero, but as the wave number grows, the phase velocity increases [28].

The combined impacts of the CNTs' mass fraction and CNTs' volume fraction inside the cluster are shown in Figure 6.11. These events have an effect on the fluctuation in the phase velocity of the beam as a function of wave number. In the illustration, a magnifier is utilized so that the overall trend of the diagrams can be better understood, and so that comparisons between the various graphics can be made more easily. It has been shown that the volume fraction of CNTs within a cluster has an impact on phase velocity that is distinct from the influence that the mass fraction of CNTs inside a cluster has on phase velocity. Thus, the phase

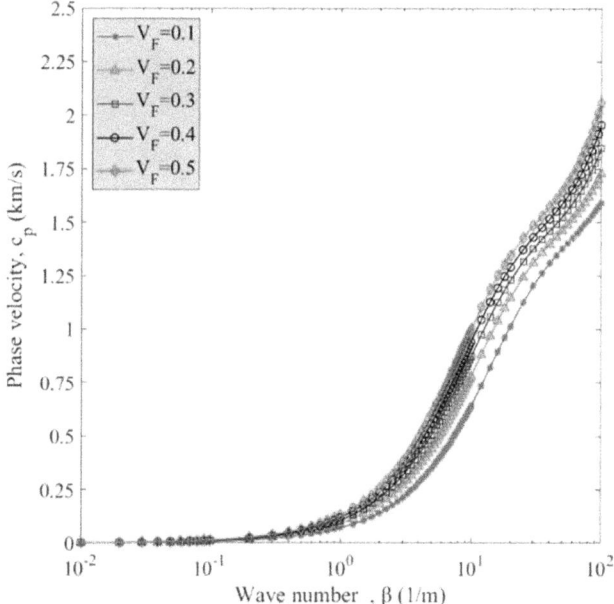

FIGURE 6.10   The effect of CF's volume fraction on the variation of the phase velocity of multi-scale nanocomposite beam versus wave number.

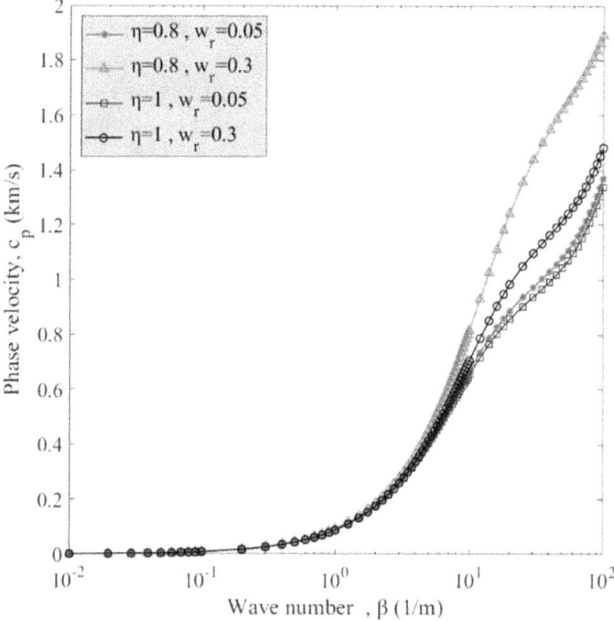

FIGURE 6.11   Both influences of volume fractions of CNT inside the cluster and CNT's mass fractions on phase velocity of multi-scale nanocomposite beam versus wave number.

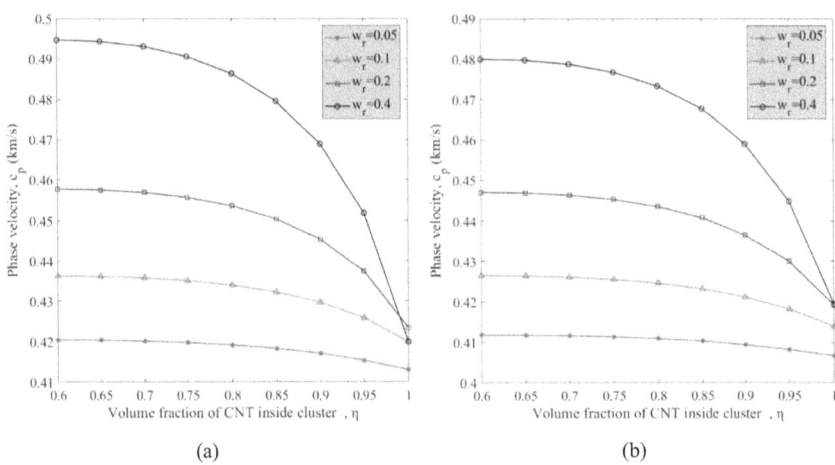

FIGURE 6.12   Variation of the phase velocity of multi-scale nanocomposite beam versus the volume fraction of CNT inside the cluster for a various mass fraction of CNT for (a) $P = 1$ and (b) $P = 2$.

velocity of a nanocomposite beam decreases with increasing volume fraction of CNTs inside the cluster but increases with increasing mass fraction of CNTs [28].

Figure 6.12 displays, for a selection of gradient indices, the results of a change in the mass fraction of CNTs relative to the volume fraction of CNTs in the cluster. This is done so in order to better understand the relationship between the two. These impacts might be seen as indicating that there is a connection between the two variables. It has been shown that a rise in the mass fraction of CNT may magnify phase velocity, while an increase in the amount of gradient index may cause phase velocity to decrease. These findings are in accordance with what has been observed. This is something that can be said, and it is also conceivable to say that whenever w_r is increased, the slope of decreasing increases bigger [28]. This is something that can be said.

It is shown that the volume fraction of the cluster and the gradient index affect the nature of the connection between the phase velocity and the volume fraction of CNT in the cluster. Figure 6.13 illustrates this correlation. Figure 6.13 shows a graph that suggests the phase velocity of a multi-scale hybrid nanocomposite beam, which falls with increasing cluster volume fraction. Similar to Figure 6.12 [28], this figure depicts the impacts of the gradient index and the volume percentage of CNTs inside the cluster.

6.2.1 THERMAL EFFECTS

The effect of temperature on the wave dispersion properties of aggregated multi-scale hybrid nanocomposites is investigated here. At $\Delta T = 10$, $K_W = K_P = 10$, $W_{CNT} = 0.02$, $V_F = 0.2$., Figure 6.14 shows the effect of various thermal loadings

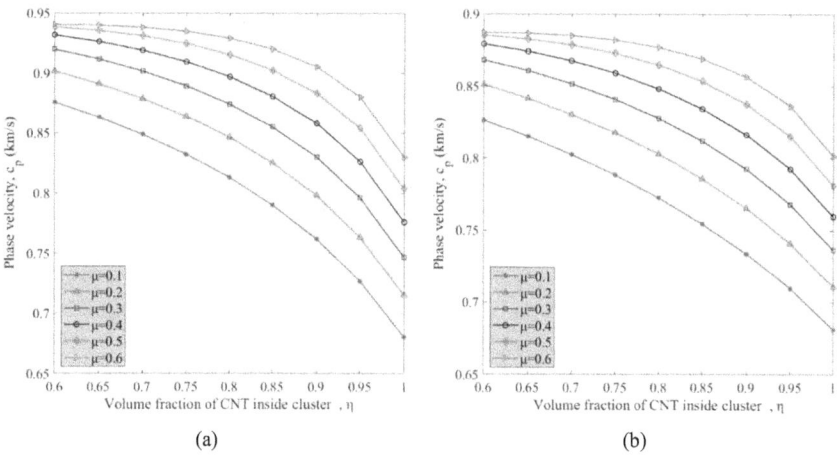

FIGURE 6.13 Variation of the phase velocity of multi-scale nanocomposite beam versus the volume fraction of CNT inside the cluster for different volume fractions of clusters for (a) $P = 1$ and (b) $P = 2$.

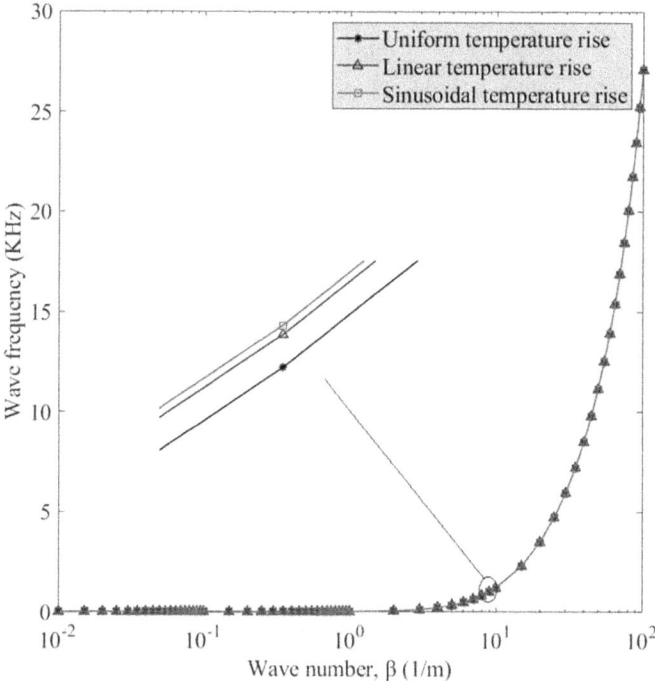

FIGURE 6.14 Variation of wave frequency versus wave number with different types of temperature rises ($\Delta T = 10$, $K_W = K_P = 10$, $W_{CNT} = 0.02$, and $V_F = 0.2$).

on the relationship between wave frequency and wave number. At smaller wave numbers, the frequency of the waves is found to be zero. However, wave frequency grows correspondingly with wave number. The maximum wave frequency is observed at the highest wave number. The different forms of temperature increase exhibit comparable quantities. However, upon closer inspection using a magnifier, it is observed that the wave frequency of the uniform temperature rise is lower than that of the other types. The linear temperature rise has a greater frequency than the uniform type, while the sinusoidal temperature rise has the highest frequency. This phenomenon arises due to the fact that uniform thermal loading proves to be more efficient than other forms of loading under specific circumstances. Homogeneous thermal loading leads to a higher reduction in wave frequency compared to other types of loading. At low wave numbers, the impact of temperature rise is greater than the influence of wave number. However, with an increase in wave number, the effect of wave number becomes dominant over the effect of temperature rise. This leads to an increase in wave frequency, as reported in reference [28].

At the specified values of $\Delta T = 10$, $K_W = K_P = 10$, $W_{CNT} = 0.02$, $V_F = 0.2$, Figure 6.15 depicts the changes in phase velocity with respect to wave number, while accounting for three distinct forms of temperature escalation. The charts illustrate that the phase velocity experiences a decrease initially as the wave

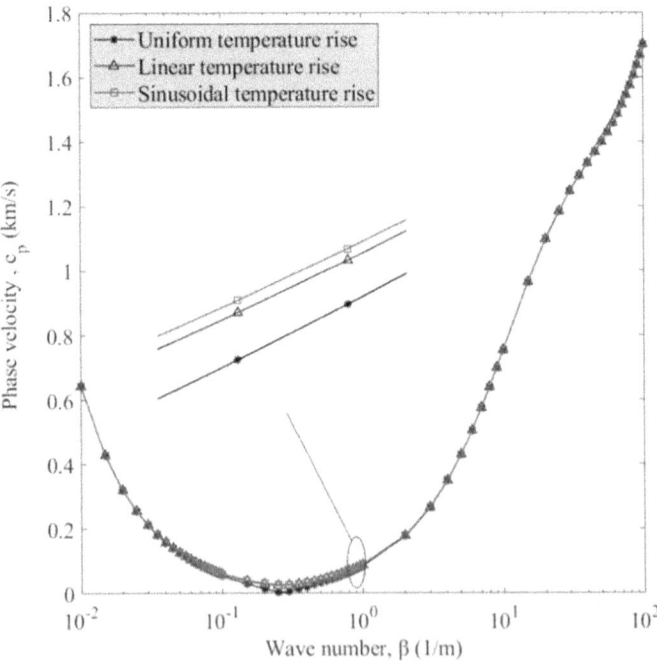

FIGURE 6.15 The effect of different types of temperature rise on phase velocity versus wave number ($\Delta T = 10$, $K_W = K_P = 10$, $W_{CNT} = 0.02$, and $V_F = 0.2$).

number increases, until it approaches zero, after which it starts to rise. The pre-dominance of the thermal effect on wave number results in a downward trend in charts with low wave numbers. However, this trend is eventually reversed, leading to an upward trend in the charts. The range of phase velocity under uniform ther-mal loading is comparatively smaller than other forms. This trend can be attrib-uted to the same cause as depicted in Figure 6.14 [28]. The observed pattern is attributed to the identical cause as depicted in Figure 6.14.

The graphical representation in Figure 6.16 illustrates the correlation between the alteration in wave frequency and the weight percentage of CNTs at various temperature increments ($\Delta T = 30$, $\beta = 10$, $K_W = K_P = 25$, $V_F = 0.2$). As expected, an elevation in the weight percentage of CNT leads to a corresponding increase in the wave frequency. This is attributed to the heightened stiffness of the material. Based on the provided diagram, the frequency values for uniform, linear, and sinusoidal types exhibit minimal variation. The sinusoidal waveform exhibits the highest frequency owing to its minimal impact on the material's stiffness, as reported in reference [28].

At the conditions shown in Figure 6.17, which are as follows: $\Delta T = 30$, $\beta = 10$, $K_W = K_P = 25$, $V_F = 0.2$, the change in phase velocity is shown against the weight fraction of CNTs. It is observable that when the weight percentage of CNT increases, the phase velocity of the multi-scale hybrid composite increases as

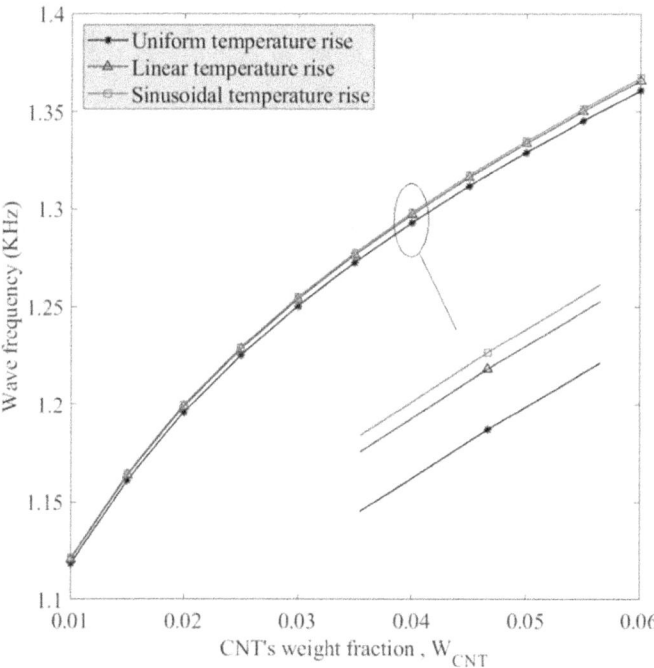

FIGURE 6.16 Both influences of different types of temperature rise and CNT's weight fraction on wave frequency ($\Delta T = 30$, $\beta = 10$, $K_W = K_P = 25$, and $V_F = 0.2$).

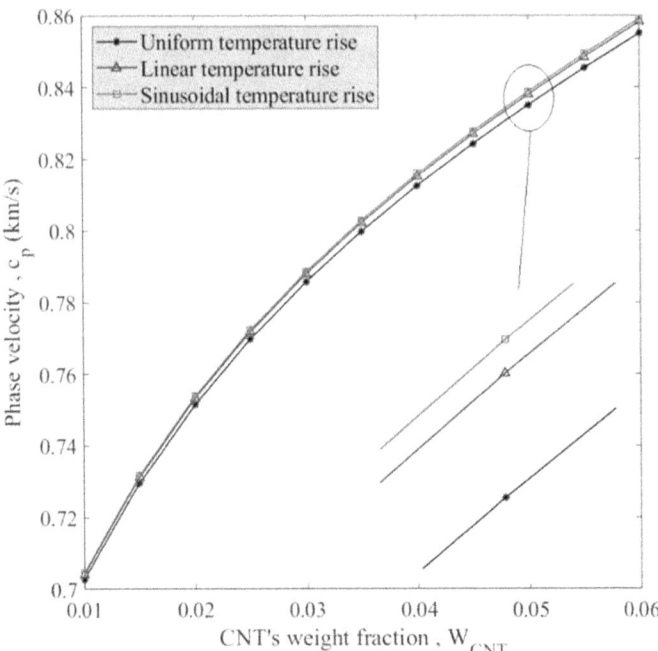

FIGURE 6.17 Variation of phase velocity versus CNT's weight fraction considering various kinds of temperature rises ($\Delta T = 30$, $\beta = 10$, $K_W = K_P = 25$, and $V_F = 0.2$).

well. This response takes place because the stiffness of the material is growing, which is the cause of the rise. The most efficient kind of temperature increase is a uniform type, followed by linear and sinusoidal varieties in that order of effectiveness. Because of this impact, the structure will have less stiffness as the temperature continues to increase [28].

Figures 6.18 and 6.19 depict the changes in wave frequency and phase velocity in relation to the Winkler coefficient, under different temperature conditions. The specific conditions for these figures are $\Delta T = 20$, $\beta = 10$, $K_P = 20$, $W_{CNT} = 0.01$, $V_F = 0.2$. This study examines the impact of varying Winkler coefficient on wave frequency and phase velocity, in relation to different types of temperature rise. The insufficient growth of phase velocity and wave frequency is directly proportional to the increase in Winkler coefficient. Based on the available evidence, it can be concluded that the Winkler coefficient does not exert a substantial influence on either the phase velocity or the wave frequency. The dissimilarities observed in the chart's visual representation can be ascribed to the influence exerted by different forms of thermal loading on the rigidity of the structure, as stated in reference [28].

The influence of the Pasternak coefficient on the wave frequency is investigated for different thermal loadings ($\Delta T = 20$, $\beta = 10$, $K_W = 30$, $W_{CNT} = 0.01$, $V_F = 0.2$) in Figure 6.20. According to the findings, a higher Pasternak coefficient increases the frequency of waves in a similar fashion. The beam's wave frequency

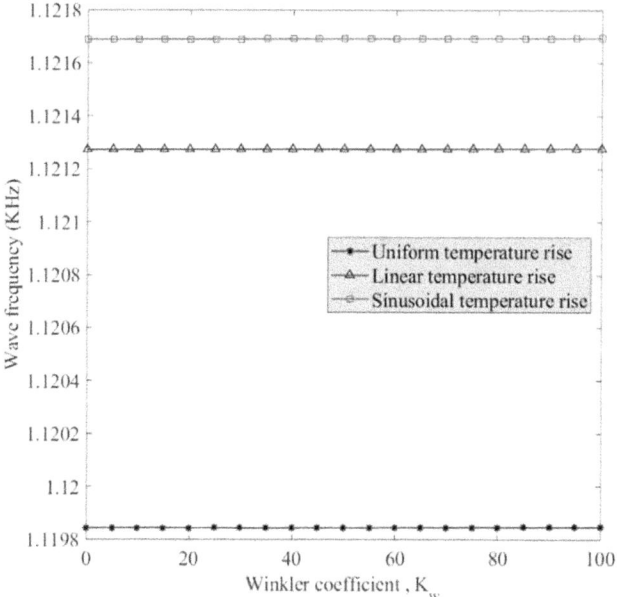

FIGURE 6.18 Illustration of the variation of wave frequency versus the Winkler coefficient with different types of temperature rises ($\Delta T = 20$, $\beta = 10$, $K_P = 20$, $W_{CNT} = 0.01$, and $V_F = 0.2$).

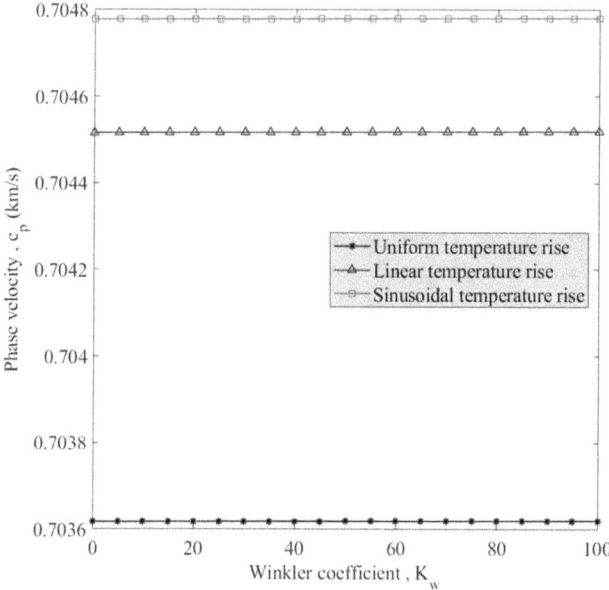

FIGURE 6.19 Changes of phase velocity versus the Winkler coefficient with different types of temperature rises ($\Delta T = 20$, $\beta = 10$, $K_P = 20$, $W_{CNT} = 0.01$, and $V_F = 0.2$).

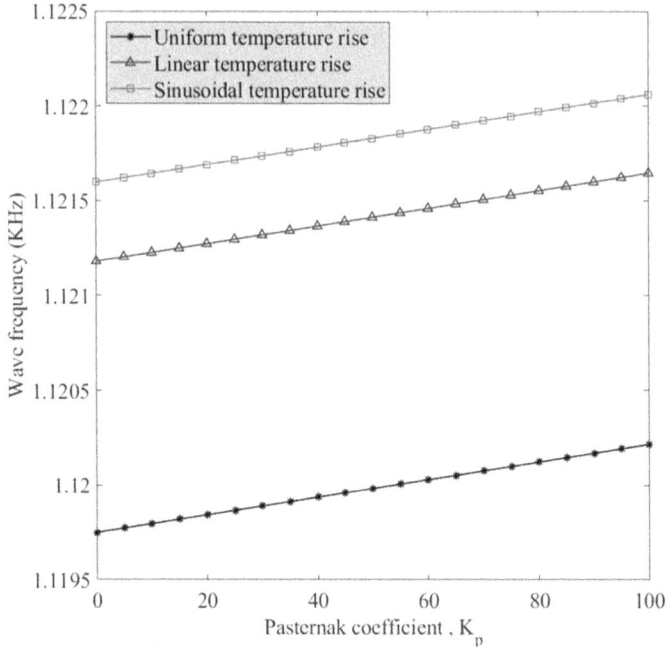

FIGURE 6.20   Coupled effects of Pasternak coefficient and different kinds of thermal loadings on the variation of wave frequency ($\Delta T = 20$, $\beta = 10$, $K_W = 30$, $W_{CNT} = 0.01$, and $V_F = 0.2$).

increases linearly with the Pasternak coefficient term included in the linear formulation's matrix component. This causes a rise in the frequency of the beam's waves. From the data in [28], it is clear that sinusoidal waves may reach greater frequencies than other wave types, whereas uniform waves can reach lower frequencies.

The link between the Pasternak coefficient and the rate of change in phase velocity under uniform, sinusoidal, and linear thermal loadings is shown in Figure 6.21. The results suggest that as the Pasternak coefficient amount grows, so does the associated rise in phase velocity. Meanwhile, at a given temperature, sinusoidal types have the highest phase velocity, followed by linear types and finally uniform types with the slowest phase velocities. However, every chart indicates an increasing trajectory [28].

The relationship between the difference in temperature ((ΔT) and the wave frequency is shown in Figure 6.22 ($\beta = 5$, $K_W = K_P = 30$, $W_{CNT} = 0.02$, $V_F = 0.2$)). This relationship holds true for a variety of thermal loadings. The amount of decrease generated by a uniform type is more than that caused by other kinds, thus it should come as no surprise that this would result in a lower wave frequency when the temperature difference between the two locations grows. This indicates that the uniform kind has a more significant impact on wave frequency than do the

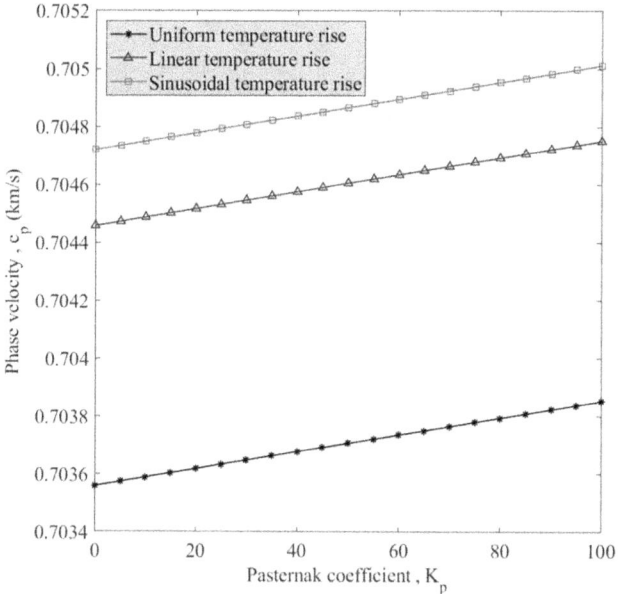

FIGURE 6.21 Phase velocity's variation versus Pasternak coefficient regarding three kinds of temperature rises ($\Delta T = 20$, $\beta = 10$, $K_W = 30$, $W_{CNT} = 0.01$, and $V_F = 0.2$).

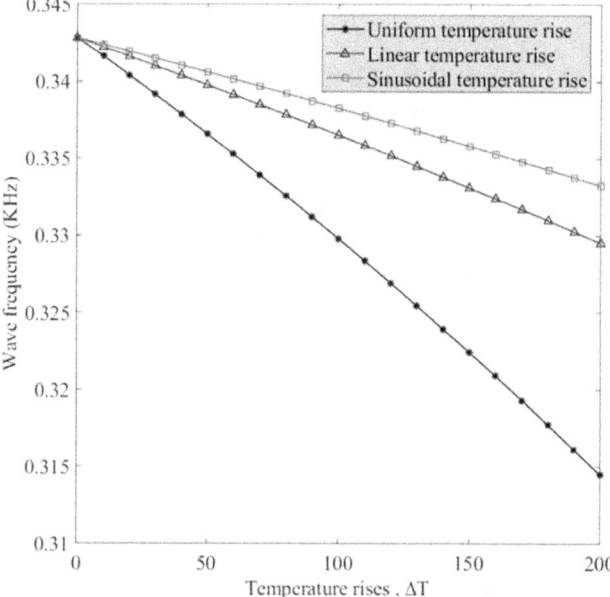

FIGURE 6.22 Variation of wave frequency versus temperature rises with respect to different types of thermal loadings ($\beta = 5$, $K_W = K_P = 30$, $W_{CNT} = 0.02$, and $V_F = 0.2$).

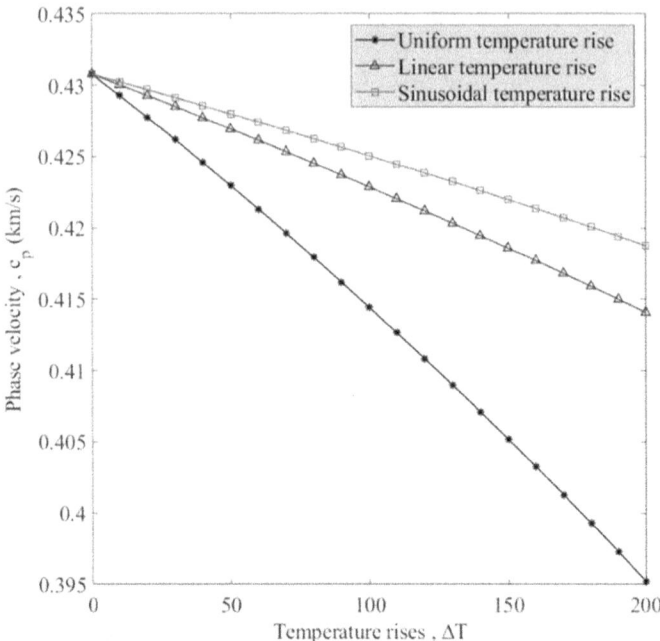

FIGURE 6.23   The influence of various kinds of thermal loadings on the variation of phase velocity versus temperature rises ($\beta = 5$, $K_W = K_P = 30$, $W_{CNT} = 0.02$, and $V_F = 0.2$).

other varieties. This behavior of the uniform type might be considered a disadvantage, which is why high frequency in high-temperature increase is not conceivable, but it is possible for the sinusoidal type [28].

Figure 6.23 shows a plot of the change in phase velocity versus temperature differential for various kinds of thermal loadings ($\beta = 5$, $K_W = K_P = 30$, $W_{CNT} = 0.02$, $V_F = 0.2$). By examining the graph in Figure 6.22, one may draw the conclusion that an increase in phase velocity occurs concurrently with an increase in temperature difference. As was also previously said, the largest and smallest decreases are shown in the uniform type and the sinusoidal kind, respectively. Actually, sinusoidal thermal loading is preferred because the structures may achieve higher frequencies than with other types of loading [28].

The changes in wave frequency and phase velocity that occur in proportion to CF's volume percentage for different temperature increases are shown in Figures 6.24 and 6.25. As can be observed in the diagram, the frequency and phase velocity of waves both rise along with an increase in CF's volume percentage. According to the information presented in the preceding paragraphs, the minimum and maximum values of all plots are uniform and sinusoidal, and they are located in close proximity to one another. It is important to draw attention to the fact that sinusoidal and linear kinds are so close to one another.

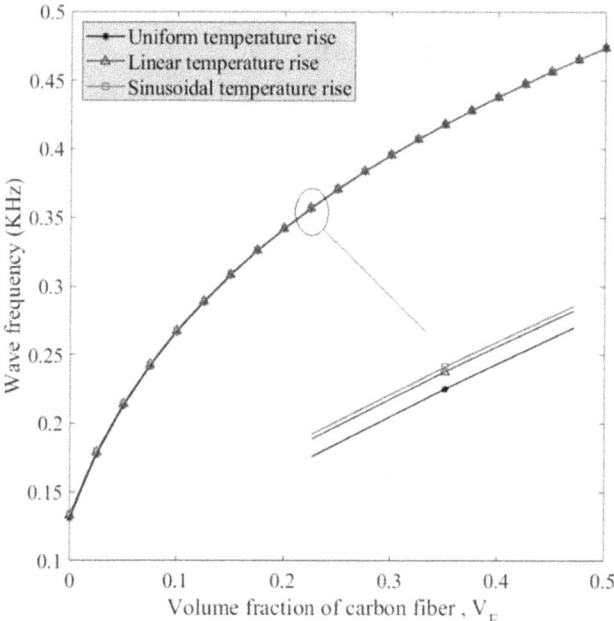

FIGURE 6.24   The effect of various kinds of thermal loadings on wave frequency versus CF's volume fraction ($\Delta T = 10$, $\beta = 5$, $K_W = K_P = 10$, $W_{CNT} = 0.02$).

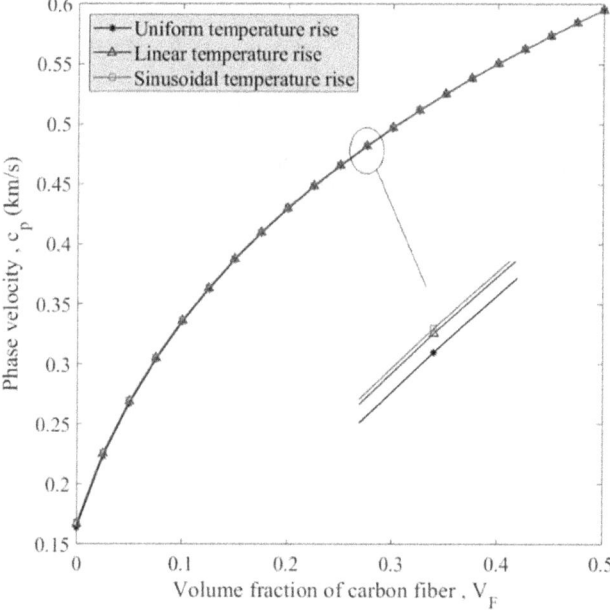

FIGURE 6.25   The illustration of both effects of volume fraction of CF and various sorts of temperature rise on changes of phase velocity ($\Delta T = 10$, $\beta = 5$, $K_W = K_P = 10$, and $W_{CNT} = 0.02$).

6.3   CNTR NANOCOMPOSITE PLATE WAVE DISPERSION CHARACTERISTICS

Figure 6.26 depicts a flat plate with dimensions (a, h). This substrate is a multi-scale nanocomposite hybrid. Epoxy resin is used as the plate's main material, with carbon fiber and carbon nanotubes (CNTs) as reinforcement [30]. To account for the influence of CNT agglomeration, the Eshelby–Mori–Tanaka model is used for the evaluation of the effective mechanical parameters of multi-scale hybrid nanocomposites. Eqs. (6.1) through (6.28) were used in Section 6.2.1 to explain this effect [26].

In addition to this, a more developed theory of the parabolic plate is presented in order to deduce the kinematic relations of the plate. An application of a shape function is made in this theory for the purpose of estimating the shear strain and stress. Following is an example of how the displacement fields of the plate might be shown here [30]:

$$u_x\left(x,y,z,t\right)=u\left(x,y,t\right)-z\frac{\partial w_b\left(x,y,t\right)}{\partial x}-f\left(z\right)\frac{\partial w_s\left(x,y,t\right)}{\partial x} \tag{6.49}$$

$$u_y\left(x,y,z,t\right)=v\left(x,y,t\right)-z\frac{\partial w_b\left(x,y,t\right)}{\partial y}-f\left(z\right)\frac{\partial w_s\left(x,y,t\right)}{\partial y} \tag{6.50}$$

$$u_z\left(x,y,z,t\right)=w_b\left(x,y,t\right)+w_s\left(x,y,t\right) \tag{6.51}$$

where, u, v, w_b, and w_s are related to longitudinal displacement and transverse displacement of the mid-surface and bending deflection and shear deflection through z-axis, respectively. Moreover, $f(z)$ is the shape function of the theorem. In the present study, this function is regarded to be $f\left(z\right)=-\frac{z}{4}+\frac{5z^3}{3h^2}$. The nonzero strains of the plate can be written as [30]:

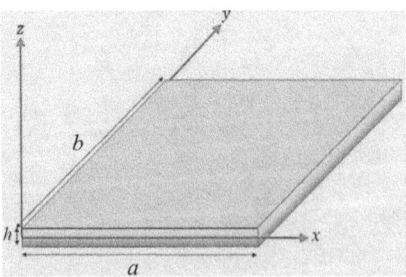

FIGURE 6.26   Schematic geometry of rectangular plate.

$$\begin{Bmatrix} \varepsilon_{xx} \\ \varepsilon_{yy} \\ \gamma_{xy} \end{Bmatrix} = \begin{Bmatrix} \varepsilon^0_{xx} \\ \varepsilon^0_{yy} \\ \gamma^0_{xy} \end{Bmatrix} + z \begin{Bmatrix} \kappa^b_{xx} \\ \kappa^b_{yy} \\ \kappa^b_{xy} \end{Bmatrix} + f(x) \begin{Bmatrix} \kappa^s_{xx} \\ \kappa^s_{yy} \\ \kappa^s_{xy} \end{Bmatrix}, \quad \begin{Bmatrix} \gamma_{xz} \\ \gamma_{yz} \end{Bmatrix} = \begin{Bmatrix} \gamma^0_{xz} \\ \gamma^0_{yz} \end{Bmatrix} \qquad (6.52a)$$

where

$$\begin{Bmatrix} \varepsilon^0_{xx} \\ \varepsilon^0_{yy} \\ \gamma^0_{xy} \end{Bmatrix} = \begin{Bmatrix} \dfrac{\partial u}{\partial x} \\[2mm] \dfrac{\partial v}{\partial y} \\[2mm] \dfrac{\partial u}{\partial y} + \dfrac{\partial v}{\partial x} \end{Bmatrix}, \quad \begin{Bmatrix} \kappa^b_{xx} \\ \kappa^b_{yy} \\ \kappa^b_{xy} \end{Bmatrix} = \begin{Bmatrix} -\dfrac{\partial^2 w_b}{\partial x^2} \\[2mm] -\dfrac{\partial^2 w_b}{\partial y^2} \\[2mm] -2\dfrac{\partial^2 w_b}{\partial x \partial y} \end{Bmatrix}, \quad \begin{Bmatrix} \kappa^s_{xx} \\ \kappa^s_{yy} \\ \kappa^s_{xy} \end{Bmatrix} = \begin{Bmatrix} -\dfrac{\partial^2 w_s}{\partial x^2} \\[2mm] -\dfrac{\partial^2 w_s}{\partial y^2} \\[2mm] -2\dfrac{\partial^2 w_s}{\partial x \partial y} \end{Bmatrix}, \quad \begin{Bmatrix} \gamma^0_{xz} \\ \gamma^0_{yz} \end{Bmatrix} = \begin{Bmatrix} \dfrac{\partial w_s}{\partial x} \\[2mm] \dfrac{\partial w_s}{\partial y} \end{Bmatrix}$$

$$(6.52b)$$

In order to expand the motion equation of the plate, Hamilton's principle is utilized. It is possible to define Hamilton's principle using the following format:

$$\int_0^t \delta(U - T)\, dt = 0 \qquad (6.53)$$

In the equation that was just presented, U and T refer, respectively, to the strain energy and the kinetic energy. The fluctuation in strain energy may be represented by the formula [30], which reads:

$$\delta U = \int_V \left(\sigma_{xx}\delta\varepsilon_{xx} + \sigma_{yy}\delta\varepsilon_{yy} + \sigma_{xy}\delta\gamma_{xy} + \sigma_{xz}\delta\gamma_{xz} + \sigma_{yz}\delta\gamma_{yz} \right) dV$$

$$= \int_0^b \int_0^a \begin{pmatrix} N_{xx}\delta\varepsilon^0_{xx} + M^b_{xx}\delta\kappa^b_{xx} + M^s_{xx}\delta\kappa^s_{xx} + N_{yy}\delta\varepsilon^0_{yy} + \\ M^b_{yy}\delta\kappa^b_{yy} + M^s_{yy}\delta\kappa^s_{yy} + N_{xy}\delta\gamma^0_{xy} + M^b_{xy}\delta\kappa^b_{xy} + \\ M^s_{xy}\delta\kappa^s_{xy} + Q_{xz}\delta\gamma^0_{xz} + Q_{yz}\delta\gamma^0_{yz} \end{pmatrix} dxdy \qquad (6.54)$$

When it is possible to represent the axial forces and bending moments in the following way:

$$\left(N_{ij}, M^b_{ij}, M^s_{ij} \right) = \int_{-h/2}^{h/2} \left(1, z, f(z) \right) \sigma_{ij} dz, \quad (i, j = x, y),$$

$$(6.55)$$

$$Q_k = \int_{-h/2}^{h/2} g(z) \sigma_k dz, \quad (k = xz, yz)$$

in which $g(z) = 1 - \dfrac{df(z)}{dz}$. Next, The first variation of kinetic energy can be written in the following form:

$$\delta T = \int_V \left(\dot{u}_x \delta \dot{u}_x + \dot{u}_y \delta \dot{u}_y + \dot{u}_z \delta \dot{u}_z \right) \rho(z) dV \qquad (6.56)$$

$$= \int_0^b \int_0^a \left\{ \begin{array}{l} I_0 \left(\dfrac{\partial u}{\partial t} \dfrac{\partial \delta u}{\partial t} + \dfrac{\partial v}{\partial t} \dfrac{\partial \delta v}{\partial t} + \dfrac{\partial w}{\partial t} \dfrac{\partial \delta w}{\partial t} \right) \\[2mm] - I_1 \left(\dfrac{\partial u}{\partial t} \dfrac{\partial^2 \delta w_b}{\partial x \partial t} + \dfrac{\partial^2 w_b}{\partial x \partial t} \dfrac{\partial \delta u}{\partial t} + \dfrac{\partial v}{\partial t} \dfrac{\partial^2 \delta w_b}{\partial y \partial t} + \dfrac{\partial^2 w_b}{\partial y \partial t} \dfrac{\partial \delta v}{\partial t} \right) \\[2mm] - J_1 \left(\dfrac{\partial u}{\partial t} \dfrac{\partial^2 \delta w_s}{\partial x \partial t} + \dfrac{\partial^2 w_s}{\partial x \partial t} \dfrac{\partial \delta u}{\partial t} + \dfrac{\partial v}{\partial t} \dfrac{\partial^2 \delta w_s}{\partial y \partial t} + \dfrac{\partial^2 w_s}{\partial y \partial t} \dfrac{\partial \delta v}{\partial t} \right) \\[2mm] + I_2 \left(\dfrac{\partial^2 w_b}{\partial x \partial t} \dfrac{\partial^2 \delta w_b}{\partial x \partial t} + \dfrac{\partial^2 w_b}{\partial y \partial t} \dfrac{\partial^2 \delta w_b}{\partial y \partial t} \right) \\[2mm] + K_2 \left(\dfrac{\partial^2 w_s}{\partial x \partial t} \dfrac{\partial^2 \delta w_s}{\partial x \partial t} + \dfrac{\partial^2 w_s}{\partial y \partial t} \dfrac{\partial^2 \delta w_s}{\partial y \partial t} \right) \\[2mm] + J_2 \left(\dfrac{\partial^2 w_b}{\partial x \partial t} \dfrac{\partial^2 \delta w_s}{\partial x \partial t} + \dfrac{\partial^2 w_s}{\partial x \partial t} \dfrac{\partial^2 \delta w_b}{\partial x \partial t} + \dfrac{\partial^2 w_b}{\partial y \partial t} \dfrac{\partial^2 \delta w_s}{\partial y \partial t} + \dfrac{\partial^2 w_s}{\partial y \partial t} \dfrac{\partial^2 \delta w_b}{\partial y \partial t} \right) \end{array} \right\} dxdy$$

The equation for the mass inertias that were employed in the previous equation looks like this [30]:

$$\left[I_0, I_1, I_2, J_1, J_2, K_2 \right] = \int_{-h/2}^{h/2} \left[1, z, z^2, f(z), zf(z), f^2(z) \right] \rho(z) dz \qquad (6.57)$$

Using equations (6.54), (6.56), and (6.53), we can get the Euler–Lagrange equations for multi-scale hybrid nanocomposite plates by setting the coefficients of δu, δv, δw_b, and δw_s equal to zero in Eq. (6.53). According to the source [30], the equations may be written as follows:

$$\frac{\partial N_{xx}}{\partial x} + \frac{\partial N_{xy}}{\partial y} = I_0 \frac{\partial^2 u}{\partial t^2} - I_1 \frac{\partial^3 w_b}{\partial x \partial t^2} - J_1 \frac{\partial^3 w_s}{\partial x \partial t^2} \qquad (6.58a)$$

$$\frac{\partial N_{xy}}{\partial x} + \frac{\partial N_{yy}}{\partial y} = I_0 \frac{\partial^2 v}{\partial t^2} - I_1 \frac{\partial^3 w_b}{\partial y \partial t^2} - J_1 \frac{\partial^3 w_s}{\partial y \partial t^2} \qquad (6.58b)$$

$$\frac{\partial^2 M_{xx}^b}{\partial x^2} + 2\frac{\partial^2 M_{xy}^b}{\partial x \partial y} + \frac{\partial^2 M_{yy}^b}{\partial y^2} = I_0 \frac{\partial^2 (w_b + w_s)}{\partial t^2} + I_1 \left(\frac{\partial^3 u}{\partial x \partial t^2} + \frac{\partial^3 v}{\partial y \partial t^2} \right)$$
$$- I_2 \left(\frac{\partial^4 w_b}{\partial x^2 \partial t^2} + \frac{\partial^4 w_b}{\partial y^2 \partial t^2} \right) - J_2 \left(\frac{\partial^4 w_s}{\partial x^2 \partial t^2} + \frac{\partial^4 w_s}{\partial y^2 \partial t^2} \right)$$

$$(6.58c)$$

$$\frac{\partial^2 M_{xx}^s}{\partial x^2} + 2\frac{\partial^2 M_{xy}^s}{\partial x \partial y} + \frac{\partial^2 M_{yy}^s}{\partial y^2} + \frac{\partial Q_{xz}}{\partial x} + \frac{\partial Q_{yz}}{\partial y} = I_0 \frac{\partial^2 (w_b + w_s)}{\partial t^2} + J_1 \left(\frac{\partial^3 u}{\partial x \partial t^2} + \frac{\partial^3 v}{\partial y \partial t^2} \right)$$
$$- J_2 \left(\frac{\partial^4 w_b}{\partial x^2 \partial t^2} + \frac{\partial^4 w_b}{\partial y^2 \partial t^2} \right) - K_2 \left(\frac{\partial^4 w_s}{\partial x^2 \partial t^2} + \frac{\partial^4 w_s}{\partial y^2 \partial t^2} \right)$$

$$(6.58d)$$

The basic elastic equations of orthotropic composites may be found by applying the elastic stress-strain relations of these materials. Because of this, it is possible to deduce the basic elastic equations of such substances. There is a way in which the constitutive equations of the plate may be presented, and it is as follows [30]:

$$\sigma_{ij} = C_{ijkl}\varepsilon_{kl} \tag{6.59}$$

In Eq. (6.30) σ_{ij}, ε_{kl}, and C_{ijkl} denote stress, strain, and elastic stiffness components, respectively. Therefore, these relations can be modified for plates in the following form [30]:

$$\begin{bmatrix} \sigma_{xx} \\ \sigma_{yy} \\ \sigma_{yz} \\ \sigma_{xz} \\ \sigma_{xy} \end{bmatrix} = \begin{bmatrix} Q_{11} & Q_{12} & 0 & 0 & 0 \\ Q_{12} & Q_{22} & 0 & 0 & 0 \\ 0 & 0 & Q_{44} & 0 & 0 \\ 0 & 0 & 0 & Q_{55} & 0 \\ 0 & 0 & 0 & 0 & Q_{66} \end{bmatrix} \begin{bmatrix} \varepsilon_{xx} \\ \varepsilon_{yy} \\ \varepsilon_{yz} \\ \varepsilon_{xz} \\ \varepsilon_{xy} \end{bmatrix} \tag{6.60}$$

in which

$$Q_{11} = \frac{E_{11}}{1-v_{12}v_{21}}, \quad Q_{12} = \frac{v_{12}E_{22}}{1-v_{12}v_{21}}, \quad Q_{22} = \frac{E_{22}}{1-v_{12}v_{21}}, \tag{6.61}$$
$$Q_{44} = G_{23}, \quad Q_{55} = G_{13}, \quad Q_{66} = G_{12}$$

When integrating from Eq. (6.58), taking into consideration the thickness direction along the z-axis, one may arrive at the consequent forces and moments shown below [30]:

$$
\begin{bmatrix} N_{xx} \\ N_{yy} \\ N_{xy} \\ M_{xx}^b \\ M_{yy}^b \\ M_{xy}^b \\ M_{xx}^s \\ M_{yy}^s \\ M_{xy}^s \end{bmatrix} = \begin{bmatrix} A_{11} & A_{12} & 0 & B_{11} & B_{12} & 0 & B_{11}^s & B_{12}^s & 0 \\ A_{12} & A_{22} & 0 & B_{12} & B_{22} & 0 & B_{12}^s & B_{22}^s & 0 \\ 0 & 0 & A_{66} & 0 & 0 & B_{66} & 0 & 0 & B_{66}^s \\ B_{11} & B_{12} & 0 & D_{11} & D_{12} & 0 & D_{11}^s & D_{12}^s & 0 \\ B_{12} & B_{22} & 0 & D_{12} & D_{22} & 0 & D_{12}^s & D_{22}^s & 0 \\ 0 & 0 & B_{66} & 0 & 0 & D_{66} & 0 & 0 & D_{66}^s \\ B_{11}^s & B_{12}^s & 0 & D_{11}^s & D_{12}^s & 0 & H_{11}^s & H_{12}^s & 0 \\ B_{12}^s & B_{22}^s & 0 & D_{12}^s & D_{22}^s & 0 & H_{12}^s & H_{22}^s & 0 \\ 0 & 0 & B_{66}^s & 0 & 0 & D_{66}^s & 0 & 0 & H_{66}^s \end{bmatrix} \begin{bmatrix} \dfrac{\partial u}{\partial x} \\[2mm] \dfrac{\partial v}{\partial y} \\[2mm] \dfrac{\partial u}{\partial y}+\dfrac{\partial v}{\partial x} \\[2mm] -\dfrac{\partial^2 w_b}{\partial x^2} \\[2mm] -\dfrac{\partial^2 w_b}{\partial y^2} \\[2mm] -2\dfrac{\partial^2 w_b}{\partial x \partial y} \\[2mm] -\dfrac{\partial^2 w_s}{\partial x^2} \\[2mm] -\dfrac{\partial^2 w_s}{\partial y^2} \\[2mm] -2\dfrac{\partial^2 w_s}{\partial x \partial y} \end{bmatrix},
$$

$$
\begin{bmatrix} Q_{xz} \\ Q_{yz} \end{bmatrix} = \begin{bmatrix} A_{44}^s & 0 \\ 0 & A_{55}^s \end{bmatrix} \begin{bmatrix} \dfrac{\partial w_s}{\partial x} \\[2mm] \dfrac{\partial w_s}{\partial y} \end{bmatrix}
$$

(6.62)

where

$$
\left[A_n, B_n, B_n^s, D_n, D_n^s, H_n^s \right] = \int_{-h/2}^{h/2} \left[1, z, f(z), z^2, zf(z), f^2(z) \right] Q_n(z) dz, \quad n = (11,12,22,66)
$$

$$
\left[A_{44}^s, A_{55}^s \right] = \int_{-h/2}^{h/2} \left[Q_{44}(z), Q_{55}(z) \right] g^2(z) dz
$$

(6.63)

It is possible to construct the governing equations for multi-scale hybrid nanocomposite plates by inserting Eq. (6.62) into Eqs. (6.53) to (6.56), as shown in [30]:

$$
A_{11}\frac{\partial^2 u}{\partial x^2} + A_{66}\frac{\partial^2 u}{\partial y^2} + \left(A_{12} + A_{66} \right)\frac{\partial^2 v}{\partial x \partial y} - B_{11}\frac{\partial^3 w_b}{\partial x^3} - \left(B_{12} + 2B_{66} \right)\frac{\partial^3 w_b}{\partial x \partial y^2}
$$

$$
-B_{11}^s\frac{\partial^3 w_s}{\partial x^3} - \left(B_{12}^s + 2B_{66}^s \right)\frac{\partial^3 w_s}{\partial x \partial y^2} - I_0\frac{\partial^2 u}{\partial t^2} + I_1\frac{\partial^3 w_b}{\partial x \partial t^2} + J_1\frac{\partial^3 w_b}{\partial x \partial t^2} = 0
$$

(6.64a)

$$\left(A_{12}+A_{66}\right)\frac{\partial^2 u}{\partial x \partial y}+A_{66}\frac{\partial^2 v}{\partial x^2}+A_{22}\frac{\partial^2 v}{\partial y^2}-\left(B_{12}+2B_{66}\right)\frac{\partial^3 w_b}{\partial x^2 \partial y}-B_{22}\frac{\partial^3 w_b}{\partial y^3}$$

$$-\left(B_{12}^s+2B_{66}^s\right)\frac{\partial^3 w_s}{\partial x^2 \partial y}-B_{22}^s\frac{\partial^3 w_s}{\partial y^3}-I_0\frac{\partial^2 v}{\partial t^2}+I_1\frac{\partial^3 w_b}{\partial y \partial t^2}+J_1\frac{\partial^3 w_b}{\partial y \partial t^2}=0 \qquad (6.64b)$$

$$B_{11}\frac{\partial^3 u}{\partial x^3}+\left(B_{12}+2B_{66}\right)\left(\frac{\partial^3 u}{\partial x \partial y^2}+\frac{\partial^3 v}{\partial x^2 \partial y}\right)+B_{22}\frac{\partial^3 v}{\partial y^3}-D_{11}\frac{\partial^4 w_b}{\partial x^4}$$

$$-2\left(D_{12}+2D_{66}\right)\frac{\partial^4 w_b}{\partial x^2 \partial y^2}-D_{22}\frac{\partial^4 w_b}{\partial y^4}-D_{11}^s\frac{\partial^4 w_s}{\partial x^4}-2\left(D_{12}^s+2D_{66}^s\right)\frac{\partial^4 w_s}{\partial x^2 \partial y^2}$$

$$-D_{22}^s\frac{\partial^4 w_s}{\partial y^4}-I_0\frac{\partial^2\left(w_b+w_s\right)}{\partial t^2}-I_1\left(\frac{\partial^3 u}{\partial x \partial t^2}+\frac{\partial^3 v}{\partial y \partial t^2}\right)+I_2\left(\frac{\partial^4 w_b}{\partial x^2 \partial t^2}+\frac{\partial^4 w_b}{\partial y^2 \partial t^2}\right) \qquad (6.64c)$$

$$+J_2\left(\frac{\partial^4 w_s}{\partial x^2 \partial t^2}+\frac{\partial^4 w_s}{\partial y^2 \partial t^2}\right)=0$$

$$B_{11}^s\frac{\partial^3 u}{\partial x^3}+\left(B_{12}^s+2B_{66}^s\right)\left(\frac{\partial^3 u}{\partial x \partial y^2}+\frac{\partial^3 v}{\partial x^2 \partial y}\right)+B_{22}^s\frac{\partial^3 v}{\partial y^3}-D_{11}^s\frac{\partial^4 w_b}{\partial x^4}$$

$$-2\left(D_{12}^s+2D_{66}^s\right)\frac{\partial^4 w_b}{\partial x^2 \partial y^2}-D_{22}^s\frac{\partial^4 w_b}{\partial y^4}-H_{11}^s\frac{\partial^4 w_s}{\partial x^4}-2\left(H_{12}^s+2H_{66}^s\right)\frac{\partial^4 w_s}{\partial x^2 \partial y^2}$$

$$-H_{22}^s\frac{\partial^4 w_s}{\partial y^4}+A_{55}^s\frac{\partial^2 w_s}{\partial x^2}+A_{44}^s\frac{\partial^2 w_s}{\partial y^2}-I_0\frac{\partial^2\left(w_b+w_s\right)}{\partial t^2}-J_1\left(\frac{\partial^3 u}{\partial x \partial t^2}+\frac{\partial^3 v}{\partial y \partial t^2}\right) \qquad (6.64d)$$

$$+J_2\left(\frac{\partial^4 w_b}{\partial x^2 \partial t^2}+\frac{\partial^4 w_b}{\partial y^2 \partial t^2}\right)+K_2\left(\frac{\partial^4 w_s}{\partial x^2 \partial t^2}+\frac{\partial^4 w_s}{\partial y^2 \partial t^2}\right)=0$$

An analytical method is used as a solution strategy in order to solve the equations that regulate the behavior of the plate. In the event that what they say is true, Eq. (6.65) [30] may be used to characterize the various components of the displacement field:

$$\begin{Bmatrix} u \\ v \\ w_b \\ w_s \end{Bmatrix}=\begin{Bmatrix} U\exp\left[i\left(\beta_1 x+\beta_2 y-\omega t\right)\right] \\ V\exp\left[i\left(\beta_1 x+\beta_2 y-\omega t\right)\right] \\ W_b\exp\left[i\left(\beta_1 x+\beta_2 y-\omega t\right)\right] \\ W_s\exp\left[i\left(\beta_1 x+\beta_2 y-\omega t\right)\right] \end{Bmatrix} \qquad (6.65)$$

in which U, V, W_b, and W_s represent wave amplitudes, respectively. β_1 is longitudinal wave number, β_2 is transverse wave number, and ω is circular frequency dispersed waves. By substituting u, v, w_b, and w_s from Eq. (6.65) in Eqs. (6.64), the following equation is gained [30]:

$$\left(\left[K \right]_{4\times4} - \omega^2 \left[M \right]_{4\times4} \right) \begin{bmatrix} U \\ V \\ W_b \\ W_s \end{bmatrix} = 0 \qquad (6.66)$$

where K and M are matrices standing for stiffness and mass, and their corresponding components are shown in Appendix 6B.

The determinant of the coefficient matrix on the left-hand side of Eq. (6.66) must be set to zero in order to solve this eigenvalue problem [30]:

$$\left| \left[K \right]_{4\times4} - \omega^2 \left[M \right]_{4\times4} \right| = 0 \qquad (6.67)$$

Following this is a series of diagrams detailing the impact of different factors on the wave propagation behavior of a multi-scale hybrid nanocomposite plate. These figures may be found in the next section. You may find these numbers farther down the page. As a means of reinforcement, the plate contains carbon fiber as well as carbon nanotubes that have been embedded in an epoxy matrix. In the order given, [11] and [12] are the references that should be consulted in order to gain information on the material properties of both the CF and the matrix. In Figure 6.27, we study how the wave frequency fluctuates in proportion to the volume fraction of the CF for various clusters with different mass percentages and different volume fractions of the clusters. We do this for a variety of different clusters with different mass percentages and different volume fractions. It has been shown that there is a direct correlation between the volume percentage of CF and the wave frequency. This tendency arises as a consequence of the fact that a rise in the volume percentage of CF results in a rise in the structure's stiffness. In addition to this, the wave frequency of a nanocomposite plate that has a higher mass percentage of CNT is higher than the wave frequency of a nanocomposite plate that has a lower mass fraction of CNT. This is because a higher mass percentage of CNT results in a larger number of CNT per unit mass. The illustration demonstrates that the volume percentage of clusters has an increasingly substantial impact on the wave frequency exhibited by multi-scale hybrid nanocomposite plates. If the cluster's volume increases by a certain percentage, the frequency of the waves will likewise rise. Those structures with a larger mass percentage of CNT and a greater volume fraction of cluster experience more frequent damage [30].

To be more specific, the goal of showing Figure 6.28 is to illustrate the impact that the volume percentage of the cluster has on the variance of wave frequency versus wave number. This may be done by looking at the relationship between the two variables. It is abundantly evident that the frequency of the waves will grow as the number of waves that exist on the nanocomposite plate will increase. In addition

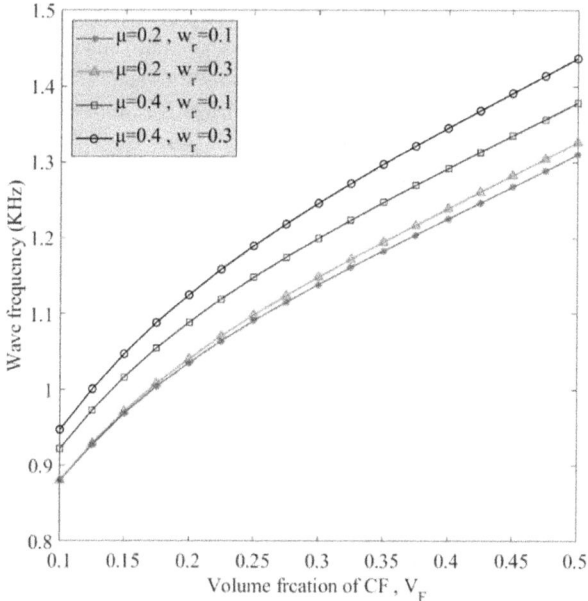

FIGURE 6.27   Illustration of the coupled effects of various volume fractions of clusters and various mass fractions of the CNTs on wave frequency versus CF's volume fraction.

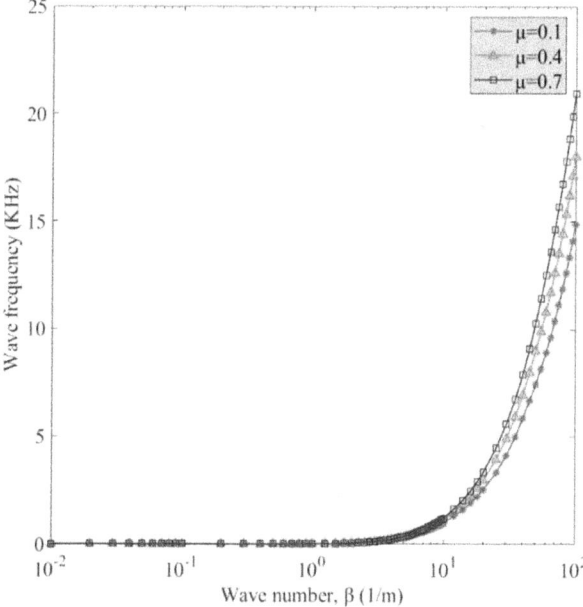

FIGURE 6.28   Variation of wave frequency of multi-scale nanocomposite plates versus wave number for different volume fractions of clusters.

to this, there is the possibility that an increase in the wave frequency will occur if the cluster's density continues to grow. Increasing the cluster volume fraction results in a higher wave frequency in multi-scale hybrid nanocomposite plates [30].

The relationship between wave frequency and wave number varies according to the volume percentage of the CFs, as seen in Figure 6.29. As can be seen, the wave frequency of the nanocomposite plate increases in a manner that is exponentially related to the rise in the wave number. While increasing the volume percentage of CF has no noticeable effect, the trend of the volume fraction of each CF can be seen more clearly when seen through a magnifying. It has been shown that the volume fraction of CF that contributes the most to the total has the highest wave frequency [30] for a given wave number.

Wave frequency and wave number variation are shown in Figure 6.30 to show the combined effects of the mass fraction and volume fraction of CNTs inside the cluster. This information may be found in the figure. It is plain to observe that the wave frequency will continue to increase even as the wave count will continue to rise. As was indicated before, the rise in the mass fraction of CNT causes an increase in the wave frequency; nevertheless, it is clear that the wave frequency will drop as the volume fraction of CNT contained inside the cluster grows. This means that the effects of the cluster volume fraction and the CNT volume fraction inside the clusters need to be considered separately. In order to increase the cluster's wave frequency, it may be necessary to reduce the amount of CNT present in the volume of the cluster [30].

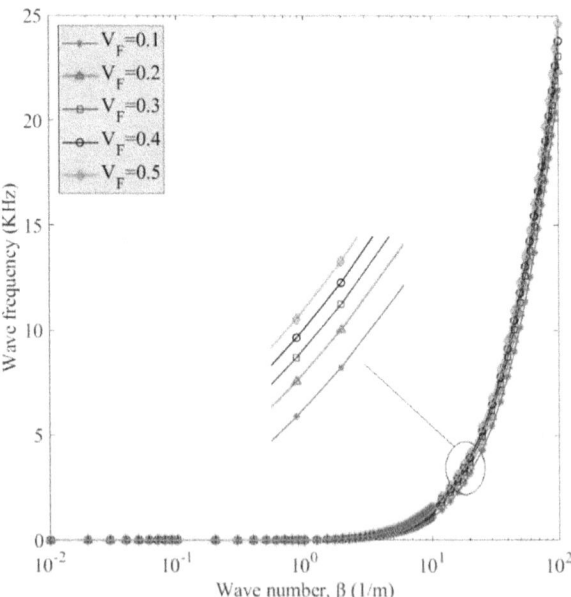

FIGURE 6.29 Variation of the wave frequency of multi-scale nanocomposite plates versus wave number for various volume fraction of CF.

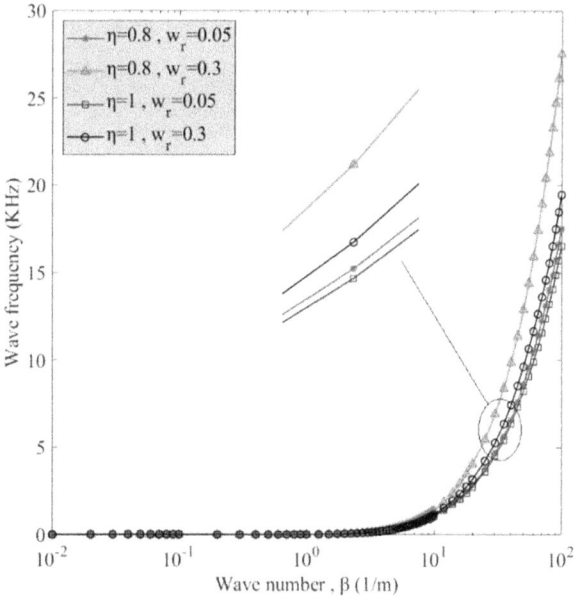

FIGURE 6.30 Variation of the wave frequency of multi-scale nanocomposite plates versus wave number for different mass fractions of CNTs and volume fractions of CNT inside cluster.

Although it is often believed that all charts have the same values for both the longitudinal and transverse wave numbers, Figure 6.31 shows how variations in the transverse wave number may have a larger impact on wave frequency than the longitudinal wave number. It is possible to draw parallels between the progression of this plot and the progression of other plots of wave frequency. It is possible to determine, via the use of a magnifier, that in the case of a fixed longitudinal wave number, wave frequency may be enhanced with the use of a bigger transverse wave number [30].

Wave frequency shifts of multi-scale hybrid nanocomposite plates are shown in Figure 6.32 as a function of the volume percentage of CNT inside the cluster and the gradient index. The correlation between the volume percentage of CNT and the change in wave frequency is also seen in this picture. This change in wave frequency is assessed in proportion to the amount of CNT that is contained inside the cluster. When schematics are compared, it is clear that an increase in the gradient index has the ability to lower the wave frequency. This is shown when the diagrams are compared. Because of this tendency, it is suggested that just a small amount of gradient index be used in order to boost the wave frequency of the structure. This is because of the propensity that gradient index has. In addition to this, each of the diagrams demonstrates a trend in a downward direction, although with different slopes. This suggests that a decrease in wave frequency may be accomplished by increasing the volume fraction of CNT contained inside the cluster.

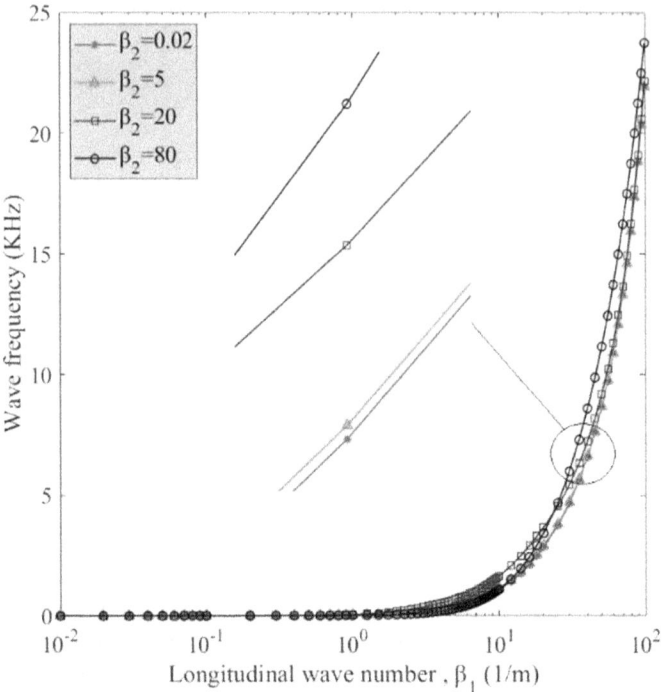

FIGURE 6.31   The influence of transverse wave number on the variation of wave frequency versus longitudinal wave number.

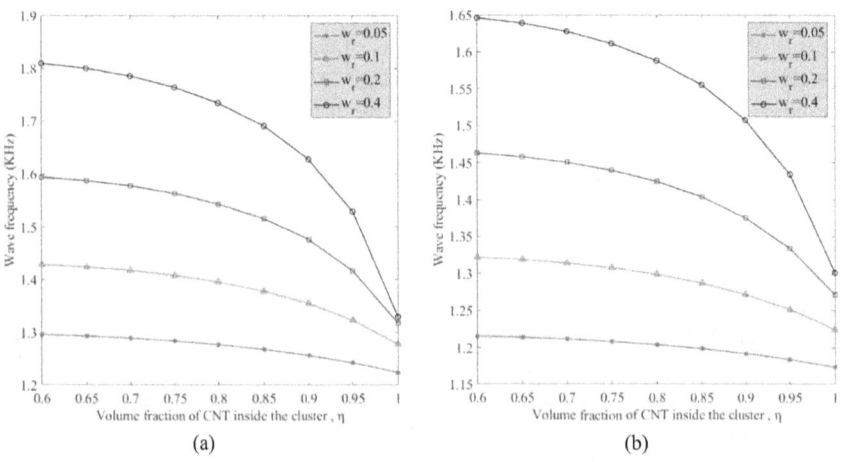

FIGURE 6.32   Change of the wave frequency of multi-scale nanocomposite plates versus the volume fraction of CNT inside the cluster for various mass fraction of CNT for (a) $P = 1$ and (b) $P = 2$.

Additionally, the mass fraction of CNT in the structure will have an effect on the frequency of waves, making them more frequent as the mass percentage of CNT increases. Additionally, it is essential to keep in mind that the structure will become more stiff as the mass percentage of CNT grows. In addition, the reason why diagrams have a negative slope is because the impact of the volume fraction of CNTs inside the cluster outweighs the influence of the mass fraction of CNTs contained within the cluster [30]. This is the reason why diagrams have a negative slope.

Figure 6.33 depicts the variation in wave frequency as a function of the volume percentage of CNT inside the cluster. This illustration is performed over a wide range of cluster volume percentages and gradient indices. The decrease in wave frequency that occurs as a direct result of an increase in the volume fraction of CNT contained inside the cluster is not hard to see at all. Once volume percentage of CNT inside the cluster = 0.6 has the greatest value of wave frequency, the value of wave frequency begins to diminish, and this reduction is caused by a decrease in the volume fraction of the cluster. The explanation for this behavior is that the structure's stiffness has been decreased. In addition to this, according to the figure, the wave frequency of the multi-scale nanocomposite plate will drop if the gradient index is raised [30]. Figure 6.34 shows the effect of varying the CNT mass fraction and cluster volume fraction on the nanocomposite plate's phase velocity. This comparison is made in relation to the CF volume fraction. It is clear that the trend of the diagrams is climbing, which indicates that the phase velocity of the multi-scale hybrid nanocomposite plate increases as the volume percentage of the CFs increases, which results in the plate having a higher degree of rigidity. In addition to this, when the cluster's volume percentage grows, the phase velocity also increases [30].

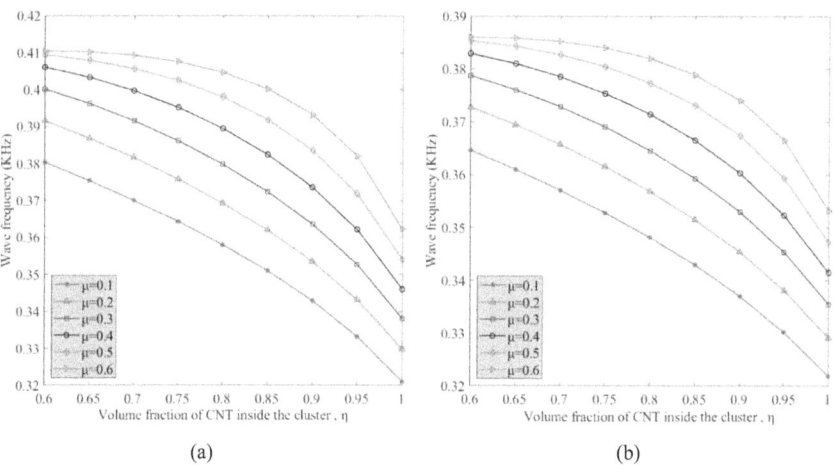

(a) (b)

FIGURE 6.33 The effect of volume fractions of clusters on variation of the wave frequency of multi-scale nanocomposite plates versus volume fractions of CNT inside the cluster for (a) $P = 1$ and (b) $P = 2$.

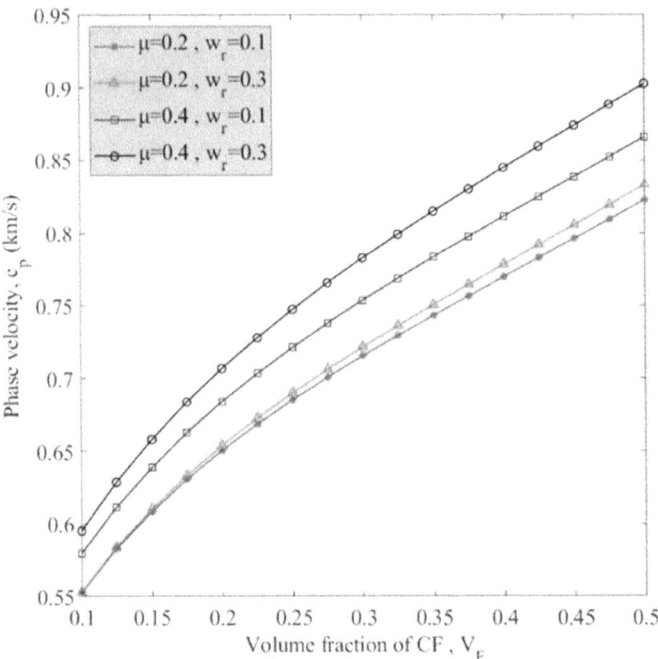

FIGURE 6.34   Variation of the phase velocity of multi-scale nanocomposite plates versus CF's volume fraction for different volume fractions of clusters and mass fractions of CNTs.

Figure 6.35 depicts how the phase velocity fluctuates in proportion to the wave number for a variety of volume fractions of the cluster. This is shown for a wide variety of cluster fractions. At the beginning of the figure, the phase velocities are all equal to zero. As the wave number grows, the velocities in each graphic increase until they reach a plateau. This is due to the fact that the phase velocity is independent of the wave number. It is possible to infer from the data [30] that an increase in the phase velocity of the plate is related to a rise in the volume fraction of a cluster.

Figure 6.36 is an illustration of the fluctuation of phase velocity versus wave number for a variety of different volume fractions of CF that have been explored. In light of the image, one may draw the conclusion that a rise in the CF volume percentage carries with it the possibility for an accompanying expansion of the phase velocity. To put it another way, while using a nanocomposite plate with a larger CF volume percentage, it is possible to get a bigger quantity of phase velocity. Similar to Figure 6.35, the phase velocity starts off as 0 and steadily increases as the wave number increases. After then, there is no further variation in phase velocity with increasing wave number [30].

We investigate the wave-number-dependent variation in the nanocomposite plate's phase velocity as a function of both the mass fraction and the volume fraction of the CNTs inside the cluster (see Figure 6.37). This is done so that we may

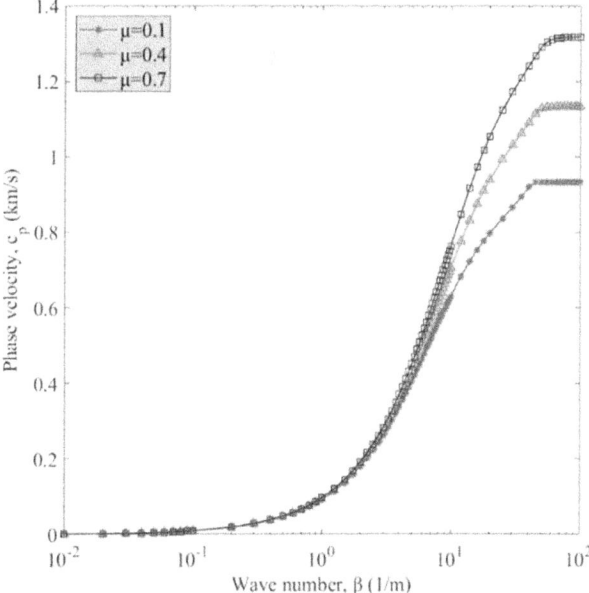

FIGURE 6.35 Illustration of phase velocity versus wave number for different volume fractions of clusters.

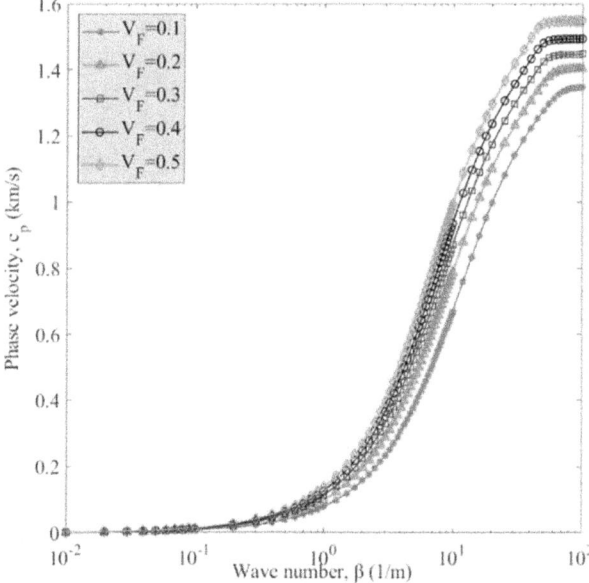

FIGURE 6.36 The effect of CF's volume fraction on the variation of phase velocity of multi-scale nanocomposite plates versus wave number.

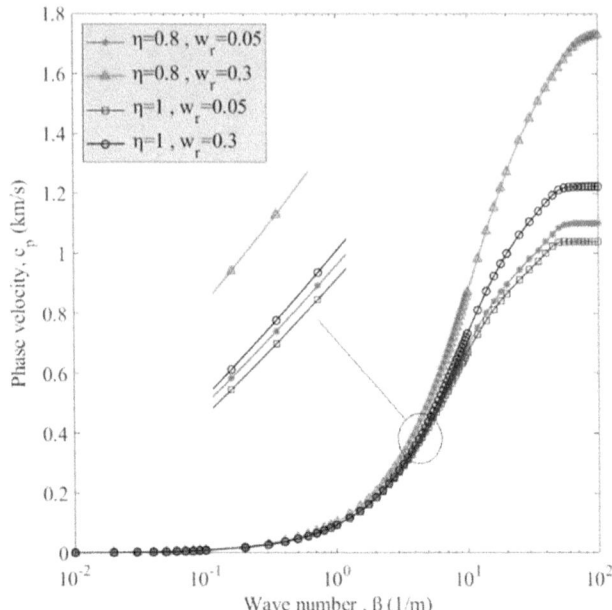

FIGURE 6.37 Both influences of volume fractions of CNT inside the cluster and CNT's mass fractions on phase velocity of multi-scale nanocomposite plates versus wave number.

answer the question, "How does the wave number influence the change in the phase velocity of the nanocomposite plate?" The graphic includes a magnifier so that the viewer can get a better look at the overall trend of the diagrams as well as the differences and similarities between each figure. It is clear that the mass fraction of CNTs contained inside a cluster and the volume fraction of CNTs contained within a cluster both have distinct influences on the phase velocity of the system. As a direct consequence of this, the phase velocity of a nanocomposite plate drops as the volume percentage of CNTs within the cluster grows, but the phase velocity of a nanocomposite plate increases when the mass fraction of CNTs inside the cluster increases [30].

For a variety of transverse wave numbers, Figure 6.38 shows how the phase velocity of a multi-scale nanocomposite plate fluctuates as a function of the number of longitudinal waves. The phase velocity is expected to rise as the number of longitudinal waves grows. If the longitudinal wave number were to be divided into three ranges, the range corresponding to a value of $\beta = 5$ would have the highest value of phase velocity, the range corresponding to a value of $\beta = 20$ would have the second-highest value of phase velocity, and the range corresponding to a value of $\beta = 80$ would have the third-highest value of phase velocity [30].

Figure 6.39 depicts the relationship between the phase velocity and the volume fraction of CNTs in the cluster over a range of gradient indices, illustrating the effects of fluctuations in the cluster's mass fraction. The results suggested that a

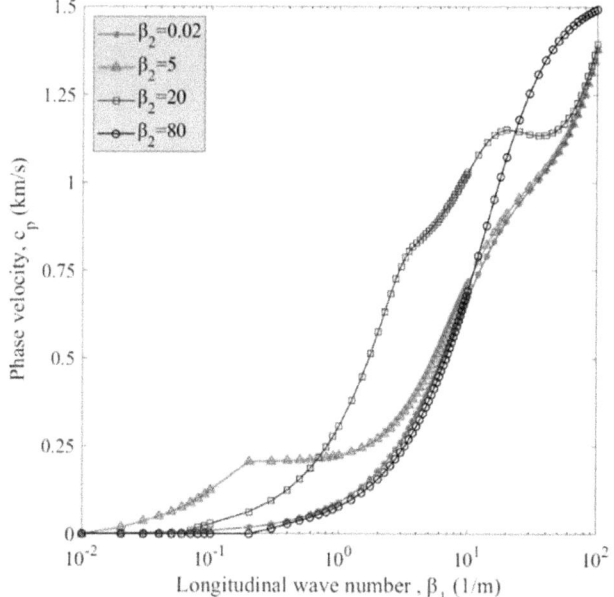

FIGURE 6.38 Variation of phase velocity versus longitudinal wave number for various transverse wave numbers.

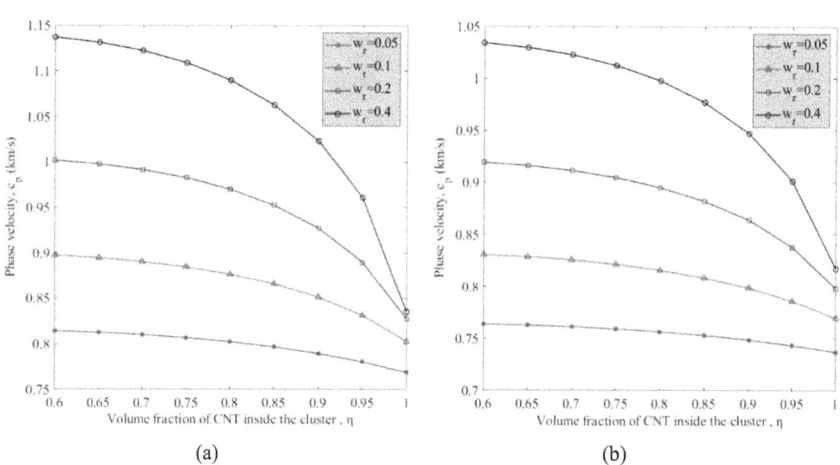

FIGURE 6.39 Variation of the phase velocity of multi-scale nanocomposite plates versus the volume fraction of CNT inside the cluster for various mass fraction of CNT for (a) $P = 1$ and (b) $P = 2$.

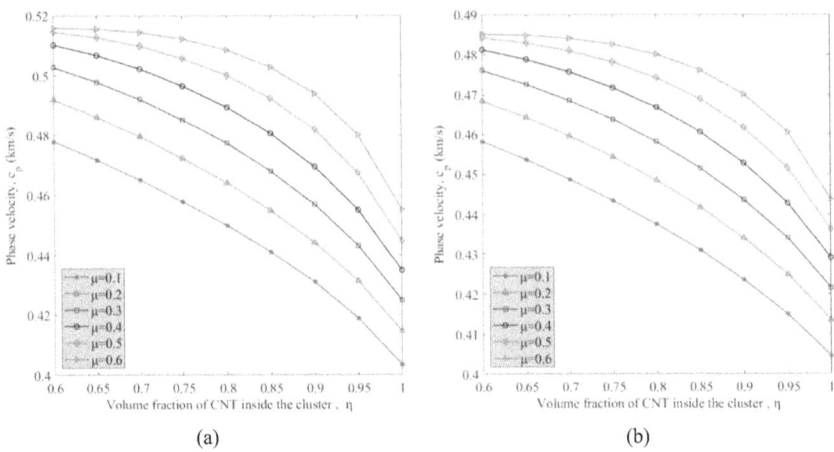

FIGURE 6.40 Variation of the phase velocity of multi-scale nanocomposite plates versus the volume fraction of CNT inside the cluster for different volume fractions of clusters for (a) $P = 1$, (b) and $P = 2$.

higher mass percentage of CNT would raise the phase velocity, whereas a higher amount of gradient index might reduce it. Additionally, it is possible to state that every time w_r is raised, the slope of decreasing gets bigger [30]; this is something that may be asserted.

Figure 6.40 illustrates the link that exists between the phase velocity and the volume percentage of CNT that is present in the cluster. This link is shown using a wide range of volume fractions of the cluster as well as a number of various gradient indices. When you look at the graph in Figure 6.40, you'll see that the phase velocity of the multi-scale hybrid nanocomposite plate falls as the cluster's volume percentage goes up. This is something that you can determine by looking at the graph. The impacts of the gradient index, as well as the volume fraction of CNTs discovered within the cluster, are precisely the same as in Figure 6.39 [30], which can be located at this location.

6.4   CNTR NANOCOMPOSITE SHELL WAVE DISPERSION CHARACTERISTICS

Figure 6.41 displays the geometry and coordinate system of the structure for the sake of clarity. In this chapter, the Eshelby–Mori–Tanaka micromechanical model and the law of mixing are used to find the corresponding mechanical characteristics of the component materials. Section 6.3 introduced Eqs. (6.1)–(6.28), which were used to get the following findings [26]. Next, the multi-scale hybrid nano-composite shell's motion equations are derived using first-order shell theory and Hamilton's principle.

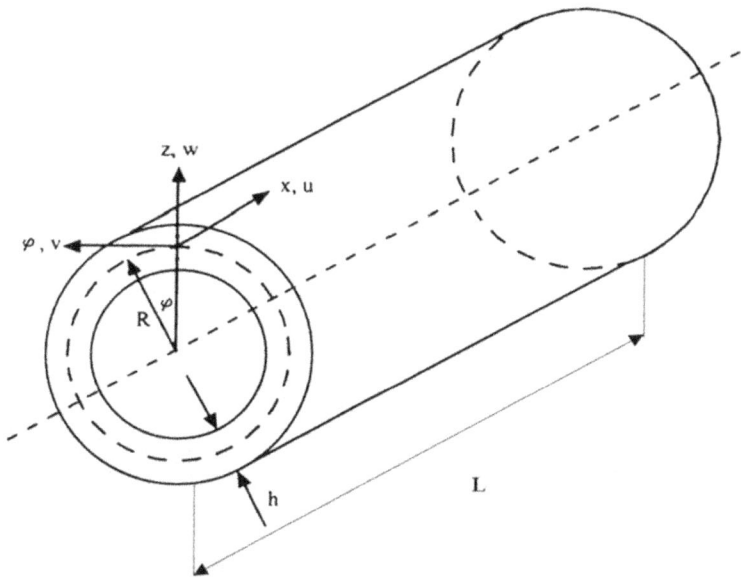

FIGURE 6.41 Geometry and coordinate systems of a CNTR nanocomposite cylindrical shell.

The nanocomposite shell's kinematic relations will be created within the framework of the first-order shear deformable shell theory, and once finished, they may be defined as follows [32]:

$$u_x\left(x,\varphi,z,t\right) = u\left(x,\varphi,t\right) + z\theta_x\left(x,\varphi,t\right) \tag{6.68}$$

$$u_\varphi\left(x,\varphi,z,t\right) = v\left(x,\varphi,t\right) + z\theta_\varphi\left(x,\varphi,t\right) \tag{6.69}$$

$$u_z\left(x,\varphi,z,t\right) = w\left(x,\varphi,t\right) \tag{6.70}$$

where u, v, and w denote axial, circumferential, and lateral displacements, respectively. Furthermore, θ_x and θ_φ indicate the rotation components about axial and circumferential directions, respectively. Henceforward, the nonzero strains of a shell-type element can be expressed as [32]:

$$\begin{Bmatrix} \varepsilon_{xx} \\ \varepsilon_{\varphi\varphi} \\ \varepsilon_{x\varphi} \\ \varepsilon_{xz} \\ \varepsilon_{\varphi z} \end{Bmatrix} = \begin{Bmatrix} \dfrac{\partial u}{\partial x} + z\dfrac{\partial \theta_x}{\partial x} \\[2ex] \dfrac{1}{R}\left(\dfrac{\partial v}{\partial \varphi} + z\dfrac{\partial \theta_\varphi}{\partial \varphi} + w \right) \\[2ex] \dfrac{1}{R}\dfrac{\partial u}{\partial \varphi} + \dfrac{\partial v}{\partial x} + \dfrac{z}{R}\dfrac{\partial \theta_x}{\partial \varphi} + z\dfrac{\partial \theta_\varphi}{\partial x} \\[2ex] \theta_x + \dfrac{\partial w}{\partial x} \\[2ex] \theta_\varphi + \dfrac{1}{R}\dfrac{\partial w}{\partial \varphi} - \dfrac{v}{R} \end{Bmatrix} \tag{6.71}$$

Now, in order to obtain Euler–Lagrange equations of a nanocomposite shell, the implementation of Hamilton's concept may be put in the following form:

$$\int_0^t \delta\left(U - T - V\right) dt = 0 \tag{6.72}$$

where U, T, and V, respectively, stand for the strain, kinetic energy, and work done by external (applied) loading. For a linear elastic solid, the variation in strain energy may be expressed using the following expression [32]:

$$\delta U = \int_{-\frac{h}{2}}^{\frac{h}{2}} \int_0^{2\pi} \int_0^L \sigma_{ij} \delta\varepsilon_{ij} R\, dx\, d\varphi\, dz \tag{6.73}$$

Additionally, the variation of kinetic energy may be shown as:

$$\delta T = \int_{-\frac{h}{2}}^{\frac{h}{2}} \int_0^{2\pi} \int_0^L \rho(z)\left[\left(\frac{\partial \delta u_x}{\partial t}\right)^2 + \left(\frac{\partial \delta u_\varphi}{\partial t}\right)^2 + \left(\frac{\partial \delta u_z}{\partial t}\right)^2 \right] R\, dx\, d\varphi\, dz \tag{6.74}$$

Last but not least, the variation in the amount of work done by external loadings may be expressed using the following form [32]:

$$\delta V = \int_{-\frac{h}{2}}^{\frac{h}{2}} \int_0^{2\pi} \int_0^L \left(N_r + N_x \frac{\partial^2 w}{\partial x^2} + \frac{N_\varphi}{R^2}\frac{\partial^2 w}{\partial \varphi^2} \right) \delta w R\, dx\, d\varphi\, dz \tag{6.75}$$

where the radial, axial, and circumferential loadings are denoted by the symbols N_r, N_x, and N_φ, respectively. Since the current study did not take any loading into account, the change in the amount of work caused by external loadings may be assumed to be zero. Consequently, in order to derive the motion equation for the cylindrical shell, Eqs. (6.73) through (6.75) are entered into Eq. (6.72), and the resulting equation may be stated as follows [32]:

$$\frac{\partial N_{xx}}{\partial x} + \frac{1}{R}\frac{\partial N_{x\varphi}}{\partial \varphi} = I_0 \frac{\partial^2 u}{\partial t^2} + I_1 \frac{\partial^2 \theta_x}{\partial t^2} \tag{6.76}$$

$$\frac{\partial N_{x\varphi}}{\partial x} + \frac{1}{R}\frac{\partial N_{\varphi\varphi}}{\partial \varphi} + \frac{Q_{z\varphi}}{R} = I_0 \frac{\partial^2 v}{\partial t^2} + I_1 \frac{\partial^2 \theta_\varphi}{\partial t^2} \tag{6.77}$$

$$\frac{\partial Q_{xz}}{\partial x} + \frac{1}{R}\frac{\partial Q_{z\varphi}}{\partial \varphi} - \frac{N_{\varphi\varphi}}{R} = I_0 \frac{\partial^2 w}{\partial t^2} \tag{6.78}$$

$$\frac{\partial M_{xx}}{\partial x} + \frac{1}{R}\frac{\partial M_{x\varphi}}{\partial \varphi} - Q_{xz} = I_1 \frac{\partial^2 u}{\partial t^2} + I_2 \frac{\partial^2 \theta_x}{\partial t^2} \tag{6.79}$$

$$\frac{\partial M_{x\varphi}}{\partial x} + \frac{1}{R}\frac{\partial M_{\varphi\varphi}}{\partial \varphi} - Q_{\varphi z} = I_1 \frac{\partial^2 v}{\partial t^2} + I_2 \frac{\partial^2 \theta_\varphi}{\partial t^2} \tag{6.80}$$

where

$$\begin{Bmatrix} N_{xx} & N_{\varphi\varphi} & N_{x\varphi} \\ M_{xx} & M_{\varphi\varphi} & M_{x\varphi} \end{Bmatrix} = \int_{-\frac{h}{2}}^{\frac{h}{2}} \left(\sigma_{xx}, \sigma_{\varphi\varphi}, \sigma_{x\varphi}\right) \begin{Bmatrix} 1 \\ z \end{Bmatrix} dz \tag{6.81}$$

$$\left[Q_{xz}, Q_{z\varphi}\right] = \kappa_s \int_{-\frac{h}{2}}^{\frac{h}{2}} \left[\sigma_{xz}, \sigma_{z\varphi}\right] dz \tag{6.82}$$

$$\left[I_0, I_1, I_2\right] = \int_{-\frac{h}{2}}^{\frac{h}{2}} \left[1, z, z^2\right] \rho(z) dz \tag{6.83}$$

in which κ_s is shear correction factor. The link between stress and strain in a multi-scale hybrid nanocomposite may be expressed as follows:

$$\sigma_{ij} = C_{ijkl}\varepsilon_{kl} \tag{6.84}$$

Here σ_{ij}, ε_{kl}, and C_{ijkl} denote components of Cauchy stress, strain, and elasticity tensors, respectively. Integrating from above equation over the shell's thickness, the following relation can be achieved [32]:

$$\begin{bmatrix} N_{xx} \\ M_{xx} \\ N_{\varphi\varphi} \\ M_{\varphi\varphi} \end{bmatrix} = \begin{bmatrix} A_{11} & B_{11} & \dfrac{A_{12}}{R} & \dfrac{B_{12}}{R} \\ B_{11} & D_{11} & \dfrac{B_{12}}{R} & \dfrac{D_{12}}{R} \\ A_{12} & B_{12} & \dfrac{A_{11}}{R} & \dfrac{B_{11}}{R} \\ B_{12} & D_{12} & \dfrac{B_{11}}{R} & \dfrac{D_{11}}{R} \end{bmatrix} \begin{bmatrix} \dfrac{\partial u}{\partial x} \\ \dfrac{\partial \theta_x}{\partial x} \\ \dfrac{\partial v}{\partial \varphi} + w \\ \dfrac{\partial \theta_\varphi}{\partial \varphi} \end{bmatrix},$$

$$\begin{bmatrix} N_{x\varphi} \\ M_{x\varphi} \end{bmatrix} = \begin{bmatrix} A_{66} & B_{66} \\ B_{66} & D_{66} \end{bmatrix} \begin{bmatrix} \dfrac{1}{R}\dfrac{\partial u}{\partial \varphi} + \dfrac{\partial v}{\partial x} \\ \dfrac{1}{R}\dfrac{\partial \theta_x}{\partial \varphi} + \dfrac{\partial \theta_\varphi}{\partial x} \end{bmatrix},$$

$$Q_{xz} = A_{55}^s\left(\theta_x + \dfrac{\partial w}{\partial x}\right), \quad Q_{\varphi z} = A_{55}^s\left(\theta_\varphi + \dfrac{1}{R}\dfrac{\partial w}{\partial \varphi} - \dfrac{v}{R}\right) \tag{6.85}$$

in which

$$\begin{bmatrix} A_{11} & B_{11} & D_{11} \\ A_{12} & B_{12} & D_{12} \\ A_{66} & B_{66} & D_{66} \end{bmatrix} = \int_{-\frac{h}{2}}^{\frac{h}{2}} \left[1, z, z^2\right] \begin{bmatrix} \dfrac{E_{11}}{1-v_{12}v_{21}} \\ \dfrac{v_{12}E_{22}}{1-v_{12}v_{21}} \\ \dfrac{E_{11}}{2(1+v_{12})} \end{bmatrix} dz \tag{6.86}$$

$$A_{55}^s = \kappa_s \int_{-\frac{h}{2}}^{\frac{h}{2}} \dfrac{E_{11}}{2(1+v_{12})}\, dz \tag{6.87}$$

As shown in reference [32], the final governing equations for the multi-scale hybrid nanocomposite shell may be derived. By connecting Eqs. (6.76)–(6.80) with Eq. (6.85), we may do this.

$$
A_{11}\frac{\partial^2 u}{\partial x^2} + B_{11}\frac{\partial^2 \theta_x}{\partial x^2} + \frac{A_{12}}{R}\left(\frac{\partial^2 v}{\partial x \partial \varphi} + \frac{\partial w}{\partial x}\right)
$$
$$
+ \frac{B_{12}}{R}\frac{\partial^2 \theta_\varphi}{\partial x \partial \varphi} + \frac{A_{66}}{R}\left(\frac{1}{R}\frac{\partial^2 u}{\partial \varphi^2} + \frac{\partial^2 v}{\partial x \partial \varphi}\right)
$$
$$
+ \frac{B_{66}}{R}\left(\frac{1}{R}\frac{\partial^2 \theta_x}{\partial \varphi^2} + \frac{\partial^2 \theta_\varphi}{\partial x \partial \varphi}\right) - I_0\frac{\partial^2 u}{\partial t^2} - I_1\frac{\partial^2 \theta_x}{\partial t^2} = 0
\tag{6.88}
$$

$$
A_{66}\left(\frac{1}{R}\frac{\partial^2 u}{\partial x \partial \varphi} + \frac{\partial^2 v}{\partial x^2}\right) + B_{66}\left(\frac{1}{R}\frac{\partial^2 \theta_x}{\partial x \partial \varphi} + \frac{\partial^2 \theta_\varphi}{\partial x^2}\right) + \frac{A_{12}}{R}\frac{\partial^2 u}{\partial x \partial \varphi}
$$
$$
+ \frac{B_{12}}{R}\frac{\partial^2 \theta_x}{\partial x \partial \varphi} + \frac{A_{11}}{R^2}\left(\frac{\partial^2 v}{\partial \varphi^2} + \frac{\partial w}{\partial \varphi}\right)
$$
$$
+ \frac{B_{11}}{R^2}\frac{\partial^2 \theta_\varphi}{\partial \varphi^2} + \frac{A_{55}^s}{R}\left(\theta_\varphi + \frac{1}{R}\frac{\partial w}{\partial \varphi} - \frac{v}{R}\right) - I_0\frac{\partial^2 v}{\partial t^2} - I_1\frac{\partial^2 \theta_\varphi}{\partial t^2} = 0
\tag{6.89}
$$

$$
A_{55}^s\left(\frac{\partial \theta_x}{\partial x} + \frac{\partial^2 w}{\partial x^2}\right) + \frac{A_{55}^s}{R}\left(\frac{\partial \theta_\varphi}{\partial \varphi} + \frac{1}{R}\frac{\partial^2 w}{\partial \varphi^2} - \frac{1}{R}\frac{\partial v}{\partial \varphi}\right)
$$
$$
- \frac{A_{12}}{R}\frac{\partial u}{\partial x} - \frac{B_{12}}{R}\frac{\partial \theta_x}{\partial x} - \frac{A_{11}}{R^2}\left(\frac{\partial v}{\partial \varphi} + w\right) - \frac{B_{11}}{R^2}\frac{\partial \theta_\varphi}{\partial \varphi} - I_0\frac{\partial^2 w}{\partial t^2} = 0
\tag{6.90}
$$

$$
B_{11}\frac{\partial^2 u}{\partial x^2} + D_{11}\frac{\partial^2 \theta_x}{\partial x^2} + \frac{B_{12}}{R}\left(\frac{\partial^2 v}{\partial x \partial \varphi} + \frac{\partial w}{\partial x}\right) + \frac{D_{12}}{R}\frac{\partial^2 \theta_\varphi}{\partial x \partial \varphi}
$$
$$
+ \frac{B_{66}}{R}\left(\frac{1}{R}\frac{\partial^2 u}{\partial \varphi^2} + \frac{\partial^2 v}{\partial x \partial \varphi}\right) + \frac{D_{66}}{R}\left(\frac{1}{R}\frac{\partial^2 \theta_x}{\partial \varphi^2} + \frac{\partial^2 \theta_\varphi}{\partial x \partial \varphi}\right)
$$
$$
- A_{55}^s\left(\theta_x + \frac{\partial w}{\partial x}\right) - I_1\frac{\partial^2 u}{\partial t^2} - I_2\frac{\partial^2 \theta_x}{\partial t^2} = 0
\tag{6.91}
$$

$$
B_{66}\left(\frac{1}{R}\frac{\partial^2 u}{\partial x \partial \varphi} + \frac{\partial^2 v}{\partial x^2}\right) + D_{66}\left(\frac{1}{R}\frac{\partial^2 \theta_x}{\partial x \partial \varphi} + \frac{\partial^2 \theta_\varphi}{\partial x^2}\right) + \frac{B_{12}}{R}\frac{\partial^2 u}{\partial x \partial \varphi}
$$
$$
+ \frac{D_{12}}{R}\frac{\partial^2 \theta_x}{\partial x \partial \varphi} + \frac{B_{11}}{R^2}\left(\frac{\partial^2 v}{\partial \varphi^2} + \frac{\partial w}{\partial \varphi}\right) + \frac{D_{11}}{R^2}\frac{\partial^2 \theta_\varphi}{\partial \varphi^2}
$$
$$
- A_{55}^s\left(\theta_\varphi + \frac{1}{R}\frac{\partial w}{\partial \varphi} - \frac{v}{R}\right) - I_1\frac{\partial^2 v}{\partial t^2} - I_2\frac{\partial^2 \theta_\varphi}{\partial t^2} = 0
\tag{6.92}
$$

In order to solve analytically the governing equations of a multi-scale hybrid nanocomposite shell, it is assumed that a solution function of exponential form exists. As a result, the relationship between the displacement field and the other variables is supposed to be as follows [32]:

$$
\begin{Bmatrix} u \\ v \\ w \\ \theta_x \\ \theta_\varphi \end{Bmatrix} = \begin{Bmatrix} U \exp\left[i\left(\beta x + n\varphi - \omega t\right)\right] \\ V \exp\left[i\left(\beta x + n\varphi - \omega t\right)\right] \\ W \exp\left[i\left(\beta x + n\varphi - \omega t\right)\right] \\ \Theta_x \exp\left[i\left(\beta x + n\varphi - \omega t\right)\right] \\ \Theta_\varphi \exp\left[i\left(\beta x + n\varphi - \omega t\right)\right] \end{Bmatrix}
\tag{6.93}
$$

In which U, V, and W represent the amplitudes of the displacement and Θ_x and Θ_φ are the amplitudes of rotation. Moreover, β and n stand for longitudinal and circumferential wave number, respectively, and finally ω is frequency. Afterward, by substituting u, v, w, θ_x, and θ_φ from Eq. (6.93) in Eqs. (6.88)–(6.92), the following equation is achieved [32]:

$$
\left(\left[K\right]_{5\times5} - \omega^2 \left[M\right]_{5\times5}\right) \begin{bmatrix} U \\ V \\ W \\ \Theta_x \\ \Theta_\varphi \end{bmatrix} = 0
\tag{6.94}
$$

where K and M are matrices representing the stiffness and mass, respectively, and the components of these matrices are listed in Appendix 6C.

In order to find a solution to this eigenvalue issue, the determinant of the coefficient matrix that is located on the left-hand side of Eq. (6.94) has to be set to zero [32]:

$$
\left| \left[K\right]_{5\times5} - \omega^2 \left[M\right]_{5\times5} \right| = 0
\tag{6.95}
$$

A collection of graphics is supplied at the end of this section in order to provide clarity on the influence that different factors have on the wave propagation of multi-scale hybrid nanocomposite shells. Epoxy is expected to be used in the construction of the cylinder's shell, and it will be strengthened with carbon fiber and carbon nanotubes. It is possible to derive the material characteristics of CFs from [11, 12].

The effect that the volume percentage of the cluster has on the change of the wave frequency of the multi-scale hybrid nanocomposite shell in relation to the circumferential wave number is shown in Figure 6.42. It is possible to deduce, on the basis of the figure, that the wave frequency would first experience a tiny fall,

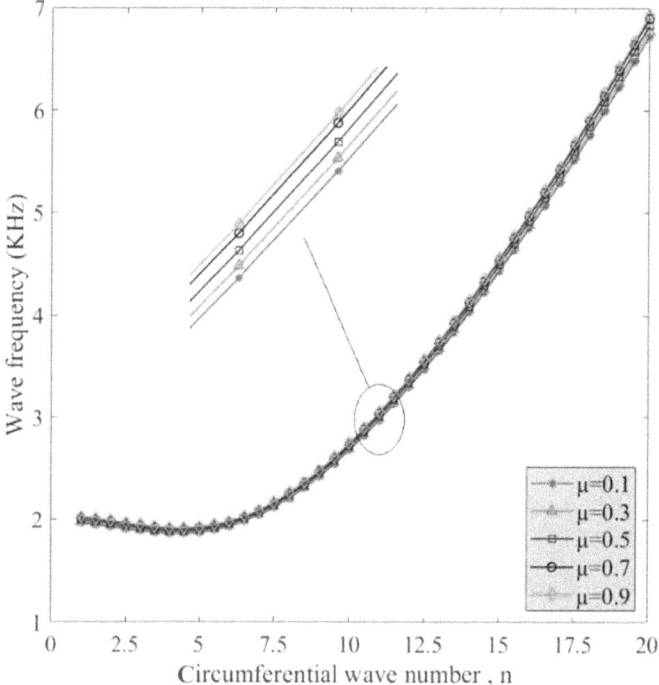

FIGURE 6.42 Illustration of variation of wave frequency versus circumferential wave number for different volume fractions of clusters ($R/h = 20$, $\eta = 0.9$, $V_F = 0.3$, $w_r = 0.2$, $P = 2$, and $\beta = 10$).

and, subsequently, it will experience an increase as a result of a rise in the circumferential wave number. In addition, by making use of the magnifier, one may discover that a growing effect is associated with an increasing volume portion of the cluster. To put it another way, when there is a rise in the volume fraction of the cluster, there is also an increase in the quantity of wave frequency. The physical definition of the phrase μ is what's behind this action, and it's the reason why. When μ equals 1, all of the nanoreinforcements are grouped together into a single cluster; as a result, the nanofillers may be dispersed equally throughout the structure [32].

Figure 6.43 depicts the way in which the wave frequency of a multi-scale hybrid nanocomposite shell varies in relation to the number of circumferential waves for a variety of gradient indices. As has been shown, a rise in the circumferential wave number results in a decrease in wave frequency, which is followed by an increase in wave frequency. By examining the enlarged figure, one may get to the conclusion that there is sufficient capacity for an increase in the gradient index to result in a drop in wave frequency. On the other hand, a reduction in wave frequency may be brought about by an increase in the gradient index [32].

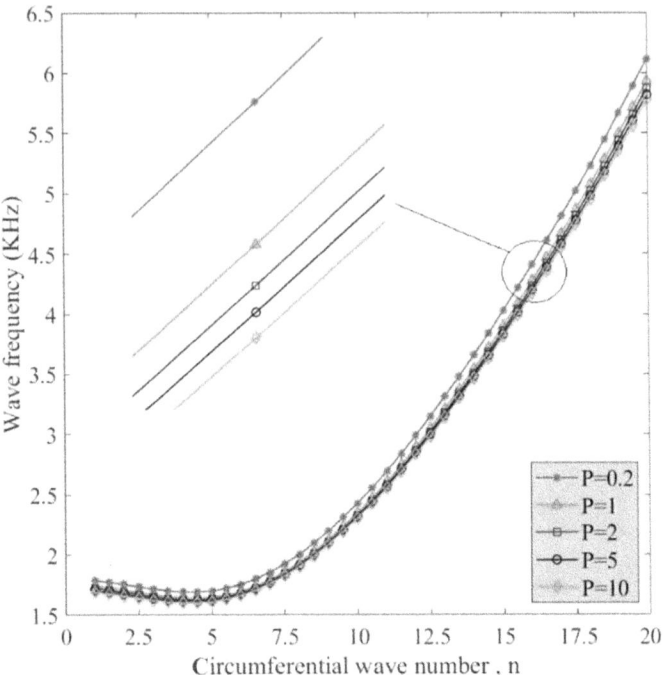

FIGURE 6.43   Variation of wave frequency versus circumferential wave number for different gradient indices ($R/h = 20$, $\eta = 0.8$, $V_F = 0.3$, $w_r = 0.2$, $\mu = 0.5$, and $\beta = 10$).

The influence of the CNT's volume percentage and radius on the frequency shift as a function of the circumferential wave number is seen in Figure 6.44. As the slope increases, the steepness of the curves with a lower radius-to-thickness ratio becomes more apparent, and the influence of the volume percentage of CNT contained inside the cluster decreases. To put it another way, a greater wave frequency occurs in a CNT cluster that has a smaller volume percentage [32].

The relationship between the wave frequency and the circumferential wave number varies depending on the volume percentage seen in Figure 6.45. It is abundantly obvious that every curve adheres to the same pattern, which states that as the number of circumferential waves increases, the wave frequency also increases. In addition, when the circumferential wave number is held constant, the curve that has a greater volume percentage of CF has a higher wave frequency when compared to the curves that have lower volume fractions of CF. The explanation for this phenomenon is that an increase in the volume percentage of CF results in a structure that is more rigid [32].

Figure 6.46 displays a graph that illustrates the correlation between the fluctuation in wave frequency and the circumferential wave number. The graph includes data for different volume fractions of CF and weight fractions of CNT. Upon visual inspection of the image, it can be inferred that the weight fraction of carbon

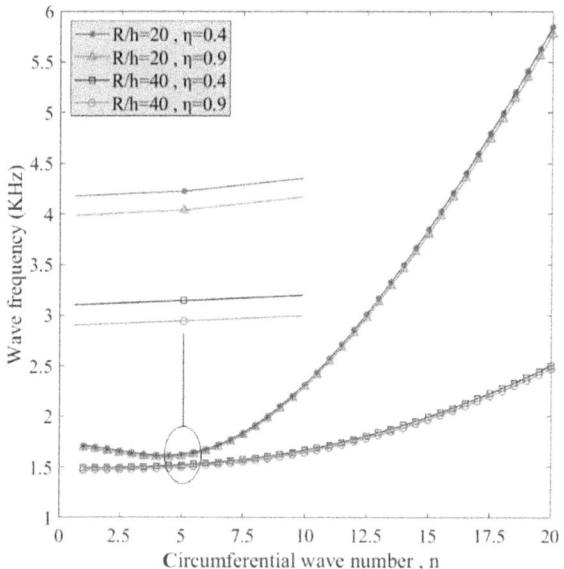

FIGURE 6.44 Variation of wave frequency versus circumferential wave number for various radius-to-thickness ratios and volume fraction of CNT inside the cluster ($V_F = 0.2$, $\eta = 0.75$, $\mu = 0.5$, $w_r = 0.1$, $P = 1$, and $\beta = 10$).

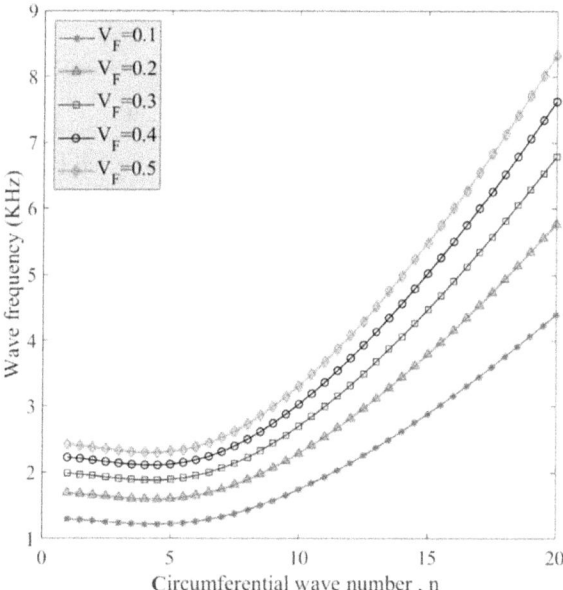

FIGURE 6.45 The effect of CF's volume fraction on wave frequency versus circumferential wave number ($R/h = 20$, $\eta = 1$, $\mu = 0.7$, $w_r = 0.2$, $P = 1$, and $\beta = 10$).

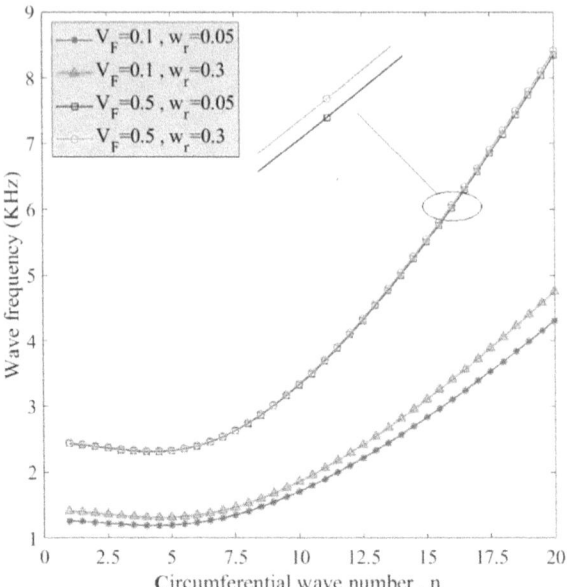

FIGURE 6.46   Variation of wave frequency versus circumferential wave number for different volume fractions of CF and weight fractions of CNT ($R/h = 20$, $\eta = 0.9$, $\mu = 0.6$, $P = 1$, and $\beta = 10$).

nanotubes (CNTs) and the volume percentage of carbon fibers (CF) exhibit a positive correlation with the observed impact. In accordance with previous research [32], it can be stated that when the circumferential wave number remains constant, an increase in the volume proportion of CF and weight percentage of CNT in a given curve will result in a higher wave frequency. On the other hand, if you reduce the CF volume percentage and the CNT weight fraction, the wave frequency will drop. Frequency is exactly related to the product of the volume fraction of CF and the weight fraction of CNT. The wave frequency grows with both the CF volume fraction and the CNT weight fraction.

Figure 6.47 depicts the results of an examination of the relationship between the variation in wave frequency and the number of longitudinal waves as a function of cluster volume. As the wave frequency of the hybrid multi-scale hybrid nanocomposite shell rises, the number of longitudinal waves increases exponentially. A magnifier is shown in the picture, allowing the volume proportion of the cluster to be determined; this fraction, like in Figure 6.42, is increasing with time. According to the data presented in reference [32], the cluster's wave frequency will rise as its volume percentage grows.

In Figure 6.48, an investigation into how the volume percentage of CF affects the relationship between the fluctuation in wave frequency and the number of longitudinal waves is carried out. The rise in the volume percentage of CF contributes to an increase in the stiffness of the multi-scale hybrid nanocomposite shell.

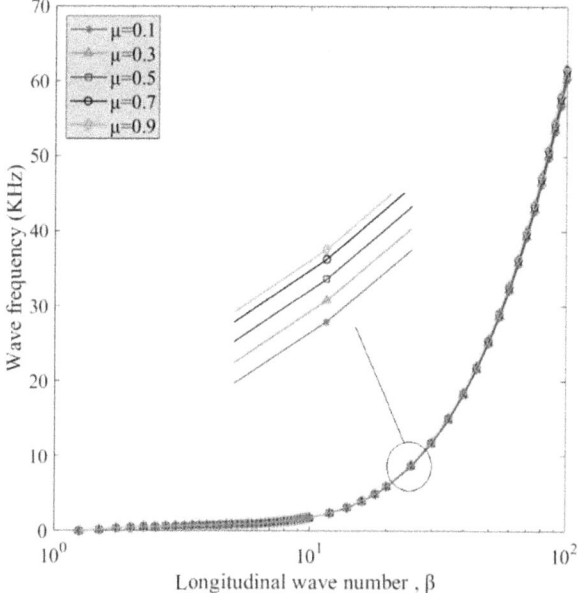

FIGURE 6.47 The influence of the volume fraction of cluster on the variation of wave frequency versus longitudinal wave number ($R/h = 30$, $\eta = 0.9$, $V_F = 0.3$, $w_r = 0.1$, $P = 1$, and $n = 1$).

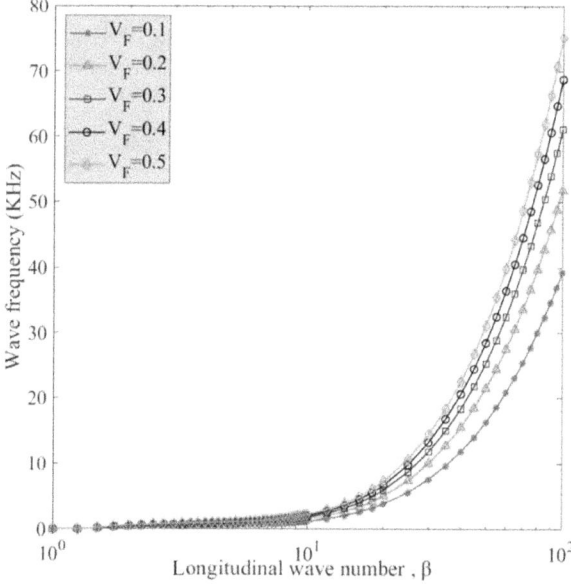

FIGURE 6.48 Variation of wave frequency versus longitudinal wave number for various CF's volume fractions ($R/h = 30$, $\eta = 0.75$, $\mu = 0.5$, $w_r = 0.1$, $P = 1$, and $n = 1$).

Therefore, an increase in the volume percentage of CF will result in a rise in the value wave frequency. Similar to what is seen in Figure.6.47, the wave frequency increases exponentially as the number of longitudinal waves increases [32].

The fluctuation of wave frequency (a) and phase velocity (b) within the cluster as a function of the volume percentage of CNT is shown in Figure 6.49 for many distinct clusters with varying volume fractions. It has been shown that increasing the volume percentage of CNT contained inside the cluster has a diminishing impact, which in turn causes a reduction in the wave frequency and phase velocity of the system. In addition, when $\mu = 0.6$, the slope of the descending line is changeable in the sense that the slope of the curve at the beginning is less than the slope of the curve at the end of the line. In the case of other curves, the variance of a slope that is falling grows smaller and eventually goes to zero [32].

Figure 6.50 depicts the impact of varying weight percentages of CNT on the alteration of wave frequency (a) and phase velocity (b) in relation to the volume fraction of CNT present in the cluster. The relationship between the volume percentage of CNT within a cluster and the weight fraction of CNT in a curve is evident. Specifically, as the volume percentage of CNT increases, the curve with a higher weight fraction of CNT is more significantly impacted compared to the curve with a lower weight fraction of CNT. Although all curves exhibit a downward trend, their degree of decline is not consistent throughout. When the value of η exceeds 0.95, the impact of the volume fraction of carbon nanotubes (CNTs) within the cluster becomes more significant than the effect of the weight fraction of CNTs. According to research [32], the curve with the lowest weight fraction of CNT has the highest wave frequency and phase velocity values.

For various cluster volume fractions, Figure 6.51 analyzes how the phase velocity varies with the circumferential wave number. All of this data is presented for the benefit of the reader. A magnifying tool is needed to show the overall

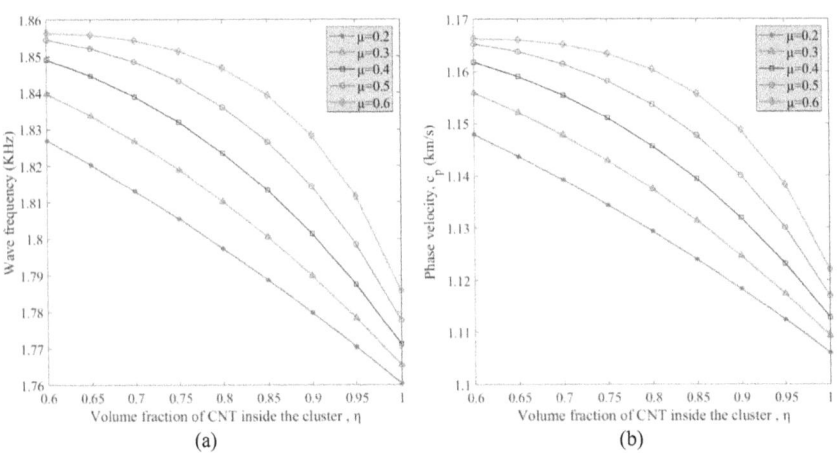

FIGURE 6.49 Illustration of the variation of wave frequency (a) and phase velocity (b) versus the volume fraction of CNT inside the cluster for various volume fractions of cluster ($R/h = 30$, $w_r = 0.2$, $V_F = 0.2$, $P = 2$, $n = 10$, and $\beta = 10$).

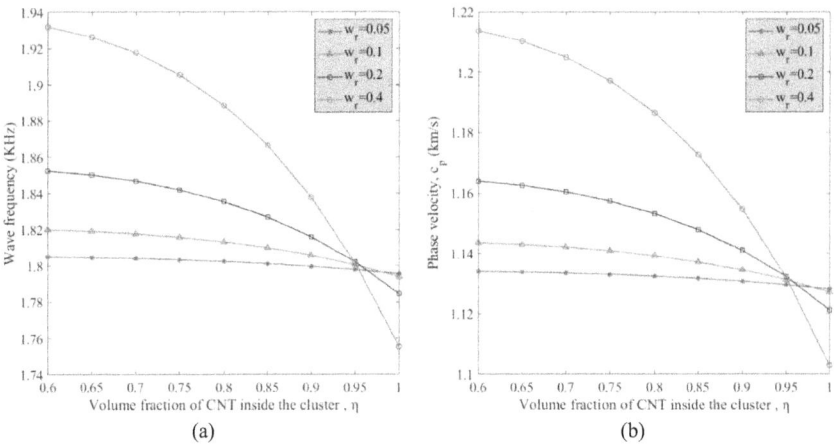

FIGURE 6.50 Variation of wave frequency (a) and phase velocity (b) versus the volume fraction of CNT inside the cluster for different weight fractions of CNT ($R/h = 30$, $\mu = 0.5$, $V_F = 0.3$, $P = 1$, $n = 1$, and $\beta = 10$).

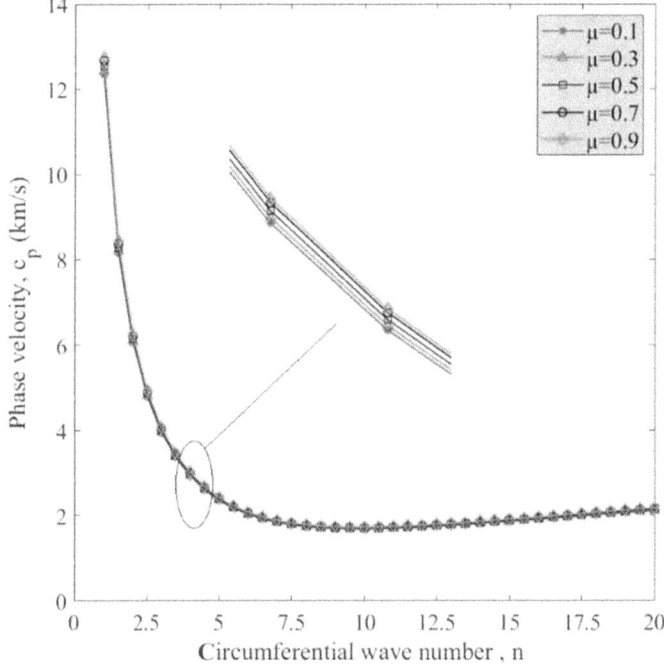

FIGURE 6.51 Variation of phase velocity versus circumferential wave number for various volume fractions of cluster ($R/h = 20$, $\eta = 0.9$, $V_F = 0.3$, $w_r = 0.2$, $P = 2$, and $\beta = 10$).

orientation of the curves in more detail. A drop in phase velocity is seen for the multi-scale hybrid nanocomposite shell, followed by an increase in phase velocity. The phase velocity is said to grow with the cluster's volume percentage, as stated in reference [32], while the effect of the cluster's volume fraction is determined to be very small.

Figure 6.52 depicts the way the phase velocity varies in relation to the circumferential wave number for a variety of different gradient indices. It can be understood that the gradient index has a diminishing impact on the variation of phase velocity in light of the fact that the magnifying curve shows this. As seen in Figure 6.51, a rise in the circumferential wave number results in a significant drop in phase velocity, which is then followed by a steady increase [32].

Carbon nanotube concentration and radius-to-thickness ratio are used to calculate the phase velocity and the circumferential wave number, respectively (see Figure 6.53). The phase velocity of a structure is predicted to decrease if the ratio of its radius to its thickness is made very great. According to reference [32], the phase velocity of the multi-scale hybrid nanocomposite shell decreases as the volume percentage of CNTs inside the cluster increases.

Figure 6.54 investigates the effect that the volume percentage of CF has on the relationship between the fluctuation in phase velocity and the circumferential wave number. As can be seen, as the total number of circumferential waves increases, the phase velocity first drops dramatically, followed by a steady ascent along an increasingly gentle slope. In addition to this, the phase velocity of the system may be increased by increasing the quantity of the CF's volume portion [32].

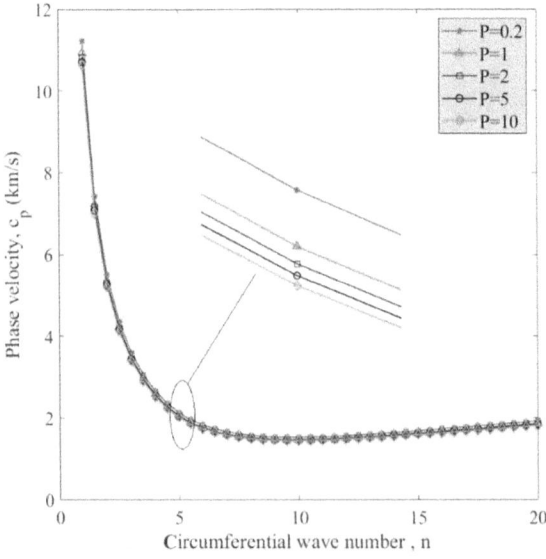

FIGURE 6.52 The effect of gradient index on variation of phase velocity versus circumferential wave number ($R/h = 20$, $\eta = 0.8$, $V_F = 0.3$, $w_r = 0.2$, $\mu = 0.5$, and $\beta = 10$).

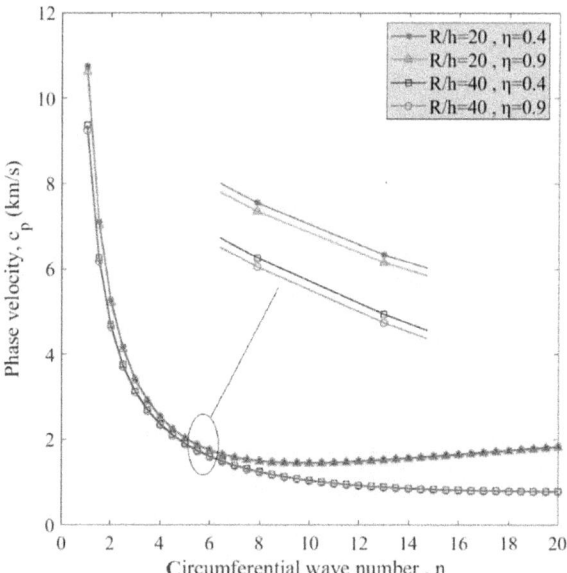

FIGURE 6.53 Both effects of radius-to-thickness ratio and the volume fraction of CNT inside the cluster on variation of phase velocity versus circumferential wave number ($V_F = 0.2$, $\eta = 0.75$, $\mu = 0.5$, $w_r = 0.1$, $P = 1$, and $\beta = 10$).

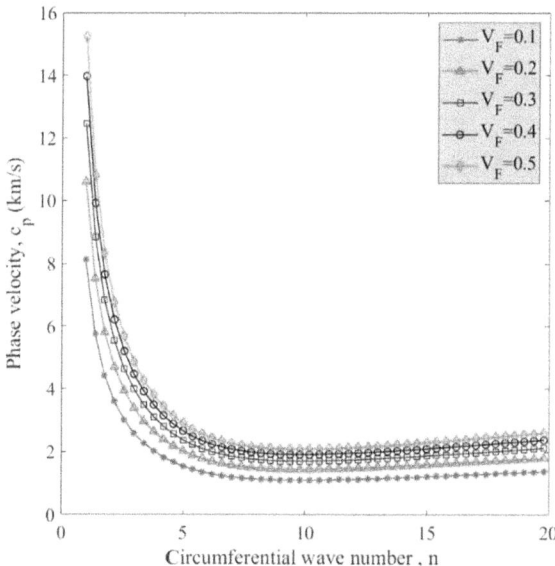

FIGURE 6.54 Variation of phase velocity versus circumferential wave number for different volume fractions of CF ($R/h = 20$, $\eta = 1$, $\mu = 0.7$, $w_r = 0.2$, $P = 1$, and $\beta = 10$).

In Figure 6.55, we see how the phase velocity varies with the circumferential wave number depending on the volume fraction of the CF and the weight fraction of the CNT. Also included in this figure is a discussion of the relationship between the two variables. Reinforcing a structure with carbon may be accomplished using either carbon fiber or carbon nanotubes. As a consequence of this, the stiffness of a multi-scale hybrid nanocomposite shell may be increased by incorporating CF and CNT into the matrix, and as a consequence of this, both the volume proportion of CF and the weight fraction of CNT have an increasing impact [32].

The influence of the volume percentage of the cluster on the fluctuation of the phase velocity versus the longitudinal wave number is seen in Figure 6.56. The curve has three distinct sections that may be distinguished from one another: $\beta < 2$, $2 < \beta < 3$, and $\beta > 3$. The trend of phase velocity is heading in an upward direction in the first and third areas. On the other hand, the trend of phase velocity is heading in the opposite direction in the second zone. In addition, the volume percentage of the cluster has an increasing role in the change of phase velocity, which is the primary cause for this pattern of behavior, as was covered in a previous section [32].

Figure 6.57 depicts how the proportion of CF in the volume affects the fluctuation in phase velocity and the number of longitudinal waves. Similar patterns may be seen in this image in Figure 6.56. It is evident that the amount of phase velocity will grow if a big value is set to CF's volume fraction, and this may be shown using [32].

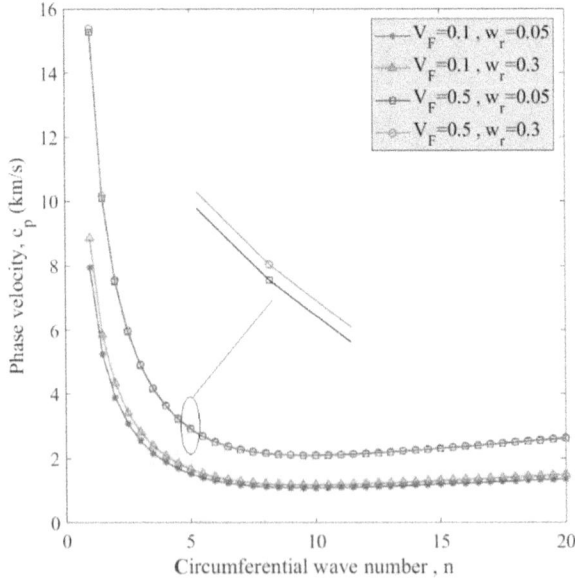

FIGURE 6.55 Variation of the phase velocity versus the circumferential wave number for various volume fractions of CF and weight fraction of CNT ($R/h = 20$, $\eta = 0.9$, $\mu = 0.6$, $P = 1$, and $\beta = 10$).

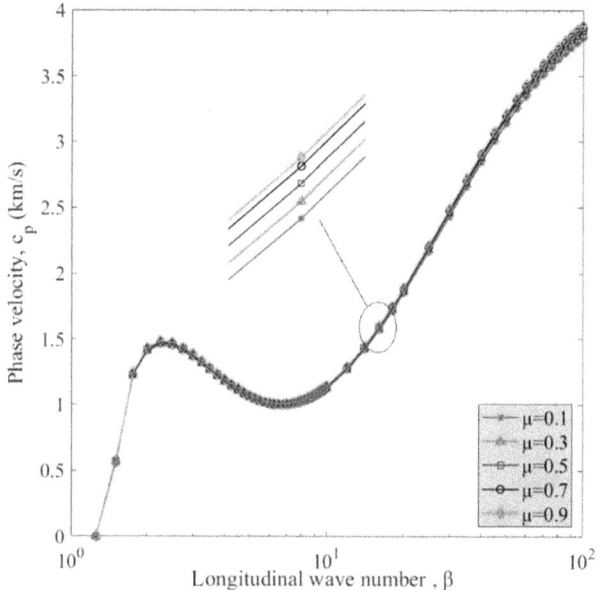

FIGURE 6.56 Illustration of the variation of phase velocity versus the longitudinal wave number for different volume fraction of cluster ($R/h = 30$, $\eta = 0.9$, $V_F = 0.3$, $w_r = 0.1$, $P = 1$, and $n = 1$).

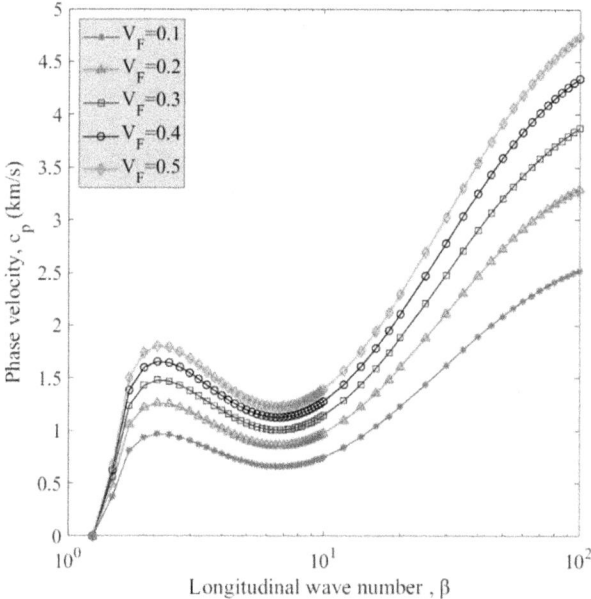

FIGURE 6.57 The influence of the volume fraction of CF on the variation of phase velocity versus longitudinal wave number ($R/h = 30$, $\eta = 0.75$, $\mu = 0.5$, $w_r = 0.1$, $P = 1$, and $n = 1$).

6.5 SUMMARY

In this chapter, the aggregation influence of CNTs is taken into consideration as the wave propagation characteristics of multi-scale hybrid nanocomposite structures are investigated. It was determined that the Eshelby–Mori–Tanaka strategy and the rule of combination were the most effective methods for accomplishing the goal of producing materials with comparable attributes. In addition, the shear deformation theory and Hamilton's principle were used in order to develop the equations that describe motion. In the next paragraphs, we discuss the concluding assertions that are the most significant in the following order:

- Either raising the volume proportion of CFs or the weight fraction of CNTs may result in an increase in the wave frequency and the phase velocity of a multi-scale hybrid nanocomposite structure.
- This can be accomplished. Structures that are made up of many different scales of hybrid nanomaterials are subject to a growing impact from the volume percentage of clusters, which in turn has an effect on the wave frequency and phase velocity.
- The gradient index and the volume percentage of CNTs that are included within the clusters have a declining impact on the fluctuation of frequency and phase velocity that is induced by the presence of a multi-scale hybrid nanocomposite structure. This is because the gradient index and the volume percentage of CNTs that are included inside the clusters are both related to the inclusion of CNTs.
- According to the results, if you want a structure to be able to tolerate greater frequency ranges, it is best to build it out of cylinders that have length-to-thickness ratios that are on the lower end.
- According to the results, the wave frequency and phase velocity values will continue to grow to a greater degree as the number of waves increases. This is because the number of waves causes a multiplicative effect on the values.
- As the values of the Winkler and Pasternak parameters, which, respectively, indicate the linear and shear layers of the elastic foundation, rise, the wave frequency and phase velocity also rise as a consequence.
- In light of the fact that the slenderness ratio has a reducing impact on the variability of the phase velocity, it is possible to reduce the phase velocity by choosing a beam with a larger slenderness ratio. Muti-walled carbon nanotubes (MWCNTs) might potentially increase the elastic modulus of the beam if they were to be included in a polymer matrix. As a consequence, wave frequency and phase velocity would both benefit from the modification. It is essential to highlight that the efficacy of this reinforcing might possibly be increased by using direct MWCNT reinforcements, and this is something that should not be overlooked.
- There is a possibility that the aggregation factor will have a dampening impact on the elastic modulus of the beam. In other words, agglomeration of MWCNTs has a negative effect not only on the wave frequency but also on the phase velocity of the nanocomposite structure.

APPENDICES

APPENDIX A

$$K_{11} = -A_{11}\beta^2$$
$$K_{12} = B_{11}i\beta^3$$
$$K_{13} = B_{11}^s i\beta^3$$
$$K_{21} = -B_{11}i\beta^3$$
$$K_{22} = -D_{11}\beta^4$$
$$K_{23} = -D_{11}^s \beta^4$$
$$K_{31} = -B_{11}^s i\beta^3$$
$$K_{32} = -D_{11}^s \beta^4$$
$$K_{33} = -H_{11}^s \beta^4 - A_{55}^s \beta^2$$
$$M_{11} = I_0$$
$$M_{12} = -I_1 i\beta$$
$$M_{13} = -J_1 i\beta$$
$$M_{21} = I_1 i\beta$$
$$M_{22} = I_0 + I_2 \beta^2$$
$$M_{23} = I_0 + J_2 \beta^2$$
$$M_{31} = J_1 i\beta$$
$$M_{32} = I_0 + J_2 \beta^2$$
$$M_{33} = I_0 + K_2 \beta^2$$

APPENDIX B

$$K_{11} = -A_{11}\beta_1^2 - A_{66}\beta_2^2$$
$$K_{12} = -\left(A_{12} + A_{66}\right)\beta_1\beta_2$$
$$K_{13} = B_{11}i\beta_1^3 + i\left(B_{12} + 2B_{66}\right)\beta_2^2\beta_1$$
$$K_{14} = B_{11}^s i\beta_1^3 + i\left(B_{12}^s + 2B_{66}^s\right)\beta_2^2\beta_1$$
$$K_{21} = -\left(A_{12} + A_{66}\right)\beta_1\beta_2$$
$$K_{22} = -A_{66}\beta_1^2 - A_{22}\beta_2^2$$
$$K_{23} = iB_{22}\beta_2^3 + i\left(B_{12} + 2B_{66}\right)\beta_1^2\beta_2$$
$$K_{24} = iB_{22}^s \beta_2^3 + i\left(B_{12}^s + 2B_{66}^s\right)\beta_1^2\beta_2$$
$$K_{31} = -B_{11}i\beta_1^3 - i\left(B_{12} + 2B_{66}\right)\beta_2^2\beta_1$$
$$K_{32} = -iB_{22}\beta_2^3 - i\left(B_{12} + 2B_{66}\right)\beta_1^2\beta_2$$
$$K_{33} = -D_{11}\beta_1^4 - 2\left(D_{12} + 2D_{66}\right)\beta_1^2\beta_2^2 - D_{22}\beta_2^4$$
$$K_{34} = -D_{11}^s \beta_1^4 - 2\left(D_{12}^s + 2D_{66}^s\right)\beta_1^2\beta_2^2 - D_{22}^s \beta_2^4$$

$$K_{41} = -B_{11}^s i\beta_1^3 - i\left(B_{12}^s + 2B_{66}^s\right)\beta_2^2\beta_1$$

$$K_{42} = -iB_{22}^s\beta_2^3 - i\left(B_{12}^s + 2B_{66}^s\right)\beta_1^2\beta_2$$

$$K_{43} = -D_{11}^s\beta_1^4 - 2\left(D_{12}^s + 2D_{66}^s\right)\beta_1^2\beta_2^2 - D_{22}^s\beta_2^4$$

$$K_{44} = -H_{11}^s\beta_1^4 - 2\left(H_{12}^s + 2H_{66}^s\right)\beta_1^2\beta_2^2 - H_{22}^s\beta_2^4 - A_{55}^s\beta_1^2 - A_{44}^s\beta_2^2$$

$$M_{11} = I_0;\ M_{13} = -I_1 i\beta_1;\ M_{14} = -J_1 i\beta_1$$

$$M_{22} = I_0;\ M_{23} = -I_1 i\beta_2;\ M_{24} = -J_1 i\beta_2;\ M_{31} = I_1 i\beta_1$$

$$M_{32} = I_1 i\beta_2;\ M_{33} = I_0 + I_2 i\beta_1^2 + I_2 i\beta_2^2$$

$$M_{34} = I_0 + J_2 i\beta_1^2 + J_2 i\beta_2^2;\ M_{41} = J_1 i\beta_1;\ M_{42} = J_1 i\beta_2$$

$$M_{43} = I_0 + J_2 i\beta_1^2 + J_2 i\beta_2^2;\ M_{44} = I_0 + K_2 i\beta_1^2 + K_2 i\beta_2^2$$

APPENDIX C

$$k_{11} = -A_{11}\beta^2 - \frac{A_{66}}{R^2}n^2,$$

$$k_{12} = -\beta n\left(\frac{A_{12} + A_{66}}{R}\right), k_{13} = \frac{A_{12}}{R}\beta i,$$

$$k_{14} = -B_{11}\beta^2 - \frac{B_{66}}{R^2}n^2, k_{15} = -\beta n\left(\frac{B_{12} + B_{66}}{R}\right),$$

$$k_{22} = -A_{66}\beta^2 - \frac{A_{11}}{R^2}n^2 - \frac{A_{55}^s}{R^2},$$

$$k_{23} = n\left(\frac{A_{11} + A_{55}^s}{R^2}\right)i, k_{24} = -\beta n\left(\frac{B_{12} + B_{66}}{R}\right),$$

$$k_{25} = -B_{66}\beta^2 - \frac{B_{11}}{R^2}n^2 - \frac{A_{55}^s}{R},$$

$$k_{33} = -A_{55}^s\beta^2 - \left[\frac{A_{11}}{R^2} + n^2\frac{A_{55}^s}{R^2}\right],$$

$$k_{34} = \left(A_{55}^s - \frac{B_{12}}{R}\right)i\beta, k_{35} = n\left(\frac{A_{55}^s}{R} - \frac{B_{11}}{R^2}\right)i,$$

$$k_{44} = -D_{11}\beta^2 - n^2\frac{D_{66}}{R^2} - A_{55}^s,$$

$$k_{45} = -\beta n\frac{D_{12} + D_{66}}{R}, k_{55} = -D_{66}\beta^2 - n^2\frac{D_{11}}{R^2} - A_{55}^s$$

$$m_{11} = m_{22} = m_{33} = I_0,$$

$$m_{14} = m_{25} = I_1,$$

$$m_{44} = m_{55} = I_2$$

REFERENCES

[1] Al-Furjan, M. S. H., Habibi, M., Ebrahimi, F., Mohammadi, K., & Safarpour, H. (2022). Wave dispersion characteristics of high-speed-rotating laminated nanocomposite cylindrical shells based on four continuum mechanics theories. *Waves in Random and Complex Media, 32*(4), 1599–1625.

[2] Ebrahimi, F., & Dabbagh, A. (2019). A comprehensive review on modeling of nanocomposite materials and structures. *Journal of Computational Applied Mechanics, 50*(1), 197–209.

[3] Ebrahimi, F., & Dabbagh, A. (2020). A brief review on the influences of nanotubes' entanglement and waviness on the mechanical behaviors of CNTR polymer nanocomposites. *Journal of Computational Applied Mechanics, 51*(1), 247–252.

[4] Nouraei, M., Haghi, P., & Ebrahimi, F. (2021). Modeling dynamic characteristics of the thermally affected embedded laminated nanocomposite beam containing multiscale hybrid reinforcement. *Waves in Random and Complex Media,* 1–30.

[5] Ebrahimi, F., Nopour, R., & Dabbagh, A. (2022). Effects of polymer's viscoelastic properties and curved shape of the CNTs on the dynamic response of hybrid nanocomposite beams. *Waves in Random and Complex Media,* 1–18.

[6] Dabbagh, A., Rastgoo, A., & Ebrahimi, F. (2019). Finite element vibration analysis of multi-scale hybrid nanocomposite beams via a refined beam theory. *Thin-Walled Structures, 140,* 304–317.

[7] Dabbagh, A., & Ebrahimi, F. (2021). Postbuckling analysis of meta-nanocomposite beams by considering the CNTs' agglomeration. *The European Physical Journal Plus, 136*(11), 1168.

[8] Beam Dabbagh, A., Rastgoo, A., & Ebrahimi, F. (2021). Static stability analysis of agglomerated multi-scale hybrid nanocomposites via a refined theory. *Engineering with Computers, 37,* 2225–2244.

[9] Dabbagh, A., Rastgoo, A., & Ebrahimi, F. (2021). Thermal buckling analysis of agglomerated multiscale hybrid nanocomposites via a refined beam theory. *Mechanics Based Design of Structures and Machines, 49*(3), 403–429.

[10] Safarpour, M., Ebrahimi, F., Habibi, M., & Safarpour, H. (2021). On the nonlinear dynamics of a multi-scale hybrid nanocomposite disk. *Engineering with Computers, 37*(3), 2369–2388.

[11] Ebrahimi, F., & Habibi, S. (2018). Nonlinear eccentric low-velocity impact response of a polymer-carbon nanotube-fiber multiscale nanocomposite plate resting on elastic foundations in hygrothermal environments. *Mechanics of Advanced Materials and Structures, 25*(5), 425–438.

[12] Ebrahimi, F., & Habibi, S. (2018). Thermal effects on nonlinear dynamic characteristics of polymer-CNT-fiber multiscale nanocomposite structures. *Structural engineering and mechanics: An international journal, 67*(4), 403–415.

[13] Ebrahimi, F., & Dabbagh, A. (2015). On thermo-mechanical vibration analysis of multi-scale hybrid composite beams. *Journal of Vibration and Control,* 1077546318806800.

[14] Ebrahimi, F., & Dabbagh, A., Rastgoo, A., & Rabczuk, T. (2020). Agglomeration effects on static stability analysis of multi-scale hybrid nanocomposite plates. *Computers, Materials & Continua, 63*(1), 41–64.

[15] Ebrahimi, F., Dabbagh, A., & Rastgoo, A. (2020). Static stability analysis of multiscale hybrid agglomerated nanocomposite shells. *Mechanics Based Design of Structures and Machines,* 1–17.

[16] Ebrahimi, F., & Dabbagh, A. (2021). An analytical solution for static stability of multi-scale hybrid nanocomposite plates. *Engineering with Computers*, *37*(1), 545–559.

[17] Ebrahimi, F., Dabbagh, A., & Rastgoo, A. (2021). Free vibration analysis of multi-scale hybrid nanocomposite plates with agglomerated nanoparticles. *Mechanics Based Design of Structures and Machines*, *49*(4), 487–510.

[18] Ebrahimi, F., & Dabbagh, A. (2019). Vibration analysis of multi-scale hybrid nano-composite plates based on a Halpin-Tsai homogenization model. *Composites Part B: Engineering*, *173*, 106955.

[19] Ebrahimi, F., & Dabbagh, A. (2022). Vibration analysis of multi-scale hybrid nano-composite shells by considering nanofillers' aggregation. *Waves in Random and Complex Media*, *32*(3), 1060–1078.

[20] Ebrahimi, F., & Dabbagh, A. (2019). Vibration analysis of multi-scale hybrid nano-composite plates based on a Halpin-Tsai homogenization model. *Composites Part B: Engineering*, *173*, 106955.

[21] Ebrahimi, F., & Mahesh, V. (2020). On nonlinear vibration of sandwiched polymer-CNT/GPL-fiber nanocomp.

[22] Nopour, R., Ebrahimi, F., Dabbagh, A., & Aghdam, M. M. (2022). Nonlinear forced vibrations of three-phase nanocomposite shells considering matrix rheological behavior and nano-fiber waviness. *Engineering with Computers*, 1–18.

[23] Dabbagh, A., Rastgoo, A., & Ebrahimi, F. (2022). Post-buckling analysis of imperfect multi-scale hybrid nanocomposite beams rested on a nonlinear stiff substrate. *Engineering with Computers*, 1–14.

[24] Ebrahimi, F. & Rostami, P. (2018). Wave propagation analysis of carbon nanotube reinforced composite beams. *The European Physical Journal Plus*, *133*, 285.

[25] Ebrahimi, F., & Dabbagh, A. (2022). *Mechanics of multiscale hybrid nanocomposites*. Elsevier.

[26] Ebrahimi, F., & Dabbagh, A. (2020). *Mechanics of nanocomposites: homogenization and analysis*. CRC Press.

[27] Ebrahimi, F. (Ed.). (2012). *Nanocomposites: New trends and developments*. BoD–Books on Demand.

[28] Ebrahimi, F., Seyfi, A., & Dabbagh, A. (2019). Wave dispersion characteristics of agglomerated multi-scale hybrid nanocomposite beams. *The Journal of Strain Analysis for Engineering Design*, *54*(4), 276–289.

[29] Ebrahimi, F., & Seyfi, A. (2021). Wave dispersion analysis of embedded MWCNTs-reinforced nanocomposite beams by considering waviness and agglomeration factors. *Waves in Random and Complex Media*, 1–20.

[30] Ebrahimi, F., & Seyfi, A. (2022). Wave propagation response of agglomerated multi-scale hybrid nanocomposite plates. *Waves in Random and Complex Media*, *32*(3), 1338–1362.

[31] Ebrahimi, F., Enferadi, A., & Dabbagh, A. (2022). Wave dispersion behaviors of multi-scale CNT/glass fiber/polymer nanocomposite laminated plates. *Polymers*, *14*(24), 5448.

[32] Ebrahimi, F., & Seyfi, A. (2021). Wave propagation response of multi-scale hybrid nanocomposite shell by considering aggregation effect of CNTs. *Mechanics Based Design of Structures and Machines*, *49*(1), 59–80.

7 Wave Dispersion Characteristics of Metal Foam Structures

7.1 BACKGROUND

The present chapter concerns the examination of wave propagation in metal foam structures that are porous, including beams, plates, and cylindrical shells. These structures are situated on an elastic substrate that undergoes deformation. The analysis is conducted within the context of shear deformation theory. This research investigates three discrete forms of temperature escalation, specifically uniform, linear, and sinusoidal. Foam-based structures are commonly composed of metallic foams that consist of magnesium, nickel, titanium, and tungsten, which are regarded as essential components. The uniform distribution of pores is observed to display both symmetrical and asymmetrical patterns across the thickness. The utilization of the Hamiltonian framework is employed in the derivation of the equations of motion for porous metal foam formations. Following the application of a shear deformation theory, the governing equations related to porous metal foam structures are derived and subsequently solved through analytical methods. This research investigates the impacts of various parameters, as demonstrated by the given illustrations. The literature has extensively examined the implications of various characteristics, including but not limited to porosity, as evidenced by previous studies [1–5].

7.1.1 A REVIEW ON WAVE DISPERSION ANALYSIS OF METAL FOAM STRUCTURES

Porosity is an inherent phenomenon that arises as a result of the manufacturing process. The augmentation of the porosity coefficient possesses the capability to reduce the rigidity of mechanical constructions, consequently influencing their efficacy. The impact of porosity on mechanical behavior has been a subject of great interest among scientists, owing to its structural imperfection. Consequently, a multitude of inquiries have been carried out to examine the mechanical characteristics of porous structures across various scales, including macro, micro, and nano levels [6–8]. A study was conducted by Ebrahimi and Zia (2019), which focused on the analysis of nonlinear vibrations in beams composed of functionally graded materials (FGMs). The study also took into account the impact of porosity. The investigation conducted by the researchers involved the utilization of both the multiple scales method and Galerkin's method. In recent decades, scholars and experts have been actively pursuing a substance possessing exceptional characteristics to

DOI: 10.1201/9781003270263-7

augment the efficacy of various systems and structures. The weight parameter is of great significance in the analysis and design of structures. The implementation of lightweight materials has been extensively embraced in diverse engineering fields due to their favorable stiffness-to-weight ratio. Porous materials, such as metal foam and ceramic foam, constitute a noteworthy class of lightweight materials that are utilized in a variety of fields, including civil construction, automotive and marine industries, and aerospace engineering. Aluminum foams have been utilized in the automotive sector with the aim of reducing the weight of vehicles and absorbing energy [9, 10]. The term "metal foam" is used to describe a cellular structure that is comprised of solid metal and contains a notable proportion of gas-filled pores within its volume. The unique characteristic of metallic foams confers upon this category of material an exceptional degree of low weight. Consequently, a multitude of scientific investigations have been carried out to scrutinize the mechanical characteristics of metal foam structures across various scales including macro, micro, and nano, as evidenced by sources [1–5, 11–16]. The nonlocal strain gradient theory (NSGT) framework was utilized by Ebrahimi et al. to conduct an analysis of wave propagation in functionally graded porous (FGP) nanoscale beams [12]. Ebrahimi et al. (2019) undertook an investigation into the influence of porosity on the transmission of waves in a nanocomposite shell that was strengthened with graphene platelets (GPL). The investigation employed the first-order shear deformation theory (FSDT) of the shell. Ebrahimi et al. [14] conducted an analytical investigation into the vibration phenomenon of a cylindrical shell that is incorporated within porous metal foam. The investigation examined the impacts of varying porosity distributions. The study examined three distinct classifications of porosity distribution across the thickness, namely uniform, symmetric, and asymmetric. The investigation encompassed the impacts of both simply supported and clamped boundary conditions. As per their findings, an increase in the porosity coefficient holds the potential to decrease the shell frequency. Selecting a symmetric porosity distribution may result in a relatively smaller decrease in the stiffness of the system. In a previous investigation, Ebrahimi and colleagues examined the matter of vibration in relation to a rectangular plate situated on a viscoelastic substrate consisting of porous metal foam [15]. Selvamani et al. (2011) proposed a finite element model for examining the properties of wave propagation in a nonlinear thermal piezoelectric hygroscopic nanoporous metal foam nanobeam. The Rayleigh–Ritz finite element method was employed to observe the dynamic response of the nanobeam. The research investigated the influence of multiple variables, such as piezoelectric strain, porosity coefficient, moisture concentration, slenderness ratio, and thickness-to-diameter ratio, on the critical buckling load of metal foam nanobeams. Ebrahimi and Seyfi [1] conducted a study on the transmission of flexural wave in beams made of porous metal foam that are supported by a Winkler–Pasternak foundation. The researchers employed a sophisticated higher-order shear deformable beam theory to carry out their investigation. The study examines different porosity distributions along the thickness direction, such as uniform, symmetric, and asymmetric distributions. The present investigation scrutinizes the impact of various parameters, such as the distribution of porosity, the coefficient of porosity, the ratio of slenderness, the number of waves, and the coefficients of elastic foundation, on the oscillations of wave frequency, escape frequency, and phase velocity. The results are conveyed via a sequence of visual representations, which are thoroughly scrutinized. The impact of hygrothermal environment on wave propagation characteristics of porous graphene foam (PGF) structures, such as beams, plates, and shells, was deliberated by the authors. These structures are supported by an elastic medium and were analyzed using the refined higher-order beam theory. This was documented in references [2] for beams, [3, 4] for plates, and [5] for shells. The beam is considered to exist in a specific state. The present investigation aims to analyze the effects of various parameters on the variability of wave frequency and phase velocity of the PGF beam. The aforementioned parameters encompass wave number, porosity coefficient, diverse forms of porosity distribution, Winkler–Pasternak coefficients, hygrothermal environment, and slenderness ratio.

7.2   GRAPHENE FOAM BEAM WAVE DISPERSION CHARACTERISTICS

The metal foam beam is assumed to be supported by an elastic substrate with a designated coordinate system, as illustrated in Figure 7.1. The beam's length, denoted by L, and its thickness, denoted by h, are postulated.

The present study investigates the impact of porosity distribution on the effective properties of porous metal foam, taking into account both uniform and non-uniform patterns. Figure 7.2 displays a schematic illustration of the porosity distribution across the beam's thickness. The computation of the effective properties of a uniformly porous metal foam beam can be expressed in the form outlined in reference [1].

$$E(z) = E_1\left(1 - e_0\xi\right) \tag{7.1a}$$

$$G(z) = G_1\left(1 - e_0\xi\right) \tag{7.1b}$$

$$\rho(z) = \rho_1\sqrt{1 - e_0\xi} \tag{7.1c}$$

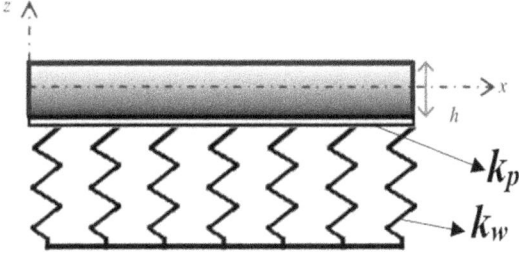

FIGURE 7.1   Schematic of metal foam beam resting on an elastic substrate.

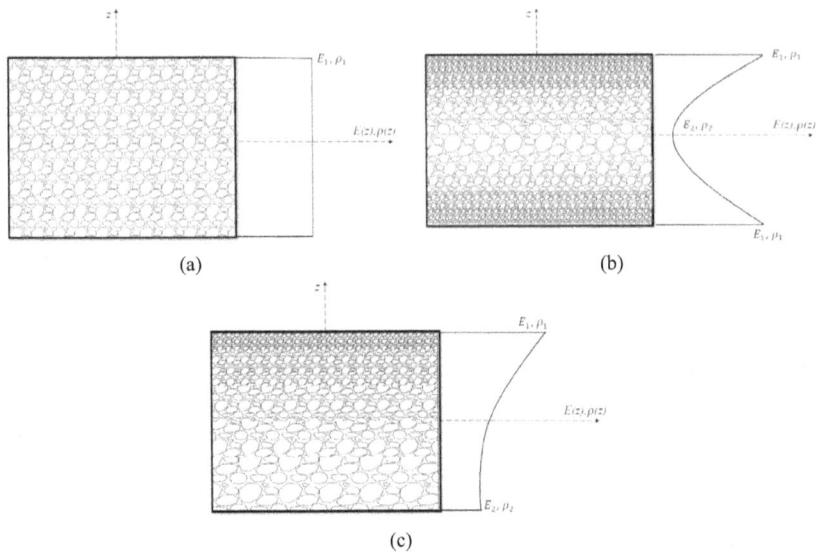

FIGURE 7.2 Schematic of different porosity distributions across the beam's thickness for (a) uniform porosity distribution, (b) symmetric porosity distribution, and (c) asymmetric porosity distribution.

In addition, there exist two distinct patterns that are deemed to be indicative of nonuniform distribution, namely symmetric and asymmetric. The effective characteristics of a metal foam beam with symmetrical porosity can be expressed as per reference [1].

$$E(z) = E_1\left(1 - e_0 \cos\left(\frac{\pi z}{h}\right)\right) \tag{7.2a}$$

$$G(z) = G_1\left(1 - e_0 \cos\left(\frac{\pi z}{h}\right)\right) \tag{7.2b}$$

$$\rho(z) = \rho_1\left(1 - e_m \cos\left(\frac{\pi z}{h}\right)\right) \tag{7.2c}$$

Furthermore, it is possible to rephrase the aforementioned equation to account for an uneven distribution of porosity, as indicated in reference [1].

$$E(z) = E_1\left(1 - e_0 \cos\left(\frac{\pi z}{2h} + \frac{\pi}{4}\right)\right) \tag{7.3a}$$

$$G(z) = G_1\left(1 - e_0 \cos\left(\frac{\pi z}{2h} + \frac{\pi}{4}\right)\right) \tag{7.3b}$$

$$\rho(z) = \rho_1\left(1 - e_m \cos\left(\frac{\pi z}{2h} + \frac{\pi}{4}\right)\right) \tag{7.3c}$$

The symbols E_1, G_1, and ρ_1 denote Young's modulus, shear modulus, and mass density of the porous metal foam at its maximum value. Furthermore, it should be noted that the variables E_2, G_2, and ρ represent the minimum values of the aforementioned parameters, respectively. In addition, the coefficients e_0 and e_m denote porosity and density, respectively, and can be precisely defined according to sources [1, 2].

$$e_0 = 1 - \frac{E_2}{E_1} = 1 - \frac{G_2}{G_1}, \quad (0 < e_0 < 1) \tag{7.4}$$

$$e_m = 1 - \frac{\rho_2}{\rho_1} \tag{7.5}$$

Furthermore, the value of *em* can be computed based on the customary mechanical characteristics of an open-cell metallic foam, as documented in references [1, 2]:

$$\frac{E_2}{E_1} = \left(\frac{\rho_2}{\rho_1}\right)^2 \tag{7.6}$$

$$e_m = 1 - \sqrt{1 - e_0} \tag{7.7}$$

In this approach, the constancy of Poisson's ratio can be assumed due to its negligible changes. The aforementioned expression, as presented in Eqs. (7.1)–(7.3), can be expressed as follows:

$$\xi = \frac{1}{e_0}\left(1 - \left[\frac{2}{\pi}\left(\sqrt{1 - e_0} - 1\right) + 1\right]^2\right) \tag{7.8}$$

The refined higher-order shear deformation beam theory is utilized to attain the kinetic relation of a metal foam beam. The displacement field of beams can be expressed in the form presented in reference [1].

$$u_x(x, z, t) = u(x, t) - z\frac{\partial w_b(x, t)}{\partial x} - f(z)\frac{\partial w_s(x, t)}{\partial x} \tag{7.9}$$

$$u_z(x,z,t) = w_b(x,t) + w_s(x,t)$$ (7.10)

The variables u, w_b, and w_s represent longitudinal displacement, bending deflection, and shear deflection, respectively. The function $f(z)$ serves as a shape function that provides an estimation of the distribution of either shear stress or strain in the direction of thickness. The model presented in this study can be utilized without the need for a shear correction factor, owing to the exerted shape function. The current study expresses the shape function as [1].

$$f(z) = -\frac{z}{4} - \frac{5z^3}{3h^2}$$ (7.11)

The nonzero strains of the beam can be identified as follows:

$$\varepsilon_{xx} = \frac{\partial u}{\partial x} - z\frac{\partial^2 w_b}{\partial x^2} - f(z)\frac{\partial^2 w_s}{\partial x^2}, \ \gamma_{xz} = g(z)\frac{\partial w_s}{\partial x}$$ (7.12)

where

$$g(z) = 1 - \frac{df(z)}{dz}$$ (7.13)

The development of the motion equations will be undertaken utilizing Hamilton's principle as the framework, which is presented as follows [1]:

$$\int_{t_0}^{t_1} \delta(U - T + V)\,dt = 0$$ (7.14)

The symbols U, T, and V represent strain energy, kinetic energy, and work done by external forces, respectively. Moreover, the symbol δ is utilized to represent variation. The definition of the variation of strain energy can be expressed as follows:

$$\delta U = \int_V \left(\sigma_{xx}\delta\varepsilon_{xx} + \sigma_{xz}\delta\gamma_{xz}\right)dV$$ (7.15)

By substituting Eqs. (7.9) through (7.13) into Eq. (7.15), the resulting equation is as follows:

$$\delta U = \int_0^L \left(N\frac{\partial\delta u}{\partial x} - M_b\frac{\partial^2\delta w_b}{\partial x^2} - M_s\frac{\partial^2\delta w_s}{\partial x^2} + Q\frac{\partial\delta w_s}{\partial x}\right)dx$$ (7.16)

The stress resultants N, M_b, M_s, and Q can be computed as follows:

$$\begin{pmatrix} N \\ M_b \\ M_s \end{pmatrix} = \int_A \begin{pmatrix} 1 \\ z \\ f(z) \end{pmatrix} \sigma_{xx} dA \tag{7.17a}$$

$$Q = \int_A g(z)\sigma_{xz} dA \tag{7.17b}$$

One possible expression for the initial derivative of kinetic energy is as follows:

$$\delta T = \int_V \left(\dot{u}_x \delta \dot{u}_x + \dot{u}_z \delta \dot{u}_z \right) \rho(z) dV$$

$$= \int_0^L \left(\begin{array}{l} I_0 \left(\dfrac{\partial u}{\partial t} \dfrac{\partial \delta u}{\partial t} + \dfrac{\partial (w_b + w_s)}{\partial t} \dfrac{\partial \delta (w_b + w_s)}{\partial t} \right) - I_1 \left(\dfrac{\partial u}{\partial t} \dfrac{\partial^2 \delta w_b}{\partial x \partial t} + \dfrac{\partial^2 w_b}{\partial x \partial t} \dfrac{\partial \delta u}{\partial t} \right) \\[2mm] -J_1 \left(\dfrac{\partial u}{\partial t} \dfrac{\partial^2 \delta w_s}{\partial x \partial t} + \dfrac{\partial^2 w_s}{\partial x \partial t} \dfrac{\partial \delta u}{\partial t} \right) + I_2 \dfrac{\partial^2 w_b}{\partial x \partial t} \dfrac{\partial^2 \delta w_b}{\partial x \partial t} + K_2 \dfrac{\partial^2 w_s}{\partial x \partial t} \dfrac{\partial^2 \delta w_s}{\partial x \partial t} \\[2mm] + J_2 \left(\dfrac{\partial^2 w_b}{\partial x \partial t} \dfrac{\partial^2 \delta w_s}{\partial x \partial t} + \dfrac{\partial^2 w_s}{\partial x \partial t} \dfrac{\partial^2 \delta w_b}{\partial x \partial t} \right) \end{array} \right) dx \tag{7.18}$$

The dot-superscript present in all equations signifies differentiation with respect to time. The mass inertias utilized in the aforementioned equation are computable.

$$\begin{pmatrix} I_0 \\ I_1 \\ I_2 \\ J_1 \\ J_2 \\ K_2 \end{pmatrix} = \int_A \begin{pmatrix} 1 \\ z \\ z^2 \\ f(z) \\ zf(z) \\ f^2(z) \end{pmatrix} \rho(z) dA \tag{7.19}$$

Subsequently, the fluctuation of the external work carried out by exogenous forces can be formulated in the subsequent manner:

$$\delta V = \int_0^L \left(k_p \frac{\partial^2 \delta (w_b + w_s)}{\partial x^2} - k_w \delta (w_b + w_s) \right) dx \tag{7.20}$$

The Euler–Lagrange equations of porous metal foam beams can be derived by inserting Eqs. (7.16), (7.18), and (7.20) into Eq. (7.14) and equating the coefficients of δu, δw_b, and δw_s to zero.

$$\frac{\partial N}{\partial x} = I_0 \frac{\partial^2 u}{\partial t^2} - I_1 \frac{\partial^3 w_b}{\partial x \partial t^2} - J_1 \frac{\partial^3 w_s}{\partial x \partial t^2} \tag{7.21}$$

$$\frac{\partial^2 M_b}{\partial x^2} = I_0 \frac{\partial^2 (w_b + w_s)}{\partial t^2} + I_1 \frac{\partial^3 u}{\partial x \partial t^2} - I_2 \frac{\partial^4 w_b}{\partial t^2 \partial x^2}$$
$$- J_2 \frac{\partial^4 w_s}{\partial t^2 \partial x^2} + k_w (w_b + w_s) - k_p \frac{\partial^2 (w_b + w_s)}{\partial x^2} \tag{7.22}$$

$$\frac{\partial^2 M_s}{\partial x^2} + \frac{\partial Q}{\partial x} = I_0 \frac{\partial^2 (w_b + w_s)}{\partial t^2} + J_1 \frac{\partial^3 u}{\partial x \partial t^2} - J_2 \frac{\partial^4 w_b}{\partial t^2 \partial x^2}$$
$$- K_2 \frac{\partial^4 w_s}{\partial t^2 \partial x^2} + k_w (w_b + w_s) - k_p \frac{\partial^2 (w_b + w_s)}{\partial x^2} \tag{7.23}$$

The present study aims to investigate the elastic stress-strain relationships of porous metal foam in order to derive the fundamental elastic equations governing the behavior of such materials. The constitutive equations can be expressed in the subsequent format, as stated in reference [1].

$$\sigma_{ij} = C_{ijkl} \varepsilon_{kl} \tag{7.24}$$

The symbols σ_{ij}, ε_{kl}, and C_{ijkl} denote the tensors for stress, strain, and elastic stiffness, respectively. Thus, it is possible to alter these relationships for beams in the following manner:

$$\sigma_{xx} = E_{11} \varepsilon_{xx} \tag{7.25}$$

$$\sigma_{xz} = G_{12} \gamma_{xz} \tag{7.26}$$

The following equations can be expressed by integrating Eqs. (7.25) and (7.26) over the cross-sectional area of the beam.

$$N = A \frac{\partial u}{\partial x} - B \frac{\partial^2 w_b}{\partial x^2} - B_s \frac{\partial^2 w_s}{\partial x^2} \tag{7.27}$$

$$M_b = B \frac{\partial u}{\partial x} - D \frac{\partial^2 w_b}{\partial x^2} - D_s \frac{\partial^2 w_s}{\partial x^2} \tag{7.28}$$

$$M_s = B_s \frac{\partial u}{\partial x} - D_s \frac{\partial^2 w_b}{\partial x^2} - H_s \frac{\partial^2 w_s}{\partial x^2} \tag{7.29}$$

$$Q = A_s \frac{\partial w_s}{\partial x} \tag{7.30}$$

in which

$$\begin{pmatrix} A \\ B \\ B_s \\ D \\ D_s \\ H_s \end{pmatrix} = \int_A \begin{pmatrix} 1 \\ z \\ f(z) \\ z^2 \\ zf(z) \\ f^2(z) \end{pmatrix} E_{11} dA \tag{7.31}$$

$$A_s = \int_A g^2(z) G_{12} dA \tag{7.32}$$

The governing equations of porous metal foam beams can be derived by substituting Eqs (7.26) through (7.30) into Eqs. (7.21) through (7.23).

$$A \frac{\partial^2 u}{\partial x^2} - B \frac{\partial^3 w_b}{\partial x^3} - B_s \frac{\partial^3 w_s}{\partial x^3} - I_0 \frac{\partial^2 u}{\partial t^2} + I_1 \frac{\partial^3 w_b}{\partial x \partial t^2} + J_1 \frac{\partial^3 w_s}{\partial x \partial t^2} = 0 \tag{7.33}$$

$$B \frac{\partial^3 u}{\partial x^3} - D \frac{\partial^4 w_b}{\partial x^4} - D_s \frac{\partial^4 w_s}{\partial x^4} - I_0 \frac{\partial^2 (w_b + w_s)}{\partial t^2} - I_1 \frac{\partial^3 u}{\partial x \partial t^2}$$
$$+ I_2 \frac{\partial^4 w_b}{\partial t^2 \partial x^2} + J_2 \frac{\partial^4 w_s}{\partial t^2 \partial x^2} + k_w (w_b + w_s) - k_p \frac{\partial^2 (w_b + w_s)}{\partial x^2} = 0 \tag{7.34}$$

$$B_s \frac{\partial^3 u}{\partial x^3} - D_s \frac{\partial^4 w_b}{\partial x^4} - H_s \frac{\partial^4 w_s}{\partial x^4} + A_s \frac{\partial^2 w_s}{\partial x^2} - I_0 \frac{\partial^2 (w_b + w_s)}{\partial t^2} - J_1 \frac{\partial^3 u}{\partial x \partial t^2}$$
$$+ J_2 \frac{\partial^4 w_b}{\partial t^2 \partial x^2} + K_2 \frac{\partial^4 w_s}{\partial t^2 \partial x^2} + k_w (w_b + w_s) - k_p \frac{\partial^2 (w_b + w_s)}{\partial x^2} = 0 \tag{7.35}$$

The application of an analytical method is utilized to resolve the governing equation of a beam composed of porous metal foam. According to the literature [1], the components of the displacement field are expected to be as follows.

$$\begin{bmatrix} u \\ w_b \\ w_s \end{bmatrix} = \begin{bmatrix} U e^{i(\beta x - \omega t)} \\ W_b e^{i(\beta x - \omega t)} \\ W_s e^{i(\beta x - \omega t)} \end{bmatrix} \tag{7.36}$$

The variables U, W_b, and W_s denote wave amplitudes. The symbols β and ω are utilized to denote the wave number and circular frequency dispersed waves, correspondingly. By substituting the variables u, w_b, and w_s from Eq. (7.36) into Eqs. (7.33) through (7.35), we obtain the resultant equation:

$$\left(\left[K - \omega^2 M\right]_{3\times3}\right)\begin{bmatrix} U \\ W_b \\ W_s \end{bmatrix} = 0 \qquad (7.37)$$

The stiffness and mass matrices are represented by K and M, respectively. The constituent elements of these matrices are provided in Appendix A [1]. The aforementioned eigenvalue problem has been resolved to obtain the value of ω, thereby enabling the determination of the wave frequency. It is necessary to equate the determinant in question to zero.

$$\left|\left[K - \omega^2 M\right]_{3\times3}\right| = 0 \qquad (7.38)$$

Furthermore, the phase velocity can be calculated by dividing the frequency of the wave by its wave number, as follows:

$$c_p = \frac{\omega}{\beta} \qquad (7.39a)$$

The derivation of the escape frequency of a beam composed of porous metal foam involves setting β to infinity. It is evident that the propagation of flexural waves ceases beyond the escape frequency.

This section presents a series of diagrams that illustrate the impact of different parameters on the wave frequency, escape frequency, and phase velocity of a metal foam beam. The current study assumes material properties of the metal foam to be $E_1 = 200$ GPa, $\rho_1 = 7850$ kg/m³, $v = 1/3$. The diagrams uniformly depict the thickness of the metal foam beam as $h = 5$ cm. The present study will proceed to introduce elastic foundation parameters for the purpose of facilitating comprehension [1].

$$K_W = \frac{k_W L^4}{X^*}, \quad K_P = \frac{k_P L^2}{X^*}, \quad X^* = \frac{E_1 h^3}{12\left(1 - v^2\right)} \qquad (7.39b)$$

Figure 7.3 illustrates the fluctuation in wave frequency of a metal foam beam in relation to wave number across various porosity distributions. The study examines the porosity distribution in three different scenarios: (a) a uniform distribution, (b) a symmetric distribution, and (c) an asymmetric distribution. The analysis is conducted across a range of porosity coefficients. It is evident that the porosity

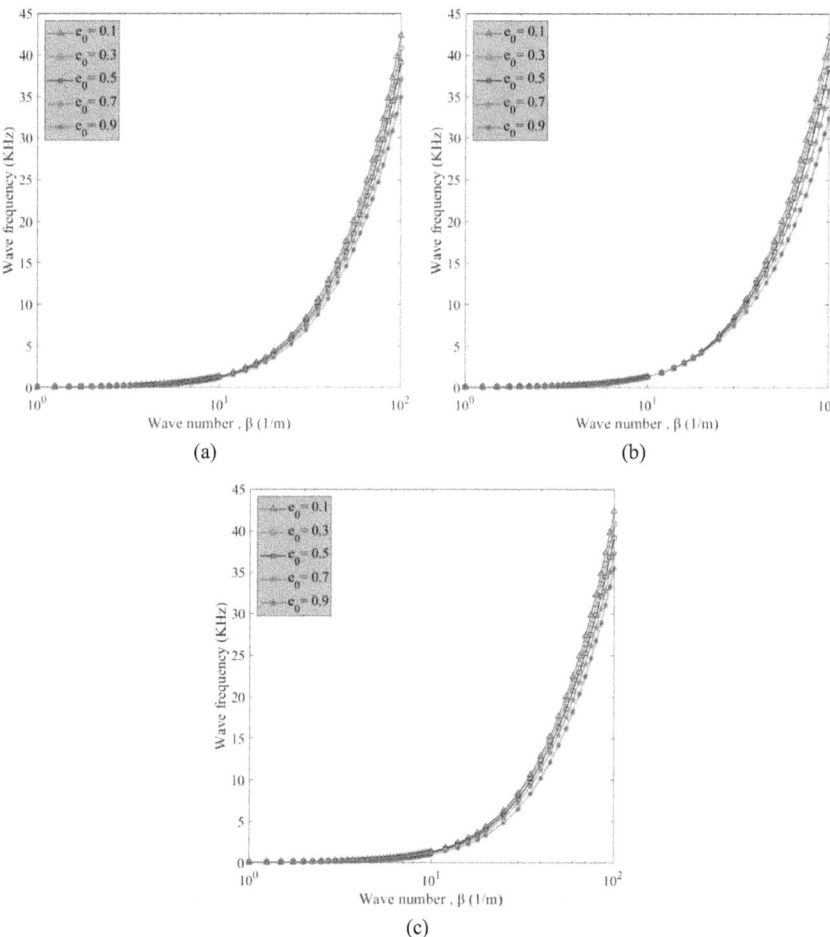

FIGURE 7.3 Variation of wave frequency of metal foam beam versus wave number according to different porosity coefficients for (a) uniform porosity distribution, (b) symmetric porosity distribution, and (c) asymmetric porosity distribution $\left(K_W = 50, K_P = 40, \dfrac{L}{h} = 20 \right)$.

coefficient exhibits a diminishing influence in all porosity distributions. Stated differently, an increase in the porosity coefficient results in a decrease in the wave frequency of a metal foam beam. The cause of this behavior can be attributed to the presence of porosity, which results in a reduction in the stiffness of the structure and a consequent weakening of its overall integrity. Moreover, the frequency of waves in a metal foam beam will increase exponentially as the wave number increases. Through a comparison of the outcomes of three distinct porosity distributions, it can be inferred that the symmetric distribution exhibits elevated levels of wave frequency [1].

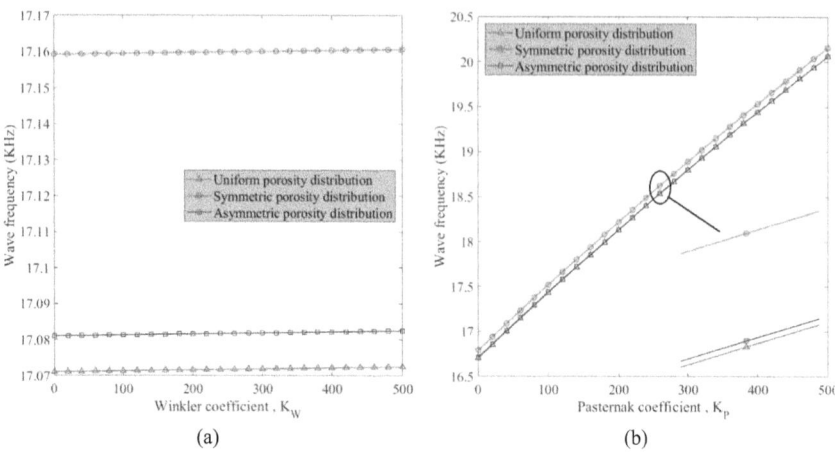

FIGURE 7.4   The effect of different types of porosity distribution on the variation of wave frequency of metal foam beam versus (a) Winkler coefficient and (b) Pasternak coefficient ($e_0 = 0.3$, $\beta = 50$, and $\dfrac{L}{h} = 20$).

Figure 7.4 illustrates the correlation between the porosity distribution and the wave frequency variation of a metal foam beam subjected to different elastic foundation parameters, namely the Winkler coefficient and the Pasternak coefficient. The present study reveals that the wave frequency of a metal foam beam is significantly influenced by both Winkler and Pasternak coefficients, which exhibit an increasing trend. The observed results are deemed reasonable owing to the hardening effect of the foundation. Moreover, it can be observed from diagram (a) that an increase in the Winkler coefficient results in a gradual rise in wave frequency. Additionally, it is noteworthy that the slope of diagram (b) is greater than that of diagram (a). Through meticulous observation of diagrams, it can be discerned that porosity distributions exhibiting symmetry are capable of withstanding higher frequencies when compared to their nonsymmetric counterparts. Conversely, it can be observed that the wave frequency is highest for materials exhibiting symmetric porosity distribution, while those with uniform porosity distribution display the lowest wave frequency. [1].

Figure 7.5 is presented to demonstrate the relationship between the escape frequency of a metal foam beam and its slenderness ratio, considering different porosity distributions. Based on the presented figure, it can be observed that the curves exhibit a relatively consistent pattern. The results indicate that the asymmetric porosity distribution exhibits the highest escape frequency, followed by the uniform porosity distribution, and finally, the symmetric porosity distribution displays the lowest escape frequency.

Figure 7.6 illustrates the impact of the porosity coefficient on the phase velocity of a metal foam beam as a function of wave number, under three distinct porosity distributions: uniform, symmetric, and asymmetric. It can be inferred that an

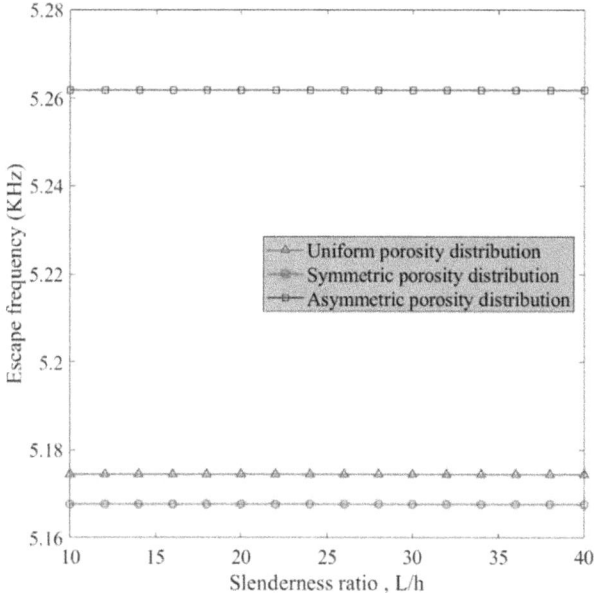

FIGURE 7.5 Variation of the escape frequency of metal foam beam versus the porosity coefficient for different types of porosity distribution ($K_W = 50$, $K_P = 20$, and $e_0 = 0.2$).

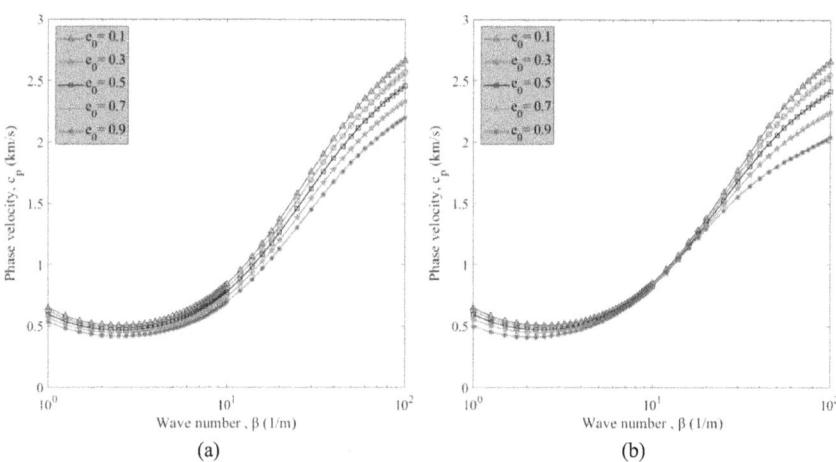

FIGURE 7.6 Variation of the phase velocity of metal foam beam versus wave number for various porosity coefficients for (a) uniform porosity distribution, (b) symmetric porosity distribution.

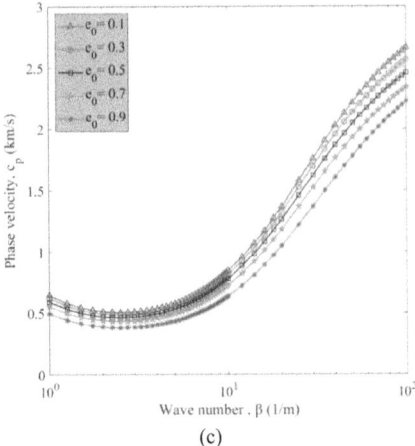

(c)

FIGURE 7.6   (Continued) (c) asymmetric porosity distribution $(K_W = 50, K_P = 40, \dfrac{L}{h} = 20)$.

increase in porosity coefficient results in a decrease in phase velocity. All curves exhibit a consistent pattern whereby an increase in wave number initially results in a decrease in phase velocity, followed by a gradual increase. The diagram depicting symmetric porosity distribution reveals a noteworthy observation. Specifically, within the range of $8 < \beta < 18$, the phase velocity value of curve $e_0 = 0.9$ surpasses that of curve $e_0 = 0.7$. Subsequently, the curves revert to their prior trajectory.

Figure 7.7 illustrates the fluctuation in phase velocity of a metal foam beam in relation to the porosity coefficient, considering different slenderness ratios for an asymmetric porosity coefficient. The current configuration typically exhibits uneven dispersion of porosity, thereby necessitating the creation of a diagram that accounts for asymmetric distribution of porosity. It is evident from the observation that the curves exhibit a descending trend, and there is a corresponding decrease in phase velocity as the slenderness ratio increases. Furthermore, an increase in the porosity coefficient results in a reduction of phase velocity.

7.3   GRAPHENE FOAM PLATE WAVE DISPERSION CHARACTERISTICS

This section delves into the wave dispersion characteristics of the metal foam plate, as depicted in Figure 7.8 [3].

The distribution of porosity in porous metal foam is taken into account when analyzing the impact of symmetric and asymmetric patterns on the effective properties of said foam. Figure 7.8 displays a schematic illustration of the porosity

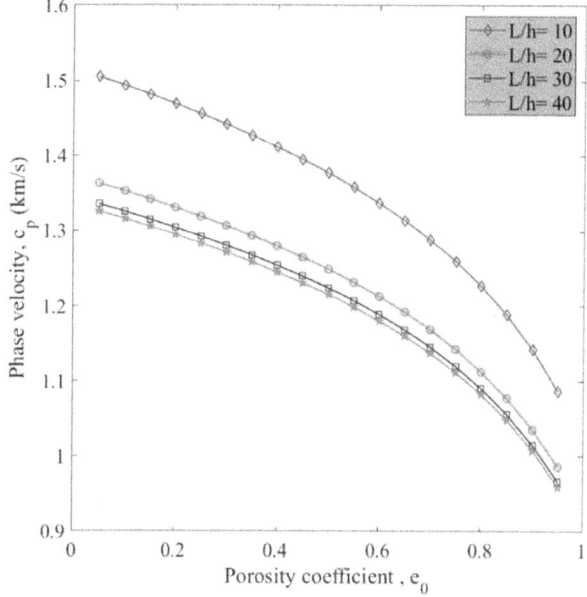

FIGURE 7.7 Variation of the phase velocity of metal foam beam with asymmetric porosity distribution versus porosity coefficient considering various slenderness ratios ($K_W = 60$, $K_P = 30$, $\beta = 20$).

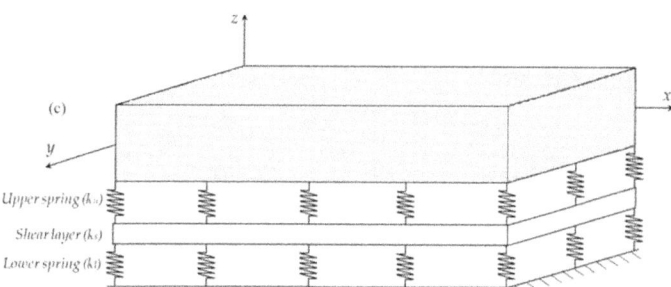

FIGURE 7.8 The configuration of rectangular metal foam plates resting on Kerr foundation.

distribution across the plate's thickness. The effective characteristics of a metal foam plate with symmetrical porosity can be mathematically represented as:

$$E(z) = E_1 \left(1 - e_0 \cos\left(\frac{\pi z}{h} \right) \right) \tag{7.40}$$

$$G(z) = G_1\left(1 - e_0 \cos\left(\frac{\pi z}{h}\right)\right) \tag{7.41}$$

$$\rho(z) = \rho_1\left(1 - e_m \cos\left(\frac{\pi z}{h}\right)\right) \tag{7.42}$$

Furthermore, the effective characteristics of a metal foam plate with asymmetric porosity can be computed using the following method:

$$E(z) = E_1\left(1 - e_0 \cos\left(\frac{\pi z}{2h} + \frac{\pi}{4}\right)\right) \tag{7.43a}$$

$$G(z) = G_1\left(1 - e_0 \cos\left(\frac{\pi z}{2h} + \frac{\pi}{4}\right)\right) \tag{7.43b}$$

$$\rho(z) = \rho_1\left(1 - e_m \cos\left(\frac{\pi z}{2h} + \frac{\pi}{4}\right)\right) \tag{7.43c}$$

The symbols E_1, G_1, and ρ_1 denote Young's modulus, shear modulus, and mass density, respectively, of the porous metal foam with the highest attainable value. The minimum values of the aforementioned variables are E_2, G_2, and ρ_2, respectively. In addition, the coefficients e_0 and e_m pertain to porosity and density and may be defined as follows:

$$e_0 = 1 - \frac{E_2}{E_1} = 1 - \frac{G_2}{G_1}, \quad (0 < e_0 < 1) \tag{7.44}$$

$$e_m = 1 - \frac{\rho_2}{\rho_1} \tag{7.45}$$

Furthermore, the value of e_m can be computed based on the standard mechanical characteristics of an open-cell metallic foam, such as:

$$\frac{E_2}{E_1} = \left(\frac{\rho_2}{\rho_1}\right)^2 \tag{7.46}$$

$$e_m = 1 - \sqrt{1 - e_0} \tag{7.47}$$

In this approach, the constancy of Poisson's ratio can be assumed due to its minimal variation.

The refined higher-order shear deformation plate theory has been utilized to establish kinetic relationships for a metal foam plate. The expression for the displacement field of plates can be represented as follows:

$$u_1\left(x, y, z, t\right) = u\left(x, y, t\right) - z\frac{\partial w_b\left(x, y, t\right)}{\partial x} - f\left(z\right)\frac{\partial w_s\left(x, y, t\right)}{\partial x} \tag{7.48}$$

$$u_2\left(x, y, z, t\right) = v\left(x, y, t\right) - z\frac{\partial w_b\left(x, y, t\right)}{\partial y} - f\left(z\right)\frac{\partial w_s\left(x, y, t\right)}{\partial y} \tag{7.49}$$

$$u_3\left(x, y, z, t\right) = w_b\left(x, y, t\right) + w_s\left(x, y, t\right) \tag{7.50}$$

The variables u, w_b, and w_s represent longitudinal displacement, bending deflection, and shear deflection, respectively. The function $f(z)$ serves as a shape function that provides an estimation of the distribution of shear stress or strain in the direction of thickness. The model presented can be utilized without the need for a shear correction factor, owing to the exerted shape function. Three different types of shape functions, specifically Ambartsumian, hyperbolic, and inverse cotangential, can be presented in the following order:

$$f\left(z\right) = \left(\frac{z}{2}\right) \times \left(\frac{h^2}{4} - \frac{z^2}{3}\right) \tag{7.51a}$$

$$f\left(z\right) = h \times \sinh\left(\frac{z}{h}\right) - z \times \cosh\left(\frac{1}{2}\right) \tag{7.51b}$$

$$f\left(z\right) = \cot^{-1}\left(\frac{rh}{z}\right) + z \times \left[\frac{-4r}{h\left(4r^2 + 1\right)}\right]; \quad r = 0.46 \tag{7.51c}$$

The current study employs the use of Integrated Circuit Substrate Design Tool (ICSDT) and conducts a comparative analysis with alternative methods. The plate's strains that are not equal to zero can be expressed as:

$$\begin{Bmatrix} \varepsilon_{xx} \\ \varepsilon_{yy} \\ \gamma_{xy} \end{Bmatrix} = \begin{Bmatrix} \varepsilon_{xx}^0 \\ \varepsilon_{yy}^0 \\ \gamma_{xy}^0 \end{Bmatrix} + z \begin{Bmatrix} \kappa_{xx}^b \\ \kappa_{yy}^b \\ \kappa_{xy}^b \end{Bmatrix} + f\left(z\right) \begin{Bmatrix} \kappa_{xx}^s \\ \kappa_{yy}^s \\ \kappa_{xy}^s \end{Bmatrix}, \quad \begin{Bmatrix} \gamma_{xz} \\ \gamma_{yz} \end{Bmatrix} = \begin{Bmatrix} \gamma_{xz}^0 \\ \gamma_{yz}^0 \end{Bmatrix} \tag{7.52}$$

where

$$
\left\{\begin{array}{c} \varepsilon_{xx}^0 \\ \varepsilon_{yy}^0 \\ \gamma_{xy}^0 \end{array}\right\} = \left\{\begin{array}{c} \dfrac{\partial u}{\partial x} \\[2mm] \dfrac{\partial v}{\partial y} \\[2mm] \dfrac{\partial u}{\partial y} + \dfrac{\partial v}{\partial x} \end{array}\right\}, \quad \left\{\begin{array}{c} \kappa_{xx}^b \\ \kappa_{yy}^b \\ \kappa_{xy}^b \end{array}\right\} = \left\{\begin{array}{c} -\dfrac{\partial^2 w_b}{\partial x^2} \\[2mm] -\dfrac{\partial^2 w_b}{\partial y^2} \\[2mm] -2\dfrac{\partial^2 w_b}{\partial x \partial y} \end{array}\right\}, \quad \left\{\begin{array}{c} \kappa_{xx}^s \\ \kappa_{yy}^s \\ \kappa_{xy}^s \end{array}\right\} = \left\{\begin{array}{c} -\dfrac{\partial^2 w_s}{\partial x^2} \\[2mm] -\dfrac{\partial^2 w_s}{\partial y^2} \\[2mm] -2\dfrac{\partial^2 w_s}{\partial x \partial y} \end{array}\right\}, \quad \left\{\begin{array}{c} \gamma_{xz}^0 \\ \gamma_{yz}^0 \end{array}\right\} = \left\{\begin{array}{c} \dfrac{\partial w_s}{\partial x} \\[2mm] \dfrac{\partial w_s}{\partial y} \end{array}\right\}
$$

$$(7.53)$$

The application of Hamilton's principle is utilized in the derivation of the motion equations governing the behavior of the plate. The principle of Hamilton can be formulated as follows according to reference [45]:

$$
\int_{t_1}^{t_2} \delta\left[\Pi_K - \left(\Pi_S + \Pi_F\right)\right] dt = 0
$$

$$(7.54)$$

The equation in question involves the variables Π_K, Π_S, and Π_F, which respectively represent kinetic energy, strain energy, and work performed by an external force. The fluctuation of kinetic energy may be expressed as follows:

$$
\delta\Pi_K = \int_V \left(\dot{u}_1 \delta\dot{u}_1 + \dot{u}_2 \delta\dot{u}_2 + \dot{u}_3 \delta\dot{u}_3\right)\rho(z)\,dV
$$

$$
= \int_0^b \int_0^a \left\{\begin{array}{c}
I_0\left(\dfrac{\partial u}{\partial t}\dfrac{\partial \delta u}{\partial t} + \dfrac{\partial v}{\partial t}\dfrac{\partial \delta v}{\partial t} + \dfrac{\partial w}{\partial t}\dfrac{\partial \delta w}{\partial t}\right) \\[3mm]
-I_1\left(\dfrac{\partial u}{\partial t}\dfrac{\partial^2 \delta w_b}{\partial x \partial t} + \dfrac{\partial^2 w_b}{\partial x \partial t}\dfrac{\partial \delta u}{\partial t} + \dfrac{\partial v}{\partial t}\dfrac{\partial^2 \delta w_b}{\partial y \partial t} + \dfrac{\partial^2 w_b}{\partial y \partial t}\dfrac{\partial \delta v}{\partial t}\right) \\[3mm]
-J_1\left(\dfrac{\partial u}{\partial t}\dfrac{\partial^2 \delta w_s}{\partial x \partial t} + \dfrac{\partial^2 w_s}{\partial x \partial t}\dfrac{\partial \delta u}{\partial t} + \dfrac{\partial v}{\partial t}\dfrac{\partial^2 \delta w_s}{\partial y \partial t} + \dfrac{\partial^2 w_s}{\partial y \partial t}\dfrac{\partial \delta v}{\partial t}\right) \\[3mm]
+I_2\left(\dfrac{\partial^2 w_b}{\partial x \partial t}\dfrac{\partial^2 \delta w_b}{\partial x \partial t} + \dfrac{\partial^2 w_b}{\partial y \partial t}\dfrac{\partial^2 \delta w_b}{\partial y \partial t}\right) \\[3mm]
+K_2\left(\dfrac{\partial^2 w_s}{\partial x \partial t}\dfrac{\partial^2 \delta w_s}{\partial x \partial t} + \dfrac{\partial^2 w_s}{\partial y \partial t}\dfrac{\partial^2 \delta w_s}{\partial y \partial t}\right) \\[3mm]
+J_2\left(\dfrac{\partial^2 w_b}{\partial x \partial t}\dfrac{\partial^2 \delta w_s}{\partial x \partial t} + \dfrac{\partial^2 w_s}{\partial x \partial t}\dfrac{\partial^2 \delta w_b}{\partial x \partial t} + \dfrac{\partial^2 w_b}{\partial y \partial t}\dfrac{\partial^2 \delta w_s}{\partial y \partial t} + \dfrac{\partial^2 w_s}{\partial y \partial t}\dfrac{\partial^2 \delta w_b}{\partial y \partial t}\right)
\end{array}\right\} dx\,dy
$$

$$(7.55)$$

The mass inertias utilized in the aforementioned equation have been formulated as follows:

$$[I_0, I_1, J_1, I_2, J_2, K_2] = \int_{-\frac{h}{2}}^{\frac{h}{2}} [1, z, f(z), z^2, zf(z), f^2(z)] \rho(z) dz \quad (7.56)$$

Subsequently, the fluctuation in strain energy can be articulated in the ensuing format:

$$\delta\Pi_S = \int_V (\sigma_{xx}\delta\varepsilon_{xx} + \sigma_{yy}\delta\varepsilon_{yy} + \sigma_{xy}\delta\gamma_{xy} + \sigma_{xz}\delta\gamma_{xz} + \sigma_{yz}\delta\gamma_{yz}) dV$$

$$= \int_0^b \int_0^a \begin{pmatrix} N_{xx}\delta\varepsilon_{xx}^0 + M_{xx}^b\delta\kappa_{xx}^b + M_{xx}^s\delta\kappa_{xx}^s + N_{yy}\delta\varepsilon_{yy}^0 + \\ M_{yy}^b\delta\kappa_{yy}^b + M_{yy}^s\delta\kappa_{yy}^s + N_{xy}\delta\gamma_{xy}^0 + M_{xy}^b\delta\kappa_{xy}^b + \\ M_{xy}^s\delta\kappa_{xy}^s + Q_{xz}\delta\gamma_{xz}^0 + Q_{yz}\delta\gamma_{yz}^0 \end{pmatrix} dxdy \quad (7.57)$$

The axial forces and bending moments can be expressed in the following manner:

$$\left(N_{ij}, M_{ij}^b, M_{ij}^s\right) = \int_{-\frac{h}{2}}^{\frac{h}{2}} \left(1, z, f(z)\right)\sigma_{ij}dz, \quad (i, j = x, y)$$

$$(7.58)$$

$$\begin{Bmatrix} Q_{xz} \\ Q_{yz} \end{Bmatrix} = \int_{-\frac{h}{2}}^{\frac{h}{2}} g(z) \begin{Bmatrix} \sigma_{xz} \\ \sigma_{yz} \end{Bmatrix} dz$$

in which $g(z) = 1 - \dfrac{df(z)}{dz}$. Subsequently, the external forces' work variation can be formulated as follows:

$$\delta\Pi_F = \int_0^L \int_A \begin{pmatrix} N^T \left(\dfrac{\partial(w_b + w_s)}{\partial x} \dfrac{\partial\delta(w_b + w_s)}{\partial x} + \dfrac{\partial(w_b + w_s)}{\partial y} \dfrac{\partial\delta(w_b + w_s)}{\partial y} \right) \\ + q_{Kerr}\delta(w_b + w_s) \end{pmatrix} dAdx$$

$$(7.59)$$

in which

$$N^T = \int_{-\frac{h}{2}}^{\frac{h}{2}} E(z)\alpha\Delta T dz \qquad (7.60)$$

This chapter investigates the effects of various types of thermal loading, namely uniform temperature rise (UTR), linear temperature rise (LTR), and sinusoidal temperature rise (STR), on the wave propagation behavior of the proposed structure. Previous works in open literature that have considered thermal environments have also been referenced [46, 47]. The temperature at the outset is regarded as the ambient temperature, denoted as $T_0 = 300$ K. The ultimate temperature (T) resulting from diverse forms of temperature increase can be expressed as:

$$\begin{cases} T = T_0 + \Delta T & \text{UTR} \\ T = T_0 + \Delta T\left(\dfrac{1}{2}+\dfrac{z}{h}\right) & \text{LTR} \\ T = T_0 + \Delta T\left(1-\cos\left(\dfrac{\pi}{4}+\dfrac{\pi z}{2h}\right)\right) & \text{STR} \end{cases} \qquad (7.61)$$

The symbol ΔT represents the difference between the final temperature T and the initial temperature T0. The Kerr substrate model comprises a shear layer with stiffness k_s that is affixed to two distinct linear layers, namely the upper and lower layers, each possessing a stiffness of k_u and k_l, respectively. The expression for the plate's deflection is given by:

$$w = w_1 + w_2 \qquad (7.62)$$

The variables w_1 and w_2 are utilized to represent the deflection of the upper springs and the shear layer, respectively. The magnitudes of the contact pressures beneath the plate and within the shear layer are denoted as q_1 and q_2, respectively, as follows:

$$q_1 = k_u w_1 = k_u\left(w - w_2\right) \qquad (7.63a)$$

$$q_2 = k_l w \qquad (7.63b)$$

The differential equation that governs the shear layer can be expressed in the following form:

$$k_l w_2 - k_s \nabla^2 w_2 = q_1 = q_{\text{Kerr}} \qquad (7.64)$$

The distributed reaction between the Kerr substrate and the lower surface of the plate can be expressed by eliminating w_2 from Eqs. (7.63b) and (7.64).

$$q_{\text{Kerr}} - \left(\frac{k_s}{k_u + k_l} \right) \nabla^2 q_{\text{Kerr}} = \left(\frac{k_l k_u}{k_u + k_l} \right) (w_b + w_s) - \left(\frac{k_s k_u}{k_u + k_l} \right) \nabla^2 (w_b + w_s) \quad (7.65)$$

The Laplacian operator, denoted by ∇^2, can be defined in the x- and y-directions.

$$\nabla^2 = \frac{\partial^2}{\partial x^2} + \frac{\partial^2}{\partial y^2} \quad (7.66)$$

The motion equations of porous metal foam plates can be derived by inserting Eqs. (7.55), (7.57), and (7.59) into Eq. (7.54) and equating the coefficients of δu, δv, δw_b, and δw_s to zero.

$$\frac{\partial N_{xx}}{\partial x} + \frac{\partial N_{xy}}{\partial y} = I_0 \frac{\partial^2 u}{\partial t^2} - I_1 \frac{\partial^3 w_b}{\partial x \partial t^2} - J_1 \frac{\partial^3 w_s}{\partial x \partial t^2} \quad (7.67)$$

$$\frac{\partial N_{xy}}{\partial x} + \frac{\partial N_{yy}}{\partial y} = I_0 \frac{\partial^2 v}{\partial t^2} - I_1 \frac{\partial^3 w_b}{\partial y \partial t^2} - J_1 \frac{\partial^3 w_s}{\partial y \partial t^2} \quad (7.68)$$

$$\frac{\partial^2 M_{xx}^b}{\partial x^2} + 2\frac{\partial^2 M_{xy}^b}{\partial x \partial y} + \frac{\partial^2 M_{yy}^b}{\partial y^2} - \left(N^T - \frac{k_s k_u}{k_u + k_l} \right) \nabla^2 (w_b + w_s) - \left(\frac{k_l k_u}{k_u + k_l} \right) (w_b + w_s)$$
$$= I_0 \frac{\partial^2 (w_b + w_s)}{\partial t^2} + I_1 \left(\frac{\partial^3 u}{\partial x \partial t^2} + \frac{\partial^3 v}{\partial y \partial t^2} \right) - I_2 \left(\frac{\partial^4 w_b}{\partial x^2 \partial t^2} + \frac{\partial^4 w_b}{\partial y^2 \partial t^2} \right) \quad (7.69)$$
$$- J_2 \left(\frac{\partial^4 w_s}{\partial x^2 \partial t^2} + \frac{\partial^4 w_s}{\partial y^2 \partial t^2} \right)$$

$$\frac{\partial^2 M_{xx}^s}{\partial x^2} + 2\frac{\partial^2 M_{xy}^s}{\partial x \partial y} + \frac{\partial^2 M_{yy}^s}{\partial y^2} + \frac{xz}{\partial x} + \frac{\partial Q_{yz}}{\partial y} - \left(N^T - \frac{k_s k_u}{k_u + k_l} \right) \nabla^2 (w_b + w_s)$$
$$- \left(\frac{k_l k_u}{k_u + k_l} \right) (w_b + w_s) = I_0 \frac{\partial^2 (w_b + w_s)}{\partial t^2} + J_1 \left(\frac{\partial^3 u}{\partial x \partial t^2} + \frac{\partial^3 v}{\partial y \partial t^2} \right) \quad (7.70)$$
$$- J_2 \left(\frac{\partial^4 w_b}{\partial x^2 \partial t^2} + \frac{\partial^4 w_b}{\partial y^2 \partial t^2} \right) - K_2 \left(\frac{\partial^4 w_s}{\partial x^2 \partial t^2} + \frac{\partial^4 w_s}{\partial y^2 \partial t^2} \right)$$

The fundamental elastic equations of porous metal foam are derived by utilizing the elastic stress-strain relations of said material. The constitutive equations pertaining to the plate can be expressed in the following manner:

$$\sigma_{ij} = C_{ijkl}\varepsilon_{kl} \tag{7.71}$$

The symbols σ_{ij}, C_{ijkl}, and ε_{kl} represent stress, elastic stiffness, and strain components, respectively. Thus, it is possible to alter these relationships for plates in the following manner:

$$\begin{bmatrix} \sigma_{xx} \\ \sigma_{yy} \\ \sigma_{yz} \\ \sigma_{xz} \\ \sigma_{xy} \end{bmatrix} = \begin{bmatrix} Q_{11} & Q_{12} & 0 & 0 & 0 \\ Q_{12} & Q_{22} & 0 & 0 & 0 \\ 0 & 0 & Q_{44} & 0 & 0 \\ 0 & 0 & 0 & Q_{55} & 0 \\ 0 & 0 & 0 & 0 & Q_{66} \end{bmatrix} \begin{bmatrix} \varepsilon_{xx} \\ \varepsilon_{yy} \\ \varepsilon_{yz} \\ \varepsilon_{xz} \\ \varepsilon_{xy} \end{bmatrix} \tag{7.72}$$

where

$$Q_{11} = Q_{22} = \frac{E(z)}{1-v^2}, \quad Q_{12} = \frac{vE}{1-v^2},$$
$$Q_{44} = Q_{55} = Q_{66} = G_{12} = \frac{E(z)}{2(1+v)} \tag{7.73}$$

The resultant forces and moments can be obtained by integrating from Eq. (7.56) with consideration of the thickness direction along the z-axis.

$$\begin{Bmatrix} N_{xx} \\ N_{yy} \\ N_{xy} \end{Bmatrix} = \begin{bmatrix} A_{11} & A_{12} & 0 \\ A_{12} & A_{22} & 0 \\ 0 & 0 & A_{66} \end{bmatrix} \begin{Bmatrix} \dfrac{\partial u}{\partial x} \\ \dfrac{\partial v}{\partial y} \\ \dfrac{\partial u}{\partial y}+\dfrac{\partial v}{\partial x} \end{Bmatrix} + \begin{bmatrix} B_{11} & B_{12} & 0 \\ B_{12} & B_{22} & 0 \\ 0 & 0 & B_{66} \end{bmatrix} \begin{Bmatrix} -\dfrac{\partial^2 w_b}{\partial x^2} \\ -\dfrac{\partial^2 w_b}{\partial y^2} \\ -2\dfrac{\partial^2 w_b}{\partial x\partial y} \end{Bmatrix}$$

$$+ \begin{bmatrix} B_{11}^s & B_{12}^s & 0 \\ B_{12}^s & B_{22}^s & 0 \\ 0 & 0 & B_{66}^s \end{bmatrix} \begin{Bmatrix} -\dfrac{\partial^2 w_s}{\partial x^2} \\ -\dfrac{\partial^2 w_s}{\partial y^2} \\ -2\dfrac{\partial^2 w_s}{\partial x\partial y} \end{Bmatrix}$$

$$\tag{7.74}$$

$$
\begin{Bmatrix} M_{xx}^b \\ M_{yy}^b \\ M_{xy}^b \end{Bmatrix} = \begin{bmatrix} B_{11} & B_{12} & 0 \\ B_{12} & B_{22} & 0 \\ 0 & 0 & B_{66} \end{bmatrix} \begin{Bmatrix} \dfrac{\partial u}{\partial x} \\ \dfrac{\partial v}{\partial y} \\ \dfrac{\partial u}{\partial y} + \dfrac{\partial v}{\partial x} \end{Bmatrix} + \begin{bmatrix} D_{11} & D_{12} & 0 \\ D_{12} & D_{22} & 0 \\ 0 & 0 & D_{66} \end{bmatrix} \begin{Bmatrix} -\dfrac{\partial^2 w_b}{\partial x^2} \\ -\dfrac{\partial^2 w_b}{\partial y^2} \\ -2\dfrac{\partial^2 w_b}{\partial x \partial y} \end{Bmatrix}
$$

$$
+ \begin{bmatrix} D_{11}^s & D_{12}^s & 0 \\ D_{12}^s & D_{22}^s & 0 \\ 0 & 0 & D_{66}^s \end{bmatrix} \begin{Bmatrix} -\dfrac{\partial^2 w_s}{\partial x^2} \\ -\dfrac{\partial^2 w_s}{\partial y^2} \\ -2\dfrac{\partial^2 w_s}{\partial x \partial y} \end{Bmatrix}
\tag{7.75}
$$

$$
\begin{Bmatrix} M_{xx}^s \\ M_{yy}^s \\ M_{xy}^s \end{Bmatrix} = \begin{bmatrix} B_{11}^s & B_{12}^s & 0 \\ B_{12}^s & B_{22}^s & 0 \\ 0 & 0 & B_{66}^s \end{bmatrix} \begin{Bmatrix} \dfrac{\partial u}{\partial x} \\ \dfrac{\partial v}{\partial y} \\ \dfrac{\partial u}{\partial y} + \dfrac{\partial v}{\partial x} \end{Bmatrix} + \begin{bmatrix} D_{11}^s & D_{12}^s & 0 \\ D_{12}^s & D_{22}^s & 0 \\ 0 & 0 & D_{66}^s \end{bmatrix} \begin{Bmatrix} -\dfrac{\partial^2 w_b}{\partial x^2} \\ -\dfrac{\partial^2 w_b}{\partial y^2} \\ -2\dfrac{\partial^2 w_b}{\partial x \partial y} \end{Bmatrix}
$$

$$
+ \begin{bmatrix} H_{11}^s & H_{12}^s & 0 \\ H_{12}^s & H_{22}^s & 0 \\ 0 & 0 & H_{66}^s \end{bmatrix} \begin{Bmatrix} -\dfrac{\partial^2 w_s}{\partial x^2} \\ -\dfrac{\partial^2 w_s}{\partial y^2} \\ -2\dfrac{\partial^2 w_s}{\partial x \partial y} \end{Bmatrix}
\tag{7.76}
$$

$$
\begin{Bmatrix} Q_{xz} \\ Q_{yz} \end{Bmatrix} = \begin{bmatrix} A_{44}^s & 0 \\ 0 & A_{55}^s \end{bmatrix} \begin{Bmatrix} \dfrac{\partial w_s}{\partial x} \\ \dfrac{\partial w_s}{\partial y} \end{Bmatrix}
\tag{7.77}
$$

in which

$$\begin{bmatrix} A_k, B_k, B_k^s \\ D_k, D_k^s, H_k^s \end{bmatrix} = \int_{-\frac{h}{2}}^{\frac{h}{2}} \begin{bmatrix} 1, z, f(z) \\ z^2, zf(z), f^2(z) \end{bmatrix} Q_k(z) dz, \quad k = (11,12,22,66)$$

(7.78)

$$\begin{bmatrix} A_{44}^s, A_{55}^s \end{bmatrix} = \int_{-\frac{h}{2}}^{\frac{h}{2}} \begin{bmatrix} Q_{44}(z), Q_{55}(z) \end{bmatrix} g^2(z) dz$$

The governing equations of porous metal foam plates can be derived by substituting Eqs. (7.74)–(7.77) into Eqs. (7.67)–(7.70). The resulting form of the equations is as follows:

$$A_{11}\frac{\partial^2 u}{\partial x^2} + A_{66}\frac{\partial^2 u}{\partial y^2} + (A_{12}+A_{66})\frac{\partial^2 v}{\partial x \partial y} - B_{11}\frac{\partial^3 w_b}{\partial x^3} - (B_{12}+2B_{66})\frac{\partial^3 w_b}{\partial x \partial y^2}$$

$$-B_{11}^s\frac{\partial^3 w_s}{\partial x^3} - (B_{12}^s+2B_{66}^s)\frac{\partial^3 w_s}{\partial x \partial y^2} - I_0\frac{\partial^2 u}{\partial t^2} + I_1\frac{\partial^3 w_b}{\partial x \partial t^2} + J_1\frac{\partial^3 w_b}{\partial x \partial t^2} = 0$$

(7.79a)

$$(A_{12}+A_{66})\frac{\partial^2 u}{\partial x \partial y} + A_{66}\frac{\partial^2 v}{\partial x^2} + A_{22}\frac{\partial^2 v}{\partial y^2} - (B_{12}+2B_{66})\frac{\partial^3 w_b}{\partial x^2 \partial y} - B_{22}\frac{\partial^3 w_b}{\partial y^3}$$

$$-(B_{12}^s+2B_{66}^s)\frac{\partial^3 w_s}{\partial x^2 \partial y} - B_{22}^s\frac{\partial^3 w_s}{\partial y^3} - I_0\frac{\partial^2 v}{\partial t^2} + I_1\frac{\partial^3 w_b}{\partial y \partial t^2} + J_1\frac{\partial^3 w_b}{\partial y \partial t^2} = 0$$

(7.79b)

$$B_{11}\frac{\partial^3 u}{\partial x^3} + (B_{12}+2B_{66})\left(\frac{\partial^3 u}{\partial x \partial y^2} + \frac{\partial^3 v}{\partial x^2 \partial y}\right) + B_{22}\frac{\partial^3 v}{\partial y^3} - D_{11}\frac{\partial^4 w_b}{\partial x^4}$$

$$-2(D_{12}+2D_{66})\frac{\partial^4 w_b}{\partial x^2 \partial y^2} - D_{22}\frac{\partial^4 w_b}{\partial y^4} - D_{11}^s\frac{\partial^4 w_s}{\partial x^4} - 2(D_{12}^s+2D_{66}^s)\frac{\partial^4 w_s}{\partial x^2 \partial y^2}$$

$$-D_{22}^s\frac{\partial^4 w_s}{\partial y^4} - \left(N^T - \frac{k_s k_u}{k_u+k_l}\right)\nabla^2(w_b+w_s) - \left(\frac{k_l k_u}{k_u+k_l}\right)(w_b+w_s)$$

(7.79c)

$$-I_0\frac{\partial^2(w_b+w_s)}{\partial t^2} - I_1\left(\frac{\partial^3 u}{\partial x \partial t^2} + \frac{\partial^3 v}{\partial y \partial t^2}\right) + I_2\left(\frac{\partial^4 w_b}{\partial x^2 \partial t^2} + \frac{\partial^4 w_b}{\partial y^2 \partial t^2}\right)$$

$$+J_2\left(\frac{\partial^4 w_s}{\partial x^2 \partial t^2} + \frac{\partial^4 w_s}{\partial y^2 \partial t^2}\right) = 0$$

$$B_{11}^s \frac{\partial^3 u}{\partial x^3} + \left(B_{12}^s + 2B_{66}^s\right)\left(\frac{\partial^3 u}{\partial x \partial y^2} + \frac{\partial^3 v}{\partial x^2 \partial y}\right) + B_{22}^s \frac{\partial^3 v}{\partial y^3} - D_{11}^s \frac{\partial^4 w_b}{\partial x^4}$$

$$-2\left(D_{12}^s + 2D_{66}^s\right)\frac{\partial^4 w_b}{\partial x^2 \partial y^2} - D_{22}^s \frac{\partial^4 w_b}{\partial y^4} - H_{11}^s \frac{\partial^4 w_s}{\partial x^4} - 2\left(H_{12}^s + 2H_{66}^s\right)\frac{\partial^4 w_s}{\partial x^2 \partial y^2}$$

$$-H_{22}^s \frac{\partial^4 w_s}{\partial y^4} + A_{55}^s \frac{\partial^2 w_s}{\partial x^2} + A_{44}^s \frac{\partial^2 w_s}{\partial y^2} - \left(N^T - \frac{k_s k_u}{k_u + k_l}\right)\nabla^2\left(w_b + w_s\right) \tag{7.79d}$$

$$-\left(\frac{k_l k_u}{k_u + k_l}\right)\left(w_b + w_s\right) - I_0 \frac{\partial^2\left(w_b + w_s\right)}{\partial t^2} - J_1\left(\frac{\partial^3 u}{\partial x \partial t^2} + \frac{\partial^3 v}{\partial y \partial t^2}\right)$$

$$+J_2\left(\frac{\partial^4 w_b}{\partial x^2 \partial t^2} + \frac{\partial^4 w_b}{\partial y^2 \partial t^2}\right) + K_2\left(\frac{\partial^4 w_s}{\partial x^2 \partial t^2} + \frac{\partial^4 w_s}{\partial y^2 \partial t^2}\right) = 0$$

An analytical solution method is utilized to solve the governing equations of porous metal foam plates. It is assumed that the components of the displacement field are as follows:

$$\begin{Bmatrix} u \\ v \\ w_b \\ w_s \end{Bmatrix} = \begin{Bmatrix} U \exp\left[i\left(\beta_1 x + \beta_2 y - \omega t\right)\right] \\ V \exp\left[i\left(\beta_1 x + \beta_2 y - \omega t\right)\right] \\ W_b \exp\left[i\left(\beta_1 x + \beta_2 y - \omega t\right)\right] \\ W_s \exp\left[i\left(\beta_1 x + \beta_2 y - \omega t\right)\right] \end{Bmatrix} \tag{7.80}$$

The wave amplitudes are denoted by U, V, W_b, and W_s, respectively. The symbols β_1 and β_2 denote the longitudinal and transverse wave numbers, respectively. Furthermore, circular frequency dispersed waves are present. The equation presented is derived by inserting the values of u, v, w_b, and w_s from Eq. (7.80) into Eqs. (7.79a) through (7.79d).

$$\left([K]_{4\times4} - \omega^2 [M]_{4\times4}\right)\begin{bmatrix} U \\ V \\ W_b \\ W_s \end{bmatrix} = 0 \tag{7.81a}$$

The stiffness and mass matrices, denoted by K and M, respectively, have their constituent elements detailed in Appendix B [3].

In order to address the eigenvalue problem at hand, it is necessary to equate the determinant of the coefficient matrix on the left-hand side of Eq. (7.81a) to zero.

$$\left|[K]_{4\times4} - \omega^2 [M]_{4\times4}\right| = 0 \tag{7.81b}$$

TABLE 7.1
The Mechanical Properties
of Rhenium Foam [3]

E (GPa)	463
ρ (kg/m³)	21020
α (μm/(m·K))	6.2
ν	0.3

Moreover, it is possible for set $\beta_1 = \beta_2 = \beta$ to determine the phase velocity by dividing the wave frequency by the wave number, as expressed below:

$$c_p = \frac{\omega}{\beta} \qquad (7.81c)$$

The current study considers Rhenium foam as a constitutive material for a plate. The mechanical properties of Rhenium foam are presented in Table 7.1.

The diagrams presented in the context take into account a uniform thickness of the metal foam plate, specifically $h = 5$ cm. Section 7.2 will present the elastic foundation parameters for the purpose of facilitating comprehension.

$$K_u = \frac{k_u L^4}{X^*}, \quad K_l = \frac{k_l L^4}{X^*}, \quad K_s = \frac{k_s L^2}{X^*}, \quad X^* = \frac{E_1 h^3}{12\left(1-v^2\right)}$$

Asymmetric porosity distribution has been a subject of greater scrutiny in the diagrams presented due to the typical occurrence of uneven porosity distribution in porous materials. Moreover, this chapter has placed greater emphasis on the sinusoidal pattern of temperature increase [3].

The impact of the porosity coefficient on the fluctuation of wave frequency of a metal foam plate in relation to wave number at ($K_u = K_l = 30$, $K_s = 10$, $\Delta T = 30$) is demonstrated in Figure 7.9. It has been observed that the porosity coefficient has a negative impact on the frequency of waves, resulting in a decrease in their values. The observed phenomenon can be attributed to the inverse relationship between the stiffness of a structure and the porosity coefficient, whereby an increase in the latter leads to a decrease in the former. Furthermore, it is evident that an increase in wave number corresponds to a proportional increase in wave frequency. The utilization of a magnifier facilitates the precise depiction of diagrammatical trends. The magnified diagram suggests that the diagram exhibiting the highest porosity coefficient is associated with the lowest wave frequency value [3].

The purpose of plotting Figure 7.10 is to demonstrate the impact of the porosity coefficient on the phase velocity variation of a metal foam plate in relation to wave number, specifically for two scenarios: (I) a symmetric porosity distribution and (II) an asymmetric porosity distribution. This analysis was conducted at

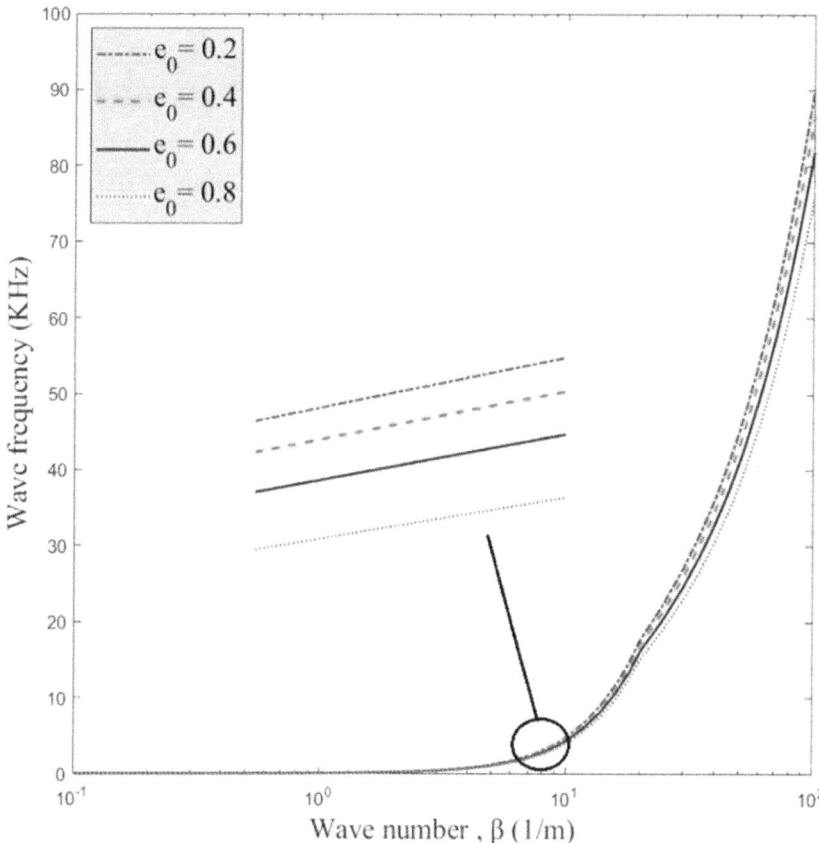

FIGURE 7.9 Variation of wave frequency versus wave number for different porosity coefficients (e_0) ($K_u = K_l = 30$, $K_s = 10$, $\Delta T = 30$).

specific values of $K_u = K_l = 30$, $K_s = 10$, $\Delta T = 30$, which are detailed within the figure. While symmetric and asymmetric distributions differ in their shape, they exhibit a similar pattern. The symmetric distribution of porosity results in a higher phase velocity value. The figure indicates that within the range of $0.1 < \beta < 1$, an increase in wave number results in a reduction of the phase velocity of the metal foam plate. Conversely, within the range of $1 < \beta < 23$, an increase in wave number leads to an increase in the phase velocity of the metal foam plate. Beyond $\beta = 23$, the phase velocity of the metal foam plate remains constant. Moreover, an increase in the porosity coefficient results in a reduction of the phase velocity value [3].

Figure 7.11 displays the frequency variation of wave propagation in a metal foam plate as a function of temperature change, for different types of temperature rise. The parameters used in the analysis are $K_u = K_l = 20$, $K_s = 5$, $e_0 = 0.2$, $\beta = 10$. It is evident that alterations in temperature have a diminishing impact, resulting in

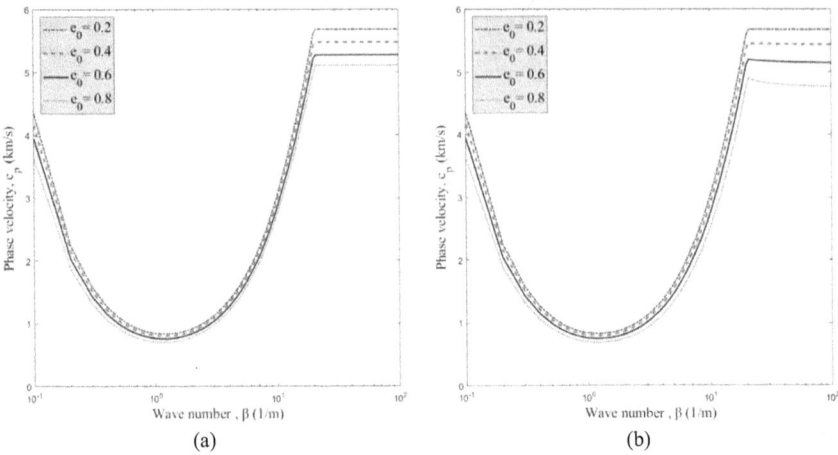

FIGURE 7.10   The effect of porosity coefficient on variation of phase velocity versus wave number for different kinds of porosity distribution (a) symmetric porosity distribution (b) asymmetric porosity distribution ($K_u = K_l = 30$, $K_s = 10$, $\Delta T = 30$).

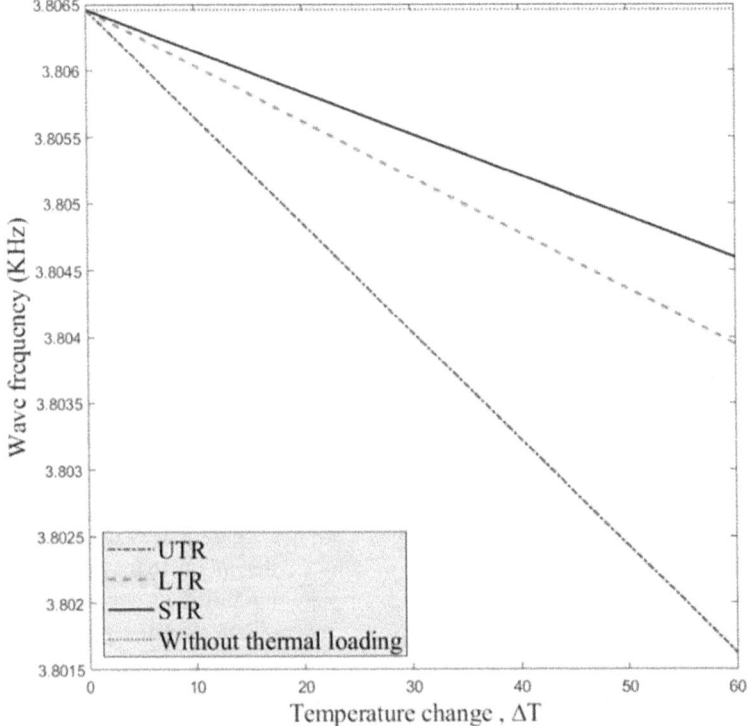

FIGURE 7.11   Variation of wave frequency versus temperature change for various types of temperature rise ($K_u = K_l = 20$, $K_s = 5$, $e_0 = 0.2$, $\beta = 10$).

a reduction of wave frequency values as temperature increases. Moreover, the curve remains unaltered in the absence of thermal loading. As the temperature increases, the stiffness of the structure decreases. The impact of temperature rise on wave frequency is more pronounced in cases where the temperature rise is uniform, as compared to other types of temperature rise. Additionally, it has been observed that the sinusoidal type of temperature rise has a relatively minor effect on the variability of wave frequency [3].

Figure 7.12 examines the impact of various temperature increases on the phase velocity variation of a metal foam plate in response to temperature changes. The investigation considers two scenarios: (I) a symmetric porosity distribution and (II) an asymmetric porosity distribution, with parameters set at ($K_u = K_l = 20$, $K_s = 5$, $e_0 = 0.2$, $\beta = 10$). Based on the depicted figure, it can be inferred that opting for a symmetric porosity distribution in the structure can result in an increase in phase velocity. As previously mentioned, alterations in temperature have a comparable impact on the reduction of phase velocity in metal foam plates. Similar to previous findings, it has been observed that the impact of temperature rise on the phase velocity is highest for uniform and sinusoidal patterns, while it is lowest for sinusoidal patterns [3].

Figure 7.13 depicts the frequency variation of wave propagation in a metal foam plate as a function of its length-to-thickness ratio, for different values of the linear layer of Kerr substrate (K_l, K_u), under the conditions of ($\Delta T = 40$, $K_s = 5$, $e_0 = 0.2$, $\beta = 5$). All curves exhibit a downward trend, whereby the frequency of waves decreases as the ratio of length to thickness increases. Furthermore, it is commonly understood that increasing the number of linear layers in a Kerr substrate leads to an increase in the corresponding wave frequency. To clarify, elevating the Kerr substrate value results in a heightened stiffness of the structure.

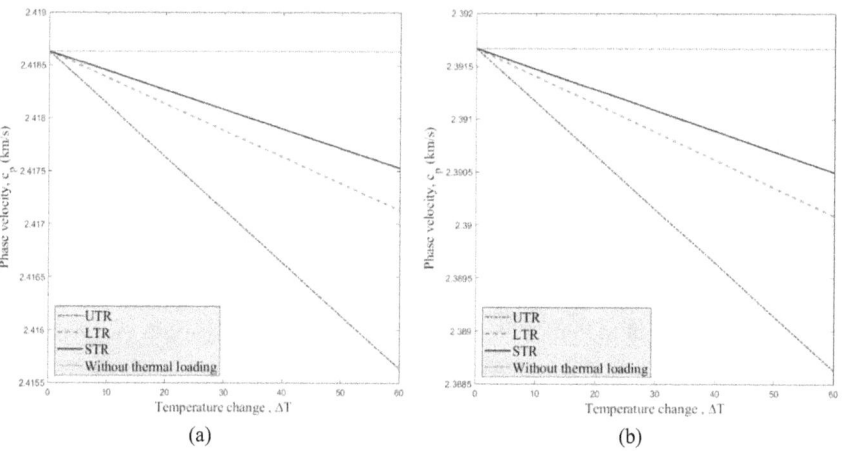

FIGURE 7.12 Variation of phase velocity versus temperature change for various types of temperature rise for different kinds of porosity distributions: (a) symmetric porosity distribution and (b) asymmetric porosity distribution ($K_u = K_l = 20$, $K_s = 5$, $e_0 = 0.2$, $\beta = 10$).

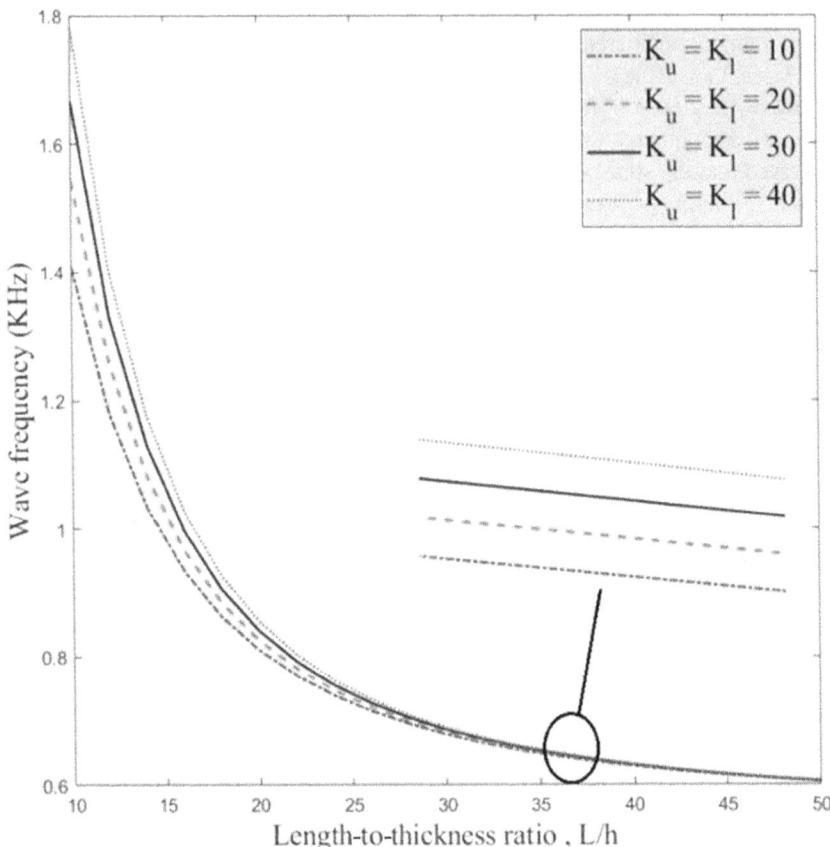

FIGURE 7.13 Variation of wave frequency versus length-to-thickness ratio for different amounts of linear layer of Kerr foundation ($\Delta T = 40$, $K_s = 5$, $e_0 = 0.2$, $\beta = 5$).

The illustration presented in Figure 7.14 exhibits the fluctuation in phase velocity of a metal foam plate in relation to wave number, while considering varying degrees of shear layer of Kerr substrate at a fixed set of parameters ($\Delta T = 40$, $K_l = K_u = 40$, $e_0 = 0.2$, $\beta = 5$). Evidently, an increase in the length-to-thickness ratio results in a reduction of phase velocity, characterized by a steep initial slope that gradually diminishes. Conversely, the shear layer of the Kerr substrate exhibits a growing impact akin to that of the linear layer of the Kerr substrate. A lower designation of the shear layer of the Kerr substrate results in a commensurate decrease in the phase velocity [3].

Figure 7.15 depicts the impact of various shape factors on the frequency of variation waves in a metal foam plate, as a function of the length-to-thickness ratio and for different wave numbers (β values of 1, 10, 20, and 100). The data was obtained under the conditions of ($\Delta T = 40$ (STR), $K_l = K_u = 40$, $K_s = 5$, $e_0 = 0.2$) The aforementioned diagrams have been plotted with meticulous attention to

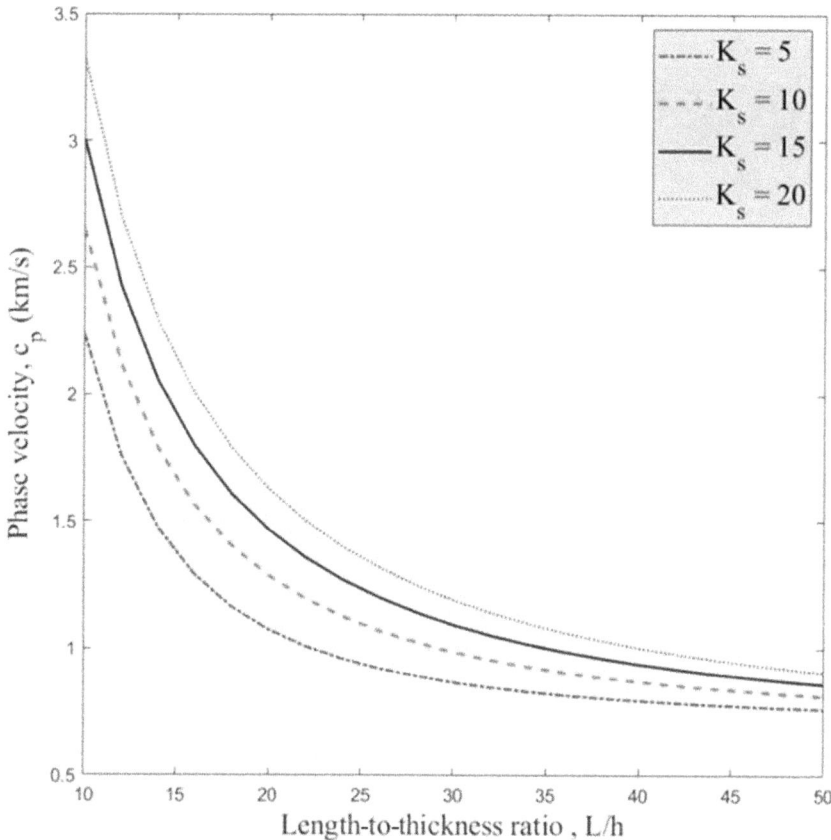

FIGURE 7.14 Illustration of the variation of phase velocity versus length-to-thickness ratio for different amounts of shear layer of Kerr foundation ($\Delta T = 40$, $K_l = K_u = 40$, $e_0 = 0.2$, $\beta = 5$).

ensure a symmetrical distribution of porosity. It is evident that when β equals 1, there are negligible variations among the different shape factors. Through the utilization of a magnifying instrument, it is discernible that the curve of the ICSDT exhibits a greater elevation than the remaining curves. Conversely, the Automated Substrate Design Tool (ASDT) displays the least amount of curvature when compared to the other aforementioned curves. In contrast to the prevailing trend, the curve of high-speed dynamic testing (HSDT) exhibits the highest value in comparison to the others. When β equals 20, the values of ICSDT, HSDT, and ASDT are higher compared to when β equals 1. At a beta value of 100, all curves exhibit identical values and the trend of the diagram remains constant. It is noteworthy to mention that in regard to diagrams, the values of HSDT and accelerated static dynamic testing curves exhibit close proximity to one another. Furthermore, as previously stated, the augmentation of the length-to-thickness ratio is associated with a reduction in effect [3].

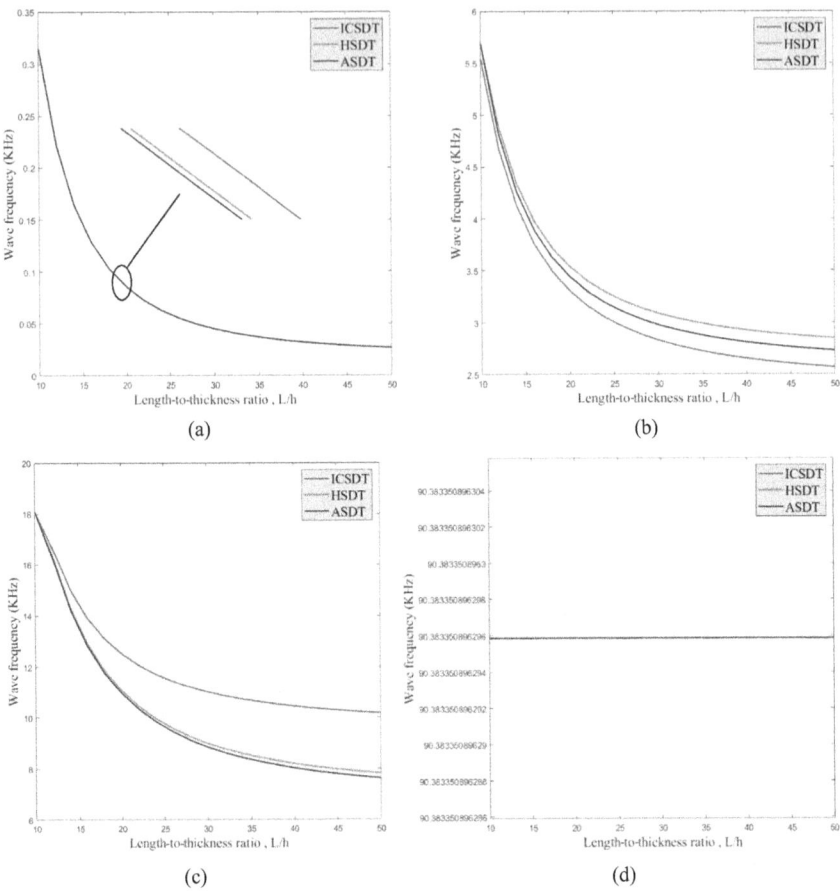

FIGURE 7.15   Variation of wave frequency versus length-to-thickness ratio for different shape functions and for different values of wave number (a) $\beta = 1$ (b) $\beta = 10$ (c) $\beta = 20$ and (d) $\beta = 100$ ($\Delta T = 40$, $K_l = K_u = 40$, $K_s = 5$, $e_0 = 0.2$).

7.4   GRAPHENE FOAM SHELL WAVE DISPERSION CHARACTERISTICS

The assumed structure depicted in Figure 7.16 is an embedded cylindrical shell characterized by its radius R, length L, and thickness h [5].

Figure 7.17 displays a schematic representation of the porosity distribution across the thickness of the cylindrical shell.

The present study considers the uniform porosity distribution (UPD), symmetric porosity distribution (SPD), and asymmetric porosity distribution (APD) as

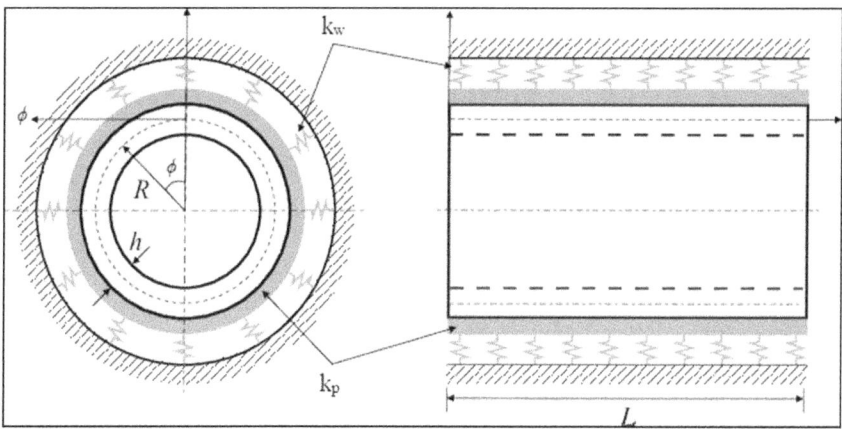

FIGURE 7.16 Geometry and coordinate systems of an embedded cylindrical shell.

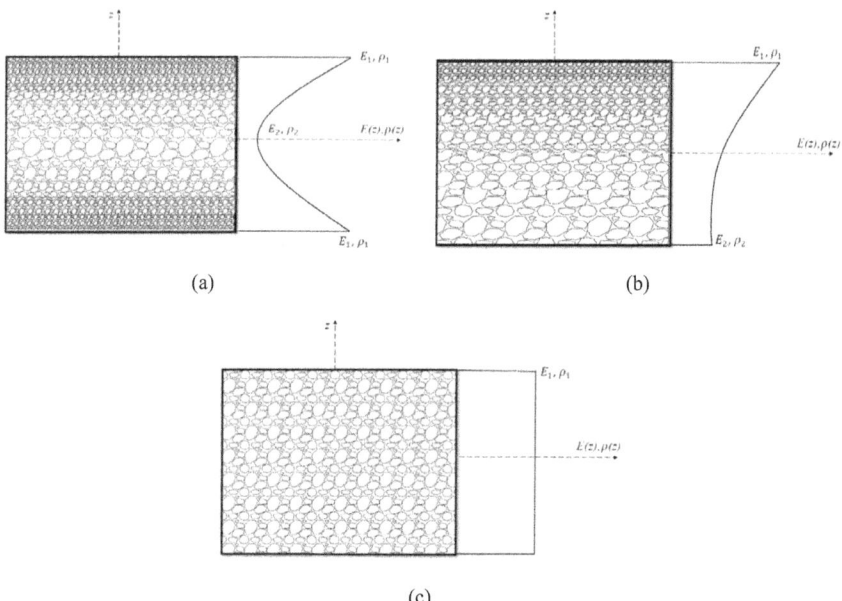

FIGURE 7.17 Schematic illustration of various porosity distributions through the thickness of shell for (a) SPD, (b) APD, and (c) UPD.

factors affecting the effective properties of porous metal foam. An investigation is conducted to analyze the impact of these factors on said properties.

$$E(z) = E_1 \left(1 - e_0 \eta \right) \tag{7.82a}$$

$$G(z) = G_1\left(1 - e_0\eta\right) \qquad (7.82b)$$

$$\rho(z) = \rho_1\sqrt{1 - e_0\eta} \qquad (7.82c)$$

The effective characteristics of a cylindrical shell made of symmetrically porous metal foam can be mathematically represented as:

$$E(z) = E_1\left(1 - e_0\cos\left(\frac{\pi z}{h}\right)\right) \qquad (7.83a)$$

$$G(z) = G_1\left(1 - e_0\cos\left(\frac{\pi z}{h}\right)\right) \qquad (7.83b)$$

$$\rho(z) = \rho_1\left(1 - e_m\cos\left(\frac{\pi z}{h}\right)\right) \qquad (7.83c)$$

Furthermore, the effective characteristics of a cylindrical shell made of metal foam with asymmetric porosity can be computed using the following method:

$$E(z) = E_1\left(1 - e_0\cos\left(\frac{\pi z}{2h} + \frac{\pi}{4}\right)\right) \qquad (7.84a)$$

$$G(z) = G_1\left(1 - e_0\cos\left(\frac{\pi z}{2h} + \frac{\pi}{4}\right)\right) \qquad (7.84b)$$

$$\rho(z) = \rho_1\left(1 - e_m\cos\left(\frac{\pi z}{2h} + \frac{\pi}{4}\right)\right) \qquad (7.84c)$$

The symbols E_1, G_1, and ρ_1 denote Young's modulus, shear modulus, and mass density, respectively, of the porous metal foam at its maximum value. The variables E_2, G_2, and ρ_2 represent the minimum values of the aforementioned parameters. Moreover, the coefficients e_0 and e_m pertain to porosity and density, respectively, and can be precisely defined as follows:

$$e_0 = 1 - \frac{E_2}{E_1} = 1 - \frac{G_2}{G_1}, \quad (0 < e_0 < 1) \qquad (7.85)$$

$$e_m = 1 - \frac{\rho_2}{\rho_1} \qquad (7.86)$$

Furthermore, the value of *em* can be computed based on the customary mechanical characteristics of an open-cell metallic foam, such as:

$$\frac{E_2}{E_1} = \left(\frac{\rho_2}{\rho_1}\right)^2 \tag{7.87}$$

$$e_m = 1 - \sqrt{1 - e_0} \tag{7.88}$$

In this approach, the constancy of Poisson's ratio can be assumed due to its negligible variations. The expression η denoted by Eqs. (7.81)–(7.83) can be expressed as per reference [5].

$$\eta = \frac{1}{e_0}\left(1 - \left[\frac{2}{\pi}\left(\sqrt{1 - e_0} - 1\right) + 1\right]^2\right) \tag{7.89}$$

According to the Finite Strip Displacement Theory, the kinematic relationships of a circular cylindrical shell can be obtained through the equations presented in reference [5].

$$u_x\left(x,\phi,z,t\right) = u\left(x,\phi,t\right) + z\theta_x\left(x,\phi,t\right) \tag{7.90}$$

$$u_\phi\left(x,\phi,z,t\right) = v\left(x,\phi,t\right) + z\theta_\phi\left(x,\phi,t\right) \tag{7.91}$$

$$u_z\left(x,\phi,z,t\right) = w\left(x,\phi,t\right) \tag{7.92}$$

The variables *u*, *v*, and *w* represent axial, circumferential, and lateral displacements, correspondingly. Additionally, it should be noted that θ_x and θ_ϕ represent the axial and circumferential rotation components, correspondingly. Therefore, the strains that are not equal to zero for an element of shell type can be expressed as:

$$\begin{Bmatrix} \varepsilon_{xx} \\ \varepsilon_{\phi\phi} \\ \varepsilon_{x\phi} \\ \varepsilon_{xz} \\ \varepsilon_{\phi z} \end{Bmatrix} = \begin{Bmatrix} \dfrac{\partial u}{\partial x} + z\dfrac{\partial \theta_x}{\partial x} \\[2mm] \dfrac{1}{R}\left(\dfrac{\partial v}{\partial \phi} + z\dfrac{\partial \theta_\phi}{\partial \phi} + w\right) \\[2mm] \dfrac{1}{R}\dfrac{\partial u}{\partial \phi} + \dfrac{\partial v}{\partial x} + \dfrac{z}{R}\dfrac{\partial \theta_x}{\partial \phi} + z\dfrac{\partial \theta_\phi}{\partial x} \\[2mm] \theta_x + \dfrac{\partial w}{\partial x} \\[2mm] \theta_\phi + \dfrac{1}{R}\dfrac{\partial w}{\partial \phi} - \dfrac{v}{R} \end{Bmatrix} \tag{7.93}$$

Subsequently, the application of Hamilton's principle was utilized to derive the Euler–Lagrange equations for a nanocomposite shell.

$$\int_0^t \delta\left(U - K - W\right) dt = 0 \tag{7.94}$$

The variables U, K, and W are utilized to denote the strain and kinetic energies, and the work executed by the external force, correspondingly. The expression for the change in strain energy of a linear elastic solid can be represented as follows:

$$\delta U = \int_{-\frac{h}{2}}^{\frac{h}{2}} \int_0^{2\pi} \int_0^L \sigma_{ij} \delta\varepsilon_{ij} R\,dx\,d\phi\,dz \tag{7.95}$$

Moreover, the alteration of kinetic energy can be formulated as:

$$\delta K = \int_{-\frac{h}{2}}^{\frac{h}{2}} \int_0^{2\pi} \int_0^L \rho(z)\left[\left(\frac{\partial \delta u_x}{\partial t}\right)^2 + \left(\frac{\partial \delta u_\phi}{\partial t}\right)^2 + \left(\frac{\partial \delta u_z}{\partial t}\right)^2\right] R\,dx\,d\phi\,dz \tag{7.96}$$

The formulation for the variation of work done by an external force can be expressed as follows:

$$\delta W = \int_{-\frac{h}{2}}^{\frac{h}{2}} \int_0^{2\pi} \int_0^L \left\{N_r + N_x \frac{\partial^2 w}{\partial x^2} + \frac{N_\phi}{R^2}\frac{\partial^2 w}{\partial \phi^2} - k_w + k_p\left(\frac{\partial^2 w}{\partial x^2} + \frac{1}{R^2}\frac{\partial^2 w}{\partial \varphi^2}\right)\right\}\delta w R\,dx\,d\phi\,dz \tag{7.97}$$

The variables N_r, N_x, and N_ϕ denote the loadings in the radial, axial, and circumferential directions, respectively. Loading was not taken into consideration in the current study. Furthermore, the symbols k_w and k_p represent the coefficients of Winkler and Pasternak, correspondingly. Thus, in order to derive the equation of motion for a circular cylindrical shell, it is necessary to substitute Eqs. (7.95) through (7.97) into Eq. (7.94).

$$\frac{\partial N_{xx}}{\partial x} + \frac{1}{R}\frac{\partial N_{x\phi}}{\partial \phi} = I_0 \frac{\partial^2 u}{\partial t^2} + I_1 \frac{\partial^2 \theta_x}{\partial t^2} \tag{7.98}$$

$$\frac{\partial N_{x\phi}}{\partial x} + \frac{1}{R}\frac{\partial N_{\phi\phi}}{\partial \phi} + \frac{Q_{z\phi}}{R} = I_0 \frac{\partial^2 v}{\partial t^2} + I_1 \frac{\partial^2 \theta_\phi}{\partial t^2} \tag{7.99}$$

$$\frac{\partial Q_{xz}}{\partial x} + \frac{1}{R}\frac{\partial Q_{z\phi}}{\partial \phi} - \frac{N_{\phi\phi}}{R} - k_w + k_p\left(\frac{\partial^2 w}{\partial x^2} + \frac{1}{R^2}\frac{\partial^2 w}{\partial \varphi^2}\right) = I_0 \frac{\partial^2 w}{\partial t^2} \tag{7.100}$$

$$\frac{\partial M_{xx}}{\partial x} + \frac{1}{R}\frac{\partial M_{x\phi}}{\partial \phi} - Q_{xz} = I_1 \frac{\partial^2 u}{\partial t^2} + I_2 \frac{\partial^2 \theta_x}{\partial t^2} \tag{7.101}$$

$$\frac{\partial M_{x\phi}}{\partial x} + \frac{1}{R}\frac{\partial M_{\phi\phi}}{\partial \phi} - Q_{\phi z} = I_1 \frac{\partial^2 v}{\partial t^2} + I_2 \frac{\partial^2 \theta_\phi}{\partial t^2} \tag{7.102}$$

where

$$\begin{bmatrix} N_{xx} & N_{\phi\phi} & N_{x\phi} \\ M_{xx} & M_{\phi\phi} & M_{x\phi} \end{bmatrix} = \int_{-\frac{h}{2}}^{\frac{h}{2}} \begin{bmatrix} 1 \\ z \end{bmatrix} \begin{bmatrix} \sigma_{xx} & \sigma_{\phi\phi} & \sigma_{x\phi} \end{bmatrix} dz \tag{7.103}$$

$$\begin{bmatrix} Q_{xz} & Q_{z\phi} \end{bmatrix} = \kappa_s \int_{-\frac{h}{2}}^{\frac{h}{2}} \begin{bmatrix} \sigma_{xz} & \sigma_{z\phi} \end{bmatrix} dz \tag{7.104}$$

$$\begin{bmatrix} I_0 & I_1 & I_2 \end{bmatrix} = \int_{-\frac{h}{2}}^{\frac{h}{2}} \rho(z) \begin{bmatrix} 1 & z & z^2 \end{bmatrix} dz \tag{7.105}$$

The value of κ_s, which represents the shear correction factor, is assumed to be equal to $\frac{\pi^2}{12}$. The stress-strain correlation of a porous metal foam may be expressed as follows:

$$\sigma_{ij} = C_{ijkl}\varepsilon_{kl} \tag{7.106}$$

The symbols σ_{ij}, ε_{kl}, and C_{ijkl} denote the constituent elements of the Cauchy stress, strain, and elasticity tensors, correspondingly. By performing integration of the aforementioned equation across the shell thickness, the subsequent relationship can be derived:

$$
\begin{bmatrix} N_{xx} \\ M_{xx} \\ N_{\phi\phi} \\ M_{\phi\phi} \end{bmatrix} = \begin{bmatrix} A_{11} & B_{11} & \dfrac{A_{12}}{R} & \dfrac{B_{12}}{R} \\[2mm] B_{11} & D_{11} & \dfrac{B_{12}}{R} & \dfrac{D_{12}}{R} \\[2mm] A_{12} & B_{12} & \dfrac{A_{11}}{R} & \dfrac{B_{11}}{R} \\[2mm] B_{12} & D_{12} & \dfrac{B_{11}}{R} & \dfrac{D_{11}}{R} \end{bmatrix} \begin{bmatrix} \dfrac{\partial u}{\partial x} \\[2mm] \dfrac{\partial \theta_x}{\partial x} \\[2mm] \dfrac{\partial v}{\partial \phi}+w \\[2mm] \dfrac{\partial \theta_\phi}{\partial \phi} \end{bmatrix}, \begin{bmatrix} N_{x\phi} \\ M_{x\phi} \end{bmatrix}
$$
$$(7.107)$$

$$
= \begin{bmatrix} A_{66} & B_{66} \\ B_{66} & D_{66} \end{bmatrix} \begin{bmatrix} \dfrac{1}{R}\dfrac{\partial u}{\partial \phi}+\dfrac{\partial v}{\partial x} \\[2mm] \dfrac{1}{R}\dfrac{\partial \theta_x}{\partial \phi}+\dfrac{\partial \theta_\phi}{\partial x} \end{bmatrix}'
$$

$$
Q_{\phi z} = A_{44}^s\left(\theta_\phi + \frac{1}{R}\frac{\partial w}{\partial \phi} - \frac{v}{R}\right), Q_{xz} = A_{55}^s\left(\theta_x + \frac{\partial w}{\partial x}\right)
$$

where

$$
\begin{bmatrix} A_{11} & B_{11} & D_{11} \\ A_{12} & B_{12} & D_{12} \\ A_{66} & B_{66} & D_{66} \end{bmatrix} = \int_{-\frac{h}{2}}^{\frac{h}{2}} \begin{bmatrix} \dfrac{E(z)}{1-v^2} \\[2mm] \dfrac{vE(z)}{1-v^2} \\[2mm] \dfrac{E(z)}{2(1+v)} \end{bmatrix} \begin{bmatrix} 1 & z & z^2 \end{bmatrix} dz
$$
$$(7.108)$$

$$
A_{44}^s = A_{55}^s = \kappa_s \int_{-\frac{h}{2}}^{\frac{h}{2}} \frac{E(z)}{2(1+v)}\, dz
$$
$$(7.109)$$

The governing equations of a metal foam circular cylindrical shell can be derived by combining Eqs. (7.98)–(7.102) with Eq. (7.107).

$$
A_{11}\frac{\partial^2 u}{\partial x^2} + B_{11}\frac{\partial^2 \theta_x}{\partial x^2} + \frac{A_{12}}{R}\left(\frac{\partial^2 v}{\partial x \partial \phi} + \frac{\partial w}{\partial x}\right) + \frac{B_{12}}{R}\frac{\partial^2 \theta_\phi}{\partial x \partial \phi} +
$$
$$
\frac{A_{66}}{R}\left(\frac{1}{R}\frac{\partial^2 u}{\partial \phi^2} + \frac{\partial^2 v}{\partial x \partial \phi}\right) + \frac{B_{66}}{R}\left(\frac{1}{R}\frac{\partial^2 \theta_x}{\partial \phi^2} + \frac{\partial^2 \theta_\phi}{\partial x \partial \phi}\right) - I_0\frac{\partial^2 u}{\partial t^2} - I_1\frac{\partial^2 \theta_x}{\partial t^2} = 0
$$
$$(7.110)$$

$$A_{66}\left(\frac{1}{R}\frac{\partial^2 u}{\partial x\partial\phi}+\frac{\partial^2 v}{\partial x^2}\right)+B_{66}\left(\frac{1}{R}\frac{\partial^2\theta_x}{\partial x\partial\phi}+\frac{\partial^2\theta_\varphi}{\partial x^2}\right)+\frac{A_{12}}{R}\frac{\partial^2 u}{\partial x\partial\phi}+\frac{B_{12}}{R}\frac{\partial^2\theta_x}{\partial x\partial\phi}$$

$$+\frac{A_{11}}{R^2}\left(\frac{\partial^2 v}{\partial\phi^2}+\frac{\partial w}{\partial\phi}\right)+\frac{B_{11}}{R^2}\frac{\partial^2\theta_\phi}{\partial\phi^2}+\frac{A_{55}^s}{R}\left(\theta_\phi+\frac{1}{R}\frac{\partial w}{\partial\phi}-\frac{v}{R}\right)-I_0\frac{\partial^2 v}{\partial t^2}-I_1\frac{\partial^2\theta_\phi}{\partial t^2}=0$$

$$(7.111)$$

$$A_{55}^s\left(\frac{\partial\theta_x}{\partial x}+\frac{\partial^2 w}{\partial x^2}\right)+\frac{A_{55}^s}{R}\left(\frac{\partial\theta_\phi}{\partial\phi}+\frac{1}{R}\frac{\partial^2 w}{\partial\phi^2}-\frac{1}{R}\frac{\partial v}{\partial\phi}\right)-\frac{A_{12}}{R}\frac{\partial u}{\partial x}$$

$$-\frac{B_{12}}{R}\frac{\partial\theta_x}{\partial x}-\frac{A_{11}}{R^2}\left(\frac{\partial v}{\partial\phi}+w\right)-\frac{B_{11}}{R^2}\frac{\partial\theta_\phi}{\partial\phi}-k_w+k_p\left(\frac{\partial^2 w}{\partial x^2}+\frac{1}{R^2}\frac{\partial^2 w}{\partial\varphi^2}\right)-I_0\frac{\partial^2 w}{\partial t^2}=0$$

$$(7.112)$$

$$B_{11}\frac{\partial^2 u}{\partial x^2}+D_{11}\frac{\partial^2\theta_x}{\partial x^2}+\frac{B_{12}}{R}\left(\frac{\partial^2 v}{\partial x\partial\phi}+\frac{\partial w}{\partial x}\right)+\frac{D_{12}}{R}\frac{\partial^2\theta_\phi}{\partial x\partial\phi}+\frac{B_{66}}{R}\left(\frac{1}{R}\frac{\partial^2 u}{\partial\phi^2}+\frac{\partial^2 v}{\partial x\partial\phi}\right)$$

$$+\frac{D_{66}}{R}\left(\frac{1}{R}\frac{\partial^2\theta_x}{\partial\phi^2}+\frac{\partial^2\theta_\phi}{\partial x\partial\phi}\right)-A_{55}^s\left(\theta_x+\frac{\partial w}{\partial x}\right)-I_1\frac{\partial^2 u}{\partial t^2}-I_2\frac{\partial^2\theta_x}{\partial t^2}=0$$

$$(7.113)$$

$$B_{66}\left(\frac{1}{R}\frac{\partial^2 u}{\partial x\partial\phi}+\frac{\partial^2 v}{\partial x^2}\right)+D_{66}\left(\frac{1}{R}\frac{\partial^2\theta_x}{\partial x\partial\phi}+\frac{\partial^2\theta_\phi}{\partial x^2}\right)+\frac{B_{12}}{R}\frac{\partial^2 u}{\partial x\partial\phi}+\frac{D_{12}}{R}\frac{\partial^2\theta_x}{\partial x\partial\phi}$$

$$+\frac{B_{11}}{R^2}\left(\frac{\partial^2 v}{\partial\phi^2}+\frac{\partial w}{\partial\phi}\right)+\frac{D_{11}}{R^2}\frac{\partial^2\theta_\phi}{\partial\phi^2}-A_{55}^s\left(\theta_\varphi+\frac{1}{R}\frac{\partial w}{\partial\phi}-\frac{v}{R}\right)-I_1\frac{\partial^2 v}{\partial t^2}-I_2\frac{\partial^2\theta_\phi}{\partial t^2}=0$$

$$(7.114)$$

The analytical solution for the governing equations of a circular cylindrical shell made of metal foam is expected to be exponential in nature. Consequently, the assumed displacement field was expressed as:

$$\begin{Bmatrix} u \\ v \\ w \\ \theta_x \\ \theta_\phi \end{Bmatrix}=\begin{Bmatrix} U\exp\left[i\left(k_x x+k_n\phi-\omega_n t\right)\right] \\ V\exp\left[i\left(k_x x+k_n\phi-\omega_n t\right)\right] \\ W\exp\left[i\left(k_x x+k_n\phi-\omega_n t\right)\right] \\ \Theta_x\exp\left[i\left(k_x x+k_n\phi-\omega_n t\right)\right] \\ \Theta_\phi\exp\left[i\left(k_x x+k_n\phi-\omega_n t\right)\right] \end{Bmatrix}\qquad(7.115)$$

The amplitudes of displacement in a given system are represented by U, V, and W, while the amplitudes of rotation are denoted by Θ_x and Θ_ϕ. Additionally, it can be observed that k_x and k_n denote the longitudinal and circumferential wave numbers, correspondingly. The symbol ω_n denotes the circular frequency. Subsequently, through the utilization of the variables u, v, w, θ_x, and θ_ϕ as replacements for Eq. (7.115) in Eqs. (7.110)–(7.114), the ensuing equation can be obtained:

$$\left(\left[K \right]_{5\times5} - \omega_n^2 \left[M \right]_{5\times5} \right) \begin{bmatrix} U \\ V \\ W \\ \Theta_x \\ \Theta_\phi \end{bmatrix} = 0 \tag{7.116}$$

The matrices K and M represent stiffness and mass, respectively. The constituent elements of said matrices are provided in Appendix C [5].

In order to obtain the natural frequency and solve the eigenvalue problem, it is necessary to equate the determinant of the coefficient matrix on the left-hand side of Eq. (7.116) to 0.

$$\left| \left[K \right]_{5\times5} - \omega_n^2 \left[M \right]_{5\times5} \right| = 0 \tag{7.117}$$

The calculation of wave frequency involves the division of circular frequency by 2π. Moreover, the phase velocity can be acquired by means of:

$$c = \frac{\omega_n}{k_n} \tag{7.118}$$

The elastic foundation parameters will be defined below for the purpose of facilitating comprehension.

$$K_W = \frac{k_w L^4}{X^*}, \quad K_p = \frac{k_p L^2}{X^*}, \quad X^* = \frac{E_1 h^3}{12\left(1-v^2\right)} \tag{7.119}$$

Furthermore, it can be observed from Figure 7.18 that the cylindrical shell composed of metal foam is in contact with a Winkler foundation that exhibits variable linear, parabolic, and sinusoidal parameters. The Winkler parameter can be introduced for various distributions in the following manner [5]:

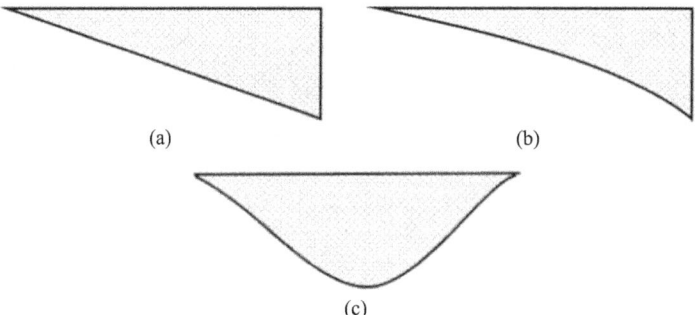

FIGURE 7.18 Variation of Winkler foundation in axial direction. (a) linear type, (b) parabolic type, and (c) sinusoidal type.

$$\bar{K}_w = \begin{cases} K_w\left(1-\xi\,\dfrac{x}{L}\right) & \text{Linear type} \\[2ex] K_w\left(1-\xi\left(\dfrac{x}{L}\right)^2\right) & \text{Parabolic type} \\[2ex] K_w\left(1-\xi\sin\left(\dfrac{\pi x}{L}\right)\right) & \text{Sinusoidal type} \end{cases} \qquad (7.120)$$

This section presents an analysis of the effects of various parameters on the wave frequency and phase velocity of a cylindrical shell made of metal foam. The analysis is presented through a series of diagrams. The present study examines the mechanical properties of four types of metal foam, namely magnesium, nickel, titanium, and tungsten, when utilized as constitutive material for cylindrical shells. Table 7.2 provides the relevant mechanical properties of the aforementioned metal foams. The diagrams presented depict a consistent thickness of 0.05 m for the cylindrical shell composed of metal foam. It is noteworthy that the diagrams pertain to four distinct types of metal foam [5]. Figure 7.19 displays the relationship between wave frequency and circumferential wave number across different porosity coefficients. It is evident that the values of wave frequency exhibit an increase as the circumferential wave number is incremented. Additionally, it can be observed that the porosity coefficient exhibits a negative correlation with wave frequency, indicating an inverse relationship between these two variables. The presence of pores and voids within a structure results in reduced strength and diminished stiffness. This is the underlying cause for the observed structural behavior. Magnesium exhibits the highest wave frequency value in comparison to other elements, while titanium displays the lowest wave frequency value [5].

TABLE 7.2

The Mechanical Properties of Metal Foams [5]

Mechanical Properties	Magnesium	Nickel	Titanium	Tungsten
E (GPa)	45	21	116	411
ρ (kg/m³)	1740	8908	4506	19300
ν	0.35	0.31	0.33	0.28

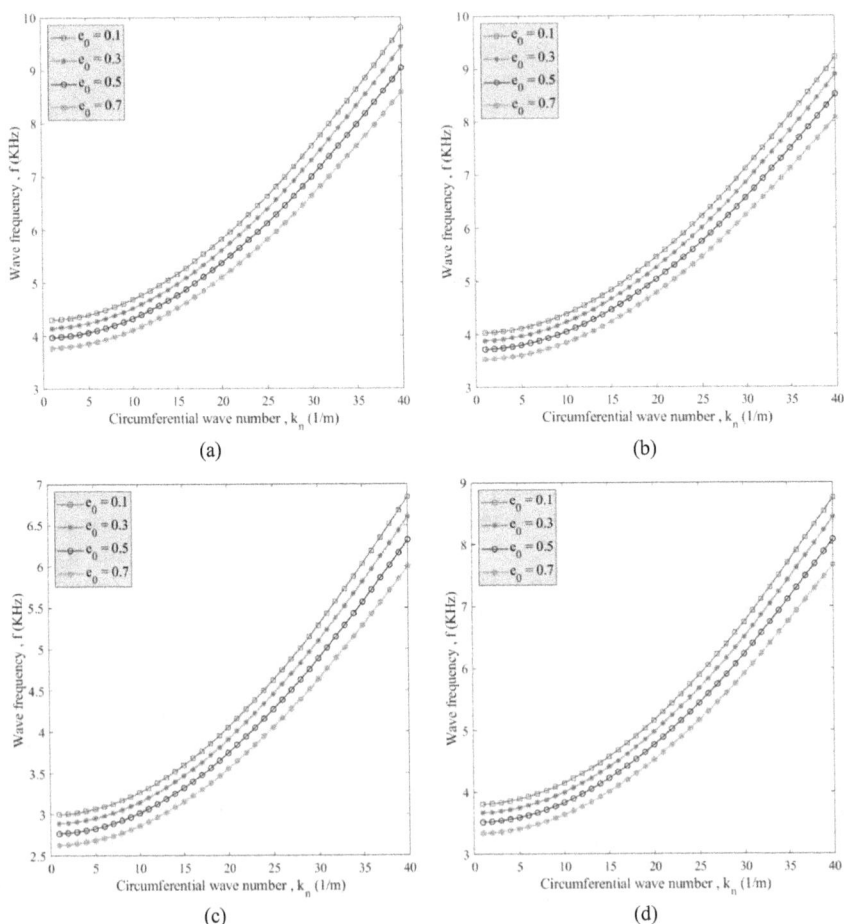

(a)   (b)

(c)   (d)

FIGURE 7.19   Variation of wave frequency versus circumferential wave number for various porosity coefficients $\left(\dfrac{L}{h} = \dfrac{R}{h} = 30, k_w = 50 \left(\text{sinusoidal} \right), k_p = 10, k_x = 20, \text{and } \xi = 0.8, \text{UPD} \right)$.

(a) Magnesium, (b) nickel, (c) titanium, and (d) tungsten.

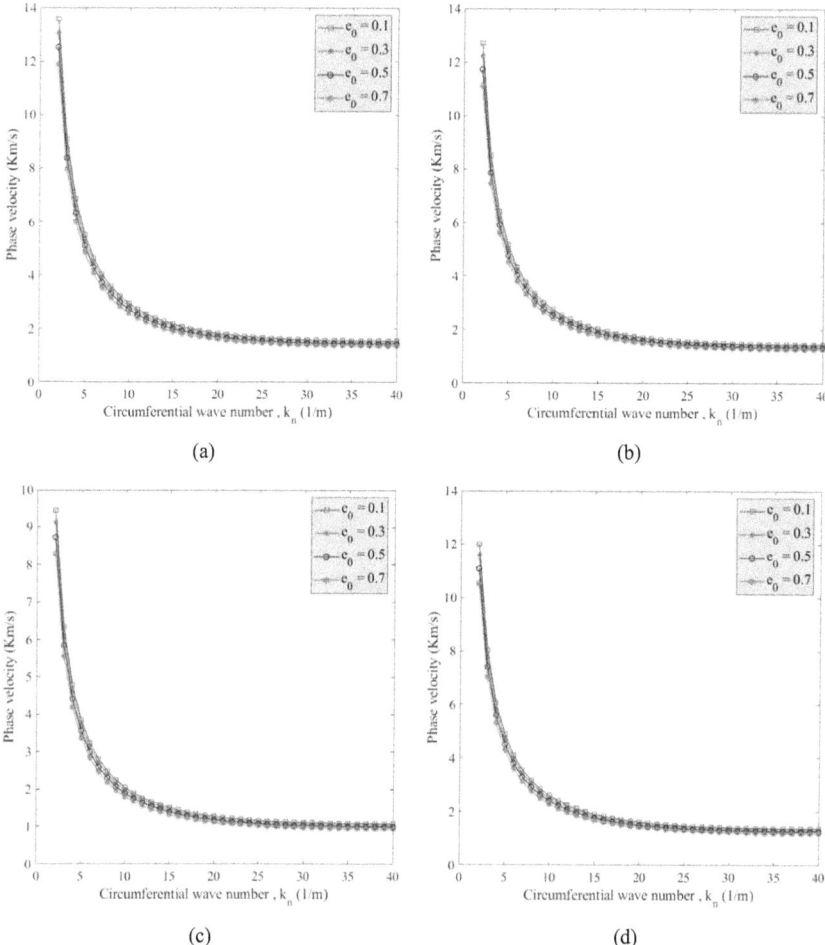

FIGURE 7.20 Variation of phase velocity versus circumferential wave number for various porosity coefficients $\left(\dfrac{L}{h} = \dfrac{R}{h} = 30, k_w = 50(\text{sinusoidal}), k_p = 10, k_x = 20, \text{and } \xi = 0.8, \text{UPD}\right)$. (a) Magnesium, (b) nickel, (c) titanium, and (d) tungsten.

Figure 7.20 illustrates the relationship between porosity coefficients and the variation of phase velocity with circumferential wave number. The diagrams indicate that as the circumferential wave number increases, there is a significant reduction in phase velocity. Subsequently, the decreasing slope gradually diminishes until the wave frequency reaches a specific value, at which point it stabilizes. Furthermore, the diagrams illustrate the adverse impact of the porosity coefficient, whereby a reduction in the porosity coefficient results in an increase in phase velocity values. Similar to the preceding illustration, it can be observed that magnesium and titanium foams exhibit the maximum and minimum phase velocity values, respectively [5].

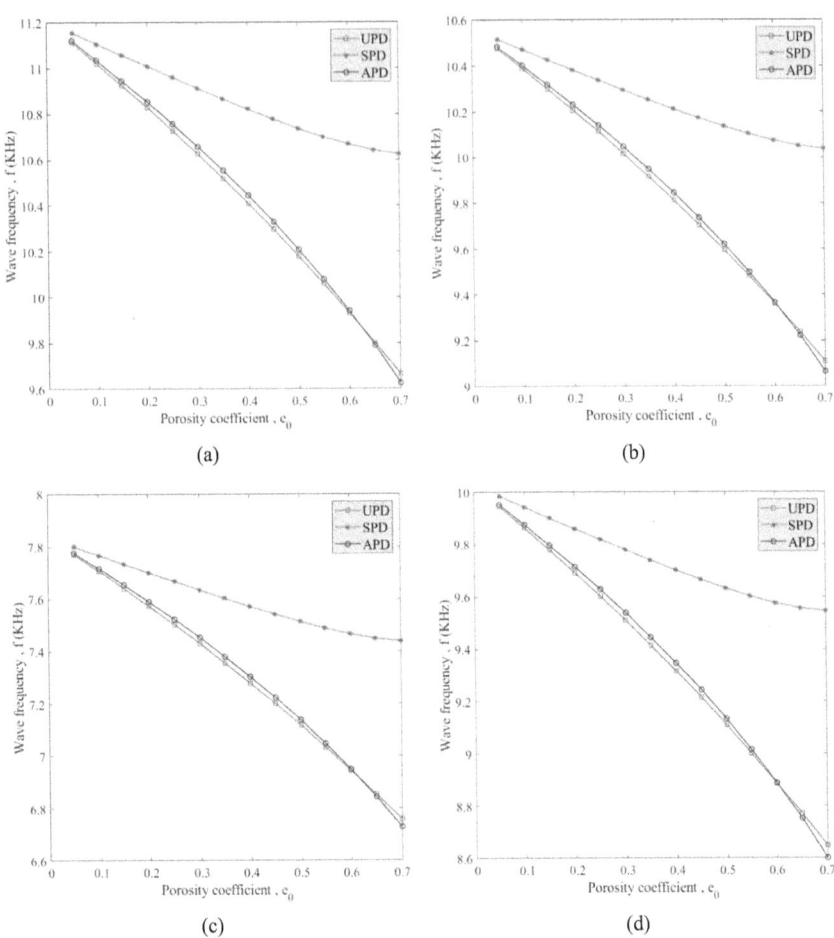

FIGURE 7.21 Variation of wave frequency versus porosity coefficient for various porosity distributions $\left(\dfrac{L}{h} = \dfrac{R}{h} = 30, k_w = 40 \left(\text{constant}\right), k_p = 10, k_x = k_n = 30, \text{and} \xi = 0.8 \right)$. (a) Magnesium, (b) nickel, (c) titanium, and (d) tungsten. [5]

Figure 7.21 has been constructed to examine the impact of porosity distribution on the fluctuation of wave frequency in relation to the porosity coefficient. Based on the presented diagrams, it can be observed that the impact of SPD on wave frequency values is comparatively lower than that of other factors. The diagrams present a noteworthy aspect. When the value of e0 falls within the range of 0.05 to 0.6, the wave frequency is observed to be at its minimum for UPD. However, beyond this range, APD takes over from UPD, indicating that for higher values of porosity coefficient, the cylindrical shell with APD exhibits the lowest wave frequency. As previously stated, the coefficient of porosity exerts a detrimental

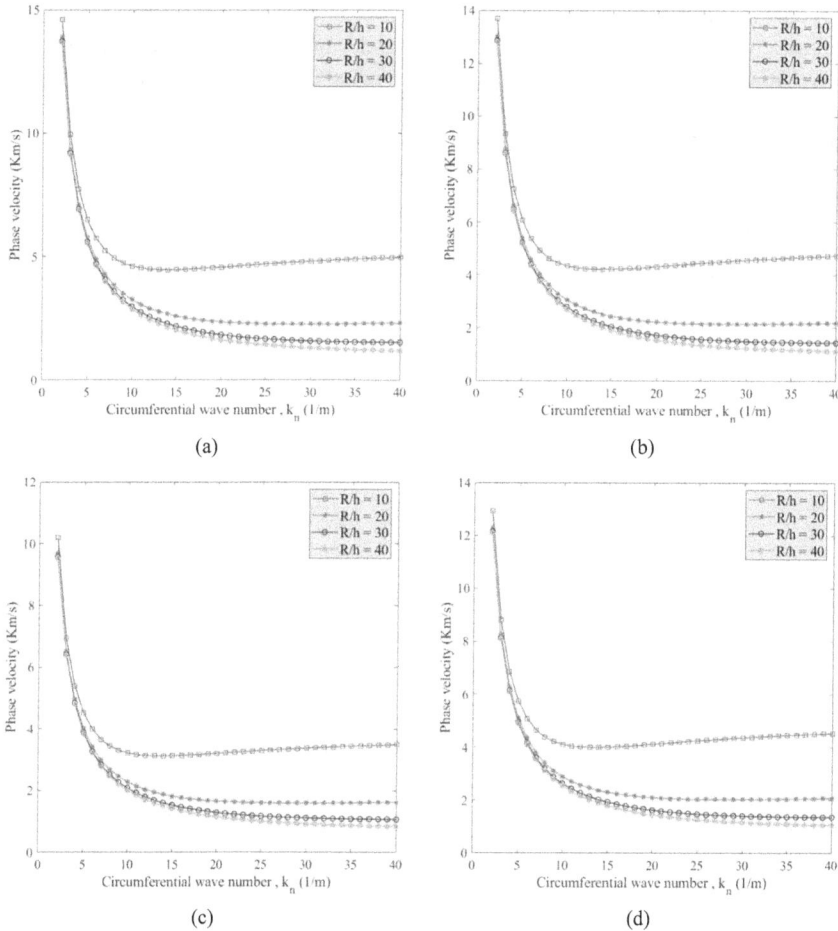

FIGURE 7.22 Variation of phase velocity versus circumferential wave number for various radius-to-thickness ratios ($\frac{L}{h} = 30, k_w = 60\,(\text{sinusoidal}), e_0 = 0.4, k_p = 30, k_x = 20$, and $\xi = 0.8, \text{SPD}$). (a) Magnesium, (b) nickel, (c) titanium, and (d) tungsten.

impact on the values of wave frequency. Moreover, magnesium exhibits the maximum wave frequency, succeeded by Nickel, tungsten, and titanium in that order.

Figure 7.22 illustrates the impact of the ratio of radius to thickness (R/h) on the fluctuation of phase velocity in relation to circumferential wave number. It can be observed that as the ratio of radius to thickness increases, there is a reduction in phase velocity. The observed behavior can be attributed to the mitigating impact of an increase in the ratio of radius to thickness. Conversely, an increase in the ratio of radius to thickness has a diminishing effect on the fluctuation of phase velocity measurements. It is noteworthy that at a Reynolds number to hydraulic diameter ratio of 10, the phase velocity experiences a reduction with a steep slope.

Beyond a certain circumferential wave number, the phase velocity reaches its minimum value. Subsequently, the phase velocity increases as the circumferential wave number increases. However, in certain instances, the trend exhibited by the curves is entirely downward. Titanium, tungsten, nickel, and magnesium exhibit the highest phase velocity in that order [5].

The influence of the length-to-thickness ratio (L/h) on the variation of phase velocity with respect to circumferential wave number is demonstrated in Figure 7.23. The phenomenon of acousto-elasticity is characterized by an increase in wave frequency values as the circumferential wave number grows and the length-to-thickness ratio amount decreases. In other terms, the increase in circumferential wave number and length-to-thickness ratio has a positive and negative

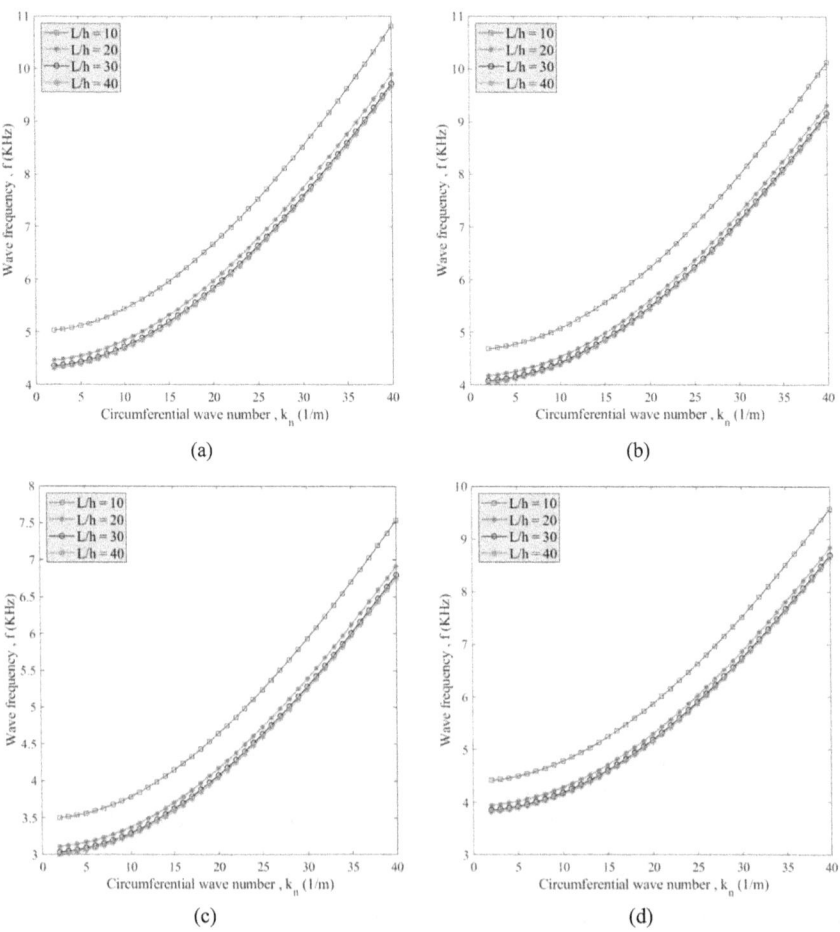

FIGURE 7.23   Variation of wave frequency versus circumferential wave number for various length-to-thickness ratios ($\frac{R}{h} = 30, k_w = 50(\text{parabolic}), e_0 = 0.4, k_p = 25, k_x = 20,$ and $\xi = 0.8$, SPD). (a) Magnesium, (b) nickel, (c) titanium, and (d) tungsten.

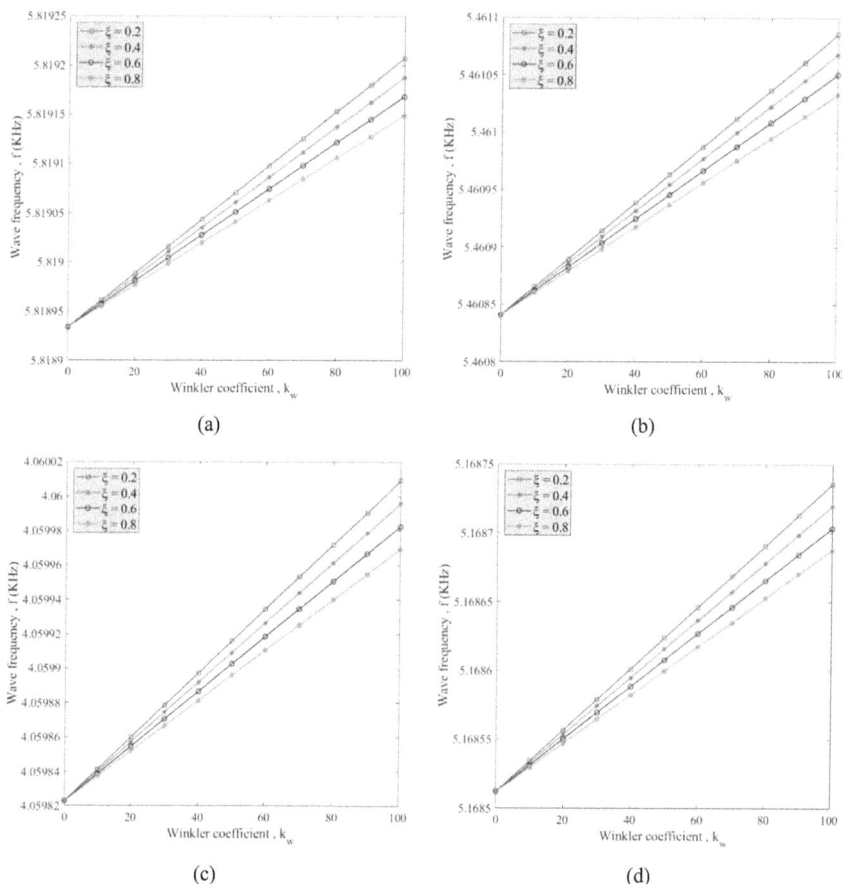

FIGURE 7.24 Variation of wave frequency versus Winkler coefficient for various ξ parameters $\left(\dfrac{L}{h} = \dfrac{R}{h} = 30, k_w \left(\text{parabolic} \right), e_0 = 0.4, k_p = 20, k_x = k_n = 20, \text{SPD} \right)$. (a) Magnesium, (b) nickel, (c) titanium, and (d) tungsten.

impact, respectively, on the variation of phase velocity. In addition, the decrease in the ratio of length to thickness results in an increase in structural rigidity, leading to a higher frequency of waves at a lower length-to-thickness ratio. The findings indicate that the cylindrical shell exhibiting a lower radius-to-thickness ratio and length-to-thickness ratio, as depicted in Figures 7.22 and 7.23, is capable of withstanding higher wave frequency and phase velocity [5].

Figures 7.24 and 7.25 demonstrate the impact of the ξ parameter on the fluctuation of wave frequency and phase velocity concerning the Winkler coefficient. According to Figures 7.9 and 7.10, it can be inferred that the Winkler foundation exhibits characteristics of both parabolic and linear types. Based on the diagrams, it can be inferred that the ξ parameter has a positive correlation with both wave frequency and phase velocity. Specifically, as the ξ parameter increases, so do the values of wave frequency and phase velocity.

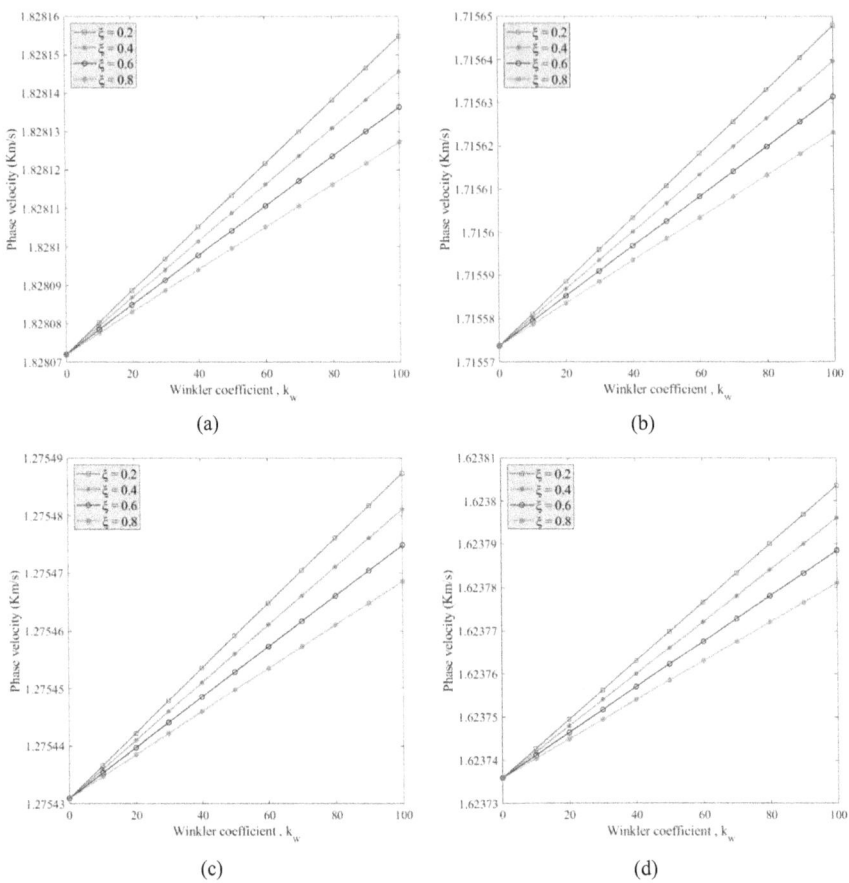

FIGURE 7.25   Variation of phase velocity versus Winkler coefficient for various ξ parameters $\left(\dfrac{L}{h} = \dfrac{R}{h} = 30, k_w \left(\text{linear} \right), e_0 = 0.4, k_p = 20, k_x = k_n = 20, \text{SPD} \right)$. (a) Magnesium, (b) nickel, (c) titanium, and (d) tungsten.

Moreover, as the Winkler coefficient increases, there is a corresponding increase in both the wave frequency and phase velocity values. The stiffness of a structure is increased when it is supported by or connected to a foundation, thereby resulting in the observed behavior. Like other diagrams presented previously, these diagrams illustrate that magnesium and titanium exhibit the highest and lowest values of wave frequency and phase velocity, respectively [5]. Figure 7.26 illustrates the impact of different types of Winkler foundation on the variation of phase velocity with respect to the Winkler coefficient. It can be observed that the sinusoidal type has the least effect on the variation of phase velocity compared to the other types. Following this, the linear type exhibits a lesser effect compared to the remaining types. Finally, the constant type and parabolic type have the most

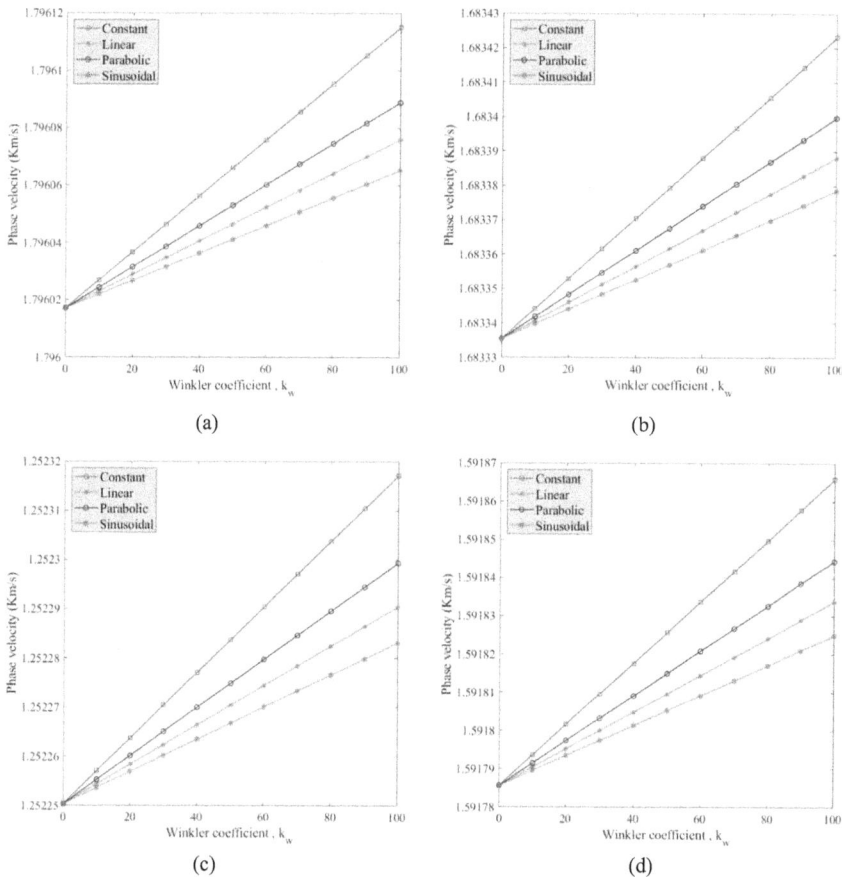

FIGURE 7.26 Variation of phase velocity versus Winkler coefficient for various types of Winkler foundations $\left(\dfrac{L}{h} = \dfrac{R}{h} = 30, e_0 = 0.3, k_p = 30, k_x = k_n = 20, \text{and} \xi = 0.8, \text{APD} \right)$. (a) Magnesium, (b) nickel, (c) titanium, and (d) tungsten.

significant impact on the variation of phase velocity, respectively. Moreover, the increase in the Winkler coefficient results in a corresponding increase in structural rigidity, leading to an elevation in phase velocity measurements [5].

The impact of the Pasternak coefficient on the fluctuation of phase velocity for different porosity coefficients is investigated in Figure 7.27. It is evident that the Pasternak coefficient, when applied to the shear layer of an elastic foundation, has a significant impact on the fluctuation of phase velocity values. To elaborate, it can be observed that the phase velocity values exhibit an increase when a higher value is selected for the Pasternak coefficient. Furthermore, when the Pasternak coefficient remains constant, an increase in the porosity coefficient results in a decrease in phase velocity values [5].

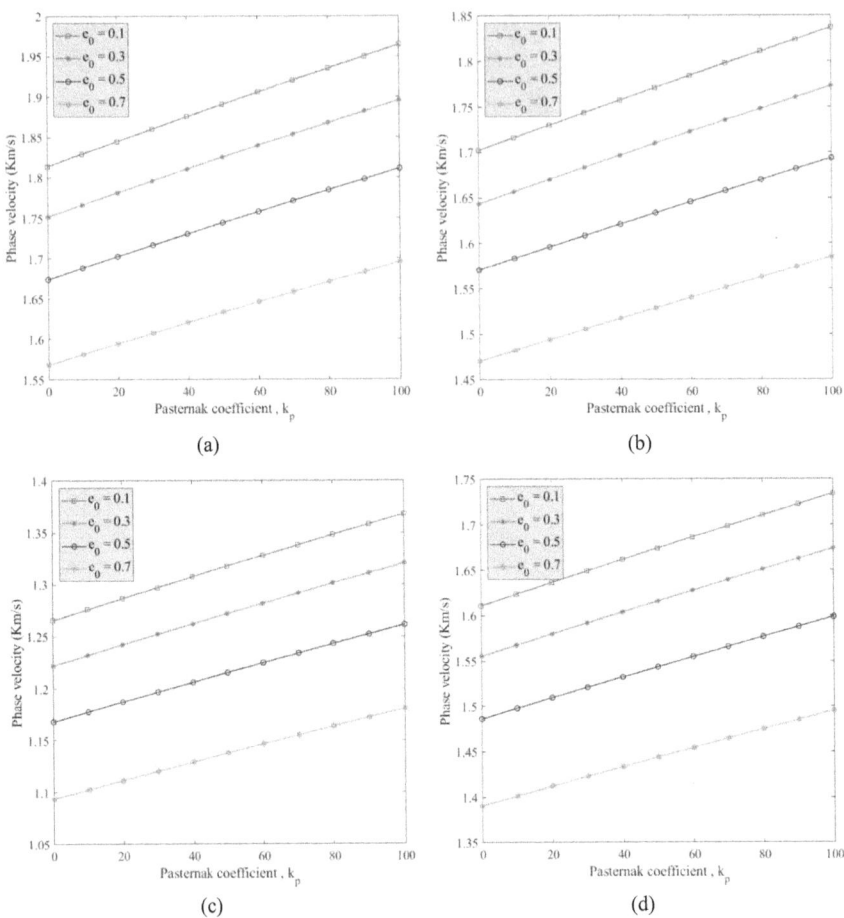

FIGURE 7.27   Variation of phase velocity versus Pasternak coefficient for various porosity coefficients $\left(\dfrac{L}{h} = \dfrac{R}{h} = 30, k_w = 50(\text{sinusoidal}), e_0 = 0.3, k_x = k_n = 20, \text{and } \xi = 0.8, \text{APD} \right)$. (a) Magnesium, (b) nickel, (c) titanium, and (d) tungsten.

7.5   SUMMARY

The current chapter investigates the thermal wave propagation of metal foam structures that are porous and are situated on an elastic foundation. The distribution of pores or voids is both symmetrical and asymmetrical. This study examines the impact of three distinct temperature increases. The Hamiltonian method was utilized to derive the equations of motion for structures composed of porous metal foam. Various patterns are taken into consideration when analyzing the distribution of porosity across the thickness of structures. Ultimately, the equations of governance are resolved through an analytical approach. This study investigates the impact of multiple parameters, such as temperature fluctuations, porosity coefficient, and Kerr substrate, on the alteration of wave frequency and phase velocity values in porous metal foam. Henceforth, the most significant highlights are delineated as follows:

- The porosity coefficient and temperature change have been observed to have a negative correlation with the wave frequency and phase velocity of porous metal foam structures.
- The impact of a two-variant elastic foundation on the wave frequency of porous metal foam is becoming more significant.
- The frequency and phase velocity of waves are significantly impacted by the Kerr substrate, as well as the Winkler and Pasternak coefficients, which exhibit an increasing influence.
- The utilization of a sinusoidal Winkler foundation results in comparatively lower alterations to phase velocity values when compared to alternative foundation types.
- The phase velocity of porous metal foam is observed to decrease with a reduction in the slenderness ratio. Specifically, the growth of the radius-to-thickness and length-to-thickness ratios is found to have a diminishing impact.
- The distribution of porosity can have an impact on the ability to withstand wave frequencies, with symmetric porosity distribution exhibiting the highest tolerance and uniform porosity distribution exhibiting the lowest tolerance.
- The asymmetrically porous metal foam structure exhibits the highest escape frequency compared to other structures.
- The structure exhibiting a symmetric porosity distribution demonstrates higher values of wave frequency and phase velocity when compared to the structure with an asymmetric porosity distribution.
- The cylindrical shell with SPD exhibits higher values of wave frequency and phase velocity when compared to the cylindrical shell with APD and UPD.

APPENDICES

APPENDIX A

$$k_{11} = -A\beta^2$$

$$k_{12} = Bi\beta^3$$

$$k_{13} = -B_s i\beta^3$$

$$k_{22} = -D\beta^4 - k_w - k_P\beta^2$$

$$k_{23} = -D_s\beta^4 - k_w - k_p\beta^2$$

$$k_{33} = -H_s\beta^4 - A_s\beta^2 - k_w - k_p\beta^2$$

$$m_{11} = I_0$$

$$m_{12} = -I_1\beta i$$

$$m_{13} = -J_1\beta i$$

$$m_{22} = I_0 + I_2\beta^2$$

$$m_{23} = I_0 + J_2\beta^2$$

$$m_{33} = I_0 + K_2\beta^2$$

APPENDIX B

$$K_{11} = -A_{11}\beta_1^2 - A_{66}\beta_2^2;$$
$$K_{12} = (A12 + A66)\beta_1\beta_2;$$
$$K_{13} = iB_{11}\beta_1^3 + i(B_{12} + 2B_{66})\beta_2^2\beta_1;$$
$$K_{14} = iB_{11}^S\beta_1^3 + i(B_{12}^S + 2B_{66}^S)\beta_2^2\beta_1;$$
$$K_{22} = -A_{22}\beta_2^2 - A_{66}\beta_1^2;$$
$$K_{23} = iB_{22}\beta_2^3 + i(B_{12} + 2B_{66})\beta_1^2\beta_2;$$
$$K_{24} = iB_{22}^S\beta_2^3 + i(B_{12}^S + 2B_{66}^S)\beta_1^2\beta_2;$$

$$K_{33} = -D_{11}\beta_1^4 - 2(D_{12} + 2D_{66})\beta_1^2\beta_2^2 - D_{22}\beta_2^4 - \frac{K_l K_u}{K_l + K_u}$$

$$-\frac{K_s K_u}{K_l + K_u}(\beta_1^2 + \beta_2^2) - \frac{K_l K_u}{K_l + K_u}(\beta_1^2 + \beta_2^2) - \frac{K_s K_u}{K_l + K_u}(\beta_1^4 + \beta_2^4) + N^T(\beta_1^2 + \beta_2^2);$$

$$K_{34} = D_{11}^S\beta_1^4 + 2(D_{12}^S + 2D_{66}^S)\beta_1^2\beta_2^2 + D_{22}^S\beta_2^4 + \frac{K_l K_u}{K_l + K_u}$$

$$+\frac{K_s K_u}{K_l + K_u}(\beta_1^2 + \beta_2^2) + \frac{K_l K_u}{K_l + K_u}(\beta_1^2 + \beta_2^2) + \frac{K_s K_u}{K_l + K_u}(\beta_1^4 + \beta_2^4) - N^T(\beta_1^2 + \beta_2^2);$$

$$K_{44} = -H_{11}^S\beta_1^4 - 2(H_{12}^S + 2H_{66}^S)\beta_1^2\beta_2^2 - H_{22}^S\beta_2^4 - A_{55}^s\beta_1^2 - A_{44}^s\beta_2^2 - \frac{K_l K_u}{K_l + K_u}$$

$$-\frac{K_s K_u}{K_l + K_u}(\beta_1^2 + \beta_2^2) - \frac{K_l K_u}{K_l + K_u}(\beta_1^2 + \beta_2^2) - \frac{K_s K_u}{K_l + K_u}(\beta_1^4 + \beta_2^4) + N^T(\beta_1^2 + \beta_2^2);$$

$$M_{11} = I_0;$$
$$M_{12} = 0;$$
$$M_{13} = -iI_1\beta_1;$$
$$M_{14} = -iJ_1\beta_1;$$

$$M_{22} = I_0;$$
$$M_{23} = -iI_1\beta_2;$$
$$M_{24} = -iJ_1\beta_2;$$
$$M_{33} = I_0 + I_2\left(\beta_1^2 + \beta_2^2\right);$$
$$M_{34} = -I_0 - J_2\left(\beta_1^2 + \beta_2^2\right);$$
$$M_{44} = I_0 + K_2\left(\beta_1^2 + \beta_2^2\right)$$

APPENDIX C

The components of stiffness and mass matrices are as follows:

$$K_{11} = -A_{11}k_x^2 - \frac{A_{66}}{R^2}k_n^2, K_{12} = -k_x k_n\left(\frac{A_{12} + A_{66}}{R}\right), K_{13} = \frac{A_{12}}{R}k_x i,$$

$$K_{14} = -B_{11}k_x^2 - \frac{B_{66}}{R^2}k_n^2, K_{15} = -k_x k_n\left(\frac{B_{12} + B_{66}}{R}\right), K_{22} = -A_{66}k_x^2 - \frac{A_{11}}{R^2}k_n^2 - \frac{A_{55}^s}{R^2},$$

$$K_{23} = k_n\left(\frac{A_{11} + A_{55}^s}{R^2}\right)i, K_{24} = -k_x k_n\left(\frac{B_{12} + B_{66}}{R}\right), K_{25} = -B_{66}k_x^2 - \frac{B_{11}}{R^2}k_n^2 - \frac{A_{55}^s}{R},$$

$$K_{33} = -A_{55}^s k_x^2 - \left(\frac{A_{11}}{R^2} + k_n^2 \frac{A_{55}^s}{R^2}\right) - k_w - k_p\left(k_x^2 + \frac{k_n^2}{R^2}\right),$$

$$K_{34} = \left(A_{55}^s - \frac{B_{12}}{R}\right)ik_x, K_{35} = k_n\left(\frac{A_{55}^s}{R} - \frac{B_{11}}{R^2}\right)i,$$

$$K_{44} = -D_{11}k_x^2 - k_n^2\frac{D_{66}}{R^2} - A_{55}^s, K_{45} = -k_x k_n\frac{D_{12} + D_{66}}{R},$$

$$K_{55} = -D_{66}k_x^2 - k_n^2\frac{D_{11}}{R^2} - A_{55}^s$$

$$M_{11} = M_{22} = M_{33} = I_0,$$
$$M_{14} = M_{25} = I_1,$$
$$M_{44} = M_{55} = I_2$$

REFERENCES

[1] Ebrahimi, F., & Seyfi, A. (2021). A wave propagation study for porous metal foam beams resting on an elastic foundation. *Waves in Random and Complex Media*, 1–15.
[2] Ebrahimi, F., & Seyfi, A. (2022). On hygrothermal wave dispersion characteristics of embedded graphene foam. *Waves in Random and Complex Media*, 1–20.
[3] Ebrahimi, F., & Seyfi, A. (2022). Studying propagation of wave of metal foam rectangular plates with graded porosities resting on Kerr substrate in thermal environment via analytical method. *Waves in Random and Complex Media*, 32, 2, 832–855.

[4] Ebrahimi, F., & Seyfi, A. (2022). On wave propagation characteristics of hygrother-mally excited graphene foam plates. *Waves in Random and Complex Media*, 1–20.

[5] Ebrahimi, F., & Seyfi, A. (2022). Studying propagation of wave in metal foam cylin-drical shells with graded porosities resting on variable elastic substrate. *Engineering with Computers*, 1–17.

[6] Ebrahimi, F., Seyfi, A., & Dabbagh, A. (2019). A novel porosity-dependent homog-enization procedure for wave dispersion in nonlocal strain gradient inhomogeneous nanobeams. *The European Physical Journal Plus*, *134*(5), 226.

[7] Ebrahimi, F, Seyfi, A, & Dabbagh, A. (2019). Dispersion of waves in FG porous nanoscale plates based on NSGT in thermal environment. *Advances in Nano Research*, *7*(5), 325–335.

[8] Ebrahimi, F., & Zia, M. (2015). Large amplitude nonlinear vibration analysis of func-tionally graded Timoshenko beams with porosities. *Acta Astronautica*, *116*, 117–125.

[9] Banhart, J. (2001). Manufacture, characterisation and application of cellular metals and metal foams. *Progress in Materials Science*, *46*(6), 559–632.

[10] Smith, B, Szyniszewski, S, Hajjar, J, et al. (2012). Steel foam for structures: A review of applications, manufacturing and material properties. *Journal of Constructional Steel Research*, *71*, 1–10.

[11] Magnucka-Blandzi, E. (2010). Non-linear analysis of dynamic stability of metal foam circular plate. *Journal of Theoretical and Applied Mechanics*, *48*(1), 207–217.

[12] Ebrahimi, F., Seyfi, A., & Dabbagh, A. (2019). A novel porosity-dependent homog-enization procedure for wave dispersion in nonlocal strain gradient inhomogeneous nanobeams. *The European Physical Journal Plus*, *134*(5), 226.

[13] Ebrahimi, F., et al. (2019). Wave dispersion characteristics of porous graphene platelet-reinforced composite shells. *Structural Engineering and Mechanics*, *71*(1), 099.

[14] Ebrahimi, F., Dabbagh, A., & Rastgoo, A. (2019). Vibration analysis of porous metal foam shells rested on an elastic substrate. *The Journal of Strain Analysis for Engineering Design*, *54*(3), 199–208.

[15] Ebrahimi, F., Dabbagh, A., & Taheri, M. (2021). Vibration analysis of porous metal foam plates rested on viscoelastic substrate. *Engineering with Computers*, *37*, 3727–3739.

[16] Selvamani, R., Rexy, J. B., & Ebrahimi, F. (2022). Finite element modeling and analysis of piezoelectric nanoporous metal foam nanobeam under hygro and nonlin-ear thermal field. *Acta Mechanica*, *233*(8), 3113–3132.

Index

Pages in *italics* refer to figures and pages in **bold** refer to tables.

For Product Safety Concerns and Information please contact our EU
representative GPSR@taylorandfrancis.com
Taylor & Francis Verlag GmbH, Kaufingerstraße 24, 80331 München, Germany

www.ingramcontent.com/pod-product-compliance
Lightning Source LLC
Chambersburg PA
CBHW060817170526
45158CB00001B/7